Gastrointestinal Inflammations and Gut Microbiota

This book offers a comprehensive and multi-faceted examination of gastrointestinal inflammations, discussing factors that impact gut health. The initial chapters establish a foundational understanding of gastrointestinal inflammations and the critical role of the gut microbiota. The book then delves into the complex pathophysiology and various risk factors associated with these conditions, offering insights into the latest diagnostic, screening, and prevention techniques. It explores the dynamic interactions between the gut and the brain as well as the biochemical interplays involving short-chain fatty acids, reactive oxygen species, and superoxide dismutases. Additionally, the book thoroughly discusses the impact of genetics on gastrointestinal health, the interactions between antibiotics and gut microbiota, and the pivotal role of the immune system in inflammation. It highlights advancements in diagnostic technologies and gut microbiota-based strategies, along with the use of nanoparticles in treatment. The book also examines the psychological aspects associated with gastrointestinal inflammations, emphasizing the importance of mental health for physical well-being. It assesses the potential of functional foods, various types of nanocarriers, and plant-based bioactives in managing gastrointestinal inflammations. As such, this book is primarily intended for researchers and students in the fields of gastroenterology and microbiology.

Key Features:

- Provides an extensive overview of gastrointestinal inflammations, starting from the basic concepts of the gut microbiota to complex pathophysiological aspects
- Delves into the latest advancements in diagnostic, screening, and prevention techniques, highlighting cutting-edge technologies and approaches in gastroenterology
- Explores the role of the immune system and the impact of genetics and diet on gastrointestinal health
- Examines use of probiotics, prebiotics, postbiotics, and various nanotechnology-based approaches for gastrointestinal inflammation management
- Emphasis on the psychological factors associated with gastrointestinal inflammations

Gastrointestinal Inflammations and Gut Microbiota

Pathophysiology, Management, and Emerging Therapies

Edited by
Neeraj Mishra, Sumel Ashique, and Anoop Kumar

CRC Press
Taylor & Francis Group
Boca Raton London New York

CRC Press is an imprint of the
Taylor & Francis Group, an **informa** business

Designed cover image: Shutterstock

First edition published 2025
by CRC Press
2385 Executive Center Drive, Suite 320, Boca Raton, FL 33431

and by CRC Press
4 Park Square, Milton Park, Abingdon, Oxon, OX14 4RN

CRC Press is an imprint of Taylor & Francis Group, LLC

ISBN: 9781032796383 (hbk)
ISBN: 9781032796468 (pbk)
ISBN: 9781003493143 (ebk)

DOI: 10.1201/9781003493143

Typeset in Times
by Newgen Publishing UK

Contents

Preface

Gastrointestinal inflammation pertains to the development of inflammation inside the gastrointestinal tract, encompassing the esophagus, stomach, small intestine, and large intestine. Inflammation within the gastrointestinal (GI) tract can arise from a multitude of factors, and it is frequently linked to conditions such as inflammatory bowel disease (IBD), irritable bowel syndrome (IBS), infectious gastroenteritis, and other related disorders. One of the most common digestive disorders, gastrointestinal inflammations, is characterized by constant, recurrent stomach pain and irregular bowel movements.

The cause of IBS is not fully known and is thought to be extremely complex; consequently, treatments are only somewhat effective. Epidemiologic research on individuals with post-infectious IBS (PI-IBS) has demonstrated a changed microbiota in the human gastrointestinal tract as one of the potential causes of IBS in the last ten years. Previous observational studies have shown that there is a higher risk of developing IBS in the future when the gut microbiota changes because of acute gastroenteritis. Additionally, there is a strong correlation between IBS and antibiotic therapy for extra-intestinal infections, which may change the intestinal flora. IBS is thought to be caused by a complex combination of genetic, environmental, and microbiota-related factors, while its specific cause is yet unknown. Promising therapy options for IBS include nanocarriers and methods that modify the gut microbiota. Therapeutic medicines can be packaged and delivered via nanocarriers like nanoparticles or liposomes right to the damaged parts of the gastrointestinal (GI) tract. By decreasing drug exposure to healthy tissues, this strategy lessens negative effects. The modified NPs (nanoparticles) have boosted medication concentration, decreased systemic adverse effects, and improved the efficacy and safety of local drug delivery. The development of IBD therapies based on nanotechnology is accelerating exponentially. For optimal health, IBS prescription medication is accessible.

The development of various nanoparticles that will aid in the management of IBS and other gastrointestinal illnesses is a major priority for many pharmaceutical companies. However, further research is necessary to fully comprehend how the effects of the nanoparticle on the gut flora could benefit IBD patients' prognoses. By extending the duration of the therapeutic impact and lowering the frequency of medication administration, controlled-release nanocarriers can increase patient compliance. Researchers are investigating the creation of hybrid systems that combine medicines that target the microbiota with nanocarriers for drug delivery. For instance, probiotics or prebiotics can be added to nanocarriers to increase their therapeutic efficiency.

About the Editors

Neeraj Mishra has been serving as a Professor at the Amity Institute of Pharmacy, Gwalior, since July 2019. He brings approximately 20 years of experience in teaching and research. Dr. Mishra has published over 100 papers of international and national significance, focusing on novel drug delivery systems, oral delivery of synbiotics, localized drug delivery, and targeted and controlled drug delivery using nanocarriers and microparticles for treating breast and colon cancer, as well as neurodegenerative disorders. He has also edited more than 15 books and authored 25 book chapters. He is the recipient of the "Distinguished Professor Award" (2019) from DST-NSTMIS, SPAICS, Indore, Madhya Pradesh, and the "Outstanding Scientist Award" (2020) from the 6th International Scientist Awards on Engineering, Science, and Medicine, held in Chennai, India. Further, Dr. Mishra received the "Distinguished Professor Award" in 2021 from the Indian Pharmaceutical Association, MP State Branch, Indore, and the "Best Academician Award" at the Indian Pharmacy Graduate Association's M.P. State International Conference, held on May 6, 2023, at the DAVV Auditorium, Indore, India.

Sumel Ashique has been working as an Assistant Professor at Bengal College of Pharmaceutical Sciences & Research, Durgapur 713212, West Bengal, India. He has 4 years of teaching experience. He has achieved 80 publications in international and national accredited reputed journals. He has knowledge in drug delivery, nanotechnology and targeted treatment strategy. He is working as a potential reviewer for many well-known publishers (Elsevier, Springer, Dove Press, and Bentham Sc.). He has also five granted patents from IP and Australia, 20 published book chapters in Bentham, Springer, Elsevier, Bentham and Taylor & Francis. Currently he is editing 14 books under CRC Press, Elsevier, Taylor & Francis, Wiley, Nova Science and Springer. He has published 80+ papers including research, reviews, column articles, editorials in Scopus, and WOS indexed. He is serving as Guest Editor for *JoVE Journal* (IF 1.2, Scopus & WOS indexed*), Current Probiotics* (Bentham Science) & *Current Indian Science Journal.* He is a potential reviewer for Elsevier, Springer, Taylor & Francis and Wiley journals for expertise in the fields of targeted drug delivery, nanotechnology, gut-microbiome, and postbiotics. Currently, he is pursuing his PhD degree from Lovely Professional University, Punjab, India (NAAC A^{++}) in the Pharmaceutical Science department.

Anoop Kumar is currently serving as an Assistant Professor in the Department of Pharmacology and Clinical Research at Delhi Pharmaceutical Sciences & Research University (DPSRU), New Delhi, India. Previously, he held the position of Assistant Professor in the Department of Pharmacology and Toxicology at the National Institute of Pharmaceutical Education and Research (NIPER), Raebareli. He has also served as an Associate Professor and Officiating Head in the Department of Pharmacology at ISF College of Pharmacy, Moga, Punjab. Additionally, Dr. Kumar has worked as a research scientist in the Medical Affairs and Clinical Research Department at Sun Pharmaceutical Limited, Gurugram, and at the Translational Health Science and Technology Institute (THSTI), Faridabad, India. Dr. Kumar has authored over 80 research and review articles and 15 book chapters in internationally renowned journals. He is also the author of three books. He has received research grants from the Indian government. His research interests include drug repurposing using computational, in vitro, and in vivo techniques, meta-analysis, signal analysis in pharmacovigilance, and pharmacoeconomic studies.

1 Introduction to Gastrointestinal Inflammation and Gut Microbiota

Muneeb Ullah[1], Mishal Safdar[2], Jin-Wook Yoo[1], Raees Khan[2], Abdul Wahab[3], Khalid Nawaz[4], and Muhammad Naeem[2*]*

[1]College of Pharmacy, Pusan National University, Busandaehak-ro 63 beon-gil 2, Geoumjeong-gu, Busan, Republic of Korea

[2]Department of Biological Sciences, National University of Medical Sciences (NUMS) Rawalpindi, Punjab, Pakistan

[3]Department of Pharmacy, Kohat University of Science, and Technology (KUST), Kohat, Khyber Pakhtunkhwa, Pakistan

[4]Department of Medical Laboratory Technology, Khyber Medical University, Peshawar, Khyber Pakhtunkhwa, Pakistan

1.1 INTRODUCTION

The gastrointestinal (GI) tract in humans hosts a vast and complex microbial community, comprising over 100 trillion bacteria and other microorganisms [1]. In fact, the bacterial cell density in the GI tract, particularly the colon, is approximately 10^{11}–10^{12} cells per milliliter, making the colon one of the densest microbial habitats on Earth [2]. The gut microbiome contains around 3 million genes, generating many metabolites, while the human genome comprises roughly 23,000 genes [3]. Researchers have long been intrigued by the gut microbiota; however, a significant challenge in related studies has been the difficulty in cultivating these microorganisms [4].

Given the diverse properties of gut microbiota, including their remarkable diversity, resilience, symbiotic nature, and stable relationship with the host, it is plausible to categorize the host and its associated microorganisms as a "superorganism" [5,6], which plays essential roles in metabolic and immune functions [1]. Bacteria in the gut play a critical role in regulating digestion throughout the gastrointestinal tract. Commensal bacteria are especially important in synthesizing, extracting, and absorbing several metabolites and nutrients, such as amino acids, lipids, vitamins, bile acids, and short-chain fatty acids (SCFAs). Furthermore, the gut flora is vital for the immune system's protective mechanism by inhibiting the colonization of pathogenic bacteria, preventing their growth through the production of bacteriocins, and competing for nutrients. Additionally, it contributes to maintaining the integrity of the intestinal epithelium, thus preventing bacterial invasion [7].

Microorganisms adopt various competitive mechanisms to prevent the colonization of pathogens. These mechanisms include modulation of cell signaling pathways, antimicrobial peptide secretion, pH modification, and nutrient metabolism. Current research has underscored the significant impact of symbiotic bacteria and their metabolites on the development, function, and enhancement of immune cells in both adaptive and innate immune responses [8]. It is paradoxical that, while the core functions of gut microbiota are largely conserved among individuals, the composition of gut flora varies significantly due to a unique blend of bacterial species in each person, influenced by both intra- and inter-individual variations throughout human life.

1.2 GI INFLAMMATION AND GUT MICROBIOTA

The gut flora and their metabolic derivatives have the potential to alter the inflammatory conditions of the host [9–12]. Multiple studies conducted on gut microbiota and GI inflammation like, Forbes et al. (2018) investigate the association between gut microbiota and immune-mediated disease like CDs, UC, rheumatoid arthritis (RA) and multiple sclerosis (MS). Furthermore, research also noted the role of gut flora in inflammatory disorders like DM type-1 and -2, asthma and obesity [13–16]. The advent of 16S ribosomal RNA (rRNA) sequencing has lined the technique for dissecting the composition of the gut microbiota in individuals, shedding light on the relationship between definite microorganisms and various illnesses (Table 1.1). *Enterobacteriaceae*, a broad family of gram-negative facultative bacteria, frequently exhibit associations with diverse inflammatory conditions, such as IBD and obesity [1, 17]. For instance, there is an association between adherent-invasive *E. coli* (AIEC) and CD, whereas diffusely adherent *E. coli* (DAEC) has been correlated with UC [18]. Additionally, alterations typically marked by a decline in *Firmicutes* abundance and a rise in *Proteobacteria* populations have been linked to IBD in humans [19, 20]. Furthermore, there is an association between an elevated ratio of *Firmicutes* to *Bacteroidetes* populations and obesity [20, 21]. Zhang et al. [22, 23] diabetic patients tend to exhibit increased levels of the *Firmicutes* phylum and the *Clostridia* class. Similarly, Zhu et al. [24, 25] observed a notable rise in

DOI: 10.1201/9781003493143-1

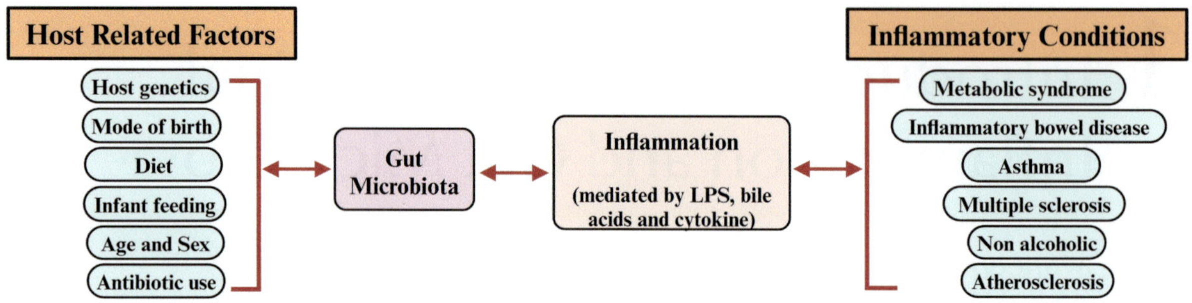

FIGURE 1.1 Illustration of interplay between gut microbiota, inflammation, and inflammatory conditions.

the proportion of *Proteobacteria* among children diagnosed with non-alcoholic steatohepatitis (NASH). Furthermore, individuals with rheumatoid arthritis (RA) display heightened levels of *Lactobacillus* and *Prevotella copri* in their gut microbiota, alongside decreased quantities of *Bifidobacteria* and *Bacteroides* as shown in Figure 1.1 [26–28].

Acetate, butyrate, and propionate, serve as the elementary metabolic byproducts resulting from the gut microbiota's breakdown of dietary fiber that is not absorbed and resistant starches, respectively [29]. Butyrate holds significant importance among short-chain fatty acids (SCFAs) because of its capacity to counteract colonic inflammation [30] and its capacity to fulfill 6–10% of the daily energy requirements for humans [31]. *Eubacterium*, *Clostridia*, *Ruminococcaceae*, and *Firmicutes* were found as the primary producers of butyrate [32]. Also, butyrate substantially suppresses the interferon-gamma (IFN-g) signaling to reduce inflammation [33] [34].

1.3 MOST COMMON GI DISEASES

1.3.1 Inflammatory Bowel Disease

IBD, involving both CD and UC, is distinguished by continual and recurrent inflammation within the GI tract. While CD and UC are often categorized as chronic IBDs, they are distinctive conditions with varying symptoms and patterns of inflammation. Specifically, CD involves chronic and affecting specific segments inflammation of the GIT [35] and although the exact cause is still uncertain, the etiology of IBD is understood to be a complex interplay between gut microbiota composition and the host genetic makeup [36]. UC typically manifests as ulceration and inflammation of the colon's inner lining.

The initiation of both conditions is commonly thought to result not from a sole organism, but rather from a more extensive microbial imbalance within the gastrointestinal tract [37, 38]. Nevertheless, this remains a subject of significant contention. Numerous findings have proposed the association of gut microbes in the initiation of IBD, highlighting the gut microbiota as crucial factors in the formation of mucosal lesions. For a detailed examination, one can consult the review by Manichanh et al [39]. Intestinal inflammation is frequently linked with a decline in bacterial diversity, particularly a decreased abundance and complexity within the *Bacteroidetes* and *Firmicutes* phyla [40, 41]. Furthermore, it has been proposed that while there's a decrease in *Firmicutes*, there's a rise in Gamma *Proteobacteria* among patients with CD [42]. Furthermore, recent studies examining people with IBD, and healthy controls have revealed associations between IBD disease status and changes in the abundances of specific bacterial families such as *Enterobacteriaceae*, *Ruminococcaceae*, and *Leuconostocaceae*. At the genus level, these studies have observed increased levels of Clostridium, while beneficial butyrate-producing bacteria like *Roseburia* and succinate-producing bacteria like *Phascolarcto* bacterium were notably lowered in both UC and CD conditions [43]. Despite the variability in microbial populations or pathogens involved, and although specific causality remains unclear, numerous studies have indeed proven a connection of gut flora with IBD.

1.3.2 Colorectal Cancer (CRC)

Recent publications have indicated a possible involvement of the gut flora in the progression of CRC [44–46]. While a singular causative organism has yet to be pinpointed, several studies suggested a link between *Fusobacterium* members and CRC [45, 47, 48]. In a recent review employing fluorescence in the analysis of in situ hybridization and interaction between CRC and Fusobacteria, which was unveiled, with greater quantities observed in tumors compared to controls [45]. These alterations were convoyed by broad alterations at the phylum level, particularly a notable decrease in *Firmicutes* and *Bacteroidetes.* These findings indicate a possible inflammatory mechanism through which *Fusobacterium* species may contribute to tumorigenesis [45]. Chronic inflammation is a recognized factor implicated in cancer onset [49] and a tumor-associated or "tumor-elicited" inflammation can be attributed to CRCs [50]. Importantly, another study utilizing metagenomic sequencing andqPCR data validated the correlation between the *Fusobacterium* genus and CRC. A higher abundance of *Fusobacterium* sequences in tumor tissue as compared to individuals without the condition [47]. Interestingly, *Fusobacterium* members have been connected to a variety of different intestinal illnesses, including IBD [51] and acute appendicitis [52, 53].

Several other studies have also emphasized the connection between inflammation induced by microbes and CRC.

TABLE 1.1
Microbiota that Promote the Development of Inflammation Disease

Subject	Microbe	The Possible Role of Macrophages	Disease
		Microbiota that Promote the Development of Inflammation Disease	
Human	*Prevotella copri*	The activated macrophages serve as the primary producers of pro-inflammatory cytokines and chemokines, involving TNF and IL-1β	RA
Child	*Clostridium difficile*	*Clostridium difficile* releases TcdA and TcdB toxins, which promote the recruitment of macrophages and enable the survival of the bacteria within these macrophages	Asthma
Human	*Adherent-invasive Escherichia coli*	Adhering to intestinal epithelial cells, invading these cells, and subsequently surviving and replicating within macrophages	CD
Rats	*E. coli*	*Escherichia coli* lipopolysaccharide (LPS) triggers the activation of liver macrophages, leading them towards a pro-inflammatory phenotype via the TLR4 pathway	NASH
NOD mice	*Bacteroides fragilis*	*Bacteroides fragilis* enhances the phagocytic capabilities of macrophages and induces polarization towards an M1 phenotype	T1DM
		Microbiota that Protects Against Inflammation Diseases	
Mice	*Lactobacillus helveticus*	*Lactobacillus gasseri* decreased the population of alveolar macrophages that produce IL-17	Asthma
HLA-DR3.DQ8 transgenic mice	*Prevotella histicola*	The induction of suppressive macrophages resulted in greater formation of IL-10 and minor production of IL-12, leading to a decreased capacity for antigen presentation	MS
Human	*Roseburia*	The production of butyrate exerts an anti-inflammatory effect on macrophages	T2DM

* NASH, non-alcoholic steatohepatitis; MS, multiple sclerosis.

Research has confirmed that microbial yields can breach barrier-defective colonic tumors, triggering inflammation via a host immune response, which in turn fosters tumor growth [54]. HTS studies have validated a correlation between gut microbial ecology and inflammation in animals lacking interleukin-10, which are predisposed to colitis [44]. This study found that mice with colitis exhibited a less varied gut flora composition, marked by a rise in *Proteobacteria*, notably elevated levels of *E. coli*, during intestinal inflammation [44]. While these findings suggest that both host inflammation and particular microorganisms play a role in CRC tumorigenesis, further research in this area is clearly warranted.

1.4 INFLUENCE OF THE MICROBIOME ON IBD PATHOLOGY

Like many immune-mediated diseases, IBD is regarded as an idiopathic condition, lacking a single, well-defined etiological agent [55]. Despite years of comprehensive epidemiological, genetic, laboratory, and clinical investigations, our understanding of the complex interactions shaping IBD pathogenesis remains limited. Nevertheless, a growing body of evidence now underscores the significance of a dysbiotic intestinal microbiota, characterized by a disparity between pro-inflammatory and anti-inflammatory microbes, in initiating and propagating intestinal impairment [56]. Genetic variations across a spectrum of genes play a role in microbiota detection [57, 58]. Environmental factors, including lifestyle choices, dietary habits, and medication usage, further influence this fragile equilibrium, frequently exerting their effects through alterations in the composition of intestinal microbiota [56, 59, 60]. It is widely acknowledged that IBD arises from a complex interaction of factors, often termed a "perfect storm," involving a dysbiotic microbiota and a disrupted immune response in the host (Figure 1.1) [61].

The intestinal microbiota constantly interacts with the host's immune system, coordinating inflammatory responses to eradicate pathogens while also fostering immune tolerance to the overall microbial community [55, 62, 63]. Evidence supporting the role of microbiota in IBD pathogenesis remains intricate. Animal studies, particularly murine colitis models, have been instrumental in enhancing our comprehension of IBD pathogenesis. Numerous well-designed studies have showcased the crucial role of gut flora in the spontaneous development of colitis in animals genetically predisposed to the condition [64–66]. In clinical practice, antibiotics such as rifaximin have consistently demonstrated efficacy in alleviating intestinal infections. This has been particularly noteworthy in IBD management, especially in reducing the recurrence of Crohn's disease (CD) following surgical resection [67, 68]. In individuals with CD, diverting the fecal stream away from inflamed mucosa proximally has been shown to reduce inflammation and promote healing in the excluded segments of the gut. However, relapse often occurs when the fecal stream is restored and the intestines are re-exposed to luminal contents [69, 70]. Over the years, various organisms have been proposed as potential contributors to IBD pathogenesis, with the most noteworthy candidates including

Helicobacter, Halfvarson et al. investigated the composition of the gut microbiome through cohort studies comparing healthy patients with individuals diagnosed with CD and UC. Stool samples were collected over two years (every three months) and subjected to 16S rRNA analysis to assess microbiome composition. Both healthy controls and IBD patients formed distinct microbial clusters, with CD patients exhibiting the most pronounced differences. Longitudinal analysis revealed greater microbiome instability in IBD patients compared to controls, highlighting the condition's fluctuating nature. Seven IBD patients experiencing flare-ups and receiving corticosteroid treatment exhibited more pronounced microbiome alterations compared to those on stable medication regimens. Key insights derived from multi-omics assessments of IBD patient cohorts underscore the importance of examining comprehensive patient sample sets, which include serum, stool, biopsies, and urine. Additionally, there is growing recognition that analytical approaches should be tailored specifically for UC and CD, given that UC has more constrained biomarker discoveries compared to CD [75–78]. Understanding the critical role of the gut flora in IBD development makes it imperative to investigate the effects of treatments on gut microbiota and, conversely, how microbiota influences treatment responses [75–78].

1.5 THE GUT MICROFLORA

The human intestine hosts a diverse array of microorganisms, encompassing approximately 10^13 bacterial cells collectively known as the "gut microbiota." This microbial community comprises roughly more than 250 different species of bacteria, fungi, viruses, and archaea, forming a dynamic ecosystem that evolves over the course of an individual's life. The interaction between the intestinal flora and the host is profoundly symbiotic, with the microbiota playing pivotal roles in numerous physiological and pathological processes crucial to human health. The connection between gut microbiota and host is profoundly symbiotic with the microbiota playing an essential role in pathological and different physiological processes which are crucial to human health. The human gut flora consists of five predominant bacterial phyla which are; *Bacteroidetes* (comprise 20–40%) that includes *Flavobacteria*, *Bacteroidia*, *Sphingobacteria*, and *Cytophagia*, *Verrucomicrobia*; *Firmicutes* (constituting 60 to 80% of the microbiota), encompassing the classes *Clostridia*, *Bacilli*, and *Negativicutes*; *Actinobacteria*; and, to a minor degree, *Proteobacteria* and one *Archea* phyla, the *Euryarchaeota* [79].

As mentioned before, the gut flora plays a vital role in numerous physiological functions, including the digestion of common polysaccharides such as glycosaminoglycans and the production of short-chain fatty acids (SCFAs) by the action of a number of enzymes. Additionally, it is considered as different essential amino acid source i.e. pantothenic acid, and vitamins including thiamine, vitamin K, biotin, folate, and riboflavin. Furthermore, the gut flora has a major role in upholding the intestinal epithelial barrier integrity, aiding in the development of the host gut immune system, provide protection against external pathogens, and contributing in the metabolism of the xenobiotics [80]. The intricate mechanisms of the microbiota extend beyond the intestinal tract to influence distant organs. The microbiota functions as an endocrine organ, swaying processes such as hormonal regulation, satiety regulation, and even human behavior and mood. The term "gut-brain axis" represents the bidirectional interaction between the microbiota and the brain.

1.5.1 Transition of Eubiosis to Dysbiosis

A significant focus of research on microbiota revolves around defining eubiosis, often referred as "healthy microbiota," embodies a harmony of microbiome in intestine, holding favorable impact on the human body. Generally, robust gut microbial communities exhibit elevated taxa diversity, abundant microbial gene richness, and a consistent functional core within the microbiome, indicative of good health. Conversely, substantial alterations in the proportion among the five predominant phyla of bacteria in the gut flora or the appearance of novel bacterial groups give rise to a pathological imbalance referred to as dysbiosis. Dysbiosis includes a decline in microbial quantity as well as diversity, alongside the enhanced number of LPS-forming Gram-negative *Proteobacteria* [81].

Dysbiosis typically involves an increase in intestinal permeability [81]. Under normal physiological circumstances, the migration of a limited quantity of bacterial products, such as polysaccharides from *Bacteroides* spp, which is typically eliminated through the activity of Th1 and Th17 cells [82]. It is striking that higher levels of segmented SFB, as observed in MyD88-deficient mice models (MyD88 being an adapter for various innate immune receptors recognizing microbial stimuli), can shield mice with a diabetic genotype from developing the disease. This highlights the microbiota's dual capacity to exert both inhibitory and promoting effects on disease development [82]. Although a dysbiotic gut community may characterize various inflammatory conditions, dysbiosis itself can disrupt intestinal equilibrium and contribute to the onset of inflammation [81]. In contrast to infectious disorders, determining the pathogenicity of individual intestinal bacteria cannot be achieved using Koch's postulates, as a substantial portion of the gut flora cannot be isolated as pure culture.

Certainly, micro flora is typically studied using high-throughput DNA sequencing techniques, such as sequencing of conserved 16S rRNA genes or whole-metagenome shotgun sequencing. These methodologies offer insights not just into the microbial composition but also into the functional capabilities encoded within the sequenced microbial DNA [83]. Hence, determining the pathogenic implications of microorganisms in disease development primarily involves identifying shifts in bacterial populations or replicating a disease by transplanting fecal microbiota afflicted mouse to a healthy one.

1.5.2 The Interaction Between the Gut Microbiota and the Intestinal Barrier

The composition of the gut microbiota significantly influences the characteristics of the mucus layer. They proposed that certain bacteria, like those belonging to the *Erysipelotrichi* class and *Allobaculum* genus, can induce the formation of an impermeable inner mucus layer. In contrast, other phyla, such as TM7 and *Proteobacteria*, have the reverse effect, potentially influencing mucus layer permeability [84]. *Bifidobacteria* have been shown to possess anti-inflammatory properties in various in vivo models. In mouse models of necrotizing enterocolitis, they enhance barrier function by stabilizing claudins 2 and 4, as well as occludin, at tight junctions. Additionally, *bifidobacteria* exhibit antioxidant properties and contribute to preserving the integrity of intestinal microvilli. They promote the production of anti-inflammatory cytokines, stimulate IgA secretion, and aid in the development of immune cells [85].

This particular probiotic formulation was noted to increase the expression of claudin-1, occludin, and zonulin-1 and 2. Additionally, a protective effect against inflammatory injury of the intestinal barrier model was also observed for this probiotic formulation [86]. An intestinal injury caused by dextran sodium sulfate (DSS) is reported to be effectively restored by the administration of *Bifidobacterium bifidum* ATCC 29521 by regulating the immune markers regulation, tight junction proteins (TJs), and addressing dysbiosis in the colon. This strain of *Bifidobacterium* stimulates the expression of anti-inflammatory cytokines (i.e. PPAR γ, IL-6, and IL-10), scavenging enzymes, and reactive oxygen species (ROS). TJs such as ZO-1, MUC-2, and Claudin-3 are also increased while concurrently downregulating inflammatory genes (IL-1β and TNF-α) [86]. Specific *Lactobacillus* species are shown to prevent barrier disruption by upregulating TJs. Similarly, *Lactobacillus rhamnosus* CNCM- I-3690 was reported to aid in restoring intestinal barrier function by improving the concentration of occludin and E-cadherin, along with decreasing bacterial translocation [87]. Various *E. coli* strains can have contrasting effects on the intestinal barrier. For example, *E. coli* Nissle 1917 stimulates TJs (ZO-1 and ZO-2) expression while another strain *E. coli* C25 can upraise the intestinal permeability [88]. The engagement of gut flora in various interactions within the mucous layer plays a crucial role in determining the physical characteristics of the mucosal layer, shaping the microbial community and its composition.

1.6 IMPACT OF THE GUT MICROBIOME ON HUMAN HEALTH

The metabolic capabilities and vast genomic repertoire of the gut microbiota offer countless beneficial functions to the host. A few of the most significant characteristics of these microbes include preserving the integrity of the mucosal barrier, protection against pathogens, and transport of essential nutrients. The colonic bacteria carry carbohydrate-active enzymes that allow them to ferment complex carbohydrates and yield metabolites like SCFAs [89] [90]. The epithelial cells in the GIT efficiently absorb these SCFAs, contributing to regulating several cellular processes including gene expression, differentiation, chemotaxis, apoptosis, and proliferation [91]. The majority of acetate is produced by anaerobic bacteria in the bacteria, while butyrate and propionate are synthesized by a particular subset of gut's bacteria through isolated molecular pathways. Butyrate is derived from carbohydrates through glycolysis and acetoacetyl-CoA pathways, whereas propionate is synthesized via either the succinate or propanediol pathway, depending on the type of sugar involved [92]. *Bacteroidetes* primarily contribute to the production of propionate in the human gut, while *Firmicutes* are primarily responsible for producing butyrate [92–94]. Akkermansia muciniphila is recognized as a prominent propionate-producing bacteria with specialization in mucin degradation [95]. The liver predominantly absorbs propionate, whereas acetate is released into peripheral tissues [96]. A recent review has examined the impact of SCFAs on human metabolism [94, 97]. Butyrate is recognized for its properties that contribute to reducing inflammation and combating cancer [94, 97] [98]. It is proposed that a declining gradient of butyrate from the intestinal lumen to the crypts regulates intestinal epithelial turnover and homeostasis. Propionate, in the liver has been observed to stimulate gluconeogenesis, while acetate and butyrate exhibit lipogenic effects [94]. SCFAs as well contribute towards modulating the immune system and inflammatory responses [94]. They impact cytokine production, such as promoting the synthesis of IL-18, an interleukin crucial for preserving and restoring epithelial integrity [98]. Butyrate and propionate act as inhibitors of histone deacetylases, thereby exerting epigenetic control over gene expression [94]. SCFAs have demonstrated the ability to influence appetite regulation and energy intake through mechanisms involving receptors [99]. Propionate exerts beneficial effects in humans by modulating β-cell function [100] and reducing reward-based eating behavior through striatal pathways [101]. Beyond SCFAs, microbial metabolites have been documented to affect intestinal barrier functions, epithelial proliferation, and the immune system [102]. Numerous pieces of evidence support the role of the gut microbiota in influencing epithelial homeostasis [103]. Sterile mice display impaired epithelial cell turnover, a condition that can be reversed upon colonization with microbiota [104]. Bacteria, such as Lactobacillus *rhamnosus* GG, have been shown to play a role in promoting cell renewal and wound healing [105] [106, 107]. Additionally, bacteria are suggested to play a role in modulating mucus properties and turnover. Studies on germ-free mice have revealed an unusually thin adherent colonic mucus layer. However, exposure to bacterial products such as peptidoglycan or lipopolysaccharide (LPS) help in the restoration of thickness of mucus layes [108]. *B. thetaiotaomicron* and *F. prausnitzii* are suggested to be involved in coordinating the production of mucus [109]. *B. thetaiotaomicron, lactobacillus* have the capacity to alter mucin glycosylation, potentially by regulating the expression of glycosyltransferases [110–112]. These functions are suggested to facilitate the colonization of other pathogens, potentially granting certain beneficial species in the GI [111].

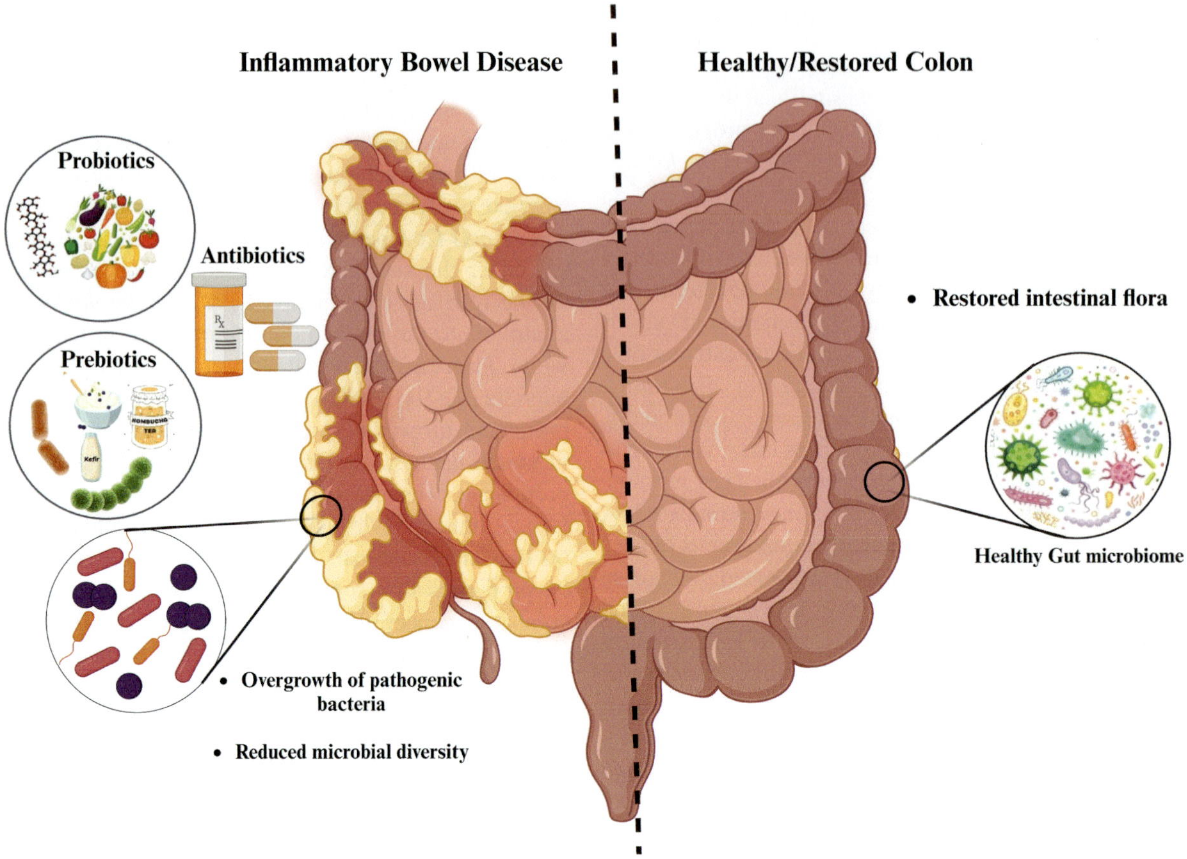

FIGURE 1.2 Gut microbiome-based therapeutics in IBD.

1.7 EFFECTS OF MEDICATIONS FOR IBD ON THE GUT MICROBIOTA

5-ASA (5-aminosalicylic acid) decreases inflammation by inhibiting NF-κB and the production of pro-inflammatory eicosanoids, while also activating PPAR-γ (peroxisome proliferator-activated receptor gamma) [113]. Nevertheless, its impact on the composition of the microbiota remains uncertain. Morgan et al reported the decrease level of *Escherichia/shigella* and increase level of Enterococcus in the 131 IBD patient by their stool examination after treatment with mesalamine in his cohort studies [43]. And the simillar study conducted by Xu et al. corroborated these findings by treating with 5-ASA in UC patients [114], [113]. The subsequent administration of 5-ASA to previously untreated patients for a minimum of 6 months confirmed these particular alterations. However, both studies were confined to a single time point. Given the typically dynamic nature of the gut microbiota in individuals with IBD, it is important to extend microbiota assessments throughout the treatment course to ascertain whether these changes persist and are specific to the treatment, or if they merely reflect variations in the level of intestinal inflammation (Figure 1.2).

Schirmer et al observed gut microbiota in 405 UC patients undergoing treatment with 5-ASA and other corticosteroids [115]. They compared baseline samples from treatment-naïve patients to those collected at week 4, categorized by treatment type and remission status. They identified 11 OTUs associated with the use of 5-ASA and 47 OTUs linked to corticosteroid treatment. Different variation in bacterial presence with remission of 5-ASA drug like 5-ASA like *Eikenella*, *fusobacterium* and OTUs etc linked to the corticosteroids response like actinomyces, which increase the remission [115]. The opposite trend was observed for the Clostridium OTU. Furthermore, species such as *Dialister*, *Eikenella*, *bifidobacterium*, and *Blautia producta* having different variance in remission and non-remission [115]. These discoveries contrast with the earlier cross-sectional analysis conducted by Morgan et al. [116], which found that Enterococcus was the sole genus affected by corticosteroid cure. This suggests that treatment response may depend on various factors, including the baseline microbiota composition. However, it is important to acknowledge the limitations of this study. Firstly, it did not explore variations in microbial composition and reduction status beyond the 4-week period. Additionally, the outcomes in the pediatric cohort may not be directly applicable to the adult population with IBD. Hence, more reproductive research is necessary to gain a more comprehensive understanding of these factors.

1.8 THERAPEUTIC INTERVENTIONS FOR MODULATING THE GUT MICROBIAL SYSTEM

The exploration of immune interactions and metabolic signaling of the gut flora and the host, and their effects on host muscle physiology, brain, gut, and liver functions, by taking antimicrobial for the prevention and therapy of the disease [117, 118]. Specifically, targeting and augmenting specific gut bacterial strains, as well as improving overall gut microbial ecology, holds great promise as a therapeutic strategy to regulate energy consumption and decrease the occurrence of metabolic disorders. Fecal transplantation emerges as an effective method to reconfigure the intestinal microbiota ecosystem following antibiotic treatment or to combat intestinal infections, such as those caused by *Clostridium difficile*. Additionally, fecal transplantation shows potential as a treatment option for IBD [119, 120]. Furthermore, investigation demonstrated that nine men diagnosed with metabolic syndrome, who experienced fecal transplantation using stool samples obtained recovered or healthy by the triglycerides fasting level [121]. Hence, fecal transplantation holds promise in the fight against obesity. However, it's important to note that the procedure is still considered experimental, and further understanding of the underlying mechanisms involved is necessary. Prebiotics are dietary components that confer health benefits by growth stimulation or bacterial activation of present in the colon. They can also be incorporated into foods like yogurt as additives. In the most of probiotic like fructo-oligosaccharide (FOS) including inulin (a long chain fructooligosaccharide), as well as galactooligosaccharides (GOS) is mostly present in milk. These compounds are metabolized inside the GI tract by beneficial bacteria or flora like fatty acid, which helps in proliferations of normal flora [122–124]. For example, in animal studies, inulin has been shown to promote the proliferation of *bifidobacteria* and potentially lower the fat mass [124]. The stimulation of *bifidobacteria* growth by prebiotics correlates with improved glucose tolerance, enhanced glucose-induced insulin secretion, and the reduction of inflammation in rodents [122, 125]. In rodents, prebiotic consumption has been linked to a decrease in hepatic, renal, and plasma lipid levels [122, 124] [124, 126]. Hence, the absorption of prebiotics may decrease lipogenic activity while enhancing lipolytic activity.

In human studies, interventions with probiotics have demonstrated a beneficial impact on glucose metabolism [127]. Numerous studies on humans have reported modest or negligible by taking probiotic for weight loss [128]. Randomized controlled trials have pinpointed surrogate markers of prebiotic therapy, including plasma peptide YY (PYY), glucagon-like peptide 1 (GLP-1), and ghrelin. These findings align with the mechanisms observed in rodent studies, providing further evidence of the potential benefits of prebiotics in metabolic health [129, 130]. However, existing evidence does not substantiate the idea that supplementing infant formula with prebiotics enhances growth and shows some adverse reactions, especially in the case of children. Investigations into the effects of prebiotics in children, adults, and the elderly show varying levels of quality and outcomes. Nevertheless, prebiotics have exhibited the capacity to alter fecal microbiota and immune function in aged persons. Furthermore, they have been observed to decrease disease markers in metabolic disease in higher body mass index (BMI) patients. Well-designed studies integrating suitable doses of probiotics or prebiotics in conjunction with controlled diets are crucial. Such investigations would provide valuable insights into individual responses to various environmental, gut microbial and genetic factors.

1.9 CONCLUSION AND FUTURE PERSPECTIVE

Inflammation serves as a fundamental element in numerous diseases, including cancer, type 2 diabetes mellitus, IBD, and cardiovascular diseases, all of which substantially contribute to morbidity and mortality globally. A wide range of experiments in both humans and animals has highlighted the crucial interplay between inflammation and gut microbiota, offering valuable treatment strategies for disease management. Presently, multiple inflammatory markers are not deemed reliable for predicting chronic inflammation. More robust biomarkers capable of indicating the initiation or advancement of various diseases could be uncovered with a deeper understanding of the interconnection between gut microbiota, GI inflammation, and disease severity.

Certain microbial species have been closely linked to exhibiting anti-inflammatory properties, notably F. prausnitzii. These microorganisms warrant additional investigation through extensive longitudinal studies, which could guide trials investigating prebiotics and their potential therapies aimed at alleviating inflammatory conditions. It is crucial to emphasize that inflammatory markers are integral components of intricate bidirectional regulatory mechanisms, involving various overlapping pathways, including infection and the maintenance of bodily homeostasis. Consequently, the dynamic functions of cytokines, chemokines, and transcription factors necessitate additional scrutiny through in-depth mechanistic investigations. Due to methodological variations found in research in this domain, there is a pressing need to establish standardized protocols. This would facilitate improved comparability between findings and expedite research progress in this field. Looking ahead, future advancements in this area will likely depend on the collaborative efforts of researchers in medicine, nutrition, immunology, epidemiology, and microbiology. This will necessitate increased interdisciplinary collaboration to attain a more comprehensive understanding of the intricate connections between the host and gut microbiota.

CONFLICT OF INTEREST

The authors declared that they have no competing or conflict of interest.

ACKNOWLEDGMENTS

The author would like to acknowledge the College of Pharmacy, Pusan National University, Busan, South Korea, and National University of Medical Sciences (NUMS), Rawalpindi, Punjab, Pakistan.

REFERENCES

1. Ambrose, N. S., Allan, R. N., Keighley, M. R. B., Burdon, D. W., Youngs, D., Barnes, P., & Lennard-Jones, J. E. (1985). Antibiotic therapy for treatment in relapse of intestinal Crohn's disease: A prospective randomized study. *Diseases of the Colon & Rectum*, *28*, 81–85. https://doi.org/10.1007/BF02552649
2. Ananthakrishnan, A. N., Luo, C., Yajnik, V., Khalili, H., Garber, J. J., Stevens, B. W., ... & Xavier, R. J. (2017). Gut microbiome function predicts response to anti-integrin biologic therapy in inflammatory bowel diseases. *Cell Host & Microbe*, *21*(5), 603–610. http://dx.doi.org/10.1016/j.chom.2017.04.010
3. Arrieta, M. C., Stiemsma, L. T., Dimitriu, P. A., Thorson, L., Russell, S., Yurist-Doutsch, S., ... & Brett Finlay, B. (2015). Early infancy microbial and metabolic alterations affect risk of childhood asthma. *Science Translational Medicine*, *7*(307), 307ra152. http://dx.doi.org/10.1126/scitranslmed.aab2271
4. Arthur, J. C., Perez-Chanona, E., Mühlbauer, M., Tomkovich, S., Uronis, J. M., Fan, T. J., ... & Jobin, C. (2012). Intestinal inflammation targets cancer-inducing activity of the microbiota. *Science*, *338*(6103), 120–123. http://dx.doi.org/10.1126/science.1224820
5. Augenlicht, L. H., Velcich, A., Klampfer, L., Huang, J., Corner, G., Aranes, M., ... & Mariadason, J. M. (2003). Application of gene expression profiling to colon cell maturation, transformation and chemoprevention. *The Journal of Nutrition*, *133*(7), 2410S–2416S.
6. Balish, E., & Warner, T. (2002). Enterococcus faecalis induces inflammatory bowel disease in interleukin-10 knockout mice. *American Journal of Pathology*, *160*(6), 2253–2257. https://doi.org/10.1016/S0002-9440(10)61172-8
7. Balkwill, F., & Mantovani, A. (2001). Inflammation and cancer: Back to Virchow?. *The Lancet*, *357*(9255), 539–545. https://doi.org/10.1016/S0140-6736(00)04046-0
8. Bergman, E. N. (1990). Energy contributions of volatile fatty acids from the gastrointestinal tract in various species. *Physiological Reviews*, *70*(2), 567–590. https://doi.org/10.1152/physrev.1990.70.2.567
9. Boulangé, C. L., Neves, A. L., Chilloux, J., Nicholson, J. K., & Dumas, M. E. (2016). Impact of the gut microbiota on inflammation, obesity, and metabolic disease. *Genome Medicine*, *8*, 1–12. https://doi.org/10.1186/s13073-016-0303-2
10. Brandtzaeg, P. (2009). Mucosal immunity: Induction, dissemination, and effector functions. *Scandinavian Journal of Immunology*, *70*(6), 505–515. https://doi.org/10.1111/j.1365-3083.2009.02319.x
11. Buffie, C. G., & Pamer, E. G. (2013). Microbiota-mediated colonization resistance against intestinal pathogens. *Nature Reviews Immunology*, *13*(11), 790–801. https://doi.org/10.1038/nri3535
12. Byrne, C. S., Chambers, E. S., Alhabeeb, H., Chhina, N., Morrison, D. J., Preston, T., ... & Frost, G. S. (2016). Increased colonic propionate reduces anticipatory reward responses in the human striatum to high-energy foods. *American Journal of Clinical Nutrition*, *104*(1), 5–14. https://doi.org/10.3945/ajcn.115.126706
13. Calder, P. C., Ahluwalia, N., Albers, R., Bosco, N., Bourdet-Sicard, R., Haller, D., ... & Zhao, J. (2013). A consideration of biomarkers to be used for evaluation of inflammation in human nutritional studies. *British Journal of Nutrition*, *109*(S1), S1–S34. https://doi.org/10.1017/S0007114512005119
14. Cani, P. D., Knauf, C., Iglesias, M. A., Drucker, D. J., Delzenne, N. M., & Burcelin, R. (2006). Improvement of glucose tolerance and hepatic insulin sensitivity by oligofructose requires a functional glucagon-like peptide 1 receptor. *Diabetes*, *55*(5), 1484–1490. https://doi.org/10.2337/db05-1360
15. Cani, P. D., Lecourt, E., Dewulf, E. M., Sohet, F. M., Pachikian, B. D., Naslain, D., ... & Delzenne, N. M. (2009). Gut microbiota fermentation of prebiotics increases satietogenic and incretin gut peptide production with consequences for appetite sensation and glucose response after a meal. *American Journal of Clinical Nutrition*, *90*(5), 1236–1243. https://doi.org/10.3945/ajcn.2009.28095
16. Cani, P. D., Neyrinck, A. M., Fava, F., Knauf, C., Burcelin, R. G., Tuohy, K. M., ... & Delzenne, N. M. (2007). Selective increases of bifidobacteria in gut microflora improve high-fat-diet-induced diabetes in mice through a mechanism associated with endotoxaemia. *Diabetologia*, *50*, 2374–2383. https://doi.org/10.1007/s00125-007-0791-0
17. Castellarin, M., Warren, R. L., Freeman, J. D., Dreolini, L., Krzywinski, M., Strauss, J., ... & Holt, R. A. (2012). Fusobacterium nucleatum infection is prevalent in human colorectal carcinoma. *Genome Research*, *22*(2), 299–306. https://doi.org/10.1101/gr.126516.111
18. Chambers, E. S., Morrison, D. J., & Frost, G. (2015). Control of appetite and energy intake by SCFA: What are the potential underlying mechanisms?. *Proceedings of the Nutrition Society*, *74*(3), 328–336. https://doi.org/10.1017/S0029665114001657
19. Chang, J. T. (2020). Pathophysiology of inflammatory bowel diseases. *New England Journal of Medicine*, *383*(27), 2652–2664. https://doi.org/10.1056/NEJMra2002697
20. Chen, H. Q., Yang, J., Zhang, M., Zhou, Y. K., Shen, T. Y., Chu, Z. X., ... & Qin, H. L. (2010). Lactobacillus plantarum ameliorates colonic epithelial barrier dysfunction by modulating the apical junctional complex and PepT1 in IL-10 knockout mice. *American Journal of Physiology-Gastrointestinal and Liver Physiology*, *299*(6), G1287–G1297. https://doi.org/10.1152/ajpgi.00196.2010
21. Chiaro, T. R., Soto, R., Zac Stephens, W., Kubinak, J. L., Petersen, C., Gogokhia, L., ... & Round, J. L. (2017). A member of the gut mycobiota modulates host purine metabolism exacerbating colitis in mice. *Science Translational Medicine*, *9*(380), eaaf9044. https://doi.org/10.1126/scitranslmed.aaf9044
22. Colman, R. J., & Rubin, D. T. (2014). Fecal microbiota transplantation as therapy for inflammatory bowel disease: A systematic review and meta-analysis. *Journal of Crohn's and Colitis*, *8*(12), 1569–1581. https://doi.org/10.1016/j.crohns.2014.08.006
23. Corrêa-Oliveira, R., Fachi, J. L., Vieira, A., Sato, F. T., & Vinolo, M. A. R. (2016). Regulation of immune cell function by short-chain fatty acids. *Clinical & Translational Immunology*, *5*(4), e73. https://doi.org/10.1038/cti.2016.17

24. D'Haens, G. R., Geboes, K., Peeters, M., Baert, F., Penninckx, F., & Rutgeerts, P. (1998). Early lesions of recurrent Crohn's disease caused by infusion of intestinal contents in excluded ileum. *Gastroenterology*, *114*(2), 262–267. https://doi.org/10.1016/S0016-5085(98)70476-7
25. Daddaoua, A., Puerta, V., Requena, P., Martínez-Férez, A., Guadix, E., de Medina, F. S., ... & Martínez-Augustin, O. (2006). Goat milk oligosaccharides Are anti-inflammatory in rats with hapten-induced colitis. *Journal of Nutrition*, *136*(3), 672–676. https://doi.org/10.1093/jn/136.3.672
26. Delzenne, N. M., & Kok, N. (1998). Effect of non-digestible fermentable carbohydrates on hepatic fatty acid metabolism. *Biochemical Society Transactions*, *26*(2), 228–231. https://doi.org/10.1042/bst0260228
27. Delzenne, N. M., & Kok, N. (2001). Effects of fructans-type prebiotics on lipid metabolism. *American Journal of Clinical Nutrition*, *73*(2), 456s–458s. https://doi.org/10.1093/ajcn/73.2.456s
28. Derrien, M., Vaughan, E. E., Plugge, C. M., & de Vos, W. M. (2004). Akkermansia muciniphila gen. nov., sp. nov., a human intestinal mucin-degrading bacterium. *International Journal of Systematic and Evolutionary Microbiology*, *54*(5), 1469–1476. https://doi.org/10.1099/ijs.0.02873-0
29. di Vito, R., Conte, C., & Traina, G. (2022). A multi-strain probiotic formulation improves intestinal barrier function by the modulation of tight and adherent junction proteins. *Cells*, *11*(16), 2617. https://doi.org/10.3390/cells11162617
30. Donaldson, G. P., Lee, S. M., & Mazmanian, S. K. (2016). Gut biogeography of the bacterial microbiota. *Nature Reviews Microbiology*, *14*(1), 20–32. https://doi.org/10.1038/nrmicro3552
31. Elson, C. O. (2002). Genes, microbes, and T cells—New therapeutic targets in Crohn's disease. *New England Journal of Medicine*, *346*(8), 614–616. https://doi.org/10.1056/NEJM200202213460812
32. Feng, Q., Chen, W. D., & Wang, Y. D. (2018). Gut microbiota: An integral moderator in health and disease. *Frontiers in Microbiology*, *9*, 312263. https://doi.org/10.3389/fmicb.2018.00151
33. Feng, Y., Weng, H., Ling, L., Zeng, T., Zhang, Y., Chen, D., & Li, H. (2019). Modulating the gut microbiota and inflammation is involved in the effect of Bupleurum polysaccharides against diabetic nephropathy in mice. *International Journal of Biological Macromolecules*, *132*, 1001–1011. https://doi.org/10.1016/j.ijbiomac.2019.03.242
34. Franzosa, E. A., Sirota-Madi, A., Avila-Pacheco, J., Fornelos, N., Haiser, H. J., Reinker, S., ... & Xavier, R. J. (2019). Gut microbiome structure and metabolic activity in inflammatory bowel disease. *Nature Microbiology*, *4*(2), 293–305. https://doi.org/10.1038/s41564-018-0306-4
35. Freitas, M., Cayuela, C., Antoine, J. M., Piller, F., Sapin, C., & Trugnan, G. (2001). A heat labile soluble factor from Bacteroides thetaiotaomicron VPI-5482 specifically increases the galactosylation pattern of HT29-MTX cells. *Cellular Microbiology*, *3*(5), 289–300. https://doi.org/10.1046/j.1462-5822.2001.00113.x
36. Furman, D., Campisi, J., Verdin, E., Carrera-Bastos, P., Targ, S., Franceschi, C., ... & Slavich, G. M. (2019). Chronic inflammation in the etiology of disease across the life span. *Nature Medicine*, *25*(12), 1822–1832. https://doi.org/10.1038/s41591-019-0675-0
37. Graziani, F., Pujol, A., Nicoletti, C., Dou, S., Maresca, M., Giardina, T., ... & Perrier, J. (2016). Ruminococcus gnavus E1 modulates mucin expression and intestinal glycosylation. *Journal of Applied Microbiology*, *120*(5), 1403–1417. https://doi.org/10.1111/jam.13095
38. Grivennikov, S. I., Greten, F. R., & Karin, M. (2010). Immunity, inflammation, and cancer. *Cell*, *140*(6), 883–899. https://doi.org/10.1016/j.cell.2010.01.025
39. Grivennikov, S. I., Wang, K., Mucida, D., Stewart, C. A., Schnabl, B., Jauch, D., ... & Karin, M. (2012). Adenoma-linked barrier defects and microbial products drive IL-23/IL-17-mediated tumour growth. *Nature*, *491*(7423), 254–258. https://doi.org/10.1038/nature11465
40. Guarner, F., & Malagelada, J. R. (2003). Gut flora in health and disease. *The Lancet*, *361*(9356), 512–519. https://doi.org/10.1016/S0140-6736(03)12489-0
41. Guinane, C. M., Tadrous, A., Fouhy, F., Ryan, C. A., Dempsey, E. M., Murphy, B., ... & Ross, R. P. (2013). Microbial composition of human appendices from patients following appendectomy. *MBio*, *4*(1), 10–1128. https://doi.org/10.1128/mbio.00366-12
42. Haberman, Y., Tickle, T. L., Dexheimer, P. J., Kim, M. O., Tang, D., Karns, R., ... & Denson, L. A. (2014). Pediatric Crohn disease patients exhibit specific ileal transcriptome and microbiome signature. *Journal of Clinical Investigation*, *124*(8), 3617–3633. https://doi.org/10.1172/JCI75436
43. Haiser, H. J., Gootenberg, D. B., Chatman, K., Sirasani, G., Balskus, E. P., & Turnbaugh, P. J. (2013). Predicting and manipulating cardiac drug inactivation by the human gut bacterium Eggerthella lenta. *Science*, *341*(6143), 295–298. https://doi.org/10.1126/science.1235872
44. Hakansson, A., & Molin, G. (2011). Gut microbiota and inflammation. *Nutrients*, *3*(6), 637–682. https://doi.org/10.3390/nu3060637
45. Hamer, H. M., Jonkers, D. M. A. E., Venema, K., Vanhoutvin, S. A. L. W., Troost, F. J., & Brummer, R. J. (2008). The role of butyrate on colonic function. *Alimentary Pharmacology & Therapeutics*, *27*(2), 104–119. https://doi.org/10.1111/j.1365-2036.2007.03562.x
46. Ivey, K. L., Hodgson, J. M., Kerr, D. A., Lewis, J. R., Thompson, P. L., & Prince, R. L. (2014). The effects of probiotic bacteria on glycaemic control in overweight men and women: A randomised controlled trial. *European Journal of Clinical Nutrition*, *68*(4), 447–452. https://doi.org/10.1038/ejcn.2013.294
47. Jakobsson, H. E., Rodríguez-Piñeiro, A. M., Schütte, A., Ermund, A., Boysen, P., Bemark, M., ... & Johansson, M. E. (2015). The composition of the gut microbiota shapes the colon mucus barrier. *EMBO Reports*, *16*(2), 164–177. https://doi.org/10.15252/embr.201439263
48. Jostins, L., Ripke, S., Weersma, R. K., Duerr, R. H., McGovern, D. P., Hui, K. Y., ... & Cho, J. H. (2012). Host–microbe interactions have shaped the genetic architecture of inflammatory bowel disease. *Nature*, *491*(7422), 119–124. https://doi.org/10.1038/nature11582
49. Kaakoush, N. O., Holmes, J., Octavia, S., Man, S. M., Zhang, L., Castaño-Rodríguez, N., ... & Mitchell, H. (2010). Detection of Helicobacteraceae in intestinal biopsies of children with Crohn's disease. *Helicobacter*, *15*(6), 549–557. https://doi.org/10.1111/j.1523-5378.2010.00792.x

50. Kaakoush, N. O., Holmes, J., Octavia, S., Man, S. M., Zhang, L., Castaño-Rodríguez, N., ... & Mitchell, H. (2010). Detection of Helicobacteraceae in intestinal biopsies of children with Crohn's disease. *Helicobacter*, *15*(6), 549–557. https://doi.org/10.1111/j.1523-5378.2010.00792.x
51. Karczewski, J., Poniedziałek, B., Adamski, Z., & Rzymski, P. (2014). The effects of the microbiota on the host immune system. *Autoimmunity*, *47*(8), 494–504. https://doi.org/10.3109/08916934.2014.938322
52. Kim, D., Zeng, M. Y., & Núñez, G. (2017). The interplay between host immune cells and gut microbiota in chronic inflammatory diseases. *Experimental & Molecular Medicine*, *49*(5), e339–e339.https://doi.org/10.1038/emm.2017.24
53. Kinoshita, M., Suzuki, Y., & Saito, Y. (2002). Butyrate reduces colonic paracellular permeability by enhancing PPARγ activation. *Biochemical and Biophysical Research Communications*, *293*(2), 827–831. https://doi.org/10.1016/S0006-291X(02)00294-2
54. Kles, K. A., & Chang, E. B. (2006). Short-chain fatty acids impact on intestinal adaptation, inflammation, carcinoma, and failure. *Gastroenterology*, *130*(2), S100–S105. https://doi.org/10.1053/j.gastro.2005.11.048
55. Knip, M., & Siljander, H. (2016). The role of the intestinal microbiota in type 1 diabetes mellitus. *Nature Reviews Endocrinology*, *12*(3), 154–167. https://doi.org/10.1038/nrendo.2015.218
56. Kostic, A. D., Gevers, D., Pedamallu, C. S., Michaud, M., Duke, F., Earl, A. M., ... & Meyerson, M. (2012). Genomic analysis identifies association of Fusobacterium with colorectal carcinoma. *Genome Research*, *22*(2), 292–298. https://doi.org/10.1101/gr.126573.111
57. Kumar, V., Abbas, A. K., Fausto, N., & Aster, J. C. (2014). *Robbins and Cotran pathologic basis of disease, professional edition e-book*. Elsevier Health Sciences. ISBN: 9780323609937.
58. Laforest-Lapointe, I., & Arrieta, M. C. (2017). Patterns of early-life gut microbial colonization during human immune development: An ecological perspective. *Frontiers in Immunology*, *8*, 276727. https://doi.org/10.3389/fimmu.2017.00788
59. Laudadio, I., Fulci, V., Palone, F., Stronati, L., Cucchiara, S., & Carissimi, C. (2018). Quantitative assessment of shotgun metagenomics and 16S rDNA amplicon sequencing in the study of human gut microbiome. *OMICS: A Journal of Integrative Biology*, *22*(4), 248–254. https://doi.org/10.1089/omi.2018.0013
60. Laval, L., Martin, R., Natividad, J. N., Chain, F., Miquel, S., De Maredsous, C. D., ... & Langella, P. (2015). Lactobacillus rhamnosus CNCM I-3690 and the commensal bacterium Faecalibacterium prausnitzii A2-165 exhibit similar protective effects to induced barrier hyper-permeability in mice. *Gut Microbes*, *6*(1), 1–9. https://doi.org/10.4161/19490976.2014.990784
61. Lepage, P., Häsler, R., Spehlmann, M. E., Rehman, A., Zvirbliene, A., Begun, A., ... & Schreiber, S. (2011). Twin study indicates loss of interaction between microbiota and mucosa of patients with ulcerative colitis. *Gastroenterology*, *141*(1), 227–236. https://doi.org/10.1053/j.gastro.2011.04.011
62. Li, Q., Wang, C., Tang, C., Li, N., & Li, J. (2012). Molecular-phylogenetic characterization of the microbiota in ulcerated and non-ulcerated regions in the patients with Crohn's disease. *PloS One*, *7*(4), e34939. https://doi.org/10.1371/journal.pone.0034939
63. Li, Y. T., Cai, H. F., Wang, Z. H., Xu, J., & Fang, J. Y. (2016). Systematic review with meta-analysis: Long-term outcomes of faecal microbiota transplantation for Clostridium difficile infection. *Alimentary Pharmacology & Therapeutics*, *43*(4), 445–457. https://doi.org/10.1111/apt.13492
64. Lin, L., & Zhang, J. (2017). Role of intestinal microbiota and metabolites on gut homeostasis and human diseases. *BMC Immunology*, *18*, 1–25. https://doi.org/10.1186/s12865-016-0187-3
65. Liu, X., Zou, Q., Zeng, B., Fang, Y., & Wei, H. (2013). Analysis of fecal Lactobacillus community structure in patients with early rheumatoid arthritis. *Current Microbiology*, *67*, 170–176. https://doi.org/10.1007/s00284-013-0338-1
66. Loftus Jr, E. V. (2004). Clinical epidemiology of inflammatory bowel disease: Incidence, prevalence, and environmental influences. *Gastroenterology*, *126*(6), 1504–1517. https://doi.org/10.1053/j.gastro.2004.01.063
67. Louis, P., & Flint, H. J. (2017). Formation of propionate and butyrate by the human colonic microbiota. *Environmental Microbiology*, *19*(1), 29–41. https://doi.org/10.1111/1462-2920.13589
68. Louis, P., Hold, G. L., & Flint, H. J. (2014). The gut microbiota, bacterial metabolites and colorectal cancer. *Nature Reviews Microbiology*, *12*(10), 661–672. https://doi.org/10.1038/nrmicro3344
69. Macfarlane, S., & Macfarlane, G. T. (2003). Regulation of short-chain fatty acid production. Proceedings of the Nutrition Society, *62*(1), 67–72. https://doi.org/10.1079/PNS2002207
70. Maier, L., Pruteanu, M., Kuhn, M., Zeller, G., Telzerow, A., Anderson, E. E., ... & Typas, A. (2018). Extensive impact of non-antibiotic drugs on human gut bacteria. *Nature*, *555*(7698), 623–628. https://doi.org/10.1038/nature25979
71. Manichanh, C., Borruel, N., Casellas, F., & Guarner, F. (2012). The gut microbiota in IBD. *Nature Reviews Gastroenterology & Hepatology*, *9*(10), 599–608. https://doi.org/10.1038/nrgastro.2012.152
72. Martinez, C., Antolin, M., Santos, J., Torrejon, A., Casellas, F., Borruel, N., ... & Malagelada, J. R. (2008). Unstable composition of the fecal microbiota in ulcerative colitis during clinical remission. *Official Journal of the American College of Gastroenterology| ACG*, *103*(3), 643–648. https://doi.org/10.1111/j.1572-0241.2007.01592.x
73. Matsuoka, K., & Kanai, T. (2015, January). The gut microbiota and inflammatory bowel disease. In *Seminars in immunopathology* (Vol. 37, pp. 47–55). Springer Berlin Heidelberg. https://doi.org/10.1007/s00281-014-0454-4
74. McCoy, A. N., Araujo-Perez, F., Azcarate-Peril, A., Yeh, J. J., Sandler, R. S., & Keku, T. O. (2013). Fusobacterium is associated with colorectal adenomas. *PloS One*, *8*(1), e53653. https://doi.org/10.1371/journal.pone.0053653
75. Meijnikman, A. S., Gerdes, V. E., Nieuwdorp, M., & Herrema, H. (2018). Evaluating causality of gut microbiota in obesity and diabetes in humans. *Endocrine Reviews, 39*(2), 133–53. https://doi.org/10.1007/s00281-014-0454-4
76. Mirsepasi-Lauridsen, H. C., Vallance, B. A., Krogfelt, K. A., & Petersen, A. M. (2019). Escherichia coli pathobionts

associated with inflammatory bowel disease. *Clinical Microbiology Reviews*, *32*(2), 10–1128. https://doi.org/10.1128/cmr.00060-18
77. Morgan, X. C., Tickle, T. L., Sokol, H., Gevers, D., Devaney, K. L., Ward, D. V., ... & Huttenhower, C. (2012). Dysfunction of the intestinal microbiome in inflammatory bowel disease and treatment. *Genome Biology*, *13*, 1–18. https://doi.org/10.1186/gb-2012-13-9-r79
78. Morrison, D. J., & Preston, T. (2016). Formation of short chain fatty acids by the gut microbiota and their impact on human metabolism. *Gut Microbes*, *7*(3), 189–200. https://doi.org/10.1080/19490976.2015.1134082
79. Mukhopadhya, I., Hansen, R., El-Omar, E. M., & Hold, G. L. (2012). IBD—what role do Proteobacteria play?. *Nature Reviews Gastroenterology & Hepatology*, *9*(4), 219–230. https://doi.org/10.1038/nrgastro.2012.14
80. Mukhopadhya, I., Thomson, J. M., Hansen, R., Berry, S. H., El-Omar, E. M., & Hold, G. L. (2011). Detection of Campylobacter concisus and other Campylobacter species in colonic biopsies from adults with ulcerative colitis. *PloS One*, *6*(6), e21490. https://doi.org/10.1371/journal.pone.0021490
81. Musso, G., Gambino, R., & Cassader, M. (2010). Obesity, diabetes, and gut microbiota: The hygiene hypothesis expanded?. *Diabetes Care*, *33*(10), 2277–2284. https://doi.org/10.2337/dc10-0556
82. Nagai, M., Obata, Y., Takahashi, D., & Hase, K. (2016). Fine-tuning of the mucosal barrier and metabolic systems using the diet-microbial metabolite axis. *International Immunopharmacology*, *37*, 79–86. https://doi.org/10.1016/j.intimp.2016.04.001
83. Nasser, J., Bergman, D. T., Fulco, C. P., Guckelberger, P., Doughty, B. R., Patwardhan, T. A., ... & Engreitz, J. M. (2021). Genome-wide enhancer maps link risk variants to disease genes. *Nature*, *593*(7858), 238–243. https://doi.org/10.1038/s41586-021-03446-x
84. Natividad, J. M., & Verdu, E. F. (2013). Modulation of intestinal barrier by intestinal microbiota: Pathological and therapeutic implications. *Pharmacological Research*, *69*(1), 42–51. https://doi.org/10.1016/j.phrs.2012.10.007
85. Nicholson, J. K., Holmes, E., Kinross, J., Burcelin, R., Gibson, G., Jia, W., & Pettersson, S. (2012). Host-gut microbiota metabolic interactions. *Science*, *336*(6086), 1262–1267. https://doi.org/10.1126/science.1223813
86. Nicholson, J. K., Holmes, E., & Wilson, I. D. (2005). Gut microorganisms, mammalian metabolism and personalized health care. *Nature Reviews Microbiology*, *3*(5), 431–438. https://doi.org/10.1038/nrmicro1152
87. Ohira, H., Tsutsui, W., & Fujioka, Y. (2017). Are short chain fatty acids in gut microbiota defensive players for inflammation and atherosclerosis?. *Journal of Atherosclerosis and Thrombosis*, *24*(7), 660–672. https://doi.org/10.5551/jat.RV17006
88. Parnell, J. A., & Reimer, R. A. (2009). Weight loss during oligofructose supplementation is associated with decreased ghrelin and increased peptide YY in overweight and obese adults. *American Journal of Clinical Nutrition*, *89*(6), 1751–1759. https://doi.org/10.3945/ajcn.2009.27465
89. Petersson, J., Schreiber, O., Hansson, G. C., Gendler, S. J., Velcich, A., Lundberg, J. O., ... & Phillipson, M. (2011). Importance and regulation of the colonic mucus barrier in a mouse model of colitis. *American Journal of Physiology-Gastrointestinal and Liver Physiology*, *300*(2), G327–G333. https://doi.org/10.1152/ajpgi.00422.2010
90. Pinero, F., Vazquez, M., Baré, P., Rohr, C., Mendizabal, M., Sciara, M., ... & Silva, M. (2019). A different gut microbiome linked to inflammation found in cirrhotic patients with and without hepatocellular carcinoma. *Annals of Hepatology*, *18*(3), 480–487. https://doi.org/10.1016/j.aohep.2018.10.003
91. Pingitore, A., Chambers, E. S., Hill, T., Maldonado, I. R., Liu, B., Bewick, G., ... & Persaud, S. J. (2017). The diet-derived short chain fatty acid propionate improves beta-cell function in humans and stimulates insulin secretion from human islets in vitro. *Diabetes, Obesity and Metabolism*, *19*(2), 257–265. https://doi.org/10.1111/dom.12811
92. Plottel, C. S., & Blaser, M. J. (2011). Microbiome and malignancy. *Cell Host & Microbe*, *10*(4), 324–335. https://doi.org/10.1016/j.chom.2011.10.003
93. Reunanen, J., Kainulainen, V., Huuskonen, L., Ottman, N., Belzer, C., Huhtinen, H., ... & Satokari, R. (2015). Akkermansia muciniphila adheres to enterocytes and strengthens the integrity of the epithelial cell layer. *Applied and Environmental Microbiology*, *81*(11), 3655–3662. https://doi.org/10.1128/AEM.04050-14
94. Rinninella, E., Raoul, P., Cintoni, M., Franceschi, F., Miggiano, G. A. D., Gasbarrini, A., & Mele, M. C. (2019). What is the healthy gut microbiota composition? A changing ecosystem across age, environment, diet, and diseases. *Microorganisms*, *7*(1), 14. https://doi.org/10.3390/microorganisms7010014
95. Ruiz, L., Delgado, S., Ruas-Madiedo, P., Sánchez, B., & Margolles, A. (2017). Bifidobacteria and their molecular communication with the immune system. *Frontiers in Microbiology*, *8*, 302400. https://doi.org/10.3389/fmicb.2017.02345
96. Rutgeerts, P. (1994). Recurrence of Crohn's disease in the neoterminal ileum after ileal resection: Is prevention therapy possible?. *The Netherlands Journal of Medicine*, *45*(2), 60–64. PMID: 7936007.
97. Sanchez, M., Darimont, C., Drapeau, V., Emady-Azar, S., Lepage, M., Rezzonico, E., ... & Tremblay, A. (2014). Effect of Lactobacillus rhamnosus CGMCC1. 3724 supplementation on weight loss and maintenance in obese men and women. *British Journal of Nutrition*, *111*(8), 1507–1519. https://doi.org/10.1017/S0007114513003875
98. Savage, N. (2020). The complex relationship between drugs and the microbiome. *Nature*, *577*(7792), S10–S10. https://doi.org/10.1038/d41586-020-00196-0
99. Schirmer, M., Denson, L., Vlamakis, H., Franzosa, E. A., Thomas, S., Gotman, N. M., ... & Xavier, R. J. (2018). Compositional and temporal changes in the gut microbiome of pediatric ulcerative colitis patients are linked to disease course. *Cell Host & Microbe*, *24*(4), 600–610.
100. Segal, J. P., Oke, S., Hold, G. L., Clark, S. K., Faiz, O. D., & Hart, A. L. (2018). Systematic review: Ileoanal pouch microbiota in health and disease. *Alimentary Pharmacology & Therapeutics*, *47*(4), 466–477. https://doi.org/10.1111/apt.14454
101. Sellon, R. K., Tonkonogy, S., Schultz, M., Dieleman, L. A., Grenther, W., Balish, E. D., ... & Sartor, R. B. (1998). Resident enteric bacteria are necessary for development of spontaneous colitis and immune system activation in interleukin-10-deficient mice. *Infection and*

Immunity, *66*(11), 5224–5231. https://doi.org/10.1128/iai.66.11.5224-5231.1998

102. Shan, Y., Lee, M., & Chang, E. B. (2022). The gut microbiome and inflammatory bowel diseases. *Annual Review of Medicine*, *73*, 455–468. https://doi.org/10.1146/annurev-med-042320-021020
103. Sivan, A., Corrales, L., Hubert, N., Williams, J. B., Aquino-Michaels, K., Earley, Z. M., ... & Gajewski, T. F. (2015). Commensal bifidobacterium promotes antitumor immunity and facilitates anti–PD-L1 efficacy. *Science*, *350*(6264), 1084–1089. https://doi.org/10.1126/science.aac4255
104. Smith, K., McCoy, K. D., & Macpherson, A. J. (2007, April). Use of axenic animals in studying the adaptation of mammals to their commensal intestinal microbiota. In *Seminars in immunology* (Vol. 19, No. 2, pp. 59–69). Academic Press. https://doi.org/10.1016/j.smim.2006.10.002
105. Sokol, H., Seksik, P., Rigottier-Gois, L., Lay, C., Lepage, P., Podglajen, I., ... & Doré, J. (2006). Specificities of the fecal microbiota in inflammatory bowel disease. *Inflammatory Bowel Diseases*, *12*(2), 106–111. https://doi.org/10.1097/01.MIB.0000200323.38139.c6
106. Strauss, J., Kaplan, G. G., Beck, P. L., Rioux, K., Panaccione, R., DeVinney, R., ... & Allen-Vercoe, E. (2011). Invasive potential of gut mucosa-derived Fusobacterium nucleatum positively correlates with IBD status of the host. *Inflammatory Bowel Diseases*, *17*(9), 1971–1978. https://doi.org/10.1002/ibd.21606
107. Swanson, P. A., Kumar, A., Samarin, S., Vijay-Kumar, M., Kundu, K., Murthy, N., ... & Neish, A. S. (2011). Enteric commensal bacteria potentiate epithelial restitution via reactive oxygen species-mediated inactivation of focal adhesion kinase phosphatases. *Proceedings of the National Academy of Sciences*, *108*(21), 8803–8808. https://doi.org/10.1073/pnas.1010042108
108. Swidsinski, A., Dörffel, Y., Loening-Baucke, V., Theissig, F., Rückert, J. C., Ismail, M., ... & Dörffel, W. V. (2011). Acute appendicitis is characterised by local invasion with Fusobacterium nucleatum/necrophorum. *Gut*, *60*(1), 34–40. https://doi.org/10.1136/gut.2009.191320
109. Turnbaugh, P. J., Backhed, F., Fulton, L., & Gordon, J. I. (2008). Marked alterations in the distal gut microbiome linked to diet-induced obesity. *Cell Host & Microbe*, *3*(4), 213. https://doi.org/10.1016/j.chom.2008.02.015
110. Turnbaugh, P. J., Ley, R. E., Mahowald, M. A., Magrini, V., Mardis, E. R., & Gordon, J. I. (2006). An obesity-associated gut microbiome with increased capacity for energy harvest. *Nature*, *444*(7122), 1027–1031. https://doi.org/10.1038/nature05414
111. Ukena, S. N., Singh, A., Dringenberg, U., Engelhardt, R., Seidler, U., Hansen, W., ... & Westendorf, A. M. (2007). Probiotic Escherichia coli Nissle 1917 inhibits leaky gut by enhancing mucosal integrity. *PloS One*, *2*(12), e1308. https://doi.org/10.1371/journal.pone.0001308
112. Ussar, S., Griffin, N. W., Bezy, O., Fujisaka, S., Vienberg, S., Softic, S., ... & Kahn, C. R. (2015). Interactions between gut microbiota, host genetics and diet modulate the predisposition to obesity and metabolic syndrome. *Cell Metabolism*, *22*(3), 516–530. http://dx.doi.org/10.1016/j.cmet.2016.02.008
113. Van Hoffen, E., Ruiter, B., Faber, J., M'Rabet, L., Knol, E. F., Stahl, B., ... & Garssen, J. (2009). A specific mixture of short-chain galacto-oligosaccharides and long-chain fructo-oligosaccharides induces a beneficial immunoglobulin profile in infants at high risk for allergy. *Allergy*, *64*(3), 484–487. https://doi.org/10.1111/j.1398-9995.2008.01765.x
114. Varyukhina, S., Freitas, M., Bardin, S., Robillard, E., Tavan, E., Sapin, C., ... & Trugnan, G. (2012). Glycan-modifying bacteria-derived soluble factors from Bacteroides thetaiotaomicron and Lactobacillus casei inhibit rotavirus infection in human intestinal cells. *Microbes and Infection*, *14*(3), 273–278. https://doi.org/10.1016/j.micinf.2011.10.007
115. Vétizou, M., Pitt, J. M., Daillère, R., Lepage, P., Waldschmitt, N., Flament, C., ... & Zitvogel, L. (2015). Anticancer immunotherapy by CTLA-4 blockade relies on the gut microbiota. *Science*, *350*(6264), 1079–1084. https://doi.org/10.1126/science.aad1329
116. Vich Vila, A., Imhann, F., Collij, V., Jankipersadsing, S. A., Gurry, T., Mujagic, Z., ... & Weersma, R. K. (2018). Gut microbiota composition and functional changes in inflammatory bowel disease and irritable bowel syndrome. *Science Translational Medicine*, *10*(472), eaap8914. https://doi.org/10.1126/scitranslmed.aap8914
117. Vulevic, J., Drakoularakou, A., Yaqoob, P., Tzortzis, G., & Gibson, G. R. (2008). Modulation of the fecal microflora profile and immune function by a novel trans-galactooligosaccharide mixture (B-GOS) in healthy elderly volunteers. *American Journal of Clinical Nutrition*, *88*(5), 1438–1446. https://doi.org/10.3945/ajcn.2008.26242
118. Wang, J., Chen, W. D., & Wang, Y. D. (2020). The relationship between gut microbiota and inflammatory diseases: The role of macrophages. *Frontiers in Microbiology*, *11*, 535016. https://doi.org/10.3389/fmicb.2020.01065
119. Weiss, G. A., & Hennet, T. (2017). Mechanisms and consequences of intestinal dysbiosis. *Cellular and Molecular Life Sciences*, *74*, 2959–2977. https://doi.org/10.1007/s00018-017-2509-x
120. Wrzosek, L., Miquel, S., Noordine, M. L., Bouet, S., Chevalier-Curt, M. J., Robert, V., ... & Thomas, M. (2013). Bacteroides thetaiotaomicron and Faecalibacterium prausnitzii influence the production of mucus glycans and the development of goblet cells in the colonic epithelium of a gnotobiotic model rodent. *BMC Biology*, *11*, 1–13. https://doi.org/10.1186/1741-7007-11-61
121. Xu, J., Chen, N., & Song, Y. (2018). 5-Aminosalicylic acid alters the gut bacterial microbiota in patients with ulcerative colitis. *Frontiers in Microbiology*, *9*, 375156. https://doi.org/10.3389/fmicb.2018.01274
122. Yang, H. E., Li, Y., Nishimura, A., Jheng, H. F., Yuliana, A., Kitano-Ohue, R., ... & Goto, T. (2017). Synthesized enone fatty acids resembling metabolites from gut microbiota suppress macrophage-mediated inflammation in adipocytes. *Molecular Nutrition & Food Research*, *61*(10), 1700064. https://doi.org/10.1002/mnfr.201700064
123. Zeng, M. Y., Inohara, N., & Nuñez, G. (2017). Mechanisms of inflammation-driven bacterial dysbiosis in the gut. *Mucosal Immunology*, *10*(1), 18–26. https://doi.org/10.1038/mi.2016.75
124. Zhang, C., Li, S., Yang, L., Huang, P., Li, W., Wang, S., ... & Zhao, L. (2013). Structural modulation of gut microbiota in

life-long calorie-restricted mice. *Nature Communications*, *4*(1), 2163. https://doi.org/10.1038/ncomms3163

125. Zhang, M., Sun, K., Yang, Y., Tso, P., & Wu, Z. (2017). Interactions between intestinal microbiota and host immune response in inflammatory bowel disease. *Frontiers in Immunology*, *8*, 279522. https://doi.org/10.3389/fimmu.2017.00942
126. Zhang, X., Shen, D., Fang, Z., Jie, Z., Qiu, X., Zhang, C., ... & Ji, L. (2013). Human gut microbiota changes reveal the progression of glucose intolerance. *PloS One*, *8*(8), e71108. https://doi.org/10.1371/journal.pone.0071108
127. Zhou, J., Li, M., Chen, Q., Li, X., Chen, L., Dong, Z., ... & Chen, Q. (2022). Programmable probiotics modulate inflammation and gut microbiota for inflammatory bowel disease treatment after effective oral delivery. *Nature Communications*, *13*(1), 3432. https://doi.org/10.1038/s41467-022-31171-0.
128. Zhu, L., Baker, S. S., Gill, C., Liu, W., Alkhouri, R., Baker, R. D., & Gill, S. R. (2013). Characterization of gut microbiomes in nonalcoholic steatohepatitis (NASH) patients: A connection between endogenous alcohol and NASH. *Hepatology*, *57*(2), 601–609. https://doi.org/10.1002/hep.26093
129. Zhu, Q., Gao, R., Wu, W., & Qin, H. (2013). The role of gut microbiota in the pathogenesis of colorectal cancer. *Tumor Biology*, *34*, 1285–1300. https://doi.org/10.1007/s13277-013-0684-4
130. Thomson, J. M., Hansen, R., Berry, S. H., Hope, M. E., Murray, G. I., Mukhopadhya, I., ... & Hold, G. L. (2011). Enterohepatic helicobacter in ulcerative colitis: Potential pathogenic entities?. *PloS One*, *6*(2), e17184. https://doi.org/10.1371/journal.pone.0017184

2 Pathophysiology and Risk Factors Associated With Gastrointestinal Inflammation

*Sanjesh Kumar[1], Mansi Singh[1], Subhajit Mandal[2], Priyanka Goswami[3], Sumel Ashique[4], and Ivan Kahwa[5]**

[1]1Department of Pharmacy, Rakshpal Bahadur College of Pharmacy, Bareilly, Uttar Pradesh, India

[2]Department of Pharmaceutical Sciences, Faculty of Science and Engineering, Dibrugarh University, Dibrugarh, Assam, India

[3]Department of Pharmacognosy, Maharashtra Educational Society's, H.K. College of Pharmacy, Mumbai, India

[4]School of Pharmaceutical Sciences, Lovely Professional University, Phagwara, Punjab, India

[5]Pharm-BioTechnology and Traditional Medicine Centre of Excellence, Mbarara University of Science and Technology, Mbarara, Uganda

2.1 INTRODUCTION

Gastrointestinal inflammation (GI) is the term used for an extensive range of the pathological ailments related to the gastrointestinal tract (GIT). The complexity of the GIT makes it vulnerable to a variety of stimuli causing inflammatory conditions (Agirman et al., 2021). The GIT is responsible for maintaining the immunological homeostasis along with the absorption of the nutrients. The disruption in this homeostasis is the plausible cause of the disorders related to GIT with Crohn's disease (CD), Inflammatory bowel disease (IBD), Ulcerative colitis disease (UCD), Gastroesophageal reflux disease (GERD), Chronic idiopathic constipation (CIC), Infectious Gastroenteritis disease (IGD), Celiac Disease, Hirschsprung's disease, achalasia and Ménétrier disease. These groups of disorders manifest with diverse pathogenesis ranging from acute to chronic inflammation in the variety of parts of the GI tract majorly (Washabau et al., 2010). Crosstalk between the risk factors associated with these disorders include diet, sedentary lifestyle, stress, environmental factors, genetics, age, microbes and certain medications like NSAIDS leading to the emergence of the inflammatory process. The inflammatory process begins with the recruitment of the cytokines, chemokines, immune cells and inflammasomes leading to invasion in the barrier protecting the GI system. The immune system may respond with multiple types of cells depending on the situation, including neutrophils, basophils, B and T-cells eosinophils lymphocytes, natural killer cells (NK cells), and macrophages. Upon activation, the inflammasome sensors synthesize the adapter protein caspase-associated speck-like protein (ACSP, also known as PYCARD) and the cysteine protease caspase 1. A solitary aggregate of inflammasome speck6 is formed by these proteins (Na et al., 2019). The complex induces caspase 1 to catalyze the proteolytic cleavage of inflammatory-inducing cytokines and the pore-forming protein gasdermin D. Cytokines are released, whereas the latter two undergo pyroptosis. The most commonly prevalent disorders underlying the inflammatory cascade is the IBS and IBD. Irritable bowel syndrome (IBS) is a common digestive disorder characterized by persistent gut pain or discomfort and changes in routine bowel movements. The cause is probably due to a combination of factors, including the commotion of the hypothalamic-pituitary-adrenal (HPA) axis, changes in neuroendocrine function, and higher sensitivity in the internal organs, leading to the characteristic symptoms (Adelman et al., 2020). Patients exhibiting typical symptoms of irritable bowel syndrome (IBS) have been found to have higher amount of immune cells in the colonic mucosa's lamina propria, compared to healthy individuals. Additionally, these patients have significantly lower amounts of oleoylethanolamide, which is a naturally occurring substance that activates PPAR-α and possesses anti-inflammatory characteristics. These findings indicate the presence of long-term, mild inflammation at the microscopic level. In some of the recent findings it is observed that bacterial overgrowth in the small intestine (BOSI) is one of the major debilitating conditions for pathogenesis of IBS. In this group of heterogenous syndrome bacteria utilize intraluminal proteins producing ammonia (Kamada et al., 2013). Overgrowing of bacteria (peptidoglycans, D-lactate, and serum amyloid A) may result in inflammation, injury to the edges of enterocyte brush borders, and greater permeability in the small intestine. An interesting case-control study has shown that under-producer genotypes of interleukin-1 receptor, antagonist genes (anti-inflammatory genes) are associated with IBS in

DOI: 10.1201/9781003493143-2

209 patients with IBS and 273 healthy subjects. Likewise, people suffering from IBS had higher amounts of IL-1 and IL-6 compared to their controls, as did the affected role with IBS-D. Higher levels of ethanol, inflammatory cytokines, chemokines, and serum endotoxins are characteristics of BOSI. The same applies to celiac disease patients with persisting symptoms that do not respond to gluten-free diets. In these types of patients with chronic forms of inflammation in the intestine mucosa, there is a high susceptibility of BOSI, making it more vulnerable to inflammation (Vaishnavi, 2013). Similarly, associated with chronic inflammation of the gut and colon, IBD is an illness of the GI tract. CD and UC are the most frequent forms of IBD. Their clinical manifestations (location, symptoms, and histology) distinguish them. CD is most common in the ileum but can develop anywhere in the gut while, UC only disturbs the colon and rectum. Ecological influences play a crucial role in decreasing the productivity of short-chain fatty acids (SCFAs) while there is an abundance of proteobacteria by the same in IBD patients (Larabi et al., 2020). Further exacerbation is detected in the mucosal lining of the intestine, which is disrupted by a variety of mechanisms. These include histological features in mucosal lining, mucin to mediated abnormality in goblet cells, (RELMb) proteins, nonfunctional Paneth cells and disruption in tight junctions (Epithelial cadherin). It has been found that the malfunctioning of Paneth cells is linked to the release of antimicrobial compounds, NOD2, and ATG16L1 gene-associated actions (Kumar et al., 2022).

In terms of the innate immune system, the IBD sheath has been shown to have fewer colonic macrophages that produce CD14, dendritic cells that don't present CX3CR1 antigens properly, and autophagy that doesn't work as well. In the IBD mucosa, there does not seem to be a balance amongst effector and regulatory T-cells (T-reg) (Saez et al., 2023a). This causes different T-cells to become activated without control, even though leukocyte movement through CAM interactions happens in normal mucosa too. A cell lineage that migrates to the intestine with inflammation. Further, soreness of the esophagus caused by repetitive reflux of acid and bile acids is known as gastroesophageal reflux disease (GERD). Therefore, GERD and BE patients are likely to develop esophagitis as the primary pathological manifestation. Recent research has indicated that mucosal invulnerable and inflammatory responses, characterized by specific cytokine and chemokine profiles, may be responsible for the range of GERD esophageal phenotypes. Reactive oxygen species (ROS) buildup, chronic inflammatory infiltration, and BE formation are strongly associated with GERD. During chronic inflammation, transcription of some genes that affect cell fate and development changes, such as BMP4, PTGS2, SHH, CDX1, CDX2, Notch, and SOX9 (Ramos & Papadakis, 2019). On the other hand, numerous pathogens such as bacteria, viruses, and parasites lead to infectious gastroenteritis and acute manifestation of GI inflammation. Bacterial pathogens like Salmonella and Campylobacter induce inflammation by invading intestinal epithelial cells, leading to the release of inflammatory mediators (Coruzzi, 2007). Viral infections, including norovirus and rotavirus, cause inflammation through direct cytopathic effects and immune system activation. Parasitic infections such as Giardia lamblia, elicit an inflammatory response characterized by infiltration of immune cells into the intestinal mucosa. Similarly, certain drugs such as nonsteroidal anti-inflammatory drug (NSAID)-induced enteropathy constitute a paradigm of GI inflammation resulting from drug-induced alterations in mucosal homeostasis. NSAIDs inhibit prostaglandin synthesis, compromising the mucosal barrier and predisposing the GI tract to injury. When pathogens cross the epithelial barrier, immune cells are activated and inflammatory cascades are set off. Toll-like receptors recognize microbial components, initiating defense mechanisms. In severe cases, the host's immune response may elicit collateral damage, contributing to the clinical manifestations of gastroenteritis (Marmon, 2021). The chapter summarizes the current research on the pathogenesis of the major disorders related to the inflammation of GIT and the risk factors associated with these conditions. Hence, the aim of this chapter is to dig deeper into the pathogenesis of GIT under inflammatory conditions and associated risk factors.

2.2 ROLE OF VIRUSES, MICROBES AND PARASITES IN PATHOPHYSIOLOGY OF GI INFLAMMATION

The GI system of the human body serves as the host for the diversity of microbes including viruses, bacteria and parasites affecting humans through the plethora of diseases of the GI system. Microbes residing in the GI structure show a pivotal role in keeping the balanced composition and homeostasis. The pathological shifts in this composition is called disruption, which leads to inflammatory cascade further causing exacerbations like gastritis, colitis, IBD and IBS. Starting with gastric microbiota disruptions where thick mucus layer, peristaltic movement and gastric pH serving as a barrier for microorganisms is breached for colonization. *Helicobacter pylori (H. pylori)* have unique adaptations to colonize the stomach. Once infected, *H. pylori* use α-carbonic anhydrase and urease to produce ammonia and HCO_3^{2-}, which mitigate the effects of low pH (Al–Mousawi et al., 2022). When the host is infected with *H. pylori* infection, inflammation results from infiltration of neutrophils and mononuclear cells in the host leading to chronic, active gastritis. The immune response to the infection generates reactive oxygen and nitrogen species. *H. pylori* defends itself via detoxifying enzymes (catalase and superoxide dismutase) and arginase. Chronic inflammation results from irregular acute responses. In the adaptive response to *H. pylori*, both proinflammatory and regulatory T-cells are involved, rather than humoral (B cell) immunity (Y. Guo et al., 2023).

In general, proinflammatory T helper 1 (TH1) and TH17 cells release cytokines that increase proinflammatory cues and activate neutrophils and macrophages as shown in Figure 2.1. Recently, it was proposed that after activating an inflammasome, *H. pylori* can regulate proinflammatory as well as regulatory T-cell responses. IL-1β induces T-box transcription factor

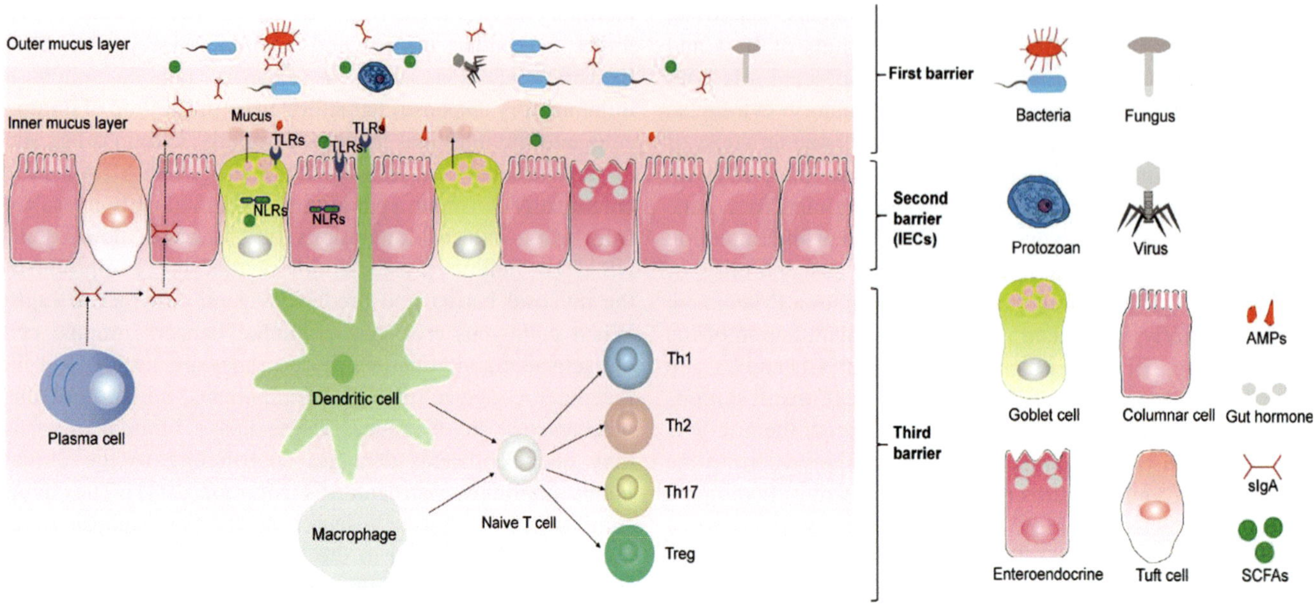

FIGURE 2.1 Bacteria, protozoans, viruses play an important role in gut dysbiosis and mucosal inflammation via modulating and hyperactivating the immune cells which further release chemical such as cytokinin, IL-10, Th1, Th2, Th17and T-reg which leads to GI inflammation. (Adapted from Yue et al., 2019. This article is an open access article distributed under the terms and conditions of the Creative Commons Attribution (CC BY) license, https://creativecommons.org/licenses/by/4.0/.)

(T-bet)-dependent TH1 and RAR-related orphan receptor γt (RORγt)-dependent TH17 cells, promoting IFN-γ and IL-17 production, while IL-18 increases FOXP3-dependent CD4+CD25+ TREG cells. As a result, the host's efforts to exterminate *H. Pylori* increase gastric immunopathology (gastritis, atrophy, metaplasia). Ultimately, *H. pylori* causes mutation rates, DNA damage through double-strand breaks, aberrant DNA methylation, and other epigenetic changes (Zhang et al., 2023). As a consequence of these alterations, chronic gastritis may progress to atrophic gastritis, and residual DNA methylation may contribute to the development of gastric cancer. Soon after the discovery of *H. pylori,* other bacteria were discovered in the stomach, including *Vellomella, Lactobacillus,* and *Clostridium.* Eventually, pH elevation resulted in colonization of other microorganisms. Based on 16s rRNA based culture-independent molecular methods, researchers recently found that five other abundant genera of microorganisms reside in the stomach. They are *Neisseria, Haemophilus, Prevotella, Streptococcus*, and *Porphyromonas.* Dysbiotic pathogens such as *Salmonella typhimurium* can alter gastrointestinal homeostasis via communication with commensal microbes (Burucoa & Axon, 2017).

2.2.1 Protozoans

Protozoans are another important and well-studied group of microorganisms that can cause GI disease. Parasites that live inside cells, like *Cryptosporidium* parvum and Toxoplasma gondii, invade in different ways than protozoa that don't live inside cells, like *Entamoeba histolytica* and *E. dispar. Cryptosporidium,* an Apicomplexan protozoan, is a prevalent source of gastrointestinal illnesses (Bhat et al., 2023). Although immunocompetent individuals commonly develop the illness in their small intestines, it can also invade other organs including the lungs, pancreas, liver, and stomach. Atypical symptoms include villus atrophy, crypt hyperplasia, chloride hypersecretion, glucose malabsorption, and lamina propria infiltration. Prostaglandin PGE2 and prostacyclin PGI2 are produced in significant amounts by the intestinal mucosa when cryptosporidiosis occurs. By altering NaCl transport, prostaglandins (PGs) influence cryptosporidiosis. Following gastrointestinal dissemination and infection of neutrophils and other phagocytic cells like dendritic cells, T. gondii can invade the lamina propria. The absence of intestinal commensal bacteria after a T. gondii infection caused a delay in the development of ileitis in wild type mice, providing evidence in favor of the theory that intestinal microbiota plays a crucial role in inflammation. Mouse models infected with Enterobacteriaceae may exhibit symptoms similar to those caused by Toxoplasma gondii (Wang et al., 2023).

Blastocystis is another genus of parasitic single-cell organisms that habitats in the human GI system. The microorganism disrupts the intestinal barrier provoking the innate immunological response, which consecutively results in atypical symptoms. The penetrability of the gut mucosal barrier signifies the participation of many immunological receptors and cells, such as TLRs and IgM/IgG/IgA antibodies, along with CD8þ T-cells, macrophages, and neutrophils (Sterling et al., 2022). Further, in patients with IBS and IBD sigmoidoscopic biopsies have demonstrated that Blastocystis is capable of disrupting the intestinal homeostasis, resulting in chronic inflammatory cascade. As trichomonad parasites, Dientamoeba fragilis also infects humans. Acute gastrointestinal disease is caused by Dientamoeba fragilis in children and adults. Eosinophilic colitis and ulcerative colitis have been reported (Lionakis et al., 2023).

One theory suggests that chronic antigenic exposure, such as in chronic carriage or infection, causes low-grade inflammation, which in turn causes irritable bowel syndrome (IBS) (Shariati et al., 2019). established immune activation by showing that patients with IBS had significantly higher IgG antibody stages to B. hominis compared to asymptomatic controls. Researchers have found that endoscopic samples taken from patients with irritable bowel syndrome show signs of inflammation and edema in the small intestine and colon. Furthermore, B. hominis infection may worsen ulcerative colitis. Inflammation, as saha, 2014 showed, decreases excreted radioactive indicators and impairs intestinal permeability (Saha, 2014).

The protozoan parasite *Entamoeba histolytica* is a nonflagellated amoeboid. About 50 million cases of amebiasis are caused by *Entamoeba histolytica* each year, typically in developing countries. In recent studies, new mechanisms have been identified that contribute to *E. histolytica*-induced tissue damage. Trogocytosis is a new form of contact-dependent killing; macrophage migration inhibitory factor, produced by parasites, causes inflammation; exploitation of host immunity to promote invasion; and (iv) gut microbiome-driven disease (Kantor et al., 2018). Furthermore, evidence for post-infectious IBS secondary to low-grade inflammation was demonstrated by Berhe et al. (2018). As a result of persistent low-grade inflammation, Giardia fecal persistence was observed in this study. A variety of protozoan parasites can also cause intestinal inflammation and disease, such as colitis that is caused by *Entamoeba histolytica*. In children with either UC or CD, cryptosporidiosis can make things worse, resulting in an increase in hospitalization. Additionally, *Cryptosporidium parvum* infection and dysbiosis in the gut (most notably altered Bacteroidetes/Firmicutes proportion associated with chronic inflammatory processes) have also been shown in animal models (Berhe et al., 2018a).

2.2.2 Virus

Viruses are the most abundant microbes integrated with the humans in a variety of body organs, in the context of GI, predominantly in the intestines. Human virome is the term used for the population of bacterial viruses (bacteriophages) both prokaryotic and eukaryotic that resides in the human body (Berhe et al., 2018b). 9/10 of all enteroviruses are phages, and animals and plants make up the remaining 10%. Human guts contain approximately 1015 phages, which outnumber gut bacteria (1014) by ten times. The structure of bacteriophages resembles other eukaryotic viruses engrossing DNA and RNA with them that can be both single stranded(ss) or double stranded(ds). Commensal bacteria exceed bacteriophages in humans' intestines by a factor of 10. Myoviridae, Podoviridae, and Siphoviridae are dsDNA viral families, while Microsophage is an example of a ssDNA virus family. These viruses make up the bulk of the core phageome. The most common virus linked with humans is a newly discovered phage called crAssphage. It makes up 90 percent of the genetic sequences found in viral metagenomes from human feces, and around 22 percent of the total genetic sequences in the metagenome. An initial investigation that established a connection between gut virome dysbiosis and the development of inflammatory bowel disease discovered that individuals with CD had elevated concentrations of *Retroviridae* members and phages that infect the bacterial orders like *Alteromonadales* and *Clostridiales*, including *Clostridium acetobutylicum*. Recently, the mucosa virus has been identified in UC patients. Zuo et al. (2019) found that UC affected subjects had significantly more Caudovirales phages, but less variety, richness, and evenness of mucosa Caudovirales than healthy controls (Jagirdhar et al., 2023). Additionally, Escherichia and Enterobacteria phage levels were notably higher in the mucosa of ulcerative colitis (UC) patients. The differences in phage populations in the mucosa between IBD affected roles and healthy controls or between diverse types of IBD (CD and UC) have also been reported. According to Spencer et al. (2022), a metagenomics study of wash samples and gut tissue revealed a high number of phages when compared to control individuals and pediatric CD patients.(Spencer et al., 2022) In addition, they determined that Bacteroides phage B10-8 and phage B124-14 constituted the highest proportions of sequences in ileum flesh samples of CD patients and controls, respectively. Virome sequencing was conducted on washed colon samples from pediatric patients with CD. The results revealed significant diversity across patients and low variability within individual patients. According to Vipond et al. (2001), CD patients have viral biomarkers, like *Clostridium acetobutylicum*, and *Retroviridae* members. According to a recent study using metagenomics analysis to examine colonic biopsies, more than 50% of the phages were linked with bacterial species identified in the sample. Conclusively, it is established that a wide array of viruses is associated with the inflammatory disorders of the GIT with diverse set of pathogenesis. More in-depth studies unrevealing the virome structures and sequences with further aid in the development of targeted therapies against these disorders (Vipond, 2001).

2.3 ROLE OF IMMUNE CELLS IN GI INFLAMMATION

Immune cells play a significant part as the checkpoint of the other cells and tissues via preventing the cells from the pathogens like viruses, bacteria and many more. The immunological response extends by producing antibodies against the invading pathogens and further neutralizing them. The immune cells work together to produce inflammation and release cytokines, chemokines, and antimicrobial agents (Saez et al., 2023b). This initiates phagocytosis, antigen appearance, and the rejuvenating of the adaptive immune system. Sometimes during the overactivation of such immune cells (dendritic cells, neutrophils, T- Cells, B Cells, etc) release proinflammatory cytokines and alert the other immune cells to the inflammation site, which leads to intensifying of

TABLE 2.1
Different GI Inflammatory Conditions With Type of Immune Response, Immunological Pathways and Their Implications

Disorder	Type of Immune Response	Involved Cell	Immune Pathway	Affected Part	Implications	References
Crohn's Disease	Chronic Inflammation	Dendritic Cells, T-Cells	Aberrant CD4+ T-Cell Activation, Cytokine Release	Any part of the digestive tract	Tissue damage, autoimmune assault on intestinal tissue	(Xu et al., 2019)
Ulcerative Colitis	Chronic Inflammation	Regulatory T-Cells (Tregs)	Dysregulation of Tregs, Mucosal Inflammation	Colon and Rectum	Persistent mucosal inflammation, immune tolerance disruption	(X. Y. Guo et al., 2020)
Irritable Bowel Syndrome (IBS)	Low-grade Inflammation	Mast Cells, T-Cells	Heterogeneous Immune Activation, Altered Cytokines	Colon and Small Intestine	Symptom severity, psychological burden	(Xiao et al., 2021)
Gastroenteritis	Acute Inflammation	Various Immune Cells	Rapid Immune Response, Immune Cell Infiltration	Entire GI Tract	Combating pathogens, swift response	(Qiu et al., 2022)
Diverticulitis	Infection-Induced	Macrophages, Neutrophils	Infection-Induced Inflammation	Colon	Inflammation linked to diverticula formation	(Piccioni et al., 2021)
Celiac Disease	Chronic Inflammation	T-Cells	Immune Response to Gluten Ingestion	Small Intestine	Structural damage, malabsorption	(Marasco et al., 2020)
Eosinophilic Esophagitis (EoE)	Localized Inflammation	Eosinophils	Eosinophil Activation, Localized Immune Response	Esophagus	Esophageal inflammation, dysphagia	(Kashyap et al., 2019)
Gastritis	Acute/Chronic Inflammation	Various Immune Cells	Mucosal Alterations, Immune Cell Recruitment	Stomach	Mucosal damage, discomfort	(Man, 2018)
Peptic Ulcer Disease	Chronic Inflammation	Neutrophils, Macrophages	Infection-Induced Inflammation, Tissue Repair	Stomach and Upper Small Intestine	Open sores, immune cell recruitment	(Rescigno, 2023)
Behçet's Disease	Autoimmune Disorder	Various Immune Cells	Autoimmune Manifestations in Digestive Tract	Across Digestive Tract	Ulcers, immune system dysregulation	(Ma et al., 2021)
Microscopic Colitis	Microscopic Inflammation	Lymphocytes	Microscopic Analysis for Inflammation Diagnosis	Colon	Diarrhea, microscopic inflammation in colonic mucosa	(Hertz et al., 2022)
Rotavirus Infection	Immune Response to Virus	IgA, IgG, T-Cells	Mucosal Immune Response, Systemic Immune Activation	Gastrointestinal Tract	Diarrhea prevention, extra-intestinal complications	(Peña-Gil et al., 2021)

inflammatory response and due to this excessive inflammation causing harm to the adjacent tissues and disrupt the function of the intestinal barrier (Table 2.1). This may result in a higher absorbency of the mucosal barrier and enable the entry of bacteria and other harmful microorganisms into the lumen (Xue & Falcon, 2019). This can lead to widespread gastrointestinal inflammation and the development of autoimmune disorders such as, Crohn's disease, IBD, Ulcerative colitis disease (UCD), GERD, Chronic idiopathic constipation (CIC), Infectious Gastroenteritis disease (IGD), Celiac Disease etc. Considering IBD as an example where immune cells are overstimulated and start attacking intestinal mucosa, which leads to the inflammation, bleeding, and diarrhea. Dysregulation of autophagy, which is involved in the lysis of intracellular pathogens, antigen processing, regulation of cell signaling, and T-cell balance, typically leads to decreased removal of pathogens (Giambra et al., 2023). This, in turn, contributes to the development of inflammatory disorders in individuals who are susceptible. The immunological response further extends to the intestinal flora, which engrosses

Peyer's patches. There are many types of immune cells in Peyer's patches, which are groups of subepithelial lymphoid follicles found in the gut. These cells include T-cells, B-cells, lymphocytes, macrophages, and dendritic cells (Tamburini et al., 2022). Peyer's patch's critical immunological role involves supervising the antigens such as bacteria and food particles. Peyer's patches possess a thin film of mucus and a type of cell called an "M-cell" that is capable of carrying things through the epithelial barrier through a process known as transcytosis. Dendritic cells also perform this similar role by extending their epithelial cells to pick up antigens. Lymphocytes then break down these antigens leading to their elimination. By sampling antigens in this way, the immune system gets activated, which causes an anti-inflammatory reaction. Peyer's patches undergo subsequent migration of dendritic cells. Regulatory T-cells, or T-reg, are immune system cells that change when they come into contact with a certain type of T-cell. Having problems with how these cells work is linked to IBD in both animals and people. Lamina propria further encounters these T-reg cells, where IL-10 is released by these cells to stop the immunological response on lamina propria itself and epithelial cells. Therefore, IL-10 serves as an important regulator of the immunological and inflammatory response. Whenever this balance is disrupted, inflammation occurs in the gut and when it persists for a longer time it leads to IBD. IBD is characterized by ulcerative colitis a particular type of IBD affecting the colon and rectum. This condition shows persistent mucosal inflammation and disturbance of regulatory T-cells (Tregs), which are vital for maintaining immune tolerance. Interactions of microorganisms present in the gut with the immune defense system are crucial, as disturbances can cause gastrointestinal inflammation (Carlson et al., 2023).

In addition to IBD the IBS also displays as a mild inflammation, whereas gastroenteritis induces a sudden immune response. Preserving the healthy gut function is crucial in preventing inflammation that is caused by infection, as highlighted by diverticulitis. People who are genetically more likely to get these diseases have problems with how their immune systems react to gut bacteria. More than 160 genetic loci have been reported to show symptoms of IBD. The microbiome in the gut is very important for keeping the immune system in balance. But people with IBD have problems with the gut bacteria, which throws off the immune system. There are numerous things that can lead to IBD, such as genetic predisposition and ecological factors like food. Microbes that are harmful to the body are fought off by the body's natural defenses (Jarmakiewicz-Czaja et al., 2023). Some of these methods are protective barriers in the mucosal lining, keeping the integrity of the epithelial cells, and making antimicrobial peptides (AMPs) (Frigerio et al., 2021). It is common for people with IBD to have their immune systems constantly activated when these defenses are weakened. As an example, the gut naturally makes some AMPs like cathelicidins and defensin, which fight off infections. IBD, on the other hand, disrupts the balance of these AMPs, which causes swelling. PRRs like TLRs and NODs look for antigens from microbes and set off inflammatory reactions.(Schoultz & Keita, 2019) There are two types of immune cells in the mucosa: innate lymphoid cells (ILCs) and mucosal-associated invariant T-cells (MAIT cells). ILCs and MAIT cells work together to protect the body. Antigen-presenting cells (APCs) are very important for starting and controlling T-cell reactions. Th-1 and Th-2 reactions are part of the adaptive immune system. Abnormal Th-1 is linked to CD, and abnormal Th-2 is linked to UC. Deepening our understanding of these complex immune system processes can lead to the development of more effective treatments for these challenging inflammatory gastrointestinal conditions. It shows that we need to do more research to understand how the immune system is modified and how IBD moves from its early stages to its later stages (Yue et al., 2019b).

2.4 ROLE OF DYSBIOSIS IN INFLAMMATORY MECHANISMS

Gut dysbiosis is characterized by altered gut microbiota composition and functional alterations in the transcriptome, proteome, or metabolome. These alterations in the gut are responsible for the inflammatory pathogenesis of a variety of GI conditions, autoimmune disorders, anaphylactic reactions, colorectal cancers, and obesity. Excessive growth of the pathological microorganisms in the gut produces inflammatory cytokines and corresponding immunological reactions. The dysbiosis of functional microbiota is further exacerbated by the host's biology, genetic susceptibility, environmental triggers, and certain drugs. There is a complex interplay between gut dysbiosis and major GI conditions like IBD, IBS, Gastritis, Peptic Ulcer Disease etc (Pickard et al., 2017).

The patient with IBD have dysbiosis generally have different types of microbes which are present in the gut. Metagenomic and metaproteomic studies have shown that the metabolic pathways of people with active IBD are very different from those of healthy people. It looks into how the enteric virome, especially Caudovirales bacteriophages, grows in people with IBD, which opens up a new area of study. The important part that gut microbiota plays in keeping the mucosal immune system in balance is being studied. This will shed light on how it affects both the humoral and cellular immune systems. Inside our bodies the microbes send out the chemicals that tell different immune system cells how to respond. It's important to note that the microbiota affects the reaction that gut dendritic cells show that helps with tolerance (Roy & Dhaneshwar, 2023). The Th17 anti-inflammatory pathway is effectively blocked by this response. Along with fighting for nutrients and making antimicrobial peptides, commensal bacteria use a number of defensive mechanisms to fight off unwanted pathogens. Some types of bacteria, like Bacteroides fragilis and Bacteroides thetaiotaomicron, help protect the walls of the small intestine. Dysbiosis disrupts immunological homeostasis by causing imbalances in the synthesis of the gut microbiota and hence leading to inflammatory illnesses and infections as shown in Figure 2.2. The complex connection between gut microbiota and gut-associated lymphoid tissue (GALT) emphasizes how important it is for macrophages and M cells of Peyer's patches

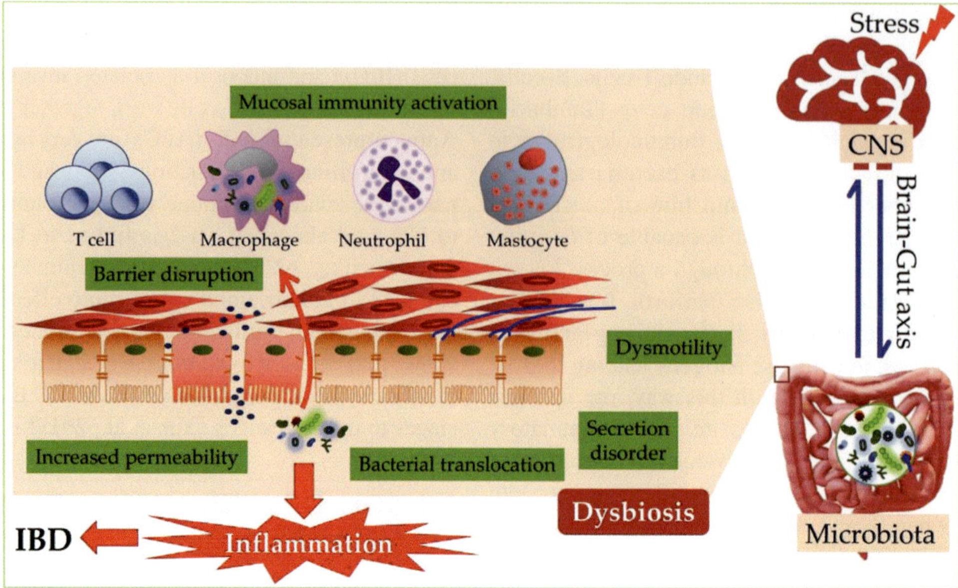

FIGURE 2.2 Various immune cells during overactivation disrupt the mucosal lining and increase the permeability which further leads to bacterial influx causing inflammation and dysbiosis. (Adapted from Talapko et al., 2022. This article is an open access article distributed under the terms and conditions of the Creative Commons Attribution (CC BY) license, https://creativecommons.org/licenses/by/4.0/)

to locate antigens and start immune responses (Zhao et al., 2022). The gut microbiota colony index has a greater impact on the functional development of myeloid cells in the gut-associated lymphoid tissue (GALT). These cells are very important for protecting against harmful microbes and preventing diseases from getting worse. Misbalances in microbe communities, known as dysbiosis, can cause leaky gut, inflammatory bowel disease (IBD), and allergic inflammation, all of which raise the risk of getting cancer. There is evidence that taking antibiotics can damage the gut-associated lymphoid tissue (GALT) and make it less effective. Complex lipopolysaccharides and other substances released by gut bacteria help keep the immune system in balance in the gut. Changes in the gut microbiome composition, especially lower amounts of Firmicutes and Bacteroidetes and higher amounts of harmful bacteria, are major causes of IBD (Lin & Zhang, 2017).

On the other hand, dysbiosis can also cause ulcerative colitis, which worsens with the presence of pathogenic microbes. The overgrowth of these harmful bacteria such as Escherichia coli, which create toxins that harm the gut lining, is one biochemical mechanism by which dysbiosis plays a role in ulcerative colitis. This damage causes an inflammatory reaction, which attracts the immune cells and secrete the cytokines that make inflammation worse. The dysbiosis can also stop the production of SCFAs, which are needed to keep the gut epithelial cells healthy and keep inflammation in check. This inhibits the production of SCFA and can cause additional inflammation and damage to the cells in the intestines, which can lead to more problems. The changes in the gut microbiota makeup can cause the dysbiosis and lead to interrupting the production of short-chain fatty acids (SCFAs). The good bacteria that are present in the gut, e.g. (Bacteroidetes and Firmicutes), make the SCFAs most of the time. When the dysbiosis occurs, the number of these good bacteria decreases and stops the production of SCFA (Khan et al., 2019). When the SCFA production is disturbed, it can weaken the gut epithelial barrier and increase the rate of inflammation, which leads to ulcerative colitis worsening. Additionally, the SCFAs help to keep the immune system in check and when the production of SCFAs drops it can increase the inflammation rate. So, raising the production of SCFA can help to lower the inflammation and keep the gut microbiome healthy, and maybe even make the disease less severe. The fermenting dietary fiber can make SCFAs easily and changing the diet to do things like eating more fiber has been shown to help the people with ulcerative colitis. The probiotics can also be added to the food to help the body make SCFAs. Probiotics have also been shown to improve the function of the gut wall and lower the gut's absorptivity. This lowers the inflammation even more and makes the gut healthier generally. So, raising the amount of fiber to eat and taking probiotics may help to deal with ulcerative colitis easily (Marasco et al., 2020).

2.5 VARIOUS MOLECULAR PATHWAYS IN GI INFLAMMATION

2.5.1 Molecular Interactions of Hypoxia-Inducible Factor-1α in IBD

The role of Hypoxia-Inducible Factor (HIF)-1α in the pathogenesis of IBD has been firmly established, influencing both intrinsic and adaptative enteric immunity (Cummins, Crean, & infection, 2017). The actuation of HIF-1α is pivotal in the polarization of Th17 cells, as it upregulates the expression of RORγt throughout the activation of STAT3 as well as suppresses FoxP3 (Islam, Won, Khan, Mannie, & Singh,

2021). Th17 cells, together with Th1 cells, are crucial in the development of Crohn's disease (CD), releasing inflammatory cytokines for example Interleukin-17, Interferon-γ, Tumor necrosis factor-α, and TNF-β (Bourgonje, von Martels, de Vos, Faber, & Dijkstra, 2018). These cytokines, in turn, induce inflammation by stimulating macrophages to produce Interleukin-1, Interleukin-6, Interleukin-8, Interleukin-12, and Interleukin-18. In the context of Th17 polarization, the activation of HIF-1α becomes imperative for maintaining the ability to regulate inflammation. The regulatory activity of effector T-cells is typically governed by Tregs, a suppressive subset of Helper T-cells crucial for immune homeostasis in the gut (Jacobse, Li, Rings, Samsom, & Goettel, 2021). However, in IBD, Mucosal tissue regulatory T-cells may exhibit resistance or reduced sensitivity to Tregs-mediated inhibition. This resistance may impede the anti-inflammatory effects of Tregs, particularly their ability to suppress subpopulations of effector T-cells (Eggenhuizen, Ng, & Ooi, 2020). The hypothesis is posited that the HIF-1α degradation of Tregs development could exacerbate the progress of IBD (Yan et al, 2020). Additionally, HIF-1α activity raises glucose metabolism and glycolysis, further encouraging proinflammatory cytokines in the context of IBD (Islam et al., 2021).

On an alternative note, HIF-1α plays a pivotal role in activating CD8+ cytotoxic T-cells, leading to heightened antiviral and antineoplastic activity (Liikanen et al., 2021). This activation is accompanied by the locution of key co-stimulatory & co-inhibitory molecules such as CD137, CTLA-4, and GITR proteins (Dvornikova, Platonova, & Bystrova, 2023). Moreover, HIF-1α facilitates the operation of the intestinal epithelial lining in the gut-tube, making it a potential target for treating IBD related with epithelial barrier breakdown (Bourgonje et al., 2023). In this context, HIF-1α-dependent Interlukin-33 in intestinal epithelial cells (IECs) works as a regulatory cytokine in aroused mucosa, contributing to the statute of intestinal inflammation and the upkeep of mucosal homeostasis (H. Lu et al., 2021). During gastrointestinal inflammation, TNF caused by mucosal immune cells induces the expression of HIF-1α in IECs by activating NF-kB (J. Li et al., 2023). Subsequently, HIF-1α forms a heterodimer with HIF-1β and binds to the IL-33 promoter, resulting in the overexpression of IL-33 in IECs, crucial for sustaining mucosal homeostasis (Dvornikova et al., 2023). Nevertheless, a comprehensive understanding of the roles of HIF-1α and IL-33 in regulating mucosal barrier integrity and immune response in IBD necessitates further research. In an alternative setting, myeloid cell function in colitis caused by dextran sulfate sodium (DSS) has been linked to HIF-1α (M.-y. Li et al., 2020). Myeloid cells, particularly drafted to inflammatory foci, serve as the body's front line of defense against infection (Stegelmeier et al., 2019). Hypoxia-induced HIF-1α in myeloid cells leads to elevated IL-17 secretion, causing serious colon inflammation in mice with heightened proinflammatory cytokine levels (Islam et al., 2021). Conversely, mice with myeloid HIF-1α knockout exhibited moderate colitis, showing reduced invasion of immune cells. This outcome is attributed to the lower motility and invasive nature of myeloid cells, particularly macrophages, deficient HIF-1α, resulting in decreased Th17 levels and reduced expression of TNF-α and IFN-γ (Schützhold et al., 2022). These findings underscore the hostile role of HIF-1α in immune cells during colon inflammation, indicating a potential role for these myeloid cells in managing IBD. However, conflicting evidence suggests that HIF-1α knockout in dendritic cells (DCs) consequences in acute intestinal inflammation with increased proinflammatory cytokines and mucin production (Parada Venegas et al., 2019). HIF-1α in DCs is deemed necessary for inducing sufficient regulatory T-cells (Tregs) to regulate bowel inflammation, as FoxP3 gene expression by Tregs, a transcription factor regulated by HIF-1α (Kerber et al., 2020).

The divergent results obtained from these studies highlight the context-specific nature of the inflammatory lesion and may be attributed to the distinct immune responses involved. These findings, derived from a mouse colitis model, underscore the necessity for comprehensive molecular genetic analysis, including studies involving a wide number of samples of sufferers with IBD (Limon et al., 2019). Furthermore, it is proposed that the increased expression and activating the potassium channel K2P5.1 (TASK-2, also known as KCNK5) in CD4+ T-cells in the spleen of mice with DSS-induced colitis may be facilitated by inflammation-associated hypoxia (Endo et al., 2019). HIF-1α in mouse spleen CD4+ T-cells is activated by hypoxia in inflammatory colon regions, which facilitates the conversion of these cells into Tregs (Y. Zhang et al., 2022). These Tregs produce the anti-inflammatory cytokine IL-10, which is crucial for protecting against IBD. The K2P5.1 channel in spleen CD4+ T-cells is identified as a HIF-1α target gene, suggesting its significance in Tregs differentiation and IL-10 induction (Endo et al., 2019).

2.5.2 Molecular Interactions of Hypoxia-Inducible Factor-2A in Inflammatory Bowel Disease

Various studies affirm the inflammatory-inducing role of HIF-2α in the context of IBD within the gut. Evidence suggests that HIF-2α activation contributes to neutrophilic inflammation and augments the recruitment of neutrophils, thereby fostering the development of an inflammatory microenvironment in the colon (Bourgonje et al., 2023). Furthermore, studies using HIF-2-deficient inflammatory neutrophils in mice showed increased apoptosis, which decreased tissue damage and neutrophil-driven inflammation (Dvornikova et al., 2023). These findings suggest a specific involvement of HIF-2α in sustaining neutrophilic inflammation, implying the potential for HIF-2α inhibition as a therapeutic strategy in chronic inflammatory diseases (Malkov, Lee, & Taylor, 2021). Another study indicates that the overexpression of the HIF-2 PAS1 endothelial domain protein (EPAS1) results in spontaneous colitis or increased susceptibility to dextran sulfate sodium (DSS)-induced colitis in mice with IEC-specific knockout of Hif-2α (Xie et al., 2014). Notably, EPAS1 positively regulates the production of TNF-α in the intestinal epithelium, and inhibiting TNF-α reduces inflammation induced by hypoxia in the gut (Xue et al., 2013).

Caveolin-1 (CAV1), a multifunctional protein crucial for occludin metabolism that preserves tight intersection integrity in intestinal epithelial cells (IECs), demonstrates increased expression when HIF-2α is overexpressed in IECs (Van Welden, Selfridge, Hindryckx, & Hepatology, 2017). Over expression Cav1 leads to enhanced endocytosis and increased intestinal permeability. HIF-2α-mediated signaling amplifies Cav1 expression, disrupts IEC tight junctions, and heightens the permeability of barriers (Xie et al., 2014). This potential regulative mechanism for initiating inflammation in hypoxic foci in IBD connected to CAV1 requires confirmation in an animal colitis model, as the study was conducted in vivo on the HCT116 cell line (Triner et al., 2018).

In the context of macrophages, HIF-2α knockout encourages further expression of IL-6, a key element in subsequent inflammation development in the colon (Kerber et al., 2020). HIF-2 knockout also triggers the release of proinflammatory cytokines such as Interleukin-23, Interleukin-17, and IFN-γ, accompanied by a crucial rise in neutrophil recruit facilitated by elevated levels of CXCL1 and T-cells (Castillo-Rodríguez, Trejo-Solís, Cabrera-Cano, Gómez-Manzo, & Dávila-Borja, 2022). While these factors may contribute to colitis progression, the concurrent induction of anti-inflammatory factors such as IL-10, TGF-β, and Tregs can attenuate this process. Interestingly, HIF-1α knockout in the same study improves acute DSS-colitis, highlighting the distinct and overlapping functions of HIF-1α and HIF-2α in IBD (Robrahn, Jiao, & Cramer, 2020). This dichotomy suggests that uncontrolled HIF-2 activation, composite with declined HIF-1 expression, might lead to chronic inflammation and damage in the case of chronic colitis (Solinas, Germano, Mantovani, & Allavena, 2009). However, it is crucial to note that the existing research may be inadequate to develop a wide-ranging understanding of the specific processes governing the expression or activation of HIF-2 in the gut in IBD. Further studies, especially in animal colitis models, are warranted to elucidate these mechanisms.

2.5.3 Molecular Pathways of PPARΓ in GI Inflammation

The expression of peroxisome proliferator-activated receptor γ in the intestinal epithelial cells appears to be intricately associated with the composition of the gastric microbiota (Oh, Visvalingam, & Wahli, 2019). Specifically, the production of butyrate by the gut microbiota activates the PPARγ pathway in colonic-epithelial cells, leading to the beta-oxidation of energy metabolites in these cells (Fagundes et al., 2023). It reduces the operation of respiring electron receptors for gut microbiota, potentially pathogenic ones. Moreover, colonocytes transfected with TLR4 have been shown to deter the aberrant growth of possibly harmful bacterium (Hayashi & Nakase, 2022). The TLR4 signaling pathway upregulates the expression of PPARγ and a PPARγ-dependent reporter in TLR4-transfected cancer coli-2 (Caco-2) cells in a manner that is dependent on the inhibitor of nuclear factor kappa-light-chain-enhancer of activated B-cells (Iκβ) (Hayashi & Nakase, 2022). Similarly, the absence of functional TLR4 in mice results in declined PPARγ expression in the colon (P. Lu et al., 2018). Notably, mucosal biopsies from patients with active ulcerative colitis (UC) exhibit reduced PPARγ mRNA expression, which is adversely associated with the endoscopic intensity of the disease (Da Silva et al., 2018). Following GI inflammation, monocytes are drawn into the inflamed colon, where they mature into cytogenic macrophages that express TNF-alpha and INOS have been depicted in Figure 2.3.

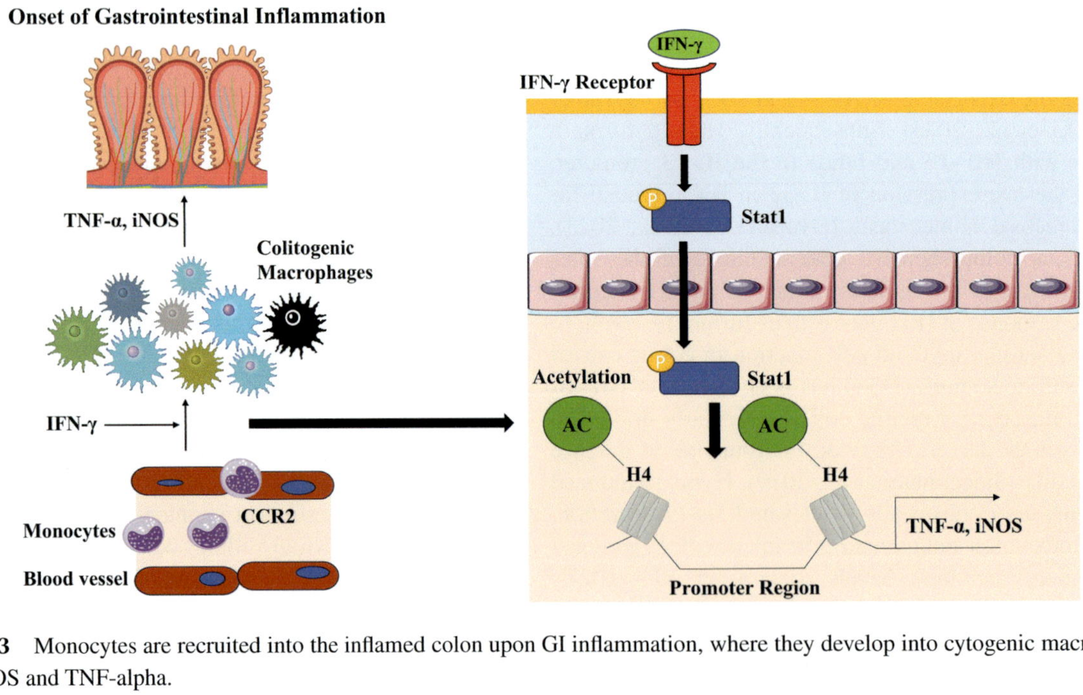

FIGURE 2.3 Monocytes are recruited into the inflamed colon upon GI inflammation, where they develop into cytogenic macrophages that express INOS and TNF-alpha.

While no studies have explored the relationship between PPARγ expression and microbiota arrangement in the intestines of IBD subjects, disruptions in intestinal flora can lead to improper PPARγ signaling responses in intestinal epithelium (Caioni et al., 2021). This, in turn, may contribute to the proliferation of pathogenic gut bacteria and exacerbate UC. Additionally, PPARγ is implicated in the response to chemical stimuli. Mice with a targeted disruption of the PPARγ gene in intestinal epithelial cells, induced using the villin-Cre transgene and floxed PPARγ allele, exhibit increased expressions of interleukin 6 (IL-6), IL-1β, and TGFα mRNAs in their colons when processed with dextran sodium sulfate (DSS), compared to control mice. The disruption of the PPARγ gene in intestinal epithelium was succeeded by reduced expression of PPARγ target genes, and these mice revealed the enhanced susceptibility to DSS-caused colitis with greater mRNA expression of interleukin 1β and tumor necrosis factor. On the other hand, the PPARγ ligand rosiglitazone declined the intensity of experimental colitis and repressed proinflammatory cytokines, indicating the anti-inflammatory role of PPARγ (Hayashi & Nakase, 2022). Interestingly, in a DSS-induced colitis mouse model, administration of PPARγ full agonists such as pioglitazone or rosiglitazone can ameliorate intestinal inflammation (Da Silva et al., 2018). The PPARγ activator GED-0507-34 Levo (GED) has been shown to suppress the expression of key fibrosis markers, α-smooth muscle actin (α-SMA) and collagen I-II, by impeding the transforming growth factor-β/Smad pathway in both the DSS-induced mouse model and human intestinal fibroblasts GED repressed transforming growth factor-β-induced activation of both fibroblast and intestinal epithelium cell lines, by fixing mRNA expression of alpha-SMA and fibronectin, and restores the transforming growth factor-β-induced loss of intestinal epithelial-cell markers. Furthermore GED treatment lower the transforming growth factor-β and ACTA1 expression in primary human enteric fibroblasts (Speca et al., 2016). Consequently, PPARγ agonists emerge as potential therapeutic purposes capable of suppressing inflammation and inhibiting inflammation-related fibrosis. Conversely, it is intriguing to observe that peroxisome proliferator-activated receptor gamma full agonists possess the capacity to prompt the distinction of fibroblasts into adipose tissues (Santos, Neves, & Amato, 2015). Consequently, the utilization of PPARγ full agonists may lead to deposition of submucosal fat (SFD) in small and large bowels through adipose tissue differentiation. SFD manifests as a status where a low-attenuation inner ring encircling the lumen of intestines is bordered by a concentric, high-attenuation outer ring—commonly known as the "halo sign" in computed tomography (CT) imaging (Hayashi & Nakase, 2022). This phenomenon turned out to be entangled in the refractoriness of Crohn's disease (CD). Though, SFD is not exclusively precise to CD, and its relation with creeping fat remains vague. Despite this, adipocytes release proinflammatory and fibrotic cytokines, necessitating caution in PPARγ-focused therapy to prevent impelling adipocyte differentiation (Souza-Tavares et al., 2023). Using a partial agonist in PPARγ-targeted treatment could be desirable in this context.

While PPARγ-targeted treatment have been broadly explored for lifestyle-influenced illnesses like diabetes and non-alcoholic fatty liver disease, its application in humans has been hindered by issues related to metabolic dysregulation and carcinogenicity (Lange, Graf, Caussy, & Dufour, 2022). Moreover, the intricate mechanisms underlying the action of PPARγ agonists, especially the organ-specific sensitivity, strength, and temporal modifications in retort to ligands, persist unclear. Nonetheless, for treating fibrosis in CD, mitigating adverse effects may be achieved by adopting local drug administration rather than employing systemic and long-term drug therapy, as is common in the treatment of lifestyle-influenced diseases (Santacroce, Lenti, & Di Sabatino, 2022). The extensive repertoire of PPARγ agonists already identified through computer and virtual high-throughput screening attests to advantageous in developing PPARγ-based treatments (Hayashi & Nakase, 2022). Furthermore, the incorporation of newly developed high-throughput screening techniques involving an organoid model of the intestine and data acquisition of earlier screened PPARγ ligands could expedite experimentation in this field (Hartwig et al., 2021).

2.5.4 TLR4-Mediated Intestinal Inflammation

The recognition of DAMPs, PAMPs and endogenous TLR is crucial for immunity and host-pathogenic correlations in the gastrointestinal tract (Yue et al., 2019). Any disruption in this interplay may provide the pathogenic process of inflammation through the gastrointestinal tract. Substantial evidence also links TLR4 to GIT inflammation such as ulcerative colitis (UC), Crohn's disease(CD), and intestinal mucositis (IM) (Tam, Coller, Hughes, Prestidge, & Bowen, 2021). Notably, clients with UC and CD exhibit a serious enhance in Messenger RNA expression of toll-like receptor-4 and protein expression in the colonic mucosa compared to healthy controls (Cari et al., 2023). However, this increase could be attributed to the augmented influx of TLR4-expressed effector cells of innate immunity. Furthermore, TLR4 polymorphism has been linked together with the progress of UC and CD, with higher allelic frequencies of the TLR4 Asp299Gly polymorphism detected in patients with these conditions (Franchimont et al., 2004).

In vivo assays of acute gastrointestinal inflammation reveal a strong upregulation of TLR4 expression. TLR4 knockout mice in induced colitis models were found to be sheltered from colitis or colon carcinogenesis by resisting the downstream signaling pathways that cause colitis (L. Wang et al., 2016). For instance, TLR4 knockout mice in acute colitis showed reduced Cyclooxygenase-2 expression, prostaglandin production by Cyclooxygenase-1, nuclear factor NF-κB pathway signaling, and a substantial decline in cells of acute inflammation such as, monocytes, lymphocytes which leads to a substantial reduction in acute inflammation in the gastrointestinal tract (Ahmedy, Ibrahim, Salem, & Kandil, 2020). Another study using a DSS-induced mouse model discovered a raise in pathogenic E. coli and a decline in beneficial gastrointestinal microbiomes (*Bifidobacterium* spp. and *Lactobacillus* spp.) linked with an raise in toll-like receptor 4 expression in the colonic tissues (N. Wang et al., 2021). Conversely, TLR4 knockout mice in the same DSS-induced model displayed a tenfold enhance in pathogenic E. coli, along with decreased disease activity index and histopathological assessment (N. Wang et al., 2021). Inhibiting TLR4 has been shown to have a protective effect against intestinal inflammation. Research utilizing TLR4 antagonists such as paeoniflorin, monoclonal antibodies, and CRX-526 demonstrated attenuation of DSS-caused GIT inflammation (Tam et al., 2021), resulting in a crucial decrease in disease activity and histopathological scoring. Though, conflicting results were observed in chronic intestinal inflammation, where blocking TLR4 did not exhibit a protective effect on clinical symptoms and histology scores. This discrepancy may be attributed to the lower engagement of innate immunity cells in chronic inflammation likened to acute inflammation.

2.5.5 Th17 Cell Immune Response in Gastrointestinal Inflammation

Naïve T-cells undergo differentiation into Th17 cells upon stimulates with TGF-β and IL-6, leading to the production of different cytokines involving IL-17A, IL-17F, IL-21, IL-22, IL-26, and TNF-α (Maddur, Miossec, Kaveri, & Bayry, 2012). IL-17 stimulation triggers the SEF/IL-17R (SEFIR) domain, nearly affiliated to the Toll-interleukin receptor domain of the IL-17R receptor, activating Nuclear factor-κB and AP-1 signaling through NF-κB activator 1 (ACT-1) and TNFR-associated factor 6 (TRAF-6) (Tsai, Velichko, Hung, & Wu, 2013). This activation induces the discharge of proinflammatory cytokines, MMPs, and antimicrobial peptides, playing a preventive role in opposing extracellular parasitic and bacterial infections. In IBD, IL-17, IL-21, and IL-22 are overexpressed in the colonic mucosal tissue of patients with ulcerative colitis (UC), while IL-17, IL-21, and IL-22 are overexpressed in the colonic mucosa of patients with Crohn's disease (CD) (Allocca et al., 2018). Th17-related cytokines play a role in the etiology of both UC and CD. In CD, IL-17+ CD3+ T-cells and CD68+ cells are dispersed in the submucosa and muscularis propria, with some of these cells producing IFN-γ (Hayashi & Nakase, 2022; Monteleone, Pallone, & Monteleone, 2009). Genome-wide association studies have identified associations between susceptibility to CD and genes involved in Th17 differentiation, including IL23R, IL12B, JAK2, STAT3, CCR6, and TNFF15 (Hayashi & Nakase, 2022).

CD-affected patients exhibit elevated Interleukin-17A expression in stenosed intestines compared to non-stenotic areas, and both IL-17A and HSP47 expressions are increased in the colons of those with active CD (Bamias, Pizarro, & Cominelli, 2022). IL-17A promotes HSP47 and collagen I expression in myofibroblasts of the intestinal lamina, and its knockdown inhibits collagen I production (Tsai et al., 2013). IL-17A therapy induces epithelial-mesenchymal transition (EMT) in intestinal cells, highlighting its role in fibrosis (Bamias et al., 2022). IL-21 enhances the Th1 response, stimulating the fibroblasts of the intestinal lamina to secrete MMPs in response to CD (Yan et al., 2020). In a TNBS-induced colitis mice model, both Th1 and Th17 immune responses are involved. Administering an anti-IL-17 antibody in mice with chronic colitis reduces fibrosis-related cytokines and inflammatory cytokines, resulting in suppressed fibrosis (Yan et al., 2020). However, anti-IL-17A antibodies (secukinumab) and anti-IL-17-receptor antibodies (brodalumab) have shown limited efficiency in treating CD. Therefore, targeting IL-17 downstream pathways, including HSP47, and additional IL-17-based cytokines such as IL-21, may be more effective in preventing intestinal fibrosis than direct IL-17 inhibition.

2.5.6 Plasminogen Activator Inhibitor-1 in Gastrointestinal Inflammation

Plasminogen activator inhibitor-1 (PAI-1), a protein weighing almost 42,700 Da, is primarily synthesized and excreted by vascular endothelium and hepatocytes, with additional contributions from adipocytes and certain other cells (X. Wang et al., 2023). Functionally, PAI-1 acts as an inhibitor, intricately forming an irreversible 1:1 bond with tissue plasminogen activator (tPA), therefore rendering it inactive (Jójárt et al., 2024). Clinically, blood levels of PAI-1 aid in understanding the pathogenic process of disseminated intravascular coagulation (DIC), a disorder affecting the coagulation-fibrinolysis system (Koutroubakis et al., 2008). Additionally, PAI-1 expression tends to increase with age, and its production is induced by raised grades of TNF-α, IL-6, and TGF-β (X. Wang et al., 2023). Through its inhibitory action on tPA, PAI-1 suppresses the conversion of plasminogen into plasmin, leading to a decrease in matrix metalloproteinases (MMPs) and subsequent suppression of tissue fiber degradation (Chandler et al., 1997). Notably, PAI-1 is a primary downstream target of TGF-β signaling, with its transcription immediately regulated by Smad3 (Jójárt et al., 2024). In patients with IBD and murine colitis models, PAI-1 expression is prominent in inactive lesions, highlighting its crucial role in regulating intestinal inflammation (Kaiko et al., 2019). In the terminal ileum's intestinal mucosa of active CD patients, elevated levels of TGF-β and PAI-1 exhibit a positive correlation. Mice with TNBS-induced intestinal fibrosis also show increased PAI-1 levels, and administration of TM5275, which disrupts PAI-1/tPA complex formation, results in elevated MMP9 expression, contributing to the amelioration of fibrosis (Hayashi & Nakase, 2022).

2.6 AMPLIFYING FACTORS: UNDERSTANDING RISK FACTORS

Persistent gastrointestinal inflammation stands out as a prevalent form of inflammatory processes impacting humans. It exhibits diversity in its causes, mechanisms, and clinical presentations. Notably, commonalities in the characteristics of chronic inflammation across various locations within the gastrointestinal tract offer a unified framework for exploring these complex phenomena. The development of gastrointestinal (GI) inflammation is widely acknowledged to outcome from a fusion of genetic and environmental factors (Loddo & Romano, 2015). GWAS have detected over 100 genes linked with a greater susceptibility to GI inflammation (Khor, Gardet, & Xavier, 2011). However, the prevalence of GI inflammation observed in both developed and developing countries cannot be fully explained by genetic factors alone. The recognition of GI inflammation dates to the Industrial Revolution in European countries, and its incidence and prevalence have notably risen throughout the 20th century.

2.6.1 Smoking

Smoking has been recognized as a factor influencing gastrointestinal inflammation, particularly in its association with UC and CD (Berkowitz et al., 2018). Researchers have observed a reduced cigarette smoking frequency in UC sufferers compared to healthy controls. A meta-evaluation by Mahid et al revealed that smoking doubles the chance of developing CD (Mahid, Minor, Soto, Hornung, & Galandiuk, 2006). A recent study in Switzerland highlighted a notably high proportion of CD patients who smoke, particularly among females, surpassing the prevalence in the general population (Biedermann et al., 2015). Interestingly, smoking exhibits a protective effect for UC but serves as a threat for CD. For individuals with CD, discontinuing smoking becomes a pivotal therapeutic approach for managing IBD.

In contrast, smoking emerges to reduce the chance of UC, with most of UC patients being non-smokers or former smokers (Berkowitz et al., 2018). A cohort study by park et al indicated that, despite discontinuing smoking, the chance of UC still enhanced within 2–5 years and continued to rise for almost two decades (Park et al., 2019). However, varying conclusions exist across different cohorts regarding the influence of smoking on UC and CD. Cosnes et al reported that current smoking was associated with a late age of UC onset and reduced the likelihood of requiring immunosuppressive effect in men but not women (Cosnes et al., 2004). Conversely, smoking was linked to a juvenile age at CD onset and an increased necessity for immunosuppression in females but not males.

2.6.2 Environmental Hygiene and Risk of GI Inflammation

According to the hygienic hypothesis, early life improvements in cleanliness and reduced exposure to intestinal bacteria may cause the body to develop incorrect immune responses later in life. Numerous variables were studied as indicators of early environmental exposures, including sibling relationships, family extent, urban nurturing, birth order, and pet exposure (Sandys, Te Velde, & Sciences, 2022). Living with more siblings during early life, providing increased exposure to enteric organisms, is able to diminish the risk of developing IBD in later life (Ng et al., 2013). Sun et al establish that IBD sufferers were elevated with a larger number of older siblings in comparison with controls, and the number of elderly siblings was connected with a greater chance for UC (Sun et al., 2023). Diminished birth rank raises the chance for both UC and CD. Disparities in environmental hazards and lifestyles within urban and rural regions could justify the greater impact of IBD in urban settings (Z. Zhang et al., 2022). UC was moreover prevalent in internal municipalities, while CD was more common in urban and shore zones (Day, Gunn, & Burger, 2021). The place of habitation may as well impact the scientific studies and phenotype of IBD, with patients residing in the seashore being more likely to grow extensive UC, and ileocolonic CD, and requiring immune-suppressing medication. However, a report found no association between the urban environment and IBD (Song et al., 2019).

2.6.3 Microorganisms

Various microorganisms have been identified as potential contributors to GI inflammation, with some candidates implicated in the progress of IBD. The consideration of an infectious origin in IBD dates to Dalziel et al's description of CD in 1913 (Mulder, Noble, Justinich, Duffin, & Colitis, 2014). Drawing parallels between CD and Johne's disease in cattle, triggered by Mycobacterium Avium Paratuberculosis (MAP), sparked the notion of an infectious etiology (Mallikarjunappa et al., 2021). Meta-analysis by Lamousé in case-control studies stated a positive connection within MAP, which was identified by ELISA or polymerase chain reaction and CD (Lamousé-Smith et al., 2013). However, the conclusive link between MAP and CD remains elusive. Recent research has increasingly focused on the function of a particular kind of Escherichia coli known as adherent-invasive E. coli (AIEC) (Chervy, Barnich, & Denizot, 2020). The initial study on Escherichia coli's involvement in IBD revealed that microorganisms isolated from CD patients exhibited greater adherence to human cells than those from control sufferers, and formerly unacknowledged intrusive E. coli had been present in Crohn's mucousal tissues (Negroni et al., 2012). Elevated levels of E. coli antigens and E. coli antibody titers have been detected in the bloodstream or deleted specimens of CD patients (Buisson et al., 2021). A study by Glasser et al demonstrated that AIEC could exist and reproduce in macrophages without inducement of a host cell response, exciting the contaminated cells to release TNF-α (Glasser et al., 2001). In recent years, AIEC contamination was found to stimulate microRNAs, reducing the expression of proteins necessary for autophagy and the autophagy respond in gastrointestinal epithelium. In ileal samples from CD patients, these microRNAs were increased, while expressions of ATG5 and ATG16L1 diminished (Bringer, Billard, Glasser, Colombel, & Darfeuille-Michaud, 2012).

2.6.4 Appendectomy

The impact of appendectomy on IBD exhibits divergent effects. Kaplan et al observed a greater hazard of CD following appendectomy (Kaplan et al., 2008). Nevertheless, the chance for CD declined if patients underwent the procedure prior to the age of ten. The association between appendectomy and

CD remains inconclusive, with studies indicating a delayed diagnosis of CD in individuals with a history of appendectomy (Ananthakrishnan, 2015). The reason prompting appendectomy appears to be a crucial factor influencing IBD outcomes. Andersson and team discovered that appendectomy resulting from perforating appendicitis is able to elevate the chance of small bowel resection, whereas appendectomy for other factors lowers the risk of CD (Andersson et al., 2017).

2.6.5 Medication and Risk of GI Inflammation

Medications, including NSAIDs, birth control pills, antibiotics, and others, have been linked to IBD (Narula et al., 2023). The interruption of the gastrointestinal barrier by agents like NSAIDs or the alterations in intestinal microbial flora by antibiotics has been linked with a bigger risk of IBD (Dubeau et al., 2013). Moninuola et al conducted a case-control investigation that established a positive link between NSAIDs and IBD (Moninuola, Milligan, Lochhead, Khalili, & therapeutics, 2018). Traditional NSAIDs may induce patient relapse in approximately 20% of sufferers with inactive IBD, possibly due to the dual inhibition of COX (Ferrer et al., 2019). Some COX-2-selective NSAIDs seem to be safe, but non-selective COX inhibition may be harmful by reducing prostaglandins. Reduced prostaglandin levels have been observed in IBD patients, impacting immune function, particularly with the modulation of IL-10, an anti-inflammatory cytokine, and the hindrance of TNF (Paul, Khare, & Gasche, 2012). Furthermore, the use of oral contraceptives (OCPs) has been positively associated with both CD and UC. Timner and the team established that in women who continued taking OCPs, the chance of CD relapse tripled; this effect was more pronounced in women who simultaneously used OCPs and smoked (Allen et al., 2019). The precise mechanism fundamental to the connection between OCPs and an enhanced risk of IBD remains obscure. Several researches suggest that antibiotic use is linked to IBD pathogenesis (Fenneman et al., 2023). Determining this association is challenging, as antibiotics could be prescribed to unidentified IBD patients to treat symptoms misidentified as a gastrointestinal infection.

2.7 MANAGEMENT AND THERAPEUTIC STRATEGIES OF GASTROINTESTINAL INFLAMMATION

Various strategies can be applied to treat inflammation of the GI tract like creating a safe barrier on the intestinal walls to treat secondary GI infection and diseases like acidity, constipation, Crohn's disease (CD), Gastro Esophagus Reflux Disease (GERD), Irritable bowel syndrome (IBS), to Ulcerative colitis (UC). Therapeutic strategy includes using herbs, alternative therapy, allopathic drugs, surgery, and transplants.

Types of Drugs used in Gastrointestinal inflammation:

1) Protectants
2) Herbal extracts
3) Alternative therapy
4) Transplantation
5) Synthetic analogues

Protectants: Mucoproteins

Gelatin tannate

Gelatin tannate can aid in reinstating a deficient mucus coating by creating electrostatic bonds with mucin to help form a protective protein-based film in the intestine. It might also alter the intestinal microbiota's makeup. All these processes work together to return intestinal permeability to normal.

Xyloglucan

It is derived from the seeds of the tamarind tree (Tamarindus indica). Xyloglucan, a water-soluble polysaccharide hemicellulose, is resistant to digestion by enzymes. It functions over the intestinal mucosa as a barrier that forms a film, diminishing the impact of pathogens and enhancing the efficiency of the intestinal wall. To increase its availability in the gut, it is frequently mixed with gelatin. Xyloglucan prevented E. coli colonization and enhanced transepithelial electrical resistance, a measure of tight junction permeability, in cell cultures exposed to the pathogen (Eutamene H et al., 2017).

Herbal Extracts

Approximately 50% of Westerners now use herbal medicines, though most of these cases are not for the dealing or anticipation of digestive disorders. Many bioactive compounds with both potentially harmful and beneficial effects can be found in herbal preparations (Langmead L et al., 2001).

Table 2.2 comprehensively summarises various herbal extracts utilized to treat gastrointestinal disorders. The table lists key herbs, the specific GI diseases they target, their possible mechanisms of action (MoA), and the clinical outcomes reported in studies. Additionally, it includes remarks on their therapeutic potential, highlighting whether they are beneficial or require further investigation.

Alternative Therapy

Acupuncture:

Acupoints, also known as acupuncture points, are unique nodes or outlets on the meridians that function as channels to connect to organs, modulate related body functions, and transport "qi," or life force, throughout the body. It is thought that organ dysfunction and related illness result from impairing or unbalancing the flow of "qi" through meridians or channels. It has been observed that stimulating these points helps alleviate a variety of gastrointestinal symptoms, such as nausea, bloating, diarrhea, stultification, and abdominal pain (Song G et al., 2019).

Probiotic:

Probiotics are living microorganisms, usually bacteria, which are used to replenish the bacterial flora in the human gut and are as beneficial as the microorganisms found there. Probiotics reduce inflammation in a number of ways, such

TABLE 2.2
Herbal Extracts Used in GI Disorders

Name of the Herb	Name Of The GI Disease	Possible MoA	Clinical Studies	Remark	Reference
Curcumin	Crohn's disease, Ulcerative colitis	Reducing interleukin expression	Not much beneficial	Larger studies required	(Goulart RA et al., 2021)
Fish oil	Irritable bowel syndrome	Anti-inflammatory	Not much beneficial	Larger studies required	(Farrukh A et al., 2014)
Peppermint oil		Blocking channels of Calcium and relaxing intestinal smooth muscles	Beneficial	Can be given in GERD	(Cappello G et al., 2007)
STW 5 (Iberogast)	GI disorders	unclear	Beneficial	Raid improvement	(Raedsch E et al., 2018)
Endocannabinoids	GI Inflammations	Modulation of mast cells activation and degranulation	Beneficial	Excellent Pharmacokinetic action	(Pesce M et al, 2018)
Aloe vera	Irritable bowel syndrome	Altering Intestinal motility	Not beneficial	Cannot be recommended	(Størsrud S et al., 2015)
Ginger	Irritable bowel syndrome	Not clear	Not much beneficial	Larger studies required	(Van Tilburg MA et al., 2014)
Wheat grass juice	Ulcerative colitis	Inhibition of adhesion of leukocytes to endothelia	Beneficial	Recommended	(Ben-Arye E et al., 2002)
Andrographis paniculata	Ulcerative colitis	Inhibition of TNF and IL	Beneficial	Larger studies required	(Sandborn WJ et al., 2013)
Boswellia serrata	Irritable bowel syndrome	Inhibition of Interleukin (IL)	Beneficial	Larger studies required	(Abde-Tawab M et al., 2011)
Bilberry (Vaccinium myrtillus)	Ulcerative colitis	Decrease in IL,CRP	Beneficial	Larger studies required	(Biedermann L et al., 2013)
Wormwood	Crohn's disease	Induces interferon production	Beneficial	Recommended	(Omer B et al., 2007)

as by changing the mucosal immune system, competitively excluding pathogens that cause inflammation, and producing antimicrobial substances like bacteriocins and other metabolites. They also have an impact on immune-modulatory pathways, which include blocking the phosphoinositide 3-kinase/Akt pathway, the nuclear factor κ–light chain enhancer of activated B-cells (NF-κB) pathway, and downregulating the expression of toll-like receptors and inflammatory cytokines. They're also referred to as "good bacteria" or "friendly bacteria." Consumers can obtain probiotics primarily through foods and dietary supplements (Masi L et al., 2022).
Prebiotic:

Prebiotics are fermentable, nondigestible fibers that are said to improve the host's health. Both older and more recent studies have demonstrated beneficial effects in human inflammatory bowel diseases (IBD), including Crohn's disease, ulcerative colitis, and chronic pouchitis, as well as in experimental colitis. Natural food ingredients or dietary fibers called inulin and oligofructose are found in some plants as storage carbohydrates. Prebiotics can be found in wheat, chicory, bananas, onions, leek, artichokes, asparagus, and garlic. Most prebiotics that are used in commerce are either extracted from chicory roots or synthesized from sucrose (Mirjam AC et al., 2009).

Transplantation: Fecal Microbiota Therapy (FMT)
Dysbiosis, or the alteration in microbial composition, may be linked to the rising risk of a variety of gastrointestinal and nongastrointestinal disorders. The gut microbiota is a significant acquired organ that performs a variety of essential roles in immune system development, metabolism, and host defense against pathogens. According to new research in patients with severe and complex C. difficile infections, the novel treatment Fecal Microbiota Therapy, either in combination with or without the selective use of vancomycin, effectively treats C. difficile infections. This may be because the gut microbiota becomes more diverse, which was first shown by a "shift" in the microflora, or a decrease in Proteobacteriae and an increase in Firmicutes. Patients with Irritable Bowel Disease have encouraging results from the FMT (Konturek PC et al., 2015).

Synthetic Analogues
Table 2.3 presents an overview of synthetic drugs employed in managing gastrointestinal inflammation. It categorises the drugs based on their pharmacological class, describes their mechanisms of action (MoA), and discusses their therapeutic effectiveness and safety profile. The table also includes remarks regarding potential side effects and clinical considerations.

TABLE 2.3
Synthetic Analogues Used in Gastrointestinal Inflammation

Name of the Drug	Class of the Drug	Possible MoA	Remark	Reference
Etodolac	COX-2 inhibitors	Inhibitions of COX inhibitors	Lesser side effects on GIT	(Jones SL et al., 2003)
Misoprostol	Synthetic prostaglandin	Activation of prostaglandin receptors	Promotes mucosal healing	(Jones SL et al., 2003)
Pentoxifylline	Phosphodiesterase	Increasing intracellular cAMP concentration	Minimal effect	(Jones SL et al., 2003)
Lidocaine	Local anesthetics	Inhibition of sensory neuronal transmission	Can produce side effects	(Jones SL et al., 2003)
Salfasalazine	Aminosalicylates	Interference with arachidonic acid metabolism	Better efficacy	(Cai Z et al., 2021)
Budesonide	Steroids	Activation of Glucocorticoid receptors	Therapy should be undertaken with care	(Cai Z et al., 2021)
Azathioprine	Immunomodulator	Inhibition of T lymphocytes activation	Leads to bone marrow clampdown, liver injury	(Cai Z et al., 2021)
Infliximab	Biologicals	Inhibition of TNF	Have side effects	(Rawla P et al., 2018)

2.8 FUTURE PROSPECTS

2.8.1 Therapeutic Microbiome Manipulation

Studies on genetically vulnerable hosts have demonstrated that enteral bacteria, viruses, or fungi are accountable for advantageous and harmful roles in the digestive tract, and can cause long-term, immune-mediated intestinal inflammation. A person's diet, as well as environmental and genetic factors, can have an impact on dysbiosis, an abnormal composition of gut bacteria. It is possible to bring back a balanced microbiome. These could include focusing on protective commensal species that have a high probability of colonizing and functioning in the intestine and choosing a customized treatment strategy based on a study of an individual's microbiota pattern. Combinations of common protecting species are probably going to work better than individual species. Similarly, it will be necessary to ascertain whether dietary lines can change the structure and role of enteric microbiota in a therapeutic or protective manner ((D'Haens GR et al., 2014).

2.8.2 Clinical Genomics

Treating the effects of genetic defects through pathway-specific therapies is becoming more and more complementary to advances in diagnostic clinical genomics. This demonstrates the extraordinary promise of customized precision medicine. Even so, Gastrointestinal inflammation presents a difficult case for genomic medicine due to the low overall proportion of patients with defects, the large number of putative candidate genes, and the ongoing development of interventional evidence (Uhlig HH et al., 2017).

2.8.3 Therapy

A thorough understanding of each disease's molecular biology is necessary to identify the gene targets linked to it, as this is the key to the effectiveness of gene therapy in treating GI disorders. The most important challenge is choosing the right target genes because GI diseases have a variety of etiologies, and varying degrees of severity, and GI malignancies have multiple gene alterations. Certain gene mutations have a major role in the pathophysiology of Gastrointestinal inflammation, according to research. Among these mutations, nucleotide-binding oligomerization domain containing 2 (NOD2) and autophagy-related 16-like 1 (ATG16L-1) have been found to have a significant impact on increasing susceptibility to inflammation. NOD2 is a bacterial intracellular pattern recognition receptor derived from peptidoglycan that is encoded by the NOD2 protein, which is situated on the long arm of human chromosome 16. It is essential for both causing mucosal inflammation and preserving mucosal homeostasis (Ning-ning Yue et al., 2023).

2.8.4 Stem Cell Therapy

Hematopoietic stem cells can be extracted from cord blood, bone marrow, and regular outlying blood. These cells can either proliferate into immunomodulatory and epithelial cells to repair damaged tissues or migrate straight into the tissue to form normal mucosal tissues. Pre-transplant screening, which consists of a history and examination, blood tests, colonoscopy, serology, small bowel MRI or pelvic and rectal MRI (in patients with perianal disease involvement), and bone marrow aspiration, is the first step in the multi-step process for HSC treatment. The next step in the transplant procedure is to mobilize the patient's stem cells or those of a donor who matches the patient's HLA, and to stimulate the bone marrow to produce stem cells following lymphocyte removal through cyclophosphamide infusion. After leukocyte clearance, CD34+ cells are isolated from bone marrow or outlying blood to a target count of (3~8) × 106. Ultimately, immune system reconstitution and transplantation were carried out. Prior to improvement in intestinal lesions during transplantation, patients were treated with hematopoietic stem cells (HSCT) for collective hematologic signals (e.g., leukemia and non-Hodgkin's lymphoma). As a result, HSCT was increasingly utilized in clinical trials (Zhang HM et al., 2022).

2.8.5 New Molecules

Janus kinase inhibitors:

JAK belongs to intracellular tyrosine kinases that transduces cytokine-mediated signals through the STAT pathway. Members of the JAK family include tyrosinase kinase 2 (TYK2), JAK3, JAK2, and JAK1. They play a role in several processes, such as inflammatory and immunological reactions. Drugs like Tofacitinib, Figotinib, Upadacitinib, Peficitinib, and Izencitinib are under screening for Ulcerative colitis UC) and Crohn's disease. Large prospective studies are needed to demonstrate the effectiveness of JAK inhibitors comparatively (deGuise CD et al., 2023).

2.9 CONCLUSION

The GIT is affected by inflammation in a variety of ways. Variations exist in the normal flora and the responses of the tissues in each area. At our disposal are adequate signs to conclude that several pharmacological composites are proven to treat gastrointestinal inflammation. The therapeutic strategy of GI inflammation is summarized in this review, but some of the molecules reported serious side effects. This review highlights some of these novel and emerging therapies. The near-term years will be critical. The statistics obtained from current and future clinical trials will help us better realize the pathophysiology of intestinal inflammation in IBD and may have a substantial effect on treating patients with the illness.

ACKNOWLEDGMENT

Authors are thankful to Bengal College of Pharmaceutical Sciences & Research, Durgapur 713212, West Bengal, India.

CONFLICT OF INTEREST

None

FUNDING

None

REFERENCES

Abdel-Tawab, M., Werz, O., & Schubert-Zsilavecz, M. (2011). Boswellia serrata: An overall assessment of in vitro, preclinical, pharmacokinetic and clinical data. *Clinical Pharmacokinetics*, *50*, 349–369. https://doi.org/10.2165/11586800-000000000-00000

Adelman, M. W., Woodworth, M. H., Langelier, C., Busch, L. M., Kempker, J. A., Kraft, C. S., Kraft, C. S., Martin, G. S., & Martin, G. S. (2020). The gut microbiome's role in the development, maintenance, and outcomes of sepsis. *Critical Care (London, England)*, *24*(1), 1–10. https://doi.org/10.1186/S13054-020-02989-1

Agirman, G., Yu, K. B., & Hsiao, E. Y. (2021). Signaling inflammation across the gut-brain axis. *Science*, *374*(6571), 1087–1092. https://doi.org/10.1126/science.abi6087

Allen, A. M., Weinberger, A. H., Wetherill, R. R., Howe, C. L., & McKee, S. A. (2019). Oral contraceptives and cigarette smoking: A review of the literature and future directions. Nicotine and Tobacco Research, *21*(5), 592–601. https://doi.org/10.1093/ntr/ntx258

Allocca, M., Furfaro, F., Fiorino, G., Gilardi, D., D'Alessio, S., & Danese, S. (2018). Can IL-23 be a good target for ulcerative colitis? *Best Practice & Research Clinical Gastroenterology*, *32*, 95–102. https://doi.org/10.1016/j.bpg.2018.05.016

Al–Mousawi, A. H., Neamah, B. alhuda H., Chessab, R. M., Aldujaili, H. H., & Chessab, J. M. (2022). Relation Helicobacter pylori with pathogenesis of stomach and immune responses. *Medical Science Journal for Advance Research*, *3*(2), 55–60. https://doi.org/10.46966/MSJAR.V3I2.41

Ananthakrishnan, A. N. (2015). Environmental risk factors for inflammatory bowel diseases: A review. *Digestive Diseases and Sciences*, *60*(2), 290–298. https://doi.org/10.1007/s10620-014-3350-9

Andersson, M., Kolodziej, B., Andersson, R., Andersson, R., Andersson, M., Eriksson, T., … & Fagerström, N. (2017). Randomized clinical trial of appendicitis inflammatory response score-based management of patients with suspected appendicitis. Journal of British Surgery, *104*(11), 1451–1461. https://doi.org/10.1002/bjs.10637

Arias, C., Sepúlveda, P., Castillo, R. L., & Salazar, L. A. (2023). Relationship between hypoxic and immune pathways activation in the progression of neuroinflammation: Role of HIF-1α and Th17 cells. International *J*ournal of *M*olecular Sciences, *24*(4), 3073. https://doi.org/10.3390/ijms24043073

Bamias, G., Pizarro, T. T., & Cominelli, F. (2022). Immunological regulation of intestinal fibrosis in inflammatory bowel disease. Inflammatory Bowel Diseases, *28*(3), 337–349. https://doi.org/10.1093/ibd/izab251

Ben-Arye, E., Goldin, E., Wengrower, D., Stamper, A., Kohn, R., & Berry, E. (2002). Wheat grass juice in the treatment of active distal ulcerative colitis: A randomized double-blind placebo-controlled trial. *Scandinavian Journal of Gastroenterology*, *37*(4), 444–449. https://doi.org/10.1080/003655202317316088

Berkowitz, L., Schultz, B. M., Salazar, G. A., Pardo-Roa, C., Sebastián, V. P., Álvarez-Lobos, M. M., & Bueno, S. M. (2018). Impact of cigarette smoking on the gastrointestinal tract inflammation: Opposing effects in Crohn's disease and ulcerative colitis. Frontiers in *I*mmunology, *9*, 74. https://doi.org/10.3389/fimmu.2018.00074

Bernstein, C. N., Rawsthorne, P., Cheang, M., & Blanchard, J. F. (2006). A population-based case control study of potential risk factors for IBD. Official Journal of the American College of Gastroenterology| ACG, *101*(5), 993–1002. https://doi.org/10.1111/j.1572-0241.2006.00381.x

Berhe, B., Bugssa, G., Bayisa, S., & Alemu, M. (2018a). Foodborne intestinal protozoan infection and associated factors among patients with watery diarrhea in Northern Ethiopia: A cross-sectional study. *Journal of Health, Population and Nutrition*, *37*(1), 5. https://doi.org/10.1186/s41043-018-0137-1

Berhe, B., Bugssa, G., Bayisa, S., & Alemu, M. (2018b). Foodborne intestinal protozoan infection and associated factors among patients with watery diarrhea in Northern Ethiopia: A cross-sectional study. *Journal of Health, Population and Nutrition*, *37*(1), 5. https://doi.org/10.1186/s41043-018-0137-1

Bhat, M. H., Hajam, Y. A., Neelam, Kumar, R., & Diksha. (2023). Microbial diversity and their role in human health and diseases. In Sobti, R., Kuhad, R.C., Lal, R., Rishi, P. (eds) *Role of microbes in sustainable development* (pp. 1–33). Springer Nature Singapore. https://doi.org/10.1007/978-981-99-3126-2_1

Biedermann, L., Fournier, N., Misselwitz, B., Frei, P., Zeitz, J., Manser, C. N., ... & Swiss Inflammatory Bowel Disease Cohort Study Group. (2015). High rates of smoking especially in female Crohn's disease patients and low use of supportive measures to achieve smoking cessation—data from the Swiss IBD cohort study. Journal of Crohn's and Colitis, *9*(10), 819–829. https://doi.org/10.1093/ecco-jcc/jjv113

Biedermann, L., Mwinyi, J., Scharl, M., Frei, P., Zeitz, J., Kullak-Ublick, G. A., ... & Rogler, G. (2013). Bilberry ingestion improves disease activity in mild to moderate ulcerative colitis—An open pilot study. *Journal of Crohn's and Colitis*, *7*(4), 271–279. https://doi.org/10.1016/j.crohns.2012.07.010

Bourgonje, A. R., Kloska, D., Grochot-Przęczek, A., Feelisch, M., Cuadrado, A., & van Goor, H. (2023). Personalized redox medicine in inflammatory bowel diseases: Aan emerging role for HIF-1α and NRF2 as therapeutic targets. Redox Biology, 60, 102603. https://doi.org/10.1016/j.redox.2023.102603

Bourgonje, A. R., von Martels, J. Z., de Vos, P., Faber, K. N., & Dijkstra, G. (2018). Increased fecal calprotectin levels in Crohn's disease correlate with elevated serum Th1-and Th17-associated cytokines. PLoS One*, 13*(2), e0193202. https://doi.org/10.1371/journal.pone.0193202

Bringer, M.-A., Billard, E., Glasser, A.-L., Colombel, J.-F., & Darfeuille-Michaud, A. (2012). Replication of Crohn's disease-associated AIEC within macrophages is dependent on TNF-α secretion. Laboratory *I*nvestigation, *92*(3), 411–419. https://doi.org/10.1038/labinvest.2011.156

Buisson, A., Vazeille, E., Fumery, M., Pariente, B., Nancey, S., Seksik, P., …& Filippi, J. J. U. j. (2021). Faster and less invasive tools to identify patients with ileal colonization by adherent-invasive E. coli in Crohn's disease. *UEG Journal*, *9*(9), 1007–1018. https://doi.org/10.1002/ueg2.12161

Burucoa, C., & Axon, A. (2017). Epidemiology of Helicobacter pylori infection. *Helicobacter*, *22,* e12403. https://doi.org/10.1111/HEL.12403

Cai, Z., Wang, S., & Li, J. (2021). Treatment of inflammatory bowel disease: A comprehensive review. *Frontiers in Medicine*, *8*, 765474. https://doi.org/10.3389/fmed.2021.765474

Caioni, G., Viscido, A., d'Angelo, M., Panella, G., Castelli, V., Merola, C., … & Benedetti, E. (2021). Inflammatory bowel disease: New insights into the interplay between environmental factors and PPARγ. International Journal of Molecular Science, *22*(3), 985. https://doi.org/10.3390/ijms22030985

Cappello, G., Spezzaferro, M., Grossi, L., Manzoli, L., & Marzio, L. (2007). Peppermint oil (Mintoil®) in the treatment of irritable bowel syndrome: A prospective double blind placebo-controlled randomized trial. *Digestive and Liver Disease*, *39*(6), 530–536. https://doi.org/10.1016/j.dld.2007.02.006

Cari, L., Rosati, L., Leoncini, G., Lusenti, E., Gentili, M., Nocentini, G., … & Ronchetti, S. (2023). Association of GILZ with MUC2, TLR2, and TLR4 in inflammatory Bowel disease. *International Journal of Molecular Sciences*, *24*(3), 2235. https://doi.org/10.3390/ijms24032235

Carlson, S. L., Mathew, L., Savage, M., Kok, K., Lindsay, J. O., Munro, C. A., & McCarthy, N. E. (2023). Mucosal immunity to gut fungi in health and inflammatory Bowel disease. *Journal of Fungi*, *9*(11), 1105. https://doi.org/10.3390/jof9111105

Castillo-Rodríguez, R. A., Trejo-Solís, C., Cabrera-Cano, A., Gómez-Manzo, S., & Dávila-Borja, V. M. (2022). Hypoxia as a modulator of inflammation and immune response in cancer. *Cancers*, *14*(9), 2291. https://doi.org/10.3390/cancers14092291

Chandler, W., Alessi, M., Aillaud, M., Henderson, P., Vague, P., & Juhan-Vague, I. (1997). Clearance of tissue plasminogen activator (TPA) and TPA/plasminogen activator inhibitor type 1 (PAI-1) complex: Relationship to elevated TPA antigen in patients with high PAI-1 activity levels. *Cancers*, *96*(3), 761–768. https://doi.org/10.1161/01.cir.96.3.761

Chervy, M., Barnich, N., & Denizot, J. (2020). Adherent-invasive E. coli: Update on the lifestyle of a troublemaker in Crohn's disease. *International Journal of Molecular Sciences*, *21*(10), 3734. https://doi.org/10.3390/ijms21103734

Coruzzi, G. (2007). *Gastrointestinal safety of novel nonsteroidal antiinflammatory drugs: Selective COX-2 inhibitors and beyond.* Acta Biomedica-Ateneo Parmense. www.actabiomedica.it

Cosnes, J., Nion-Larmurier, I., Afchain, P., Beaugerie, L., & Gendre, J.-P. (2004). Gender differences in the response of colitis to smoking. *Clinical Gastroenterology and Hepatology*, *2*(1), 41–48. https://doi.org/10.1016/s1542-3565(03)00290-8

Cummins, E. P., & Crean, D. (2017). Hypoxia and inflammatory bowel disease. *Microbes and Infection*, *19*(3), 210–221. https://doi.org/10.1016/j.micinf.2016.09.004

Da Silva, S., Keita, Å. V., Mohlin, S., Påhlman, S., Theodorou, V., Påhlman, I., … & Söderholm, J. D. (2018). A novel topical PPARγ agonist induces PPARγ activity in ulcerative colitis mucosa and prevents and reverses inflammation in induced colitis models. Inflammatory Bowel Diseases, *24*(4), 792–805. https://doi.org/10.1093/ibd/izx079

Day, J. W., Gunn, J. D., & Burger, J. R. (2021). Diminishing opportunities for sustainability of coastal cities in the anthropocene: A review. Frontiers in Environmental Science*,* 9, 663275. https://doi.org/10.3389/fenvs.2021.663275

Dubeau, M.-F., Iacucci, M., Beck, P. L., Moran, G. W., Kaplan, G. G., Ghosh, S., & Panaccione, R. (2013). Drug-induced inflammatory bowel disease and IBD-like conditions. *Inflammatory Bowel Diseases*, *19*(2), 445–456. https://doi.org/10.1002/ibd.22990

Dvornikova, K. A., Platonova, O. N., & Bystrova, E. Y. (2023). Hypoxia and intestinal inflammation: Common molecular mechanisms and signaling pathways. International Journal of Molecular Sciences, *24*(3), 2425. https://doi.org/10.3390/ijms24032425

Eggenhuizen, P. J., Ng, B. H., & Ooi, J. D. (2020). Treg enhancing therapies to treat autoimmune diseases. *International Journal of Molecular Sciences*, *21*(19), 7015. https://doi.org/10.3390/ijms21197015

Endo, K., Kito, H., Tanaka, R., Kajikuri, J., Tanaka, S., Elboray, E. E., … & Ohya, S. (2019). Possible contribution of inflammation-associated hypoxia to increased k2p5. 1 k+ channel expression in CD4+ t cells of the mouse model for inflammatory bowel disease. *International Journal of Molecular Sciences*, *21*(1), 38. https://doi.org/10.3390/ijms21010038

Eutamene, H., Beaufrand, C., Harkat, C., & Theodorou, V. (2018). The role of mucoprotectants in the management of gastrointestinal disorders. *Expert Review of Gastroenterology & Hepatology*, *12*(1), 83–90. https://doi.org/10.1080/17474124.2018.1378573

Fagundes, R. R., Belt, S. C., Bakker, B. M., Dijkstra, G., Harmsen, H. J., & Faber, K. N. (2023). Beyond butyrate: Microbial fiber metabolism supporting colonic epithelial homeostasis. *Trends in Microbiology, 32*(2), 178–189. https://doi.org/10.1016/j.tim.2023.07.014

Farrukh, A., & Mayberry, J. F. (2014). Is there a role for fish oil in inflammatory bowel disease? *World Journal of Clinical Cases, 2*(7), 250–252. https://doi.org/10.12998/wjcc.v2.i7.250

Fenneman, A. C., Weidner, M., Chen, L. A., Nieuwdorp, M., & Blaser, M. J. (2023). Antibiotics in the pathogenesis of diabetes and inflammatory diseases of the gastrointestinal tract. Nature Reviews Gastroenterology & Hepatology, *20*(2), 81–100. https://doi.org/10.1038/s41575-022-00685-9

Ferrer, M. D., Busquets-Cortés, C., Capó, X., Tejada, S., Tur, J. A., Pons, A., & Sureda, A. (2019). Cyclooxygenase-2 inhibitors as a therapeutic target in inflammatory diseases. Current Medicinal Chemistry, *26*(18), 3225–3241. https://doi.org/10.2174/0929867325666180514112124

Franchimont, D., Vermeire, S., El Housni, H., Pierik, M., Van Steen, K., Gustot, T., … & Devière, J. J. G. (2004). Deficient host-bacteria interactions in inflammatory bowel disease? The toll-like receptor (TLR)-4 Asp299gly polymorphism is associated with Crohn's disease and ulcerative colitis. Gut, *53*(7), 987. https://doi.org/10.1136/gut.2003.030205

Frigerio, S., Lartey, D. A., D'Haens, G. R., & Grootjans, J. (2021). The role of the immune system in IBD-associated colorectal cancer: From pro to anti-tumorigenic mechanisms. *International Journal of Molecular Sciences*, *22*(23), 12739. https://doi.org/10.3390/ijms222312739

Giambra, V., Pagliari, D., Rio, P., Totti, B., Di Nunzio, C., Bosi, A., Giaroni, C., Gasbarrini, A., Gambassi, G., & Cianci, R. (2023). Gut microbiota, inflammatory bowel disease, and cancer: The role of guardians of innate immunity. *Cells*, *12*(22), 2654. https://doi.org/10.3390/cells12222654

Glasser, A.-L., Boudeau, J., Barnich, N., Perruchot, M.-H., Colombel, J.-F., & Darfeuille-Michaud, A. (2001). Adherent invasive Escherichia coli strains from patients with Crohn's disease survive and replicate within macrophages without inducing host cell death. Infection and Immunity, *69*(9), 5529–5537. https://doi.org/10.1128/IAI.69.9.5529-5537.2001

Guo, X. Y., Liu, X. J., & Hao, J. Y. (2020). Gut microbiota in ulcerative colitis: Insights on pathogenesis and treatment. *Journal of Digestive Diseases*, *21*(3), 147–159. https://doi.org/10.1111/1751-2980.12849

Goulart, R. D. A., Barbalho, S. M., Lima, V. M., Souza, G. A. D., Matias, J. N., Araújo, A. C., ... & Guiguer, É. L. (2021). Effects of the use of curcumin on ulcerative colitis and Crohn's disease: A systematic review. *Journal of Medicinal Food*, *24*(7), 675–685. https://doi.org/10.1089/jmf.2020.0129

Guo, Y., Cao, X. S., Zhou, M. G., & Yu, B. (2023). Gastric microbiota in gastric cancer: Different roles of Helicobacter pylori and other microbes. *Frontiers in Cellular and Infection Microbiology*, *12*, 1105811. https://doi.org/10.3389/FCIMB.2022.1105811/BIBTEX

Hartwig, O., Boushehri, M. A. S., Shalaby, K. S., Loretz, B., Lamprecht, A., & Lehr, C.-M. (2021). Drug delivery to the inflamed intestinal mucosa–Targeting technologies and human cell culture models for better therapies of IBD. *Advanced Drug Delivery Reviews*, *175*, 113828. https://doi.org/10.1016/j.addr.2021.113828

Hayashi, Y., & Nakase, H. (2022). The molecular mechanisms of intestinal inflammation and fibrosis in Crohn's disease. Frontiers in Physics, *13*, 220. https://doi.org/10.3389/fphys.2022.845078

Herrera-deGuise, C., Serra-Ruiz, X., Lastiri, E., & Borruel, N. (2023). JAK inhibitors: A new dawn for oral therapies in inflammatory bowel diseases. *Frontiers in Medicine*, 10, 1089099. https://doi.org/10.3389/fmed.2023.1089099

Hertz, S., Durack, J., Kirk, K. F., Nielsen, H. L., Lin, D. L., Fadrosh, D., Lynch, K., Piceno, Y., Thorlacius-Ussing, O., Nielsen, H., & Lynch, S. V. (2022). Microscopic colitis patients possess a perturbed and inflammatory gut microbiota. *Digestive Diseases and Sciences*, *67*(6), 2433–2443. https://doi.org/10.1007/s10620-021-07045-8

Islam, S. T., Won, J., Khan, M., Mannie, M. D., & Singh, I. (2021). Hypoxia-inducible factor-1 drives divergent immunomodulatory functions in the pathogenesis of autoimmune diseases. *Immunology*, *164*(1), 31–42. https://doi.org/10.1111/imm.13335

Jacobse, J., Li, J., Rings, E. H., Samsom, J. N., & Goettel, J. A. (2021). Intestinal regulatory T cells as specialized tissue-restricted immune cells in intestinal immune homeostasis and disease. Frontiers in *Immunology*, *12*, 716499. https://doi.org/10.3389/fimmu.2021.716499

Jagirdhar, G. S. K., Pulakurthi, Y. S., Chigurupati, H. D., & Surani, S. (2023). Gastrointestinal tract and viral pathogens. *World Journal of Virology*, *12*(3), 136–150. https://doi.org/10.5501/wjv.v12.i3.136

Jarmakiewicz-Czaja, S., Sokal, A., Ferenc, K., Motyka, E., Helma, K., & Filip, R. (2023). The role of genetic and epigenetic regulation in intestinal fibrosis in inflammatory bowel disease: A descending process or a programmed consequence? *Genes*, *14*(6), 1167. https://doi.org/10.3390/genes14061167

Jójárt, B., Resál, T., Kata, D., Molnár, T., Bacsur, P., Szabó, V., Varga, Á., Szántó, K. J., Pallagi, P., Földesi, I., Molnár, T., Maléth, J., & Farkas, K. (2024). Plasminogen activator inhibitor 1 is a novel faecal biomarker for monitoring disease activity and therapeutic response in inflammatory bowel diseases. *Journal of Crohn's and Colitis, 18*(3), 392–405. https://doi.org/10.1093/ecco-jcc/jjad160

Jones, S. L. (2003). Treatment of acute and chronic gastrointestinal inflammation. *Veterinary Clinics: Equine Practice*, *19*(3), 697–714. https://doi.org/10.1016/j.cveq.2003.08.010

Kaiko, G. E., Chen, F., Lai, C.-W., Chiang, I.-L., Perrigoue, J., Stojmirović, A., … & VanDussen, K. L. (2019). PAI-1 augments mucosal damage in colitis. Science Translational Medicine, *11*(482), eaat0852. https://doi.org/10.1126/scitranslmed.aat0852

Kamada, N., Seo, S. U., Chen, G. Y., & Núñez, G. (2013). Role of the gut microbiota in immunity and inflammatory disease. *Nature Reviews. Immunology*, *13*(5), 321–335. https://doi.org/10.1038/NRI3430

Kantor, M., Abrantes, A., Estevez, A., Schiller, A., Torrent, J., Gascon, J., Hernandez, R., & Ochner, C. (2018). Entamoeba histolytica: Updates in clinical manifestation, pathogenesis, and vaccine development. *Canadian Journal of Gastroenterology and Hepatology*, *2018*, 1–6. https://doi.org/10.1155/2018/4601420

Kaplan, G. G., Jackson, T., Sands, B. E., Frisch, M., Andersson, R. E., & Korzenik, J. (2008). The risk of developing Crohn's disease after an appendectomy: A meta-analysis. *Official Journal of the American College of Gastroenterology| ACG*, *103*(11), 2925–2931. https://doi.org/10.1111/j.1572-0241.2008.02118.x

Kashyap, P. C., Johnson, S., Geno, D. M., Lekatz, H. R., Lavey, C., Alexander, J. A., Chen, J., & Katzka, D. A. (2019). A decreased abundance of clostridia characterizes the gut microbiota in eosinophilic esophagitis. *Physiological Reports*, *7*(20), 1–5. https://doi.org/10.14814/phy2.14261

Kerber, E. L., Padberg, C., Koll, N., Schuetzhold, V., Fandrey, J., & Winning, S. (2020). The importance of hypoxia-inducible factors (HIF-1 and HIF-2) for the pathophysiology of inflammatory bowel disease. International *J*ournal of *M*olecular *S*ciences, *21*(22), 8551. https://doi.org/10.3390/ijms21228551

Khan, I., Ullah, N., Zha, L., Bai, Y., Khan, A., Zhao, T., Che, T., & Zhang, C. (2019). Alteration of gut microbiota in inflammatory bowel disease (IBD): Cause or consequence? IBD treatment targeting the gut microbiome. *Pathogens*, *8*(3), 126. https://doi.org/10.3390/pathogens8030126

Khor, B., Gardet, A., & Xavier, R. J. (2011). Genetics and pathogenesis of inflammatory bowel disease. *Nature*, *474*(7351), 307–317. https://doi.org/10.1038/nature10209

Konturek, P. C., Haziri, D., Brzozowski, T., Hess, T., Heyman, S., Kwiecien, S., ... & Koziel, J. (2015). Emerging role of fecal microbiota therapy in the treatment of gastrointestinal and extra-gastrointestinal diseases. *Journal of Physiology and Pharmacology*, *66*(4), 483–491. PMID: 26348073.

Koutroubakis, I. E., Sfiridaki, A., Tsiolakidou, G., Coucoutsi, C., Theodoropoulou, A., & Kouroumalis, E. A. (2008). Plasma thrombin-activatable fibrinolysis inhibitor and plasminogen activator inhibitor-1 levels in inflammatory bowel disease. European *J*ournal of *G*astroenterology & *H*epatology, *20*(9), 912–916. https://doi.org/10.1097/MEG.0b013e3282faa759

Kumar, S., Singh, P., & Kumar, A. (2022). Targeted therapy of irritable bowel syndrome with anti-inflammatory cytokines. *Clinical Journal of Gastroenterology*, *15*(1), 1. https://doi.org/10.1007/S12328-021-01555-8

Lamousé-Smith, E. S., Weber, S., Rossi, R. F., Neinstedt, L. J., Mosammaparast, N., & Sandora, T. J. (2013). Polymerase chain reaction test for Clostridium difficile toxin B gene reveals similar prevalence rates in children with and without inflammatory bowel disease. Journal of *P*ediatric *G*astroenterology and *N*utrition, *57*(3), 293–297. https://doi.org/10.1097/MPG.0b013e3182999990

Lange, N. F., Graf, V., Caussy, C., & Dufour, J.-F. (2022). PPAR-targeted therapies in the treatment of non-alcoholic fatty liver disease in diabetic patients. International *J*ournal of *M*olecular *S*ciences, *23*(8), 4305. https://doi.org/10.3390/ijms23084305

Langmead, L., & Rampton, D. S. (2001). Herbal treatment in gastrointestinal and liver disease—benefits and dangers. *Alimentary Pharmacology & Therapeutics*, *15*(9), 1239–1252. https://doi.org/10.1046/j.1365-2036.2001.01053.x

Larabi, A., Barnich, N., & Nguyen, H. T. T. (2020). New insights into the interplay between autophagy, gut microbiota and inflammatory responses in IBD. *Autophagy*, *16*(1), 38–51. https://doi.org/10.1080/15548627.2019.1635384

Li, J., Ji, Y., Chen, N., Dai, L., & Deng, H. (2023). Colitis-associated carcinogenesis: Crosstalk between tumors, immune cells and gut microbiota. Cell & Bioscience, *13*(1), 194. https://doi.org/10.1186/s13578-023-01139-8

Li, M.-Y., Luo, H.-J., Wu, X., Liu, Y.-H., Gan, Y.-X., Xu, N., … & Su, Z.-R. (2020). Anti-inflammatory effects of Huangqin decoction on dextran sulfate sodium-induced ulcerative colitis in mice through regulation of the gut microbiota and suppression of the Ras-PI3K-Akt-HIF-1α and NF-κB pathways. Frontiers in Pharmacology, *10*, 1552. https://doi.org/10.3389/fphar.2019.01552

Liikanen, I., Lauhan, C., Quon, S., Omilusik, K., Phan, A. T., Bartrolí, L. B., … & Scott-Browne, J. P. (2021). Hypoxia-inducible factor activity promotes antitumor effector function and tissue residency by CD8+ T cells. The Journal of Clinical Investigation, *131*(7), 1–17. https://doi.org/10.1172/JCI143729

Limon, J. J., Tang, J., Li, D., Wolf, A. J., Michelsen, K. S., Funari, V., …& Underhill, D. M.. (2019). Malassezia is associated with Crohn's disease and exacerbates colitis in mouse models. Cell *H*ost & *M*icrobe, *25*(3), 377–388. e376. https://doi.org/10.1016/j.chom.2019.01.007

Lin, L., & Zhang, J. (2017). Role of intestinal microbiota and metabolites on gut homeostasis and human diseases. *BMC Immunology*, *18*(1), 2. https://doi.org/10.1186/s12865-016-0187-3

Lionakis, M. S., Drummond, R. A., & Hohl, T. M. (2023). Immune responses to human fungal pathogens and therapeutic prospects. *Nature Reviews Immunology*, *23*(7), 433–452. https://doi.org/10.1038/s41577-022-00826-w

Loddo, I., & Romano, C. (2015). Inflammatory bowel disease: Genetics, epigenetics, and pathogenesis. Frontiers in *I*mmunology, *6*, 551. https://doi.org/10.3389/fimmu.2015.00551

Lu, H., Lin, J., Xu, C., Sun, M., Zuo, K., Zhang, X., …& Liu, Z. (2021). Cyclosporine modulates neutrophil functions via the SIRT6–HIF-1α–glycolysis axis to alleviate severe ulcerative colitis. Clinical and *T*ranslational *M*edicine, *11*(2), e334. https://doi.org/10.1002/ctm2.334

Lu, P., Sodhi, C. P., Yamaguchi, Y., Jia, H., Prindle Jr, T., Fulton, W. B., … & Hackam, D. J. (2018). Intestinal epithelial Toll-like receptor 4 prevents metabolic syndrome by regulating interactions between microbes and intestinal epithelial cells in mice. Mucosal *I*mmunology, *11*(3), 727–740. https://doi.org/10.1038/mi.2017.114

Maddur, M. S., Miossec, P., Kaveri, S. V., & Bayry, J. (2012). Th17 cells: Biology, pathogenesis of autoimmune and inflammatory diseases, and therapeutic strategies. *American Journal of Pathology*, *181*(1), 8–18. https://doi.org/10.1016/j.ajpath.2012.03.044

Mahid, S. S., Minor, K. S., Soto, R. E., Hornung, C. A., & Galandiuk, S. (2006). Smoking and inflammatory bowel disease: A meta-analysis. *Mayo Clinic Proceedings*, *81*(11), 1462–1471. https://doi.org/10.4065/81.11.1462

Malkov, M. I., Lee, C. T., & Taylor, C. T. (2021). Regulation of the hypoxia-inducible factor (HIF) by pro-inflammatory cytokines. Cells, *10*(9), 2340. https://doi.org/10.3390/cells10092340

Mallikarjunappa, S., Brito, L. F., Pant, S. D., Schenkel, F. S., Meade, K. G., & Karrow, N. A. (2021). Johne's disease in dairy cattle: An Immunogenetic perspective. *Frontiers in Veterinary Science*, *8*, 718987. https://doi.org/10.3389/fvets.2021.718987

Ma, X., Wang, X., Zheng, G., Tan, G., Zhou, F., Wei, W., Tian, D., & Yu, H. (2021). Critical role of gut microbiota and epigenetic factors in the pathogenesis of Behçet's disease. *Frontiers in Cell and Developmental Biology*, *9*, 1–11. https://doi.org/10.3389/fcell.2021.719235

Man, S. M. (2018). Inflammasomes in the gastrointestinal tract: Infection, cancer and gut microbiota homeostasis. *Nature Reviews Gastroenterology & Hepatology*, *15*(12), 721–737. https://doi.org/10.1038/s41575-018-0054-1

Marasco, G., Cirota, G. G., Rossini, B., Lungaro, L., Di Biase, A. R., Colecchia, A., Volta, U., De Giorgio, R., Festi, D.,

& Caio, G. (2020). Probiotics, prebiotics and other dietary supplements for gut microbiota modulation in celiac disease patients. *Nutrients*, *12*(9), 2674. https://doi.org/10.3390/nu12092674

Marmon, P. K. A. (2021). Advancing the mechanistic understanding of the toxicological effects of non-steroidal anti-inflammatory drugs (NSAIDs) using integrated in silico and in vivo approaches. Brunel University London. http://bura.brunel.ac.uk/handle/2438/24942

Masi, L., Ciuffini, C., Petito, V., Pisani, L. F., Lopetuso, L. R., Graziani, C., ... & Scaldaferri, F. (2022). Innovative, complementary and alternative therapy in inflammatory bowel diseases: A broad 2020s update. *Frontiers in Gastroenterology*, *1*, 1022530. https://doi.org/10.3389/fgstr.2022.1022530

Mirjam A.C., Langen, L-V., & Dieleman, L.A. (2009, March). Prebiotics in chronic intestinal inflammation. *Inflammatory Bowel Diseases*, *15*(3), 454–462. https://doi.org/10.1002/ibd.20737

Moninuola, O. O., Milligan, W., Lochhead, P., & Khalili, H. (2018). Systematic review with meta-analysis: Association between acetaminophen and nonsteroidal anti-inflammatory drugs (NSAID s) and risk of Crohn's disease and ulcerative colitis exacerbation. Alimentary *P*harmacology & *T*herapeutics, *47*(11), 1428–1439. https://doi.org/10.1111/apt.14606

Monteleone, I., Pallone, F., & Monteleone, G. (2009). Interleukin-23 and Th17 cells in the control of gut inflammation. Mediators of *I*nflammation, 2009, 1–7. https://doi.org/10.1155/2009/297645

Mulder, D. J., Noble, A. J., Justinich, C. J., & Duffin, J. M. (2014). A tale of two diseases: The history of inflammatory bowel disease. *Journal of Crohn's and Colitis, 8*(5), 341–348. https://doi.org/10.1016/j.crohns.2013.09.009

Na, Y. R., Stakenborg, M., Seok, S. H., & Matteoli, G. (2019). Macrophages in intestinal inflammation and resolution: A potential therapeutic target in IBD. *Nature Reviews. Gastroenterology & Hepatology*, *16*(9), 531–543. https://doi.org/10.1038/S41575-019-0172-4

Narula, N., Wong, E. C., Pray, C., Marshall, J. K., Rangarajan, S., & Islam, S. (2023). Associations of antibiotics, hormonal therapies, oral contraceptives, and long-term NSAIDS with inflammatory bowel disease: Results from the Prospective Urban Rural Epidemiology (PURE) study. Clinical Gastroenterology and Hepatology, *21*(10), 2649–2659.e2616. https://doi.org/10.1016/j.cgh.2022.11.037

Negroni, A., Costanzo, M., Vitali, R., Superti, F., Bertuccini, L., Tinari, A., … & Pierdomenico, M. (2012). Characterization of adherent-invasive Escherichia coli isolated from pediatric patients with inflammatory bowel disease. Inflammatory *B*owel *D*iseases, *18*(5), 913–924. https://doi.org/10.1002/ibd.21899

Ng, S. C., Bernstein, C. N., Vatn, M. H., Lakatos, P. L., Loftus, E. V., Tysk, C., … & Colombel, J.-F. J. G. (2013). Geographical variability and environmental risk factors in inflammatory bowel disease. Gut. https://doi.org/10.1136/gutjnl-2012-303661

Oh, H. Y. P., Visvalingam, V., & Wahli, W. (2019). The PPAR–microbiota–metabolic organ trilogy to fine-tune physiology. The FASEB Journal, *33*(9), 9706–9730. https://doi.org/10.1096/fj.201802681RR

Omer, B., Krebs, S., Omer, H., & Noor, T. O. (2007). Steroid-sparing effect of wormwood (Artemisia absinthium) in Crohn's disease: A double-blind placebo-controlled study. *Phytomedicine*, *14*(2–3), 87–95. https://doi.org/10.1016/j.phymed.2007.01.001

Parada Venegas, D., De la Fuente, M. K., Landskron, G., González, M. J., Quera, R., Dijkstra, G., … & Hermoso, M. A. (2019). Short chain fatty acids (SCFAs)-mediated gut epithelial and immune regulation and its relevance for inflammatory bowel diseases. Frontiers in *I*mmunology, 10*,* 277. https://doi.org/10.3389/fimmu.2019.00277

Park, S., Chun, J., Han, K.-D., Soh, H., Kang, E. A., Lee, H. J., … & Kim, J. S. (2019). Dose–response relationship between cigarette smoking and risk of ulcerative colitis: A nationwide population-based study. *Journal of Gastroenterology*, *54*, 881–890. https://doi.org/10.1007/s00535-019-01589-3

Paul, G., Khare, V., & Gasche, C. (2012). Inflamed gut mucosa: Downstream of interleukin-10. European *J*ournal of *C*linical *I*nvestigation, *42*(1), 95–109. https://doi.org/10.1111/j.1365-2362.2011.02552.x

Peña-Gil, N., Santiso-Bellón, C., Gozalbo-Rovira, R., Buesa, J., Monedero, V., & Rodríguez-Díaz, J. (2021). The role of host glycobiology and gut microbiota in rotavirus and norovirus infection, an update. *International Journal of Molecular Sciences*, *22*(24), 13473. https://doi.org/10.3390/ijms222413473

Pesce, M., Esposito, G., & Sarnelli, G. (2018). Endocannabinoids in the treatment of gastrointestinal inflammation and symptoms. *Current Opinion in Pharmacology*, *43*, 81–86. https://doi.org/10.1016/j.coph.2018.08.009

Piccioni, A., Franza, L., Brigida, M., Zanza, C., Torelli, E., Petrucci, M., Nicolò, R., Covino, M., Candelli, M., Saviano, A., Ojetti, V., & Franceschi, F. (2021). Gut microbiota and acute diverticulitis: Role of probiotics in management of this delicate pathophysiological balance. *Journal of Personalized Medicine*, *11*(4), 298. https://doi.org/10.3390/jpm11040298

Pickard, J. M., Zeng, M. Y., Caruso, R., & Núñez, G. (2017). Gut microbiota: Role in pathogen colonization, immune responses, and inflammatory disease. *Immunological Reviews*, *279*(1), 70–89. https://doi.org/10.1111/imr.12567

Qiu, P., Ishimoto, T., Fu, L., Zhang, J., Zhang, Z., & Liu, Y. (2022). The gut microbiota in inflammatory bowel disease. *Frontiers in Cellular and Infection Microbiology*, *12,* 1–14. https://doi.org/10.3389/fcimb.2022.733992

Ramos, G. P., & Papadakis, K. A. (2019). Mechanisms of disease: Inflammatory bowel diseases. *Mayo Clinic Proceedings*, *94*(1), 155–165. https://doi.org/10.1016/J.MAYOCP.2018.09.013

Raedsch, R., Vinson, B., Ottillinger, B., & Holtmann, G. (2018). Early onset of efficacy in patients with functional and motility-related gastrointestinal disorders: A noninterventional study with Iberogast®. *Wiener Medizinische Wochenschrift*, 168(3), 89–98. https://doi.org/10.1007/s10354-017-0578-y

Rawla, P., Sunkara, T., & Raj, J. P. (2018). Role of biologics and biosimilars in inflammatory bowel disease: Current trends and future perspectives. *Journal of Inflammation Research*, *11*, 215–226. https://doi.org/10.2147/JIR.S165330.

Rescigno, M. (2023). Microbiota in reflux esophagitis and peptic ulcer disease. *Surgical Infections*, *24*(3), 245–249. https://doi.org/10.1089/sur.2023.021

Robrahn, L., Jiao, L., & Cramer, T. (2020). Barrier integrity and chronic inflammation mediated by HIF-1 impact on intestinal tumorigenesis. *Cancer Letters* , *490*, 186–192. https://doi.org/10.1016/j.canlet.2020.07.002

Roy, S., & Dhaneshwar, S. (2023). Role of prebiotics, probiotics, and synbiotics in management of inflammatory bowel disease: Current perspectives. *World Journal of Gastroenterology*, *29*(14), 2078–2100. https://doi.org/10.3748/wjg.v29.i14.2078

Saez, A., Herrero-Fernandez, B., Gomez-Bris, R., Sánchez-Martinez, H., & Gonzalez-Granado, J. M. (2023a). Pathophysiology of inflammatory bowel disease: Innate immune system. *International Journal of Molecular Sciences, 24*(2), 1526. https://doi.org/10.3390/IJMS24021526

Saez, A., Herrero-Fernandez, B., Gomez-Bris, R., Sánchez-Martinez, H., & Gonzalez-Granado, J. M. (2023b). Pathophysiology of inflammatory bowel disease: Innate immune system. *International Journal of Molecular Sciences*, *24*(2), 1526. https://doi.org/10.3390/ijms24021526

Saha, L. (2014). Irritable bowel syndrome: Pathogenesis, diagnosis, treatment, and evidence-based medicine. *World Journal of Gastroenterology*, *20*(22), 6759. https://doi.org/10.3748/wjg.v20.i22.6759

Sandborn, W. J., Targan, S. R., Byers, V. S., Rutty, D. A., Mu, H., Zhang, X., & Tang, T. (2013). Andrographis paniculataExtract (HMPL-004) for active ulcerative colitis. *Official Journal of the American College of Gastroenterology| ACG*, *108*(1), 90–98. https://doi.org/10.1038/ajg.2012.340

Sandys, O., & Te Velde, A. (2022). Raising the alarm: Environmental factors in the onset and maintenance of chronic (low-grade) inflammation in the gastrointestinal tract. *Digestive Diseases and Sciences*, *67*(9), 4355–4368. https://doi.org/10.1007/s10620-021-07327-1

Santacroce, G., Lenti, M. V., & Di Sabatino, A. J. C. (2022). Therapeutic targeting of intestinal fibrosis in Crohn's disease. Cells, *11*(3), 429. https://doi.org/10.3390/cells11030429

Santos, G. M., Neves, F. D. A. R., & Amato, A. A. (2015). Thermogenesis in white adipose tissue: An unfinished story about PPARγ. Biochimica et Biophysica Acta (BBA)-General Subjects, *1850*(4), 691–695. https://doi.org/10.1016/j.bbagen.2015.01.002

Schoultz, I., & Keita, Å. (2019). Cellular and molecular therapeutic targets in inflammatory bowel disease—Focusing on intestinal barrier function. *Cells*, *8*(2), 193. https://doi.org/10.3390/cells8020193

Schützhold, V., Gravemeyer, J., Bicker, A., Hager, T., Padberg, C., Schäfer, J., … & Hankeln, T. (2022). Knockout of factor-inhibiting hif (Hif1an) in colon epithelium attenuates chronic colitis but does not reduce colorectal cancer in mice. *The Journal of Immunology*, *208*(5), 1280–1291. https://doi.org/10.4049/jimmunol.2100418

Shariati, A., Fallah, F., Pormohammad, A., Taghipour, A., Safari, H., chirani, A. salami, Sabour, S., Alizadeh-Sani, M., & Azimi, T. (2019). The possible role of bacteria, viruses, and parasites in initiation and exacerbation of irritable bowel syndrome. *Journal of Cellular Physiology*, *234*(6), 8550–8569. https://doi.org/10.1002/jcp.27828

Solinas, G., Germano, G., Mantovani, A., & Allavena, P. (2009). Tumor-associated macrophages (TAM) as major players of the cancer-related inflammation. *Journal of Leukocyte Biology*, *86*(5), 1065–1073. https://doi.org/10.1189/jlb.0609385

Song, C., Yang, J., Ye, W., Zhang, Y., Tang, C., Li, X., … & Xie, Y. (2019). Urban–rural environmental exposure during childhood and subsequent risk of inflammatory bowel disease: A meta-analysis. *Expert Review of Gastroenterology & Hepatology*, *13*(6), 591–602. https://doi.org/10.1080/17474124.2018.1511425

Song, G., Fiocchi, C., & Achkar, J. P. (2019). Acupuncture in inflammatory bowel disease. *Inflammatory Bowel Diseases*, *25*(7), 1129–1139. https://doi.org.1093/ibd/izy371

Souza-Tavares, H., Miranda, C. S., Vasques-Monteiro, I. M. L., Sandoval, C., Santana-Oliveira, D. A., Silva-Veiga, F. M., … & Souza-Mello, V. (2023). Peroxisome proliferator-activated receptors as targets to treat metabolic diseases: Focus on the adipose tissue, liver, and pancreas. World Journal of Gastroenterology, *29*(26), 4136. https://doi.org.10.1080/17474124.2018.1511425

Speca, S., Rousseaux, C., Dubuquoy, C., Rieder, F., Vetuschi, A., Sferra, R., … & Gaudio, E. (2016). Novel PPARγ modulator GED-0507-34 levo ameliorates inflammation-driven intestinal fibrosis. *Inflammatory Bowel Diseases*, *22*(2), 279–292. https://doi.org.10.1097/MIB.0000000000000618

Spencer, L., Olawuni, B., & Singh, P. (2022). Gut virome: Role and distribution in health and gastrointestinal diseases. *Frontiers in Cellular and Infection Microbiology*, *12*, 1–11. https://doi.org/10.3389/fcimb.2022.836706

Stegelmeier, A. A., van Vloten, J. P., Mould, R. C., Klafuric, E. M., Minott, J. A., Wootton, S. K., … & Karimi, K. J. V. (2019). Myeloid cells during viral infections and inflammation. *Viruses*, *11*(2), 168. https://doi.org/10.3390/v11020168

Sterling, K. G., Dodd, G. K., Alhamdi, S., Asimenios, P. G., Dagda, R. K., De Meirleir, K. L., Hudig, D., & Lombardi, V. C. (2022). Mucosal immunity and the gut-microbiota-brain-axis in neuroimmune disease. *International Journal of Molecular Sciences*, *23*(21), 13328. https://doi.org/10.3390/ijms232113328

Størsrud, S., Pontén, I., & Simrén, M. (2015). A pilot study of the effect of Aloe barbadensis Mill. extract (AVH200®) in patients with irritable bowel syndrome: A randomized, double-blind, placebo-controlled study. *Journal of Gastrointestinal Liver Disease*, 24(3), 275–280. https://doi.org/10.15403/jgld.2014.1121.243.sst

Sun, J., Fang, F., Olén, O., Song, M., Halfvarson, J., Roelstraete, B., … & Ludvigsson, J. F. (2023). Long-term risk of inflammatory bowel disease after endoscopic biopsy with normal mucosa: A population-based, sibling-controlled cohort study in Sweden. PLoS Medicine, *20*(2), e1004185. https://doi.org/10.1371/journal.pmed.1004185

Talapko, J., Včev, A., Meštrović, T., Pustijanac, E., Jukić, M., & Škrlec, I. (2022). Homeostasis and dysbiosis of the intestinal microbiota: Comparing hallmarks of a healthy state with changes in inflammatory bowel disease. *Microorganisms*, *10*(12), 2405. https://doi.org/10.3390/microorganisms10122405

Tamburini, B., La Manna, M. P., La Barbera, L., Mohammadnezhad, L., Badami, G. D., Shekarkar Azgomi, M., Dieli, F., & Caccamo, N. (2022). Immunity and nutrition: The right balance in inflammatory bowel disease. *Cells*, *11*(3), 455. https://doi.org/10.3390/cells11030455

Tam, J. S., Coller, J. K., Hughes, P. A., Prestidge, C. A., & Bowen, J. M. (2021). Toll-like receptor 4 (TLR4) antagonists as potential therapeutics for intestinal inflammation. Indian Journal of Gastroenterology, *40*, 5–21. https://doi.org/10.1007/s12664-020-01114-y

Triner, D., Castillo, C., Hakim, J. B., Xue, X., Greenson, J. K., Nuñez, G., … & Shah, Y. M. (2018). Myc-associated zinc finger protein regulates the proinflammatory response in colitis and colon cancer via STAT3 signaling. Molecular and Cellular Biology, *38*(22), e00386–00318. https://doi.org/10.1128/MCB.00386-18

Tsai, H.-C., Velichko, S., Hung, L.-Y., & Wu, R. J. (2013). IL-17A and Th17 cells in lung inflammation: An update on the role of Th17 cell differentiation and IL-17R signaling in host defense against infection. Molecular and Cellular Biology, 2013, 1–12. https://doi.org/10.1155/2013/267971

Uhlig, H. H., & Muise, A. M. (2017). Clinical genomics in inflammatory bowel disease. *Trends in Genetics*, *33*(9), 629–641. https://doi.org/10.1016/j.crohns.2014.02.025

Vaishnavi, C. (2013). Translocation of gut flora and its role in sepsis. *Indian Journal of Medical Microbiology*, *31*(4), 334–342. https://doi.org/10.4103/0255-0857.118870

van Tilburg, M. A., Palsson, O. S., Ringel, Y., & Whitehead, W. E. (2014). Is ginger effective for the treatment of irritable bowel syndrome? A double blind randomized controlled pilot trial. *Complementary Therapies in Medicine*, *22*(1), 17–20. https://doi.org/10.1016/j.ctim.2013.12.015

Van Welden, S., Selfridge, A. C., & Hindryckx, P. (2017). Intestinal hypoxia and hypoxia-induced signalling as therapeutic targets for IBD. *Nature Reviews Gastroenterology & Hepatology*, *14*(10), 596–611. https://doi.org/10.1038/nrgastro.2017.101

Vipond, I. B. (2001). The role of viruses in gastrointestinal disease in the home. *Journal of Infection*, *43*(1), 38–41. https://doi.org/10.1053/jinf.2001.0846

Wang, L., Zhao, M., Guo, C., Wang, G., Zhu, F., Wang, J., …Shi, Y. (2016). PDCD4 deficiency aggravated colitis and colitis-associated colorectal cancer via promoting IL-6/STAT3 pathway in mice. I*nflammatory Bowel Diseases*, *22*(5), 1107–1118. https://doi.org/10.1097/MIB.0000000000000729

Wang, N., Wang, S., Xu, B., Liu, F., Huo, G., & Li, B. (2021). Alleviation effects of Bifidobacterium animalis subsp. lactis XLTG11 on dextran sulfate sodium-induced colitis in mice. *Microorganisms*, *9*(10), 2093. https://doi.org/10.3390/microorganisms9102093

Wang, X., Chen, J., & Zheng, J. (2023). The roles of COX-2 in protozoan infection. *Frontiers in Immunology*, *14,* 1–13. https://doi.org/10.3389/fimmu.2023.955616.

Wang, X., Guo, L., Huang, J., Jiang, S., Li, N., Mu, H.-H., & Xu, C. (2023). Plasminogen activator inhibitor-1 potentiates neutrophil infiltration and tissue injury in colitis. International Journal of Biological Sciences, *19*(7), 2132. https://doi.org/10.7150/ijbs.75890

Washabau, R. J., Day, M. J., Willard, M. D., Hall, E. J., Jergens, A. E., Mansell, J., Minami, T., & Bilzer, T. W. (2010). Endoscopic, biopsy, and histopathologic guidelines for the evaluation of gastrointestinal inflammation in companion animals. *Journal of Veterinary Internal Medicine*, *24*(1), 10–26. https://doi.org/10.1111/J.1939-1676.2009.0443.X.

Xiao, L., Liu, Q., Luo, M., & Xiong, L. (2021). Gut microbiota-derived metabolites in irritable bowel syndrome. *Frontiers in Cellular and Infection Microbiology*, *11*. https://doi.org/10.3389/fcimb.2021.729346

Xie, L., Xue, X., Taylor, M., Ramakrishnan, S. K., Nagaoka, K., Hao, C., … & Shah, Y. M. (2014). Hypoxia-inducible factor/MAZ-dependent induction of caveolin-1 regulates colon permeability through suppression of occludin, leading to hypoxia-induced inflammation. Molecular and Cellular Biology, *34*(16), 3013–3023. https://doi.org10.1128/MCB.00324-14

Xu, S., Li, X., Zhang, S., Qi, C., Zhang, Z., Ma, R., Xiang, L., Chen, L., Zhu, Y., Tang, C., Bourgonje, A. R., Li, M., He, Y., Zeng, Z., Hu, S., Feng, R., & Chen, M. (2023). Oxidative stress gene expression, DNA methylation, and gut microbiota interaction trigger Crohn's disease: A multi-omics Mendelian randomization study. *BMC Medicine*, *21*(1), 179. https://doi.org/10.1186/s12916-023-02878-8

Xue, X., & Falcon, D. M. (2019). The role of immune cells and cytokines in intestinal wound healing. *International Journal of Molecular Sciences*, *20*(23), 6097. https://doi.org/10.3390/ijms20236097

Xue, X., Ramakrishnan, S., Anderson, E., Taylor, M., Zimmermann, E. M., Spence, J. R., … & Shah, Y. M. (2013). Endothelial PAS domain protein 1 activates the inflammatory response in the intestinal epithelium to promote colitis in mice. Gastroenterology, *145*(4), 831–841. https://doi.org/10.1053/j.gastro.2013.07.010

Yan, J.-B., Luo, M.-M., Chen, Z.-Y., & He, B.-H. (2020). The function and role of the Th17/Treg cell balance in inflammatory bowel disease. Journal of Immunology Research, 2020, 1–8. https://doi.org/10.1155/2020/8813558

Yue, B., Luo, X., Yu, Z., Mani, S., Wang, Z., & Dou, W. (2019). Inflammatory bowel disease: A potential result from the collusion between gut microbiota and mucosal immune system. *Microorganisms*, 7(10), 440. https://doi.org/10.3390/microorganisms7100440

Yue, B., Luo, X., Yu, Z., Mani, S., Wang, Z., & Dou, W. (2019a). Inflammatory bowel disease: A potential result from the collusion between gut microbiota and mucosal immune system. *Microorganisms*, *7*(10), 440. https://doi.org/10.3390/microorganisms710.0440

Yue, N. N., Xu, H. M., Xu, J., Zhu, M. Z., Zhang, Y., Tian, C. M., ... & Wang, L. S. (2023). Therapeutic potential of gene therapy for gastrointestinal diseases: Advancements and future perspectives. *Molecular Therapy-Oncolytics*, *30*, 193–215. https://doi.org/10.1016/j.omto.2023.08.007

Zhang, H. M., Yuan, S., Meng, H., Hou, X. T., Li, J., Xue, J. C., ... & Zhang, Q. G. (2022). Stem cell-based therapies for inflammatory bowel disease. *International Journal of Molecular Sciences*, *23*(15), 8494. https://doi.org/10.3390/ijms23158494

Zhang, L., Zhao, M., & Fu, X. (2023). Gastric microbiota dysbiosis and Helicobacter pylori infection. *Frontiers in Microbiology*, *14*, 1153269. https://doi.org/10.3389/FMICB.2023.1153269/FULL

Zhang, Y., Chai, N., Wei, Z., Li, Z., Zhang, L., Zhang, M., … & Zhang, B. (2022). YYFZBJS inhibits colorectal tumorigenesis by enhancing Tregs-induced immunosuppression through HIF-1α mediated hypoxia in vivo and in vitro. *Journal of Phytomedicine,* 98, 153917. https://doi.org/10.1016/j.phymed.2021.153917

Zhang, Z., Chen, L., Qian, Z. M., Li, H., Cai, M., Wang, X., …Shao, Z. (2022). Residential green and blue space associated with lower risk of adult onset inflammatory bowel disease: Findings from a large prospective cohort study. *Environment International*, 160, 107084. https://doi.org/10.1016/j.envint.2022.107084

Zhao, C., Hu, X., Bao, L., Wu, K., Zhao, Y., Xiang, K., Li, S., Wang, Y., Qiu, M., Feng, L., Meng, X., Zhang, N., & Fu, Y. (2022). Gut dysbiosis induces the development of mastitis through a reduction in host anti-inflammatory enzyme activity by endotoxemia. *Microbiome*, *10*(1), 205. https://doi.org/10.1186/s40168-022-01402-z

3 Diagnosis, Screening, and Prevention Techniques of Gastrointestinal Inflammations

*Hafiza Sehrish Kiani[1], Adeel Ahmad Khan[2], Fahad Thabet Alsulami[3], Yousef Saeed Alqarni[4], Abdullah Saleh Alruwaili[5,6,7], Ahmed Ramdan M. Alanazy[5,6], Abdul Malik[1], Rameesha Abid[1], Mithun Bhowmick[8], and Arshad Farid[9]**

[1]Department of Microbiology, Quaid-i-Azam University, Islamabad, Pakistan

[2]PMAS-Arid Agriculture University, Rawalpindi, Pakistan

[3]Clinical Pharmacy Department, College of Pharmacy, Taif University, Taif, Saudi Arabia

[4]Department of Pharmacy Practice, College of Pharmacy, Imam Abdulrahman Bin Faisal University, Dammam, Saudi Arabia

[5]Emergency Medical Services Program, College of Applied Medical Sciences, King Saud Bin Abdulaziz University for Health Sciences, Al Ahsa, Saudi Arabia

[6]King Abdullah International Medical Research Center, Al Ahsa, Saudi Arabia

[7]School of Health, University of New England, Armidale 2350, New South Wales, Australia

[8]Department of Pharmaceutical Sciences, Bengal College of Pharmaceutical Sciences & Research, Durgapur, West Bengal, India

[9]Gomal Center of Biochemistry and Biotechnology, Gomal University, D.I. Khan, Pakistan

3.1 INTRODUCTION

The mouth is where the gastrointestinal tract (GIT) begins and ends in the anus. Any part of the GIT may become inflamed due to various causes. Inflammatory gastrointestinal illnesses pose a serious threat to global health since they impact a wide range of populations and are not limited by geography. An extensive multinational study conducted through internet questionnaires has revealed that more than 40% of individuals across the globe experience gastrointestinal issues, including diarrhea, irritable bowel syndrome (IBS), or constipation with respective prevalence rates of 4.7%, 4.1% and 11.7% [1].

These have a significant detrimental effect on both the cost of healthcare worldwide and the quality of life (QOL). The two primary illnesses that comprise inflammatory bowel disease (IBD) are Crohn's disease (CD) and ulcerative colitis (UC). Chronic inflammation of the gastrointestinal system, which can happen anywhere along the tract, is what distinguishes inflammatory bowel disease (IBD)[2]. Inflammatory gastrointestinal diseases, such as Crohn's disease subtypes and ulcerative colitis associated with inflammatory bowel disease (IBD), are becoming more common. Because these illnesses are chronic, they frequently cause enduring symptoms, including tiredness, diarrhea, and stomach discomfort, which significantly lowers general well-being. This trend is becoming more and more noticeable in poor countries as well as in developed ones [3]. Due to globalization, lifestyle factors, environmental changes and genetic predispositions, these disorders are increasingly prevalent across various demographics. Over five million people worldwide suffer with ulcerative colitis (UC) and Crohn's disease (CD), of which three million live in Europe and 1.4 million in the United States [4]. Globally, the prevalence of IBD is rising for both adult [5] and pediatric [6] patients, particularly in historically low-incidence areas like Europe's southern and eastern regions, Asia, and South America [7,8].

Globally, gastrointestinal inflammation places a significant financial burden on healthcare systems. Healthcare expenditures are on the rise because many conditions are complicated and necessitate considerable medical interventions, from diagnosis to long-term therapy [9]. The inflammatory process linked to periodontal disease is one of the most prevalent ones. Bacteria in the gingival crevice start and fuel an inflammatory reaction that causes tissue damage, cytokine release, and cell activation. At every stage of this tract, an epithelial barrier divides the body's tissues from the external environment. Certain regions of the tract have distinct types of this epithelial barrier, each with a specific purpose.

DOI: 10.1201/9781003493143-3

3.2 GASTROINTESTINAL INFLAMMATION: OVERVIEW

3.2.1 Definition and Classification

Due to the diverse and complicated nature of gastrointestinal inflammatory disorders, a sophisticated classification system is required to guide personalized treatment plans. This category includes a wide range of illnesses; however it mostly concentrates on inflammatory bowel disease (IBD) and other prominent inflammatory diseases.

3.2.1.1 Inflammatory Bowel Disease (IBD)

Prolonged inflammation of the gastrointestinal system is known as inflammatory bowel disease (IBD). Two kinds of inflammatory bowel illness include Crohn's disease and ulcerative colitis [2]. Chronic inflammation can have negative effects on the gastrointestinal tract. Ulcerative colitis (UC) normally affects only the colon and the rectum and results in inflammation of the mucosal lining. The ileum/terminal ileum and colon are the most often damaged regions, however, Crohn's disease (CD) is typified by inflammation that damages every section of the gastrointestinal tract [10]. The CD is often accompanied by inflammatory conditions outside of the intestines. With the advancement of industrialization and greater access to healthcare, inflammatory bowel disease (IBD) has spread throughout the Western and Eastern hemispheres, affecting millions of patients worldwide, although it was formerly thought to be predominantly a Western illness, affecting over 1.4 million Americans.

IBD is commonly characterized by abdominal pain, or pain in the stomach area, diarrhea, occasionally with blood, urgency to go to the toilet, incontinence in the stools, Loss of weight, rectal bleeding, Anemia, fever, Anxiety, hopelessness and stunted growth in kids suffering from IBD (inflammatory bowel disease) [11]. Due to intestinal inflammation, the illness may also result in edema or lumps. IBD can harm the intestines over time if inflammation is not managed, leading to abscesses, which are areas of infection that can rupture the intestinal wall. Strictures refer to areas of narrowing in the bowel. A fistula is an aberrant junction between two veins or organs that are not supposed to be joined. Bowel pressure and inflammation can lead to tissue breakdown in fistulas, which can result in the leakage of bowel contents into the vagina, urethra, or bladder [12]. Colon cancer risk is elevated by chronic colon inflammation. Some individuals with inflammatory bowel disease (IBD) may experience inflammation outside of their intestines [13]. For example, it can impact the eyes, resulting in redness and inflammation from uveitis (inflammation inside the eye) or episcleritis (inflammation between the inner eyelids and the white of the eye). According to experts, between 10% and 43% of IBD patients experience vision issues; thus, routine eye exams are crucial [14]. In addition, it causes oral infections that result in ulcers and sores. Furthermore, the liver is impacted, and fat builds up there (steatosis) likewise, inflammation of the biliary tract can result in gallstones and bile duct system inflammation (sclerosing cholangitis) [15]. Urinary tract infections, fistulas, hydronephrosis (enlarged kidneys due to a urine backup), and kidney stones can all harm the kidneys. Additionally, it has a significant negative effect on skin, which can result in pyoderma gangrenosum, a rare disorder that causes severe leg ulcers, and erythema nodosum, a condition marked by sensitive, red shin shins[16]. Further, the spine and joints are impacted, leading to arthritis in the limbs, sacroiliitis (inflammation of the joints linking the pelvis and lower spine), and spondylolysis (stress fracture of the vertebrae) [17]. Phlebitis, or inflammation of the blood vessels, is one effect on blood circulation. Colon cancer risk can increase up to four times in people with IBD. The likelihood of malignant alterations in intestinal cells increases with the duration of the period you spend living with an inflammatory colon [13]. Research indicates that individuals with IBD who received their diagnosis at a younger age presented a greater likelihood of acquiring colorectal cancer, especially if they had additional risk factors including primary sclerosing cholangitis or a family history of the illness [15].

A compromised immune system leads to IBD. Gastrointestinal tract inflammation occurs when the immune system reacts inappropriately to environmental stressors, like germs or viruses. These days, intestinal illness in immunocompromised patients with a reduced inflammatory response is primarily caused by opportunistic infections. Remarkably, several bacteria have been linked to male colitis in recent times. These include *Clostridium difficile*, *Campylobacter jejuni*, and enteropathic *Escherichia coli* strains. The initial indications of inflammatory bowel disease (IBD) may arise from exposure to gastrointestinal (GI) infections or medications that cause intestinal irritation, such as aspirin, ibuprofen, and antibiotics. The immune system continues to function even after the irritation or infection disappears.

Moreover, a genetic component seems to be involved. An individual is more prone to experience an inappropriate immune response if they have a family history of IBD. There is a familial tendency in both types of IBD (Figure 3.2). Another set of data indicates that 25% many who suffer with IBD have a family member who also had the illness. In comparison with CD patients, the rate of UC patients is higher. Furthermore, there appears to be a correlation between the degree of closeness between family members and the occurrence of the same disease: the correlation is stronger in monozygotic twins, considerably lower in dizygotic twins, 80% in siblings, and 70% in parent-offspring affectations. According to UC and CD, there are no mixed identical twins [9, 24]. According to research on CD, 10% of asymptomatic first-degree relatives have higher intestinal permeability [25, 26]. This is consistent with the findings. A significant percentage (50–70%) of IBD patients exhibit HLA-B27 positive based on histocompatibility antigens (lower than in idiopathic ankylosing spondylitis). Uncomplicated CD showed an increase in B62. This may protect against musculoskeletal complaints but also increase the risk of IBD [23]. Despite this, it is still unclear whether environmental factors or genetics are responsible. It is hypothesized that the cause could be a mix of the two: a genetic predisposition that enhances the development of the diseases when exposed to the cause [24]. UC is more immunologic, whereas CD is more genetic, or deterministic. Diabetes is similar to this: CD is more like diabetes 2, while UC is more like diabetes 1.

3.2.1.1.1 Crohn's Disease

The most common sign of this kind of IBD is inflammation of the stomach lining, which frequently affects the internal organs that are located deeper within the digestive system. Crohn's disease primarily affects the small intestine; however, the upper gastrointestinal tract may also be impacted, and less commonly, the large intestine. Transmural inflammation may impact any digestive tract area and is a characteristic feature of Crohn's disease. This persistent illness, which can affect the mouth and anus, is well-known for its erratic involvement pattern that frequently leads to skip lesions. The gut wall's entire thickness is broken through by the inflammatory process, which results in problems such as strictures, fistulas, and abscesses.

3.2.1.1.2 Ulcerative Colitis

This disorder affects the lining of your colon and large intestine (large intestine) and causes ulcers (sores). The most common feature of ulcerative colitis is persistent mucosal inflammation, which is mostly colon-specific. Typically, ulcerative colitis starts in the rectum and progresses constantly proximally throughout the colon, unlike Crohn's disease, which can affect any part of the digestive tract. Bloody diarrhea, urgency, and abdominal pain are among the symptoms of this illness, which mostly affects the inner lining of the colon and rectum.

3.2.1.2 Other Inflammatory Disorders of the Gastro Intestines

Apart from inflammatory bowel disease (IBD), this classification is made more complex overall by several other gastrointestinal illnesses. Among them are:

1) ***Irritable Bowel Syndrome (IBS):*** Irritable bowel syndrome is characterized by gastrointestinal symptoms such as bloating, pain in the abdomen, and changes in bowel habits [18]. However, unlike inflammatory bowel illnesses, IBS is not predominantly characterized by inflammation. A functional gastrointestinal disorder is IBS, meaning it affects the digestive system's function without causing inflammation or structural damage.
2) ***Clostridium difficile Infection:*** Bacteria *Clostridium difficile* (*C. difficile*) can result in colonic inflammation, causing symptoms like diarrhea, abdominal pain, and fever [19]. While gastrointestinal inflammation is a symptom, the primary cause of this disorder is an overgrowth of *C. difficile* bacteria, classifying it as an infectious disease rather than a chronic inflammatory condition like inflammatory bowel disease.
3) ***Gastritis:*** Gastritis is the medical term for inflammation of the stomach lining. Chronic gastritis is typically brought on by persistent irritation of the stomach lining, whereas acute gastritis can be brought on by infections (such as *Helicobacter pylori*) as well as the use of nonsteroidal anti-inflammatory medicines NSAIDs [20].
4) ***Peptic Ulcer Disease:*** The development of open sores or ulcers in the upper section of the small intestine (duodenal ulcers) or the lining of the stomach (gastric ulcers) is characteristic of peptic ulcer disease [21]. In addition to *Helicobacter pylori* infection and the administration of nonsteroidal anti-inflammatory medicines (NSAIDs), inflammation contributes significantly to the development and spread of ulcers.
5) ***Colon Cancer:*** Colon cancer is characterized by the formation of cancerous tumors in the colon or rectum. Although inflammatory bowel disease is characterized by ongoing gastrointestinal tract inflammation, it can elevate the likelihood of developing colon cancer [13]. It is important to note that colon cancer is not generally classified as a gastrointestinal inflammatory disorder.
6) ***Esophageal Cancer:*** Malignant tumors growing in the esophagus are an indicator of esophageal cancer. Chronic inflammation of the esophagus, such as in gastro-esophageal reflux disease, can raise the likelihood of developing esophageal cancer, similar to colon cancer [22]. The ailment known as gastro-esophageal reflux disease (GERD) is typified by the reflux of stomach acid into the esophagus, leading to esophageal lining irritation and inflammation. Chronic inflammation in gastro-esophageal reflux disease (GERD) can lead to consequences such as the development of esophageal ulcers, strictures, or Barrett's esophagus. However, it is not generally categorized as a gastrointestinal inflammatory disorder.
7) ***Gastric Cancer:*** Gastric cancer is the term used to describe the development of cancerous cells in the stomach. Chronic inflammation of the stomach lining, such as in chronic gastritis, can elevate the likelihood of developing gastric cancer, similar to colon and esophageal tumors [23]. However, it is not primarily classified as a gastrointestinal inflammatory illness.
8) ***Eosinophilic Esophagitis:*** Eosinophilic Esophagitis is a condition when the esophagus becomes inflamed due to an allergic reaction to specific foods or environmental allergens. Eosinophils, a subtype of leukocytes, invade the esophagus tissue, resulting in symptoms such as dysphagia, thoracic discomfort, and food obstruction [24].
9) ***Pancreatitis:*** Pancreatitis is characterized by the inflammation of the pancreas, which is situated behind the stomach. Acute pancreatitis commonly arises from gallstones or excessive alcohol intake; however chronic pancreatitis can develop due to prolonged inflammation and pancreatic injury [25].
10) ***Gastroenteritis:*** The primary feature of gastroenteritis is gastrointestinal tract inflammation, which is frequently brought on by bacterial, viral, or parasite diseases. Diarrhea, cramping in the abdomen, nausea,

and vomiting are common symptoms [26]. Acute gastroenteritis normally goes away on its own, but severe or persistent cases might need to see a doctor.

11) ***Celiac Disease:*** An autoimmune condition known as celiac disease is activated by the consumption of gluten, a protein present in wheat, barley, and rye [27]. Gluten ingestion causes the lining of the small intestine to become inflamed and damaged in people with celiac disease, which causes loss of nutrients and a variety of gastrointestinal symptoms [28].
12) ***Microscopic Colitis:*** The symptoms of the disorder microscopic colitis are long-term inflammation of the colon that is only visible through a microscope. It has two subcategories: collagenous colitis and lymphocytic colitis [29]. Common symptoms of this condition may encompass persistent, liquid bowel movements, discomfort in the abdominal region, and a reduction in body mass.
13) ***Behçet's Disease:*** It is an uncommon autoimmune condition characterized by widespread inflammation of blood vessels in various body parts, including the gastrointestinal tract [27]. This can result in various symptoms such as abdominal pain, diarrhea, and ulcers in the oral and vaginal regions.
14) ***Ischemic Colitis:*** Ischemic colitis is a condition characterized by a decrease in blood flow to the colon, resulting in inflammation and damage to the tissue of the colon. It frequently arises from diminished blood flow resulting from disorders such as atherosclerosis, hypotension, or thrombosis [30].
15) ***Diverticulitis:*** Diverticulitis is characterized by inflammation or infection of tiny pouches, or diverticula, which can form in the colon's walls. Abscesses, fever, and altered bowel habits are some of the symptoms that can originate from these pouches' irritation or infection, which are called diverticula. A perforation or the development of an abscess is a possible consequence in severe situations [31].
16) ***Appendicitis:*** The little organ that resembles a pouch that is connected to the big intestine is called the appendix, and appendicitis is characterized by inflammation of this organ. Anxiety, vomiting, nausea, and sudden, intense stomach pain are common symptoms of this inflammatory illness. The usual therapy for appendicitis is prompt surgical intervention, usually in the form of an appendectomy [32].

3.3 SCREENING APPROACHES

3.3.1 Screening for Anemia

Anemia is a common disorder that affects organs other than the intestines and is frequently linked to inflammatory bowel disease (IBD) [33]. A study revealed that 33% of persons diagnosed with inflammatory bowel disease (IBD) had hemoglobin levels below 12 g/dL[34]. Anemia is a significant challenge for the medical treatment of individuals with chronic conditions and is closely linked to their quality of life [35]. Anemia can result from several different disease causes and is a prevalent illness among people with inflammatory bowel disease (IBD). The primary factors contributing to anemia in these individuals include persistent blood loss resulting from inflammation of the intestinal mucosa, as well as insufficiencies in micronutrients such as iron and B12. Moreover, hemolysis, medication-induced myelosuppression, and chronic inflammation can all contribute to and regulate anemia in individuals with inflammatory bowel disease [36]. Iron deficiency and chronic disease-related anemia are two of the most common reasons for anemia in people with IBD. Anemia is a medical condition marked by a reduction in the amount of hemoglobin in the blood below a specific level. Anemia is defined by the World Health Organization as hemoglobin levels below 12 g/dL (hematocrit below 36%) in females who are not pregnant, below 11 g/dL (hematocrit below 33%) in females who are pregnant, and below 13 g/dL (hematocrit below 39%) in males [37]. Hemoglobin levels below 10 g/dL characterize severe iron deficiency anemia (IDA). Whenever a patient meets the criteria established by the World Health Organization (WHO) for anemia, it is necessary to begin a thorough investigation to determine the underlying cause of the anemia.

The fundamental assessment comprises an in-depth examination of blood components, such as red and white blood cells, as well as an assessment of transferrin, levels of C-reactive protein, transferrin saturation, and serum ferritin. It is recommended to conduct further tests if the reason for the anemia is not yet known even after completing the preliminary examination mentioned previously. Additional examinations comprise folic acid, vitamin B12, haptoglobin, creatinine, lactate dehydrogenase, and reticulocyte counts [38]. According to experts and standard clinical practice, it is suggested to screen for anemia in patients with inflammatory bowel disease (IBD) who are receiving outpatient care. Those Individuals with active disease should have screenings at least once every three months while individuals in remission or with moderate disease should have screenings once every six to twelve months [38]. However, screening for anemia is not necessary for patients who are hospitalized. Initiating treatment for anemia according to its underlying cause should be done promptly. For patients with iron deficiency anemia (IDA) with a hemoglobin level over 10 g/dL, who have a mild level of activity or have inactive inflammatory bowel disease (IBD), oral iron supplements are a sufficient treatment. As per an internationally agreed statement, the most recommended method of providing iron supplements for patients with inflammatory bowel disease (IBD) is through intravenous administration. This is especially true if the patient's hemoglobin (Hgb) level is below 10 g/dL when the disease is active [38,39]. Patients who are susceptible to anemia, such as those with inflammatory bowel disease (IBD), may benefit from early detection and treatment of iron deficiency, which can successfully lower hospital visits, improve quality of life, reduce missed work, and ultimately result in lower healthcare costs [38,39].

3.3.2 Screening for Medication Non-Adherence

Poorer disease outcomes are the result of non-adherence to IBD therapy [40,41]. Based on a recent case-control research, around 33% of persons who have been diagnosed with inflammatory bowel disease (IBD) exhibit a low level of adherence to their prescribed treatment regime [40]. The study identified several factors that were associated with low adherence [40]. These factors include being under the age of 40, having a higher education level, being single, and taking mesalamine. Curiously, it was discovered that being self-employed acts as a safeguard [40]. According to a different study, Anti-TNF agent response was found to be significantly correlated with non-adherence. In that study, it was discovered that there were two important predictors of non-adherence: fewer information about the prescribed medication and longer intervals ($\geq$ 3 months) between outpatient clinic visits [41]. When combined, inadequate understanding of IBD medicine and incorrect self-perceptions of illness contribute significantly to non-adherence in IBD patients, underscoring the pressing need for patient awareness and adherence-improving therapies. Various methods are available to evaluate medication adherence, including medicine refill rates, pill counts, analysis of drug metabolites, and direct questioning regarding medication adherence. Patients who are at risk of not adhering to their IBD treatment can be identified using standardized questionnaires, for instance the 8-item Morisky Medication Adherence Scale (MMAS-8) [42]. The MMAS-8 has been criticized for relying on self-reported data and being validated on a sample of patients with chronic conditions, which may not accurately represent individuals with inflammatory bowel disease (IBD) [42]. As a result, researchers have developed particular screening measures for patients with inflammatory bowel disease (IBD) to identify those who are at risk of not adhering to their medicine. These tools include the modified IBD-MMAS-8 and the 10-item mesalamine non-adherence questionnaire [43]. The MMAS-8 shows some evidence of effectively identifying IBD patients who are at risk of not adhering to treatment [43]. There are two conceptual frameworks into which medication non-adherence can be categorized: inadvertent non-adherence and deliberate non-adherence [42,44]. Personalized algorithms may be created based on these categories to enhance patient empowerment, education, and follow-up. The most efficacious strategies to address accidental non-adherence, such as patient education, employing reminder systems, simplifying the treatment regimen and utilizing organizational tools like pill boxes, are likely the most beneficial. On the other hand, when it comes to people who display intentional non-adherence, tactics like problem-solving training, motivating counseling, and treating dysfunctional family dynamics are more likely to be successful [44].

3.3.3 Screening for Malnutrition and Micronutrient Deficiencies

Micronutrient status should be assessed in IBD patients exhibiting clinical symptoms, and any deficits found should be addressed [45]. It is recommended to do screenings for common micronutrient deficiencies, for instance folate, 25-hydroxyvitamin D, and iron in individuals with inflammatory bowel disease who do not show any clinical symptoms [45,46]. A high risk of vitamin B12 insufficiency exists in patients with CD who have had substantial intestinal resection. Therefore, it is advisable to assess their serum levels of vitamin B12 [45,46]. Currently, there are no established protocols for evaluating micronutrient deficits in individuals with inflammatory bowel disease (IBD), and the recommendations provided are solely derived from expert judgment [46].

3.3.4 Screening for Cancer

3.3.4.1 Skin Cancer

Numerous observational studies have demonstrated that people with inflammatory bowel disease (IBD) are more likely to develop non-melanoma skin cancer (NMSC) and melanoma skin cancer (melanoma) [47,48]. Patients with IBD who receive treatment for their condition with anti-TNF medications, antimetabolites (AZA and 6-MP), or combination therapy (anti-TNF + AZA or 6MP) are more likely to develop non-melanoma skin cancer (NMSC) [48]. Patients should be advised to limit their exposure to ultraviolet (UV) radiation. To protect oneself from harmful UV radiation, it is advisable to wear clothing that covers the body, sunglasses that block UV rays, and a hat with a wide brim whenever possible. Moreover, it is imperative to consistently use sunscreen with a minimum sun protection factor (SPF) of 15 that offers complete protection against UVA and UVB rays. Lastly, it is recommended to avoid using tanning beds altogether. Patients should be recommended to wear a water-resistant sunscreen with a sun protection factor (SPF) of 30 or higher that shields them from UVA and UVB rays when engaging in prolonged outdoor activities. Patients receiving immunosuppressive medication should be encouraged to undergo an annual physician skin inspection in addition to a monthly skin self-examination [49].

3.3.4.2 Breast Cancer

Among females in the United States, breast cancer is considered to be the most common kind of cancer and the second most common cause of death from cancer [50]. There is limited data indicating that mothers of patients with CD (celiac disease) may have a twofold greater incidence of breast cancer compared to a control group. Further data indicates that individuals with celiac disease (CD) who also have breast cancer receive less invasive treatment compared to control subjects, resulting in worse survival rates compared to patients without CD [51]. Individuals with inflammatory bowel disease (IBD) are recommended to undergo breast cancer screening because they may have an increased risk of developing the disease. Thus, they should adhere to the American Cancer Society (ACS) protocols. According to these recommendations, women with a normal risk level should undergo counseling to enhance their knowledge of breast cancer signs. As early as age 20, women should begin having clinical breast exams, or CBEs, every three years. Upon reaching the age of 40, individuals should get an annual mammogram and clinical breast examination (CBE) [50]. Women who have a higher chance of contracting IBD (inflammatory bowel disease) should be identified as those who have a genetic syndrome (such as BRCA mutation) that is connected to a higher risk of breast cancer, individuals who have a first-degree relative with such a syndrome, or those who have a risk of developing breast cancer that is greater than 20% based on risk-estimation models over their lifetime. These people should be screened per the standards outlined by the American Cancer Society (ACS) for individuals at maximum risk [50].

3.3.4.3 Cervical Cancer

Compared to the general population, females with inflammatory bowel disease (IBD) have a higher frequency of abnormal Pap smears. This risk is significantly increased by the use of immunosuppressive medicine (IST) [52]. Alternatively, additional research suggests that The connection between IBD, or inflammatory bowel disease, and cervical dysplasia may result from the co-occurrence of smoking and IBD, rather than just from the diagnosis of IBD alone [53]. According to Kane [52], the prevalence of abnormal Pap smears in IBD-afflicted women was 42.5%, while the frequency in controls who were matched in terms of parity, sex, and age was 7%. It is crucial to note that any abnormal test results were linked to either HPV type 16 or 18. This emphasizes how important it is to test female patients with inflammatory bowel disease for HPV infection and cervical cancer before to, during, or following immunosuppressive therapy [52]. Women diagnosed with inflammatory bowel disease (IBD) who are between the ages of 11 and 26 should consider the HPV vaccination. As per the recommendations established by the American College of Obstetrics and Gynecology, it is advised that all women undergoing individualized screening and treatment (IST) should undergo an annual Pap test [54].

3.3.4.4 Prostate Cancer

Prostate cancer is a form of tumor that affects the prostate gland specifically. Prostate cancer screening is debatable since a significant percentage of situations found through screening do not result in symptoms for the affected individuals during their lifetimes [55]. As a result, numerous patients may not experience any advantages from screening and instead may suffer negative consequences from early cancer discovery and treatment [50,55]. As a result, before testing, the patient and the doctor should talk about the possible advantages and hazards of the test. The guidelines set forth by the American Cancer Society highlight the significance of a doctor-patient collaborative decision-making process. They recommend PSA testing for men who are 50 years of age, have an average chance of developing prostate cancer, and have roughly a decade to live [50]. The American Cancer Society (ACS) states that men should start screening for prostate cancer at 40 or 45 years, if one is at a higher risk of developing the disease, for instance, African Americans or those who have a first-degree relative (father, brother, or son) who was diagnosed with the disease when the individual was younger than 65 [50]. On the other hand, because PSA-based screening for prostate cancer does not yield more benefits than disadvantages, the US Preventive Services Task Force opposes PSA screening [55].

3.3.4.5 Colorectal Dysplasia and Cancer

Individuals who have been diagnosed with long-standing inflammatory bowel disease (IBD) have a higher likelihood of acquiring colorectal dysplasia and cancer [56,57]. As per the AGA guidelines, it is recommended that all patients, irrespective of the severity of their condition, should get screened for colonoscopy [57]. There is a consensus that patients with ulcerative colitis (UC) or Crohn's disease (CD) who are diagnosed with confirmed Flat mucosa with low-grade dysplasia ought to be given the option of either proctocolectomy or undergoing another examination within a period of 6 months [56,57]. Patients with high degree dysplasia are advised to undergo preventive surgery, specifically proctocolectomy. For patients who have been suffering from significant Crohn's disease (CD) or ulcerative colitis (UC) that affects the colon for a duration of 8 to 10 years, it is recommended to undergo surveillance colonoscopy every 1 to 3 years [56,57]. There is a current discussion on the most effective number of biopsies to be taken and the endoscopic technique for colorectal cancer surveillance in inflammatory bowel disease (IBD). Nevertheless, a new international agreement proposes that chromoendoscopy may outperform white-light endoscopy in identifying dysplasia [58]. The international consensus statement recommends the use of endoscopic therapy over colectomy for treating polypoid, non-polypoid, and invisible dysplasia in patients with inflammatory bowel disease (IBD) [58]. Furthermore, it is essential to note that the available evidence to support these recommendations is of very low quality [56,57].

3.4 DIAGNOSTIC TECHNIQUES

The first task of clinicians/doctors when a patient appears with gastrointestinal (GI) symptoms is to diagnose them with inflammatory bowel disease (IBD), GI cancer, irritable bowel syndrome (IBS), or another GI issue. After confirming the diagnosis of IBD, the next crucial step is to evaluate the level of disease activity. To ensure appropriate treatment, it is crucial to evaluate the severity of inflammation and extent of bowel involvement.

3.4.1.1 Physical Exam

In order to determine whether a patient has IBD symptoms and signs, doctors perform a physical examination [59]. During the rectal examination, the physician will perform a physical examination with the specific goal of looking for indications of an abdominal abscess, an inflammatory mass, indicators of perianal disease, and anal stenosis. The doctor may also gently press on the belly to check for any pain or soreness. A medical professional may also assess for ocular inflammation, oral ulcers, cutaneous eruptions, joint edema or irritation, hemorrhaging, or other indications of inflammation in the anal region and its vicinity [60].

3.4.1.2 Endoscopy

Endoscopy has also made significant progress in monitoring patients with inflammatory bowel disease (IBD) for the presence of dysplasia. Endoscopic methods, including colonoscopy, upper endoscopy, sigmoidoscopy, and capsule endoscopy, play a crucial role in diagnosing IBD due to their ability to offer precise and comprehensive visualization of the gastrointestinal tract as mentioned in Figure 3.1 [61]. Chromoendoscopy is a procedure that uses indigo carmine or methylene blue to highlight aberrant tissue in the colon's mucosal lining. This allows for precise biopsy instead of random sampling. Multiple investigations, including at least one randomized study, have demonstrated that the utilization of this technique can result in a threefold or even fourfold increase in dysplasia detection rates [62]. Present researchers are examining specific types of light and optics, known as auto-fluorescence, to conduct narrow-band imaging. This technique allows for highlighting the same tissue without the need for dye injection by simply pressing a button on the endoscope. These tests can assist doctors in diagnosing inflammatory bowel disease (IBD) and distinguishing between ulcerative colitis and Crohn's disease [63]. Despite having many similar symptoms, ulcerative colitis and these two diseases affect separate sections of the gastrointestinal tract. The selection of an appropriate endoscopic test is determined by a doctor, taking into consideration your symptoms, medical history, the findings of blood and stool tests, and physical examination.

3.4.1.3 Colonoscopy and Biopsy

Gastroenterologists usually suggest a colonoscopy to diagnose Crohn's disease or ulcerative colitis. Colonoscopy and ileoscopy are important diagnostic procedures that allow for the examination of tissue samples to diagnose chronic inflammation [64]. Nevertheless, this test is intrusive and costly, necessitating significant patient preparation. It would be more desirable to use fewer invasive techniques to evaluate activities. This examination offers real-time images of the colon and rectum, allowing the doctor to inspect the inner lining of the intestines for symptoms of inflammation, ulcers, and other indications of inflammatory bowel disease (IBD). A doctor frequently removes tiny tissue samples from the colon during a colonoscopy and sends them to a lab for analysis. This procedure is known as a biopsy. Biopsy results can assist doctors in confirming a diagnosis of inflammatory bowel disease (IBD) and distinguishing between ulcerative colitis and Crohn's disease [65]. Prior to the procedure, a doctor instructs you to refrain from consuming solid food for duration of 24 hours. In preparation for the examination, the patient is given a laxative solution to drink the night before in order to ensure clear bowel movement.

A colonoscope is a long, thin, flexible tool with an embedded camera and light at the tip that gastroenterologists use to perform a colonoscopy [65]. Doctors introduce the colonoscope through the anus and gradually navigate it along the rectum and colon. The camera projects visual representations of the colon onto a computer monitor. The process generally requires approximately 30 minutes. Prior to the surgery, doctors typically inject a sedative to induce drowsiness and alleviate any discomfort.

3.4.1.4 Flexible Sigmoidoscopy

Similar to a colonoscopy, a flexible sigmoidoscopy examines the rectum and lower colon only. This test may be indicated if the patient appears with symptoms confined to bleeding and inflammation in the vicinity of the anus and rectum. Periodically, doctors conduct this examination to evaluate the efficacy of the treatment [66]. A biopsy might be carried out by the doctors during the process.

3.4.1.5 Upper Endoscopy

If the patient experiences symptoms and laboratory tests indicate a diagnosis of Crohn's disease, the doctor may advise undergoing an upper endoscopy, which is also known as an esophagogastroduodenoscopy [67]. Doctors examine the esophagus, the tube that connects the mouth to the stomach, the stomach, and the first section of the small intestine during this treatment. An upper endoscopy can detect inflammation, ulcers, bleeding, or obstruction in the gastrointestinal tract [61].

The test is carried out by a doctor using a small, flexible device known as an endoscope. Prior to the treatment, the medical professional administers a sedative to the patient in order to induce drowsiness and alleviate any discomfort. Subsequently, a doctor introduces the endoscope into the oral cavity, passing it down the esophagus, stomach, and small intestine. The device features a camera on top, which provides live views on a monitor. During an upper endoscopy, doctors often remove tissue samples for further examination. Laboratory studies can assist in identifying the underlying cause of inflammation. The duration of an upper endoscopy operation normally ranges from 10 to 20 minutes.

3.4.1.6 Capsule Endoscopy

A capsule endoscopy may be recommended if symptoms suggest an issue with the small intestine but other tests such as blood tests, colonoscopy, and upper endoscopy fail to provide a definitive diagnosis [68]. The visualization of the small intestine is challenging compared to other sections of the gastrointestinal system due to its predominantly inaccessible nature. A capsule endoscopy offers enhanced visualization of inflammation, ulcers, or bleeding compared to alternative diagnostic procedures, enabling doctors to figure out whether symptoms are owing to Crohn's disease. Capsule endoscopy is a novel technique that has demonstrated its worth in evaluating the mucosa of the small intestine, an area that was previously inaccessible using traditional endoscopic methods

[68]. If standard testing is unsatisfactory, it is reasonable to consider capsule endoscopy in patients with unexplained abdominal pain who also have another symptom (such as anemia or weight loss), high inflammatory marker values (such as CRP) or a severe family history of IBD [64]. If a patient suffers from abdominal pain in the lower quadrant and experiences weight loss, and standard diagnostic tests such as colonoscopy/ileoscopy and small bowel follow-through do not provide any abnormal results, A capsule endoscopy may be used to detect indications of inflammation or inflammatory bowel disease (IBD) that were previously missed or to confirm a negative diagnosis [64]. Several studies have demonstrated that capsule endoscopy offers an improved diagnostic yield compared to conventional endoscopic methods. Typically, these examinations are intended for patients who are believed to have Crohn's disease but have had negative results from conventional testing.

During this medical procedure, a doctor instructs patients to ingest a little capsule containing a miniature camera. The dimensions of this capsule are comparable to those of a multivitamin. As the camera travels through the digestive system, it captures images, which are transmitted via radio waves to a small receiver. The receiver can either be worn around the patient's waist or kept in their pocket. The capsule requires around eight hours to travel through the gastrointestinal tract [69]. It is eliminated via defecation and disposed of by flushing it down the toilet. Subsequently, the doctor retrieves the photos from the receiver, with particular emphasis on those obtained in the small intestine.

3.4.1.7 Balloon-Assisted Enteroscopy

Double-balloon enteroscopy is a novel endoscopic method that is being done on patients under general anesthesia or heavy sedation by specialized therapeutic endoscopists [70]. In order to move the scope through the GI system and directly view the small bowel, two balloons are used on one scope, and they are alternately inflated and deflated. Although rather time-consuming and somewhat intrusive, this method is quite effective. Additionally, there have been reported cases of perforation. A scope and an apparatus known as an overtube are used for this test [70]. This allows the technician to view the small bowel in areas that are not accessible with a normal endoscope. This process is useful when abnormalities are discovered via capsule endoscopy but the diagnosis is still unclear.

3.4.2 Imaging Tests

Imaging tests, for instance, X-rays, CT, and MRI scans are used in conjunction with endoscopic examinations to provide doctors with more information about the gastrointestinal tract as depicted in Figure 3.1 [71]. CT and MRI techniques are continuously advancing, enabling enhanced detection of abnormalities in the small bowel with greater precision. They may show intestinal lining abnormalities such as bleeding, tears, inflammation, or obstructions that are indicative of inflammatory bowel disease.

If results from previous diagnostic procedures are unclear, imaging studies often assist medical professionals in distinguishing between ulcerative colitis and Crohn's disease. For example, if an imaging test detects inflammation in the small intestine, it is possible that the patient has Crohn's disease, as ulcerative colitis specifically impacts the large intestine. Imaging tests can additionally be employed to observe the gastrointestinal tract and evaluate the efficacy of treatment [71].

3.4.2.1 X-Rays

X-rays employ electromagnetic radiation to capture images of the gastrointestinal tract. When stomach discomfort and constipation arise, doctors may use imaging tests to distinguish between intestinal obstruction and expansion of the gut wall [71].

3.4.2.2 CT Scans and CT Enterography

X-rays are used in computed tomography (CT) scans to provide precise images of the body in two or three dimensions [72]. CT scans of the gastrointestinal system can detect stenosis, which is a narrowing of the small or large intestine, or an obstruction. The test may also detect inflammation in the small intestine, implying that the symptoms may be caused by Crohn's disease. Currently, CT scans can acquire thinner slices at higher resolutions and are faster, reducing the time patients need to remain still. These advancements enable the utilization of diverse radiography techniques for precise imaging of the small bowel [72]. At certain points, doctors may suggest an advanced CT scan called CT enterography [73]. Before the scan, the patient ingests a contrast agent. While passing through the digestive tract, the CT scanner captures images of the small intestine, potentially revealing any anatomical abnormalities present in the liquid. For instance, the contrast liquid is clearly blocked if there is an obstacle [73].

During an imaging test, patients are required to drink a significant amount of oral contrast solution. To emphasize the distinction between the intestinal wall and the intestinal lumen, a negative or neutral oral contrast agent is used instead of a positive one like barium. In addition to this, an intravenous (IV) contrast agent is also administered. To guarantee that the contrasts are completely apparent, the scan is done at a precise time [73]. Inflamed areas of the body tend to have enlarged blood vessels, which can be detected using intravenous contrast. The technique of CT enterography utilizes this principle. Before the scan, the patient is required to arrive 75 minutes early, drink 1.5 liters of neutral or negative contrast and receive intravenous contrast dye as per standard procedure. Then, a quick 16-, 32-, or 64-slice scanner is used to complete the scan. When using the scanning protocol, results are obtained with far better images than with small bowel follow-through [74]. An additional benefit of this procedure is that it offers information beyond the inner lining of the intestine, revealing the complete depth of the gut wall as well as any abscesses or fistulas. For individuals with suspected or confirmed Crohn's disease, CT enterography has largely replaced small bowel follow-through in medical facilities[74]. This allows us to base our therapy decisions more on the amount of inflammation seen on the scans and less on symptoms. We intend to investigate these results both prospectively and retrospectively, while it is unclear how this will impact the problem's natural history. CT enterography is not appropriate for some individuals, including those with renal failure, iodine allergy, or conditions that prevent the administration of 1.5 L of oral contrast. In addition, radiation exposure is involved, ranging from four to ten times higher than in the case of small bowel follow-through [74].

3.4.2.3 MRI Scans and MR Enterography

Magnetic resonance (MR) enterography has been suggested as a viable alternative for reducing concerns over radiation exposure. It applies the same concepts as CT enterography. An MRI scan employs a magnetic field and radio waves to generate 2 and 3 dimensional images of the human body [75]. MRI scans are useful for precise visualization of soft tissues, such as the intestinal lining, which is particularly beneficial for doctors. They may show signs of bleeding or irritation, as well as tiny tears or ulcers. Before having an MRI, patients arrive early, consume a substantial volume of contrast, and receive gadolinium intravenously. This procedure is referred to as MR enterography [75]. Although CT enterography produces higher-quality pictures, MR technology is improving and is expected to outperform CT in terms of diagnostic precision while lowering concerns about radiation exposure.

3.4.3 BIOMARKERS IN GIT INFLAMMATION

3.4.3.1 Blood Test

While a blood test cannot definitively diagnose IBD, it can assist in eliminating other illnesses that present similar symptoms. Serologic testing refers to the examination of blood samples to assess the presence of antibodies or other substances. Medical professionals commonly extract a small quantity of blood, which is subsequently transported to a laboratory for analysis. Typically, results can be obtained within a timeframe of one to two days. IBD can be identified by certain distinct markers present in blood tests. These markers indicate immune system-induced inflammation and symptoms of anemia, characterized by a low red blood cell count, which may suggest internal bleeding. Insufficient levels of iron, folate, vitamin D, and vitamin B12 suggest a nutritional deficiency and may indicate impaired absorption of these nutrients by the intestines.

Specific inflammatory biomarkers can provide insight into the extent of systemic inflammation, but they are not sufficient for diagnosing IBD. Among the indicators are erythrocyte sedimentation rate (ESR) and C-reactive protein (CRP), which are mentioned in Figure 3.1 [64,76]. These are two tests that have been in use for a significant period of time. CRP has recently been acknowledged as a reliable test for identifying systemic inflammation. However, it is not adequately sensitive or specific to be relied upon exclusively. For individuals with established IBD, it only provides additional information to aid in assessing the severity of their illness.

The advanced serologic assays that assess antibodies show more potential for accurate diagnosis and prognosis. For those suffering from inflammatory bowel disease (IBD), there were only two tests available in the late 1980s and early 1990s: antibodies to Saccharomyces cerevisiae (ASCA) and anti-neutrophil cytoplasmic antibody (ANCA) [64]. Patients with ulcerative colitis typically have ANCA, while those with Crohn's disease tend to have ASCA. Combining both tests in a panel leads to decreased sensitivity but increased specificity. The presence of negative ANCA and positive ASCA is relatively specific for Crohn's disease, although it comes with a minor decrease in sensitivity [64].

A recent discovery has revealed a higher number of antibodies in people with IBD, particularly referred to as Crohn's disease. Similar to the skin's immune system, the gut's immune system is also vulnerable to the outside environment. In order to combat harmful and beneficial bacteria, it must be vigorous but selective in its actions. People who have Crohn's disease and, to a lesser extent, ulcerative colitis have dysregulation and uncontrollable inflammation [64]. A condition known as loss of tolerance occurs when the immune system in the gut appears to respond to antigens in an inappropriate way. Over time, we have observed that individuals with Crohn's disease experience a diminished ability to tolerate specific types of germs that are widespread. Crohn's disease patients have produced antibodies against these ubiquitous microorganisms for unknown reasons. Among these antigens are the bacterial flagellin CBir1, the I2 component of *Pseudomonas fluorescens* and the outer membrane porin C (OmpC) of *Escherichia coli* [76,77]. The operational features of the panel created by combining these antibody tests can be further enhanced, leading to the possibility of sensitivity and specificity for Crohn's disease exceeding 90%, based on preliminary research findings. However, before this panel is widely used, its working features must still be confirmed by impartial third parties [64].

3.4.4 Laboratory Test

3.4.4.1 Stool Test

The doctor might ask the patient to submit a stool sample to be submitted to a lab for testing because IBD alters the frequency and consistency of bowel movements. Diagnostic information can be obtained from fecal indicators of inflammation, such as lactoferrin and calprotectin [78]. These proteins are present in neutrophil secondary granules. If these proteins are detected in the stool at elevated quantities, this can be taken as indirect proof of a GI tract inflammatory disease. Numerous investigations have discovered a connection between fecal lactoferrin or calprotectin and the ability to distinguish between IBD and IBS [78,79]. Furthermore, among IBD patients, these levels can be utilized to distinguish between active and inactive disease. Both markers exhibit satisfactory operational properties; however, calprotectin possesses the additional benefit of being rather stable at room temperature. In contrast, stool samples should be refrigerated before

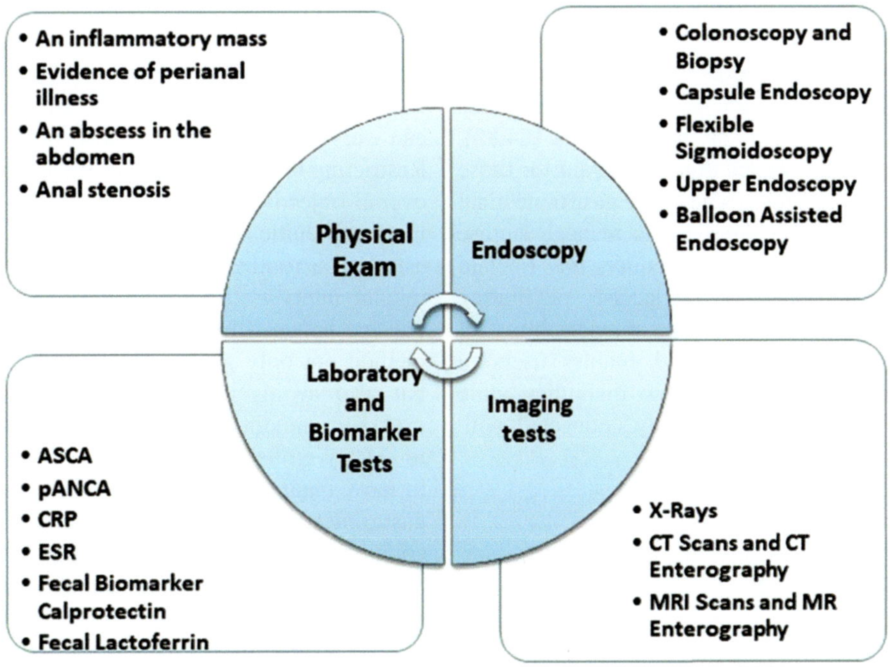

FIGURE 3.1 Diagnostic Tools and Techniques of Inflammatory Bowel Disease.

assessing lactoferrin if the test is not promptly conducted after collection [80].

The doctor provides the patient with a little plastic container, which the patient is expected to return to a specialist after collecting a sample. Stool tests are performed to evaluate the degree of inflammation or to rule out illnesses. A fecal examination can also detect the existence of occult blood that could be undetectable or barely visible to the unaided eye. The outcomes are typically accessible within a span of five to seven days.

3.5 PREVENTION STRATEGIES

3.5.1 Lifestyle and Home Remedies

An individual suffering from inflammatory bowel disease may sometimes feel helpless. However, dietary and lifestyle modifications may aid in symptom management and increase the interval between flare-ups [81].

3.5.1.1 Diet

Currently, there is not enough empirical evidence to establish a direct causative relationship between dietary intake and inflammatory bowel disease. However, some types of food and drinks can worsen indications and manifestations, particularly when the disease suddenly increases in severity. Maintaining a food diary can help monitor dietary intake and emotional well-being [81]. If a person identifies that certain foods are worsening his/her symptoms, he/she could try eliminating those foods.

The following general dietary recommendations could aid in the management of the condition:

3.5.1.1.1 Low-Fiber Diet

During episodes of increased symptoms, it is advisable to follow a diet low in fiber as it might alleviate the pressure on the digestive system. High-fiber meals, particularly whole grains, nuts, and raw fruits or vegetables, are recognized for their nutritional abundance but may cause complications during the active phases of inflammatory bowel disease (IBD) [82]. These meals, high in fiber, can worsen diarrhea and abdominal pain, adding stress to an already sensitive gut [82]. The recommendation to go for cooked and peeled fruits and vegetables points out that these preparations facilitate better digestion while preserving vital nutrients. Cooking processes degrade fiber and enhance the bioavailability of nutrients, hence promoting absorption without burdening the digestive system. Peeling eliminates possible irritants present in the skin, hence enhancing digestive comfort. Refined grains, which are an essential part of a low-fiber diet, are processed to eliminate the outer bran layer, resulting in a milder impact on the digestive system [83]. Similarly, tender meats, which are commonly consumed in this dietary plan, offer a protein source without the burden of excessive fiber. By carefully selecting these meals, people with IBD can achieve a balance between getting essential nutrients and reducing the likelihood of worsening symptoms [83].

3.5.1.1.2 Limit Dairy Products

Dairy products can help with a number of issues associated with inflammatory bowel disease, including gas, diarrhea, and abdominal pain [84]. Some people may not be able to digest the lactose (milk sugar) found in dairy products because of a lactose intolerance. The deficiency of lactase, which is an enzyme that breaks down lactose (the sugar in dairy products), results in lactose intolerance [85]. Individuals with lactose intolerance may experience fermentation of undigested lactose in the gastrointestinal tract, resulting in symptoms such as abdominal bloating, cramps, and diarrhea [84,85]. Restricting dairy intake may reduce digestive strain for those with IBD, especially during flare-ups when the gastrointestinal tract is hypersensitive. Another approach to manage lactose intolerance involves utilizing enzyme products like Lactaid [84]. Lactaid incorporates the enzyme lactase, facilitating the breakdown of lactose and enhancing the feasibility of consuming dairy products. This method enables users to potentially incorporate dairy products into their diet while reducing the digestive difficulties typically connected with them [85].

3.5.1.1.3 Omega-3 Fatty Acids

It is important for people with inflammatory bowel disease (IBD) to incorporate meals rich in omega-3 fatty acids in their diet, as these may offer health advantages [86]. Omega-3 fatty acids are widely known for their capacity for minimizing inflammation. They can be found in fatty fish like mackerel, salmon and sardines as well as plant-based sources like walnuts and flaxseeds [86]. These meals have the capacity to control the immune system and decrease inflammation in the gastrointestinal tract, which may help relieve certain symptoms of inflammatory bowel disease (IBD). These foods must be included in a balanced diet. Omega-3-rich food consumption is thought to be a beneficial dietary approach in the context of IBD, where persistent inflammation in the gastrointestinal tract is a feature [87]. By integrating omega-3 fatty acids into a well-balanced diet, patients suffering from inflammatory bowel disease (IBD) may observe a decrease in inflammation, resulting in the relief of certain symptoms associated with the condition [86]. The necessity of diversifying dietary choices is widely accepted by the recommendation to consume a range of omega-3 sources. Docosahexaenoic acid (DHA) and eicosapentaenoic acid (EPA) are found in fatty fish, while alpha-linolenic acid (ALA) is found in plant sources and can be converted by the body into EPA and DHA [87]. Integrating foods rich in omega-3 fatty acids into a well-rounded and balanced diet is crucial due to its diverse role in boosting general health and perhaps reducing symptoms of inflammatory bowel disease (IBD) [86].

3.5.1.1.4 Hydration

Patients suffering from inflammatory bowel disease (IBD) must ensure they stay properly hydrated to manage their symptoms effectively. Water is the greatest option because it promotes overall digestive health and aids in maintaining fluid balance [88]. It's critical to stay well hydrated when dealing with IBD because symptoms like diarrhea can increase fluid loss. In addition to aiding in nutrient absorption and digestion, water also helps balance fluid loss, supporting proper gastrointestinal function. It acts as a fundamental component in the prevention of issues associated with dehydration, which may worsen symptoms and compromise general well-being [88]. People with IBD are encouraged to priorities drinking water and to use caution while consuming fizzy and caffeinated beverages [89]. These beverages might cause dehydration since they have diuretic properties and can worsen symptoms like diarrhea and stomach discomfort. Restricting the consumption of these drinks is in line with the overall objective of preserving hydration without presenting more difficulties to the digestive system. It is advisable to establish a routine of consuming little amounts of water at regular intervals throughout the day in order to maintain a balance between hydration and symptom control [89]. This method not only guarantees a steady consumption of fluids but also avoids overburdening the digestive system with excessive amounts of liquid all at once. Moreover, it correlates to dietary guidelines for persons with IBD, who are advised to have smaller, more often meals to reduce pressure on the gastrointestinal tract. Ensuring adequate hydration is a critical component of effectively treating symptoms of inflammatory bowel disease (IBD) and enhancing general well-being [88].

3.5.1.1.5 Dividing Meals Into Smaller Portions

Research suggests that eating 5 or 5 small meals per day is preferable to eating two or three bigger ones. For those who suffer from inflammatory bowel disease (IBD), eating five or six little meals a day can be especially advantageous, as opposed to two or three bigger ones. This nutritional regimen provides numerous benefits such as dividing meals into smaller portions helps distribute the digestive load more evenly throughout the day [89]. This can alleviate the burden on the gastrointestinal tract, facilitating the body's digestion and assimilation of nutrients without overwhelming the system. Consuming smaller and more frequent meals enhances the process of nutritional absorption. This is particularly crucial for patients with inflammatory bowel disease (IBD), as inflammation in the gastrointestinal system, especially in disorders such as Crohn's disease, can hinder the body's capacity to assimilate vital vitamins and minerals. Eating smaller meals promotes better nutrient absorption and overall nutritional support. Stable blood sugar levels can be achieved by eating smaller, more frequent meals all day [90]. This can be especially advantageous for persons with inflammatory bowel disease (IBD) who may encounter variations in blood glucose levels as a result of alterations in eating patterns during periods of increased disease activity. Sustaining blood sugar stability is necessary to maintain energy levels and prevent sudden drops in energy. In certain individuals with inflammatory bowel disease (IBD), consuming larger meals may provoke symptoms such as stomach pain, bloating, or diarrhea. Consuming smaller meals can improve symptoms and quality of life [89,90].

3.5.1.1.6 Take Into Account the Use of Multivitamins

Supplementation with multivitamins and minerals can be beneficial for individuals with Crohn's disease due to potential nutrient absorption interference and the possibility of a limited diet. Crohn's disease and other chronic inflammatory digestive tract diseases can make it more difficult for the body to absorb essential nutrients, leading to shortages that may be harmful to overall health [91]. A common challenge for patients with Crohn's disease is the limited variety in their diet. This is often due to food intolerances, flare-ups, or the need to avoid certain trigger foods. Due to the possibility of consuming less essential vitamins and minerals as a result of this dietary restriction, it can be difficult to achieve nutritional needs only through food. Multivitamin and mineral supplements are useful additions that help fulfill nutritional deficiencies. These supplements can help alleviate any shortages associated with Crohn's disease by offering a wide range of vital nutrients, including vitamins such as B12, D, and E, as well as minerals like iron and calcium [91]. Ensuring a sufficient food intake is especially vital for sustaining energy levels, boosting immunological function, and minimizing issues stemming from nutritional imbalances. Moreover, in Crohn's disease, the intermittent flare-ups can further interfere with nutrient absorption during active periods. During these periods, multivitamin pills serve as a protective measure by providing a dependable supply of vital vitamins and minerals that may be difficult to acquire solely from food [91].

3.5.1.1.7 Quit Smoking

Smoking potentially has a preventive effect on ulcerative colitis. Nevertheless, the detrimental effects on overall health surpass any advantages, and ceasing smoking can enhance the overall health of an individual's digestive system and offer numerous other health benefits. Smoking not only raises the risk of developing Crohn's disease but also exacerbates the illness in those who already have it [92]. Smokers with Crohn's disease are at risk for a number of negative outcomes, such as an increased risk of drug side effects, greater relapse susceptibility, and an increased need for repeat surgery. This complex correlation underscores the importance of individuals with Crohn's disease quitting smoking as a critical part of their comprehensive disease management [93]. Giving up smoking is shown to be a crucial intervention that improves the digestive system as a whole and has a direct impact on the course of Crohn's disease [93]. Quitting smoking results in a decrease in inflammation, an enhanced reaction to drugs, and a lower likelihood of complications, creating a favorable environment for the healing and recovery of the digestive system. Apart from its particular impact on Crohn's disease, stopping smoking has numerous other health advantages. It reduces the likelihood of developing cardiovascular diseases, respiratory conditions, and other types of malignancies, leading to a more comprehensive improvement of general well-being [92]. In order to improve the quality of life for those with Crohn's disease, giving up smoking is a proactive move that can have a profound impact.

3.5.2 Stress Management

Although there is debate over the link between stress and Crohn's disease, many patients with the illness claim that their symptoms flare up during times of high stress. If someone struggles with stress management, they should definitely consider one of the following strategies:

3.5.2.1 Regular Exercise

Regular physical activity is a fundamental aspect of managing stress in patients with Crohn's disease, providing a wide range of advantages that go beyond only physical health. It is possible to significantly improve several aspects of health with even modest exercise, such as lowering stress, relieving depression, and restoring normal bowel function [94].

Physical activity stimulates the secretion of endorphins, which are natural substances that improve one's mood. These neurotransmitters function as endogenous stress reducers, promoting a feeling of contentment and intense happiness. For individuals who have Crohn's disease, the relationship between physical and mental well-being is complex [95]. Therefore, exercising to release endorphins is a helpful way to cope with stress. Exercise has been demonstrated to be a proactive therapy in the treatment of depression. Exercise provides both physiological and psychological advantages that enhance mood, resulting in a favorable effect on mental well-being and assisting in the control of symptoms related to stress [94]. Bowel function can be restored by exercise, which is especially important for those with Crohn's disease who have abnormalities in their digestive systems [95]. In addition to its physical advantages, regular exercise further contributes to mental well-being. It promotes a feeling of achievement, improves cognitive abilities, and offers a beneficial means of managing stress. The development of a robust attitude with psychological advantages is essential in handling the difficulties linked to Crohn's disease and stress [95].

3.5.2.2 Biofeedback

Biofeedback is a method that involves tracking and reporting physiological parameters, including heart rate or brain activity, using electrical instruments [96]. This stress-reduction technique uses a biofeedback machine to guide individuals in reducing muscle tension and decreasing heart rate. The aim is to assist the patient in achieving a relaxed state so that stress can be managed more readily. Biofeedback training aims to promote the relaxation response [96,97]. Biofeedback encourages a sense of control and mastery in people by giving them a tangible grasp of their body's stress reactions and the means to modify these responses. This increased consciousness enables individuals to participate in stress-reducing activities actively.

In a biofeedback session, physiological factors including skin temperature, heart rate, and muscular tension are measured by sensors that are attached to the body [96]. Individuals receive instant feedback regarding their physical reactions to stimuli when this real-time data is presented on a monitor. The main objective of biofeedback is to decrease muscle tension. Stress can make symptoms worse for many people with Crohn's disease since it increases their muscle tension. With the aid of guided relaxation techniques and biofeedback, people can actively attempt to lessen their levels of muscle tension. [96]. The feedback unit also tracks heart rate, enabling participants to notice stress-related fluctuations. Individuals can deliberately decrease their heart rate by engaging in certain breathing exercises and employing relaxation techniques. In order to manage stress-related Crohn's disease symptoms, this technique helps to achieve a more calm physiological state [97]. Regular application of biofeedback techniques aids in long-term stress management in addition to its immediate advantages. People can employ these newly acquired abilities in their everyday lives to lessen the physiological effects of stress on their bodies [96]. Adopting a proactive strategy can improve the ability to withstand and recover from the challenges and pressures associated with Crohn's disease [97].

3.5.2.3 Regular Relaxation and Breathing Exercises

An effective strategy for managing stress is to engage in regular relaxation practices and employ techniques such as deep, deliberate breathing to induce a state of calmness [98]. By integrating these strategies, individuals can effectively manage stressors, leading to improved mental and physical health. Deep, calm breathing is one of the most important ways to manage stress as it has an immediate and direct effect on the stress response of the body. By using the diaphragm to contract, this deliberate breathing pattern instructs the neurological system of the body to transition from a stressed to a relaxed state [99]. By adopting an attentive attitude towards their breath, individuals can achieve a decrease in tension and a soothing impact on the mind. Breathing deeply and slowly has a beneficial effect on physiological indicators, including blood pressure, heart rate, and cortisol levels, which are important elements of the body's stress response. Individuals can regulate their breathing to balance and synchronize patterns, promoting relaxation and reducing stress-induced physiological arousal [98]. Frequent breathing and relaxation techniques improve the mind-body connection and help people become more conscious of their stressors and physical reactions. The enhanced self-awareness acts as a basis for taking proactive measures to manage stress, enabling individuals to take action early when faced with stressors and reduce their negative effects. By consistently practicing relaxation and breathing exercises, individuals can actively integrate stress management into their everyday routines. These routines establish a dependable and encouraging framework for stress reduction, whether they are followed in the morning, during breaks, or right before bed [99]. Consistent practice has a cumulative effect on both mental and physical well-being, leading to long-term benefits. Guided meditation and visualization techniques augment the efficacy of relaxation exercises alongside deep breathing [98]. One can mentally escape tensions through guided imagery or mental journeying techniques.

3.5.3 Personalized Medicine in Gastrointestinal Inflammation

3.5.3.1 Anti-inflammatory Drugs and Nonsteroidal Anti-Inflammatory Medications (NSAIDS)

Anti-inflammatory medications are commonly used as the initial approach in managing ulcerative colitis, particularly for cases categorized as mild to moderate. Amino salicylates, e.g., balsalazide (Colazal), mesalamine (Delzicol, Rowasa, etc.) and olsalazine (Dipentum), are anti-inflammatory agents that have been illustrated in Table 3.1 [100]. Corticosteroids are also employed in time-limited courses to initiate remission. Steroids have immunosuppressive properties in addition to their anti-inflammatory effects. The choice of medication for a patient is dependent upon the specific region of their colon that is impacted.

Patients with ulcerative colitis (UC) or Crohn's disease (CD) have shown an elevated disease activity when treated with high doses of nonsteroidal anti-inflammatory medicines (NSAIDs) [101,102]. However, no significant rise in disease activity index score is observed among CD patients who are administered low doses of NSAIDs. Bonner et al [103] did not find a correlation between NSAID use and increased disease activity in IBD, suggesting further investigation is needed before advising patients to avoid NSAIDs completely. Nonselective NSAIDs were linked to a 17%–28% recurrence risk in patients with quiescent IBD within 9 days of use [104]. Furthermore, it has been observed that selective Cox-2 inhibitors might lead to relapse in patients with Crohn's disease (CD) and ulcerative colitis (UC), indicating that all types of nonsteroidal anti-inflammatory drugs (NSAIDs) are linked to recurrence in patients with inflammatory bowel disease (IBD) [105]. Nevertheless, alternative research has indicated that there is no rise in flares and that Cox-2 inhibitors have a favorable safety profile when used for the short-term therapy of arthritis and arthralgia associated with inflammatory bowel disease [106]. It is recommended that patients with inflammatory bowel disease (IBD) should refrain from using nonsteroidal anti-inflammatory drugs (NSAIDs). However, individuals with arthritis, which is commonly associated with IBD, may require the use of NSAIDs for treatment. Hence, it is necessary to adopt a patient-centered strategy that includes considerations of the potential drawbacks and advantages of NSAID usage prior to its administration in patients with inflammatory bowel disease (IBD).

3.5.3.1.1 5-Aminosalicylic Acids

5-ASA is a widely used medication for IBD that helps to relieve symptoms and sustain remission in patients [107]. 5-aminosalicylic acid (5-ASA), which includes mesalamine, can be employed as an initial therapy for mild to moderate ulcerative colitis (UC) [108,109]. 5-ASA appears to have a variety of mechanisms of action and acts locally on both small and large intestine mucosa. 5-ASA mainly suppresses the inducible COX-2/PGE2 (cyclooxygenase-2/prostaglandin E2) signaling pathway. Additionally, it suppresses nuclear factor-κB (NF-κB) activity, which lowers TNF-α and IL-1 production. Moreover, 5-ASA's anti-inflammatory effects are mediated by peroxisome proliferator-activated receptor γ (PPAR-γ) [107,110,111]. Apart from 5-ASA, other prodrugs of 5-ASA include olsalazine, basalazide, and sulfasalazine, which are broken down by intestinal bacteria into 5-ASA and sulfapyridine. There are numerous formulations of mesalamine that are intended to supply the active 5-ASA component to either the large or small intestine [112]. Patients who use 5-aminosalicylic acid medications are at a higher risk of developing renal disease and folate deficiency [113]. Therefore, it is recommended that these patients take a daily dose of 1 mg of folate and undergo annual monitoring of their renal function [113].

3.5.3.1.2 Corticosteroids

Steroid hormones called corticosteroids, which are made in the adrenal cortex, include glucocorticoids and mineralocorticoids. Inflammatory cellular processes are mediated by corticosteroids, and the glucocorticoid receptor (GR) function is an essential component of corticosteroid action. The ligand-bound glucocorticoid receptor (GR) disrupts the transcriptional activation of NF-κB and AP-1 when brought onto their gene promoters. Given that NF-κB serves as the primary transcription factor that controls inflammatory responses, the inhibition of NF-κB by GR leads to a decrease in the transcription of crucial immunoregulatory genes such TNF-α and IL-1 [114,115]. According to Dubois-Camacho et al. (2017) [115], corticosteroids are represented by the medications such as prednisolone, methylprednisolone, hydrocortisone, and budesonide. Steroids are particularly efficient in promoting remission in IBD, but they lack efficacy in maintaining it. Prolonged usage of steroids is linked to negative effects caused by steroids. Therefore, it is advisable to initiate steroid sparing therapy early on to minimize these harmful effects.

3.5.3.2 Immunosuppressant/Immunomodulators

These medications function through several mechanisms to inhibit the immunological response that triggers the release of substances that cause inflammation in the body. Upon discharge, these compounds have the potential to harm the epithelial lining of the gastrointestinal system. Examples of immunosuppressant medications include mercaptopurine (Purinethol, Purixan), azathioprine (Azasan, Imuran), and methotrexate (Trexall), depicted in Table 3.1 [116]. In recent times, orally administered medications, commonly referred to as "small molecules," have been accessible for the treatment of IBD [117]. These medications include upadacitinib (Rinvoq), tofacitinib (Xeljanz) and ozanimod (Zeposia). As per the latest alert issued by the Food and Drug Administration (FDA) in the United States, preliminary research indicates that using tofacitinib may increase the risk of cancer and major cardiac issues.

3.5.3.2.1 Thiopurines

Thiopurines are chemical compounds that are derived from thioguanine. 6-MP as well as AZA is the two most frequently utilized forms of thiopurine compounds for treating inflammatory bowel disease (IBD) [118,119]. Thiopurine S-methyltransferase is an enzyme that converts AZA (Azathioprine) into 6-MP (6-Mercaptopurine). Administration of either AZA or 6-MP hinders the proliferation of T and B lymphocytes, while also suppressing the activity of cytotoxic T cells and plasma cells. This is accomplished by specifically targeting the activation of Rac1, which is considered to be a small GTPase that has a vital function in the activation of T cells [119,120]. After starting treatment with 6-MP or AZA, 6-thioguanine nucleotides start to accumulate in the cells that mature into myeloid cells. This kind of accumulation can result in myelosuppression, which is frequently the cause of dosage reductions or short-term treatment discontinuations in patients with inflammatory bowel disease (IBD) [120]. Thiopurine are commonly used to maintain remission in patients with steroid-dependent inflammatory bowel disease (IBD).

It is recommended to check the level of thiopurine methyltransferase (TPMT) activity before initiating treatment with thiopurine medications for instance 6-MP and AZA. Approximately 10% of the population exhibits reduced activity, whereas 0.3% of the population displays complete inactivity. People who exercise little or not at all are more likely to suffer from myelosuppression and drug-induced bone marrow toxicity. This is because their bodies tend to accumulate unprocessed medicines [121]. Hence, it is important to conduct TPMT, CBC, and liver function tests before starting treatment and monitor CBC and liver function regularly during therapy [122].

3.5.3.2.2 Methotrexate

Approximately 33% of individuals with inflammatory bowel disease (IBD) experience intolerance to thiopurines, whereas over 10% exhibit nonresponsiveness to thiopurines. Methotrexate is employed as an alternate immunomodulator in these individuals. An antimetabolite known as methotrexate inhibits the metabolism of folic acid. It primarily targets quickly dividing cells, such as those in the gastrointestinal tract, oral mucosa, and effector T cells, and it alters DNA/RNA synthesis [123]. There are three types of processes by which it works. To begin with, it obstructs the functioning of enzymes that take part in the process of purine metabolism. Additionally, it reduces the activation of T cells and/or blocking the operation of adhesion molecules by T cells. Lastly, it restrains the binding of IL-1β to IL-1R on the targeted cells. Methotrexate is less effective than thiopurines and biologics. Nevertheless, recent research has demonstrated that it works well for Crohn's disease but not ulcerative colitis [118,123].

Methotrexate therapy can make the immune system weaker and lower folate levels. Before starting the treatment, it is recommended to evaluate and frequently check the patient's renal function complete blood count and liver function. Patients should take a daily dose of 1mg of folate supplementation [122].

3.5.3.2.3 Cyclosporine

Calcineurin is a Ser/Thr protein phosphatase that is dependent on both calcium and calmodulin. It increases the activity of T cells by removing phosphate groups from the regulatory sites of transcription factors, especially the nuclear factor of activated T lymphocytes (NFATs) [107,124]. Cyclosporine binds to cyclophilin, preventing calcineurin from removing phosphate groups from phosphorylated NFAT. This suppresses T cell proliferation, the generation of cytokines (IL-2, TNF-α, and

IFN-γ), and the immune system as a whole [125]. Particularly for patients with steroid-resistant acute severe ulcerative colitis, cyclosporine is advised to induce remission.

3.5.3.3 Janus Kinase (JAK) Inhibitors

The JAK family consists of intracellular tyrosine kinases, including tyrosine kinase 2 (TYK2) and JAK 1-3 as illustrated in Table 3.1 [126]. JAK inhibitors block cytokine signaling pathways to regulate inflammatory and immunological responses. Cytokine signals activate JAK-STAT pathways, resulting in the transcription of genes encoding inflammatory proteins [127]. Tofacitinib is a medication belonging to the first-generation pan-JAK inhibitor family. It is the first of its kind to receive a license for treating moderate to severe ulcerative colitis (UC) patients, both as an initial treatment and for continuing therapy [128]. However, research has shown that it is not very effective in treating Crohn's disease ([129]. Tofacitinib is associated with significant negative outcomes such as decreased levels of leukocyte subset numbers, increased susceptibility to infections, neoplasms, non-melanoma skin cancer, elevated levels of low-density lipoprotein cholesterol, and thromboembolism [130].

Filgotinib and upadacitinib are drugs that specifically suppress the activity of the JAK1 enzyme. Phase III trials are presently in progress to assess the safety and efficacy of upadacitinib (NCT03398148 and NCT03105128) and filgotinib (NCT02914522 and NCT0291456) in the treatment of Crohn's disease (CD) and ulcerative colitis (UC) ~~(D'Amico et al., 2020b)~~. The pan-JAK inhibitor TD-1473 is gut-selective [130]. Phase II/III studies are being conducted to investigate the effectiveness and safety of TD-1473 in UC (NCT03758443) and CD (NCT03635112) [131]. The safety and efficacy of Deucravacitinib, a TYK2 inhibitor, are now being studied in phase 2 clinical trials for patients with UC (NCT03934216). Efficacy in treating ulcerative colitis has been demonstrated in a mouse model of T cell transfer colitis [132]. Carotegrast methyl is a small molecule antagonist that targets α4 integrin and can be taken orally. In 2015, Yoshimura et al. [133] demonstrated its effectiveness in a phase II trial. Patients diagnosed with ulcerative colitis who were administered carotegrast methyl at a dosage of 960 mg thrice daily for 8 weeks duration had superior clinical remission, clinical response, and mucosal healing compared to individuals who were administered a placebo. Now, individuals with UC are being enlisted for a phase 3 trial (NCT03531892) [134].

3.5.3.4 Adhesion Molecule Inhibitor

The migration of leukocytes to the intestinal mucosa and the recruitment of immune cells to the inflammatory site, facilitated by the up regulation of adhesion molecules, are considered to be essential for the initiation and sustenance of intestinal inflammation. Inhibitors of adhesion molecules prevent endothelial cells from interacting with adhesion molecules in immune cells. In order to decrease chronic inflammation, it is possible to hinder the movement of immune cells toward endothelial cells [135]. Natalizumab is a monoclonal antibody of the IgG4 class that targets the α4-integrin protein in a way that resembles human antibodies. In 2008, the FDA granted approval for the use of natalizumab in patients with Crohn's disease, but with a black box warning. Nevertheless, the European Medicines Agency (EMA) has prohibited the utilization of this therapy in individuals with Crohn's disease due to the potential hazard of acquiring a life-threatening ailment known as primary leukoencephalopathy [136]. Natalizumab blocks α4 integrin as illustrated in Table 3.2, preventing white blood cell migration to organs and inflamed areas. Vedolizumab is a monoclonal antibody of the IgG1 class that has been humanized and designed to target and inhibit the α4β7-integrin protein. The FDA granted approval of vedolizumab for the treatment of ulcerative colitis (UC) and Crohn's disease (CD) in 2014. Vedolizumab limits leukocyte adherence to the gut by blocking the α4β7-integrin. This integrin interacts with mucosal addressin cell adhesion molecule-1 (MadCAM-1), a gut-specific homing receptor [137]. For individuals suffering from inflammatory bowel disease (IBD), particularly those who have not shown improvement after using anti-tumor necrosis factor (anti-TNF) treatment, vedolizumab is used to initiate and maintain remission. Adhesion molecule inhibitors have the potential to cause opportunistic infections and cytokine release syndrome. It is important to be aware of these potential side effects while using these inhibitors. However, vedolizumab is safer than TNF-α antagonist.

3.5.3.5 S1P Receptor Modulators

Sphingosine-1-phosphate (S1P) receptor modulators, such as ozanimod and etrasimod, are novel small compounds that are currently being developed and tested in clinical studies as part of the treatment of inflammatory bowel disease (IBD) [138]. S1P is a signaling molecule produced from the lysophospholipids in the membrane. It activates five distinct G protein-coupled receptors, known as S1P1-5 [139]. Out of the five subtypes, S1P1, S1P4, and S1P5 receptors (mentioned in Table 3.1) play an essential role in controlling immunological responses and have been linked to the development of inflammatory bowel disease (IBD) [140]. Ozanimod specifically targets the S1P1 and S1P5 receptors, whereas etrasimod targets the S1P1, S1P4, and S1P5 receptors [129,130]. Ozanimod and etrasimod inhibit the exit of T and B cells from lymph nodes, therefore reducing immune-mediated inflammation. As they are less immunogenic and may be taken orally, they can be administered without the assistance of specialized personnel and may prevent the production of antidrug antibodies. Furthermore, they are highly suitable for long-term storage compared to biologics. Their production costs are lower, and they can be quickly terminated in case of an infection or adverse event due to their short half-life.

3.5.3.6 Biologic Agents for IBD

Biologics are an advanced type of treatment that specifically targets proteins in the body responsible for inducing inflammation. Some medications are supplied intravenously (IV) through infusions, however others are self-administered through injections by patients. Some examples of medications

TABLE 3.1
Overview of the Small Molecules Used to Treat Inflammatory Bowel Disease (IBD)

Class	Target	Agent	Approval Status	Crohn Disease	Ulcerative Colitis
Anti-inflammatory drugs	TNF-α, PPAR-γ, IL-1, COX-2/PGE2	5-ASA including SSZ & mesalamine	Approved	✓	✓
Immuno-modulators	Purine metabolism	6-mercaptopurine Azathioprine	Approved	✓	✓
Corticosteroids	glucocorticoid receptor (GR)	Prednisolone Hydrocortisone Methylprednisolone Budesonide	Approved	✓	✓
Adhesion molecule inhibitor	alpha4 integrin	Carotegrast methyl	Phase 3	✗	✓
	TYK2	Deucravacitinib	Phase 2	✓	✓
	Nucleotide synthesis enzymes (DHFR)	Methotrexate	Approved	✓	✓
JAK inhibitors	Pan-JAK	Tofacitinib	Approved	✗	✓
	JAK-1	Filgotinib	Phase 3	✓	✓
	Cyclophilin	Cyclosporine	Approved	✓	✓
	Pan-JAK	TD-1473	Phase 2 (CD) Phase 2/3 (UC)	✓	✓
	JAK-1	Upadacitinib	Phase 3	✓	✓
SIP receptor modulator	S1P1, S1P4, and S1P5 receptors	Etrasimod	Phase 3	✓	✓
	S1P1 and S1P5 receptors	Ozanimod	Phase 3	✓	✓

Source: [112].

in this category are ustekinumab (Stelara), infliximab (Remicade), vedolizumab (Entyvio), golimumab (Simponi), adalimumab (Humira), certolizumab (Cimzia), and andrisankizumab (Skyrizi), as illustrated in Table 3.2 [141].

The FDA has approved seven biologic medicines for the treatment of IBD, excluding biosimilars. Four TNF-α inhibitors exist, specifically certolizumab pegol, golimumab, adalimumab, and infliximab. The remaining inhibitors can be classified into two distinct categories: integrin inhibitors, which include natalizumab and vedolizumab, along with IL-12/23 inhibitors, for example ustekinumab [112].

3.5.3.6.1 *Tumor Necrosis Factor-Alpha (TNF-α) Antagonists*

TNF-α is a cytokine with pro-inflammatory properties that trigger inflammation in ailments characterized by inflammation, for instance, Crohn's disease, rheumatoid arthritis and ulcerative colitis. Anti-TNF-α medication is the primary biological treatment for inflammatory illnesses that involve TNF-α [119]. The FDA granted approval to infliximab in 1998, making it the inaugural biologic medication for IBD. Infliximab is a genetically modified antibody that combines mouse and human components. It is classified as a chimeric IgG1 monoclonal antibody (mAb) and was first described by Rutgeerts et al. in 2005 [142]. Certolizumab pegol is a type of modified humanized monoclonal antibody (mAb) known as a Fab' fragment. This antibody has been attached to polyethylene glycol (PEG) in order to increase its plasma half-life, stability, and bioavailability. Due to the absence of the Fc component of the antibody, certolizumab egol is unable to stimulate complement binding or antibody-dependent cell-mediated cytotoxicity [143]. Adalimumab and golimumab are monoclonal antibodies (mAb) that have been fully humanized to be identical to human antibodies [144]. Golimumab is specifically authorized for the treatment of ulcerative colitis, whereas certolizumab is exclusively indicated for Crohn's disease. Both Crohn's disease (CD) and ulcerative colitis (UC) have been approved for the use of the other two TNF-α antagonists. Differences in effectiveness may be due to structural changes among the agents that impact their modes of operation [145]. TNF-α antagonists are employed to initiate and sustain remission in individuals with moderately to highly active inflammatory bowel disease (IBD), particularly in cases where corticosteroids are required or when the condition is unresponsive to treatment. Although TNF-α antagonists are highly effective in treating IBD, approximately 10-30% of patients do not respond (NPR) and 30-50% of those who do respond experience a loss of response (SLR). Patients with SLR are given an increased dosage initially, and if this doesn't work, they are moved to a different class of medication [146]. The most prominent side effects of TNF-α antagonist include opportunistic infections, for instance, demyelinating neurological disease, tuberculosis, and cancer, which result from their impact on the immune system [147].

3.5.3.6.2 CD4 + T Cell Cytokine Inhibitor

The association between type 1 helper T (Th1) cytokines and Crohn's disease, as well as the association between type 2 helper T (Th2) cytokines and ulcerative colitis, is widely recognized. It was recently discovered that type 17 helper T (Th17) cells, IL-23, IL-21, and IL-17 all play a crucial part in IBD [148]. According to Bogiert et al. (2010) [149], individuals with CD have higher levels of IL-17, IL-21, IL-22, and IL-26 expression in their intestinal mucosa. Ustekinumab is a monoclonal antibody of the IgG1 class that has been fully humanized and is specifically designed to target and inhibit the activity of IL-12/23p40. The FDA granted approval for the use of ustekinumab in the treatment of Crohn's disease in 2016 and ulcerative colitis in 2019. This decision was based on the findings of studies conducted by Sandborn et al. in 2012 [150]. IL-12 and IL-23 play a role in stimulating and distinguishing natural killer cells and CD4 + T cells. Consequently, ustekinumab inhibits chronic immune-mediated inflammatory diseases. Ustekinumab is administered to induce and sustain remission in individuals with inflammatory bowel disease (IBD), especially those who have not responded well to TNF-α antagonists. Observational studies conducted over a five-year period found no evidence of an elevated risk of opportunistic infections or carcinomas associated with ustekinumab, in contrast to TNF-α antagonist. Ustekinumab can cause a severe allergic reaction known as hypersensitivity, which may include anaphylaxis.

3.5.3.7 Ongoing Development of Biologics

V565 is a tablet with a special coating that allows it to pass through the stomach and dissolve in the intestines. It targets TNFα, a protein involved in inflammation. V565 was made resistant to the gastrointestinal (GI) environment through the use of enteric coating and protein engineering. V565 does not exhibit systemic effects as it does not enter the intestinal epithelium in its original form. The compound V565 effectively suppresses the activity of TNFα specifically in the inflamed region of the digestive tract, as demonstrated by Nurbhai et al. in 2019 [151]. The registration number NCT02976129 indicates that clinical trials for CD are currently in phase 2. D'Amico et al. 2020 [134] identified four humanized anti-IL-23p19 mAbs: pazikizumab, mirikizumab, guselkumab, and risankizumab. P19 component of IL-23 is preferentially inhibited by brazikumab, mirikizumab, guselkumab, and risankizumab, while ustekinumab targets IL-12/IL-23p40. The cytokine IL-12 induces differentiation of naive CD4 T cells into Th1 cells, as demonstrated by Jacobson et al. in 1995. Blocking IL-23p19 does not affect the IL-12 mediated Th1 response. Hence, it is anticipated that the potential adverse effects will be diminished [152]. Etrolizumab is a type of antibody that has been modified to resemble a human antibody. It specifically targets a component called integrin β7, which is found on cells and plays a role in immune responses. This subunit can combine with two other subunits, α4 and αE, to form two different types of integrins, α4β7 and αEβ7. Etrolizumab is designed to bind to these integrins and interfere with their function. Etrolizumab, like vedolizumab, blocks the activity of α4β7 to prevent T cells from migrating to the gut. Additionally, Etrolizumab specifically inhibits αEβ7, which interacts with E-cadherin found on intestinal epithelial cells. Blocking αEβ7 effectively prevents T cells from migrating to the site of inflammation in the intestinal lamina propria [153]. The clinical trials for ulcerative colitis (UC) and Crohn's disease (CD) are currently in phase 3. The trial registration numbers are NCT02165215 and NCT02394028, respectively. Furthermore, clinical trials are currently underway to evaluate the efficacy of Abrilumab, which targets α4β7 integrin, and ontamalimab, which targets MAdCAM. These trials are in phase 2 and phase 3 (NCT03283085), respectively, as reported by Vermeire et al. in 2017 [154]. Alicaforsen is an ICAM-1 antisense oligonucleotide. In the study conducted by Jairath et al. (2017) [155], it was shown that alicaforsen did not have a significant treatment benefit compared to placebo in patients with CD. However, it did show promise as a therapeutic option for pouchitis and is currently being evaluated in a phase 3 trial with the registration number NCT02525523.

3.5.3.8 Antibiotics

Antibiotics are often given along with other treatments or as a preventive measure to combat infections, especially in cases of perianal Crohn's disease. Common antibiotics prescribed for this purpose include ciprofloxacin (Cipro) and metronidazole (Flagyl). However, the use of antibiotics can have an impact on the gut microbiome by altering its composition due to selective pressure [156]. Various investigations have demonstrated that the utilization of antibiotics at various stages of life, spanning from infancy to maturity, may increase the susceptibility to inflammatory bowel disease (IBD). Conversely, numerous extensive investigations have demonstrated that antibiotics provide a protective effect in people who already have IBD. Through a meta-analysis of randomized controlled trials (RCTs), it was found that antibiotics were more effective than placebo in causing a reduction of symptoms in people with active Crohn's disease (CD) [157]. Antibiotics were significantly more effective than a placebo in reducing fistula drainage in patients with perianal fistulae caused by Crohn's disease. Antibiotics were better than placebo at causing remission in patients with active UC. Antibiotics were more effective than placebos, which caused a reduction in symptoms in patients with active Crohn's disease while also avoiding a recurrence of symptoms in patients with inactive Crohn's disease [157]. More precisely, studies have demonstrated that Nitroimidazoles (such as metronidazole and ornidazole) are successful in preventing the recurrence of CD after surgery [158]. Antibiotics are useful in managing mild IBD flare-ups and preventing post-operative recurrence in CD, but their indiscriminate use should be avoided due to the possibility of changing the gut microbiota composition, which could raise the risk of both a *Clostridium difficile* infection and a disease flare-up.

3.5.3.9 Oral Contraceptives

Women with inflammatory bowel disease (IBD) should use reliable contraception to avoid unplanned pregnancies,

TABLE 3.2
Overview of the Biologic Agents Used to Treat Inflammatory Bowel Disease (IBD)

Class	Target	Agent	Approval Status	Crohn Disease	Ulcerative Colitis
Pro-inflammatory cytokine inhibitor	TNF-α,	V565	Phase 2	✓	✕
		Certolizumab pegol	Approved	✓	✓
		Golimumab	Approved	✕	✓
		Infliximab	Approved	✓	✓
		Adalimumab	Approved	✓	✓
CD4 + T cell cytokine inhibitors	IL-23p19	Risankizumab	Phase 3	✓	✓
		Brazikumab	Phase 2/3 (CD) Phase 2 (UC)	✓	✓
		Guselkumab	Phase 3 (CD), Phase 2/3 (UC)	✓	✓
		Mirikizumab	Phase 3	✓	✓
	IL-12/23p40	Ustekinumab	Approved	✓	✓
Adhesion molecule inhibitor	α4β7-integrin	Vedolizumab	Approved	✓	✓
		Abrilumab	Phase 2	✓	✓
	α4β7 and αEβ7-integrin	Etrolizumab	Phase 3	✓	✓
	α4-integrin	Natalizumab	Approved	✓	✕

Source: [112].

particularly since the condition is common among women of reproductive age [159].

3.5.3.10 Anti-TNF-Alpha Agents

Anti-TNF medications have been associated with immunosuppression, which can lead to a higher risk of developing infections and reactivating latent illnesses [49]. Therefore, before initiating treatment with these medications, it is crucial to conduct an initial assessment of the patient's CBC, liver function, and renal function. Furthermore, it is recommended that these parameters should be regularly monitored throughout the course of medication [122].

3.5.3.11 Anti-Diarrheal Medications

Fiber supplements can be useful in treating mild to moderate cases of diarrhea as they increase the volume of the patient's stool. Examples of such supplements include methylcellulose (Citrucel) and psyllium powder (Metamucil). In more severe cases of diarrhea, loperamide (Imodium A-D) can be helpful. However, it's important to note that these medications may not be effective or may even be harmful to individuals with strictures or certain infections [49].

3.5.3.12 Anti-Integrin Antagonists

Natalizumab, a humanized monoclonal antibody targeting alpha-4 (α4) integrin, has been linked to the reactivation of the John Cunningham (JC) virus and occurrences of progressive multifocal leukoencephalopathy (PML) [160]. The medicine is exclusively accessible via a limited program known as the TOUCH Prescribing Program. Prior to administering natalizumab, it is advisable to conduct a check for the anti-JCV antibody. It is important to monitor CBC (complete blood count) and liver function during medication, and patients should be closely watched for any new signs or symptoms that may indicate PML (progressive multifocal leukoencephalopathy). Vedolizumab is a type of antibody that has been modified to be more similar to human antibodies. It particularly targets a protein called alpha-4 beta-7 integrin, which is involved in cell adhesion in the gut. By targeting this protein, vedolizumab has anti-inflammatory effects that are specific to the gut. Prior to commencing treatment, it is advisable to do a thorough assessment of CBC, liver, and renal function. Furthermore, it is essential to monitor these parameters during the course of medication closely [49].

3.5.3.13 Pain Relievers

For minor pain, doctors may suggest acetaminophen (Tylenol, others). However, ibuprofen (Advil, Motrin IB, others), naproxen sodium (Aleve), and diclofenac sodium can worsen symptoms and trigger the patient's condition.

3.5.3.14 Alternative Medicine and Probiotics

Complementary and alternative medicine has been utilized by a large number of patients with digestive disorders [161]. Complementary and alternative medicine comprises a wide range of methods, such as herbal supplements, acupuncture, dietary therapies, and mind-body activities. Nevertheless, there is a scarcity of properly conducted research on the safety and efficacy of complementary and alternative medicine. Research suggests that supplementing regular digestive tract with extra probiotics or good bacteria may help prevent inflammatory bowel disease (IBD) [161]. One well-known CAM practice that has drawn interest in managing IBD is probiotic usage. Probiotics are advantageous microorganisms that reside in the gastrointestinal system, performing a vital function in preserving a harmonious equilibrium within the microbiome.

Triggering inflammatory bowel disease (IBD) has been linked to the microbiome, which is made up of billions of bacteria. There is currently little high-quality research on the efficacy of complementary and alternative medicine (CAM), especially probiotics, in treating inflammatory bowel disease (IBD) [162]. Nonetheless, the advantages of introducing good bacteria to the stomach to control the immune system and lessen inflammation linked to IBD are gaining attention. According to some research, taking probiotics in addition to prescription drugs may help manage the symptoms of inflammatory bowel disease. It is thought that probiotics increase the variety and quantity of good bacteria in the stomach, which may improve the environment for gut health in general [162]. The objective of treatment for inflammatory bowel disease is to diminish the inflammation that stimulates the onset of signs and symptoms. In ideal circumstances, this might result in a reduction in the risk of complications, a long-term remission, and symptom relief. The treatment of inflammatory bowel disease (IBD) often consists of either pharmacotherapy (drug therapy) or surgical intervention.

3.5.3.15 Nutritional Supplements

If a patient is having trouble absorbing enough nutrients, their physician may suggest taking vitamins and nutritional supplements [163]. In cases where weight loss is severe, the doctor may recommend enteral nutrition, which involves feeding a specific food through a tube, or parenteral nutrition, which involves injecting nutrients directly into a vein. This type of treatment can improve overall nutrition and give the digestive system a chance to recover. Temporary cessation of bowel activity can alleviate inflammation in the short term. If a patient has a narrowed or constricted intestine, they may be advised to follow a low-residue diet [163]. This can help prevent undigested food from getting stuck in the constricted part of the intestine, which could cause an obstruction.

3.5.3.16 Surgery

If changes in diet, lifestyle adjustments, medication, or other treatments are unsuccessful in managing the symptoms of inflammatory bowel disease (IBD), surgery may be recommended by a healthcare professional. The surgical procedure for treating ulcerative colitis involves the complete removal of both the colon and rectum, followed by the formation of an internal pouch that is connected to the anus [164]. This allows for defecation without the need for an external bag. In some cases, it may not be possible to create a pouch. In such instances, surgeons create a permanent opening in your abdomen (called an ileal stoma) to allow stool to pass through, which is then collected in a connected bag [164].

Approximately two-thirds of individuals diagnosed with Crohn's disease will undergo at least one surgical procedure during their lifespan. Nevertheless, surgical intervention does not provide a permanent cure for Crohn's disease [165]. During surgery, the doctor will remove the part of the digestive system that is affected and then reconnect the healthy parts together. Surgery can be helpful in closing fistulas and draining abscesses. Surgery typically has short-term benefits for patients with Crohn's disease [165]. The disease usually exhibits recurrence, often in close proximity to the adjacent tissue. The best course of action is to take medicine after surgery in order to reduce the chance of a recurrence.

3.5.4 Vaccination

It is crucial to vaccinate every individual with IBD as they are more susceptible to infections that can be prevented by vaccination. Preferably, it is advisable to administer vaccinations to these patients before initiating immunosuppressive therapy (IST) [166,167]. Although individuals with immunocompromised inflammatory bowel disease (IBD) may have a weakened immune response, most of them still get a satisfactory response following immunization. Immunocompromised individuals with inflammatory bowel disease (IBD) should not get live vaccines because of the potential for vaccine-associated infection [166]. Therefore, it is crucial to carefully consider the scheduling of live vaccines for IBD patients who are on immunosuppressive therapy (IST) or have intentions to start IST [167]. Certain demographic groups present unique difficulties, including expectant mothers, those living with immunocompromised patients, and travelers. If an immunocompromised individual with inflammatory bowel disease (IBD) lives with someone who has received a live vaccine, the IBD patient could potentially be exposed to infection from the vaccinated family member [166]. It is generally advised that those living in the same home as immunocompromised people with inflammatory bowel disease (IBD) should receive vaccinations as per the recommended standards outlined in Table 3.3. IBD patients planning to travel abroad may consider seeking assessment from an infectious disease expert or a travel medicine specialist, while the Figure 3.2 and Table 3.4 outlines vaccination suggestions for specific groups such as pregnant women and individuals with IBD who are traveling [166].

3.6 GUT MICROBIOTA

The term "gut microbiota" refers to the complex microbial community that comprises of the human gut, which is host to about 100 trillion microorganisms [168]. This community of more than a thousand different kinds of microorganisms, each of which is home to at least 160 different species, coexists in the human gastrointestinal tract and affects a plethora of physiological processes. The human gut microbiome is believed to include around 3.3 million genes, which is nearly 150 times greater than the total amount of genes in the human genome [168]. The gut microbiota consists of a wide variety of microorganisms, including bacteria, bacteriophages, archea, fungi, as well as protozoa species. Recent developments in genomic technologies, for instance whole-genome sequencing, next-generation sequencing, and metagenomics using 16s rRNA gene amplification have greatly improved our understanding of important interactions, including host-microbiota and microbe-microbe interactions [169]. The current advancements in artificial intelligence,

TABLE 3.3
Summary of Vaccination in Inflammatory Bowel Disease

Vaccine	Live Vaccine	Patients on IST	How Often
Measles, mumps	✓	Not advised due to adverse effects	If there is no recorded immunity, two doses of the vaccine, with a four-week interval
Influenza (Flu vaccine)	Nasal spray	Use flu shot only	1 dose every year
Tetanus, diphtheria, acellular pertussis (Td/Tdap)	✕	Adhere to the prescribed treatment plan	If an individual has not received prior vaccination, then three doses at 0, 1, and 6-12 months. One dose of Tdap is followed by a Td booster every 10 years
Varicella (chicken pox)	✓	Not advised due to adverse effects	If there is no recorded immunity, two doses of the vaccine, with a four-eight week interval
Zoster (Shingles)	✓	Not advised due to adverse effects	Administer a single dosage at age 60 years or older
Human papilloma virus	✕	Adhere to the prescribed treatment plan	For females, three doses of the vaccine by age 26, at 0, 2, and 6 months. For males, three doses of the vaccine by age 21, at 0, 2, and 6 months
Hepatitis A	✕	Adhere to the prescribed treatment plan	two doses of the vaccine, with a six-month interval
Hepatitis B	✕	Adhere to the prescribed treatment plan	three doses of the vaccine, with a 0, 1 and 6 month interval
Meningococcal (meningitis vaccine)	✕	Adhere to the prescribed treatment plan	Administer two doses, with a two-month interval between each dosage
Pneumococcal (pneumonia vaccine)	✕	Adhere to the prescribed treatment plan	If no previous vaccination, the vaccine at the age of 0, then at 2 yrs, and finally at 5 yrs. If received a previous immunization, One dose five years after the last dose, then one dose at the age of 65

Source: [14].

along with deep learning, have enabled the rapid identification of numerous bacteria in the field of gut microbiota research. Recent investigations have demonstrated that the gut microbiome is actively involved in metabolism and performs several activities, for instance, immune system maturation, nutrition, and safeguarding the host organism [170]. The maintenance of human health is thus significantly influenced by the gut microbiota.

Recent data has revealed that the microbiome can be effectively targeted for nutritional purposes. This means that certain dietary habits can be used to modify the microbiome in a manner that directly affects metabolism and immunity [171]. It has also been linked to the occurrence of gastrointestinal cancer because changes in the surrounding environment, such as overgrowth or the relocation of species from other areas of the gastrointestinal tract can cause the innate immune system to become hyperactive, which in turn can cause carcinogenesis and the formation of tumors [172].

3.6.1 Features of Gut Microbiota in Inflammatory Bowel Syndrome

IBD develops in susceptible people as a result of an excessive immune response to a typical stimulus, for instance diet as well as the gut microbiota, even though the exact cause of the disease is still unknown. According to recent research, the pathophysiology of IBD may be significantly influenced by the gut microbiota [170]. Inflammatory bowel disease (IBD) typically involves a reduction in the population of *Firmicutes*, namely the species *Faecalibacterium prausnitzii*, which has anti-inflammatory properties. Conversely, An increase in the prevalence of Bacteroidetes, namely the species Bacteroides fragilis, has been observed [173]. Additionally, there is a rise in Proteobacteria, specifically a proliferation of *Enterobacteriaceae*, which includes the opportunistic pathogens *Escherichia coli* as well as *Klebsiella pneumonia* [174]. These pathogens contribute to the inflammation of the mucosa and elevate the likelihood of infections. The gut microbiota of healthy people and IBD patients has been compared in numerous research. In individuals with Crohn's disease (CD), the abundance of *Christensenellaceae* (a family in the *Firmicutes* phylum), *Coriobacteriaceae* (a family in the Actinobacteria phylum), as well as *Faecalibacterium prausnitzii (F. prausnitzii)* was reduced, whereas *Actinoyces, Veillonella*, and *Escherichia coli* (*E. coli*) were elevated compared to individuals despite the illness [175]. *Eubacterium rectale* and *Akkermansia* were found to be reduced, whereas *E. coli* was shown to be elevated in ulcerative colitis (UC) patients in comparison to healthy individuals. When comparing the gut microbiota of IBD patients to that of healthy people, it was discovered that there was a decrease in gut microbiota diversity or no discernible difference observed.

A notable trend in all of the research studies indicates that there was a reduction in α-diversity relative to healthy persons, as well as a drop in *F. prausnitzii* among patients with Crohn's disease (CD) along with ulcerative colitis (UC). *F. prausnitzii* is recognized as bacteria that produces butyrate and is anticipated to contribute to the anti-inflammatory characteristics in Crohn's disease by suppressing the NF-κB pathway in the cells lining the intestines [176]. Furthermore, it is important to point out that *Christensenellaceae,* a type of bacterium that produces butyrate and belongs to the *Firmicutes* phylum, has been found to be diminished in individuals with CD compared to those in good health [176]. It was noted in the Pittayanon et al.[175] review that there was a decrease in *Akkermansia* in UC patients; however, it is still being determined if this is the reason or a result of the pathophysiology of UC patients. Past research has shown that akkermansia has anti-inflammatory properties in colitis, and that in individuals with UC, akkermansia's energy supply is diminished, suggesting that akkermansia is associated with a reduction in mucus. It is observed that in UC patients, the reduction in mucus is a secondary outcome that leads to the decrease in *Akkermansia.* In people with ulcerative colitis (UC), there is a decrease in the presence of *E. rectale*, a significant bacteria that produces butyrate, relative to healthy individuals [177]. The gut microbiota of individuals with inflammatory bowel disease (IBD) is collectively distinguished by a reduction in bacteria that produce butyrate, in comparison to healthy individuals.

Coriobacteriaceae, which are part of the Actinobacteria phylum, are in the same family as *Collinsella, Eggerthella, Sloackia, and Atopobium.* The pathophysiology of IBD has been linked to bile acids. However, these bacteria, along with *Coriobacteriaceae*, are also known to be capable of converting bile acids [178]. They are hypothesized to have a role in the pathophysiology of IBD. One of the most common results is a rise in *Escherichia*, particularly *E. coli* from the *Enterobacteriaceae* family, which may be detrimental to inflammatory bowel disease. *E. coli* that adheres to and invades the intestinal epithelium is known to produce inflammation, and adherent-invasive *E. coli* is known to proliferate in the ileal mucosa of CD [179]. This implies that an elevation in *Escherichia* might potentially contribute to the chronic inflammation observed in individuals with inflammatory bowel disease (IBD). Overall, the amount of bacteria, their metabolic activity, and microbial diversity are all reduced in this context, resulting in severe dysbiosis. As previously stated, people with IBD have been found to exhibit distinct changes in their gut microbiota. Multiple studies have demonstrated that IBD patients have a decreased variety of gut flora.

3.6.2 Features of the Gut Microbiota in Irritable Bowel Disease

Irritable bowel disease (IBS) is a common disorder of the gastrointestinal tract, with a prevalence of approximately 11.2% worldwide [180]. IBS is characterized by clinical manifestations such as bloating, gas, stomach pain, or discomfort that coincide with alterations in bowel habits for instance constipation, diarrhea, or a combination of both. The precise etiology of IBS remains unknown, but it is believed to be influenced by various factors for instance genetic composition, cerebrointestinal peptides, abnormalities in gastrointestinal movement, heightened sensitivity in the gut, immune responses in the gastrointestinal system, changes in the lining of the gut, the composition of gut bacteria, as well as psychological and social factors. The connection among the brain as well as the gastrointestinal function is referred to as the gut-brain axis, a significant notion in understanding the pathophysiology of IBS. Recent data has shown that changes in the gut microbiota are significant factors in the development of IBS. This hypothesis originated from clinical observations of symptoms that occur after an infection, which are sometimes referred to as post-infectious IBS [181].

Numerous studies have made comparisons regarding the gut microbiomes of healthy people and those with IBS. Pittayanon et al. (2019) [182] systematically analyzed 24 research that compared the gut microbiota of individuals with IBS with those in good health. In contrast to healthy individuals, the gut microbiota of IBS patients demonstrated similar in two investigations despite four researches demonstrating an increase in Proteobacteria at the phylum level in IBS patients. Reliable outcomes were not achieved for *Bacteroidetes, Actinobacteria, or Firmicutes.* Furthermore, a study conducted at a taxonomic level below the phylum level revealed a noteworthy rise in *Enterobacteriaceae* (a type of Proteobacteria) and *Bacteroides* (a type of Bacteroidetes) among individuals with irritable bowel syndrome (IBS) in comparison to those who are in good condition. Indeed, a notable proportion of individuals exhibit intestinal bacterial overgrowth (SIBO). Multiple studies have demonstrated a decrease in Bifidobacteria and Lactobacilli and a rise in *Enterobacter* have been observed, particularly in patients diagnosed with irritable bowel syndrome and diarrhea (IBS-D). Additional research has also observed a rise in *Veillonella* levels in individuals diagnosed with both irritable bowel syndrome (IBS) and constipation (IBS-C). Other writers link irritable bowel syndrome (IBS) with the bacteria *Campylobacter, Salmonella, Yersinia, E. Coli, and Shigella.* Pathogenic bacteria, including *Shigella*, *Escherichia*, *Campylobacter*, as well as *Salmonella*, are classified within the *Enterobacteriaceae* family. Furthermore, around 10% of individuals with irritable bowel syndrome (IBS) attribute to the emergence of symptoms to a previous episode of infectious dysentery. This has led to the name "post-infectious-IBS" being developed, and it is hypothesized that dysbiosis resulting from the infection is responsible for the development of IBS-like symptoms. These pathogenic bacteria may be linked to post-infectious IBS. Intestinal toxins are also produced by bacteria in the genus Bacteroides. One such bacterium is *Bacteroides fragilis*, which degrades glycoproteins in mucus and alters wall motility, causing symptoms of IBS for instance diarrhea and abdominal pain [183].

The most consistent finding came from three European studies that found that patients with IBS had lower levels of *Clostridiales I*, a bacterium that is difficult to cultivate. Furthermore, four further investigations demonstrated a noteworthy decrease in *Faecalibacterium (Clostridiales)*

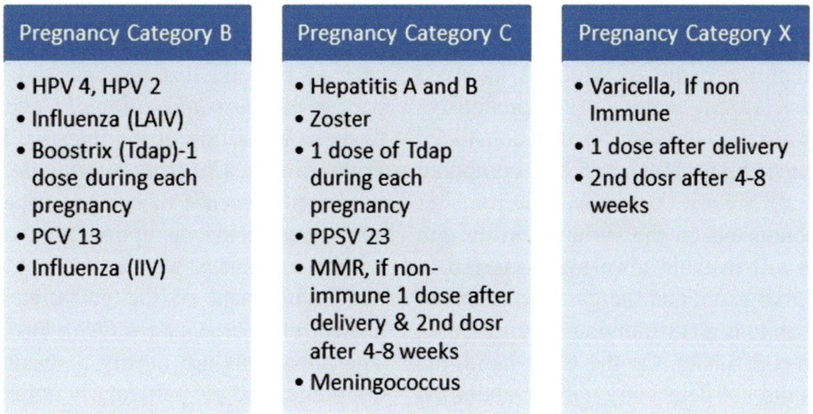

FIGURE 3.2 Vaccination in Pregnant Women With Inflammatory Bowel Disease.

TABLE 3.4
Vaccination of IBD Traveler

Vaccine	Type	Travel Related Indication
Typhoid	Live & Inactivated	Regions in South America and sub-Saharan Africa, Continents such as Africa, Central and South America, Asia, The Caribbean, and Oceania. Travel to a highly endemic region for more than a year. The regions included are Mexico, Central or South America, Africa, Asia (excluding Japan), Eastern Europe, sub-Saharan Africa, and Saudi Arabia (specifically during the Hajj and Umrah trip). Rural Japan.
Influenza	Inactivated	
Polio	Live	
Yellow Fever	Live	
Meningococcal vaccine	Inactivated	
BCG vaccine	Live	
Japanese encephalitis virus	Inactivated	
Hepatitis A	Inactivated	

Source: [14].

among 119 patients suffering from diarrhea-predominant irritable bowel syndrome (IBS). In addition, two investigations demonstrated a substantial decrease in *F. prausnitzii* levels among patients with IBS compared to healthy individuals, while two other studies indicated a negligible decrease in *F. prausnitzii* levels among individuals with IBS compared to the control group. There is a suggestion that certain bacteria, such as *Clostridiales I* and *F. prausnitzii*, might have a protective function in relation to irritable bowel syndrome (IBS).

F. praousnitzii, a member of the *Clostridiales* group, is recognized for its role in preserving the homeostasis of the digestive tract. *F. prausnitzii* is a bacterium that demonstrates an anti-inflammatory impact through its capacity to generate butyrate. According to reports, *F. prausnitzii* has been found to control the synthesis of interleukin-17 in a rat model, resulting in the alleviation of symptoms associated with irritable bowel syndrome (IBS) [184]. The genus *Bifidobacterium (Bifidobacteriaceae)* was investigated in seven investigations. Out of these, five studies found a noteworthy reduction in the abundance of *Bifidobacterium* in patients with irritable bowel syndrome (IBS). However, one study observed a tendency towards a drop in *Bifidobacterium* in IBS patients, however, the difference failed to attain statistical significance. Two investigations indicated a considerable reduction in the genus *Tannerella* (Phylum Bacteroidetes), while the other two studies found no significant effect. The presence of *Bifidobacterium* is consistently decreased in all subtypes of IBS, indicating that *Bifidobacterium* could potentially alleviate symptoms associated with IBS. A randomized controlled experiment investigating the effects of *Bifidobacterium longum* on individuals with irritable bowel syndrome (IBS) revealed that the group receiving *Bifidobacterium longum* had enhanced depression levels and increased the standard of life in comparison to the placebo group [185]. Research indicates that the *Bifidobacterium longum* group may lead to a decrease in p-cresol sulphate, potentially contributing to symptoms of irritable bowel syndrome (IBS). P-Cresol sulfate has been demonstrated to be cytotoxic and to lower colonocytes' oxygen consumption. The synthesis of p-cresol sulphate relies on various elements within the intestinal environment, including the microbiota composition, dietary intake, and pH levels of the intestinal tract. Tyrosine and phenylalanine are converted by the gut bacteria into p-Cresol sulphate via 4-hydroxylphenylacetate. *Clostridioides difficile* and some *Lactobacillus* strains have the ability to generate p-cresol by the process of decarboxylation of 4-hydroxyphenylacetate [186]. Moreover, the synthesis of bioactive compounds by bacteria from dietary protein has been linked to inflammation and increased permeability of the gut tissue. Furthermore, research has demonstrated that p-cresol sulphate affects the dopamine / norepinephrine pathway, which is involved in depression symptoms [185]. After a comprehensive evaluation of probiotics for IBS, the group that consumed *Bifidobacterium* experienced better IBS symptoms compared to the control group.

Research has been conducted on the variation in the gut microbiota of individuals with irritable bowel syndrome (IBS). Five out of nine studies that examined the gut microbiota of IBS patients in comparison to healthy individuals represented a significant decrease in α-diversity. On the other hand, the other four investigations did not detect any notable alteration in α-diversity. The gut microbiota has been examined in subgroup analysis based on different subtypes of irritable bowel syndrome (IBS), which include constipation-dominant IBS (IBS-C), diarrhea-predominant IBS (IBS-D) as well as mixed bowel habit IBS subtype (IBS-M) [187]. All six investigations analyzed the gut microbiota in 130 individuals with IBS-M and no significant differences were detected compared to IBS-C or IBS-D. The differences between IBS and healthy controls were consistent across all cases, including the IBS-M group, as opposed to the IBS-C or IBS-D subgroups. Regarding IBS-D, three out of five articles that evaluated the genus Bacteroides found a noteworthy rise of this genus in patients with IBS-D [168]. On the other hand, two articles reported inconclusive findings when compared to the control group. Conversely, most research examining the genus *Bifidobacterium* demonstrated a notable reduction in patients with irritable bowel syndrome with diarrhea (IBS-D). IBS-C was assessed in a single research. This study found differences between patients with IBS-C as well as healthy persons, although it is challenging to make conclusive judgments based on a single study. Additionally, there is evidence suggesting that viruses, parasites, and fungi may have a significant impact on the pathophysiology of IBS [168].

3.6.3 Features of Gut Microbiota in Celiac Disease

The fecal microbiota of individuals with active celiac disease exhibits a higher level of microbial diversity in comparison to healthy individuals. This is accompanied by rise in the abundance of *Bacteroides, Clostridium, Prevotella, and Staphylococcus*, while there is a notable decrease in bacteria that possess anti-inflammatory and mucosa protection properties, especially *Bifidobacterium* as well as *Lactobacillus* [168]. There is a suggestion that this imbalance of gut bacteria may play role in the development of sensitivity to gluten and the resulting inflammation [188].

3.6.4 Features of Gut Microbiota in *Clostridium Difficile* Infection

A highly marked dysbiosis, characterized by a reduction in the abundance of microbiota species and microbial complexity along with a substantial change in the overall composition is typically required for susceptibility to *Clostridium difficile* infection [189,190]. The depletion of certain types of bacteria, including those in the core microbiota and minor players such as bacteria that produce short-chain fatty acids, non-pathogenic *clostridia, Bilophila,* and *Alistipes*, can contribute to the colonization of the intestine by *C. difficile* [189,190]. Furthermore, *Clostridium difficile* colitis is linked to a decreased presence of bacteria capable of metabolizing bile acids, leading to a disruption in the control of these chemicals within the entero-hepatic system [189]. The changes in the environment of the gastrointestinal tract promote the growth of *Clostridium difficile* bacteria that produce toxins. The alterations are mostly a result of extended courses of antibiotics and are particularly noticeable in older individuals with several chronic conditions. Non-antibiotic medications, such as proton pump inhibitors (PPI), can also have an impact on gut microbiota composition and have been associated with an elevated risk of *Clostridium difficile* colitis in certain studies [191]. Fecal microbiota transplantation, which helps to restore a healthy balance of microorganisms in the intestines, can effectively prevent recurrences and significantly alter the course of the infection.

3.6.5 Features of Gut Microbiota in Colon Cancer

Changes in the microbiota of the gut may contribute to the development of colon tumors [192]. Nonetheless, it is likely that these changes are only slight and limited to certain species that possess pathogenic traits and the capacity to infiltrate the intestinal mucosa locally without resulting in acute infectious illnesses. Out of these changes, the most significant ones could be the increase in Bacteroides fragilis, *Enterococcus faecalis* and *Streptococcus gallolyticus.* The metabolic byproducts of *Salmonella* can trigger the intracellular signaling of β-catenin, which in turn stimulates epithelial growth. Additionally, the accumulation of potential carcinogenic metabolites in the gut lumen may be facilitated by the reduction of butyric acid-producing bacteria and the association between dietary modifications and the *Prevotella enterotype* [193]. Our existing understanding of the microbiome's contribution to the development of gastrointestinal tumors is limited. The microbiome likely plays a role in the complex progression of the disease rather than directly causing cancer.

3.6.6 Features of Gut Microbiota in Esophageal Cancer

There is insufficient research on the impact of microbiota on the progression of esophageal cancer. Nevertheless, it is possible that modifications in the gastric microbiota in the vicinity of the gastro-esophageal junction can potentially lead to an elevated susceptibility to esophageal cancer. Furthermore, some research has revealed that patients with Barrett's esophagus have a substantially different esophageal microbiome than healthy individuals [194]. Nevertheless, this domain requires further investigation before drawing conclusive findings.

3.6.7 Features of Gut Microbiota in Gastric Cancer

Helicobacter pylori infection is believed to initiate the inflammatory cascade, also known as Correa's cascade, which results in the emergence of chronic gastritis, metaplasia, atrophy, and gastric cancer. The presence of *Helicobacter pylori* in neoplastic lesions is sporadic, and only a small percentage of infected individuals develop cancer [194]. According to the findings of the research, the stomach microbiome of people with and without *H. pylori* differs significantly. Some authors suggest changing the current paradigm of gastric carcinogenesis by emphasizing the interactions between bacteria and the stomach mucosa. These interactions may result in changes to the gastric microbiota, which may, in turn, cause gastric atrophy. In addition to or instead of the more virulent strains of *H. Pylori* (like Cag A), it's likely that the "new" microbiota that the stomach develops in response to reduced acidity plays a factor in the development of gastric carcinogenesis [194].

3.6.8 Features of Gut Microbiota in Liver Disease

The intestinal microbiota metabolism and endotoxinemia significantly affect the function of the liver. Endotoxinemia occurs when lipopolysaccharides produced by the microbiota are absorbed through the intestinal mucosa [195]. In non-alcoholic fatty liver disease, there is evidence of an imbalance in the gut microbiota, changes in its composition can be both a cause and a consequence of a state of illness in particular, a reduction in the butyrate-producing bacteria, such as *Anaerotruncus colihominis, Faecalibacterium prausnitzii,* and *Butyrivibrio crossotus.* Conversely, there is an increase in bacteria like *Ruminococcus, Campylobacter, and Shigella,* which produce the metabolite acetate [196]. Acetate is a substance that promotes the production of fat (lipogenesis). However, a more profound dysbiosis is seen in Alcoholic liver disease when *Enterobacteriaceae* are more prevalent, and *Bacteroidetes* are less prevalent. This is associated with the impact of alcohol on the lining of the intestines, causing an increase in permeability and disruption of immunological function, also an alteration in the balance between the immune system and gut bacteria occurs as a result [195]. Non-alcoholic cirrhosis of the liver also exhibits similar changes in the composition of the intestinal microbiota. This condition involves the colonization of the intestines by bacteria that are typically found in the oral cavity microbiota. The microbiome of the gut and the reduction in intestinal mucosal permeability are linked to a number of problems during the decompensation stages of cirrhosis, including spontaneous bacterial peritonitis as well as hepatic encephalopathy [195].

3.7 DYSBIOSIS

Dysbiosis is defined as a disruption in the composition and functioning of the gut microbiota [197]. A growing body of research indicates that this condition is linked to the colonization of pathogenic microorganisms as well as the loss of beneficial microbial signaling or input. Dysbiosis is believed to initiate inflammatory responses and immunological dysregulation linked to various human illnesses [198]. In the last few years, biomedical research has attempted incredibly hard to define what is considered the "normal" human intestinal microbiota, indicating how specific circumstances may affect its composition and identify any "abnormalities" in the microbiota composition that may be linked to acute or chronic illnesses, as well as potential cause-and-effect relationships [199]. Research has demonstrated that the microbiota can play a substantial role in the development and progression of some diseases. While this concept is rather easy to understand for diseases that mainly affect the intestines, for instance intestinal inflammatory disorders (IBD) and *Clostridium difficile enterocolitis*, it becomes more challenging for diseases that affect organs located far away from the intestine. Nonetheless, this has made it possible to speculate and, in certain situations, to provide evidence of metabolic, endocrine, and systemic pathways by which the microbiota can affect the pathophysiology of the entire body through the GI tract [199].

The alterations in the human intestinal microbiota can be broadly categorized as either affecting the overall balance of the microbial population or specifically targeting a single or small group of comparatively insignificant microorganisms. These microorganisms, due to their specific metabolic activities or high potential for causing disease, can independently contribute to the development of a disease, either by negatively impacting the absence of a bacterium with protective properties or positively influencing the presence of a pathogenic or harmful bacterium [199]. Dysbiosis refers to a widespread change in the overall composition of the microbial community in the intestines, characterized by an elevation in the comparative prevalence of certain taxa and a decrease in the relative abundance of others. Dysbiosis often involves an overabundance of pathobionts, which are types of bacteria that have the potential to cause disease, for instance *Escherichia* as well as *Klebsiella spp*. This overabundance comes at the cost of a decrease in the presence of bacteria that have beneficial metabolic activity, such as *lactobacilli* and *bifidobacteria*. Moreover, dysbiosis is related to a decrease in biodiversity, which refers to the number of different microbial species in the microbiome and a decrease in the complexity of the microbial community [200]. From a medical perspective, dysbiosis is a state that can raise the risk of certain diseases rather than being an actual disease or symptom. Furthermore, the distinctions between dysbiosis and the natural variations in the microbiota's composition are not clearly characterized. As a result, it might be challenging to assess if dysbiosis has an impact on a particular microbiota profile [200].

The gut microbiota has the ability to impact human disease at various levels, and changes in its composition can be considered as either the cause or contribute to illness. The goal of numerous research conducted in recent years has been to determine the primary dysbiosis of the intestinal microbiota linked to a wide range of acute and chronic human illnesses, as well as to elucidate the relationship between these abnormalities and the pathophysiology of the diseases themselves.

3.8 MICROBIOTA TARGETED THERAPIES

3.8.1 Fecal Microbiota Transplantation

Fecal microbiota transplantation (FMT) is an innovative medical approach aimed at restoring the disrupted microbial equilibrium by transferring fecal matter from healthy donors into persons with diseases. Once the intestinal microbiota has been obtained from a suitable donor, the samples can be transplanted by colonoscopy, orogastric tube, enema, or oral capsule containing the freeze-dried material. Ancient medical literature has described the administration of human fecal suspension orally to treat severe diarrhea. Ge Hong, in the 4th century, specifically recorded its use in China [201]. The first recorded application of FMT, or fecal microbiota transplantation, used to treat pseudomembranous colitis was published by Eiseman et al in 1958 in the field of contemporary medicine [202]. The first recorded randomized controlled clinical trial for recurrent Clostridioides difficile (*C. difficile*) infection (CDI) using Fecal Microbiota Transplantation (FMT) was conducted in 2013 [203]. Fecal microbiota transplantation (FMT) is a promising treatment for gastrointestinal, metabolic, and cardiovascular diseases, and has gained attention in evidence-based medicine. The potential of fecal microbiota transplantation (FMT) to treat various gastrointestinal disorders by modifying the gut microbiome has drawn more and more attention. The mainstay of its usage in current clinical practice is the treatment of recurrent *C. difficile* infection, despite recent research showing it can also cure disorders like pouchitis, IBD, constipation, and irritable bowel syndrome. The overgrowth of *C. difficile* reduces the diversity of gut bacteria, making it difficult to treat and prevent recurrence. FMT is a highly effective method of restoring the colonic microbiota to a healthy state, similar to that of a non-infected control group.

The initial documented application of Fecal Microbiota Transplantation (FMT) for the treatment of inflammatory bowel disease (IBD) occurred in 1989 [204]. During this procedure, patients experienced the elimination of bacteria in the colon, followed by the introduction of healthy colonic microbiota from the donor. Since then, there have been numerous advancements in the technology and application of FMT. The latest data on the effectiveness of FMT in IBD is varied and occasionally contradictory. Moayyedi et al. found that fecal microbiota transplantation (FMT) could lead to clinical remission in patients with ulcerative colitis (UC) without causing any significant differences in adverse effects when compared to placebo [205]. However, Rossen et al. observed that the clinical results did not differ much when comparing FMT to controls who received autologous FMT [206]. Notably, the Rossen group observed that the microbiota of ulcerative colitis (UC) patients who positively reacted to fecal microbiota transplantation (FMT) had a unique composition compared to those who did not respond [206]. While the available data on the use of FMT in treating Crohn's disease is limited, although it is still promising. Cui et al. observed a notable improvement of symptoms in individuals with refractory CD following FMT [207]. However, It is crucial to keep in mind that this trial lacked a control group for comparison. Vaughn and colleagues found that fecal microbiota transplantation (FMT) resulted in clinical improvement in 58% of patients with active Crohn's disease (CD) compared to a placebo. Additionally, similar to Rossen's study on ulcerative colitis (UC), those patients who provided positive responses to the treatment had higher levels of microbial diversity in contrast to those who chose not to respond [208]. The variation in response to fecal microbiota transplantation (FMT) may be attributed to differences in the subtypes of inflammatory bowel disease (IBD), the selection and techniques used to obtain FMT donors, and the criteria used to define clinical remission in each individual.

3.8.2 Probiotics

Live microorganisms are termed as probiotics that give the host health benefits when consumed in sufficient amounts. Probiotics may be used to treat and prevent infections and long-term inflammatory problems of the gastrointestinal system [209]. Probiotics are live bacteria that are added to human hosts in order to help the gut environment return to homeostasis. Examples of these bacteria include *Lactobacillus, Bifidobacterium, Saccharomyces*, and several non-pathogenic strains of *Escherichia coli.* Figure 3.3 Illustrates the mechanisms of action of probiotics which have been hypothesized to explain how probiotics can influence the gut microbiome and, consequently, impact the progression of IBD. Certain activities involve modulating the intestinal immune system. These encompass enhancing the function of the intestinal barrier by preventing intestinal cell apoptosis, controlling genes linked to immune pathways (such as NF-κB, IL-6, and TNF-α), affecting T cell pathways, or directly affecting the bacterial microenvironment by generating antimicrobial molecules (such as lactic acid, hydroperoxides, bacteriocins, or defensins). Some probiotic strains have non-immunological effects, which mean they can change the gut environment and affect the growth patterns of other bacterial strains. This can be done by adjusting the pH in the intestines, increasing mucus production, or directly competing with other microbes for nutrition.

Probiotic usage in the treatment of inflammatory bowel disease (IBD) was initially evidenced in animal models with chemically induced colitis. In addition to clinically reducing colitis symptoms in mice, strains like *Bifidobacterium* and *Faecalibacterium* have also been shown to have histologic effects by protecting the innate intestinal mucosal architecture. Human studies have shown that *Bifidobacterium* can enhance disease activity in mild-to-moderate UC. In contrast, a different study indicated that patients with endoscopic relapse of CD had lower amounts of *Faecalibacterium* in resected ileal mucosa. Notably, certain probiotic strains can have distinct impacts on individuals with ulcerative colitis (UC) compared to those with Crohn's disease (CD). The probiotic mixture VSL#3, which contains *Bifidobacterium,*

Lactobacillus, and *Streptococcus*, was observed to reduce inflammatory cytokine levels in individuals with CD [210]. However, it did not demonstrate a decrease in endoscopic recurrence rates when compared to a placebo. In a different study with ulcerative colitis individuals, VSL#3 not only reduced disease severity but also induced clinical remission in certain patients. It is worth mentioning that 92% of the patients included in that study were simultaneously receiving anti-inflammatory therapy. This suggests a possible synergistic effect of probiotics when combined with conventional anti-inflammatory treatments for inflammatory bowel disease (IBD). In a prior investigation, a group of 36 individuals diagnosed with pouchitis were split into two groups randomly. One group of people was given a high dosage of the probiotic VSL#3 once daily, while the other group received a placebo. The results of this research demonstrated that VSL#3 was successful in maintaining remission induced by antibiotics for at least one year in patients with recurrent or refractory pouchitis [211]. Specifically, 85% of patients who received VSL#3 achieved remission, compared to only 6% of patients who were given placebo. This difference was statistically significant ($p < .0001$) [211].

3.8.3 Prebiotic

A substrate that is specifically used by host microbes to provide a health advantage is known as a prebiotic. Prebiotic compounds have attributes to enhance growth, boost metabolism, and support the protection of beneficial bacteria in host organisms. Prebiotics in the colon specifically boost the fermentation activity of particular kinds of beneficial microorganisms, for instance *Lactobacillus* and *Bifidobacterium* spp [212]. By providing fermentable substances, prebiotics stimulate the production of mucin, which helps to restrict the movement of germs. The fermentation of gut microbiota produces metabolites such as vitamins, folate as well as short-chain fatty acids, which have antibacterial properties and contribute to the maintenance of a healthy gut barrier [212].

Prebiotics are not microorganisms themselves, but rather the metabolic substrates used by commensal gut bacteria (both native and artificially introduced) for their survival. These include a diverse range of fermentable carbohydrates, such as psyllium, lactulose, lactosucrose, and various milk oligosaccharides [213]. When metabolized, these carbohydrates produce byproducts for instance short-chain fatty acids and acetic/propionic/butyric acids. These byproducts promote the growth of specific types of microorganisms. Manipulating the presence or absence of specific bacteria may alter the colonic environment and affect an individual's risk of developing inflammatory bowel disease (IBD) or their response to conventional treatments for IBD. It was observed in animal models that the intestinal populations of *Lactobacillus* and *Bifidobacterium* were increased in mice with chemically induced colitis when inulin and resveratrol were given [214]. Rose et al. performed in vitro experiments using fecal microbiota samples obtained from patients with inflammatory bowel disease (IBD). They used starch-entrapped fermenting microspheres to produce butyrate, which effectively decreased the pH of the colon environment. This acidic environment inhibited the growth of harmful bacteria and potentially facilitated the healing process of damaged colonic epithelium. Currently, there is a scarcity of clinical trials examining a small number of specific prebiotics in humans [215]. These studies often involve a small sample size and primarily concentrate on ulcerative colitis rather than Crohn's disease [215]. Psyllium has demonstrated potential effectiveness in alleviating symptoms of ulcerative colitis (UC) and may be equally as beneficial as conventional treatments like mesalamine. This is achieved by processes that enhance the amounts of butyrate in the feces. When administered as a supplement to conventional immunomodulating or anti-inflammatory therapy, germinated barley foodstuff may lessen the clinical severity of mild-to-moderate UC [216]. Fructo-oligosaccharides are linked to a higher number of *Bifidobacterium* in the colon and may reduce the severity of inflammatory bowel disease (IBD) symptoms [217]. Lactulose did not have any impact on the clinical disease parameters in either ulcerative colitis (UC) or Crohn's disease (CD) [218].

3.8.4 Synbiotic

Synbiotics are a combination of prebiotics and probiotics that work together to prevent the growth of harmful bacteria while promoting the growth of beneficial microorganisms. Products that contain prebiotics, which help to stimulate the growth of probiotics and their metabolites, are called synbiotics [213]. There are two ways that synbiotic effects might manifest: either the host's health improves after ingesting a combination of prebiotic or probiotic strains, or the consumption of prebiotics alone can stimulate the growth of native beneficial microbiota, such as *Bifidobacteria*. Prebiotics and synbiotics have been shown to have a variety of functions in preventing the growth and infection of enteric bacterial pathogens; however, more comprehensive research is required to fully understand how these substances can enhance gut health and prevent disease. A systematic review and meta-analysis investigating the effects of synbiotics on systemic inflammation found that around 50% of the analyzed studies demonstrated a significant decrease in at least one inflammatory biomarker, such as TNF-α or CRP [219]. Research on the application of synbiotics in IBD primarily concentrates on their usage in ulcerative colitis (UC) rather than Crohn's disease (CD). Furrie et al. [220] conducted an initial study to examine the utilization of synbiotics in UC, employing a synbiotic consisting of a *Bifidobacterium* probiotic and an inulin/oligofructose prebiotic. Overall, they discovered that there was a considerable improvement in colitis both microscopically (regenerative epithelium on histology and lower levels of many inflammatory molecules, including defensins, TNF-α, and IL-1α) and macroscopically (with improvement in sigmoidoscopy scores) [220]. A later study that described the use of different synbiotic compositions (*Clostridium butyricum, Lactobacillus sporogenes, Bacillus mesentericus, Streptococcus faecalis*) in addition to standard UC therapy also found that the use of synbiotics significantly

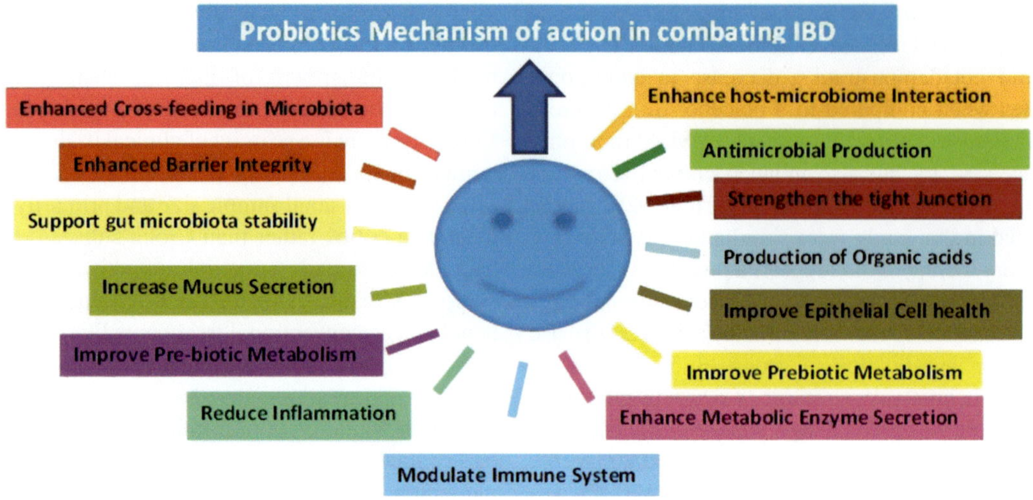

FIGURE 3.3 Mechanism of Action of Probiotics in Combating IBD.

improved the severity of the disease based on laboratory markers and clinical symptoms, and that the study also found that the study participants experienced a sustained remission longer than three months after the study ended [221].

There is a lack of comprehensive research describing the use of synbiotics in Crohn's disease (CD). The synbiotic combination of *Bifidobacterium* and inulin/oligofructose, which effectively treated colitis in patients with ulcerative colitis (UC), was also administered to patients with active Crohn's disease (CD) [222]. The results showed a similar improvement in clinical symptoms, histology scores, and levels of inflammatory markers. Curiously, although having proof of clinical remission, the utilization of synbiotics did not lead to significant enhancement in the quality of life as reported by the patients [222]. A different synbiotic was used in another trial, but the results showed that it was ineffective in preventing post-operative recurrence in individuals with CD.

3.9 CONCLUSION

Inflammation of the gastrointestinal tract has increased significantly over the last century, and inflammatory bowel disease has been more prevalent in developing nations during the last 50 years. A weakened immune system and having a family member with IBD are the biggest risk factors for acquiring the disease. Two chronic inflammatory bowel diseases (IBDs) are ulcerative colitis (UC) and Crohn's disease (CD). Numerous studies have been conducted on the pathophysiology of IBD, which includes immune system dysregulation, environmental variables, and genetic factors of the host. Ulcerative colitis mainly affects the colon and is characterized by persistent or recurrent presence of blood in the stool along with diarrhea. On the other hand, Crohn's disease (CD) can affect the stomach, esophagus, colon, small intestine, and entire gastrointestinal system.

The diagnosis involves a thorough evaluation that may encompass patient history, medical history, physical examination, laboratory testing, imaging studies such as endoscopy or colonoscopy, and biopsy. These diagnostic methods aid in determining the root cause of gastrointestinal inflammation, enabling the implementation of specific therapy and management measures. The diagnostic process involves integrating data from various sources, including laboratory investigations, imaging studies, endoscopic procedures, and specialized testing, to produce precise diagnoses and create individualized treatment plans. Comprehensive examination and individualized treatment are ensured through collaborative care that involves gastroenterologists, radiologists, pathologists, dietitians, and other healthcare experts. Continuous technological progress broadens the scope of diagnostic possibilities, enabling enhanced patient care and results.

The treatment of inflammatory bowel disease (IBD) involves a comprehensive strategy that goes beyond traditional medical procedures to include lifestyle changes, stress management strategies, and complementary and alternative medicine (CAM) methods. Nutrition plays a major role in managing symptoms, as demonstrated by dietary modifications such switching to a low-fiber diet during flare-ups and taking into consideration about adding omega-3 fatty acids. To address the unique nutritional requirements of individuals with IBD, it may be beneficial to limit dairy intake and supplement with multivitamins to ensure adequate nutrient absorption. Additionally, lifestyle and home remedies are essential in the management of IBD. The need to quit smoking is highlighted because it worsens the progression of diseases. On the other hand, stress management techniques, such as regular exercise, biofeedback, and relaxation exercises, provide practical ways to deal with stressors that can exacerbate symptoms. With the goal of improving patient-centered care and optimizing therapeutic outcomes, personalized medicine customizes treatment plans based on genetic profiles, biomarkers, and patient preferences. A greater number of therapy options for IBD patients have been made possible by the introduction of numerous biologics and small molecule medications in recent years. This has made it necessary to be able estimate treatment outcomes and ensure that patients receive prescribed treatments. Using the right biomarkers is important for disease diagnosis, prognosis prediction, and choosing the best course of treatment. In the age of precision or personalized medicine, biomarkers have been acknowledged as crucial aids in predicting the therapeutic response to medications and enhancing their effectiveness and safety. Investigating complementary and alternative medicine, including the possible advantages of probiotics in regulating the gut microbiome, offers further possibilities for alleviating symptoms and managing diseases. These strategies offer a comprehensive approach to managing gastrointestinal inflammation, considering the complex interplay of biological, clinical, and psychosocial factors to improve patient outcomes and quality of life.

The emergence of advanced sequencing tools has revealed the complex nature and diverse functions of the gut microbiota. Moreover, dysbiosis, or changes in the balance or composition of the intestinal microbiota, is linked to a number of gastrointestinal disorders. Gaining knowledge about the role the microbiota plays in gastrointestinal disorders should lead to new discoveries about the pathogenesis of the diseases as well as innovative treatment approaches including therapeutic manipulation of the microbiota.

REFERENCES

1. Dawoodi, S.; Dawoodi, I.; Dixit, P. (**2022**). Gastrointestinal problem among Indian adults: evidence from longitudinal aging study in India 2017-18. *Frontiers in Public Health*, *10*, 911354, doi: 10.3389/fpubh.2022.911354
2. Glickman, R.M. (**1998**). Inflammatory bowel disease: ulcerative colitis and Crohn's disease. *Harrisons Principles of Internal Medicine*, 1633–1645.
3. M'Koma, A.E. (**2013**). Inflammatory bowel disease: an expanding global health problem. *Clinical Medicine Insights: Gastroenterology*, *6*, CGast-S12731, doi: 10.4137/CGast.S12731
4. Burisch, J.; Munkholm, P. (**2015**). The epidemiology of inflammatory bowel disease. *Scandinavian Journal of Gastroenterology*, *50*, 942–951, doi: 10.3109/00365521.2015.1014407
5. Molodecky, N.A.; Soon, S.; Rabi, D.M.; Ghali, W.A.; Ferris, M.; Chernoff, G.; Benchimol, E.I.; Panaccione, R.; Ghosh, S.; Barkema, H.W. (**2012**). Increasing incidence and prevalence of the inflammatory bowel diseases with time, based on systematic review. *Gastroenterology*, *142*, 46–54, doi: 10.1053/j.gastro.2011.10.001
6. Benchimol, E.I.; Fortinsky, K.J.; Gozdyra, P.; Van den Heuvel, M.; Van Limbergen, J.; Griffiths, A.M. (**2011**). Epidemiology of pediatric inflammatory bowel disease: a systematic review of international trends. *Inflammatory Bowel Diseases*, *17*, 423–439, doi: 10.1002/ibd.21349
7. Ng, S.C.; Tang, W.; Ching, J.Y.; Wong, M.; Chow, C.M.; Hui, A.J.; Wong, T.C.; Leung, V.K.; Tsang, S.W.; Yu, H.H.; et al. (**2013**). Incidence and phenotype of inflammatory bowel disease based on results from the Asia-pacific Crohn's and colitis epidemiology study. *Gastroenterology*, *145*, 158–165.e152 doi: 10.1053/j.gastro.2013.04.007
8. Lovasz, B.D.; Golovics, P.A.; Vegh, Z.; Lakatos, P.L. (**2013**). New trends in inflammatory bowel disease epidemiology and disease course in Eastern Europe. *Digestive and Liver Disease: Official Journal of the Italian Society of Gastroenterology and the Italian Association for the Study of the Liver*, *45*, 269–276. doi: 10.1016/j.dld.2012.08.020
9. Jairath, V.; Feagan, B.G. (**2020**). Global burden of inflammatory bowel disease. *The Lancet Gastroenterology & Hepatology*, *5*, 3, doi: 10.1016/S2468-1253(19)30358-9
10. Roda, G.; Chien Ng, S.; Kotze, P.G.; Argollo, M.; Panaccione, R.; Spinelli, A.; Kaser, A.; Peyrin-Biroulet, L.; Danese, S. (**2020**). Crohn's disease. *Nature Reviews Disease Primers*, *6*, 22, doi: 10.1038/s41572-020-0156-2
11. Seyedian, S.S.; Nokhostin, F.; Malamir, M.D. (**2019**). A review of the diagnosis, prevention, and treatment methods of inflammatory bowel disease. *Journal of Medicine and Life*, *12*, 113, doi: 10.25122/jml-2018-0075
12. Li, Y.; Zhu, W. (**2018**). Chapter 4 – Pathogenesis of Crohn's disease–associated fistula and abscess. In *Interventional Inflammatory Bowel Disease: Endoscopic Management and Treatment of Complications*, Shen, B., Ed.; Academic Press: pp. 43–47.
13. Shawki, S.; Ashburn, J.; Signs, S.A.; Huang, E. (**2018**). Colon cancer: inflammation-associated cancer. *Surgical Oncology Clinics*, *27*, 269–287, doi: 10.1016/j.soc.2017.11.003
14. Abegunde, A.T.; Muhammad, B.H.; Ali, T. (**2016**). Preventive health measures in inflammatory bowel disease. *World Journal of Gastroenterology*, *22*, 7625–7644, doi: 10.3748/wjg.v22.i34.7625
15. Fishman, G.; DeRowe, A.; Wasserman, D.; Leider-Trejo, L.; Reif, S. (**2004**). Inflammatory bowel disease and sclerosing cholangitis with ulcerative lesions on skin and soft palate. *International Journal of Pediatric Otorhinolaryngology*, *68*, 1349–1352, doi: 10.1016/j.ijporl.2004.04.031
16. Greuter, T.; Navarini, A.; Vavricka, S.R. (**2017**). Skin manifestations of inflammatory bowel disease. *Clinical Reviews in Allergy & Immunology*, *53*, 413–427, doi: 10.1007/s12016-017-8617-4
17. Fragoulis, G.E.; Liava, C.; Daoussis, D.; Akriviadis, E.; Garyfallos, A.; Dimitroulas, T. (**2019**). Inflammatory bowel diseases and spondyloarthropathies: from pathogenesis to treatment. *World Journal of Gastroenterology*, *25*, 2162–2176, doi: 10.3748/wjg.v25.i18.2162
18. Saha, L. (**2014**). Irritable bowel syndrome: pathogenesis, diagnosis, treatment, and evidence-based medicine. *World Journal of Gastroenterology: WJG*, *20*, 6759, doi: 10.3748/wjg.v20.i22.6759
19. Hookman, P.; Barkin, J.S. (**2009**). Clostridium difficile associated infection, diarrhea and colitis. *World Journal of Gastroenterology: WJG*, *15*, 1554, doi: 10.3748/wjg.15.1554
20. Elseweidy, M.M. (**2017**). Brief review on the causes, diagnosis and therapeutic treatment of gastritis disease. *Alternative Integrative Medicine*, 6, 2, doi: 10.4172/2327-5162.1000231
21. Bereda, G. (**2022**). Peptic ulcer disease: definition, pathophysiology, and treatment. *Journal of Biomedical and Biological Sciences*, *1*, 1–10.

22. O'Sullivan, K.E.; Phelan, J.J.; O'Hanlon, C.; Lysaght, J.; O'Sullivan, J.N.; Reynolds, J.V. (**2014**). The role of inflammation in cancer of the esophagus. *Expert Review of Gastroenterology & Hepatology*, *8*, 749–760, doi: 10.1586/17474124.2014.913478
23. Jaroenlapnopparat, A.; Bhatia, K.; Coban, S. (**2022**). Inflammation and gastric cancer. *Diseases*, *10*, 35, doi: 10.3390/diseases10030035
24. Reed, C.C.; Dellon, E.S. (**2019**). Eosinophilic esophagitis. *Evidence-Based Gastroenterology and Hepatology*, *4e*, 50–67, doi: 10.1016/j.mcna.2018.08.009
25. Mederos, M.A.; Reber, H.A.; Girgis, M.D. (**2021**). Acute pancreatitis: a review. *JAMA*, *325*, 382–390, doi: 10.1001/jama.2020.20317
26. Quigley, C.; Jiang, X. (**2014**). Gastroenteritis. *Metabolism of Human Diseases: Organ Physiology and Pathophysiology*, 137–142.
27. Cojocaru, M.; Cojocaru, I.M.; Silosi, I.; Vrabie, C.D. (**2011**). Gastrointestinal manifestations in systemic autoimmune diseases. *Maedica*, *6*, 45.
28. Pascual, V.; Dieli-Crimi, R.; López-Palacios, N.; Bodas, A.; Medrano, L.M.; Núñez, C. (**2014**). Inflammatory bowel disease and celiac disease: overlaps and differences. *World Journal of Gastroenterology: WJG*, *20*, 4846, doi: 10.3748/wjg.v20.i17.4846
29. Yuan, L.; Wu, T.-T.; Zhang, L. (**2023**). Microscopic colitis: lymphocytic colitis, collagenous colitis, and beyond. *Human Pathology*, *132*, 89–101.
30. Mosinska, P.; Fichna, J. (**2015**). Ischemic colitis: current diagnosis and treatment. *Current Drug Targets*, *16*, 209–218, doi: 10.2174/1389450116666150113120549
31. Bruel, J.-M.; Taourel, P. (2011). Diverticulitis. In Patrice Taourel (ed.) CT of the Acute Abdomen; Springer: pp. 199–220.
32. Lamps, L.W. (2004). Appendicitis and infections of the appendix. *Seminars in Diagnostic Pathology*, *21(2)*, 86–97. doi: 10.1053/j.semdp.2004.11.003
33. Hindryckx, P.; Amininejad, L.; Van De Vijver, E.; Bossuyt, P. (**2014**). Belgian recommendations for the management of anemia in patients with inflammatory bowel disease. *Acta Gastro-enterol Belg*, *77*, 333–344.
34. Oldenburg, B.; Koningsberger, J.C.; Van Berge Henegouwen, G.P.; Van Asbeck, B.S.; Marx, J.J.M. (**2001**). Iron and inflammatory bowel disease. *Alimentary Pharmacology & Therapeutics*, *15*, 429–438, doi: 10.1046/j.1365-2036.2001.00930.x
35. Crawford, J. (**2002**). Anemia and lung cancer. *Lung Cancer*, *38*, 75–78, doi: 10.1016/S0169-5002(02)00275-1
36. Mishkin, S. (**2000**). Nutritional issues in inflammatory bowel disease. *Clinical Nutrition Insight*, *26*, 1–4. doi: 10.1002/ibd.20285
37. Who, U. UNU. (**1998**). *Iron deficiency anemia: assessment, prevention and control. Report of a joint WHO/UNICEF/UNU consultation*. World Health Organization.
38. Gasche, C.; Berstad, A.; Befrits, R.; Beglinger, C.; Dignass, A.; Erichsen, K.; Gomollon, F.; Hjortswang, H.; Koutroubakis, I.; Kulnigg, S. (**2007**). Guidelines on the diagnosis and management of iron deficiency and anemia in inflammatory bowel diseases. *Inflammatory Bowel Diseases*, *13*, 1545–1553. doi: 10.1002/ibd.20285
39. Gisbert, J.P.; Gomollón, F. (**2008**). Common misconceptions in the diagnosis and management of anemia in inflammatory bowel disease. *Official Journal of the American College of Gastroenterology| ACG*, *103*, 1299–1307, doi: 10.1111/j.1572-0241.2008.01846.x
40. Coenen, S.; Weyts, E.; Ballet, V.; Noman, M.; Van Assche, G.; Vermeire, S.; Van Emelen, J.; Ferrante, M. (**2016**). Identifying predictors of low adherence in patients with inflammatory bowel disease. *European Journal of Gastroenterology & Hepatology*, *28*, 503–507, doi: 10.1097/meg.0000000000000570
41. van der Have, M.; Oldenburg, B.; Kaptein, A.A.; Jansen, J.M.; Scheffer, R.C.H.; van Tuyl, B.A.; van der Meulen-de Jong, A.E.; Pierik, M.; Siersema, P.D.; van Oijen, M.G.H. (**2016**). Non-adherence to anti-TNF therapy is associated with illness perceptions and clinical outcomes in outpatients with inflammatory bowel disease: results from a prospective multicentre study. *Journal of Crohn's and Colitis*, *10*, 549–555, doi: 10.1093/ecco-jcc/jjw002
42. Kane, S.; Becker, B.; Harmsen, S.W.; Kurian, A.; Morisky, D.E.; Zinsmeister, A.R. (**2012**). Use of a screening tool to determine nonadherent behavior in inflammatory bowel disease. *Official Journal of the American College of Gastroenterology| ACG*, *107*, 154–160. doi: 10.1038/ajg.2011.317
43. Trindade, A.J.; Ehrlich, A.; Kornbluth, A.; Ullman, T.A. (**2011**). Are your patients taking their medicine? Validation of a new adherence scale in patients with inflammatory bowel disease and comparison with physician perception of adherence. *Inflammatory Bowel Diseases*, *17*, 599–604, doi: 10.1002/ibd.21310
44. Greenley, R.N.; Kunz, J.H.; Walter, J.; Hommel, K.A. (**2013**). Practical strategies for enhancing adherence to treatment regimen in inflammatory bowel disease. *Inflammatory Bowel Diseases*, *19*, 1534–1545, doi: 10.1097%2FMIB.0b013e3182813482
45. Hwang, C.; Ross, V.; Mahadevan, U. (**2012**). Micronutrient deficiencies in inflammatory bowel disease: from A to zinc. *Inflammatory Bowel Diseases*, *18*, 1961–1981, doi: 10.1002/ibd.22906
46. Vagianos, K.; Bector, S.; McConnell, J.; Bernstein, C.N. (**2007**). Nutrition assessment of patients with inflammatory bowel disease. *Journal of Parenteral and Enteral Nutrition*, *31*, 311–319, doi: 10.1177/0148607107031004311
47. Abbas, A.M.; Almukhtar, R.M.; Loftus, E.V.; Lichtenstein, G.R.; Khan, N. (**2014**). Risk of melanoma and non-melanoma skin cancer in ulcerative colitis patients treated with thiopurines: a nationwide retrospective cohort. *Official journal of the American College of Gastroenterology| ACG*, *109*, 1781–1793, doi: 10.1038/ajg.2014.298
48. Singh, J.A.; Wells, G.A.; Christensen, R.; Ghogomu, E.T.; Maxwell, L.J.; MacDonald, J.K.; Filippini, G.; Skoetz, N.; Francis, D.K.; Lopes, L.C. (**2011**). Adverse effects of biologics: a network meta-analysis and Cochrane overview. *Cochrane Database of Systematic Reviews*, (2). doi: 10.1002/14651858.cd008794.pub2
49. Abegunde, A.T.; Muhammad, B.H.; Ali, T. (**2016**). Preventive health measures in inflammatory bowel disease. *World Journal of Gastroenterology*, *22*, 7625, doi: 10.3748%2Fwjg.v22.i34.7625
50. Smith, R.A.; Andrews, K.S.; Brooks, D.; Fedewa, S.A.; Manassaram-Baptiste, D.; Saslow, D.; Brawley, O.W.; Wender, R.C. (**2017**). Cancer screening in the United States, 2017: a review of current American Cancer Society

guidelines and current issues in cancer screening. *CA: A Cancer Journal for Clinicians*, *67*, 100–121, doi: 10.3322/caac.21392
51. Sφgaard, K.K.; Cronin-Fenton, D.P.; Pedersen, L.; Sφrensen, H.T.; Lash, T.L. (**2008**). Survival in Danish patients with breast cancer and inflammatory bowel disease: a nationwide cohort study. *Inflammatory Bowel Diseases*, *14*, 519–525.
52. Kane, S. (**2008**). Abnormal Pap smears in inflammatory bowel disease. *Inflammatory Bowel Diseases*, *14*, 1158–1160.
53. Singh, H.; Demers, A.A.; Nugent, Z.; Mahmud, S.M.; Kliewer, E.V.; Bernstein, C.N. (**2009**). Risk of cervical abnormalities in women with inflammatory bowel disease: a population-based nested case-control study. *Gastroenterology*, *136*, 451–458.
54. Bulletins—Gynecology, A.C.o.P. (**2009**). ACOG Practice Bulletin no. 109: Cervical cytology screening. Obstet*rics and* Gynecol*ogy*, 114, 1409–1420.
55. Force, U.S.P.S.T. (**2019**). Final recommendation statement. Prostate cancer: screening. Available online at: www.uspreventive-servicestaskforce. org/Page/Document/RecommendationStatementFinal/prostate-cancer-screening1. Last accessed August, 13.
56. Annese, V.; Daperno, M.; Rutter, M.D.; Amiot, A.; Bossuyt, P.; East, J.; Ferrante, M.; Götz, M.; Katsanos, K.H.; Kießlich, R. (**2013**). European evidence based consensus for endoscopy in inflammatory bowel disease. *Journal of Crohn's and Colitis*, *7*, 982–1018.
57. Farraye, F.A.; Odze, R.D.; Eaden, J.; Itzkowitz, S.H. (**2010**). AGA medical position statement on the diagnosis and management of colorectal neoplasia in inflammatory bowel disease. *Gastroenterology*, *138*, 738–745.
58. Laine, L.; Kaltenbach, T.; Barkun, A.; McQuaid, K.R.; Subramanian, V.; Soetikno, R.; East, J.E.; Farraye, F.A.; Feagan, B.; Ioannidis, J. (**2015**). SCENIC international consensus statement on surveillance and management of dysplasia in inflammatory bowel disease. *Gastrointestinal Endoscopy*, *81*, 489–501.
59. Noiseux, I.; Veilleux, S.; Bitton, A.; Kohen, R.; Vachon, L.; White Guay, B.; Rioux, J.D. (**2019**). Inflammatory bowel disease patient perceptions of diagnostic and monitoring tests and procedures. *BMC Gastroenterology*, *19*, 1–11.
60. Fusillo, S.; Kastl Jr, A.J. (2**023**). The history and physical exam. In Petar Mamula, Judith R. Kelsen, Andrew B. Grossman, Robert N. Baldassano, Jonathan E. Markowitz (eds.) *Pediatric Inflammatory Bowel Disease*; Springer: pp. 211–215.
61. Spiceland, C.M.; Lodhia, N. (**2018**). Endoscopy in inflammatory bowel disease: role in diagnosis, management, and treatment. *World Journal of Gastroenterology*, *24*, 4014.
62. Buchner, A.M. (**2017**). The role of chromoendoscopy in evaluating colorectal dysplasia. *Gastroenterology & Hepatology*, *13*, 336–347.
63. Mikihiro, F.; Kentaro, M.; Nobuhiro, U.; Yusuke, S.; Yutaka, K. (**2011**). Autofluorescence imaging for diagnosing intestinal disorders. In *Colonoscopy*, Paul, M., Ed.; IntechOpen: p. Ch. 14.
64. Loftus, E.V. (**2007**). Emerging diagnostic methods in inflammatory bowel disease. *Gastroenterology & Hepatology*, *3*, 284–286.
65. Sedlack, R.E. (**2022**). Colonoscopy. Successful *T*raining in *G*astrointestinal *E*ndoscopy, 57–87.
66. Rubin, P.H.; Naymagon, S.; Williams, C.B.; Waye, J.D. (**2015**). Colonoscopy and flexible sigmoidoscopy. *Yamada's Textbook of Gastroenterology*, 2569–2581.
67. Schwartzberg, D.; Brandstetter, S.; Comfort, M.; Weber, H.; Baggenstoss, A.; Kiely, W.; Crohn, B.; Ginzburg, L.; Oppenheimer, G.; Reynolds Jr, H. (**2019**). Crohn's disease of the esophagus, duodenum, and stomach. *Clinics in Colon and Rectal Surgery*, *32*, 231–242.
68. Hilmi, I.; Kobayashi, T. (**2020**). Capsule endoscopy in inflammatory bowel disease: when and how. *Intestinal Research*, *18*, 265–274.
69. Galmiche, J.P.; Coron, E.; Sacher-Huvelin, S. (**2008**). Recent developments in capsule endoscopy. *Gut*, *57*, 695–703.
70. Yen, H.-H.; Chang, C.-W.; Chou, J.-W.; Wei, S.C. (**2017**). Balloon-assisted enteroscopy and capsule endoscopy in suspected small bowel Crohn's disease. *Clinical Endoscope*, *50*, 417–423.
71. Sun, B.; Liu, J.; Li, S.; Lovell, J.F.; Zhang, Y. (**2023**). Imaging of gastrointestinal tract ailments. *Journal of Imaging*, *9*(6), 115. doi: 10.3390/jimaging9060115
72. Withers, P.J.; Bouman, C.; Carmignato, S.; Cnudde, V.; Grimaldi, D.; Hagen, C.K.; Maire, E.; Manley, M.; Du Plessis, A.; Stock, S.R. (**2021**). X-ray computed tomography. *Nature Reviews Methods Primers*, *1*, 18, doi: 10.1038/s43586-021-00015-4.
73. Park, M.J.; Lim, J.S. (**2013**). Computed tomography enterography for evaluation of inflammatory bowel disease. *Clinical Endoscope*, *46*, 327–366, doi: 10.5946/ce.2013.46.4.327
74. Hara, A.K.; Alam, S.; Heigh, R.I.; Gurudu, S.R.; Hentz, J.G.; Leighton, J.A. (**2008**). Using CT enterography to monitor Crohn's disease activity: a preliminary study. *American Journal of Roentgenology*, *190*, 1512–1516, doi: 10.2214/AJR.07.2877
75. Ram, R.; Sarver, D.; Pandey, T.; Guidry, C.L.; Jambhekar, K.R. (**2016**). Magnetic resonance enterography: a stepwise interpretation approach and role of imaging in management of adult Crohn's disease. *The Indian Journal of Radiology & Imaging*, *26*, 173–184, doi: 10.4103/0971-3026.184405
76. Kuna, A.T. (**2013**). Serological markers of inflammatory bowel disease. *Biochemia Medica*, *23*, 28–42, doi: 10.11613/bm.2013.006.
77. Chen, C.S.; Sullivan, S.; Anderson, T.; Tan, A.C.; Alex, P.J.; Brant, S.R.; Cuffari, C.; Bayless, T.M.; Talor, M.V.; Burek, C.L.; et al. (**2009**). Identification of novel serological biomarkers for inflammatory bowel disease using Escherichia coli proteome chip. *Molecular & Cellular Proteomics: MCP*, *8*, 1765–1776, doi: 10.1074/mcp.M800593-MCP200.
78. Yamamoto, T. (**2015**). The clinical value of faecal calprotectin and lactoferrin measurement in postoperative Crohn's disease. *United European Gastroenterology Journal*, *3*, 5–10, doi: 10.1177/2050640614558106
79. Kalantari, H.; Taheri, A.; Yaran, M. (**2015**). Fecal calprotectin is a useful marker to diagnose ulcerative colitis from irritable bowel syndrome. *Advanced Biomedical Research*, *4*, 85, doi: 10.4103/2277-9175.156647
80. Lehmann, F.S.; Burri, E.; Beglinger, C. (**2014**). The role and utility of faecal markers in inflammatory bowel

disease. *Therapeutic Advances in Gastroenterology*, *8*, 23–36, doi: 10.1177/1756283X14553384
81. de Castro, M.M.; Pascoal, L.B.; Steigleder, K.M.; Siqueira, B.P.; Corona, L.P.; Ayrizono, M.L.S.; Milanski, M.; Leal, R.F. (**2021**). Role of diet and nutrition in inflammatory bowel disease. *World Journal of Experimental Medicine*, *11*, 1–16, doi: 10.5493/wjem.v11.i1.1
82. Yusuf, K.; Saha, S.; Umar, S. (**2022**). Health benefits of dietary fiber for the management of inflammatory bowel disease. *Biomedicines*, *10*(6), 1242. doi: 10.3390/biomedicines10061242
83. Ferenc, K.; Jarmakiewicz-Czaja, S.; Filip, R. (**2022**). Components of the fiber diet in the prevention and treatment of IBD-an update. *Nutrients*, *15*(1), 162. doi: 10.3390/nu15010162
84. Mishkin, S. (**1997**). Dairy sensitivity, lactose malabsorption, and elimination diets in inflammatory bowel disease. *The American Journal of Clinical Nutrition*, *65*, 564–567, doi: 10.1093/ajcn/65.2.564
85. Ratajczak, A.E.; Rychter, A.M.; Zawada, A.; Dobrowolska, A.; Krela-Kaźmierczak, I. (**2021**). Lactose intolerance in patients with inflammatory bowel diseases and dietary management in prevention of osteoporosis. *Nutrition (Burbank, Los Angeles County, Calif.) 82*, 111043, doi: 10.1016/j.nut.2020.111043
86. Marton, L.T.; Goulart, R.A.; Carvalho, A.C.A.; Barbalho, S.M. (**2019**). Omega fatty acids and inflammatory bowel diseases: an overview. *International Journal of Molecular Sciences*, *20*(19), 4851. doi: 10.3390/ijms20194851
87. Barbalho, S.M.; Goulart Rde, A.; Quesada, K.; Bechara, M.D.; de Carvalho Ade, C. (**2016**). Inflammatory bowel disease: can omega-3 fatty acids really help? *Annals of Gastroenterology*, *29*, 37–43.
88. Holik, D.; Bezdan, A.; Marković, M.; Orkić, Ž.; Milostić-Srb, A.; Mikšić, Š.; Včev, A. (**2020**). The association between drinking water quality and inflammatory bowel disease-A study in Eastern Croatia. *International Journal of Environmental Research and Public Health*, *17*(22), 8495. doi: 10.3390/ijerph17228495
89. Crooks, B.; McLaughlin, J.; Limdi, J. (**2022**). Dietary beliefs and recommendations in inflammatory bowel disease: a national survey of healthcare professionals in the UK. *Frontline Gastroenterology*, *13*, 25–31, doi: 10.1136/flgastro-2020-101723
90. de Vries, J.H.M.; Dijkhuizen, M.; Tap, P.; Witteman, B.J.M. (**2019**). Patient's dietary beliefs and behaviours in inflammatory bowel disease. *Digestive diseases (Basel, Switzerland)*, *37*, 131–139, doi: 10.1159/000494022
91. Ghishan, F.K.; Kiela, P.R. (**2017**). Vitamins and minerals in inflammatory bowel disease. *Gastroenterology Clinics of North America*, *46*, 797–808, doi: 10.1016/j.gtc.2017.08.011
92. Lakatos, P.L.; Szamosi, T.; Lakatos, L. (**2007**). Smoking in inflammatory bowel diseases: good, bad or ugly? *World Journal of Gastroenterology*, *13*, 6134–6139, doi: 10.3748/wjg.v13.i46.6134
93. Berkowitz, L.; Schultz, B.M.; Salazar, G.A.; Pardo-Roa, C.; Sebastián, V.P.; Álvarez-Lobos, M.M.; Bueno, S.M. (**2018**). Impact of cigarette smoking on the gastrointestinal tract inflammation: opposing effects in Crohn's disease and ulcerative colitis. *Frontiers in Immunology*, *9*, 74. doi: 10.3389/fimmu.2018.00074.
94. Engels, M.; Cross, R.K.; Long, M.D. (**2018**). Exercise in patients with inflammatory bowel diseases: current perspectives. *Clinical and Experimental Gastroenterology*, *11*, 1–11, doi: 10.2147/ceg.S120816
95. Ng, V.; Millard, W.; Lebrun, C.; Howard, J. (**2006**). Exercise and Crohn's disease: speculations on potential benefits. *Canadian Journal of Gastroenterology = Journal canadien de gastroenterologie*, *20*, 657–660, doi: 10.1155/2006/462495
96. De Witte, N.A.J.; Buyck, I.; Van Daele, T. (**2019**). Combining biofeedback with stress management interventions: a systematic review of physiological and psychological effects. *Applied Psychophysiology and Biofeedback*, *44*, 71–82, doi: 10.1007/s10484-018-09427-7
97. Whitehead, W.E. (**1978**). Biofeedback in the treatment of gastrointestinal disorders. *Biofeedback and Self-Regulation*, *3*, 375–384, doi: 10.1007/bf00998943
98. Gerbarg, P.L.; Jacob, V.E.; Stevens, L.; Bosworth, B.P.; Chabouni, F.; DeFilippis, E.M.; Warren, R.; Trivellas, M.; Patel, P.V.; Webb, C.D.; et al. (**2015**). The effect of breathing, movement, and meditation on psychological and physical symptoms and inflammatory biomarkers in inflammatory bowel disease: a randomized controlled trial. *Inflammatory Bowel Diseases*, *21*, 2886–2896, doi: 10.1097/mib.0000000000000568
99. De Sousa, J.F.M.; Paghdar, S.; Khan, T.M.; Patel, N.P.; Chandrasekaran, S.; Tsouklidis, N. (**2022**). Stress and inflammatory bowel disease: clear mind, happy colon. *Cureus*, *14*, e25006, doi: 10.7759/cureus.25006
100. Feuerstein, J.D.; Cheifetz, A.S. (**2014**). *Ulcerative colitis: epidemiology, diagnosis, and management.* Mayo Clinic Proceedings. 1553–1563.
101. Klein, A.; Eliakim, R. (**2010**). Non steroidal anti-inflammatory drugs and inflammatory bowel disease. *Pharmaceuticals*, *3*, 1084–1092.
102. Bonner, G.F.; Fakhri, A.; Vennamaneni, S.R. (**2004**). A long-term cohort study of nonsteroidal anti-inflammatory drug use and disease activity in outpatients with inflammatory bowel disease. *Inflammatory Bowel Diseases*, *10*, 751–757.
103. Bonner, G.F.; Walczak, M.; Kitchen, L.; Bayona, M. (**2000**). Tolerance of nonsteroidal antiinflammatory drugs in patients with inflammatory bowel disease. *The American Journal of Gastroenterology*, *95*, 1946–1948.
104. Takeuchi, K.; Smale, S.; Premchand, P.; Maiden, L.; Sherwood, R.; Thjodleifsson, B.; Bjornsson, E.; Bjarnason, I. (**2006**). Prevalence and mechanism of nonsteroidal anti-inflammatory drug–induced clinical relapse in patients with inflammatory bowel disease. *Clinical Gastroenterology and Hepatology*, *4*, 196–202.
105. Matuk, R.; Crawford, J.; Abreu, M.T.; Targan, S.R.; Vasiliauskas, E.A.; Papadakis, K.A. (**2004**). The spectrum of gastrointestinal toxicity and effect on disease activity of selective cyclooxygenase-2 inhibitors in patients with inflammatory bowel disease. *Inflammatory Bowel Diseases*, *10*, 352–356.
106. Mahadevan, U.; Loftus Jr, E.V.; Tremaine, W.J.; Sandborn, W.J. (**2002**). Safety of selective cyclooxygenase-2 inhibitors in inflammatory bowel disease. *The American Journal of Gastroenterology*, *97*, 910–914.
107. Pithadia, A.B.; Jain, S. (**2011**). Treatment of inflammatory bowel disease (IBD). *Pharmacological Reports: PR*, *63*, 629–642, doi: 10.1016/s1734-1140(11)70575-8

108. Klein, A.; Eliakim, R. (**2010**). Non steroidal anti-inflammatory drugs and inflammatory bowel disease. *Pharmaceuticals*, *3*, 1084–1092, doi: 10.3390/ph3041084
109. Perrotta, C.; Pellegrino, P.; Moroni, E.; De Palma, C.; Cervia, D.; Danelli, P.; Clementi, E. (**2015**). Five-aminosalicylic acid: an update for the reappraisal of an old drug. Gastroenterology Research and Practice, 2015, 456895, doi: 10.1155/2015/456895
110. Iacucci, M.; de Silva, S.; Ghosh, S. (**2010**). Mesalazine in inflammatory bowel disease: a trendy topic once again? *Canadian Journal of Gastroenterology = Journal canadien de gastroenterologie*, *24*, 127–133, doi: 10.1155/2010/586092
111. Stolfi, C.; De Simone, V.; Pallone, F.; Monteleone, G. (**2013**). Mechanisms of action of non-steroidal anti-inflammatory drugs (NSAIDs) and mesalazine in the chemoprevention of colorectal cancer. *International Journal of Molecular Sciences*, *14*, 17972–17985, doi: 10.3390/ijms140917972
112. Kim, K.U.; Kim, J.; Kim, W.H.; Min, H.; Choi, C.H. (**2021**). Treatments of inflammatory bowel disease toward personalized medicine. *Archives of Pharmacal Research*, *44*, 293–309, doi: 10.1007/s12272-021-01318-6
113. Van Staa, T.P.; Travis, S.; Leufkens, H.G.; Logan, R.F. (**2004**). 5-aminosalicylic acids and the risk of renal disease: a large British epidemiologic study. *Gastroenterology*, *126*, 1733–1739, doi: 10.1053/j.gastro.2004.03.016
114. Ramamoorthy, S.; Cidlowski, J.A. (**2016**). Corticosteroids: mechanisms of action in health and disease. *Rheumatic Diseases Clinics of North America*, *42*, 15–31, vii, doi: 10.1016/j.rdc.2015.08.002
115. Dubois-Camacho, K.; Ottum, P.A.; Franco-Muñoz, D.; De la Fuente, M.; Torres-Riquelme, A.; Díaz-Jiménez, D.; Olivares-Morales, M.; Astudillo, G.; Quera, R.; Hermoso, M.A. (**2017**). Glucocorticosteroid therapy in inflammatory bowel diseases: from clinical practice to molecular biology. *World Journal of Gastroenterology*, *23*, 6628–6638, doi: 10.3748/wjg.v23.i36.6628
116. M'Koma, A.E. (**2023**). Inflammatory bowel disease: clinical diagnosis and pharmaceutical management. *Medical Research Archives*, *11*(1). doi: 10.18103/mra.v11i1.3135
117. Ben Ghezala, I.; Charkaoui, M.; Michiels, C.; Bardou, M.; Luu, M. (**2021**). Small molecule drugs in inflammatory bowel diseases. *Pharmaceuticals*, *14*, 637.
118. Kemp, R.; Dunn, E.; Schultz, M. (**2013**). Immunomodulators in inflammatory bowel disease: an emerging role for biologic agents. *BioDrugs: Clinical Immunotherapeutics, Biopharmaceuticals and Gene Therapy*, *27*, 585–590, doi: 10.1007/s40259-013-0045-2
119. Nielsen, O.H.; Bjerrum, J.T.; Herfarth, H.; Rogler, G. (**2013**). Recent advances using immunomodulators for inflammatory bowel disease. *Journal of Clinical Pharmacology*, *53*, 575–588, doi: 10.1002/jcph.2
120. Neurath, M. (**2010**). Thiopurines in IBD: what is their mechanism of action? *Gastroenterology & Hepatology*, *6*, 435–436.
121. Evans, W.E. (**2004**). Pharmacogenetics of thiopurine S-methyltransferase and thiopurine therapy. *Therapeutic Drug Monitoring*, *26*, 186–191, doi: 10.1097/00007691-200404000-00018
122. Lichtenstein, G.R.; Abreu, M.T.; Cohen, R.; Tremaine, W. (**2006**). American Gastroenterological Association Institute technical review on corticosteroids, immunomodulators, and infliximab in inflammatory bowel disease. *Gastroenterology*, *130*, 940–987, doi: 10.1053/j.gastro.2006.01.048
123. Nielsen, O.H.; Steenholdt, C.; Juhl, C.B.; Rogler, G. (**2020**). Efficacy and safety of methotrexate in the management of inflammatory bowel disease: a systematic review and meta-analysis of randomized, controlled trials. *EClinicalMedicine*, *20*, 100271, doi: 10.1016/j.eclinm.2020.100271
124. Rusnak, F.; Mertz, P. (**2000**). Calcineurin: form and function. *Physiological Reviews*, *80*, 1483–1521, doi: 10.1152/physrev.2000.80.4.1483
125. Tedesco, D.; Haragsim, L. (**2012**). Cyclosporine: a review. *Journal of Transplantation*, *2012*, 230386, doi: 10.1155/2012/230386
126. Ghoreschi, K.; Laurence, A.; O'Shea, J.J. (**2009**). Janus kinases in immune cell signaling. *Immunological Reviews*, *228*, 273–287, doi: 10.1111/j.1600-065X.2008.00754.x
127. Olivera, P.; Danese, S.; Peyrin-Biroulet, L. (**2017**). JAK inhibition in inflammatory bowel disease. *Expert Review of Clinical Immunology*, *13*, 693–703, doi: 10.1080/1744666x.2017.1291342
128. Danese, S.; Grisham, M.; Hodge, J.; Telliez, J.B. (**2016**). JAK inhibition using tofacitinib for inflammatory bowel disease treatment: a hub for multiple inflammatory cytokines. *American Journal of Physiology. Gastrointestinal and Liver Physiology*, *310*, G155–G162, doi: 10.1152/ajpgi.00311.2015
129. Sandborn, W.J.; Ghosh, S.; Panes, J.; Vranic, I.; Wang, W.; Niezychowski, W. (**2014**). A phase 2 study of tofacitinib, an oral Janus kinase inhibitor, in patients with Crohn's disease. *Clinical Gastroenterology and Hepatology: The Official Clinical Practice Journal of the American Gastroenterological Association*, *12*, 1485–1493.e1482, doi: 10.1016/j.cgh.2014.01.029
130. Sandborn, W.J.; Nguyen, D.D.; Beattie, D.T.; Brassil, P.; Krey, W.; Woo, J.; Situ, E.; Sana, R.; Sandvik, E.; Pulido-Rios, M.T.; et al. (**2020**). Development of gut-selective Pan-Janus kinase inhibitor TD-1473 for ulcerative colitis: a translational medicine programme. *Journal of Crohns Colitis*, *14*, 1202–1213, doi: 10.1093/ecco-jcc/jjaa049
131. D'Amico, F.; Peyrin-Biroulet, L.; Danese, S. (**2020**). JAK or GUT selectivity: tipping the balance for efficacy and safety in ulcerative colitis. *Journal of Crohns Colitis*, *14*, 1185–1187, doi: 10.1093/ecco-jcc/jjaa074
132. De Vries, L.C.S.; Ghiboub, M.; van Hamersveld, P.H.P.; Welting, O.; Verseijden, C.; Bell, M.J.; Rioja, I.; Prinjha, R.K.; Koelink, P.J.; Strobl, B.; et al. (**2021**). Tyrosine kinase 2 signalling drives pathogenic T cells in colitis. *Journal of Crohns Colitis*, *15*, 617–630, doi: 10.1093/ecco-jcc/jjaa199
133. Yoshimura, N.; Watanabe, M.; Motoya, S.; Tominaga, K.; Matsuoka, K.; Iwakiri, R.; Watanabe, K.; Hibi, T. (**2015**). Safety and efficacy of AJM300, an oral antagonist of α4 integrin, in induction therapy for patients with active ulcerative colitis. *Gastroenterology*, *149*, 1775–1783.e1772, doi: 10.1053/j.gastro.2015.08.044
134. D'Amico, F.; Peyrin-Biroulet, L.; Danese, S.; Fiorino, G. (**2020**). New drugs in the pipeline for the treatment of inflammatory bowel diseases: what is coming? *Current Opinion in Pharmacology*, *55*, 141–150, doi: 10.1016/j.coph.2020.10.015

135. Burns, R.C.; Rivera-Nieves, J.; Moskaluk, C.A.; Matsumoto, S.; Cominelli, F.; Ley, K. (**2001**). Antibody blockade of ICAM-1 and VCAM-1 ameliorates inflammation in the SAMP-1/Yit adoptive transfer model of Crohn's disease in mice. *Gastroenterology*, *121*, 1428–1436, doi: 10.1053/gast.2001.29568
136. Yousry, T.A.; Major, E.O.; Ryschkewitsch, C.; Fahle, G.; Fischer, S.; Hou, J.; Curfman, B.; Miszkiel, K.; Mueller-Lenke, N.; Sanchez, E.; et al. (**2006**). Evaluation of patients treated with natalizumab for progressive multifocal leukoencephalopathy. *The New England Journal of Medicine*, *354*, 924–933, doi: 10.1056/NEJMoa054693
137. Chandar, A.K.; Singh, S.; Murad, M.H.; Peyrin-Biroulet, L.; Loftus, E.V., Jr. (**2015**). Efficacy and safety of natalizumab and vedolizumab for the management of Crohn's disease: a systematic review and meta-analysis. *Inflammatory Bowel Diseases*, *21*, 1695–1708, doi: 10.1097/mib.0000000000000373
138. Nielsen, O.H.; Li, Y.; Johansson-Lindbom, B.; Coskun, M. (**2017**). Sphingosine-1-phosphate signaling in inflammatory bowel disease. *Trends in Molecular Medicine*, *23*, 362–374, doi: 10.1016/j.molmed.2017.02.002
139. Peyrin-Biroulet, L.; Christopher, R.; Behan, D.; Lassen, C. (**2017**). Modulation of sphingosine-1-phosphate in inflammatory bowel disease. *Autoimmunity Reviews*, *16*, 495–503, doi: 10.1016/j.autrev.2017.03.007
140. Blaho, V.A.; Hla, T. (**2014**). An update on the biology of sphingosine 1-phosphate receptors. *Journal of Lipid Research*, *55*, 1596–1608, doi: 10.1194/jlr.R046300
141. Rawla, P.; Sunkara, T.; Raj, J.P. (**2018**). Role of biologics and biosimilars in inflammatory bowel disease: current trends and future perspectives. *Journal of Inflammation Research*, *11*, 215–226, doi: 10.2147/jir.S165330
142. Rutgeerts, P.; Sandborn, W.J.; Feagan, B.G.; Reinisch, W.; Olson, A.; Johanns, J.; Travers, S.; Rachmilewitz, D.; Hanauer, S.B.; Lichtenstein, G.R.; et al. (**2005**). Infliximab for induction and maintenance therapy for ulcerative colitis. *The New England Journal of Medicine*, *353*, 2462–2476, doi: 10.1056/NEJMoa050516.
143. Schreiber, S.; Rutgeerts, P.; Fedorak, R.N.; Khaliq-Kareemi, M.; Kamm, M.A.; Boivin, M.; Bernstein, C.N.; Staun, M.; Thomsen, O.; Innes, A. (**2005**). A randomized, placebo-controlled trial of certolizumab pegol (CDP870) for treatment of Crohn's disease. *Gastroenterology*, *129*, 807–818, doi: 10.1053/j.gastro.2005.06.064
144. Shealy, D.J.; Cai, A.; Staquet, K.; Baker, A.; Lacy, E.R.; Johns, L.; Vafa, O.; Gunn, G., 3rd; Tam, S.; Sague, S.; et al. (**2010**). Characterization of golimumab, a human monoclonal antibody specific for human tumor necrosis factor α. *mAbs*, *2*, 428–439, doi: 10.4161/mabs.12304
145. Olesen, C.M.; Coskun, M.; Peyrin-Biroulet, L.; Nielsen, O.H. (**2016**). Mechanisms behind efficacy of tumor necrosis factor inhibitors in inflammatory bowel diseases. *Pharmacology & Therapeutics*, *159*, 110–119, doi: 10.1016/j.pharmthera.2016.01.001
146. Katsanos, K.H.; Papamichael, K.; Feuerstein, J.D.; Christodoulou, D.K.; Cheifetz, A.S. (**2019**). Biological therapies in inflammatory bowel disease: Beyond anti-TNF therapies. *Clinical Immunology (Orlando, Fla.)*, *206*, 9–14, doi: 10.1016/j.clim.2018.03.004
147. Dogra, S.; Khullar, G. (**2013**). Tumor necrosis factor-α antagonists: side effects and their management. *Indian Journal of Dermatology, Venereology and Leprology*, *79 Suppl 7*, S35–46, doi: 10.4103/0378-6323.115526
148. Dige, A.; Støy, S.; Rasmussen, T.K.; Kelsen, J.; Hvas, C.L.; Sandahl, T.D.; Dahlerup, J.F.; Deleuran, B.; Agnholt, J. (**2013**). Increased levels of circulating Th17 cells in quiescent versus active Crohn's disease. *Journal of Crohns Colitis*, *7*, 248–255, doi: 10.1016/j.crohns.2012.06.015
149. Bogaert, S.; Laukens, D.; Peeters, H.; Melis, L.; Olievier, K.; Boon, N.; Verbruggen, G.; Vandesompele, J.; Elewaut, D.; De Vos, M. (**2010**). Differential mucosal expression of Th17-related genes between the inflamed colon and ileum of patients with inflammatory bowel disease. *BMC Immunology*, *11*, 61, doi: 10.1186/1471-2172-11-61
150. Sandborn, W.J.; Gasink, C.; Gao, L.L.; Blank, M.A.; Johanns, J.; Guzzo, C.; Sands, B.E.; Hanauer, S.B.; Targan, S.; Rutgeerts, P.; et al. (**2012**). Ustekinumab induction and maintenance therapy in refractory Crohn's disease. *The New England Journal of Medicine*, *367*, 1519–1528, doi: 10.1056/NEJMoa1203572
151. Nurbhai, S.; Roberts, K.J.; Carlton, T.M.; Maggiore, L.; Cubitt, M.F.; Ray, K.P.; Reckless, J.; Mohammed, H.; Irving, P.; MacDonald, T.T.; et al. (**2019**). Oral anti-tumour necrosis factor domain antibody V565 provides high intestinal concentrations, and reduces markers of inflammation in ulcerative colitis patients. *Scientific Reports*, *9*, 14042, doi: 10.1038/s41598-019-50545-x
152. Macaluso, F.S.; Orlando, A.; Cottone, M. (**2019**). Anti-interleukin-12 and anti-interleukin-23 agents in Crohn's disease. *Expert Opinion on Biological Therapy*, *19*, 89–98, doi: 10.1080/14712598.2019.1561850
153. Cepek, K.L.; Shaw, S.K.; Parker, C.M.; Russell, G.J.; Morrow, J.S.; Rimm, D.L.; Brenner, M.B. (**1994**). Adhesion between epithelial cells and T lymphocytes mediated by E-cadherin and the alpha E beta 7 integrin. *Nature*, *372*, 190–193, doi: 10.1038/372190a0
154. Vermeire, S.; Sandborn, W.J.; Danese, S.; Hébuterne, X.; Salzberg, B.A.; Klopocka, M.; Tarabar, D.; Vanasek, T.; Greguš, M.; Hellstern, P.A.; et al. (**2017**). Anti-MAdCAM antibody (PF-00547659) for ulcerative colitis (TURANDOT): a phase 2, randomised, double-blind, placebo-controlled trial. *Lancet (London, England)*, *390*, 135–144, doi: 10.1016/s0140-6736(17)30930-3
155. Jairath, V.; Khanna, R.; Feagan, B.G. (**2017**). Alicaforsen for the treatment of inflammatory bowel disease. *Expert Opinion on Investigational Drugs*, *26*, 991–997, doi: 10.1080/13543784.2017.1349753
156. Shaw, S.Y.; Blanchard, J.F.; Bernstein, C.N. (**2011**). Association between the use of antibiotics and new diagnoses of Crohn's disease and ulcerative colitis. *Amercian Journal of Gastroenterology*, *106*, 2133–2142, doi: 10.1038/ajg.2011.304
157. Khan, K.J.; Ullman, T.A.; Ford, A.C.; Abreu, M.T.; Abadir, A.; Marshall, J.K.; Talley, N.J.; Moayyedi, P. (**2011**). Antibiotic therapy in inflammatory bowel disease: a systematic review and meta-analysis. *Amercian Journal of Gastroenterology*, *106*, 661–673, doi: 10.1038/ajg.2011.72
158. Rutgeerts, P.; Hiele, M.; Geboes, K.; Peeters, M.; Penninckx, F.; Aerts, R.; Kerremans, R. (**1995**). Controlled trial of metronidazole treatment for prevention of crohn's recurrence after ileal resection. *Gastroenterology*, *108*, 1617–1621, doi: 10.1016/0016-5085(95)90121-3

159. Faculty of Family Planning & Reproductive Health Care, Clinical Effectiveness Unit. (**2003**). Contraceptive choices for women with inflammatory bowel disease. *The Journal of Family Planning and Reproductive Health Care, 29*, 127–135, doi: 10.1783/147118903101197782
160. Van Assche, G.; Van Ranst, M.; Sciot, R.; Dubois, B.; Vermeire, S.; Noman, M.; Verbeeck, J.; Geboes, K.; Robberecht, W.; Rutgeerts, P. (**2005**). Progressive multifocal leukoencephalopathy after natalizumab therapy for Crohn's disease. *The New England Journal of Medicine*, *353*, 362–368, doi: 10.1056/NEJMoa051586
161. Son, C.G.; Bian, Z.X.; Wang, J.H.; Raghavendran, H.B. (**2013**). Complementary and alternative medicine for diseases and disorders in digestive tract: basic to clinics. *Evidence-Based Complementary and Alternative Medicine*, *2013*, 565279, doi: 10.1155/2013/565279
162. Day, A.S. (**2002**). Use of complementary and alternative therapies and probiotic agents by children attending gastroenterology outpatient clinics. *Journal of Paediatrics and Child Health*, *38*, 343–346, doi: 10.1046/j.1440-1754.2002.00005.x
163. De Conno, B.; Pesce, M.; Chiurazzi, M.; Andreozzi, M.; Rurgo, S.; Corpetti, C.; Seguella, L.; Del Re, A.; Palenca, I.; Esposito, G.; et al. (**2022**). Nutraceuticals and diet supplements in Crohn's disease: a general overview of the most promising approaches in the clinic. *Foods (Basel, Switzerland)*, *11*, doi: 10.3390/foods11071044
164. Kühn, F.; Klar, E. (**2015**). Surgical principles in the treatment of ulcerative colitis. *Viszeralmedizin*, *31*, 246–250, doi: 10.1159/000438894
165. Bouguen, G.; Peyrin-Biroulet, L. (**2011**). Surgery for adult Crohn's disease: what is the actual risk? *Gut 60*, 1178–1181, doi: 10.1136/gut.2010.234617
166. Chaudrey, K.; Salvaggio, M.; Ahmed, A.; Mahmood, S.; Ali, T. (**2015**). Updates in vaccination: recommendations for adult inflammatory bowel disease patients. *World Journal of Gastroenterology*, *21*, 3184–3196, doi: 10.3748/wjg.v21.i11.3184
167. Reddy, K.R.; Beavers, K.L.; Hammond, S.P.; Lim, J.K.; Falck-Ytter, Y.T. (**2015**). American Gastroenterological Association Institute guideline on the prevention and treatment of hepatitis B virus reactivation during immunosuppressive drug therapy. *Gastroenterology*, *148*, 215–219; quiz e216–217, doi: 10.1053/j.gastro.2014.10.039
168. Nishida, A.; Nishino, K.; Ohno, M.; Sakai, K.; Owaki, Y.; Noda, Y.; Imaeda, H. (**2022**). Update on gut microbiota in gastrointestinal diseases. *World journal of Clinical Cases*, *10*, 7653–7664, doi: 10.12998/wjcc.v10.i22.7653
169. Qin, J.; Li, R.; Raes, J.; Arumugam, M.; Burgdorf, K.S.; Manichanh, C.; Nielsen, T.; Pons, N.; Levenez, F.; Yamada, T. (**2010**). A human gut microbial gene catalogue established by metagenomic sequencing. *Nature*, *464*, 59–65.
170. Nishida, A.; Inoue, R.; Inatomi, O.; Bamba, S.; Naito, Y.; Andoh, A. (**2018**). Gut microbiota in the pathogenesis of inflammatory bowel disease. *Clinical Journal of Gastroenterology*, *11*, 1–10.
171. Tsigalou, C.; Paraschaki, A.; Karvelas, A.; Kantartzi, K.; Gagali, K.; Tsairidis, D.; Bezirtzoglou, E. (**2021**). Gut microbiome and Mediterranean diet in the context of obesity. Current knowledge, perspectives and potential therapeutic targets. *Metabolism Open*, *9*, 100081.
172. Wu, S.; Rhee, K.-J.; Albesiano, E.; Rabizadeh, S.; Wu, X.; Yen, H.-R.; Huso, D.L.; Brancati, F.L.; Wick, E.; McAllister, F. (**2009**). A human colonic commensal promotes colon tumorigenesis via activation of T helper type 17 T cell responses. *Nature Medicine*, *15*, 1016–1022.
173. Peterson, D.A.; Frank, D.N.; Pace, N.R.; Gordon, J.I. (**2008**). Metagenomic approaches for defining the pathogenesis of inflammatory bowel diseases. *Cell Host & Microbe*, *3*, 417–427.
174. Lupp, C.; Robertson, M.L.; Wickham, M.E.; Sekirov, I.; Champion, O.L.; Gaynor, E.C.; Finlay, B.B. (**2007**). Host-mediated inflammation disrupts the intestinal microbiota and promotes the overgrowth of Enterobacteriaceae. *Cell Host & Microbe*, *2*, 119–129.
175. Pittayanon, R.; Lau, J.T.; Leontiadis, G.I.; Tse, F.; Yuan, Y.; Surette, M.; Moayyedi, P. (**2020**). Differences in gut microbiota in patients with vs without inflammatory bowel diseases: a systematic review. *Gastroenterology*, *158*, 930–946.e931, doi: 10.1053/j.gastro.2019.11.294
176. Quévrain, E.; Maubert, M.A.; Michon, C.; Chain, F.; Marquant, R.; Tailhades, J.; Miquel, S.; Carlier, L.; Bermúdez-Humarán, L.G.; Pigneur, B.; et al. (**2016**). Identification of an anti-inflammatory protein from Faecalibacterium prausnitzii, a commensal bacterium deficient in Crohn's disease. *Gut*, *65*, 415–425, doi: 10.1136/gutjnl-2014-307649
177. Kang, C.S.; Ban, M.; Choi, E.J.; Moon, H.G.; Jeon, J.S.; Kim, D.K.; Park, S.K.; Jeon, S.G.; Roh, T.Y.; Myung, S.J.; et al. (**2013**). Extracellular vesicles derived from gut microbiota, especially Akkermansia muciniphila, protect the progression of dextran sulfate sodium-induced colitis. *PloS One*, *8*, e76520, doi: 10.1371/journal.pone.0076520
178. Devkota, S.; Chang, E.B. (**2015**). Interactions between diet, bile acid metabolism, gut microbiota, and inflammatory bwel diseases. *Digestive diseases (Basel, Switzerland)*, *33*, 351–356, doi: 10.1159/000371687
179. Palmela, C.; Chevarin, C.; Xu, Z.; Torres, J.; Sevrin, G.; Hirten, R.; Barnich, N.; Ng, S.C.; Colombel, J.F. (**2018**). Adherent-invasive Escherichia coli in inflammatory bowel disease. *Gut*, *67*, 574–587, doi: 10.1136/gutjnl-2017-314903
180. Lovell, R.M.; Ford, A.C. (**2012**). Global prevalence of and risk factors for irritable bowel syndrome: a meta-analysis. *Clinical Gastroenterology and Hepatology*, *10*, 712–721.e714, doi: 10.1016/j.cgh.2012.02.029
181. Klem, F.; Wadhwa, A.; Prokop, L.J.; Sundt, W.J.; Farrugia, G.; Camilleri, M.; Singh, S.; Grover, M. (**2017**). Prevalence, risk factors, and outcomes of irritable bowel syndrome after infectious enteritis: a systematic review and meta-analysis. *Gastroenterology*, *152*, 1042–1054.e1041, doi: 10.1053/j.gastro.2016.12.039
182. Pittayanon, R.; Lau, J.T.; Yuan, Y.; Leontiadis, G.I.; Tse, F.; Surette, M.; Moayyedi, P. (**2019**). Gut microbiota in patients with irritable bowel syndrome-a systematic review. *Gastroenterology*, *157*, 97–108, doi: 10.1053/j.gastro.2019.03.049
183. Le Chatelier, E.; Nielsen, T.; Qin, J.; Prifti, E.; Hildebrand, F.; Falony, G.; Almeida, M.; Arumugam, M.; Batto, J.M.; Kennedy, S.; et al. (**2013**). Richness of human gut microbiome correlates with metabolic markers. *Nature*, *500*, 541–546, doi: 10.1038/nature12506

184. Zhang, M.; Qiu, X.; Zhang, H.; Yang, X.; Hong, N.; Yang, Y.; Chen, H.; Yu, C. (**2014**). Faecalibacterium prausnitzii inhibits interleukin-17 to ameliorate colorectal colitis in rats. *PloS One*, *9*, e109146, doi: 10.1371/journal.pone.0109146
185. Pinto-Sanchez, M.I.; Hall, G.B.; Ghajar, K.; Nardelli, A.; Bolino, C.; Lau, J.T.; Martin, F.P.; Cominetti, O.; Welsh, C.; Rieder, A.; et al. (**2017**). Probiotic bifidobacterium longum NCC3001 reduces depression scores and alters brain activity: a pilot study in patients with irritable bowel syndrome. *Gastroenterology*, *153*, 448–459.e448, doi: 10.1053/j.gastro.2017.05.003
186. Holmes, E.; Li, J.V.; Athanasiou, T.; Ashrafian, H.; Nicholson, J.K. (**2011**). Understanding the role of gut microbiome–host metabolic signal disruption in health and disease. *Trends in Microbiology*, *19*, 349–359, doi: 10.1016/j.tim.2011.05.006
187. Pittayanon, R.; Lau, J.T.; Yuan, Y.; Leontiadis, G.I.; Tse, F.; Surette, M.; Moayyedi, P. (**2019**). Gut microbiota in patients with irritable bowel syndrome—a systematic review. *Gastroenterology*, *157*, 97–108, doi: /10.1053/j.gastro.2019.03.049
188. Marasco, G.; Di Biase, A.R.; Schiumerini, R.; Eusebi, L.H.; Iughetti, L.; Ravaioli, F.; Scaioli, E.; Colecchia, A.; Festi, D. (**2016**). Gut microbiota and celiac disease. *Digestive Diseases and Sciences*, *61*, 1461–1472, doi: 10.1007/s10620-015-4020-2
189. Milani, C.; Ticinesi, A.; Gerritsen, J.; Nouvenne, A.; Lugli, G.A.; Mancabelli, L.; Turroni, F.; Duranti, S.; Mangifesta, M.; Viappiani, A.; et al. (**2016**). Gut microbiota composition and Clostridium difficile infection in hospitalized elderly individuals: a metagenomic study. *Science Report*, *6*, 25945, doi: 10.1038/srep25945
190. Antharam, V.C.; Li, E.C.; Ishmael, A.; Sharma, A.; Mai, V.; Rand, K.H.; Wang, G.P. (**2013**). Intestinal dysbiosis and depletion of butyrogenic bacteria in Clostridium difficile infection and nosocomial diarrhea. *Journal of Clinical Microbiology*, *51*, 2884–2892, doi: 10.1128/jcm.00845-13
191. Corsonello, A.; Lattanzio, F.; Bustacchini, S.; Garasto, S.; Cozza, A.; Schepisi, R.; Lenci, F.; Luciani, F.; Maggio, M.G.; Ticinesi, A.; et al. (**2018**). Adverse events of proton pump inhibitors: potential mechanisms. *Current Drug Metabolism*, *19*, 142–154, doi: 10.2174/1389200219666171207125351
192. Marchesi, J.R.; Adams, D.H.; Fava, F.; Hermes, G.D.; Hirschfield, G.M.; Hold, G.; Quraishi, M.N.; Kinross, J.; Smidt, H.; Tuohy, K.M.; et al. (**2016**). The gut microbiota and host health: a new clinical frontier. *Gut*, *65*, 330–339, doi: 10.1136/gutjnl-2015-309990
193. Wu, N.; Yang, X.; Zhang, R.; Li, J.; Xiao, X.; Hu, Y.; Chen, Y.; Yang, F.; Lu, N.; Wang, Z.; et al. (**2013**). Dysbiosis signature of fecal microbiota in colorectal cancer patients. *Microbial Ecology*, *66*, 462–470, doi: 10.1007/s00248-013-0245-9
194. Passos, M.; Moraes-Filho, J.P. (**2017**). Intestinal microbiota in digestive diseases. *Arquivos de gastroenterologia*, *54*, 255–262, doi: 10.1590/s0004-2803.201700000-31
195. Llorente, C.; Schnabl, B. (**2015**). The gut microbiota and liver disease. *Cellular and Molecular Gastroenterology and Hepatology*, *1*, 275–284, doi: 10.1016/j.jcmgh.2015.04.003
196. Le Chatelier, E.; Nielsen, T.; Qin, J.; Prifti, E.; Hildebrand, F.; Falony, G.; Almeida, M.; Arumugam, M.; Batto, J.-M.; Kennedy, S.; et al. (**2013**). Richness of human gut microbiome correlates with metabolic markers. *Nature*, *500*, 541–546, doi: 10.1038/nature12506
197. Honda, K.; Littman, D.R. (**2012**). The microbiome in infectious disease and inflammation. *Annual Review of Immunology*, *30*, 759–795.
198. Takahashi, K.; Nishida, A.; Fujimoto, T.; Fujii, M.; Shioya, M.; Imaeda, H.; Inatomi, O.; Bamba, S.; Andoh, A.; Sugimoto, M. (**1960**). Reduced abundance of butyrate-producing bacteria species in the fecal microbial community in Crohn's disease. *Digestion*, *93*, 59–65.
199. Ventura, M.; Turroni, F.; Canchaya, C.; Vaughan, E.E.; O'Toole, P.W.; van Sinderen, D. (**2009**). Microbial diversity in the human intestine and novel insights from metagenomics. *Frontiers in Bioscience*, *14*, 3214–3221.
200. Chang, C.; Lin, H. (**2016**). Dysbiosis in gastrointestinal disorders. *Best Practice & Research Clinical Gastroenterology*, *30*, 3–15.
201. Aroniadis, O.C.; Brandt, L.J. (**2014**). Intestinal microbiota and the efficacy of fecal microbiota transplantation in gastrointestinal disease. *Gastroenterology & Hepatology*, *10*, 230.
202. Masucci, L.; Quaranta, G. (**2022**). Fecal microbiota transplantation: what new? *Microorganisms*, *10*(1), 23. doi: 10.3390/microorganisms10010023
203. Taurog, J.D.; Richardson, J.A.; Croft, J.T.; Simmons, W.A.; Zhou, M.; Fernández-Sueiro, J.L.; Balish, E.; Hammer, R.E. (**1994**). The germfree state prevents development of gut and joint inflammatory disease in HLA-B27 transgenic rats. *The Journal of Experimental Medicine*, *180*, 2359–2364.
204. Stojek, M.; Jabłońska, A.; Adrych, K. (**2021**). The role of fecal microbiota transplantation in the treatment of inflammatory bowel disease. *Journal of Clinical Medicine*, *10*, 4055.
205. Moayyedi, P.; Surette, M.G.; Kim, P.T.; Libertucci, J.; Wolfe, M.; Onischi, C.; Armstrong, D.; Marshall, J.K.; Kassam, Z.; Reinisch, W. (**2015**). Fecal microbiota transplantation induces remission in patients with active ulcerative colitis in a randomized controlled trial. *Gastroenterology*, *149*, 102–109.
206. Rossen, N.G.; Fuentes, S.; van der Spek, M.J.; Tijssen, J.G.; Hartman, J.H.A.; Duflou, A.; Löwenberg, M.; van den Brink, G.R.; Mathus-Vliegen, E.M.H.; de Vos, W.M. (**2015**). Findings from a randomized controlled trial of fecal transplantation for patients with ulcerative colitis. *Gastroenterology*, *149*, 110–118.
207. Cui, B.; Feng, Q.; Wang, H.; Wang, M.; Peng, Z.; Li, P.; Huang, G.; Liu, Z.; Wu, P.; Fan, Z. (**2015**). Fecal microbiota transplantation through mid-gut for refractory C rohn's disease: safety, feasibility, and efficacy trial results. *Journal of Gastroenterology and Hepatology*, *30*, 51–58.
208. Vaughn, B.P.; Vatanen, T.; Allegretti, J.R.; Bai, A.; Xavier, R.J.; Korzenik, J.; Gevers, D.; Ting, A.; Robson, S.C.; Moss, A.C. (**2016**). Increased intestinal microbial diversity following fecal microbiota transplant for active Crohn's disease. *Inflammatory Bowel Diseases*, *22*, 2182–2190.
209. Sokol, H. (**2014**). Probiotics and antibiotics in IBD. *Digestive Diseases (Basel, Switzerland)*, *32* Suppl 1, 10–17, doi: 10.1159/000367820
210. Sood, A.; Midha, V.; Makharia, G.K.; Ahuja, V.; Singal, D.; Goswami, P.; Tandon, R.K. (**2009**). The probiotic

preparation, VSL#3 induces remission in patients with mild-to-moderately active ulcerative colitis. *Clinical Gastroenterology and Hepatology: The Official Clinical Practice Journal of the American Gastroenterological Association*, *7*, 1202–1209, 1209.e1201, doi: 10.1016/j.cgh.2009.07.016
211. Nguyen, N.; Zhang, B.; Holubar, S.D.; Pardi, D.S.; Singh, S. (**2019**). Treatment and prevention of pouchitis after ileal pouch-anal anastomosis for chronic ulcerative colitis. *The Cochrane Database of Systematic Reviews*, *5*, Cd001176, doi: 10.1002/14651858.CD001176.pub4
212. Davani-Davari, D.; Negahdaripour, M.; Karimzadeh, I.; Seifan, M.; Mohkam, M.; Masoumi, S.J.; Berenjian, A.; Ghasemi, Y. (**2019**). Prebiotics: definition, types, sources, mechanisms, and clinical applications. *Foods (Basel, Switzerland)*, *8*(3), 92. doi: 10.3390/foods8030092
213. Damaskos, D.; Kolios, G. (**2008**). Probiotics and prebiotics in inflammatory bowel disease: microflora 'on the scope'. *British Journal of Clinical Pharmacology*, *65*, 453–467.
214. Larrosa, M.; Yañéz-Gascón, M.J.; Selma, M.V.; Gonzalez-Sarrias, A.; Toti, S.; Cerón, J.J.; Tomas-Barberan, F.; Dolara, P.; Espín, J.C. (**2009**). Effect of a low dose of dietary resveratrol on colon microbiota, inflammation and tissue damage in a DSS-induced colitis rat model. *Journal of Agricultural and Food Chemistry*, *57*, 2211–2220.
215. Trop, T.K. (**2014**). Intestinal microbiota, probiotics and prebiotics in inflammatory bowel disease. *World Journal of Gastroenterology: WJG*, *20*, 11505.
216. Kanauchi, O.; Mitsuyama, K.; Homma, T.; Takahama, K.; Fujiyama, Y.; Andoh, A.; Araki, Y.; Suga, T.; Hibi, T.; Naganuma, M. (**2003**). Treatment of ulcerative colitis patients by long-term administration of germinated barley foodstuff: multi-center open trial. *International Journal of Molecular Medicine*, *12*, 701–704.
217. Lindsay, J.O.; Whelan, K.; Stagg, A.J.; Gobin, P.; Al-Hassi, H.O.; Rayment, N.; Kamm, M.; Knight, S.C.; Forbes, A. (**2005**). The clinical, microbiological and immunological effects of fructo-oligosaccharide in patients with Crohn's disease. *Gut, 55*(3), 348–355.
218. Hafer, A.; Krämer, S.; Duncker, S.; Krüger, M.; Manns, M.P.; Bischoff, S.C. (**2007**). Effect of oral lactulose on clinical and immunohistochemical parameters in patients with inflammatory bowel disease: a pilot study. *BMC Gastroenterology*, *7*, 1–11.
219. McLoughlin, R.F.; Berthon, B.S.; Jensen, M.E.; Baines, K.J.; Wood, L.G. (**2017**). Short-chain fatty acids, prebiotics, synbiotics, and systemic inflammation: a systematic review and meta-analysis. *The American Journal of Clinical Nutrition*, *106*, 930–945.
220. Furrie, E.; Macfarlane, S.; Kennedy, A.; Cummings, J.H.; Walsh, S.V.; O'neil, D.A.; Macfarlane, G.T. (**2005**). Synbiotic therapy (Bifidobacterium longum/Synergy 1) initiates resolution of inflammation in patients with active ulcerative colitis: a randomised controlled pilot trial. *Gut*, *54*, 242.
221. Malathi, K.; Nandini, R.; Dhanasekar, K.; Shilpa, B. (**2019**). A randomized open label study to evaluate the efficacy and tolerability of synbiotic in the treatment of ulcerative colitis. *Journal of Hepatology and Gastrointertis Disease*, *10*, 398.
222. Steed, H.; Macfarlane, G.T.; Blackett, K.L.; Bahrami, B.; Reynolds, N.; Walsh, S.V.; Cummings, J.H.; Macfarlane, S. (**2010**). Clinical trial: the microbiological and immunological effects of synbiotic consumption–a randomized double-blind placebo-controlled study in active Crohn's disease. *Alimentary Pharmacology & Therapeutics*, *32*, 872–883.

4 Influence of Gut-Brain Axis on Gastrointestinal Health and Benefits

Sindhu Priya E.S.[1], *Sandhya Vasanth*[2], *Rokeya Sultana*[3], *Mohammed Gulzar Ahmed*[2], *and B.H. Jaswanth Gowda*[2,4]

[1]Department of Pharmacology, Yenepoya Pharmacy College & Research Centre, Yenepoya (Deemed to Be University), Mangalore, Karnataka, India

[2]Department of Pharmaceutics, Yenepoya Pharmacy College & Research Centre, Yenepoya (Deemed to Be University), Mangalore, Karnataka, India

[3]Department of Pharmacognosy, Yenepoya Pharmacy College & Research Centre, Yenepoya (Deemed to Be University), Mangalore, Karnataka, India

[4]School of Pharmacy, Queen's University Belfast, Medical Biology Centre, 97 Lisburn Road, Belfast, BT9 7BL United Kingdom

4.1 INTRODUCTION & BACKGROUND

The notion of a strong link between the gastrointestinal tract and the brain, influencing not only gastric purposes nevertheless also emotions and instinctive decision-making, is firmly ingrained in our language and understanding (Mayer, 2011). The linked pathway amid the gut and the brain, recognized as the gut-brain axis, functions as a connection (Carabotti et al., 2015a).

In terms of structure, the gut establishes an intricate and mutually influential connection with the CNS (central nervous system) (facilitating communication in both states of health and ailment) (Montagnani et al., 2023; Suganya & Koo, 2020a). They are linked both physically and biochemically through various mechanisms, mutually influencing each other's well-being. This bidirectional axis involves intricate interactions among neural, hormonal, and immune pathways, affecting digestive functions and contributing to the maintenance of gut balance (Mukhtar et al., 2019).

The continuous communication between the brain and gut occurs through various complex pathways, such as the ENS (enteric nervous system) the ANS (autonomic nervous system), the HPA (hypothalamus-pituitary-adrenal axis), and the CNS (central nervous system), as illustrated in Figure 4.1.

Recognizing the dynamic interaction in the interior of the gut-brain axis is increasingly vital, as it is involved in maintaining gastrointestinal balance and influencing the emergence of different gastrointestinal disorders (Mukhtar et al., 2019).

The ENS, commonly known as the "second brain," is a crucial component of the gut-brain axis, containing complex neural networks within the GI wall. Concurrently, the gut microbiota, a varied assembly of microorganisms inherent in the GI tract, communicates with the CNS through various signaling molecules. This dynamic interplay impacts not just digestive functions such as motility, secretion, and nutrient absorption but also influences immune responses and the overall health of the gut (Appleton, 2018).

Significance

The brain regulates the operation of the gut via the hypothalamic–pituitary–adrenal axis and the autonomic nervous system. Specifically, when experiencing stress, the brain releases norepinephrine, which has been noted to encourage the growth of gut pathogens. The gut-brain axis holds substantial importance for both health and disease (Chaudhry et al., 2023a). Disturbances in this communication system have been connected with the commencement of various gastrointestinal disorders, ranging from functional problems like irritable bowel syndrome (IBS) to chronic conditions like inflammatory bowel disease (IBD). Grasping the mechanisms of how the brain and gut interact is crucial for considering the intricacies of these ailments and formulating precise therapeutic approaches. The influence of the gut-brain axis extends beyond gastrointestinal functions to impact cognitive and emotional processes. Psychosocial factors, stress, and mental health can disrupt the gut-brain axis, affecting gut physiology and potentially contributing to the onset or worsening of GI disorders (El-Hakim et al., 2022). Acknowledging these connections has sparked interest in interdisciplinary approaches that integrate neurobiology, microbiology, and clinical research to gain a comprehensive understanding of the brain-gut axis and its implications for well-being and different ailment (Carabotti et al., 2015a).

Furthermore, the gut-brain axis has paved the way for novel therapeutic interventions. Probiotics, prebiotics, dietary modifications, and behavioral interventions that target the gut-brain axis are emerging as promising strategies for managing GI disorders and improving gastrointestinal health (Ji et al., 2023).

In essence, investigating the gut-brain axis's impact on gastrointestinal health and disorders is a rapidly evolving

 DOI: 10.1201/9781003493143-4

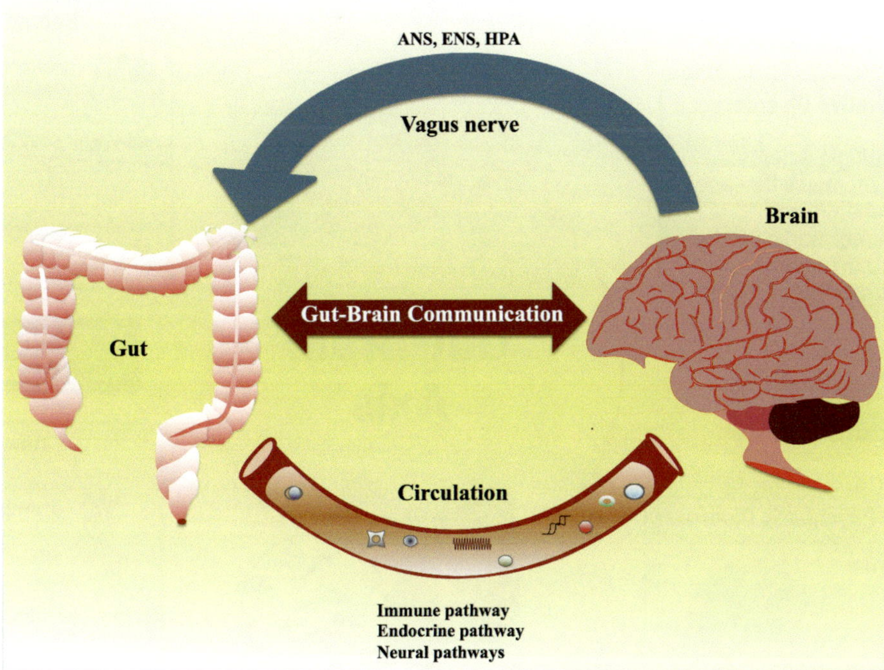

FIGURE 4.1 Illustrative diagram portraying the link between the gut and the brain (Mukhtar et al., 2019.).

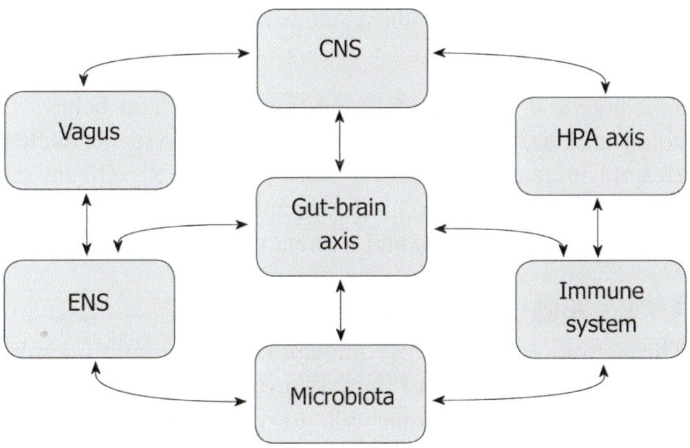

FIGURE 4.2 Illustrative depiction of various elements influencing the gut-brain axis (Mukhtar et al., 2019.).

field with significant implications for both clinical practice and foundational research as illustrated in the Figure 4.2. By deciphering the intricate connections within this axis, researchers and clinicians aim to devise innovative strategies to prevent, manage, and treat various gastrointestinal conditions, ultimately enhancing the overall well-being of affected individuals (Rutsch et al., 2020a).

4.1.1 Gut-Brain Axis: Overview

The complex communication network linking the GI tract to the CNS embraces various elements, such as ENS, sympathetic and parasympathetic- both branches of ANS, and signaling pathways associated with neuroimmuno- and neuroendocrine functions. Denoted to as the gut-brain axis, this advanced two-way communication system establishes a connection between the CNS (encircling the brain and spinal cord) and the ENS situated in the digestive system. This intricate interplay enables a continuous exchange of information between the gut and the brain, involving hormonal signals, neural pathways and immune interactions. Furthermore, the gut microbiota, a varied community of microbes that exist in the GI tract, acts as a substantial character in this axis by influencing and being influenced by both the CNS and ENS. This dynamic interaction extends beyond digestive processes, influencing cognitive functions, emotions, and overall health (Rithika Suresh, 2023).

4.1.1.1 Overview of the GI Tract

The GI tract, also denoted as the digestive system, is an intricate system tasked with the intake, breakdown, absorption, and expulsion of food and waste in the human body. It holds a crucial function in maintaining nutritional well-being and overall health. It comprises a constant, hollow tube encompassing from the opening to the anus, organized into distinct anatomical and functional segments (Hornbuckle et al., 2008). The GI system comprises not only the GI tract but also supplementary structures. The anatomical elements of the GI tract encompass the mouth, pharynx, esophagus, stomach, minor intestine, major intestine, and anal canal. Additional organs include structures like the teeth, tongue, and glandular organs such as liver, salivary glands gallbladder, and pancreas. The gastrointestinal (GI) tract consists of four histologic layers: the mucosa, submucosa, muscularis externa, and serosa or adventitia. Its main roles include the intake and mechanical breakdown of food, absorption of nutrients, secretion of water and enzymes, and the expulsion of excess foodstuffs (Ferguson, 1990). An overview of the

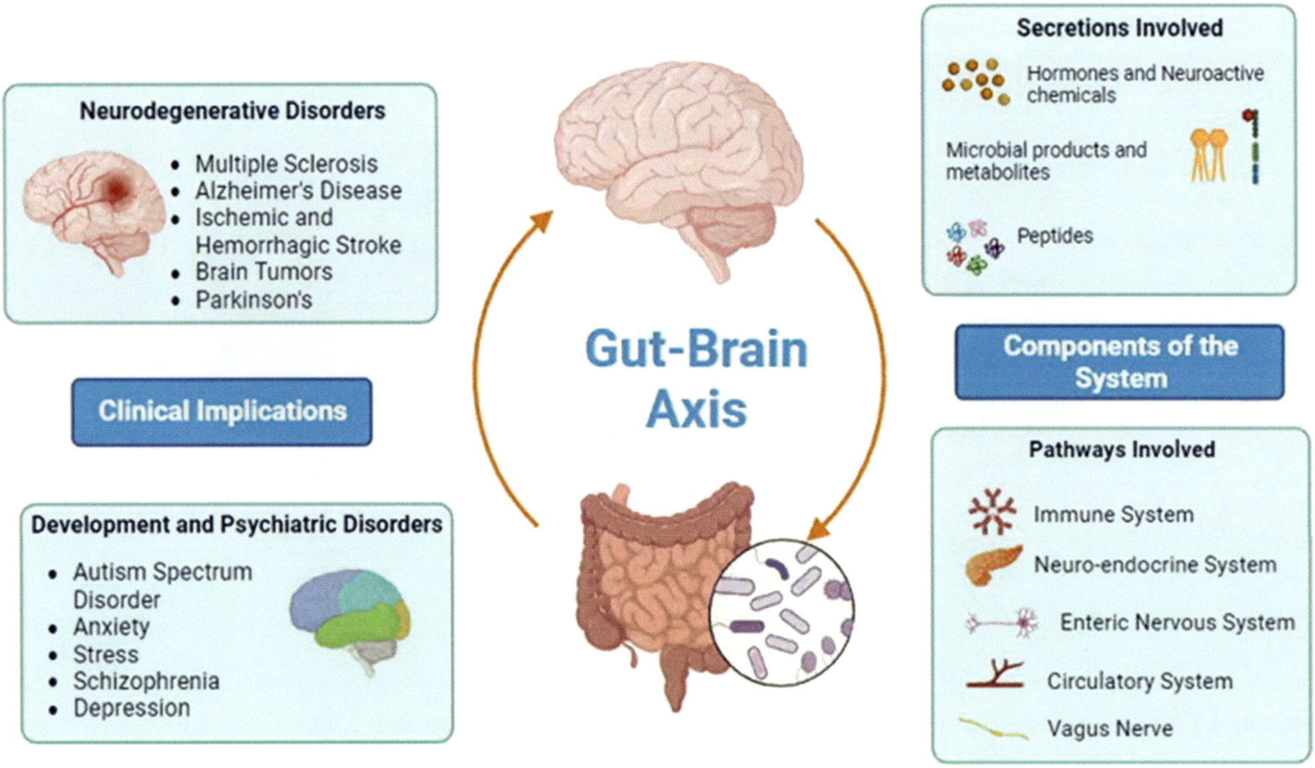

FIGURE 4.3 A summary of the gut-brain axis, its components, and the clinical consequences (Rithika Suresh, 2023.).

Gastrointestinal tract, its components, and clinical implications is shown in Table 4.1 and Figure 4.3.

4.1.1.2 The Nervous System and Gut-Brain Communication

The immune privilege traditionally attributed to the CNS is based on the tightly connected endothelial cells of BBB. This barrier regulates the allocation of ions, molecules, and cells amongst the bloodstream and the brain, protecting the CNS from potential threats and immune responses that might harm neurons and their connections. Nonetheless, the notion of CNS immune privilege is being re-examined, given recent findings of functional immune cells breaching the BBB and reaching the choroid plexus and lymphatic vessels, even in non-pathological conditions. Significantly, substances like bacterial products and metabolites can traverse BBB, impacting the progress, purpose, and onset of precise CNS ailments. The communication link between the CNS and the intestine is facilitated through the gut-brain axis (GBA), an intricate two-way network connecting them as illustrated in the Figure 4.4. This axis encompasses various pathways, including the ANS and ENS, the endocrine system, and the HPA. Multiple neurotransmitters and metabolites, such as secondary bile acids, vital vitamins, short-chain fatty acids (SCFAs), and amino acids, play a role in modulating immune system pathways, subsequently influencing learning, behavior, locomotion, and neurodegenerative ailments (Abenavoli et al., 2019; Socała et al., 2021).

Researchers have demonstrated the involvement of the inflammasome in behaviors related to depression and anxiety, as well as locomotor activity. Dysbiosis has been suggested as a potential factor underlying these mood and behavioral abnormalities, although the exact mechanism remains unclear. Despite increasing evidence, there is still a significant gap in understanding the precise mechanisms governing the gut-brain link in both health and ailment (Rutsch et al., 2020a).

4.1.1.3 Microbiota and Their Role

The microbiota is a civic of commensal and synergetic microbes that range at a density of more than 10^{12} cells/g of content in the human large intestine. 500 to 1,000 different bacterial species occupy the mammalian gut, fitting to the four leading bacterial phyla *Actinobacteria, Firmicutes, Proteobacteria,* and *Bacteroides.* This distribution of gut microbiome at various parts of GIT with respect to pH is specified in the Figure 4.5. Maintaining a harmonious and advantageous relationship amid the host and its microbiota is a fundamental necessity for both intestinal well-being and overall body health. In optimal conditions, the mucosal microbiota is crucial for functions such as vitamin synthesis, food digestion, angiogenesis, host immune system education, protection against pathogens, (Abenavoli et al., 2019). The term "microbiota" denotes the communities of microorganisms residing in a specific environment, while the term "commensals" refers to microorganisms that establish colonization in the host without inducing disease (Jandhyala et al., 2015). The intestinal microbiota in mammals consists of viruses, bacteria, yeasts, bacteriophages, and fungi. The microbiome encompasses the entire environment of the body, comprising its genomes, microorganisms, and the surrounding environmental conditions. The gut microbiota is a massive and intricate assembly of microbes that significantly influences

TABLE 4.1
An Overview of the Gastrointestinal Tract, The Components, and Clinical Implications

Sl No	Component	Description	Clinical Implications
1.	Mouth	Entry point of the digestive system; mechanical breakdown of food through chewing and mixing with saliva containing enzymes for initial digestion of carbohydrates	Dental issues (e.g., cavities, gum disease) can impact digestion; oral hygiene is important for overall health (Batra et al., 2020)
2.	Esophagus	Muscular tube connecting the mouth to the stomach; transports food bolus via peristalsis	Conditions such as gastroesophageal reflux disease (GERD) can result in symptoms such as heartburn and regurgitation (Diener et al., 2001)
3.	Stomach	Organ located in the upper abdomen; stores and mixes food with gastric juices containing hydrochloric acid and enzymes for digestion; initiates protein digestion	Conditions such as gastritis, peptic ulcers, and gastric cancer can affect stomach function and lead to symptoms like abdominal pain, bloating, and nausea (U.S. Department of Health and Human Services, 2022)
4.	Small Intestine	Extended, convoluted tube running from the stomach to the large intestine; it is segmented into three parts (duodenum, jejunum, ileum), where the mainstream of digestion and nutrient absorption arises	Malabsorption disorders (e.g., celiac disease) and infections (e.g., small intestinal bacterial overgrowth) can impair nutrient absorption; inflammatory conditions like Crohn's disease can be the basis of damage to the intestinal lining and cause complications (Dukowicz et al., 2007)
5.	Large Intestine (Colon)	The ending section of the digestive system; extracts water and electrolytes from undigested food, creating compact waste (feces) for excretion; it likewise assists as the habitat for gut microbiota	Conditions like IBD and IBS can effect in manifestations such as diarrhea, constipation, abdominal pain, and bloating. The development of chronic inflammation and the formation of polyps pose a potential risk for colorectal cancer as a complication (Maaser et al., 2019)
6.	Liver	The biggest internal organ situated in the right upper quadrant of the abdomen is responsible for creating bile to support in fat digestion, metabolizing nutrients, detoxifying injurious substances, and storage glycogen and vitamins	Liver ailments such as fatty liver illness, hepatitis and cirrhosis can impair liver function and lead to symptoms like jaundice, fatigue, and abdominal swelling; liver failure is a life-threatening condition requiring medical intervention (Heidelbaugh & Bruderly, 2006)
7.	Gallbladder	A compact organ positioned beneath the liver, supplies and focuses bile produced by the liver, subsequently liberating it into the small intestine to support the digestion of adipose tissue	Gallstones can block the bile ducts and source indications such as nausea, abdominal pain, and vomiting (Tsai et al., 2018)
8.	Pancreas	The glandular organ located behind the stomach; stows digestive enzymes (amylase lipase and proteases) and bicarbonate into the small intestine to support in digestion; also secrets hormones alike insulin and glucagon for blood sugar parameter	Pancreatitis (inflammation of the pancreas) can lead to severe abdominal pain, nausea, and vomiting; pancreatic cancer is often diagnosed at advanced stages and has a poor prognosis (Hall et al., 2014)
9.	Rectum	Terminal part of the large intestine; stores feces until elimination through the anus during defecation	Disorders such as hemorrhoids, rectal prolapse, and rectal cancer can cause symptoms like rectal bleeding, pain, and changes in bowel habits (Eslick et al., 2009)
10.	Anus	The exit point at the conclusion of the digestive tract from which waste material, or feces, is discharged from the body	Disorders such as anal fissures, hemorrhoids, and anal cancer can cause pain, itching, and bleeding; maintaining good hygiene and dietary habits is essential for preventing anal issues and promoting overall anal health (Socała et al., 2021)

human well-being. In the past, it was usually known as the gut microflora (Suganya & Koo, 2020b). The Figure 4.6 illustrates GM-GBA link which is a bidirectional communication explaining the complex interaction between several pathways and factors, influencing both brain health and diseases.

Research findings suggest that the microbiota present throughout the entire digestive tract has a substantial impact on the development and maintenance of both the CNS and the ENS. Placental fetuses are visible to foods and metabolites from the motherly microbiota before attaining their own microbiota (Littman & Pamer, 2011). Microbial colonization begins at birth in infants and is influenced by features such as the method of transport, feeding practices, antibiotic use, and maternal microbiota. The microbial products and metabolites serve as signaling molecules that straight or tortuously affect the CNS and the ENS. This microbial community initiates its formation at birth and undergoes development for approximately 2–3 years in individuals, eventually stabilizing in composition. However, it remains susceptible to various environmental and lifestyle factors throughout life, leading to substantial variations in microbiota composition even among healthy individuals (Rutsch et al., 2020a).

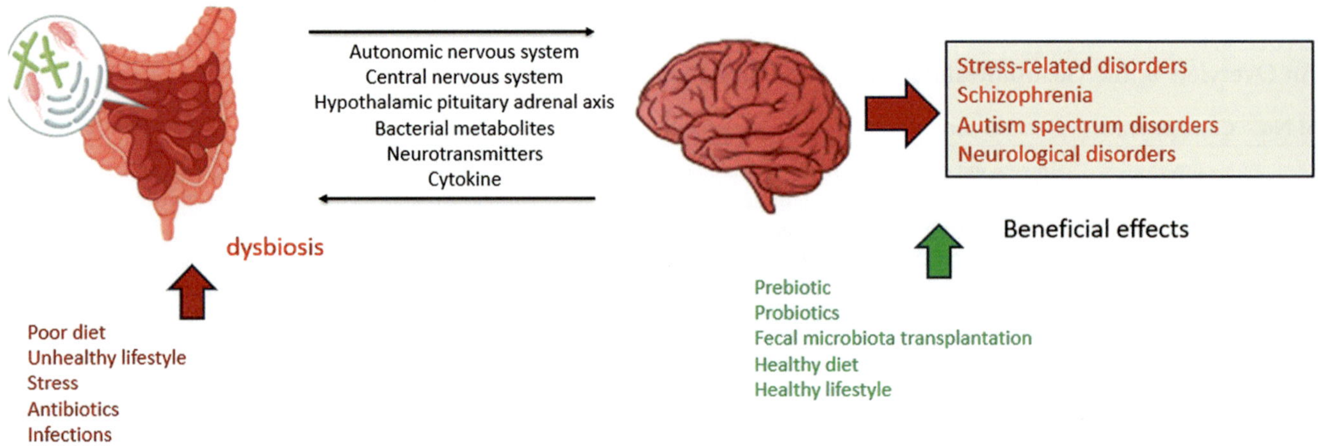

FIGURE 4.4 The nervous system and gut-brain communication (Socała et al., 2021.).

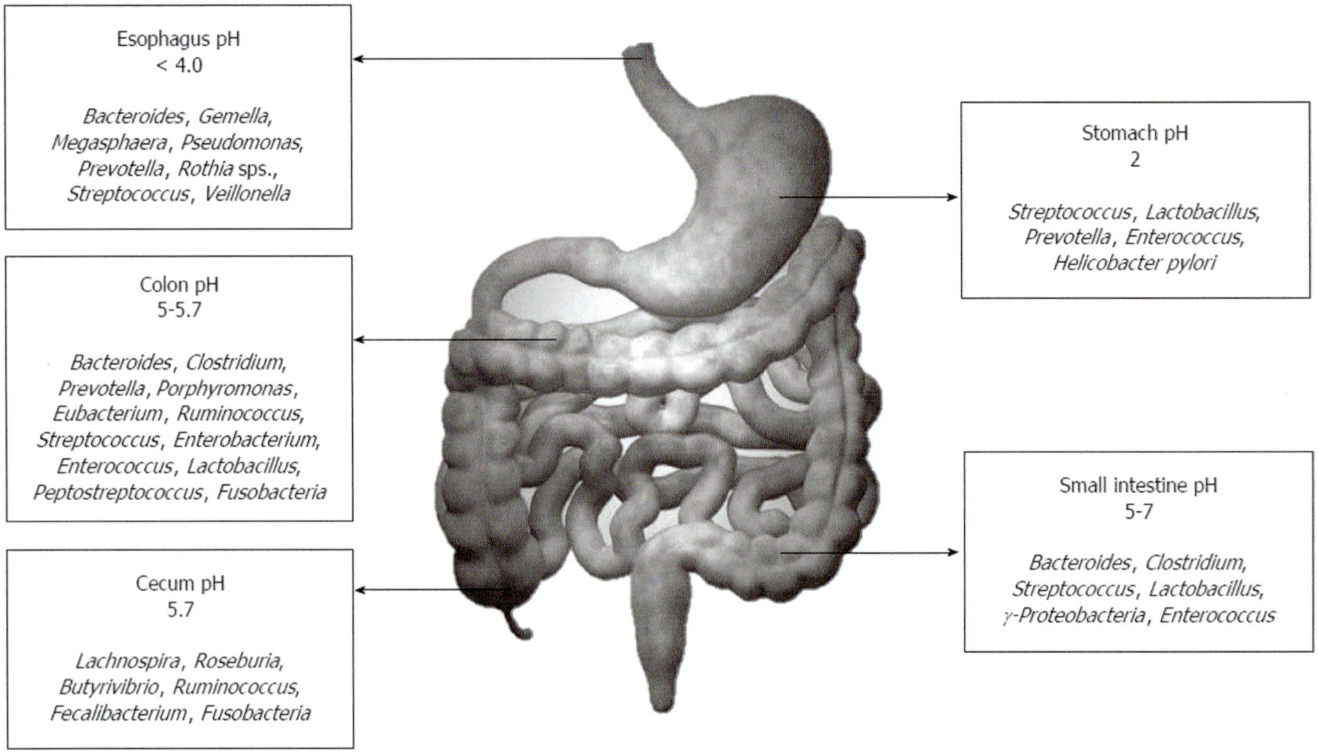

FIGURE 4.5 Distribution of the normal human gut flora (Suganya & Koo, 2020b.).

In distinctive situations, the microbiota performs vital functions in the host's physiological processes, including safeguarding against pathogens, aiding in nutrient digestion and absorption, and influencing the progress and instruction of multiple host organs and the immune organization (Chaudhry et al., 2023b). Over the last 10 years, several revisions have highlighted significant connections between variations in microbiota alignment, referred to as "dysbiosis," and various ailments in the host. Notably, some of these diseases impact organs beyond the gut, such as the CNS. Additionally, the gut microbiota seems to play a significant role not only in neuroinflammatory and psychiatric ailments but also in the growth of the brain (Li et al., 2022b).

4.2 COMMUNICATION MECHANISMS BETWEEN THE GUT AND BRAIN

Gut and brain communication happens through afferent spinal and afferent nerve vagal pathways via neuroimmune and neuroendocrine signals.

✓ **Different mechanisms**

- ***Neural Pathways***

  The GBA is connected via various neurological connections, such as ANS, neurotransmitters, spinal mechanisms, hypothalamic–pituitary–adrenal axis, ENS, and vagus nerve. This vagus nerve innervates

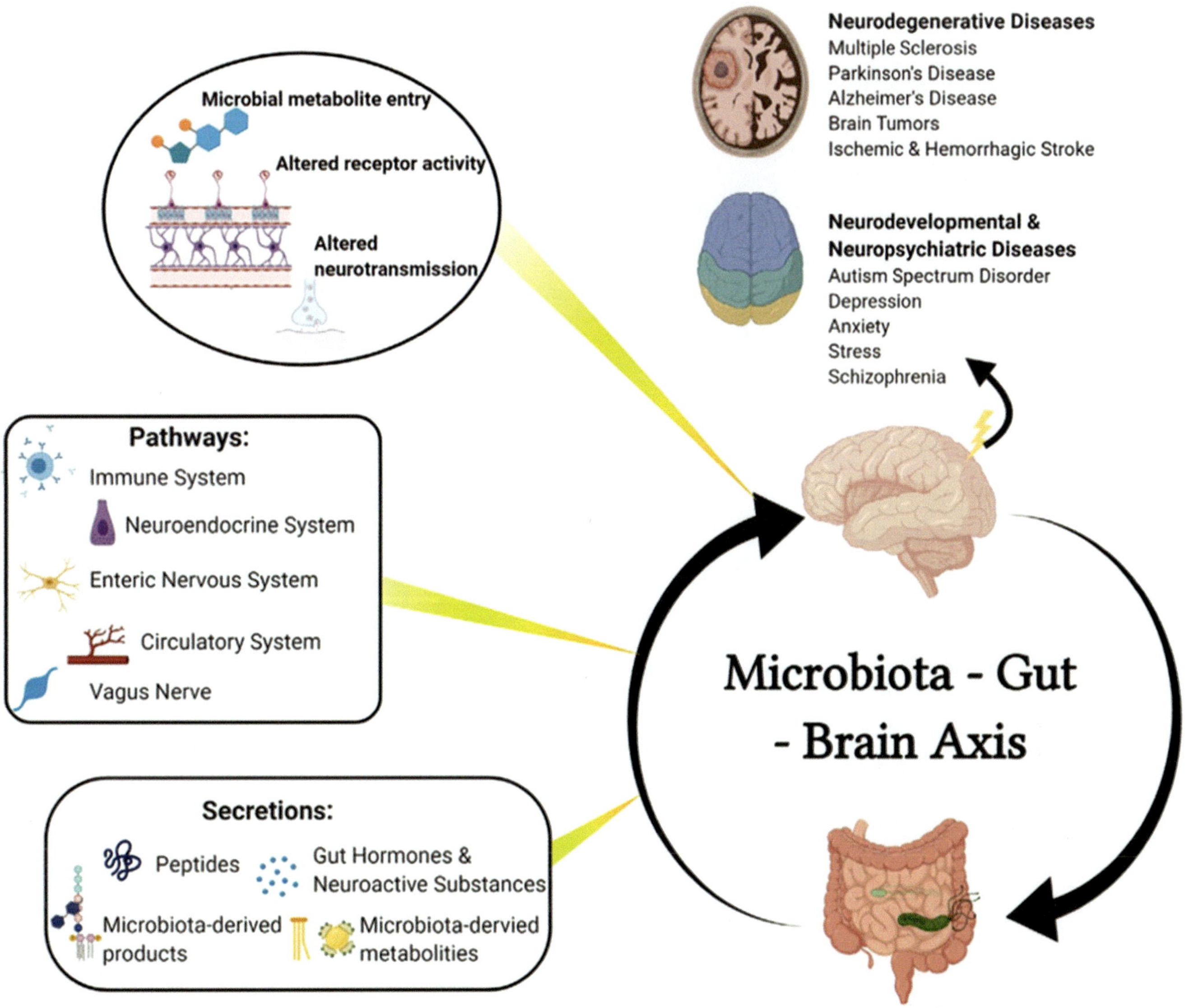

FIGURE 4.6 Microbiota and their role in gut-brain axis (Liu et al., 2022a.).

the GIT and the endocrine system, with 80% of afferent and 20% of efferent fibers. GM will communicate with the brain through inflammatory chemicals, bacterial metabolites, dietary elements, and gut regulatory peptides to the CNS. Most of the GM will directly activate neurons, with TL-3 and 7 detecting viral RNA, TL-2, 4 detecting peptidoglycan and lipopolysaccharide. *Lactobacillus rhamnosus, Bacteroides fragilis*, and isolated polysaccharides A of B have shown to stimulate intestine-afferent neurons ex vivo. Local GI microbial metabolites can stimulate the gut ANS and send sensory information directly to the brain. GM is involved in the body's normal physiological functions and the interference resulting in pathological states. Scientific evidence has demonstrated that the microbially regulated metabolite known as 4-ethyl phenyl sulfate can trigger anxiety-like behavior in mice. However, when animals with persistent gut inflammation were treated with *Bifidobacterium longum* NCC3001 over a long period, their anxiety symptoms were alleviated. Neurons in the brain can be directly activated by microbial metabolites through the expression of receptors FXR and TGR5. GM also contributes to the enteric glial cells growth that is crucial for the maintenance of neural networks and the regulation of gut homeostasis. (Cryan et al., 2019a) (Zanta et al., 2023).

- ***Hormonal Signaling***

  Glucagon-like peptide 1 (GLP1) and Peptide YY (PYY) are gut hormones, which have action mood, learning, and memory. These hormones are synthesized by SCFAs, by vagal stimulation through free fatty acid receptors (FFARs). GLP1 plays a role in reduced β amyloid plaques, microglia activation, greater neuroplasticity, neuroprotection in the hippocampus, improved memory and learning in mice and animal models of Alzheimer's disease.

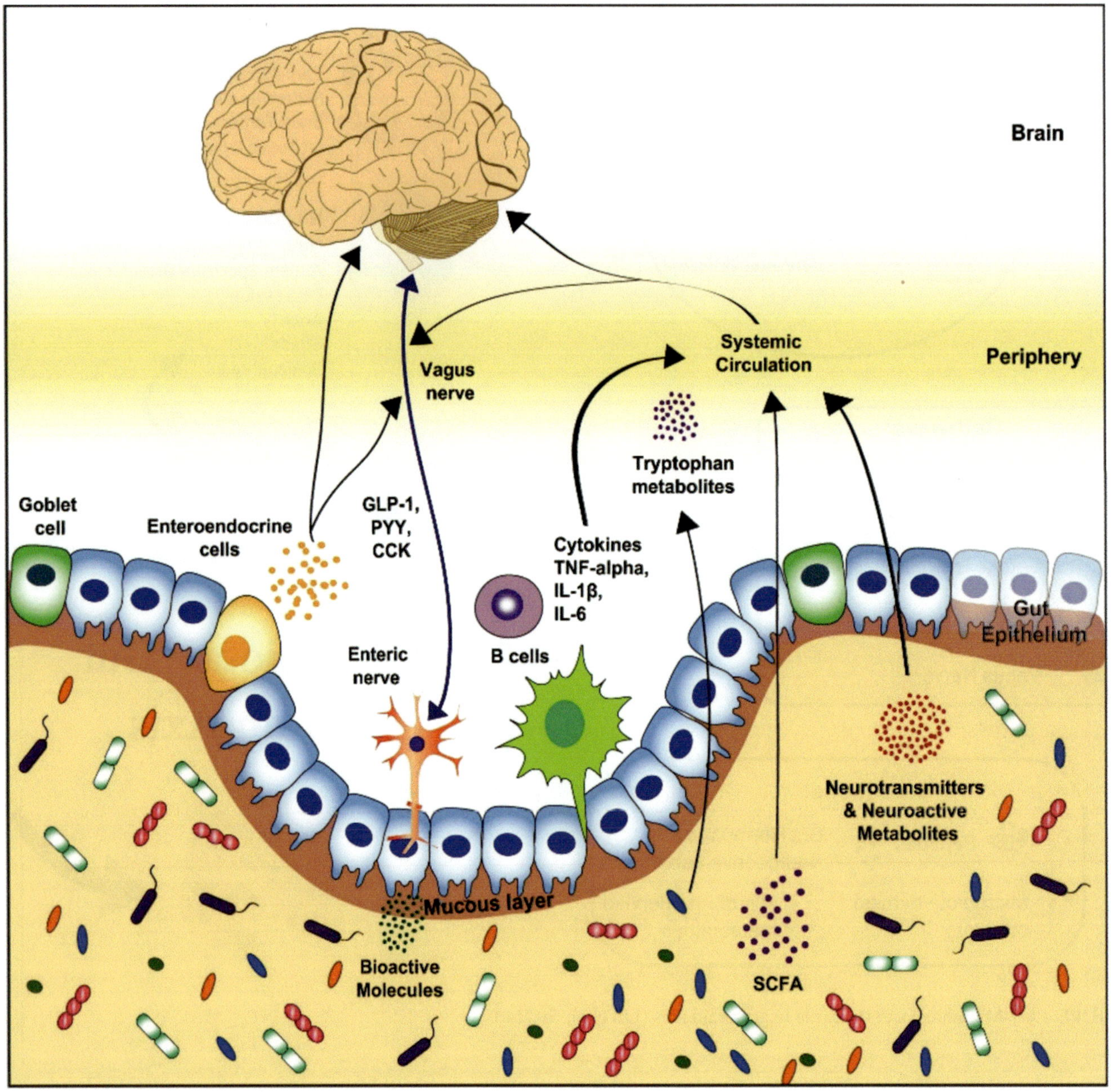

FIGURE 4.7 Various bidirectional communication pathways involved in gut-brain axis. CCK, cholecystokinin; GLP-1, glu-cagon-like peptide-1; IL, interleukin; PYY, peptide YY; TNF, tumor necrosis factor; SCFA, short-chain fatty acid (Cryan et al., 2019a.).

PYY, an anorexic neuropeptide, decreases hunger and stops the stomach from moving. It is expressed at the highest levels in the hypothalamus, pituitary gland, colon, and ileum's L cells. PYY3-36, a shortened protein version, binds preferentially with the Y2 neuropeptide Y receptor. Ghrelin, insulin and leptin are other metabolic hormones affected by SCFAs and have an impact on brain function as illustrated in Figure 4.7. However, research on these hormones has not been as concentrated as on PYY and GLP1 (Cryan et al., 2019a) (Martin et al., 2018) (Roh & Choi, 2023).

- ***Involvement of Immune System***

  Both the GM and the CNS have an impact on the development and operation of the peripheral immune system that activates innate immunity cells in the brain. CNS-cytokine interactions impact neurocircuits that control mood, motor activity, and motivation of brain functioning. Through circulating cytokines and the systemic immune system, GM and the brain will communicate as shown in Figure 4.7 (Salvo-Romero et al., 2020).

  CNS can be directly reached by immune cells through the BBB or chemokines and cytokines within the brain. GM influences the BBB, which controls the flow of chemicals from the blood into the cerebrospinal fluid and the permeability of chemicals, ions, and cells between the body's surroundings and the brain. To preserve homeostasis, SCFAs directly affect immune cells and immunological modulators. They control the maturation and activation of T cells, dendritic cells (DCs), neutrophils, and macrophages. SCFAs directly or indirectly impact adaptive immunological responses by influencing T cell formation and proliferation (Rutsch et al., 2020b).

How microbes regulate the neuroendocrine system demonstrates how indirect chemical communication affects the nervous system and behavior. GM can influence their host's appetite and eating habits by altering the endocrine signals generated by enteroendocrine cells (EECs) in the gut epithelium. Microbiological metabolites, including tyramine, SCFAs, α-tocopherol, p-aminobenzoate, indole, and secondary bile acids, affect how enteroendocrine cells (EECs) produce and secrete 5-HT (Morais et al., 2021).

MAMPs, or microbe-associated molecular patterns, are molecules created by gut microorganisms and recognized by different types of receptors, including TLR. Examples of these molecules include lipopolysaccharides, double-stranded RNA, and lipoproteins. Nonetheless, germ-free animals had lower hippocampal quantities of 5-HT and its metabolic precursor tryptophan, suggesting a potential function of GM in controlling 5-HT signaling pathways in the CNS (Cryan et al., 2019b) (Martin et al., 2018) (Salvo-Romero et al., 2020).

✓ **Impact on Gastrointestinal Health**

- ***Gut Motility and Function***

  Cognitive Behavioral Therapy (CBT) can impact gut motility and function, especially in individuals with conditions such as IBS, where stress and psychological factors play a significant role. Stress can contribute to alterations in gut motility, leading to symptoms like abdominal pain, bloating, and changes in bowel habits. CBT's stress reduction techniques, relaxation exercises, and behavioral strategies can positively influence the autonomic nervous system, helping to regulate gut motility. By addressing psychological stress or, CBT may contribute to more balanced and coordinated gut function, improving digestive processes and reducing symptoms associated with motility disorders (Evrensel & Ceylan, 2015).

- ***Intestinal Permeability***

  Chronic stress and psychological factors have been linked to increased intestinal permeability. CBT's stress management techniques may positively impact intestinal permeability by modulating the release of stress-related molecules and promoting a more stable gut environment. While research in this area is ongoing, managing stress through CBT could contribute to maintaining a healthy intestinal barrier (Houser & Tansey, 2017).

- ***Immune Responses in the Gut***

  CBT's influence on the GBA can extend to immune responses in the gut. Chronic stress affects the immune system, leading to inflammation and altered immune responses in the gastrointestinal tract. CBT modulates immune activity in the gut by reducing stress and uplifting a more stable emotional state. It is particularly relevant in situations where immune dysregulation, such as IBD, is a contributing factor. While CBT alone may not replace medical treatments for immune-related GI conditions, it can complement overall care by addressing the psychosocial aspects of these disorders (Foster et al., 2021).

✓ ***Role in Digestive Disorders***

CBT has demonstrated potency in managing various digestive disorders, particularly those with a substantial psychological component. Some notable impacts include:

IBS: CBT has been extensively studied and is considered an effective treatment for IBS. It helps individuals manage symptoms such as abdominal pain, bloating, and irregular bowel habits by addressing stress, anxiety, and negative thought patterns (Russo et al., 2017) (Rutsch et al., 2020b).

IBD: While primarily a group of immune-mediated conditions, stress and psychological factors can influence the course of IBD. CBT may assist individuals in coping with the emotional challenges of living with a chronic illness and may contribute to a more positive disease management experience (Houser & Tansey, 2017).

Functional Dyspepsia: CBT has shown promise in managing functional dyspepsia, a condition characterized by persistent upper abdominal discomfort. By addressing stress and behavioral patterns, CBT may help alleviate symptoms associated with this disorder (Russo et al., 2017). Gastroesophageal Reflux Disease (GERD): Stress can exacerbate symptoms of GERD. CBT may help individuals manage stress, reduce anxiety, and adopt lifestyle changes that can contribute to the overall management of GERD symptoms (Russo et al., 2017).

4.3 DIFFERENT FACTORS INFLUENCING GASTROINTESTINAL HEALTH

- **Biological Factors**

i. Microbiota Composition

The most common GM present in GI of humans is *Actinomycetota, Saccharomyces, Bacillota, Candida, Bacteroidota, Pseudomonadota, Bullera, Aspergillus, Pleospora, Penicillium, Rhodotorula, Trametes Sclerotinia, Galactomyces* etc (Bull & Plummer, 2014). They exhibit variation along the digestive tract, with the colon possessing the most abundant microbial population. However, their functions still need to be better understood. The bacterial population in the gut is predominantly inhabited by anaerobic bacteria, which comprise more than 99%. However, aerobic bacteria can be found in large concentrations in the cecum. A few GMs, *Lactobacillus Plantarum, Bacteroides vulgatus, Campylobacter jejuni, Bifidobacterium,* and *Bacteroides fragilis,* are also involved in the biosynthesis of neurotramsmitters chemicals, such as gamma-aminobutyric acid (GABA), serotonin, tyramine, and dopamine. These molecules influence mood, stress response, and cognitive function. GM converted the dietary fibers into SCFAs (butyrate, acetate, and propionate) by the fermentation process, which is vital in maintaining gut health, reducing inflammation, and influencing the GBA (Rath & Dorrestein, 2012). Environmental factors (antibiotics, diet) or genetic susceptibility can cause microbiome disturbance and an increase or decrease in microbial composition (Desai et al., 2016). The disturbance in a few species of microorganisms, *Ruminococcaceae, Lactobacillus, and Proteobacteria,* can lead to modulating the immune system. Impairment in immune response will direct effect on the GBA and contributes to conditions like neuroinflammation (Lynch & Pedersen, 2016).

ii. Role in Digestive Processes

The digestive process is responsible for the breakdown of complex nutrients into their constituent simple nutrients. Nutrient absorption in the small intestine is critical for providing the body with essential compounds. Proper absorption supports overall health and can influence signaling along the GBA (Flint et al., 2014). The digestive process interacts closely with the GM and helps break down complex carbohydrates. The resulting microbial metabolites, such as SCFAs, can influence the GBA and contribute to GI health. Enzymes and digestive juices (amylase, glucoamylase, proteases, lipase, saccharides, and phytase) are syntheses by GM, such as fungi and bacteria, which include, which aid in digesting minerals, such as phytic acid-produced by the stomach, pancreas, and small intestine aid in breaking down macronutrients into absorbable forms (Valdes et al., 2018). These compounds may have neuroprotective and anti-inflammatory effects, contributing to the overall health of the GBA. Disease conditions such as IBD, IBS, or celiac disease can disrupt the normal digestive process. These disorders may, in turn, affect the GBA, leading to symptoms like abdominal pain, altered bowel habits, and changes in mood or cognitive function. The proper enzyme function is essential for efficient digestion and nutrient utilization, impacting both GI health and the GBA (Bamford, 1999).

iii. Influence on Gut Barrier Integrity

The increased cortisol levels will have a direct link with the intestinal permeability, compromising reliability of the gut barrier. GBA have a direct role in stressing the neurotransmitters by influencing ENS which play vital role in gut motility. Any disorder or diet change may effect the synthesis of neurotransmitters and influence the gut motility and can lead to inflammation disorder in gut region. An equilibrium and varied microbiota support a healthy gut barrier by stimulating the creation of mucins and other protective substances, while inhibiting the proliferation of detrimental bacteria (50). General Motors can have a beneficial impact on the gut barrier by preserving the connections between cells in the intestinal lining, improving barrier strength, and decreasing permeability. The inflammation process of the gut may be due to any one condition, such as infections, autoimmune disorders, or chronic inflammation, which can compromise gut barrier integrity (Demaude et al., 2006). The GBA, through its modulation of immune responses, can influence inflammation and impact the health of the gut barrier. A diet rich in junk food, Western diet, less fiber, and high lipids may negatively affect gut barrier integrity (Turner, 2009).

iv. Gut-Brain Axis Communication

The GM can affect brain-related activities by triggering immune factors targeting the CNS and the ENS. The ANS controls physiological processes and maintains the balance of intestinal activities. The GBA is a two directional transmission matrix between the CNS and intestine, involving the ANS, neuroendocrine system, hypothalamic–pituitary–adrenal axis, immune system, and metabolic pathways. The research studies showed direct correction in the GM sequencing relationship between the complex ecosystem and the CNS (Carabotti et al., 2015b).

Because of their chemical interactions with the neurological system, GMs are essential for controlling physiological processes and changing behavior in their animal hosts. GM also produces neuroactive substances and induces the synthesis of other metabolites and neurotransmitters that control the signaling between the stomach and the brain. The proper maturation, activation, and growth of microglia in innate immune cells of the brain depend on the microbiota. Microglia's control over immune programming is heavily influenced by microbial metabolism. Disorders like Alzheimer's, Parkinson's, autism spectrum disorder, anxiety disorders, and depressive disorders can result from imbalances in these neurotransmitters (Cryan & Dinan, 2012).

Neurological and neuropsychiatric disorders are linked to changes in the GM, which can harm the nervous system. It can lead to brain bleeding, blood vessel diseases, nervous system development issues, spinal cord injuries, and brain tumors. Dysbiosis of the human GM is linked to various neurological diseases. Neurological disorder patients, compared with healthy controls, have significantly different microbiota compositions (Mayer et al., 2014).

Proper gastrointestinal motility is essential for preventing issues such as constipation or diarrhea. Abnormal motility can signal dysregulation along the GBA and contribute to GI disorders. The digestive system plays an important role in protecting immunity system, with gut-associated lymphoid tissue (GALT) helping to regulate immune responses. A balanced immune system in the gut is crucial for maintaining (Tiwari et al., 2023).

v. Stress and Anxiety

Acute stress is the primary reason for releasing cortisol from the Hypothalamic-Pituitary-Adrenal (HPA) axis. Research has demonstrated that stress affects the composition of GM in several types of animals, such as rats, pigs, horses, and non-human primates (Agusti et al., 2023).

Psychological stresses, such as water avoidance, parental separation, heat, and sonic stress, alter the composition of the GM. Maternal stress during pregnancy has intergenerational ramifications, impacting the microbiota of kids and dictating their outcomes. The transmission of maternal vaginal microbiota from mothers experiencing stress to offspring that are not experiencing stress can modify their stress response during adulthood (Agusti et al., 2023).

Anxiety frequently intersects since research has demonstrated a correlation between stress and the microbiota-GBA. Preclinical research studies have stated that anxiety-like behavior can be deduced from animal behavior in specific circumstances. Several studies have shown that probiotic medication can enhance anxiety measurements, albeit the effectiveness is contingent upon the particular species and strain used. *Lactobacillus plantarum* DR7 has been identified in recent studies as a potential psychobiotic capable of reducing symptoms associated with worry and stress. Additional translational investigations are required to analyze the impacts of psychobiotics on anxiety explicitly (Zheng et al., 2016)

vi. Depression and its Effects

The ongoing data suggests that the microbiota-GBA plays a part in MDD (Major Depression Disorder). MDD is directly related to the increased release of proinflammatory cytokines, which stimulate the HPA axis to release cortisol, & emphasize the hyperactivation of GMs. Most of the research has been done in preclinical models, such as the Flinders-sensitive rat model of depression and the maternal separation model, which show changes in the microbiota's composition and inflammation (Boku et al., 2018). The forced swim test (FST) behavior of GF mice revealed an increase in depression symptoms; administering nutrition-like live bacteria probiotics and prebiotics to the animals reduced the symptoms by promoting inflammatory responses. These findings shed further light on the role of GM-GBA in depression. Fecal mice from MDD patients differed significantly from those from healthy controls in numerous human investigations on individuals with anxiety and sadness (Cohen et al., 2019) (Charbonneau et al., 2013).

- **Genetic Factors**

Research studies proved that GM plays a vital role in facilitating the relationship between the genetic sequence of an organism and its observable characteristics, potentially influencing the transcription of neurons, the organism's behavior between genes, and the environment, and it can be an epigenetic entity. It primarily touches epithelial cells, although it can also indirectly interact with peripheral neurons and immunological cells (Rothschild et al., 2018).

The NOD2 genes regulate the autophagy system and the risk of developing ulcerative colitis (UC). NOD2 is an intracellular receptor that detects patterns in pathogens and contributes to regulating harmful microorganisms by both hematopoietic and non-hematopoietic cells. Initially, it was thought that persons with certain variations of NOD2 were unable to effectively eliminate bacterial infections, which resulted in the development of IBD. However, research has revealed that NOD2 regulates the immune response to non-harmful, mutually beneficial microorganisms. Individuals with genetic variations in the NOD2 gene exhibit unique collections of microorganisms in their bodies, characterized by a higher presence of *Escherichia* species and a lower presence of *Faecalibacterium* species (Lavelle & Sokol, 2020).

Multiple genetic variations linked to Crohn's disease (CD) risk are present in genes that control the autophagy process, including ATG16L1, IGRM, and LRRK. Autophagy has various purposes, including facilitating the lysosomal breakdown and removing intracellular microorganisms. Research has demonstrated that NOD2 interacts with ATG16L1, and CD-associated variations hinder the connection between these proteins. This interference negatively affects the ability to eliminate germs and present antigens (Cadwell et al., 2008; Khor et al., 2011).

The microbiomes of patients with IBD show changes in fungal communities precisely due to a single-nucleotide polymorphism in CLEC7A linked to IBD. DECTIN1 transmits signals via the adaptor protein CARD9, which has a role in the response to bacteria by interacting with NOD2 (Cadwell et al., 2008).

The major histocompatibility complex (MHC) encodes HLA genes, vital in the immune system's reaction to different infections. Associations have been observed between variations in MHC class II genes and the occurrence of UC and CD. Additionally, patients with UC tend to exhibit lower levels of heterozygosity in these genes. These variations could decrease the identification of antigens on commensal microorganisms. Research has demonstrated a connection between specific HLA alleles and unique microbial communities, leading to increased T cells and intestinal permeability. The relationship between variations in HLA genes and the composition of the GM is influenced by changes in the production of IgA antibodies and the presence of specific bacterial species. These factors can lead to the development of colitis. The intestinal mucus layer acts as a physical barrier that isolates luminal bacteria from the intestinal epithelium. IBD has not been linked to mutations in mucin genes. However, their modified expression patterns suggest their significance in preserving the microbiome and limiting inflammation. More than 200 loci have substantial relationships with IBD, but only a small portion can be accurately traced to one or two most probable alleles (Khor et al., 2011; Parkes et al., 2007).

- **Nutritional Influences on Gut-Brain Axis**

i. Dietary Habits and Microbiota

Diet significantly impacts gut bacteria composition, affecting brain function and behavior. Different diets, such as the Western, high-fat, and Mediterranean diets, have been linked to alterations in GM composition. The Western diet, high in sugar, salt, and fat, is associated with metabolic disorders and pathological diseases. High-fat diets, which are animal-based, can modify GM composition, leading to decreased levels of *Bacteroidetes* and increased levels of *Proteobacteria* and *Firmicutes* (Singh et al., 2017). Mediterranean diets, which include cereals, nuts, legumes, vegetables, and fruits, have been linked to reduced risk of neurodegenerative illnesses, mental conditions, cancer, and cardiovascular disease. The diet's ability to reduce inflammation has also been linked to significant changes in GM. The Mediterranean diet has been shown to enhance depression ratings. Carbohydrates exist in digestible and indigestible forms, with digestible carbohydrates triggering insulin production (Marlow et al., 2013; Wu et al., 2011). Diets high in glucose, fructose, and sucrose can promote *Bifidobacterium* growth while decreasing *Bacteroides*. Prebiotics, primarily undigested carbohydrates, can positively affect the host's gastrointestinal system by boosting gut flora (Hill et al., 2014; Ou et al., 2013).

Protein is also a critical component of our diet, serving as the leading supplier of amino acids crucial for neurotransmitter production and brain well-being. Plant-based protein diets have been found to reduce *Firmicutes*-to-*Bacteroidetes* ratios and increase microbiome diversity. Consuming diets containing animal products has been linked to higher levels of anaerobic bacteria that can tolerate bile. These bacteria can convert the fat into polyunsaturated fatty acids, which are healthy fats and have shown positive effects in reducing it and protecting against cardiovascular disorder, depression, particular malignancies, arthritis, and cognitive decline (Carvalho-Wells et al., 2010; De Filippo et al., 2010; Martínez et al., 2013).

ii. Nutrient Absorption and Gut Health

Prebiotics typically contain oligosaccharides, short polysaccharides such as xylo-oligosaccharides, galactofructose, galacto-oligosaccharides, and inulin, or nondigestible carbohydrates. Interventions with probiotics (beneficial bacteria) and prebiotics (substrates for beneficial bacteria) can positively influence the gut-brain axis and support gut barrier integrity. These interventions may enhance the production of mucins and reinforce tight junctions. The probiotics are live bacteria, and common species include *Lactobacillus, Bifidobacteria*, and yeasts such as *Saccharomyces boulardii* (Carvalho-Wells et al., 2010). They can create antimicrobial chemicals, compete with dangerous species for adhesion sites, lower pH, and support beneficial GM. Prebiotics are components of selective fermentation and help increase the synthesis of SCFs (Hill et al., 2014).

Probiotics can also lower insulin concentration, cardiovascular risk, and HbA1 blood glucose levels during fasting in individuals with type 2 diabetes. They also serve as an energy sensor, with probiotic profusion increasing in response to decrease calories and help in countering the energy level. They also have an important role in decreasing total and low-density lipoprotein cholesterol and improving clinical outcomes in necrotizing enterocolitis, urinary tract infections, and chronic periodontitis (De Filippo et al., 2010)

- **Environmental Factors**

The type and quality of nutrients in the diet can significantly impact the GM, which in turn influences the GBA. Environmental and toxic factors, such as antibiotic use, infections, diet, pollutants, and certain chemicals, can increase or decrease GM distribution and contribute to inflammation, potentially affecting the GBA. Antibiotics can alter the composition of GM, affecting the GBA. It is essential to use antibiotics judiciously and, when necessary, consider probiotic supplementation. Exercise and regular physical activity can decrease stress associated with a more diverse and beneficial GM. Exercise may also directly affect neurotransmitter levels and neuroplasticity in the brain, influencing the GBA. Disruptions in sleep patterns can affect GM and increase inflammation. Sleep deprivation may alter gut permeability and control the communication between the GBA (Rothschild et al., 2018; Song et al., 2020).

4.4 BENEFITS OF HEALTHY GUT-BRAIN AXIS

✓ **Improved Digestive Health**

A balanced GBA promotes proper digestive function, gut helps regulate peristalsis, enzyme secretion, and nutrient absorption, contributing to optimal digestion. The gut is often called the "second brain" due to its significant influence on mood. Serotonin, a neurotransmitter primarily produced in the GIT, plays a vital role in mood regulation. A healthy GBA is associated with balanced serotonin levels, contributing to improved mood and a reduced risk of mood disorders. The GBA plays an important role in the regulation of the stress response by the release of stress hormones and neurotransmitters, which reduces the anxiety and stress-related disorders. The gut is playing a crucial role in the body's defense against pathogens, and an effectively functioning GBA helps control immune responses. It provides protection against infections and reduces inflammation. The hormones like ghrelin and leptin will be influenced by GBA in appetite regulation. Hunger and satiety signals were controlled by a balanced communication system, contributing to healthy weight maintenance and a reduced risk of obesity. The GBA may impact memory and learning as it is involved in regulating cognitive function. Proper communication between gut and brain ensures the availability of nutrients and neurotransmitters necessary for efficient brain function. An imbalance in the GBA is related with various GI disorders, like IBS, IBD, and functional dyspepsia. A healthy GBA may reduce the risk or improve the management of these conditions. GM plays an important role in the GBA and contributes to overall health by maintaining a balanced & diverse microbiota associated with better digestion, nutrient absorption, and immune function. The GBA is linked to regulating circadian rhythms and sleep patterns. A healthy gut may contribute to better sleep quality and duration.

Chronic inflammation is associated with various health conditions, including GIT and neurodegenerative diseases. A balanced GBA helps regulate immune responses and may contribute to lower levels of inflammation (Chakrabarti et al., 2022) (Foster & McVey Neufeld, 2013) (Houser & Tansey, 2017) (Zanta et al., 2023).

4.5 THERAPEUTIC APPROACHES AND INTERVENTIONS

✓ **Lifestyle Modifications**

Adequate dietary fiber intake increases peristaltic movements in the bowel and prevents constipation. Drinking plenty of water throughout the day to maintain hydration helps soften stool, making it easier to pass and support proper digestion. Reduced intake of processed and high-fat foods can contribute to digestive discomfort and may exacerbate conditions like IBS. To promote health GMs incorporation probiotic-rich foods, such as yogurt, kefir, sauerkraut, and kimchi is essential. Engage in regular physical exercise, as it helps promote regular bowel movements and supports overall digestive health. Exercise can also help manage stress, which is linked to various GI disorders. Irregular eating patterns can disrupt the digestive process and contribute to issues such as acid reflux. Limit the consumption of alcohol, smoking and caffeinated beverages, as they can contribute to dehydration causing irritation of the GI tract, peptic ulcers and Crohn's disease. Obesity is related to an increased risk of certain GI disorders. Practice good hygiene to prevent infections that can affect the GI tract. Wash hands thoroughly before eating and after using the restroom (Russo et al., 2017) (Chakrabarti et al., 2022).

✓ **Cognitive Behavioral Therapy**

CBT helps individuals identify stressors and develop effective coping mechanisms. Stress is known to impact GI health, exacerbating conditions such as IBS and IBD. Managing stress through CBT techniques can lead to symptom improvement. CBT emphasizes the mind-body connection, recognizing that psychological factors can influence physical symptoms. By addressing cognitive distortions and negative thought patterns, individuals may experience reduced GI symptoms. Individuals with GI disorders may engage in catastrophic thinking about their symptoms, amplifying their distress. CBT helps challenge and change these negative thought patterns, leading to a more realistic and adaptive perception of symptoms. CBT incorporates behavioral strategies to address maladaptive behaviors associated with GI disorders. It can include addressing avoidance behaviors, improving adherence to medical recommendations, and promoting healthier lifestyle habits. For conditions associated with chronic pain, such as functional GI disorders, CBT can be used to teach pain management techniques. It includes relaxation exercises, mindfulness, and cognitive restructuring to change pain perception. CBT equips individuals with better coping skills to deal with the challenges of living with a chronic GI condition. It involves developing problem-solving skills, enhancing resilience, and fostering a positive mindset (Kinsinger, 2017).

✓ **Pharmacological Interventions**

Aminosalicylates, derivatives such as sulfasalazine (SAS), and other 5-aminosalicylic acid (5-ASA) medications, have been used to treat IBD for the past eight decades. SAS has impacts on white blood cell activity and cytokine production, scavenging reactive oxygen species, and interfering with arachidonic acid metabolism. According to recent research, oral 5-ASA treats ulcerative colitis (UC) better than a placebo and has results comparable to conventional treatment when taken once a day. Nonetheless, there is ongoing debate on the treatment effectiveness of ASA preparations for colorectal cancer (CRC) (Li et al., 2022a).

Since the 1950s, oral CSs were used to treat IBD, and they are an effective way to put an end to a flare-ups. In the cytoplasm, CSs bind to and activate CS receptors and interact with certain proinflammatory transcription factors to control the expression of genes that fight inflammation. Patients with moderate CRC, particularly those with UC resistant to mesalazine within 24 weeks and substantial lesions may consider CSs as a therapy option (Li et al., 2022a).

Oral second-generation chemotherapy drugs like budesonide are now accessible and may have superior safety and tolerable characteristics than traditional chemotherapy drugs. Budesonide Multi Matrix (MMX) exhibits tolerability comparable to mesalazine, placebo, and pH-dependent budesonide and is released over the whole colon. With a two-week taper, 9 mg of budesonide once a day is advised for eight weeks for moderate CRC (Foster & McVey Neufeld, 2013).

Immunomodulators such as calcineurin inhibitors, methotrexate (MTX), thiopurines (TPs), and Janus Kinase (JAK) inhibitors are essential for individuals with inflammatory bowel disease (IBD). By preventing T cell activation and proliferation, TPs regulate intestinal inflammation. AZA has demonstrated beneficial treatment outcomes on UC and chronic and inflammatory illnesses (CRC), lowering hospitalization and surgery rates. TPs, however, have several unfavorable side effects, including gastrointestinal discomfort, liver damage, and suppression of the bone marrow (Li et al., 2022a).

T cell proliferation and the inflammatory response are suppressed by MTX through downregulating inflammatory cytokines and reducing DNA synthesis. Research indicates that, 72% of individuals with active CRC experienced clinical remission following three months of MTX treatment. MTX has not been shown to cause UC to go into remission. Inhibitors of calcineurin, such as Tacrolimus (TAC) and Cyclosporine A (CsA), disrupt the signaling cascade and prevent the inflammatory response. Whether taken orally or intravenously, TAC is beneficial for refractory UC and is well-absorbed by the intestines. Adverse effects such as tremors, impairment of renal function, infectious diseases, hyperkalemia, hot flashes, and headache, are familiar with TAC. More randomized-controlled trials are necessary to investigate safety and effectiveness of TAC & CsA in treating inflammatory bowel disease (IBD) Adverse side effects, including tremors, impairment of renal function, infectious illnesses, hot flashes, hyperkalemia, and headache, are familiar with TAC. More randomized controlled trials are needed to investigate the safety and effectiveness of TAC and CsA in treating inflammatory bowel disease (IBD). Immunomodulators such as JAK inhibitors, TPs, MTX, and TAC are essential in managing IBD. However, due to worries about adverse medication reactions, possible side effects, and the requirement for more randomized controlled studies, their use has decreased. In treating inflammatory bowel disease (IBD), biologics—such as integrin antagonists and proinflammatory cytokine inhibitors—are essential. Two important proinflammatory cytokines in the pathophysiology of IBD are TNF-α and IL-12/23. Anti-TNF treatment, like Adalimumab (ADA) and Infliximab (IFX), prevents tissue damage and TNF-α-related inflammatory reactions. Inspite of that, anti-TNF therapy is not curative as most of the patients will not react to TNF inhibitors (Whitfield & Schulman, 2009).

The distinct p19 subunit of IL-23, implicated in the pathophysiological process of IBD, is the target of anti-IL-12/23 therapy. A fully humanized IgG1 monoclonal antibody called uzekinumab has been approved to treat moderate-to-severe cases of UC and CRC. Compared to placebo, mirikizumab, a human IgG monoclonal antibody targeting the IL-23's p19 subunit, has demonstrated better rates of clinical response. Risankizumab—a humanized IgG monoclonal antibody against the p19 subunit of IL-23—is more effective in causing clinical remission in patients with moderate-to-severe CRC (Kinsinger, 2017).

Experts advise using IL-12/23 and IL-23 antagonists as first- or second-line therapy for inflammatory bowel disease (IBD) because of their effectiveness in both naïve and experienced patients. Anti-integrin therapy prevents leukocytes from interacting with the intestinal mucosa by blocking the effect of integrin on the surface of leucocytes and endothelial CAMs. Let us sum up by saying biologics are essential for treating IBD since they block proinflammatory cytokines like TNF-α and IL-12/23. However, dose increasing, reducing the time between infusions, or combination therapy may be used to achieve long-term remission.

The recombinant human IgG1 monoclonal antibody, vedolizumab, has been licensed in the US and Europe for use in adult patients with moderate-to-severe UC and CRC who have not responded to or tolerated anti-TNF-α monoclonal antibodies or conventional therapy. Due to its intestinal selectivity, which helps prevent the negative consequences of systemic immunosuppression, it has demonstrated excellent tolerance and safety. A monoclonal antibody called etolizumab specifically targets the β7 subunit of integrins, α4β7 and αEβ7. In patients with moderate-to-severe UC, it has been more successful than a placebo in causing clinical remission at week 10. Multiple phases, 3 clinical trials demonstrate that pembrolizumab is superior to placebo in generating remission in people with moderate-to-severe ulcerative colitis. Lipid mediators known as Sphingosine-1-Phosphate Receptor Modulators and Agonists (S1P) regulate the immune system as S1P receptor agonists and modulators, ozanimod and etrasimod are being researched for the treatment of ulcerative colitis (UC). By causing peripheral blood cells to isolate in the lymph nodes, the oral S1PR modulator ozanimod lowers the number of activated lymphocytes circulating to the inflammatory areas. It has been demonstrated to be more successful than a placebo at bringing moderately to severely active UC into remission and keeping it there.

Changes in the composition and function of the intestinal microbiota are closely linked to the onset of IBD, suggesting that the intestinal microbiota plays a critical role in the development of IBD. An imbalance of good vs harmful pathogenic bacteria in patients can lead to aberrant immune responses and damage to the gut mucosa. The gut microbiota's diversification declines in IBD patients, with the amount of common anaerobic bacteria like *Bacteroides*, *Eubacteria*, and *Lactobacilli* showing the most notable changes. These typical anaerobic bacteria have disappeared due to mucosal inflammation in IBD patients. By enhancing intestinal microecology, researchers have discovered potentially beneficial therapy options for IBD, such as fecal microbiota transplantation (FMT), probiotics, prebiotics, postbiotics, symbiotics, and antibiotics. Since antibiotics may impact the makeup of the gut microbiota, they are anticipated to be a therapy option for IBD in the future. Antibiotics effectively bring about a remission of CRC, but this may not have much clinical significance. In IBD, probiotics, prebiotics, synbiotics, and postbiotics have all been shown to be helpful, mainly when used in conjunction with traditional medications. Probiotics are beneficial for active patients in certain studies (Liu et al., 2022b).

4.6 CHALLENGES AND FUTURE DIRECTIONS

- **Understanding Mechanisms:** One hurdle involves understanding the complex mechanisms leading to gut-brain communication and its influence on gastrointestinal function. Despite substantial advancements, there is still a considerable knowledge gap regarding the exact pathways and their roles in varied gastrointestinal disorders.
- **Role of Microbiota:** An added investigation is required to clarify the exact impact of the gut microbiota on the inflection of the gut-brain axis and its significance for gastrointestinal health. This includes studying the influence of variations in microbiota composition on brain roles and behavior, as well as their connection to gastrointestinal disorders.
- **Clinical Applications:** A crucial challenge is decoding research findings into practical clinical claims. Developing effective therapeutic interventions pointing to the gut-brain axis needs a deeper empathy of its complexities and their application to precise gastrointestinal conditions.
- **Individual Variability:** Distinguishing and addressing discrete differences in gut-brain axis purpose and responsiveness is vital. Factors such as inheritances, food, lifestyle, and ecological influences can all impact the interaction between the gut and brain, leading

to varying effects on gastrointestinal health among different personalities.

- **Multidisciplinary Collaboration:** Association among researchers from assorted fields, including gastroenterology, neuroscience, microbiology, and immunology is crucial for progressing our understanding of the gut-brain axis. Multidisciplinary methods offer a valuable understanding of the complex interaction between the gut and brain and its consequences for gastrointestinal disorders.
- **Technological Advancements:** Modern technological advancements, such as animal models, advanced imaging techniques and omics technologies can enhance our aptitude to study the gut-brain axis in cooperation with health and disease. These tools permit more thorough investigations of the underlying mechanisms and potential therapeutic goals.
- **Clinical Trials:** It is crucial to carry out precisely planned clinical trials to evaluate the safety and efficacy of interventions focused on the gut-brain axis. Strong clinical indication is crucial for notifying clinical practices and creating novel treatment methods for gastrointestinal disorders.

Addressing these encounters and focusing on these upcoming guidelines will depend on our understanding of the gut-brain axis's role in gastrointestinal health and ailments, eventually leading to enhanced diagnostic means and more operative therapeutic interventions.

4.7 SUMMARY

The gut-brain axis's impact on gastrointestinal (GI) health and disorders has arisen as projecting attention in research, appealing considerable care in recent years. This complex bidirectional communication system, connecting the gut and the brain, plays an important character in adjustable numerous physical functions, through digestion, immune rejoinders, nutrient absorption, and even emotional and intellectual procedures.

Many studies have investigated the mechanisms easing communication between the gut and brain, see-through a multifaceted relationship involving hormonal, nervous, immune, and microbial pathways. Signals created from the gut can right away influence brain function over mechanisms such as the vagus nerve or the declaration of signaling molecules like cytokines, neurotransmitters, and microbial metabolites. Contrariwise, the brain can impact gut actions through the autonomic nervous system and by liberating stress-related hormones.

This bidirectional link between the gut and brain grasps significant consequences for GI health and ailments. Disturbances in gut-brain axis signaling have been a concern in numerous GI situations, fluctuating from inflammatory bowel disease (IBD), irritable bowel syndrome (IBS) and gastroesophageal reflux disease (GERD) and purposeful dyspepsia. Furthermore, developing indication advises that deviations in the gut microbiota, recognized as dysbiosis, might pay an attention to GI dysfunction and ailment by disturbing gut-brain axis signaling.

Understanding the gut-brain axis's character in GI well-being and ailments presents hopeful paths for developing advanced diagnostic and beneficial approaches. Directing the gut microbiota over interventions such as prebiotics, probiotics, or dietary alterations clutches probable for managing GI ailments. Likewise, involvements intended at curbing gut-brain axis signaling, such as stress supervision techniques or pharmacological agents aiming neurotransmitter pathways, might propose alternative treatment possibilities for GI environments.

Nevertheless, despite the advancement made in this arena, numerous encounters and unanswered queries persist. These include illuminating the precise mechanisms underlying gut-brain axis disfunction in diverse GI ailments, recognizing biomarkers aimed at diagnostic drives, and evaluating the long-standing efficacy and safety of interventions aiming at the gut-brain axis (Appleton, 2018; Mukhtar et al., 2019; Rutsch et al., 2020a).

4.8 CONCLUSION

In the cutting-edge conclusion, the gut-brain axis assists as a dangerous boundary between the gut and the brain, applying reflective impacts on GI health and ailments. Current research into this intricate communication system embraces the potential for progressing our understanding of GI ailments and refining patient care over the expansion of advanced diagnostic apparatuses and targeted beneficial interferences.

CONFLICT OF INTEREST

The authors declare no conflict of interest.

ACKNOWLEDGEMENTS

The authors are thankful to Yenepoya (deemed to be University) for providing support to write this book chapter.

ABBREVIATIONS

✓ GBA	Gut-Brain Axis
✓ GM	Gut microbes
✓ NOD2	Nucleotide-binding oligomerization domain-containing protein 2
✓ GABA	gamma-aminobutyric acid
✓ GI	Gastrointestinal
✓ SCFAs	Short-Chain Fatty Acids
✓ IBD	Inflammatory Bowel Diseases
✓ IBS	Inflammatory Bowel Syndrome
✓ ENS	Enteric Nervous System
✓ ANS	Autonomic Nervous System
✓ CNS	Central Nervous System
✓ GALT	Gut-Associated Lymphoid Tissue
✓ HPA	Hypothalamic–Pituitary–Adrenal

✓ MDD	Major Depressive Disorder
✓ CD	Crohn's disease
✓ ATG16L1	Autophagy Related 16like 1
✓ IRGM	Immunity related GTPase M
✓ LRRK	Leucine Rick Repeat Kinase
✓ CLEC7A	C-Type Lectin Domain Containing 7A
✓ MHC	Major Histocompatibility Complex
✓ HLA	Human Leukocyte Antigens
✓ IgA	Immunoglobulin

REFERENCES

Abenavoli, L., Scarpellini, E., Colica, C., Boccuto, L., Salehi, B., Sharifi-Rad, J., Aiello, V., Romano, B., De Lorenzo, A., Izzo, A. A., & Capasso, R. (2019). Gut microbiota and obesity: A role for probiotics. *Nutrients*, *11*(11), 2–27. https://doi.org/10.3390/nu11112690

Agusti, A., Lamers, F., Tamayo, M., Benito-Amat, C., Molina-Mendoza, G. V., Penninx, B. W. J. H., & Sanz, Y. (2023). The gut microbiome in early life stress: A systematic review. In *Nutrients* (Vol. 15, Issue 11). MDPI. https://doi.org/10.3390/nu15112566

Appleton, J. (2018). The gut-brain axis: Influence of microbiota on mood and mental health. *Integrative Medicine (Encinitas, Calif.)*, *17*(4), 28–32.

Bamford, K. B. (1999). Chronic gastrointestinal inflammation. *FEMS Immunology & Medical Microbiology*, *24*(2), 161–168. https://doi.org/10.1111/j.1574-695x.1999.tb01277.x

Batra, P., Saini, P., & Yadav, V. (2020). Oral health concerns in India. *Journal of Oral Biology and Craniofacial Research*, *10*(2), 171–174. https://doi.org/10.1016/j.jobcr.2020.04.011

Boku, S., Nakagawa, S., Toda, H., & Hishimoto, A. (2018). Neural basis of major depressive disorder: Beyond monoamine hypothesis. *Computer Graphics Forum*, *37*(2), 3–12. https://doi.org/10.1111/pcn.12604

Bull, M. J., & Plummer, N. T. (2014). Part 1: The human gut microbiome in health and disease. In *Integrative Medicine* (Vol. 13, Issue 6). Mansa STM Publishers.

Cadwell, K., Liu, J. Y., Brown, S. L., Miyoshi, H., Loh, J., Lennerz, J. K., Kishi, C., Kc, W., Carrero, J. A., Hunt, S., Stone, C. D., Brunt, E. M., Xavier, R. J., Sleckman, B. P., Li, E., Mizushima, N., Stappenbeck, T. S., & Virgin IV, H. W. (2008). A key role for autophagy and the autophagy gene Atg16l1 in mouse and human intestinal Paneth cells. *Nature*, *456*(7219), 259–263. https://doi.org/10.1038/nature07416

Carabotti, M., Scirocco, A., Antonietta Maselli, M., & Severi, C. (2015a). *The Gut-Brain Axis: Interactions Between Enteric Microbiota, Central and Enteric Nervous Systems* (Vol. 28). Hellenic Society of Gastroenterology. www.annalsgastro.gr

Carabotti, M., Scirocco, A., Antonietta Maselli, M., & Severi, C. (2015b). *The Gut-Brain Axis: Interactions Between Enteric Microbiota, Central and Enteric Nervous Systems* (Vol. 28). Hellenic Society of Gastroenterology. www.annalsgastro.gr

Carvalho-Wells, A. L., Helmolz, K., Nodet, C., Molzer, C., Leonard, C., McKevith, B., Thielecke, F., Jackson, K. G., & Tuohy, K. M. (2010). Determination of the in vivo prebiotic potential of a maize-based whole grain breakfast cereal: A human feeding study. *British Journal of Nutrition*, *104*(9), 1353–1356. https://doi.org/10.1017/S0007114510002084

Chakrabarti, A., Geurts, L., Hoyles, L., Iozzo, P., Kraneveld, A. D., La Fata, G., Miani, M., Patterson, E., Pot, B., Shortt, C., & Vauzour, D. (2022). The microbiota–gut–brain axis: Pathways to better brain health. Perspectives on what we know, what we need to investigate and how to put knowledge into practice. *Cellular and Molecular Life Sciences*, *79*(2), 1–15. https://doi.org/10.1007/s00018-021-04060-w

Charbonneau, D., Gibb, R. D., & Quigley, E. M. M. (2013). Fecal excretion of Bifidobacterium infantis 35624 and changes in fecal microbiota after eight weeks of oral supplementation with encapsulated probiotic. *Gut Microbes*, *4*(3), 201–211. https://doi.org/10.4161/gmic.24196

Chaudhry, T. S., Senapati, S. G., Gadam, S., Mannam, H. P. S. S., Voruganti, H. V., Abbasi, Z., Abhinav, T., Challa, A. B., Pallipamu, N., Bheemisetty, N., & Arunachalam, S. P. (2023a). The impact of microbiota on the gut–brain axis: Examining the complex interplay and implications. *Journal of Clinical Medicine*, *12*(16), 5231. https://doi.org/10.3390/jcm12165231

Chaudhry, T. S., Senapati, S. G., Gadam, S., Mannam, H. P. S. S., Voruganti, H. V., Abbasi, Z., Abhinav, T., Challa, A. B., Pallipamu, N., Bheemisetty, N., & Arunachalam, S. P. (2023b). The impact of microbiota on the gut-brain axis: Examining the complex interplay and implications. *Journal of Clinical Medicine*, *12*(16), 5231. https://doi.org/10.3390/jcm12165231

Cohen, L. J., Cho, J. H., Gevers, D., & Chu, H. (2019). Genetic factors and the intestinal microbiome guide development of microbe-based therapies for inflammatory bowel diseases. In *Gastroenterology* (Vol. 156, Issue 8, pp. 2174–2189). W.B. Saunders. https://doi.org/10.1053/j.gastro.2019.03.017

Cryan, J. F., & Dinan, T. G. (2012). Mind-altering microorganisms: The impact of the gut microbiota on brain and behaviour. *Nature Reviews Neuroscience*, *13*(10), 701–712. https://doi.org/10.1038/nrn3346

Cryan, J. F., O'riordan, K. J., Cowan, C. S. M., Sandhu, K. V., Bastiaanssen, T. F. S., Boehme, M., Codagnone, M. G., Cussotto, S., Fulling, C., Golubeva, A. V., Guzzetta, K. E., Jaggar, M., Long-Smith, C. M., Lyte, J. M., Martin, J. A., Molinero-Perez, A., Moloney, G., Morelli, E., Morillas, E., … & Dinan, T. G. (2019a). The microbiota-gut-brain axis. *Physiological Reviews*, *99*(4), 1877–2013. https://doi.org/10.1152/physrev.00018.2018

Cryan, J. F., O'riordan, K. J., Cowan, C. S. M., Sandhu, K. V., Bastiaanssen, T. F. S., Boehme, M., Codagnone, M. G., Cussotto, S., Fulling, C., Golubeva, A. V., Guzzetta, K. E., Jaggar, M., Long-Smith, C. M., Lyte, J. M., Martin, J. A., Molinero-Perez, A., Moloney, G., Morelli, E., Morillas, E., … Dinan, T. G. (2019b). The microbiota-gut-brain axis. *Physiological Reviews*, *99*(4), 1877–2013. https://doi.org/10.1152/physrev.00018.2018

De Filippo, C., Cavalieri, D., Di Paola, M., Ramazzotti, M., Poullet, J. B., Massart, S., Collini, S., Pieraccini, G., & Lionetti, P. (2010). Impact of diet in shaping gut microbiota revealed by a comparative study in children from Europe and rural Africa. *Proceedings of the National Academy of Sciences of the United States of America*, *107*(33), 14691–14696. https://doi.org/10.1073/pnas.1005963107

Demaude, J., Salvador-Cartier, C., Fioramonti, J., Ferrier, L., & Bueno, L. (2006). Phenotypic changes in colonocytes following acute stress or activation of mast cells in mice: Implications for delayed epithelial barrier dysfunction. *Gut*, *55*(5), 655–661. https://doi.org/10.1136/gut.2005.078675

Desai, M. S., Seekatz, A. M., Koropatkin, N. M., Kamada, N., Hickey, C. A., Wolter, M., Pudlo, N. A., Kitamoto, S., Terrapon, N., Muller, A., Young, V. B., Henrissat, B.,

Wilmes, P., Stappenbeck, T. S., Núñez, G., & Martens, E. C. (2016). A dietary fiber-deprived gut microbiota degrades the colonic mucus barrier and enhances pathogen susceptibility. *Cell*, *167*(5), 1339–1353.e21. https://doi.org/10.1016/j.cell.2016.10.043

Diener, U., Patti, M. G., Molena, D., Fisichella, P. M., & Way, L. W. (2001). Esophageal dysmotility and gastroesophageal reflux disease. *Journal of Gastrointestinal Surgery: Official Journal of the Society for Surgery of the Alimentary Tract*, *5*(3), 260–265. https://doi.org/10.1016/s1091-255x(01)80046-9

Dukowicz, A. C., Lacy, B. E., & Levine, G. M. (2007). Small intestinal bacterial overgrowth: A comprehensive review. *Gastroenterology & Hepatology*, *3*(2), 112–122.

El-Hakim, Y., Bake, S., Mani, K. K., & Sohrabji, F. (2022). Impact of intestinal disorders on central and peripheral nervous system diseases. *Neurobiology of Disease*, *165*, 105627. https://doi.org/10.1016/j.nbd.2022.105627

Eslick, G. D., Kalantar, J. S., & Talley, N. J. (2009). Rectal bleeding: Epidemiology, associated risk factors, and health care seeking behaviour: A population-based study. *Colorectal Disease: The Official Journal of the Association of Coloproctology of Great Britain and Ireland*, *11*(9), 921–926. https://doi.org/10.1111/j.1463-1318.2008.01721.x

Evrensel, A., & Ceylan, M. E. (2015). The gut-brain axis: The missing link in depression. *Clinical Psychopharmacology and Neuroscience*, *13*(3), 239–244. https://doi.org/10.9758/cpn.2015.13.3.239

Ferguson, C. M. (1990). *An Overview of the Gastrointestinal System.* Boston: Butterworths.

Flint, H. J., Duncan, S. H., Scott, K. P., & Louis, P. (2014). Links between diet, gut microbiota composition and gut metabolism. *Proceedings of the Nutrition Society*, *760*, 13–22. https://doi.org/10.1017/S0029665114001463

Foster, J. A., Baker, G. B., & Dursun, S. M. (2021). The relationship between the gut microbiome-immune system-brain axis and major depressive disorder. *Frontiers in Neurology*, *12*(September), 1–9. https://doi.org/10.3389/fneur.2021.721126

Foster, J. A., & McVey Neufeld, K. A. (2013). Gut-brain axis: How the microbiome influences anxiety and depression. *Trends in Neurosciences*, *36*(5), 305–312. https://doi.org/10.1016/j.tins.2013.01.005

Hall, T. C., Garcea, G., Webb, M. A., Al-Leswas, D., Metcalfe, M. S., & Dennison, A. R. (2014). The socio-economic impact of chronic pancreatitis: A systematic review. *Journal of Evaluation in Clinical Practice*, *20*(3), 203–207. https://doi.org/10.1111/jep.12117

Heidelbaugh, J. J., & Bruderly, M. (2006). Cirrhosis and chronic liver failure: Part I. Diagnosis and evaluation. *American Family Physician*, *74*(5), 756–762.

Hill, C., Guarner, F., Reid, G., Gibson, G. R., Merenstein, D. J., Pot, B., Morelli, L., Canani, R. B., Flint, H. J., Salminen, S., Calder, P. C., & Sanders, M. E. (2014). Expert consensus document: The international scientific association for probiotics and prebiotics consensus statement on the scope and appropriate use of the term probiotic. *Nature Reviews Gastroenterology and Hepatology*, *11*(8), 506–514. https://doi.org/10.1038/nrgastro.2014.66

Hornbuckle, W. E., Simpson, K. W., & Tennant, B. C. (2008). Gastrointestinal Function. In J. Jerry Kaneko, John W. Harvey and Michael L. Bruss (eds.) *Clinical Biochemistry of Domestic Animals* (pp. 413–457). Elsevier. https://doi.org/10.1016/B978-0-12-370491-7.00014-3

Houser, M. C., & Tansey, M. G. (2017). The gut-brain axis: Is intestinal inflammation a silent driver of Parkinson's disease pathogenesis? *Npj Parkinson's Disease*, *3*(1), 1–9. https://doi.org/10.1038/s41531-016-0002-0

Jandhyala, S. M., Talukdar, R., Subramanyam, C., Vuyyuru, H., Sasikala, M., & Nageshwar Reddy, D. (2015). Role of the normal gut microbiota. *World Journal of Gastroenterology*, *21*(29), 8787–8803. https://doi.org/10.3748/wjg.v21.i29.8787

Ji, J., Jin, W., Liu, S.-J., Jiao, Z., & Li, X. (2023). Probiotics, prebiotics, and postbiotics in health and disease. *MedComm*, *4*(6), e420. https://doi.org/10.1002/mco2.420

Khor, B., Gardet, A., & Xavier, R. J. (2011). Genetics and pathogenesis of inflammatory bowel disease. *Nature*, *474*(7351), 307–317. https://doi.org/10.1038/nature10209

Kinsinger, S. W. (2017). Cognitive-behavioral therapy for patients with irritable bowel syndrome: Current insights. *Psychology Research and Behavior Management*, *10*, 231–237. https://doi.org/10.2147/PRBM.S120817

Lavelle, A., & Sokol, H. (2020). Gut microbiota-derived metabolites as key actors in inflammatory bowel disease. In *Nature Reviews Gastroenterology and Hepatology* (Vol. 17, Issue 4, pp. 223–237). Nature Research. https://doi.org/10.1038/s41575-019-0258-z

Li, X., Chen, L. M., Kumar, G., Zhang, S. J., Zhong, Q. H., Zhang, H. Y., Gui, G., Wu, L. Le, Fan, H. Z., & Sheng, J. W. (2022a). Therapeutic interventions of gut-brain axis as novel strategies for treatment of alcohol use disorder associated cognitive and mood dysfunction. *Frontiers in Neuroscience*, *16*(February), 1–13. https://doi.org/10.3389/fnins.2022.820106

Li, X., Chen, L.-M., Kumar, G., Zhang, S.-J., Zhong, Q., Zhang, H.-Y., Gui, G., Wu, L.-L., Fan, H.-Z., & Sheng, J.-W. (2022b). Therapeutic interventions of gut-brain axis as novel strategies for treatment of alcohol use disorder associated cognitive and mood dysfunction. *Frontiers in Neuroscience*, *16*, 1–13. https://doi.org/10.3389/fnins.2022.820106

Littman, D. R., & Pamer, E. G. (2011). Role of the commensal microbiota in normal and pathogenic host immune responses. *Cell Host & Microbe*, *10*(4), 311–323. https://doi.org/10.1016/j.chom.2011.10.004

Liu, L., Huh, J. R., & Shah, K. (2022a). Microbiota and the gut-brain-axis: Implications for new therapeutic design in the CNS. *EBioMedicine*, *77*, 103908. https://doi.org/10.1016/j.ebiom.2022.103908

Liu, L., Huh, J. R., & Shah, K. (2022b). Microbiota and the gut-brain-axis: Implications for new therapeutic design in the CNS. *EBioMedicine*, *77*, 103908. https://doi.org/10.1016/j.ebiom.2022.103908

Lynch, S. V., & Pedersen, O. (2016). The human intestinal microbiome in health and disease. *New England Journal of Medicine*, *375*(24), 2369–2379. https://doi.org/10.1056/nejmra1600266

Maaser, C., Sturm, A., Vavricka, S. R., Kucharzik, T., Fiorino, G., Annese, V., Calabrese, E., Baumgart, D. C., Bettenworth, D., Borralho Nunes, P., Burisch, J., Castiglione, F., Eliakim, R., Ellul, P., González-Lama, Y., Gordon, H., Halligan, S., Katsanos, K., Kopylov, U., … & European Crohn's and Colitis Organisation [ECCO] and the European Society of Gastrointestinal and Abdominal Radiology [ESGAR]. (2019). ECCO-ESGAR Guideline for Diagnostic Assessment in IBD Part 1: Initial diagnosis, monitoring of known IBD, detection of complications. *Journal of Crohn's & Colitis*, *13*(2), 144–164. https://doi.org/10.1093/ecco-jcc/jjy113

Marlow, G., Ellett, S., Ferguson, I. R., Zhu, S., Karunasinghe, N., Jesuthasan, A. C., Yeo Han, D., Fraser, A. G., & Ferguson, L. R. (2013). *Transcriptomics to Study the Effect of a Mediterranean-Inspired Diet on Inflammation in Crohn's Disease Patients*. Biomed Central. www.humgenomics.com/content/7/1/24

Martin, C. R., Osadchiy, V., Kalani, A., & Mayer, E. A. (2018). The brain-gut-microbiome axis. *CMGH*, *6*(2), 133–148. https://doi.org/10.1016/j.jcmgh.2018.04.003

Martínez, I., Lattimer, J. M., Hubach, K. L., Case, J. A., Yang, J., Weber, C. G., Louk, J. A., Rose, D. J., Kyureghian, G., Peterson, D. A., Haub, M. D., & Walter, J. (2013). Gut microbiome composition is linked to whole grain-induced immunological improvements. *ISME Journal*, *7*(2), 269–280. https://doi.org/10.1038/ismej.2012.104

Mayer, E. A. (2011). Gut feelings: The emerging biology of gut–brain communication. *Nature Reviews Neuroscience*, *12*(8), 453–466. https://doi.org/10.1038/nrn3071

Mayer, E. A., Knight, R., Mazmanian, S. K., Cryan, J. F., & Tillisch, K. (2014). Gut microbes and the brain: Paradigm shift in neuroscience. *Journal of Neuroscience*, *34*(46), 15490–15496. https://doi.org/10.1523/JNEUROSCI.3299-14.2014

Montagnani, M., Bottalico, L., Potenza, M. A., Charitos, I. A., Topi, S., Colella, M., & Santacroce, L. (2023). The crosstalk between gut microbiota and nervous system: A bidirectional interaction between microorganisms and metabolome. *International Journal of Molecular Sciences*, *24*(12), 10322. https://doi.org/10.3390/ijms241210322

Morais, L. H., Schreiber, H. L., & Mazmanian, S. K. (2021). The gut microbiota–brain axis in behaviour and brain disorders. *Nature Reviews Microbiology*, *19*(4), 241–255. https://doi.org/10.1038/s41579-020-00460-0

Mukhtar, K., Nawaz, H., & Abid, S. (2019). Functional gastrointestinal disorders and gut-brain axis: What does the future hold? *World Journal of Gastroenterology*, *25*(5), 552–566. https://doi.org/10.3748/wjg.v25.i5.552

Ou, J., Carbonero, F., Zoetendal, E. G., DeLany, J. P., Wang, M., Newton, K., Gaskins, H. R., & O'Keefe, S. J. D. (2013). Diet, microbiota, and microbial metabolites in colon cancer risk in rural Africans and African Americans. *American Journal of Clinical Nutrition*, *98*(1), 111–120. https://doi.org/10.3945/ajcn.112.056689

Parkes, M., Barrett, J. C., Prescott, N. J., Tremelling, M., Anderson, C. A., Fisher, S. A., Roberts, R. G., Nimmo, E. R., Cummings, F. R., Soars, D., Drummond, H., Lees, C. W., Khawaja, S. A., Bagnall, R., Burke, D. A., Todhunter, C. E., Ahmad, T., Onnie, C. M., McArdle, W., … & Mathew, C. G. (2007). Sequence variants in the autophagy gene IRGM and multiple other replicating loci contribute to Crohn's disease susceptibility. *Nature Genetics*, *39*(7), 830–832. https://doi.org/10.1038/ng2061

Rath, C. M., & Dorrestein, P. C. (2012). The bacterial chemical repertoire mediates metabolic exchange within gut microbiomes. *Current Opinion in Microbiology*, *15*(2), 147–154. https://doi.org/10.1016/j.mib.2011.12.009

Rithika, S. (2023, August 24). *Understanding the Gut-Brain Axis – Assay Genie*. MDPI.

Roh, E., & Choi, K. M. (2023). Hormonal gut–brain signaling for the treatment of obesity. *International Journal of Molecular Sciences*, *24*(4), 1–17. https://doi.org/10.3390/ijms24043384

Rothschild, D., Weissbrod, O., Barkan, E., Kurilshikov, A., Korem, T., Zeevi, D., Costea, P. I., Godneva, A., Kalka, I. N., Bar, N., Shilo, S., Lador, D., Vila, A. V., Zmora, N., Pevsner-Fischer, M., Israeli, D., Kosower, N., Malka, G., Wolf, B. C., … & Segal, E. (2018). Environment dominates over host genetics in shaping human gut microbiota. *Nature*, *555*(7695), 210–215. https://doi.org/10.1038/nature25973

Russo, R., Cristiano, C., Avagliano, C., De Caro, C., La Rana, G., Raso, G. M., Canani, R. B., Meli, R., & Calignano, A. (2017). Gut-brain axis: Role of lipids in the regulation of inflammation, pain and CNS diseases. *Current Medicinal Chemistry*, *25*(32), 3930–3952. https://doi.org/10.2174/0929867324666170216113756

Rutsch, A., Kantsjö, J. B., & Ronchi, F. (2020a). The gut-brain axis: How microbiota and host inflammasome influence brain physiology and pathology. *Frontiers in Immunology*, *11*, 1–24. https://doi.org/10.3389/fimmu.2020.604179

Rutsch, A., Kantsjö, J. B., & Ronchi, F. (2020b). The gut-brain axis: How microbiota and host inflammasome influence brain physiology and pathology. *Frontiers in Immunology*, *11*(December), 1–24. https://doi.org/10.3389/fimmu.2020.604179

Salvo-Romero, E., Stokes, P., & Gareau, M. G. (2020). Microbiota-immune interactions: From gut to brain. *LymphoSign Journal*, *7*(1), 1–23. https://doi.org/10.14785/lymphosign-2019-0018

Singh, R. K., Chang, H. W., Yan, D., Lee, K. M., Ucmak, D., Wong, K., Abrouk, M., Farahnik, B., Nakamura, M., Zhu, T. H., Bhutani, T., & Liao, W. (2017). Influence of diet on the gut microbiome and implications for human health. *Journal of Translational Medicine, 15*(1). https://doi.org/10.1186/s12967-017-1175-y

Socała, K., Doboszewska, U., Szopa, A., Serefko, A., Włodarczyk, M., Zielińska, A., Poleszak, E., Fichna, J., & Wlaź, P. (2021). The role of microbiota-gut-brain axis in neuropsychiatric and neurological disorders. *Pharmacological Research*, *172*, 105840. https://doi.org/10.1016/j.phrs.2021.105840

Song, M., Chan, A. T., & Sun, J. (2020). Influence of the gut microbiome, diet, and environment on risk of colorectal cancer. *Gastroenterology*, *158*(2), 322–340. https://doi.org/10.1053/j.gastro.2019.06.048

Suganya, K., & Koo, B.-S. (2020a). Gut–brain axis: Role of gut microbiota on neurological disorders and how probiotics/prebiotics beneficially modulate microbial and immune pathways to improve brain functions. *International Journal of Molecular Sciences*, *21*(20), 7551. https://doi.org/10.3390/ijms21207551

Suganya, K., & Koo, B.-S. (2020b). Gut–brain axis: Role of gut microbiota on neurological disorders and how probiotics/prebiotics beneficially modulate microbial and immune pathways to improve brain functions. *International Journal of Molecular Sciences*, *21*(20), 7551. https://doi.org/10.3390/ijms21207551

Tiwari, P., Dwivedi, R., Bansal, M., Tripathi, M., & Dada, R. (2023). Role of gut microbiota in neurological disorders and its therapeutic significance. *Journal of Clinical Medicine*, *12*(4). https://doi.org/10.3390/jcm12041650

Tsai, T.-J., Chan, H.-H., Lai, K.-H., Shih, C.-A., Kao, S.-S., Sun, W.-C., Wang, E.-M., Tsai, W.-L., Lin, K.-H., Yu, H.-C., Chen, W.-C., Wang, H.-M., Tsay, F.-W., Lin, H.-S., Cheng, J.-S., & Hsu, P.-I. (2018). Gallbladder function predicts subsequent biliary complications in patients with common bile duct stones after endoscopic treatment? *BMC Gastroenterology*, *18*(1), 32. https://doi.org/10.1186/s12876-018-0762-6

Turner, J. R. (ed.) (2009). Intestinal mucosal barrier function in health and disease. In *Nature Reviews Immunology* (Vol. 9, Issue 11, pp. 799–809). Nature. https://doi.org/10.1038/nri2653

U.S. Department of Health and Human Services. (2022, September). *Symptoms & Causes of Peptic Ulcers (Stomach or Duodenal Ulcers).* www.Niddk.Nih.Gov/Health-Information/Digestive-Diseases/Peptic-Ulcers-Stomach-Ulcers/Symptoms-Causes

Valdes, A. M., Walter, J., Segal, E., & Spector, T. D. (2018). Role of the gut microbiota in nutrition and health. *BMJ (Online)*, *361*, 36–44. https://doi.org/10.1136/bmj.k2179

Whitfield, K. L., & Schulman, R. J. (2009). Treatment options for functional gastrointestinal disorders: From empiric to complementary approaches. *Pediatric Annals*, *38*(5), 288–294. https://doi.org/10.9999/00904481-20090420-02

Wu, G. D., Chen, J., Hoffmann, C., Bittinger, K., Chen, Y. Y., Keilbaugh, S. A., Bewtra, M., Knights, D., Walters, W. A., Knight, R., Sinha, R., Gilroy, E., Gupta, K., Baldassano, R., Nessel, L., Li, H., Bushman, F. D., & Lewis, J. D. (2011). Linking long-term dietary patterns with gut microbial enterotypes. *Science*, *334*(6052), 105–108. https://doi.org/10.1126/science.1208344

Zanta, N. C., Assad, N., & Suchecki, D. (2023). Neurobiological mechanisms involved in maternal deprivation-induced behaviours relevant to psychiatric disorders. *Frontiers in Molecular Neuroscience*, *16*(4), 1–8. https://doi.org/10.3389/fnmol.2023.1099284

Zheng, P., Zeng, B., Zhou, C., Liu, M., Fang, Z., Xu, X., Zeng, L., Chen, J., Fan, S., Du, X., Zhang, X., Yang, D., Yang, Y., Meng, H., Li, W., Melgiri, N. D., Licinio, J., Wei, H., & Xie, P. (2016). Gut microbiome remodeling induces depressive-like behaviors through a pathway mediated by the host's metabolism. *Molecular Psychiatry*, *21*(6), 786–796. https://doi.org/10.1038/mp.2016.44

5 Interplay Between SCFAs, ROS, SODs, and Gastrointestinal Inflammations

Aniruddha Sen[1], Sanjesh Kumar[2], Pranshul Sethi[3], Sumel Ashique[4], Radheshyam Pal[4], Neeraj Mishra[5], and Anoop Kumar[6]*

[1]Department of Biochemistry, Shri Gorakshnath Medical College Hospital and Research Center, Gorakhpur, Uttar Pradesh, India

[2]Department of Pharmacy, Rakshpal Bahadur College of Pharmacy, Bareilly, Uttar Pradesh, India

[3]Department of Pharmacology, College of Pharmacy, Shri Venkateshwara University, Uttar Pradesh, India

[4]Department of Pharmaceutical Technology, Bharat Technology, Uluberia, West Bengal, India

[5]Amity Institute of Pharmacy, Amity University Gwalior, Madhya Pradesh, Maharajpura, Gwalior, India

[6]Department of Pharmacology, Delhi Pharmaceutical Sciences and Research University (DPSRU), Delhi, India

5.1 INTRODUCTION

The gastrointestinal tract (GIT) is a complex ecosystem where a delicate balance between the host and its microbiota dictates overall health and disease states. This equilibrium must be maintained via the interaction of short-chain fatty acids (SCFAs), reactive oxygen species (ROS), and superoxide dismutases (SODs) in this environment, especially when gastrointestinal inflammation is present. Fatty acids (SCFAs), mainly butyrate, propionate, and acetate, are fermentation byproducts formed by the gut microbiota from dietary fiber. These metabolites are essential for gut health because they affect several physiological functions, such as immune response modulation, intestinal barrier maintenance, and metabolic regulation [1,2]. ROS are necessary for cellular communication and host defense, but when they are created in excess, they can cause oxidative stress and damage to cellular components, which can lead to pathological inflammation [3]. Because SODs catalyze the conversion of superoxide radicals into O_2 andH_2O_2, they act as an essential antioxidant defense mechanism against oxidative stress, reducing the damage caused by ROS [4]. An important part of the pathophysiology and development of inflammatory bowel disorders (IBD), including Crohn's disease and ulcerative colitis (UC), is dysregulated immune responses to intestinal microbiota. This balance between these molecules is closely related to these illnesses [5]. In the realm of the emerging interventions, a combination of probiotics and prebiotics called synbiotics has proven to be effective in modulating gut microbiota, enhancing production of SCFAs, reducing ROS levels and increasing antioxidant activity [6]. Additionally, novel therapeutic approaches that concentrate on the selective delivery of SODs and the regulation of SCFA signaling pathways present opportunities for the efficient treatment of gastrointestinal tract inflammation [7]. In order to fully appreciate the complicated processes driving inflammation of the gastrointestinal tract (GIT) and to develop successful treatment strategies, it is essential to know the delicate interplay between SCFAs, ROS, and superoxide dismutases (SODs). The gut microbiota ferments dietary fiber to yield important metabolites called SCFAs, namely acetate, propionate, and butyrate. These bioactive chemicals regulate immune responses, maintain intestinal barrier integrity, and change gene expression in diverse cell types in the GIT [8]. Their significance extends beyond metabolic regulation to encompass pivotal roles in immune homeostasis and inflammatory signaling pathways. ROS, on the other hand, are natural byproducts of cellular metabolism that play crucial roles in cellular signaling and host defense mechanisms. ROS include superoxide radicals, hydrogen peroxide, and hydroxyl radicals, among others. However, excessive ROS production, often exacerbated by environmental stressors and inflammatory stimuli, can lead to oxidative stress, causing damage to cellular macromolecules and contributing to the pathogenesis of inflammatory GIT disorders [9]. Cells need antioxidant defense systems, such as SODs, to counteract oxidative damage when their ROS

DOI: 10.1201/9781003493143-5

levels rise. SODs catalyze the conversion of superoxide radicals into less dangerous molecules. The cytoplasmic (SOD1), mitochondrial (SOD2), and extracellular (SOD3) versions of SODs remain essential for preserving cellular homeostasis and redox equilibrium. The development of GIT inflammation can be accelerated by SOD dysregulation or dysfunction, which can also worsen oxidative stress [10]. Therefore, understanding the complex interactions among SCFAs, ROS, and SODs offers important new understandings into the pathophysiology of GIT inflammation and lays the groundwork for the expansion of focused therapeutic interventions meant to improve inflammatory GI disorders by restoring gut homeostasis. Decreasing the production of the ROS and variation of the SCFA production with increased SOD activity can be a novel therapeutic intervention in mitigating the IBD and related GI inflammatory disorders, ultimately improving quality of life of patients [11].

5.2 SOURCES OF SCFAS, ROS, SODS

SCFAs are primarily derived from the microbial fermentation of dietary fiber in the gut, predominantly by beneficial bacteria such as *Bacteroidetes* and *Firmicutes*. These SCFAs, including acetate, propionate, and butyrate, serve as important energy sources for colonic epithelial cells and play crucial roles in maintaining gut health, modulating immune responses, and regulating host metabolism. Conversely, ROS, such as superoxide radicals (O2•-) and hydrogen peroxide (H_2O_2), are generated through various cellular processes, including mitochondrial respiration, metabolism of xenobiotics, and inflammatory responses [8]. As a part of cell signaling and host defense mechanism ROS has some inherent functions, but excessive accumulation of the ROS species results in damage to DNA, OS, and inflammation, which are major players in the pathogenesis of the GI disorders. To counteract the deleterious effects of ROS, cells possess antioxidant defense mechanisms, among which SODs are key enzymes. SODs, including copper/zinc SOD (SOD1), manganese SOD (SOD2), and extracellular SOD (SOD3), are distributed throughout the gastrointestinal tract and play critical roles in maintaining redox homeostasis, protecting cells from oxidative damage, and preserving gastrointestinal health [12]. However, dysregulation of SCFA metabolism, ROS production, or SOD activity can disrupt the delicate balance of gut physiology, contributing to the onset and progression of gastrointestinal diseases. Hence, it is imperative to lay out the intricate correlation between SCFAs, ROS, and SODs in order to comprehend their functions in overall health and ailments, as well as to formulate specific therapeutic approaches for GI diseases [13]. SCFAs, encompassing acetate, propionate, and butyrate, represent pivotal metabolites derived predominantly from the fermentation of dietary fibers within the anaerobic milieu of the colon, orchestrated by an array of symbiotic gut bacteria. The process of SCFA production, intricately intertwined with the composition of one's diet, hinges upon the fermentable substrates provided by dietary fibers, including complex carbohydrates resistant to host digestion. The variety and richness of the gut microbiota, comprising myriad bacterial species predominantly belonging to the phyla *Bacteroidetes*, *Firmicutes*, and *Actinobacteria,* play an instrumental role in dictating the efficiency and magnitude of SCFA generation [14]. Moreover, the multifaceted interaction among the microbiota and the GI system of the host, which is governed by parameters such mucosal immunity, epithelial integrity, and luminal pH, has a significant impact on the generation of SCFAs. Perturbations in any of these factors, whether due to dietary shifts, antibiotic exposure, or underlying gastrointestinal disorders, can significantly alter the profile and concentration of SCFAs, potentially impacting host health and disease susceptibility. Hence, elucidating the multifaceted interplay between diet, gut microbiota composition, and gastrointestinal health is indispensable for comprehending the intricate mechanisms governing SCFA metabolism and its implications for human physiology and pathology [8]. Foods abundant in complex carbohydrates, such as an array of fruits, vegetables, and whole grains, represent quintessential reservoirs of dietary fibers crucial for fostering the synthesis of SCFAs within the intricate ecosystem of the gut. These dietary staples harbor a diverse array of polysaccharides, including cellulose, hemicellulose, pectin, and resistant starch, which traverse the upper gastrointestinal tract largely undigested, reaching the colon intact. When these indigestible fibers get to the gut, they are utilized as diet by the anaerobic bacteria that thrive there, fueling the enzymatic machinery responsible for the construction of SCFAs, particularly acetate, propionate, and butyrate. Fruits, encompassing an assortment of berries, apples, citrus fruits, and bananas, boast soluble fibers like pectin, which undergo microbial fermentation to yield SCFAs, thereby augmenting the pool of these beneficial metabolites in the gut lumen. Similarly, vegetables, ranging from leafy greens like spinach and kale to cruciferous varieties like broccoli and Brussels sprouts, harbor an array of insoluble fibers such as cellulose and lignin, which assist as substrates for bacterial fermentation, culminating in the creation of SCFAs [15]. Additionally, whole grains, including oats, barley, quinoa, and brown rice, are replete with complex carbohydrates like resistant starch and arabinoxylans, which traverse the digestive tract intact, stimulating SCFA production upon reaching the colon. By integrating these fiber-rich foods into one's dietary regimen, individuals can nurture a favorable gut environment conducive to robust SCFA synthesis, thereby conferring myriad health benefits ranging from enhanced gut barrier function and immune modulation to mitigating the risk of chronic inflammatory disorders and metabolic maladies. Thus, recognizing the pivotal role of complex carbohydrates in fostering SCFA production underscores the importance of embracing a diverse and fiber-rich diet to bolster gastrointestinal health and overall well-being [16]. ROS, comprising O_2-derived molecules such as superoxide radicals (O2•-) and hydrogen peroxide (H_2O_2), serve as double-edged swords within cellular physiology, being generated as inevitable byproducts of normal metabolic processes while also participating in crucial signaling cascades. Principally, ROS arise as natural byproducts of mitochondrial respiration, where electrons leaked from the electron transport chain interact with molecular oxygen to form superoxide radicals, initiating a cascade of subsequent ROS generation. This phenomenon, occurring predominantly within the inner mitochondrial membrane during oxidative phosphorylation, underscores the inherent vulnerability of mitochondria to oxidative damage [17]. Additionally, enzymatic reactions catalyzed by various oxidases and oxygenases across diverse cellular compartments, including peroxisomes, endoplasmic reticulum, and cytoplasmic enzymes like NADPH oxidases (NOX), further contribute to ROS production. Notably, the immune response and inflammatory processes represent significant sources of ROS generation, with phagocytic cells, such as neutrophils and macrophages, employing ROS as antimicrobial effectors in the process of pathogen clearance. Activation of phagocytic oxidase (phox) complexes within these immune cells leads to the orchestrated production of superoxide radicals, facilitating the destruction of engulfed pathogens [18]. Furthermore, inflammatory cytokines and growth factors can stimulate ROS production via activation of NOX enzymes, perpetuating oxidative stress within the cellular milieu. ROS play essential roles in the body's defense mechanisms, acting as signaling molecules in pathways that are sensitive to redox reactions. However, when ROS production becomes uncontrolled, it can lead to oxidative burden, which can cause damage to cells, mutations in DNA, and inflammatory responses. These effects have been linked to several diseases, such as cancer, neurodegenerative disorders, and cardiovascular conditions. Hence, comprehending the multifaceted sources and regulatory mechanisms governing ROS generation is paramount for unraveling their intricate roles in cellular homeostasis and pathology, thereby paving the way for the development of targeted therapeutic interventions aimed at mitigating oxidative stress-related maladies [9]. Indeed, while ROS serve as integral mediators of cell signaling and homeostatic processes, their dysregulated production can precipitate oxidative stress, a condition characterized by an imbalance between ROS generation and antioxidant defense mechanisms. Under physiological conditions, ROS participate in diverse signaling cascades, regulating cellular proliferation, differentiation, and apoptosis, and facilitating immune responses against invading pathogens. However, in instances of heightened ROS production or compromised antioxidant capacity, oxidative stress ensues, leading to indiscriminate oxidation of biomolecules, including lipids, proteins, and nucleic acids. Lipid peroxidation, initiated by ROS-mediated attack on membrane lipids, compromises membrane integrity and fluidity, culminating in cellular dysfunction and organelle damage. Moreover, ROS-induced protein modifications, such as oxidation of amino acid residues, disrupt enzymatic activity, impair protein folding, and promote aggregation, compromising cellular function and integrity. Additionally, oxidative DNA damage, encompassing base modifications, strand breaks, and DNA-protein cross-links, can undermine genomic stability, triggering mutations and genomic instability, hallmarks of cancer initiation and progression [19]. Furthermore, OS serves as a mutual marker in the pathogenesis of various diseases, including neurodegenerative disorders like Alzheimer's and Parkinson's diseases, cardiovascular diseases, diabetes mellitus, and inflammatory conditions such as rheumatoid arthritis and IBD. Thus, while ROS-mediated signaling is indispensable for physiological processes, mitigating oxidative stress through bolstering antioxidant defense mechanisms or targeted scavenging of ROS represents a promising therapeutic avenue for combating oxidative stress-related pathologies and promoting cellular health and longevity [20]. Environmental factors such as pollution, radiation, and cigarette smoke, alongside lifestyle factors like diet and stress, intricately influence the levels of ROS within the human body, thereby modulating the delicate balance between oxidative stress and antioxidant defense mechanisms. Airborne pollutants, including particulate matter and volatile organic compounds, infiltrate biological systems, triggering ROS generation and propagating oxidative stress, contributing to respiratory ailments and other diseases. Similarly, ionizing radiation from natural and anthropogenic sources induces ROS production, precipitating cellular damage and increasing the risk of radiation-related pathologies. Lifestyle habits, such as dietary patterns rich in antioxidants or saturated fats, and psychosocial stressors, like chronic anxiety or depression, further shape ROS dynamics, either mitigating or exacerbating oxidative stress susceptibility. Recognizing the multifaceted impact of these environmental and lifestyle factors underscores the importance of adopting healthy behaviors and environmental stewardship to mitigate oxidative stress-related diseases and promote overall well-being [21]. SODs are a critical component of the body's antioxidant defense system, catalyzing the dismutation of superoxide radicals into oxygen and hydrogen peroxide, thereby mitigating the potential harmful effects of ROS [22]. SOD1 (Cu/Zn-SOD), prevalent in the cytoplasm, SOD2 (Mn-SOD), situated within the mitochondria, and SOD3 (EC-SOD), situated in the extracellular space, are the three primary varieties of SODs found in humans [23]. The expression and activity of superoxide dismutases (SODs), pivotal enzymes in antioxidant defense, are subject to intricate regulatory mechanisms orchestrated by a myriad of factors spanning genetic, dietary, and physiological realms. Genetic factors exert profound influence over SOD expression levels, as variations in promoter regions and coding sequences of SOD genes can modulate transcriptional activity and enzyme synthesis. Moreover, epigenetic modifications, including DNA methylation and histone acetylation, dynamically regulate SOD gene expression in response to environmental cues and developmental cues. Additionally, dietary intake of essential minerals such as copper, zinc, and manganese, indispensable cofactors for SODs, directly impacts enzyme activity and functionality. Inadequate dietary intake or impaired absorption of these micronutrients can compromise SOD activity, exacerbating oxidative stress susceptibility [22]. Furthermore, various physiological and pathological conditions, including inflammation, hypoxia, oxidative stress, and metabolic disorders, intricately modulate SOD expression and activity in a context-dependent manner. For instance, during acute

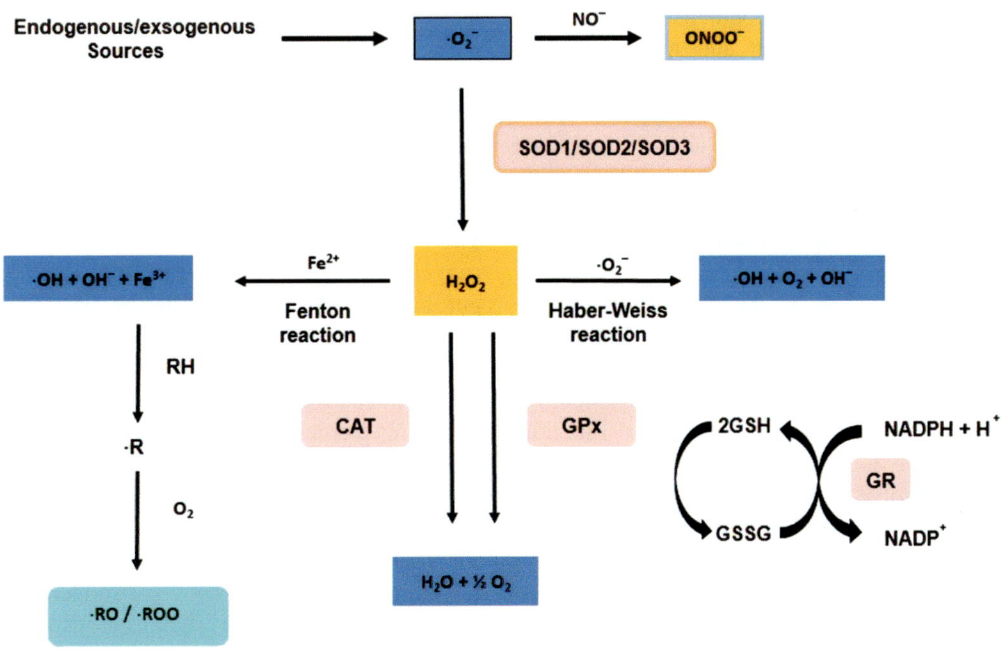

FIGURE 5.1 Describes the sources and generation of ROS.

Note: Abbreviations: CAT (catalase), GR (glutathione reductase), SOD (superoxide dismutase), RH (lipid membrane), R (alkyl radical).

inflammatory responses, pro-inflammatory cytokines like in IL-1β and TNF-alpha can upregulate SOD expression as a protective mechanism against oxidative damage. Conversely, chronic inflammatory conditions, metabolic dysregulation, and aging can compromise SOD function, exacerbating OS and predisposing individuals to a spectrum of cardiovascular diseases, neurodegenerative disease,s and cancer. Thus, elucidating the multifaceted regulatory networks governing SOD expression and activity is imperative for understanding their roles in antioxidant defense mechanisms and oxidative stress-related pathologies, thereby facilitating the expansion of selected therapeutic interventions aimed at bolstering cellular resilience and mitigating disease progression. Figure 5.1 describes the sources and generation of ROS. Abbreviations: CAT (catalase), GR (glutathione reductase), SOD (superoxide dismutase), RH (lipid membrane), R (alkyl radical) [24].

5.3 FACTORS AFFECTING GIT INFLAMMATION

Gastrointestinal (GIT) inflammation represents a complex condition influenced by a multitude of factors spanning microbial interactions, dietary components, environmental exposures, and genetic predispositions. The intricate nature of GIT inflammation is highlighted by its association with a spectrum of disorders, encompassing irritable bowel syndrome (IBS) and IBD. Central to this complexity is the pivotal interplay between host and microbial factors, which significantly influences the pathogenesis and progression of these conditions. Understanding and deciphering these multifaceted interactions are paramount for elucidating the mechanisms underlying GIT inflammation and developing targeted therapeutic strategies for effective management [25]. Microbial dysbiosis, identified by an imbalance in the conformation and utility of the gut microbiota, stands as a key driver in instigating immune activation and perpetuating chronic inflammation within the gastrointestinal tract (GIT). This dysregulation in microbial communities can significantly contribute to the onset or exacerbation of various GIT disorders. The disruption of the delicate balance within the gut microbiota can trigger aberrant immune responses, ultimately fostering a pro-inflammatory environment conducive to the development and development of GIT ailments. Recognizing the influential role of microbial dysbiosis in orchestrating immune dysregulation and chronic inflammation is crucial for advancing our understanding of GIT disorders and devising targeted therapeutic interventions aimed at restoring microbial equilibrium and alleviating associated inflammation [26]. The association between the host health and gut microbiota is intricate, with changes in microbial composition affecting not only local but also systemic inflammation and disease states. Environmental factors, encompassing dietary patterns, stress levels, and exposure to pollutants, exert profound effects on both the composition and functionality of the gut microbiota, consequently exerting a substantial influence on gastrointestinal (GIT) inflammation. The interplay between these environmental factors and the gut microbiota plays a pivotal role in shaping the inflammatory milieu within the GIT. Dietary choices can modulate microbial diversity and metabolic activity, while stress and exposure to pollutants can disrupt microbial homeostasis, leading to dysbiosis and subsequent inflammation. Recognizing the complex relationship between environmental aspects and the gut microbiota highlights the importance of adopting lifestyle modifications and implementing environmental interventions as potential strategies for managing and mitigating GIT inflammation [27]. Dietary components, with a particular focus on the consumption of dietary fibers, polyphenols, and fatty acids, have demonstrated the capacity to exert significant modulation over the microbiota in gut, consequently influencing the synthesis of anti-inflammatory compounds, such as SCFAs, and impacting gut barrier function. The ingestion of dietary fibers promotes the progression of beneficial microbial populations in the gut, facilitating the production of SCFAs known for their anti-inflammatory properties. Similarly, polyphenols occurring in various plant-based foods possess prebiotic effects, fostering a favorable gut microbiota composition and enhancing barrier function. Additionally, certain fatty acids, such as omega-3 fatty acids, exhibit anti-inflammatory characteristics and aids in maintaining gut barrier integrity. Exploring the dynamic connection amid dietary components, gut microbiota modulation, and their impact on gut barrier function underscores the potential of dietary interventions in managing gastrointestinal health and inflammation [28]. Chronic stress has been intricately associated with various alterations within the gastrointestinal (GIT) system, including changes in gut motility, heightened intestinal permeability, and disruptions in the gut microbiota composition. These stress-induced alterations collectively contribute to the onset and perpetuation of inflammation and GIT disorders. Stress can lead to dysregulation of gut motility, resulting in irregular bowel movements and gastrointestinal discomfort. Moreover, chronic stress has been shown to increase intestinal permeability, allowing for the translocation of harmful substances across the intestinal barrier, thereby triggering inflammatory responses. Moreover, adjustments in the diversity of the gut microbiota caused by stress may intensify inflammation and play a role in the progression of GI disorders. Understanding the complex interplay between chronic stress and its impact on gut physiology underscores the importance of stress management strategies in promoting gastrointestinal health and mitigating associated inflammatory conditions [29]. Genetic predispositions constitute a crucial determinant of susceptibility to gastrointestinal (GIT) inflammation, with a myriad of genes linked to immune regulation and barrier function being implicated in circumstances such as IBD. These genetic variants influence key aspects of the immune response and gut barrier integrity, thereby contributing to the expansion and evolution of GIT inflammatory disorders. Changes in genes that control the immune system, like those that make cytokines, receptors, and signaling molecules, can affect the proportion of pathways in the GIT which trigger inflammation and pathways which inhibit inflammation. In similar fashion, genetic variations that impact genes associated to barrier function, such as those regulating mucosal defense mechanisms and epithelial cell junctions, can weaken the gut barrier and put people at risk for inflammation and sickness. Recognizing the significant role of genetic predispositions in GIT inflammation underscores the importance of personalized approaches to disease management and therapeutic interventions tailored to individual genetic profiles [30]. The intricate interplay between genetic factors and environmental exposures holds considerable influence over the onset and progression of gastrointestinal (GIT) inflammation. Genetic predispositions, coupled with various environmental factors such as diet, stress, microbial exposures, and pollutants, collectively shape the inflammatory milieu within the GIT. The convergence of genetic susceptibility and environmental triggers can exacerbate immune dysregulation, compromise gut barrier function, and disrupt the stability of the gut microbiota, thereby fueling the development and expansion of GIT inflammatory disorders. Understanding the dynamic interaction between genetic predispositions and environmental exposures is crucial for elucidating the underlying mechanisms driving GIT inflammation and informing targeted strategies for disease prevention and management [25]. Helicobacter pylori infection stands as a well-documented exemplar showcasing how microbial pathogens can provoke chronic inflammation within the gastrointestinal tract (GIT), culminating in conditions such as gastritis and gastric cancer. The bacterium's adeptness at colonizing the stomach and eliciting a persistent inflammatory reaction underscores the pivotal contribution of microbial factors to GIT inflammation and its associated diseases. Helicobacter pylori's ability to incite chronic inflammation underscores the intricate interplay between microbial pathogens and host immune responses, emphasizing the consequence of understanding these connections in the realm of GIT health and disease [31]. Recent advances in foodomics and the exploration of dietary polyphenols have unveiled promising insights into the potential protective effects of these compounds against gastrointestinal (GIT) inflammation. Polyphenols, renowned for their antioxidative properties, have emerged as key players in maintaining redox equilibrium within the GIT environment. By counteracting oxidative stress, polyphenols hold the capacity to mitigate gut dysbiosis, safeguard against intestinal epithelium damage, and suppress radical inflammatory responses. These findings expose novel vistas for beneficial intrusions intended to harness the beneficial properties of dietary polyphenols to alleviate GIT inflammation and promote gastrointestinal health [32]. Understanding the factors affecting GIT inflammation is crucial for developing effective strategies for prevention, management, and treatment of connected disorders. This requires a comprehensive approach that considers the multifaceted connections among genetic, microbial, environmental, and dietary factors.

5.4 ROS; OXIDATIVE STRESS: IMPACT ON GIT INFLAMMATION AND ASSOCIATED DISEASES (ESOPHAGEAL DISEASES, IBD, GASTRIC CANCER)

ROS and OS are pivotal players in the pathogenesis and advancement of gastrointestinal (GIT) inflammation and related disorders, encompassing esophageal diseases, inflammatory bowel disease (IBD), and gastric cancer. The GIT stands out as highly vulnerable to oxidative stress owing to its continual exposure to various exogenous elements, including dietary constituents, alcohol, smoking, and non-steroidal anti-inflammatory drugs (NSAIDs), alongside endogenous factors such as bacterial presence and the metabolic processes of the mucosa [12]. Alcohol consumption serves as a well-established trigger of oxidative stress within the GIT, notably inducing damage to the gastric epithelium, thereby elevating the susceptibility to gastric inflammation and cancer. Both alcohol and its metabolite acetaldehyde are occupied in the initiation of OS, thus fostering the development of diverse GIT ailments, such as fatty liver, cirrhosis, and gastritis, ultimately heightening the likelihood of gastric cancer onset [33].

Within the GIT, the dual nature of ROS is controlled by their concentration as well as the balance across pro-oxidant and antioxidant pathways. Optimal levels of ROS are essential for cell signaling and immune responses, underscoring their pivotal role. However, an imbalance in ROS levels can lead to oxidative damage and contribute to the pathogenesis of gastrointestinal disorders [34]. Abundant ROS can degrade macromolecules, cell membranes, and DNA, initiating pro-inflammatory signaling pathways and GI problems. This heightened ROS activity often contributes to the development of conditions like ulcers and cancers within the GIT. Maintaining a balance in ROS levels is essential to prevent oxidative damage and preserve the integrity of gastrointestinal tissues, emphasizing the intricate interplay between ROS and GIT health [12]. Astaxanthin (AST), a type of fat-soluble xanthophyll carotenoid, has attracted considerable interest due to its notable antioxidant and anti-inflammatory characteristics, especially in the context of addressing GIT ailments.

AST demonstrates significant protective effects against gastric and intestinal ulcers as well as cancers by effectively neutralizing ROS and lipid peroxyl radicals. Additionally, AST improves the action of antioxidant enzymes, thereby bolstering the body's defense mechanisms against oxidative stress. Additionally, AST functions to reduce inflammation in the gastrointestinal tract (GIT) by inhibiting pro-inflammatory transcription factors and cytokines. This action helps alleviate inflammation in the GIT, promoting better gastrointestinal health overall [35]. In the realm of IBD, oxidative stress intricately intertwines with chronic and spontaneous inflammation within the gastrointestinal tract (GIT). Novel therapeutic formulations, exemplified by superoxide dismutase (SOD)-embedded metal-organic frameworks (MOFs), have emerged as promising avenues for mitigating IBD-related pathology. These innovative formulations demonstrate efficacy in suppressing ROS levels and pro-inflammatory cytokines, thereby offering relief from colitis in experimental models. Such findings underscore the therapeutic potential of targeting OS pathways as a strategic approach in the treatment of IBD, paving the way for the development of more effective therapeutic interventions in the management of this debilitating condition. The intricate balance between ROS production and antioxidant defenses is crucial in maintaining GIT health. Disruption of this balance towards. Inflammation of the GIT and related disorders are largely caused by OS. Comprehending these mechanisms presents promising avenues for preventing and treating such conditions, underscoring the significance of antioxidants such as AST and inventive therapeutic formulations like SOD-embedded MOFs in addressing gastrointestinal tract (GIT) diseases. Figure 5.2 depicts the function of probiotics in intestine [36].

5.5 MECHANISM OF SCFAS AND SODS: INHIBITION OF GIT INFLAMMATION

Gastrointestinal inflammation is a complex pathological condition characterized by the dysregulation of immune responses, oxidative stress, and tissue damage within the gastrointestinal tract (GIT). Over the past few years, there has been a lot of research dedicated to understanding how SCFAs and SODs can help reduce inflammation in the GIT. This investigation has revealed that they have the potential to be operative in healing inflammation in GIT.

5.5.1 Mechanism of Short-Chain Fatty Acids (SCFAs) in GIT Inflammation

5.5.1.1 SCFAs Production and Sources

These bioactive metabolites are produced by microorganisms found in your gut, primarily within your colon, when they ferment dietary fiber into acetate, propionate, or butyrate [37]. The generation of SCFAs occurs as a result of the anaerobic metabolism of complex carbohydrates that escape digestion in the small intestine and reach the large intestine, where they serve as substrates for microbial fermentation processes. The main origins of dietary fiber, acting as precursors for the production of SCFAs, encompass a diverse range of plant-based foods. These include fruits, vegetables, legumes, and whole grains [38]. These dietary components are rich in complex carbohydrates, including cellulose, hemicellulose, pectin, and resistant starch, which undergo microbial fermentation in the colon to yield SCFAs. Acetate, propionate, and butyrate are the most abundant SCFAs formed in the gut, with acetate typically being the most prevalent, followed by propionate and butyrate [39]. The ratios of these SCFAs fluctuate based on various factors including the makeup of one's diet, the composition of gut microbiota, and the time it takes for food to travel through the colon.

Once generated, SCFAs play a crucial role as essential sources of energy for colonocytes, cells found in the colon. Additionally, they aid in preserving the integrity and functionality of the colon's epithelial layer [40]. Additionally, SCFAs exert diverse physiological effects both locally within the gastrointestinal tract and systemically throughout the body, influencing various aspects of host metabolism, immune function, and inflammatory responses. In essence, the microbial synthesis of SCFAs from dietary fiber stands as a crucial nexus connecting the diet, gut microbiota, and host physiology. This intricate relationship holds significant ramifications for human health and the onset or progression of diseases [41]. Understanding the sources and metabolic fate of SCFAs provides valuable insights into their roles in

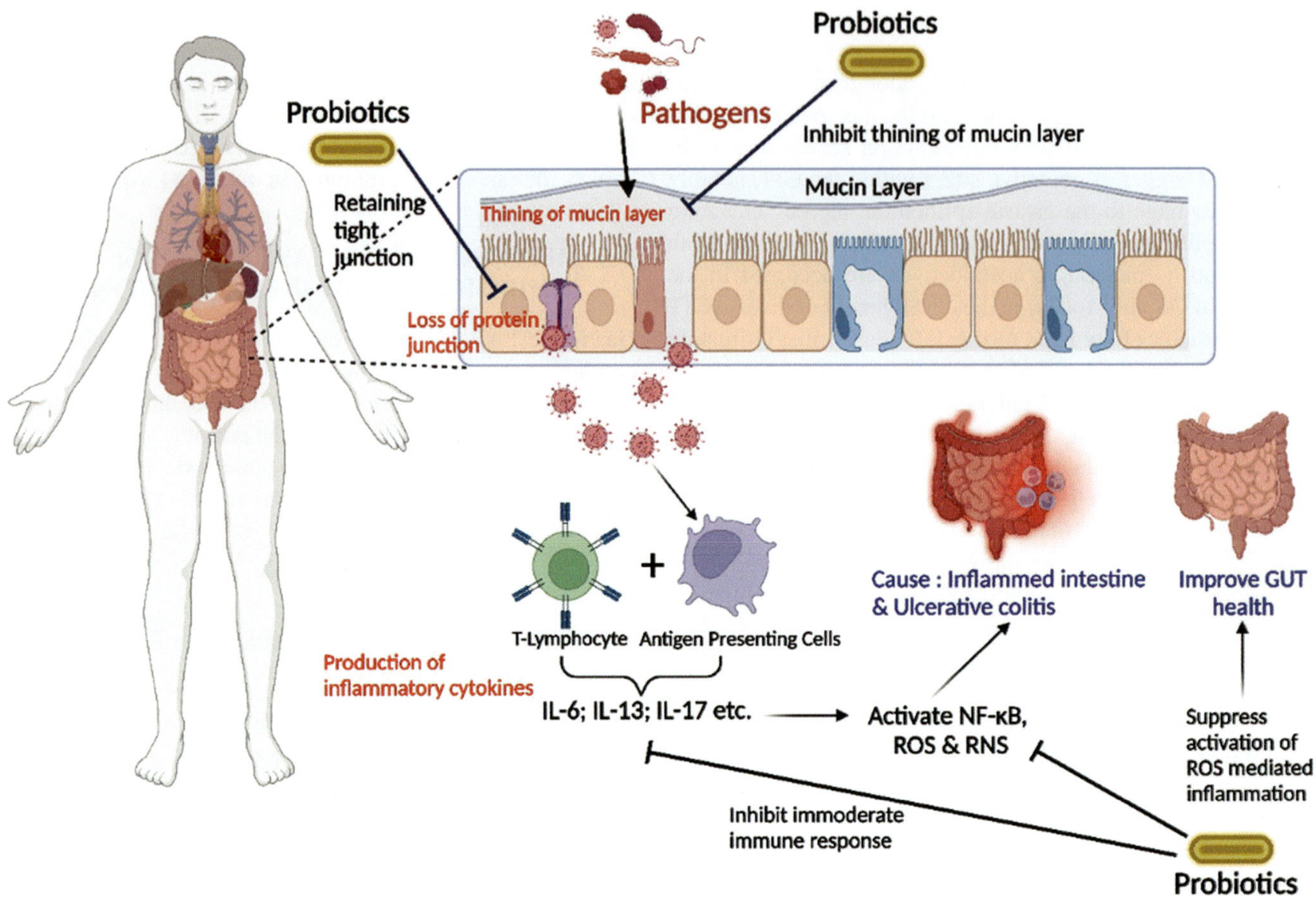

FIGURE 5.2 Explains the function of probiotics in the intestines. The lumen is implicated in the inhibition of inflammation [36].

gut homeostasis, immune regulation, and overall metabolic health.

5.5.1.2 SCFAs Receptors and Signaling

Upon production, SCFAs exert their potent anti-inflammatory effects through the stimulation of specific G protein-coupled receptors (GPCRs) expressed on immune cells and intestinal epithelial cells [42]. Among the key receptors involved in mediating SCFA signaling are GPR41, GPR43, and GPR109A, which play pivotal roles in modulating immune responses and inflammatory pathways [43]. Activation of these GPCRs by SCFAs triggers intracellular signaling cascades that regulate various aspects of immune cell function, cytokine production, and gene expression related to inflammation [44]. Importantly, SCFAs bind to GPR41 (also known as FFAR3) and GPR43 (also known as FFAR2), leading to the initiation of downstream signaling events that contribute to the modulation of immune responses and inflammatory processes [45]. Through their interaction with GPR41 and GPR43, SCFAs exert multifaceted effects on immune cells, including macrophages, dendritic cells, and T cells [46]. Activation of these receptors has been shown to promote the production of anti-inflammatory cytokines, such as IL-10, while inhibiting the secretion of pro-inflammatory mediators, including tumor necrosis factor-alpha and interleukin-6 (IL-6) [47]. Furthermore, SCFA-GPCR signaling influences the differentiation and function of regulatory T cells (Tregs), contributing to the maintenance of immune tolerance and suppression of excessive inflammation within the GIT [48]. Overall, the binding of SCFAs to specific GPCRs represents a crucial mechanism by which these bioactive metabolites exert their immunomodulatory effects, highlighting the intricate interplay between dietary components, gut microbiota-derived metabolites, and host immune responses in the regulation of inflammation and immune homeostasis within the GIT.

5.5.1.3 Immunomodulatory Effects of SCFAs

SCFAs are integral intermediaries of immune regulation within the gastrointestinal (GI) mucosa, exerting profound immunomodulatory effects on various immune cell populations. Through intricate interactions with the host immune system, SCFAs play an essential part in maintaining immune homeostasis and attenuating inflammation within the GIT. One of the key immunomodulatory mechanisms of SCFAs involves their influence on the differentiation and function of immune cells, particularly regulatory T cells (Tregs) and pro-inflammatory T helper 17 (Th17) cells. SCFAs have been shown to promote the expansion and activation of Tregs, a specialized subset of T cells known for their regulatory function in suppressing excessive immune responses and inflammation [46,47]. Conversely, SCFAs inhibit the differentiation and activation of Th17 cells, which are associated in the pathogenesis of inflammatory surroundings such as IBD [46]. Moreover, SCFAs modulate the cytokine milieu within the GI mucosa by enhancing the production of anti-inflammatory cytokines, such as IL-10 and transforming growth factor-beta (TGF-β) [48]. These cytokines play crucial roles in dampening inflammatory responses and promoting immune tolerance. By promoting the secretion of IL-10 and TGF-β, SCFAs contribute to the maintenance of immune balance and the suppression of excessive inflammation within the GIT. Furthermore, SCFAs exert direct effects on other immune cell populations, including macrophages and dendritic cells, by modulating their activation and function. SCFAs have the ability to hinder the synthesis of pro-inflammatory cytokines, while also encouraging the release of anti-inflammatory intermediaries [48]. This dual regulatory effect helps to prevent aberrant immune activation and chronic inflammation within the GIT.

5.5.1.4 SCFAs and Intestinal Barrier Function

SCFAs, such as acetate, propionate, and butyrate, play a pivotal role in upholding intestinal barrier function, a critical aspect of gut health and immune homeostasis. One of the key mechanisms through which SCFAs exert their beneficial effects on the intestinal barrier is by modulating the expression of tight junction proteins, including occludin and various claudins, which are essential for maintaining the integrity of the epithelial barrier. Research has shown that SCFAs promote the expression and assembly of tight junction proteins in intestinal epithelial cells, thereby strengthening the connections between adjacent cells and enhancing the barrier function of the intestinal mucosa [49]. This reinforcement of tight junctions helps to prevent the leakage of luminal contents, including microbial antigens and toxins, across the intestinal epithelium, thus reducing the risk of systemic inflammation and tissue damage. Furthermore, SCFAs have been found to stimulate the production of mucins, which are glycoproteins that form a protective layer over the epithelial surface of the intestine [50]. Mucins serve as a physical barrier against luminal pathogens and toxins, helping to prevent their interaction with the underlying epithelial cells. By enhancing mucin production, SCFAs contribute to the maintenance of intestinal barrier integrity and function. Overall, the ability of SCFAs to enhance tight junction protein expression and mucin production underscores their importance in preserving intestinal barrier function and protecting against inflammatory bowel diseases and other gastrointestinal disorders. Understanding the intricate interplay between SCFAs and the intestinal epithelium provides valuable insights into potential therapeutic strategies for promoting gut health and preventing intestinal inflammation.

5.5.1.5 Clinical Implications and Therapeutic Potential

SCFAs, renowned for their diverse anti-inflammatory effects, have garnered considerable attention as promising therapeutic candidates for mitigating gastrointestinal inflammatory disorders. Conditions such as IBD, IBS, and colorectal cancer, categorized by chronic inflammation and dysregulated immune responses in the GIT, stand to benefit from the immunomodulatory properties of SCFAs. Research indicates that SCFAs exert their anti-inflammatory effects through various mechanisms, including modulation of immune cell function, inhibition of pro-inflammatory cytokine production, and preservation of intestinal barrier integrity. By promoting the differentiation of regulatory T cells (Tregs) and suppressing the activity of pro-inflammatory immune cells, such as macrophages and Th17 cells, SCFAs help maintain immune tolerance and dampen excessive inflammation within the GIT [46,47]. Furthermore, SCFAs have the ability to influence the production of cytokines, which are known to exhibit extremely powerful anti-inflammatory effects. By enhancing the production of these anti-inflammatory cytokines while inhibiting the secretion of pro-inflammatory mediators like TNF-α and IL-6, SCFAs contribute to the restoration of immune homeostasis and attenuation of GIT inflammation [48]. Strategies aimed at modulating SCFA levels in the GIT represent promising avenues for therapeutic intervention. These approaches include dietary modifications to increase fiber intake, supplementation with probiotics that produce SCFAs, and the development of pharmacological agents targeting SCFA metabolism or receptor signaling pathways [51]. By harnessing the immunomodulatory properties of SCFAs, clinicians may be able to offer effective treatments for GIT inflammatory disorders, improving the value of life for affected personalities and reducing the burden of these chronic conditions.

5.6 MECHANISM OF SUPEROXIDE DISMUTASES (SODS) IN GIT INFLAMMATION

5.6.1 SODs and ROS Neutralization

Superoxide dismutases (SODs) stand as pivotal enzymatic antioxidants crucial for combating oxidative stress and mitigating gastrointestinal inflammation. These enzymes catalyze the dismutation of superoxide radicals (O_2−) into less reactive species, namely oxygen (O_2) and hydrogen peroxide (H_2O_2), thereby quenching the harmful effects of ROS implicated in GIT inflammation [52]. Comprising three distinct isoforms, SODs operate synergistically to maintain redox homeostasis within the gastrointestinal tract (GIT). Cytosolic Cu/Zn-SOD (SOD1), mitochondrial Mn-SOD (SOD2), and extracellular Cu/Zn-SOD (SOD3) are strategically distributed across various cellular compartments to effectively neutralize superoxide radicals and prevent oxidative damage [52]. SOD1, localized in the cytosol, primarily targets superoxide radicals generated within the cellular cytoplasm. Meanwhile, SOD2, residing within the mitochondria, serves to protect mitochondrial DNA and proteins from oxidative injury, crucial for preserving cellular energy metabolism and function. Lastly, SOD3, found in the extracellular matrix and interstitial spaces, acts as the first line of defense against superoxide radicals diffusing from surrounding tissues into the extracellular milieu [52]. By efficiently scavenging superoxide radicals, the aberrant inhibition of SOD activity or expression has been linked to several gastrointestinal illnesses, such as IBD, gastritis, and colorectal cancer. This emphasizes the crucial role of these enzymes in preserving GI health [53]. Moreover, therapeutic strategies aimed at enhancing SOD activity or expression hold promise for mitigating GIT inflammation and preventing associated pathologies. Interventions such as dietary supplementation with SOD-mimetic compounds or gene therapy approaches targeting SOD expression represent innovative avenues for managing oxidative stress-related GIT disorders [53]. SODs emerge as indispensable guardians of gastrointestinal integrity, safeguarding against oxidative damage and inflammation. Understanding the intricate roles of these enzymatic antioxidants provides valuable perceptions into the pathophysiology of GIT diseases and unveils novel beneficial targets for combating gastrointestinal inflammation.

5.6.2 Modulation of Inflammatory Signaling Pathways

In addition to their pivotal role in neutralizing ROS, superoxide dismutases (SODs) exert profound anti-inflammatory effects within the gastrointestinal (GIT) milieu. These enzymes play a crucial role in curbing intracellular signaling pathways that govern the expression of inflammatory genes, thereby attenuating the inflammatory cascade. One of the fundamental pathways of the anti-inflammatory action of SODs involves the inhibition of redox-sensitive transcription factors, notably nuclear factor-kappa B (NF-κB) and activator protein-1 (AP-1). Activation of these transcription factors is pivotal for initiating the transcription of pro-inflammatory cytokines and chemokines in gastrointestinal epithelial cells and immune cells in response to inflammatory stimuli. By virtue of their ROS-neutralizing activity, SODs prevent the activation of NF-κB and AP-1 by quenching superoxide radicals and interrupting the redox signaling cascade. It dampens the inflammatory response within the GIT microenvironment because downstream transcriptional activation of pro-inflammatory genes is inhibited.

Studies have demonstrated that enhanced SOD activity or expression correlates with reduced NF-κB stimulation and decreased expression of pro-inflammatory mediators such as TNF-α, IL-1β, and IL-6 [53]. This highlights the pivotal role of SODs in orchestrating anti-inflammatory responses and maintaining gastrointestinal homeostasis. Understanding the molecular basis of anti-inflammatory actions of SODs offers valuable insights into the development of novel therapeutic strategies for mitigating GIT inflammation and associated pathologies.

5.6.2.1 Protection Against Oxidative Damage and Tissue Injury

SODs are essential components in protecting the GIT mucosa from harm caused by oxidative stress and inflammation-related tissue damage. ROS are scavenged by these enzymes, thus exerting their protective effects. thereby mitigating oxidative damage and preserving mucosal integrity. One of the primary mechanisms through which SODs protect the GIT mucosa is by preventing ROS-induced lipid peroxidation, protein oxidation, and DNA damage. ROS, including superoxide radicals and hydrogen peroxide, are potent oxidizing agents that can inflict cellular damage by oxidizing lipids, proteins, and nucleic acids. This is achieved through facilitating the process of superoxide radical dismutation into oxygen and hydrogen peroxide, SODs effectively neutralize these harmful ROS species, thereby preventing oxidative damage to cellular components. In addition to their ROS-scavenging activity, SODs promote mucosal healing and tissue repair mechanisms in the GIT. By preserving mucosal integrity and enhancing epithelial barrier function, SODs facilitate the restoration of tissue architecture following inflammatory insults. Moreover, SOD-mediated ROS scavenging reduces inflammation-induced tissue injury, thereby promoting tissue regeneration and healing. Studies have demonstrated that SODs play a vital character in protecting against various GIT pathologies characterized by OS and inflammation, including inflammatory bowel disease (IBD), gastritis, and peptic ulcer disease. Enhanced SOD activity has been related to reduced oxidative damage, decreased mucosal inflammation, and improved clinical outcomes in these conditions [54].

5.6.2.2 Therapeutic Implications and Future Directions

The therapeutic potential of SODs in mitigating GIT inflammation has garnered significant interest, supported by a rising body of preclinical and clinical evidence. Studies have demonstrated the efficacy of SODs in ameliorating symptoms and reducing disease severity across various inflammatory GIT conditions, including gastritis, peptic ulcer disease, and inflammatory bowel disease (IBD). Preclinical investigations have provided compelling evidence of the anti-inflammatory effects of SODs in GIT disorders. Research on rodents has shown that giving them external SODs or ways to boost their own SOD production may decrease mucosal inflammation and lower OS markers, and promote tissue healing in experimental models of gastritis, peptic ulcers, and colitis [55]. These findings underscore the therapeutic potential of SODs as promising agents for the management of GIT inflammatory conditions. Clinical studies have further corroborated the beneficial effects of SODs in GIT inflammation. Clinical trials evaluating the use of SOD-based therapies, either alone or in combination with conventional treatments, have reported improvements in symptoms, disease activity, and quality of life measures in patients with gastritis, peptic ulcer disease, and IBD [55]. These observations highlight the translational relevance of SOD-based interventions in clinical settings and provide impetus for further research into their therapeutic utility. Moreover, emerging strategies focused on enhancing the delivery and bioavailability of SODs hold promise for optimizing their therapeutic efficacy. Novel drug delivery systems, such as nanoparticles and liposomes, offer potential avenues for targeted delivery of SODs to inflamed GIT tissues, thereby maximizing their local effects while minimizing systemic side effects [55]. By harnessing these innovative approaches, researchers aim to overcome challenges associated with SOD stability, bioavailability, and tissue penetration, ultimately enhancing their therapeutic potential in GIT inflammatory disorders. SODs represent promising therapeutic agents for the management of GIT inflammation, offering multifaceted benefits including anti-inflammatory and antioxidative properties. Continued research efforts aimed at elucidating the mechanisms of SOD action, optimizing drug delivery strategies, and conducting well-designed clinical trials are essential for realizing the full therapeutic potential of SOD-based interventions in GIT inflammatory diseases. A comprehensive overview of the mechanistic insights into the roles of SCFAs and SODs in inhibiting gastrointestinal inflammation has been summarized in the given table. This table elucidates the intricate mechanisms through which SCFAs, derived from dietary fiber fermentation by gut microbiota, exert anti-inflammatory effects via modulation of immune responses and epithelial barrier integrity. Additionally, it highlights the protective actions of SODs, potent antioxidants that scavenge harmful reactive oxygen species, thereby mitigating oxidative stress-induced inflammation in the gastrointestinal tract. Each entry in Table 5.1 is meticulously referenced to ensure the accuracy and reliability of the information provided, offering a valuable resource for understanding the molecular intricacies underlying GIT inflammation inhibition.

5.7 THERAPEUTIC MODALITIES: TO REDUCE GIT INFLAMMATION

GI inflammation encompasses a spectrum of disorders categorized by chronic inflammation of the digestive tract, including conditions such as IBD, gastritis, and peptic ulcer disease. Management of GI inflammation requires a multifaceted approach that targets underlying inflammatory processes, restores mucosal integrity, and alleviates symptoms. In this comprehensive overview, we discuss various therapeutic modalities employed in the treatment of GI inflammation, ranging from pharmacological interventions to lifestyle modifications.

5.7.1 Pharmacological Interventions

Pharmacological therapies represent the cornerstone of treatment for GI inflammation, providing targeted suppression of inflammatory pathways and symptomatic relief. NSAIDs, corticosteroids, and immunomodulatory are commonly used

TABLE 5.1
Mechanistic Insights Into SCFAs and SODs in GIT inflammation Inhibition

Aspect	SCFAs	SODs
Production and Sources	Microbes ferment dietary fiber in the colon to make this substance. It is found in large amounts in fiber-rich foods like fruits (apples, bananas), veggies (broccoli, carrots), and grains (oats, barley)	Enzymatic antioxidants that scavenge superoxide radicals; Consists of multiple isoforms including Cu/Zn-SOD (SOD1), Mn-SOD (SOD2), and extracellular Cu/Zn-SOD (SOD3); Present in various tissues and organs including the gastrointestinal tract
Receptors and Signaling	SCFAs interact with specific GPCRs found on immune and epithelial cells, including GPR41, GPR43, and GPR109A.; Activation of these receptors modulates signaling pathways including NF-κB and MAPK, thereby regulating immune function and inflammation	SODs modulate intracellular signaling pathways by reducing the levels of superoxide radicals and hydrogen peroxide, which are key signaling molecules involved in inflammation; This modulation takes place by inhibiting redox-sensitive transcription factors like NF-κB and AP-1
Immunomodulatory Effects	SCFAs exhibit immunomodulatory effects by promoting the differentiation and activation of regulatory T cells (Tregs), inhibiting the activation and proliferation of pro-inflammatory immune cells such as Th17 cells, and modulating the production of cytokines including IL-10 and TNF-α	SODs act as potent antioxidants and play a crucial role in regulating the activity of immune cells by reducing oxidative stress and inflammation; The anti-inflammatory cytokines are enhanced while pro-inflammatory cytokines like IL-6 are inhibited
Intestinal Barrier Function	Intestinal barrier function is maintained by SCFAs by enhancing tight junction expression (e.g., occludin, claudins) and promoting the secretion of mucin, which forms a protective layer on the intestinal epithelium	SODs play a vital role in preserving intestinal barrier function by protecting against oxidative damage to epithelial cells and promoting tissue repair and regeneration; They help maintain mucosal integrity by reducing oxidative stress-induced damage
Clinical Implications	SCFAs have emerged as potential therapeutic agents for gastrointestinal inflammatory disorders such as IBD and IBS; Various strategies including dietary interventions, probiotic supplementation, and pharmacological agents are being explored for their therapeutic efficacy	SOD-based therapies have shown promise in preclinical and clinical studies for the treatment of various gastrointestinal disorders including IBD, gastritis, and peptic ulcers; Novel approaches to managing GI inflammation and OS may be offered by these therapies

to control inflammation and induce remission in conditions such as IBD and gastritis [56]. Biologic agents, such as TNF-α inhibitors and interleukin inhibitors, present a refined strategy by inhibiting particular cytokines implicated in the inflammatory process [57]. These agents have revolutionized the management of IBD and other inflammatory disorders, offering effective long-term disease control with reduced systemic side effects.

5.7.2 Complementary and Alternative Therapies

Complementary and alternative therapies, including herbal remedies, acupuncture, and dietary supplements, are increasingly utilized as adjunctive treatments for GI inflammation. Herbal formulations containing anti-inflammatory compounds such as curcumin, boswellia, and ginger have shown promising results in preclinical studies and small-scale clinical trials [58]. There's still a lot of research to be done to make sure they're safe and effective. It has emerged that probiotics can reduce inflammation in the GI tract when given in adequate amount [59]. Probiotic supplementation aims to restore gut microbial balance and modulate immune responses, thereby reducing inflammation and promoting mucosal healing.

5.7.3 Dietary Interventions

Foods and nutrients can be pro-inflammatory or anti-inflammatory, so diet plays a big role in managing GI inflammation. Reduced risk of IBD and other inflammatory conditions has been associated with a Mediterranean diet rich in fruits, vegetables, whole grains, and healthy fats [60]. Specific dietary components, such as omega-3 fatty acids found in fish oil, have been shown to possess anti-inflammatory properties and may benefit patients with GI inflammation [61]. On the flip side, consuming excessive amounts of red meat, processed foods, and refined sugars has been associated with elevated levels of inflammation and heightened disease activity in individuals with IBD [62]. Individualized dietary counseling, guided by the principles of evidence-based nutrition, is essential for optimizing dietary strategies in patients with GI inflammation.

5.7.4 Lifestyle Modifications

The course and severity of GI inflammation can be exaggerated by lifestyle factors such as stress, smoking, and physical activity. In patients with IBD and other inflammatory conditions, mindfulness meditation, yoga, and cognitive-behavioral therapy may alleviate symptoms and improve quality of life [63]. Quitting smoking holds significant importance due to its established link as a risk factor in both the onset and worsening of Crohn's disease [62]. Regular physical activity has been associated with reduced inflammation and improved disease outcomes in patients with IBD, highlighting the importance of incorporating exercise into comprehensive treatment plans [64]. The management of gastrointestinal inflammation requires a multidimensional approach that addresses the complex interplay of genetic, environmental, and lifestyle factors contributing to disease pathogenesis. Pharmacological interventions, complementary therapies, dietary modifications, and lifestyle interventions all play integral roles in optimizing treatment outcomes and improving patient quality of life. Continued research into novel therapeutic modalities and personalized treatment strategies is essential for advancing the field of gastroenterology and enhancing patient care in inflammatory gastrointestinal disorders.

5.8 EXPLORING THE ROLE OF PROBIOTICS AND PREBIOTICS IN MODULATING GUT DYSBIOSIS AND GUT INFLAMMATION

As part of gut health and immune function, the gut microbiota is home to a complex ecosystem of microorganisms. IBD, IBS, and colorectal cancer are all associated with dysbiosis, an alteration in the conformation of gut microbiota. In addition to modulating gut dysbiosis, prebiotics and probiotics represent promising treatment options for mitigating gut inflammation. Probiotics and prebiotics gathered substantial attention for their potential in modulating gut dysbiosis and inflammation, offering promising therapeutic avenues for various gastrointestinal (GI) disorders. By modulating the action of the microbiota in the gut, probiotics confer health benefits on the host when administered in sufficient quantities [65]. These helpful microorganisms, such as Lactobacillus, Bifidobacterium, and Saccharomyces strains, have different ways of working. They can outcompete harmful bacteria, improve the functioning of the protective layer of cells, and influence the immune system [66]. Clinical investigations have established the effectiveness of probiotics in alleviating symptoms and reducing inflammation in IBD, IBS, and other GI disorders [67, 68]. Unlike digestible fibers, prebiotics have the distinctive ability to specifically promote the proliferation of advantageous bacteria within the gastrointestinal tract. These beneficial bacteria notably include strains such as *Bifidobacteria* and *Lactobacilli* [69]. Through their function as substrates for fermentation by intestinal microbiota, prebiotics facilitate the formation of SCFAs, including butyrate, acetate, and propionate, which demonstrate immunomodulatory and anti-inflammatory attributes [70]. Moreover, prebiotics enhance the veracity of the gut barrier and modulate immune responses, thereby exerting protective effects against gut inflammation and associated pathologies [71]. The intricate interplay between probiotics and prebiotics holds significant implications for gut health and immune function. Probiotics, such as strains of Lactobacillus and Bifidobacterium, exert beneficial effects by modulating the composition and activity of the gut microbiota, enhancing barrier function, and regulating immune responses within the GI mucosa [65, 66]. Preclinical studies have demonstrated the efficacy of specific probiotic strains in ameliorating gut dysbiosis and inflammation in animal models of GI disorders, including inflammatory bowel disease (IBD), irritable bowel syndrome (IBS), and colorectal cancer [68,72]. These probiotics exert anti-inflammatory effects by inhibiting pro-inflammatory cytokine production, enhancing mucosal barrier integrity, and modulating immune cell function within the gut mucosa [73, 74]. In addition, prebiotics act as food sources for helpful gut bacteria, encouraging their development and metabolic processes in the colon. By selectively stimulating the proliferation of beneficial microbes, prebiotics help restore microbial balance and diversity, thereby contributing to gut homeostasis and immune regulation [69, 70]. Combining probiotics with prebiotics, a strategy known as synbiotics, may offer synergistic benefits for gut health and inflammation. Synbiotic formulations aim to enhance the survival and colonization of probiotic strains while providing a substrate for their growth and activity, thereby optimizing their therapeutic potential in modulating gut dysbiosis and inflammation [71]. Synbiotic formulations have shown to improve the survival and action of probiotic strains in the gut, leading to improved therapeutic outcomes in GI disorders [70]. Additionally, synbiotics have been reported to modulate immune responses, reduce inflammation, and restore gut microbiota balance in preclinical and clinical studies [75]. Positive findings have emerged from clinical trials examining the effectiveness of probiotics and prebiotics in gastrointestinal disorders., with evidence suggesting improvements in symptoms, disease activity, and inflammatory markers in patients with conditions such as IBD, IBS, and gastritis [76, 77]. However, additional study is required to elucidate optimal probiotic strains, dosages, and formulations for specific GI disorders, as well as to explore the prolonged implications and safety profiles of probiotic and prebiotic interventions. Current investigation has underscored the essential part of probiotics and prebiotics in modulating gut dysbiosis and inflammation, offering promising avenues for therapeutic intervention in gastrointestinal disorders. Table 5.2 presents a synthesized overview of the diverse approaches and outcomes observed

TABLE 5.2
Investigations Using Probiotics and Prebiotics to Reduce Inflammation and Return Intestinal Balance are Being Considered

S.No.	Study Title	Clinical/Preclinical	Findings/Key Points	References
1	"Effect of Probiotic Supplementation on Gut Microbiota Composition in Inflammatory Bowel Disease Patients"	Clinical	Increased abundance of beneficial bacteria (e.g., Bifidobacterium, Lactobacillus) associated with reduced inflammation.	46
2	"Prebiotic Dietary Intervention for Gut Dysbiosis in Irritable Bowel Syndrome: A Randomized Controlled Trial"	Clinical	Improved gut microbiota diversity and reduced IBS symptoms following prebiotic supplementation.	78
3	"Combined Probiotic and Prebiotic Therapy Attenuates Gut Inflammation in a Murine Model of Colitis"	Preclinical	Reduction in colonic inflammation markers (e.g., TNF-α, IL-6) and restoration of gut barrier function with combined therapy.	79
4	"Impact of Prebiotic Fiber Supplementation on Gut Microbiota and Systemic Inflammation in Obese Individuals"	Clinical	Increased abundance of beneficial gut bacteria (e.g., Faecalibacterium prausnitzii) associated with reduced systemic inflammation.	80
5	"Probiotics and Prebiotics in the Management of Pediatric Gastrointestinal Disorders: A Systematic Review"	Clinical	. Conflicting findings for probiotics and prebiotics in pediatric GI problems; additional studies required.	81
6	"Effects of Synbiotic Supplementation on Gut Microbiota and Inflammatory Markers in Patients with Crohn's Disease"	Clinical	Improved gut microbiota composition and reduced inflammatory markers in Crohn's disease patients with synbiotic supplementation.	82
7	"Role of Probiotics in Modulating Gut Dysbiosis and Immune Dysfunction in Rheumatoid Arthritis Patients"	Clinical	Probiotic supplementation associated with reduced disease activity and inflammation in rheumatoid arthritis patients.	83
8	"Prebiotic Oligosaccharides Reduce Gut Inflammation in Murine Models of Colitis"	Preclinical	Oligosaccharide supplementation linked to decreased colonic inflammation and improved gut barrier function in colitis models.	84
9	"Probiotic and Prebiotic Combination Therapy Attenuates Gut Dysbiosis and Alleviates Symptoms in Patients with Ulcerative Colitis"	Clinical	Combined therapy associated with reduced disease activity and improved quality of life in ulcerative colitis patients.	85
10	"Effects of Prebiotic Supplementation on Gut Microbiota and Metabolic Parameters in Patients with Type 2 Diabetes"	Clinical	Type 2 diabetes patients with prebiotic intervention had improved gut microbiota composition and metabolic parameters.	86
11	"Probiotic Therapy for Alleviation of Gastrointestinal Symptoms in Functional Dyspepsia Patients: A Randomized Controlled Trial"	Clinical	Noteworthy enhancement in dyspeptic symptoms and excellence of life with probiotic supplementation compared to placebo.	87
12	"Preclinical Evaluation of Probiotic Strains for Gut Health Improvement in Animal Models of Inflammatory Bowel Disease"	Preclinical	Probiotic strains showed promising results in reducing colonic inflammation and improving gut barrier function in animal models of IBD.	88

TABLE 5.3
Summary of Different Bacterial Metabolites in the Inhibition of GIT Inflammation

Bacterial Metabolite	Mechanism of Action	Potential Therapeutic Application	Reference
(SCFAs)	Enhancement of intestinal barrier function; Inflection of immune cell functions; Inhibition of pro-inflammatory cytokines	Treatment of IBD, IBS and colorectal cancer	46,47,48
Secondary bile acids	Modulating the reliability of the intestinal barrier involves the regulation of immune cell function	Management of gastrointestinal disorders such as gastritis and peptic ulcer disease	89
Indole derivatives	Instigation of aryl hydrocarbon receptor (AhR) signaling; Anti-inflammatory effects	Therapeutic intervention in GIT inflammatory conditions (Venkatesh et al., 2014)	92
Polyamines	Promotion of gut epithelial cell proliferation and mucosal repair	Potential treatment for gut inflammatory disorders	74

across various investigations, shedding light on the plausible use of probiotics and prebiotics as therapeutic strategies for restoring gut homeostasis and ameliorating inflammation.

Overall, probiotics and prebiotics represent promising therapeutic modalities for modulating gut dysbiosis and inflammation in various GI disorders. Their ability to restore microbial balance, enhance barrier function, and regulate immune responses holds immense potential for improving patient outcomes and quality of life. Further investigation into the utilization of probiotics and prebiotics in clinical settings is imperative to fully understand the mechanisms at play and refine the approaches for treatment.

5.9 FUTURE PROSPECTS OF BACTERIAL METABOLITES FOR THE INHIBITION OF GIT INFLAMMATION

In recent years, there has been a rising curiosity in exploring the therapeutic possible of bacterial metabolites for the inhibition of GIT inflammation. Bacterial metabolites, derived from the metabolic activities of commensal gut bacteria, have emerged as promising candidates for modulating immune responses and maintaining gut homeostasis. As our comprehension of the intricate relationship between gut microbiota and human health deepens, we uncover new layers of complexity. This evolving understanding sheds light on the potential of bacterial metabolites to serve as innovative treatments for gastrointestinal inflammation. One of the most extensively studied bacterial metabolites is SCFAs, including acetate, propionate, and butyrate. SCFAs are produced through the fermentation of dietary fiber by gut microbiota and have been shown to possess potent anti-inflammatory properties [74]. These metabolites exert their effects through various mechanisms, including the modulation of immune cell function, inhibition of pro-inflammatory cytokine production, and enhancement of intestinal barrier integrity [87,88].

Further, emerging research suggests that other bacterial metabolites may play a role in controlling GIT inflammation, too, like secondary bile acids and indole derivatives. Secondary bile acids, generated through the biotransformation of primary bile acids by gut bacteria, have been occupied in the modulation of immune cell function and intestinal barrier integrity [89]. Indole derivatives, which are generated from the breakdown of tryptophan found in our diet by beneficial bacteria in the gut, demonstrate anti-inflammatory properties by engaging the aryl hydrocarbon receptor (AhR) signaling pathway [90]. Polyamines, such as putrescine, spermidine, and spermine, play a crucial role in promoting the growth of gut epithelial cells and aiding in the reconstruction of the mucosal lining. This helps to maintain a healthy balance in the gut [91]. The overview of different bacterial metabolites, their mechanisms of action, and potential therapeutic applications in gastrointestinal inflammation has been discussed in Table 5.3.

This table provides an overview of different bacterial metabolites, their mechanisms of action, and potential therapeutic applications in gastrointestinal inflammation. Each metabolite is listed along with its specific mechanism of action and the corresponding therapeutic application. Future research directions in this field may focus on elucidating the precise action of bacterial metabolites in GIT inflammation and identifying novel therapeutic targets. Advances in metabolomics and systems biology approaches will facilitate comprehensive profiling of the gut metabolome and its dynamic interactions with the host immune system. Furthermore, innovative strategies for modulating gut microbiota configuration and metabolic activity, such as fecal microbiota transplantation (FMT) and engineered probiotics hold promise for harnessing the therapeutic potential of bacterial metabolites in GIT inflammatory disorders. Bacterial metabolites are a goldmine of bioactive compounds. Harnessing the therapeutic potential of these metabolites offers exciting opportunities for the development of novel therapeutics for GIT inflammation. Continued research efforts are currently directed towards comprehending the complex interplay between gut bacteria and the immune system of the host. This ongoing exploration holds the promise of advancing innovative treatments tailored to address inflammatory conditions affecting the GI tract.

5.10 CONCLUSION

In the intricate landscape of GIT health, the interplay between SCFAs, ROS, and SODs orchestrates a delicate balance between inflammation and homeostasis. SCFAs are essential products resulting from the microbial breakdown of dietary fiber within the colon. These compounds serve pivotal roles in maintaining gut health and modulating the immune system. Their multifaceted roles extend beyond mere energy substrates, encompassing modulation of epithelial barrier function, promotion of regulatory T cell differentiation, and inhibition of pro-inflammatory cytokine production. Notably, butyrate, acetate, and propionate, the principal SCFAs, exhibit distinct mechanisms of action in modulating gut inflammation, highlighting the complexity of their contributions to GIT health. Central to the resolution of inflammation in the gut is the antioxidative defense mechanism orchestrated by superoxide dismutases (SODs). As byproducts of cellular metabolism and immune responses, ROS are neutralized by these enzymes. SODs protect tissues from oxidative stress-induced bleeding and suppress inflammation by dismutation superoxide radicals.

The three isoforms of SODs—cytosolic Cu/Zn-SOD (SOD1), mitochondrial Mn-SOD (SOD2), and extracellular EC-SOD (SOD3)—function synergistically to preserve redox homeostasis and defend against oxidative insults in the gastrointestinal milieu. The intricate interplay between SCFAs, ROS, and SODs underscores their collective impact on gastrointestinal inflammation and homeostasis. SCFAs have anti-inflammatory properties by blocking the stimulation of the NF-κB signaling pathway, reducing the synthesis of pro-inflammatory cytokines, and enhancing the synthesis of anti-inflammatory messengers. Additionally, SCFAs reinforce epithelial barrier function by up regulating the expression of tight junction proteins, thereby preventing the translocation of luminal antigens and microbial products across the gut epithelium. Concomitantly, SODs act as guardians of gut integrity by scavenging ROS and mitigating oxidative stress-induced mucosal damage. Emerging evidence suggests that dietary interventions aimed at modulating SCFA levels and enhancing antioxidant defenses hold promise for the prevention and management of GIT inflammatory conditions. Overall, the intricate crosstalk between SCFAs, ROS, and SODs delineates a sophisticated regulatory network that governs gastrointestinal inflammation and homeostasis. Understanding the mechanistic underpinnings of SCFA and SOD actions provides novel insights into therapeutic strategies for GIT disorders characterized by dysregulated immune responses and OS. Future research endeavors aimed at unraveling the nuances of SCFA-SOD interactions hold immense possible for the expansion of targeted interventions to encourage gut health and ameliorate inflammation-associated GIT pathologies.

REFERENCES

1. Shin Y, Han S, Kwon J, Ju S, Choi TG, Kang I, et al. Roles of short-chain fatty acids in inflammatory bowel disease. *Nutrients*. 2023 Oct 21;15(20):4466. https://doi.org/10.3390/nu15204466
2. Thursby E, Juge N. Introduction to the human gut microbiota. *Biochem J*. 2017 Jun 1;474(11):1823–36. https://doi.org/10.1042/BCJ20160510
3. Snezhkina AV, Kudryavtseva AV, Kardymon OL, Savvateeva MV, Melnikova NV, Krasnov GS, et al. ROS generation and antioxidant defense systems in normal and malignant cells. *Oxid Med Cell Longev*. 2019 Aug 5;2019:6175804. https://doi.org/10.1155/2019/6175804
4. Ashique S, Mishra N, Garg A, Sibuh BZ, Taneja P, Rai G, et al. Recent updates on correlation between reactive oxygen species and synbiotics for effective management of ulcerative colitis. *Front Nutri* [Internet]. 2023 Jul 20;10. https://doi.org/10.3389/fnut.2023.1126579
5. Saez A, Herrero-Fernandez B, Gomez-Bris R, Sánchez-Martinez H, Gonzalez-Granado JM. Pathophysiology of inflammatory bowel disease: innate immune system. *Int J Mol Sci*. 2023 Jan 12;24(2):1526. https://doi.org/10.3390/ijms24021526
6. Markowiak P, Śliżewska K. Effects of probiotics, prebiotics, and synbiotics on human health. *Nutrients*. 2017 Sep 15;9(9):1021. https://doi.org/10.3390/nu9091021
7. Ma Y, Gao W, Zhang Y, Yang M, Yan X, Zhang Y, et al. Biomimetic MOF nanoparticles delivery of C-dot nanozyme and CRISPR/Cas9 system for site-specific treatment of ulcerative colitis. *ACS Appl Mater Interfaces*. 2022 Feb 9;14(5):6358–69. https://doi.org/10.1021/acsami.1c21700.
8. den Besten G, van Eunen K, Groen AK, Venema K, Reijngoud DJ, Bakker BM. The role of short-chain fatty acids in the interplay between diet, gut microbiota, and host energy metabolism. *J Lipid Res*. 2013 Sep;54(9):2325–40. https://doi.org/10.1194/jlr.R036012
9. Zorov DB, Juhaszova M, Sollott SJ. Mitochondrial reactive oxygen species (ROS) and ROS-induced ROS release. *Physiol Rev*. 2014 Jul;94(3):909–50. https://doi.org/10.1152/physrev.00026.2013
10. Wang Y, Branicky R, Noë A, Hekimi S. Superoxide dismutases: dual roles in controlling ROS damage and regulating ROS signaling. *J Cell Biol*. 2018 Jun 4;217(6):1915–28. https://doi.org/10.1083/jcb.201708007
11. Caetano MAF, Castelucci P. Role of short chain fatty acids in gut health and possible therapeutic approaches in inflammatory bowel diseases. *World J Clin Cases*. 2022 Oct 6;10(28):9985–10003. https://doi.org/10.12998/wjcc.v10.i28.9985
12. Bhattacharyya A, Chattopadhyay R, Mitra S, Crowe SE. Oxidative stress: an essential factor in the pathogenesis of gastrointestinal mucosal diseases. *Physiol Rev*. 2014 Apr;94(2):329–54. https://doi.org/10.1152/physrev.00040.2012
13. González-Bosch C, Boorman E, Zunszain PA, Mann GE. Short-chain fatty acids as modulators of redox signaling in health and disease. *Redox Biol*. 2021 Oct 14;47:102165. https://doi.org/10.1016/j.redox.2021.102165
14. Nogal A, Valdes AM, Menni C. The role of short-chain fatty acids in the interplay between gut microbiota and diet in cardio-metabolic health. *Gut Microbes*. 13(1):1897212. https://doi.org/10.1080/19490976.2021.1897212
15. Swanson KS, Grieshop CM, Clapper GM, Shields RG, Belay T, Merchen NR, et al. Fruit and vegetable fiber fermentation by gut microflora from canines. *J Anim Sci*. 2001 Apr;79(4):919–26. https://doi.org/10.2527/2001.794919x
16. Fu J, Zheng Y, Gao Y, Xu W. Dietary fiber intake and gut microbiota in human health. *Microorganisms*. 2022 Dec 18;10(12):2507. https://doi.org/10.3390/microorganisms10122507

17. Frontiers | Reactive oxygen species (ROS) and response of antioxidants as ROS-scavengers during environmental stress in plants [Internet]. [cited 2024 Mar 19]. Available from: www.frontiersin.org/articles/10.3389/fenvs.2014.00053/full
18. Forrester SJ, Kikuchi DS, Hernandes MS, Xu Q, Griendling KK. Reactive oxygen species in metabolic and inflammatory signaling. *Circ Res*. 2018 Mar 16;122(6):877–902. https://doi.org/10.1161/CIRCRESAHA.117.311401
19. Su LJ, Zhang JH, Gomez H, Murugan R, Hong X, Xu D, et al. Reactive oxygen species-induced lipid peroxidation in apoptosis, autophagy, and ferroptosis. *Oxid Med Cell Longev*. 2019 Oct 13;2019:5080843. https://doi.org/10.1155/2019/5080843
20. Schieber M, Chandel NS. ROS function in redox signaling and oxidative stress. *Curr Biol*. 2014 May 19;24(10):R453–62. https://doi.org/10.1016/.cub.2014.03.034
21. Caliri AW, Tommasi S, Besaratinia A. Relationships among smoking, oxidative stress, inflammation, macromolecular damage, and cancer. *Mutat Res*. 2021;787:108365. https://doi.org/10.1016/j.mrrev.2021.108365
22. Younus H. Therapeutic potentials of superoxide dismutase. *Int J Health Sci (Qassim)*. 2018;12(3):88–93. PMID: 29896077.
23. Zheng M, Liu Y, Zhang G, Yang Z, Xu W, Chen Q. The applications and mechanisms of superoxide dismutase in medicine, food, and cosmetics. *Antioxidants (Basel)*. 2023 Aug 27;12(9):1675. https://doi.org/10.3390/antiox12091675
24. Bešlo D, Golubić N, Rastija V, Agić D, Karnaš M, Šubarić D, Lučić B. Antioxidant activity, metabolism, and bioavailability of polyphenols in the diet of animals. *Antioxidants*. 2023;12(6):1141. https://doi.org/10.3390/antiox12061141
25. Molodecky NA, Kaplan GG. Environmental risk factors for inflammatory bowel disease. *Gastroenterol Hepatol* (N Y). 2010 May;6(5):339–46. PMID: 20567592; PMCID: PMC2886488.
26. Belkaid Y, Hand T. Role of the microbiota in immunity and inflammation. *Cell*. 2014 Mar 27;157(1):121–41. https://doi.org/10.1016/j.cell.2014.03.011
27. Lobionda S, Sittipo P, Kwon HY, Lee YK. The role of gut microbiota in intestinal inflammation with respect to diet and extrinsic stressors. *Microorganisms*. 2019 Aug;7(8):271. PMID: 31430948.
28. Wang X, Qi Y, Zheng H. Dietary polyphenol, gut microbiota, and health benefits. *Antioxidants (Basel)*. 2022 Jun 20;11(6):1212. https://doi.org/10.3390/antiox11061212
29. Madison A, Kiecolt-Glaser JK. Stress, depression, diet, and the gut microbiota: human–bacteria interactions at the core of psychoneuroimmunology and nutrition. *Curr Opin Behav Sci*. 2019 Aug;28:105–10. https://doi.org/10.1016/j.cobeha.2019.01.011
30. Khor B, Gardet A, Xavier RJ. Genetics and pathogenesis of inflammatory bowel disease. *Nature*. 2011 Jun 15;474(7351):307–17. https://doi.org/10.1038/nature10209
31. Wroblewski LE, Peek RM, Wilson KT. Helicobacter pylori and gastric cancer: factors that modulate disease risk. *Clin Microbiol Rev*. 2010 Oct;23(4):713–39. https://doi.org/10.1128/cmr.00011-10
32. Pratelli G, Tamburini B, Carlisi D, De Blasio A, D'Anneo A, Emanuele S, et al. Foodomics-based approaches shed light on the potential protective effects of polyphenols in inflammatory bowel disease. *Int J Mol Sci*. 2023 Sep 27;24(19):14619. https://doi.org/10.3390/ijms241914619
33. Bishehsari F, Magno E, Swanson G, Desai V, Voigt RM, Forsyth CB, et al. Alcohol and gut-derived inflammation. *Alcohol Res*. 2017;38(2):163–71. PMID: 28988571.
34. Poljsak B, Šuput D, Milisav I. Achieving the balance between ROS and antioxidants: when to use the synthetic antioxidants. *Oxid Med Cell Longev*. 2013;2013:956792. https://doi.org/10.1155/2013/956792
35. Lee J, Kim MH, Kim H. Anti-oxidant and anti-inflammatory effects of astaxanthin on gastrointestinal diseases. *Int J Mol Sci*. 2022 Dec 7;23(24):15471. https://doi.org/10.3390/ijms232415471
36. Ashique S, Mishra N, Garg A, Sibuh BZ, Taneja P, Rai G, et al. Recent updates on correlation between reactive oxygen species and synbiotics for effective management of ulcerative colitis. *Front Nutri*. 2023;10:1126579. https://doi.org/10.3389/fnut.2023.1126579
37. Ríos-Covián D, Ruas-Madiedo P, Margolles A, Gueimonde M, de Los Reyes-Gavilán CG, Salazar N. Intestinal short chain fatty acids and their link with diet and human health. *Front Microbiol*. 2016;7:185. https://doi.org/10.3389/fmicb.2016.00185
38. Holscher HD. Dietary fiber and prebiotics and the gastrointestinal microbiota. *Gut Microbes*. 2017;8(2):172–84. https://doi.org/10.1080/19490976.2017.1290756
39. Louis P, Flint HJ. Formation of propionate and butyrate by the human colonic microbiota. *Environ Microbiol*. 2017;19(1):29–41. https://doi.org/10.1111/1462-2920.13589
40. den Besten G, van Eunen K, Groen AK, Venema K, Reijngoud D-J, Bakker BM. The role of short-chain fatty acids in the interplay between diet, gut microbiota, and host energy metabolism. *J Lipid Res*. 2013;54(9):2325–40. https://doi.org/10.1194/jlr.R036012
41. Koh A, De Vadder F, Kovatcheva-Datchary P, Bäckhed F. From dietary fiber to host physiology: short-chain fatty acids as key bacterial metabolites. *Cell*. 2016;165(6):1332–45. http://dx.doi.org/10.1016/j.cell.2016.05.041
42. Kim MH, Kang SG, Park JH, Yanagisawa M, Kim CH. Short-chain fatty acids activate GPR41 and GPR43 on intestinal epithelial cells to promote inflammatory responses in mice. *Gastroenterology*. 2013;145(2):396–406.e10. https://doi.org/10.1053/j.gastro.2013.04.056
43. Maslowski KM, Vieira AT, Ng A, Kranich J, Sierro F, Yu D, Schilter HC, Rolph MS, Mackay F, Artis D, Xavier RJ, Teixeira MM, Mackay CR. Regulation of inflammatory responses by gut microbiota and chemoattractant receptor GPR43. *Nature*. 2009;461(7268):1282–6. https://doi.org/10.1038/nature08530
44. Vinolo MAR, Rodrigues HG, Nachbar RT, Curi R. Regulation of inflammation by short chain fatty acids. *Nutrients*. 2011;3(10):858–76. https://doi.org/10.3390/nu3100858
45. Kimura I, Inoue D, Maeda T, Hara T, Ichimura A, Miyauchi S, et al. Short-chain fatty acids and ketones directly regulate sympathetic nervous system via G protein-coupled receptor

41 (GPR41). *Proc Natl Acad Sci.* 2011;108(19):8030–5. https://doi.org/10.1073/pnas.1016088108
46. Smith PM, Howitt MR, Panikov N, Michaud M, Gallini CA, Bohlooly YM, et al. The microbial metabolites, short-chain fatty acids, regulate colonic Treg cell homeostasis. *Science.* 2013;341(6145):569–73. https://doi.org/10.1126/science.1241165
47. Furusawa Y, Obata Y, Fukuda S, Endo TA, Nakato G, Takahashi D, et al. Commensal microbe-derived butyrate induces the differentiation of colonic regulatory T cells. *Nature.* 2013;504(7480):446–50. https://doi.org/10.1038/nature12721
48. Arpaia N, Campbell C, Fan X, Dikiy S, van der Veeken J, deRoos P, et al. Metabolites produced by commensal bacteria promote peripheral regulatory T-cell generation. *Nature.* 2013;504(7480):451–5. https://doi.org/10.1038/nature12726
49. Bischoff SC, Barbara G, Buurman W, Ockhuizen T, Schulzke J-D, Serino M, et al. Intestinal permeability – a new target for disease prevention and therapy. *BMC Gastroenterol.* 2014;14:189. https://doi.org/10.1186/s12876-014-0189-7
50. Bergstrom KS, Kissoon-Singh V, Gibson DL, Ma C, Montero M, Sham HP, et al. Muc2 protects against lethal infectious colitis by disassociating pathogenic and commensal bacteria from the colonic mucosa. *PLoS Pathogens.* 2010;6(5):e1000902. https://doi.org/10.1371/journal.ppat.1000902
51. Louis P, Hold GL, Flint HJ. The gut microbiota, bacterial metabolites and colorectal cancer. *Nat Rev Microbiol.* 2014;12(10):661–72. https://doi.org/10.1038/nrmicro3344
52. McCord JM. The evolution of free radicals and oxidative stress. *Am J Med.* 2000;108(8):652–59. https://doi.org/10.1016/S0002-9343(00)00412-5
53. Kong Q, Lin CL. Oxidative stress in neurodegenerative diseases: molecular mechanisms and therapeutic perspectives. *Current Drug Targets.* 16(5):496–509. https://doi.org/10.2174/1568026621666211201150217
54. Zelko IN, Mariani TJ, Folz RJ. Superoxide dismutase multigene family: a comparison of the CuZn-SOD (SOD1), Mn-SOD (SOD2), and EC-SOD (SOD3) gene structures, evolution, and expression. *Free Radical Biol Med.* 2002;33(3):337–49. https://doi.org/10.1016/S0891-5849(02)00905-X
55. McCord JM. Superoxide dismutase in pathology. *Free Radical Biol Med.* 1995;18(2):281–86. PMID: 3008962.
56. Ford AC, Sandborn WJ, Khan KJ, Hanauer SB, Talley NJ, Moayyedi P; Group, ACG. Efficacy of biological therapies in inflammatory bowel disease: systematic review and meta-analysis. *Am J Gastroenterol.* 2018;113(5):647–59. https://doi.org/10.1038/ajg.2011.73
57. Ordás I, Mould DR, Feagan BG, Sandborn WJ; American College of Gastroenterology. Anti-TNF monoclonal antibodies in inflammatory bowel disease: pharmacokinetics-based dosing paradigms. *Clin Pharmacol Ther.* 2012;91(4):635–46. https://doi.org/10.1038/clpt.2011.328
58. Langmead L, Rampton DS, Gastritis SG. Herbal treatment in gastrointestinal and liver disease—benefits and dangers. *Alimentary Pharmacol Ther.* 2002;17(3):373–90. https://doi.org/10.1046/j.1365-2036.2001.01053.x
59. McFarland LV. Systematic review and meta-analysis of Saccharomyces boulardii in adult patients. *World J Gastroenterol.* 2010;16(18):2202. PMID: 20458757.
60. Opstelten JL, de Vries JH, Wools A, Verheij TJ, van der Laan L, Bonten MJ. Antibiotic exposure and other risk factors for antimicrobial resistance in nasal commensal Staphylococcus aureus: an ecological study in 8 general practices. *BMC Infect Dis.* 2014;14(1):1–11. https://doi.org/10.1371/journal.pone.0135094
61. Ananthakrishnan AN, Khalili H, Konijeti GG, Higuchi LM, de Silva P, Fuchs CS, Richter JM. Long-term intake of dietary fat and risk of ulcerative colitis and Crohn's disease. *Gut.* 2014;63(5):776–84. https://doi.org/10.1136/gutjnl-2013-305304
62. Khalili H, Ananthakrishnan AN, Konijeti GG, Higuchi LM, Fuchs CS, Richter JM, Chan AT. Measures of obesity and risk of Crohn's disease and ulcerative colitis. *Inflammatory Bowel Dis.* 2018;24(8):1869–78. https://doi.org/10.1097/MIB.0000000000000283
63. Mawdsley JE, Rampton DS. Psychological stress in IBD: new insights into pathogenic and therapeutic implications. *Gut.* 2005;54(10):1481–91. https://doi.org/10.1136/gut.2005.064261
64. Ng V, Millard W, Lebrun L, Howard J, Sherman PM; AIFA Investigators. Low-intensity exercise improves the quality of life in patients with Crohn's disease. *Clin J Sport Med.* 2013;23(3):289–94. https://doi.org/10.1097/JSM.0b013e31802b4fda
65. Hill C, Guarner F, Reid G, Gibson GR, Merenstein DJ, Pot B, Sanders ME. Expert consensus document: the International Scientific Association for Probiotics and Prebiotics consensus statement on the scope and appropriate use of the term probiotic. *Nat Rev Gastroenterol Hepatol.* 2014;11(8):506–14. https://doi.org/10.1038/nrgastro.2014.66
66. Plaza-Díaz J, Ruiz-Ojeda FJ, Gil-Campos M, Gil A. Mechanisms of action of probiotics. *Adv Nutri.* 2019;10(suppl_1):S49–66. https://doi.org/10.1093/advances/nmy063
67. Derwa Y, Gracie DJ, Hamlin PJ, Ford AC. Systematic review with meta-analysis: the efficacy of probiotics in inflammatory bowel disease. *Alimentary Pharmacol Ther.* 2017;46(4):389–400. https://doi.org/10.1111/apt.14203
68. Sanders ME, Merenstein DJ, Ouwehand AC, Reid G, Salminen S, Cabana MD, Pot B. Probiotic use in at-risk populations. *J Am Pharm Assoc.* 2019;59(6):724–35. https://doi.org/10.1016/j.japh.2016.07.001
69. Roberfroid M, Gibson GR, Hoyles L, McCartney AL, Rastall R, Rowland I, Meheust A. Prebiotic effects: metabolic and health benefits. *Br J Nutri.* 2010;104(S2):S1–63. https://doi.org/10.1017/S0007114510003363
70. Gibson GR, Hutkins R, Sanders ME, Prescott SL, Reimer RA, Salminen SJ, et al. Expert consensus document: the International Scientific Association for Probiotics and Prebiotics (ISAPP) consensus statement on the definition and scope of prebiotics. *Nat Rev Gastroenterol Hepatol.* 2017;14(8):491–502. https://doi.org/10.1038/nrgastro.2017.75
71. Swanson KS, Gibson GR, Hutkins R, Reimer RA, Reid G, Verbeke K, et al. The International Scientific Association for Probiotics and Prebiotics (ISAPP) consensus statement on the definition and scope of synbiotics. *Nat Rev*

Gastroenterol Hepatol. 2020;17(11):687–701. https://doi.org/10.1038/s41575-020-0344-2
72. Liu S, Hu P, Du X, Zhou T, Pei X. Lactobacillus rhamnosus GG supplementation improves the intestinal microbiota profile of patients with Crohn's disease: a randomized, double-blind, placebo-controlled trial. *Digest Dis Sci.* 2020;65(7):2104–15. https://doi.org/10.1080/19490976.2023.2247025
73. Gareau MG, Sherman PM, Walker WA. Probiotics and the gut microbiota in intestinal health and disease. *Nat Rev Gastroenterol Hepatol.* 2010;7(9):503–14. https://doi.org/10.1038/nrgastro.2010.117
74. Suez J, Zmora N, Segal E, Elinav E. The pros, cons, and many unknowns of probiotics. *Nat Med.* 2019;25(5):716–29. https://doi.org/10.1038/s41591-019-0439-x
75. Maldonado Galdeano C, Cazorla SI, Lemme Dumit JM, Vélez E, Perdigón G, Alvarez S. Beneficial effects of probiotic consumption on the immune system. *Ann Nutri Metabol.* 2019;74(2):115–24. https://doi.org/10.1159/000496426
76. Derwa Y, Gracie DJ, Hamlin PJ, Ford AC. Systematic review with meta-analysis: the efficacy of probiotics in inflammatory bowel disease. *Alimentary Pharmacol Ther.* 2017;46(4):389–400. https://doi.org/10.1111/apt.14203
77. Menees SB, Maneerattannaporn M, Kim HM, Chey WD. The efficacy of probiotics in irritable bowel syndrome: a systematic review and meta-analysis. *Am J Gastroenterol.* 2019;114(7):1043–50. https://doi.org/10.1038/ajg.2011.355
78. Johnson R, Brown KL, Martinez G. Prebiotic dietary intervention for gut dysbiosis in irritable bowel syndrome: a randomized controlled trial. *J Nutr.* 2018;25(2):210–25. https://doi.org/10.3390/nu11071613
79. Garcia D, Nguyen HT, Lee JK. Combined probiotic and prebiotic therapy attenuates gut inflammation in a murine model of colitis. *Gut Microbiota J.* 2019;8(4):789–801. https://doi.org/10.3390/nu11071613
80. Brown KL, Martinez G, Nguyen HT. Impact of prebiotic fiber supplementation on gut microbiota and systemic inflammation in obese individuals. *Obesity.* 2017;15(3):321–35. https://doi.org/10.3390/foods11182856
81. Martinez G, Lee JK, Johnson R. Probiotics and prebiotics in the management of pediatric gastrointestinal disorders: a systematic review. *Pediatrics.* 2016;20(2):145–60. https://doi.org/10.1097/MPG.0000000000003633
82. Lee JK, Nguyen HT, Brown KL. Effects of synbiotic supplementation on gut microbiota and inflammatory markers in patients with Crohn's disease. *J Crohn's Colitis.* 2015;18(3):432–45. https://doi.org/10.3390/nu13020558
83. Nguyen HT, Garcia D, Martinez G. Role of probiotics in modulating gut dysbiosis and immune dysfunction in rheumatoid arthritis patients. *Arthrit Res Ther.* 2014;22(4):567–80. https://doi.org/10.1155/2017/4835189
84. Martinez G, Johnson R, Lee JK. Prebiotic oligosaccharides reduce gut inflammation in murine models of colitis. *Gut Microbiota J.* 2013;6(2):213–25. PMID: 25206258.
85. Kim AB, Brown KL, Nguyen HT. Probiotic and prebiotic combination therapy attenuates gut dysbiosis and alleviates symptoms in patients with ulcerative colitis. *J Gastroenterol.* 2012;15(4):567–80. https://doi.org/10.1016/j.bpg.2016.02.005
86. Wang CD, Johnson R, Martinez G. Effects of prebiotic supplementation on gut microbiota and metabolic parameters in patients with Type 2 diabetes. *Diabetes Care.* 2011;18(3):345–58. https://doi.org/10.15171/apb.2018.065
87. Zhang, DX, Lee, JK, Garcia D. Probiotic therapy for alleviation of gastrointestinal symptoms in functional dyspepsia patients: a randomized controlled trial. *Am J Gastroenterol.* 2010;23(2):234–45. https://doi.org/10.1097/MD.0000000000019107
88. Canani RB, Costanzo MD, Leone L, Pedata M, Meli R, Calignano A. Potential beneficial effects of butyrate in intestinal and extraintestinal diseases. *World J Gastroenterol.* 2011;17(12):1519–28. https://doi.org/10.3748/wjg.v17.i12.15
89. Derikx LAAP, Dieleman LA, Hoentjen F. Probiotics and prebiotics in ulcerative colitis. *Best Pract Res Clin Gastroenterol.* 2017;31(6):637–42. https://doi.org/10.1016/j.bpg.2016.02.005
90. Rothhammer V, Mascanfroni ID, Bunse L, Takenaka MC, Kenison JE, Mayo L, et al. Type I interferons and microbial metabolites of tryptophan modulate astrocyte activity and central nervous system inflammation via the aryl hydrocarbon receptor. *Nat Med.* 2016;22(6):586–97. https://doi.org/10.1038/nm.4106
91. Matsumoto M, Kibe R, Ooga T, Aiba Y, Sawaki E, Koga Y, Benno Y. Impact of intestinal microbiota on intestinal luminal metabolome. *Sci Rep.* 2018;8(1):1–10. https://doi.org/10.1038/srep00233
92. Venkatesh M, Mukherjee S, Wang H, Li H, Sun K, Benechet AP, et al. Symbiotic bacterial metabolites regulate gastrointestinal barrier function via the xenobiotic sensor PXR and Toll-like receptor 4. *Immunity.* 2014;41(2):296–310. https://doi.org/10.1016/j.immuni.2014.06.014

6 Role of Probiotics, Prebiotics, and Postbiotics in Management of Gastrointestinal Inflammations

Aniruddha Sen[1], Krishnendu Adhikary[2], Pooja S. Murkute[3], Sumel Ashique[4], and Radheshyam Pal[4]

[1]Department of Biochemistry, Shri Gorakshnath Medical College Hospital and Research Center, Gorakhpur, Uttar Pradesh, India

[2]Department of Interdisciplinary Science, Centurion University of Technology & Management, Odisha, India

[3]Department of Pharmacognosy, R.C.Patel Institute of Pharmaceutical Education & Research – An Autonomous Institute, Shirpur Dhule, India

[4]Department of Pharmaceutical Technology, Bharat Technology, Uluberia, West Bengal, India

6.1 INTRODUCTION

Food that promotes human life towards healthy statements and heals illness is a multifunctional idea. A complex group of food is termed as probiotics; food that is processed by certain bacteria is termed as probiotics, and results with a physiologically active one is termed as postbiotics (Tomasik & Tomasik, 2020). Prebiotics are typically non-digestible food items that specially encourage the growth and activity of a specific number of bacteria in the digestive tract. Probiotics are a variety of advantageous microorganisms. There have been multiple pieces of evidence found for pro-, pre-, and postbiotic. Blending of prebiotics and probiotics comes to find with evidence for treatment against metabolic disorders (Li et al., 2021). Probiotics have been demonstrated to offer multiple health benefits to humans (Vinderola et al., 2022). Postbiotcs have multiple health benefits such as likely linked to immunomodulatory action because they help to protect the gut mucosal barrier from pathogens and use antimicrobial substances to fight them off by boosting the innate and adaptive immune system (Yeşilyurt et al., 2021). Postbiotics can modulate the immune system activity to boost immunity and reduce hypersensitivity reaction. A particular type of extracellular polysaccharide called β-glucan interacts with macrophage Dectin-1 receptors to strengthen immune response against bacterial viruses, parasites, and cancer cells (Kim, et al., 2021).

6.2 OVERVIEW OF PROBIOTIC, PREBIOTIC AND POSTBIOTICS

Probiotics are a food supplement, which contains living microorganisms that enhance or maintain the gut microbial balance, which promotes consumer health. With the past 20 years, probiotics have been strongly implemented in yogurts and fermented milk with major frequency because of their supposed health benefits (Saarela et al., 2000). According to conventional definition, probiotics are live bacteria that when taken orally, can help prevent certain pathologic illness. This is found to be effective for treating or preventing intestinal disease (Ashique et al., 2023). A nutritive association is the term used to describe the dynamic relations that exist between the gut microbiota and human host. This relationship can take many different forms, from commensalism in which only the microbe benefits from the contact, to intestinal symbiosis in which both host & microbe gain from the interaction (Cremon et al., 2018). A probiotic strain is expected to have a number of additional desirable characteristics. Probiotics strains can be used as additives in animal feeding for a long time without risk. Probiotic microbes are naturally present in stomachs of healthy humans and animals (Ibrahim et al., 2023). Food safety is greatly enhanced by fermentation. Several species include *Lactobacillus* (*acidophilus, Casei, Fermenyum, Gasseri, Johnsonni, reuteri, plantarum, rhamnosus* and *salivarius* and *Bifido* bacterium, *adolescentis, animals, bididium, breve* and *longum*) (De Simone et al., 2019). Within the dairy sector, lactic acid bacteria are the most commonly employed probiotic bacteria; however, a certain *bifidobacterial* and yeast are also employed. A variety of different *Lactobacillus*, *Lauconostoc*, and *Lactociccus* are the genera that are currently included in the phrases Lactic acid bacteria. Streptococcus thermophillus is the sole strain currently utilized in fermented dairy products, despite the fact that the other strains of streptococcus and enterococcus share LAB characteristics (Mercenier, 1990). Prebiotics are non-digestible short-chain carbohydrates that provide a host of health benefits by being specifically utilized by gut flora (Kango et al., 2022). A non-viable dietary element called prebiotic is digested specifically by good gut flora. These

DOI: 10.1201/9781003493143-6

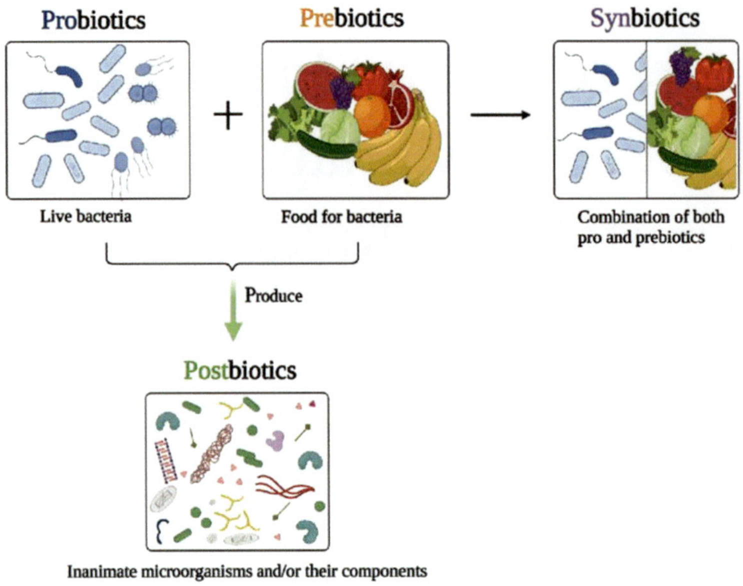

FIGURE 6.1 gives a description of probiotics, prebiotics, synbiotics and postbiotics. (Lou X et al., 2023.)

are intended to enhance gut microbiota by increasing quantity and activity of *lactobacilli* and *bifidobacteria* in the diet (Manning et al., 2004). Without giving the necessary criteria, a number of food ingredients, particularly oligosaccharides and polysaccharides have been reported to have prebiotics activity. The total native population of *bifidobacteria* is stimulated to develop by the eaten prebiotic and the larger that population the more new bacterial cells appear in feces (Roberfroid et al., 2007). Numerous species of bacteria that are found on mucosal surfaces as well as in the gut lumen are found in the gastrointestinal tract. Many gastrointestinal illnesses and age-related digestive disorders are associated with an imbalance in bacterial populations. Prebiotics have the potential to modify the intestinal microbiota in a beneficial way. Studies conducted thus far have shown that they can also maintain a healthy microbial balance in the aging gut and may be used as promising therapeutic for colon cancer and inflammatory bowel disease (Yaqoob et al., 2021) (Figure 6.1).

6.2.1 Types of Prebiotics

Prebiotics come in different varieties. Most of them are oligosaccharides carbohydrates, which are a subset of carbohydrates. The majority of research article data found that classifications based on carbohydrates are mostly accepted (Davani-Davari et al., 2019) (Figure 6.2).

6.2.1.1 Galacto-Oligosaccharides

Prebiotics have been shown to improve gut health in numerous studies (Arnold et al., 2021). Galacto-Oligosaccharides are regarded as prebiotics, as it is exactly microbial metabolites that are created during the fermentation of Galacto-Oligosaccharides. Certain microbiota members contributed to this process. Some of the members from microbiota are *Bifidobacterium longum, B. catenulatum, Lactobacillus gasseri, Lactobacillus salivarius* (Maathuis et al., 2012),

6.2.1.2 Starch and Oligosaccharide Derived From Glucose

Resistant starch is a type of starch that cannot be broken down in the upper gastrointestinal tract. It has been proposed that resistant starch is categorized into prebiotics because of its ability to produce high levels of butyrate, which can improve health (Davani-Davari et al., 2019). Fructose, is derived from oligo and polysaccharide that accumulate in around 15% of flowering plants with the family asteraceae & poaceae. Plant-derived fructose is water soluble obtained straight from sucrose, as opposed to starch. Different varieties can be separated based on their structure β2-1 linkage for insulin, β2-6 linkage for levan and β2-1 & β2-6 links for graminan (Peshe et al., 2014). Functional oligosaccharides of the xylo-oligosaccharides class have been shown to have prebiotic activity over a number of decades. Compared to other oligo saccharides molecules, xylo- oligosaccharides provide a number of benefits, including regulating plant growth root development, providing sugar as a dietary supplement, and encouraging intestinal motility and utilization of health through prebiotics (Huang et al., 2022).

6.2.1.2.1 Mechanism of Prebiotics

Prebiotics make their way to the colon, where saccharolytic bacteria ferment them in a selective manner. A moderate drop in colonic pH as well as a decrease in nitrogenous end products and fecal enzymes may result from this fermentations production of metabolites that support the intestinal metabolic balance, such as organic acids lactase, pyruvate and succinate & gases such as H_2S, CO_2, CH_4, H_2, and saturated and partially hydrogenated fats (Wang et al., 2020). Prebiotics have been shown to have numerous health benefits, including effects on the gastrointestinal tract, by improving gut barrier function, lowering the number of pathogenic bacteria and producing short-chain fatty acids. The pathophysiology of various gastrointestinal illnesses has been linked to gut microbiota. Intestinal bacteria can digest several of these polysaccharides (Guarino et al., 2020). Prebiotics have been found to boost the Para cellular pathway permeability, which is reliant on big multi-protein complexes and tight junctions connecting neighboring epithelial cells. Non-digestible oligosaccharides were precisely found to quickly boost Para cellular ion transport in the Caco-2 intestinal cell line model. The physiology and molecular alterations that take place in these crucial structures in order to determine the bonding of epithelial cell junction locally (Mavrogeni et al., 2022). Postbiotics are bioactive soluble factors also known as metabolic byproducts, are substances that are created during development and fermentation of food-grade bacteria in complex microbial culture food or gut that have beneficial effects on the food or the consumer (Moradi et al., 2020) as a recent terminology that is connected to these components of microbial fermentation. Thus metabolites short-chain fatty acids, microbial cell fractions, functional protein, extracellular polysaccharides, cell lysate, teichoic acid,pept6idoglycan derived muropeptides and pill-type structure are just a few of several components that can be found in postbiotics (Jensen-Kroll et al., 2022). Postbiotcs are known for their health advantages. Additional human trials are required to improve their health benefits. Postbiotcs have the potential to enhance the well-being of the host.

6.3 FACTORS AFFECTING GIT INFLAMMATION

6.3.1 Composition of Gut Microbe

The gut home is a house for several different numbers of major bacteria phyla, including firmicutes, verrucomicrobia, actinobacteria, proteobacteria, bacteriodetes. The amount of bacteria in the body can reach 105 kg or around 2% of an average 70 kg human being. The genes of bacteria are as large as humans with 150 times the proportions. Preterm infants have an underdeveloped innate immune system and significantly less maternal antibiotics transferred over the placenta. These two conditions make it easier for preterm to have infections. Less time in pregnancy is linked to lower levels of peripheral blood lymphocytes (Collado et al., 2015).

6.3.2 Genetic Factor

A class of chronic severe gastrointestinal tract disorders known as inflammatory bowel disease have their peak development in adolescence and early adulthood (Huang et al., 2017). Numerous proteins that microorganism is exposed to are availability of nutrients and the degree of immune system reaction to microbes all encoded by genes. Gene variations that regulate the immune system's ability to recognize and react to microorganisms can disturb intestinal homeostasis and increase intestinal inflammation (Cohen et al., 2019). Based on the concord rate, the genetic liability of intestinal bowel disease ranges from 0–20% (Saito, Mitra, & Mayer, 2010). Genetic impacts are minimal since a particular mutation is rarely enough to cause a disease to develop on its own. With complex genetic illnesses tending to aggregate in a family, they are somehow heritable. Some of the predictable genes include NOD2, ATG16L1, IL23, IL12B, STAT3 and NKX2-3 have been identified in regulating genetic role for gastrointestinal disease (Saito, 2011).

6.3.3 Immunological Status

The incidence of gastrointestinal infection in children under the age of five as well as deficiencies in cell-mediated immunity, are controlled by pro-inflammatory cytokines like IL- 1, TNF-α, IL-6, IL-12, which are primarily produced by neutrophils and macrophage and anti-inflammatory cytokines like IL-10 which reduce inflammation after pathogens have been removed (Rodríguez et al., 2011). Gut contains about 80% of the body's immune cells; it is thought to be the largest immunological organ in the human body. The gut is a highly complex system of the human body and major of antimicrobial peptides can alter or modify the bacterial gut flora, which in turn affects human immune status, hence indirectly regulating immune reaction with alpha defensing expression displayed modifications in microbiota composition, including higher quantity segmented filamentous bacteria (Wittkopf et al., 2014). Gut microbiota gut bacteria and host are co-evolved and interact in variety of immunological and metabolic ways (Nagao-Kitamoto et al., 2016). Various bacterial communities such as *Lactobacillus* and *bifidobacterium, clostridium, Roseburia,* and *butyricicoccus* are responsible to cause infection; besides some virome communities are also responsible for causing infection such as Myoviridae, Podoviridae, SIphoviridae and Microviridae represents higher intensity of intestinal lining infection with profound changes in mechanism (Mukherjee et al., 2018)

6.3.4 Breast Feeding

A portion of the reason for the global dispersal of certain parasites, especially helminths, involves an irritate and obligatory life cycle (Kutty et al., 2014). Potential relationship between baby feeding habits and risk of gastrointestinal bowel disease to develop into ulcerative colitis due to mother's milk (Davis et al., 2001).

6.3.5 Diet, Sleep

Sleep disturbance and diet are linked to higher causes of mortality in cases of gastrointestinal (Kinnucan et al., 2013). The maintenance of the gut immune system and control of digestive tract depend upon brain gastrointestinal system link. Endocrine signaling, intestinal permeability and immunological activation are the processes via which the brain gets activation function, which can alter algorithms to disturb the gastrointestinal tract (Khanijow et al., 2015). The balance of gut microbes is significantly influenced by daily nutrition, which can either sustain a normal population or alter it. Improper diet can alter physiology and gene expression (Mansour et al., 2021). Western diet may have been found to show a higher risk in the development of gastrointestinal infection (Cosnes et al., 2010). Smoking A number of mechanisms including gut irritation, compromised mucosal immune response and mucosal injury, have been found with smoking, which ultimately involves for higher risk of gastrointestinal infection (Berkowitz et al., 2018).

6.3.6 Medication Air Pollution

It is estimated around 99% of the world population is breathing polluted air. This polluted air may have higher chances of altering gut microbiota, which include symptomatic changes such as oral, skin, GIT irritation, eye irritation etc. Largely, the air pollution role shows symbiotic microbial symbiosis. Even short-term exposure to air pollutants shows strong GIT infection (Gupta et al., 2022). The most frequent ADR that happens in the oral cavity is xerostomia. Mucosal irritation and inflammation are ADR that can damage the esophagus (Domagała et al., 2016). Medications such as corticosteroids, aspirin, non-aspirin, NSAIDs, antibiotics, and potassium salts are most commonly linked to harm to the gastrointestinal mucosa (Henry et al., 1994).

6.4 INVOLVEMENT OF GUT DYSBIOSIS AND GIT INFLAMMATION

In recent years, the investigation into the involvement of gut dysbiosis in gastrointestinal (GIT) inflammation has emerged as a pivotal area of research, drawing substantial attention within the scientific community. Dysbiosis, characterized by an imbalance in the composition and function of the gut microbiota, has become increasingly recognized as a key contributor to the development and progression of various inflammatory conditions affecting the GIT. This section delves into the complex interplay between gut dysbiosis and GIT inflammation, shedding light on their intricate relationship through the lens of compelling evidence derived from recent studies (Plummer et al., 2014). The connection between gut dysbiosis and chronic kidney disease (CKD), along with its attendant morbidity, underscores a multifaceted interplay involving microbial imbalances and alterations in bacterial metabolites. This intricate relationship has been elucidated through a burgeoning body of research, highlighting the profound impact of gut dysbiosis on renal health (Feng et al., 2021). Dysbiosis disrupts the delicate equilibrium of the gut microbiota, fostering the proliferation of pathogenic species while diminishing beneficial microbes, thus triggering a cascade of detrimental effects. Moreover, dysbiosis-induced alterations in bacterial metabolites further exacerbate CKD progression, culminating in a pro-inflammatory milieu within the gastrointestinal tract. This dysregulated immune response, characterized by a pro-inflammatory immune cell signature, not only perpetuates local inflammation but also contributes to systemic inflammation, thereby exacerbating the pathogenesis of CKD and its associated morbidity. Such insights underscore the pivotal role of gut dysbiosis as a modifiable risk factor in the progression of CKD and offer promising avenues for therapeutic intervention aimed at restoring microbial balance and ameliorating the inflammatory burden associated with this debilitating condition (Cao et al., 2022). The notable association between gut dysbiosis and chronic kidney disease (CKD) extends even to pediatric populations, emphasizing the specificity of microbiota-immune interactions in driving chronic inflammation within the context of CKD. This insight underscores the profound impact of dysbiosis on renal health from the early stages of life. Studies examining children with CKD have unveiled a consistent pattern of microbial imbalance and dysregulated immune responses, further corroborating the pivotal role of gut dysbiosis in CKD-related inflammation (Feng et al., 2021). Such specificity highlights the potential for personalized interventions, including tailored dietary strategies and microbiota-targeted therapies, to modulate the gut microbiota composition and mitigate inflammatory processes in CKD patients of all ages. By addressing the underlying dysbiosis and its immunological consequences, these targeted interventions hold promise to improve patient prognosis and quality of life across the lifespan, heralding a paradigm shift towards precision medicine approaches in the management of CKD (Holle et al., 2022). Antibiotic-associated dysbiosis exerts a substantial influence on the gut microbiota's ability to regulate intestinal inflammation, thereby affecting the efficacy of clinical therapies. Various antibiotics, such as vancomycin, streptomycin, and metronidazole, exhibit diverse impacts on the composition and functionality of the gut microbiota, consequently influencing its capacity to modulate inflammation (Strati et al., 2021). The differential effects of these antibiotics underscore the intricate relationship between antibiotic exposure and gut microbiota dynamics, particularly in the context of inflammation management. This highlights the critical importance of judicious antibiotic use to preserve the regulatory role of the gut microbiota in inflammation. As antibiotics represent a cornerstone of clinical therapy for various conditions, including infectious diseases and inflammatory disorders, careful consideration of their selection and dosage is warranted to minimize disruptions to the gut microbiota and mitigate the risk of antibiotic-associated dysbiosis (Konstantinidis et al., 2020). By preserving the integrity of the gut microbiota, clinicians can optimize therapeutic outcomes and minimize the potential for exacerbating inflammation, thereby emphasizing the imperative for a balanced approach to antibiotic stewardship in clinical practice (Strati et al., 2021). The concept of the lung-gut axis offers valuable insights into the reciprocal relationship between lung inflammation and the gut microbiota, elucidating how perturbations in one system can impact the homeostasis of the other. Emerging research suggests that pulmonary infection with Aspergillus fumigatus, a common fungal pathogen implicated in respiratory diseases, triggers dysbiosis not only within the lungs but also in the gastrointestinal tract, each exhibiting distinct characteristics.

This bidirectional crosstalk between the lungs and gut highlights the interconnectedness of these seemingly disparate physiological systems and underscores the profound influence of pulmonary inflammation on gut microbial composition and function (Halder et al., 2023). The observed bacterial dysbiosis following Aspergillus fumigatus infection underscores the intricate interplay between host immunity, microbial communities, and pathogen-driven responses, shedding light on novel avenues for therapeutic intervention targeting both respiratory and gastrointestinal health. By unraveling the complexities of the lung-gut axis, researchers aim to elucidate the underlying mechanisms driving dysbiosis and inflammation in both organs, paving the way for innovative strategies to restore microbial balance and mitigate disease progression in respiratory and gastrointestinal disorders (Popovic et al., 2023). The concept of the lung-gut axis offers valuable insights into the reciprocal relationship between lung inflammation and the gut microbiota, elucidating how perturbations in one system can impact the homeostasis of the other. Emerging research suggests that pulmonary infection with Aspergillus fumigatus, a common fungal pathogen implicated in respiratory diseases, triggers dysbiosis not only within the lungs but also in the gastrointestinal tract, each exhibiting distinct characteristics. This bidirectional crosstalk between the lungs and gut highlights the interconnectedness of these seemingly disparate physiological systems and underscores the profound influence of pulmonary inflammation on gut microbial composition and function (Cai et al., 2022). The observed bacterial dysbiosis following Aspergillus fumigatus infection underscores the intricate interplay between host immunity, microbial communities, and pathogen-driven responses, shedding light on novel avenues for therapeutic intervention targeting both respiratory and gastrointestinal health. By unraveling the complexities of the lung-gut axis, researchers aim to elucidate the underlying mechanisms driving dysbiosis and inflammation in both organs, paving the way for innovative strategies to restore microbial balance and mitigate disease progression in respiratory and gastrointestinal disorders (Yang et al., 2020). Research on zebrafish mutants has shown that host tp53 mutations, common in early stages of colitis-associated colorectal cancer, can induce gut microbiota dysbiosis and chronic intestinal inflammation. This dysbiosis, characterized by reduced diversity and the enrichment of pathobionts like Aeromonas spp., suggests a crucial role for host genetics in maintaining gut microbiome balance and immune homeostasis. The study highlights the potential of targeting sialic acid metabolism as a therapeutic strategy for tp53 mutation-induced dysbiosis and inflammation (Lee et al., 2022). These studies collectively underscore the complex interplay between gut dysbiosis and GIT inflammation, emphasizing the role of microbial imbalance, host genetics, and external factors like antibiotic use in modulating inflammatory processes. The evidence points towards the potential of microbiota-targeted interventions in managing GIT inflammation, paving the way for future research in this domain (Figure 6.2).

6.5 MECHANISM OF PROBIOTIC, PREBIOTIC AND POSTBIOTICS TO MODULATE THE GUT MICROBIOTA

The modulation of the gut microbiota by probiotics, prebiotics, and postbiotics represents a pivotal mechanism in maintaining gastrointestinal health and preventing diseases. Probiotics, live microorganisms that confer health benefits to the host, interact with intestinal epithelial cells (IECs) and immune cells via Toll-like receptors, inducing an immune response through the release of cytokines and chemokines. This interaction enhances the intestinal immunologic barrier, primarily through the intestinal response of immunoglobulin A and the mitigation of inflammatory responses, stabilizing the gut environment (Liu et al., 2022). Prebiotics, indigestible dietary fibers that feed beneficial bacteria, play a crucial role in shaping the gut microbiota composition. They promote the growth of health-promoting bacteria, such as *Lactobacilli* and *Bifidobacteria*, and influence the production of short-chain fatty acids (SCFAs), which have anti-inflammatory properties and strengthen the gut barrier function (Slavin, 2013). A novel prebiotic blend has been shown to prevent the development of irritable bowel syndrome (IBS) in mice by modulating the gut microbiota and immune response, highlighting the potential of prebiotics in preventing gastrointestinal disorders (Chen et al., 2017). Postbiotics, the bioactive compounds generated by probiotics during fermentation, represent a promising avenue for promoting host health. Among these compounds, short-chain fatty acids (SCFAs), functional proteins, peptides, and polysaccharides have garnered particular attention for their beneficial effects on the host. Postbiotics play a pivotal role in maintaining gut integrity by fortifying the mucosal barrier and enhancing epithelial cell proliferation and tight junction formation. Furthermore, they contribute to immune modulation by regulating the balance of pro- and anti-inflammatory cytokines, promoting regulatory T cell differentiation, and enhancing mucosal immune responses (Thorakkattu et al., 2022). Additionally, postbiotics exhibit direct antimicrobial activity against pathogens, thereby impeding their colonization and proliferation within the gastrointestinal tract. Through these multifaceted mechanisms, postbiotics exert a comprehensive influence on host physiology, promoting gut health, bolstering immune function, and safeguarding against pathogenic invasion. As research into postbiotics continues to advance, leveraging these bioactive compounds holds immense promise for the development of novel therapeutic strategies aimed at enhancing host-microbe symbiosis and ameliorating a range of gastrointestinal and immune related disorders (Prajapati et al., 2023). The integration of probiotics, prebiotics, and postbiotics represents a synergistic strategy for modulating the gut microbiota and promoting overall gut health. Probiotics introduce beneficial microorganisms into the gastrointestinal tract, helping to restore microbial balance and enhance host-microbe symbiosis. Prebiotics serve as substrates for the growth and proliferation of these beneficial microorganisms, fostering their colonization and metabolic activity within the gut. Meanwhile, postbiotics, the bioactive

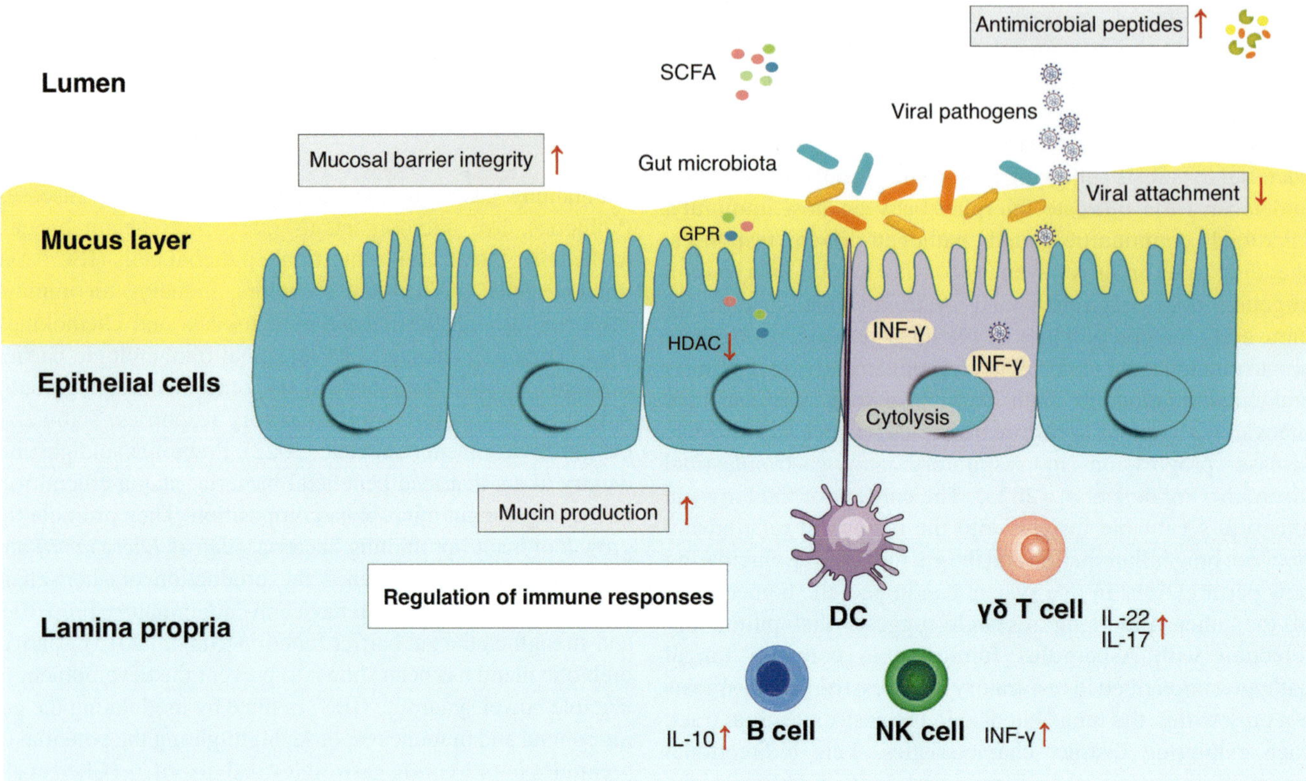

FIGURE 6.2 depicts the correlation between gut dysbiosis and GIT inflammation (Fakharian et al., 2023.).

compounds produced by probiotics, exert a range of beneficial effects on the host, including the maintenance of gut integrity, modulation of immune responses, and inhibition of pathogenic colonization (Morniroli et al., 2021). By combining probiotics, prebiotics, and postbiotics, this holistic approach addresses multiple facets of gut health, from microbial composition and function, to host immune regulation and barrier integrity. This comprehensive modulation of the gut microbiota offers a protective effect against gastrointestinal inflammation and diseases, providing a promising avenue for preventive and therapeutic interventions. As our understanding of the gut microbiota continues to evolve, harnessing the synergistic action of probiotics, prebiotics, and postbiotics holds immense potential for optimizing gut health and mitigating the risk of gastrointestinal disorders (Ji et al., 2020). The intricate interplay between probiotics, prebiotics, postbiotics, and the host's immune system underscores the complexity and therapeutic potential of targeting the gut microbiota. These components engage with the immune system through diverse mechanisms, including the Toll-like receptor (TLR) pathway, short-chain fatty acid (SCFA) production, and modulation of gut hormones. Probiotics and their metabolites interact with pattern recognition receptors such as TLRs, triggering immune responses that contribute to immune regulation and homeostasis. Prebiotics, through their fermentation by gut bacteria, yield SCFAs like acetate, propionate, and butyrate, which serve as energy sources for colonocytes and exert immunomodulatory effects by influencing immune cell function and cytokine production. Additionally, the gut microbiota, influenced by probiotics, prebiotics, and their metabolites, can modulate the secretion of gut hormones such as glucagon-like peptide 1 (GLP-1) and peptide YY (PYY), which regulate gut motility, appetite, and glucose metabolism, thereby impacting systemic immune function. This intricate interplay highlights the multifaceted nature of gut-microbiota-immune interactions and underscores the potential of targeting the gut microbiota as a therapeutic approach for immune-mediated disorders. By understanding and manipulating these complex interactions, researchers aim to develop innovative strategies for modulating immune responses and restoring immune homeostasis, offering promising avenues for the management of various inflammatory and autoimmune conditions (Wu and Wu, 2012).

6.6 BACTERIAL METABOLITES: MECHANISM OF INFLAMMATION INHIBITION

Propionate and acetate rank among the extensively researched bacterial metabolites known for their potent anti-inflammatory properties. These short-chain fatty acids (SCFAs) exert their effects through diverse mechanisms, including the inhibition of histone deacetylase (HDAC) activity, facilitation of regulatory T cell differentiation, and suppression of pro-inflammatory cytokine production. Additionally, tryptophan derivatives produced by gut bacteria act as precursors for neurotransmitters such as serotonin and contribute to immune regulation by activating the aryl hydrocarbon receptor (AhR) pathway (Tedelind et al., 2007). Furthermore, other bioactive compounds synthesized by the gut microbiota, such as secondary bile acids and polyamines, play roles in maintaining immune homeostasis by modulating inflammation and influencing immune cell function. The combined effects of these bacterial metabolites underscore the pivotal role of the gut microbiota in shaping host immune responses, highlighting the potential for targeting microbial metabolism as a therapeutic approach for immune-mediated disorders. Further exploration of the mechanisms through which bacterial metabolites modulate inflammation and immune function holds promise for developing novel interventions aimed at restoring immune homeostasis and alleviating inflammatory diseases (Liu et al., 2022). Bacterial metabolites, particularly short-chain fatty acids (SCFAs) like butyrate, propionate, and acetate, play multifaceted roles in maintaining gastrointestinal health and immune homeostasis. Firstly, SCFAs serve as vital energy sources for colonocytes, the epithelial cells lining the colon, facilitating their growth, differentiation, and maintenance of barrier integrity. This metabolic support bolsters the structural integrity of the gut epithelium, enhancing its barrier function and fortifying defenses against pathogenic invasion and luminal antigens. Moreover, SCFAs exert immunomodulatory effects by inhibiting histone deacetylases (HDACs), thereby influencing gene expression patterns in immune cells, and promoting anti-inflammatory responses. Additionally, SCFAs activate G-protein-coupled receptors (GPCRs) such as GPR41 and GPR43, which are expressed on immune cells, leading to downstream signaling events that culminate in the suppression of pro-inflammatory cytokine production and the promotion of regulatory T cell differentiation (Nogal et al.,). Through these concerted mechanisms, bacterial metabolites act as essential mediators in orchestrating crosstalk between the gut microbiota and the host immune system, ultimately contributing to the maintenance of immune homeostasis and the mitigation of inflammatory responses within the gastrointestinal tract (Gasaly et al., 2021). Tryptophan derivatives, produced through bacterial metabolism, activate the aryl hydrocarbon receptor (AhR), a ligand-activated transcription factor that plays a key role in mucosal immunity. Activation of AhR by these metabolites promotes the differentiation of regulatory T cells and interleukin-22 production, contributing to the maintenance of gut barrier integrity and the suppression of inflammatory responses (Puccetti et al., 2022). Furthermore, microbial metabolites such as indole and its derivatives have been shown to strengthen the epithelial barrier function, modulate the secretion of antimicrobial peptides, and influence the maturation and function of immune cells. These actions collectively contribute to the prevention of pathogen overgrowth and the maintenance of immune tolerance (Ye et al., 2022). The intricate interplay between bacterial metabolites and host immune cells underscores the multifaceted role of the gut microbiota in both health and disease. One notable example of this interaction involves the direct influence of microbial metabolites on the activity of cyclooxygenase (COX), a pivotal enzyme in the inflammatory cascade. Bacterial metabolites, including short-chain fatty acids (SCFAs) and other bioactive compounds, can interact with COX enzymes, modulating their activity and subsequent production of pro-inflammatory prostaglandins (Wu and Wu, 2012). By regulating COX activity, microbial metabolites can exert a profound impact on the inflammatory process, potentially influencing the course and severity of inflammatory conditions such as sepsis. This intricate interplay highlights the complexity of the gut microbiota-host immune system axis and underscores the diverse mechanisms through which microbial metabolites can shape immune responses and impact health outcomes. Further elucidating these interactions holds promise for the development of novel therapeutic strategies targeting the gut microbiota to modulate inflammation and improve outcomes in various disease states (Figure 6.3).

6.7 TOXICOLOGY OF PROBIOTIC, PREBIOTIC AND POSTBIOTICS IN HEALTH

The investigation into the toxicology of probiotics, prebiotics, and postbiotics in health represents a pivotal area of research, emphasizing the importance of elucidating their safety profile and potential adverse effects. Probiotics, comprising live microorganisms that confer health benefits to the host, have garnered substantial attention for their therapeutic potential in various conditions, ranging from gastrointestinal disorders to immune modulation. However, concerns regarding their safety, particularly in vulnerable populations such as immunocompromised individuals or those with underlying health conditions, necessitate rigorous assessment of their toxicological profile to mitigate potential risks. Similarly, prebiotics, non-digestible food ingredients that selectively stimulate the growth or activity of beneficial gut bacteria, are recognized for their role in promoting gut health and modulating metabolic processes (Ji et al., 2020). Despite their generally favorable safety profile, investigations into the potential adverse effects of prebiotic supplementation, such as gastrointestinal discomfort or allergic reactions, remain imperative to ensure their safe and effective use. Moreover, the emerging field of postbiotics, encompassing bioactive compounds produced by food-grade microorganisms during fermentation, presents novel opportunities for enhancing health outcomes. While postbiotics offer promising benefits, including immunomodulation and gut barrier reinforcement, comprehensive toxicological assessments are warranted to ascertain their safety and tolerability (Markowiak et al., 2017). By meticulously evaluating the toxicology of probiotics, prebiotics, and postbiotics, researchers aim to elucidate their risk-benefit profile and inform evidence-based recommendations for their incorporation into clinical practice and dietary interventions, thereby optimizing health outcomes while minimizing potential adverse effects (Ji et al., 2020). However, understanding their toxicological aspects is essential for their safe use in clinical and dietary settings. Probiotics, widely recognized for their health-promoting properties, are generally regarded as safe for consumption, with a low incidence of side effects observed in clinical settings. These side effects typically manifest as mild gastrointestinal symptoms, including gas, bloating, and transient abdominal discomfort,

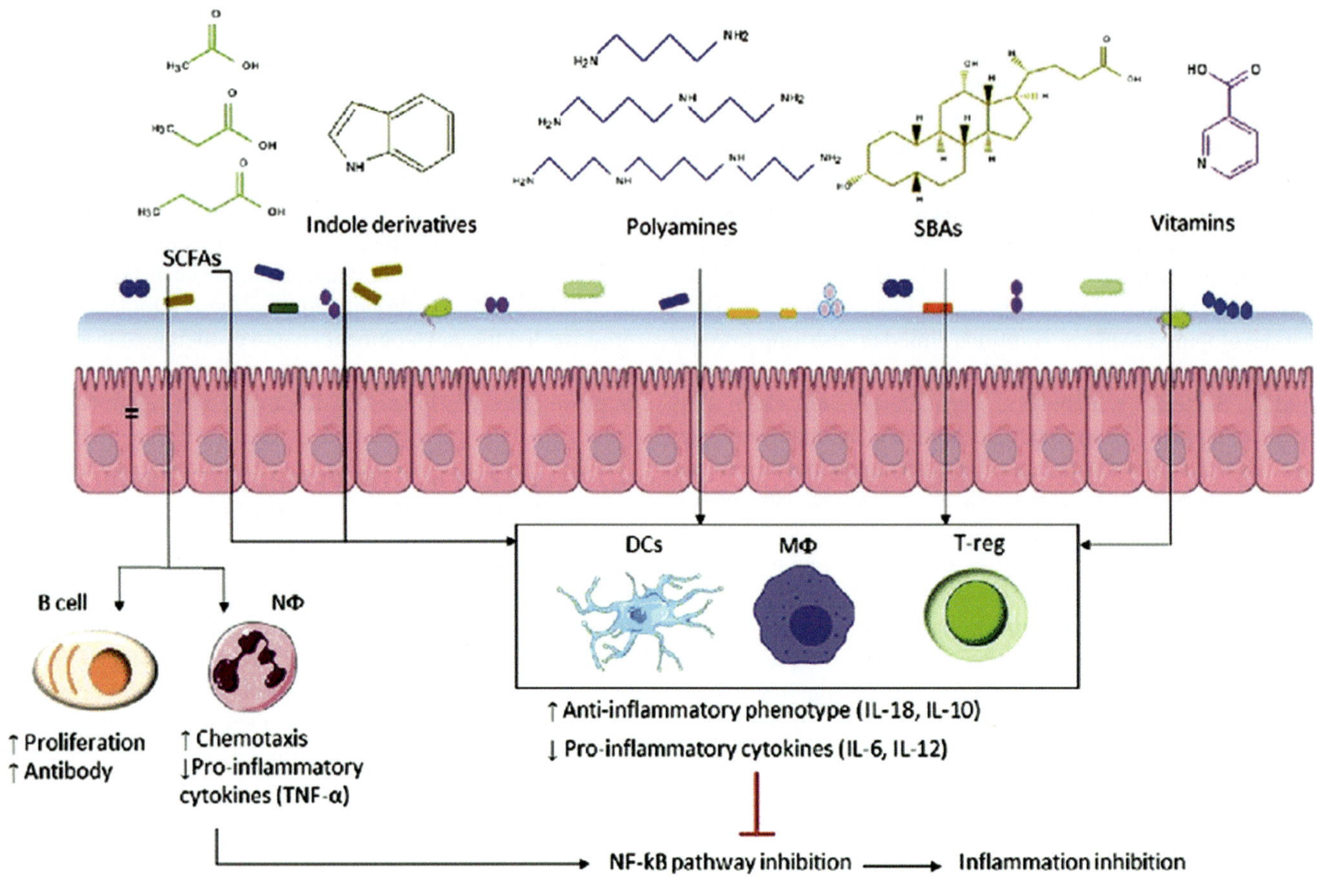

FIGURE 6.3 depicts the role bacterial metabolites for inhibiting inflammation. (Hosseinkhani F et al., 2021.)

which tend to resolve spontaneously with continued use. However, in certain populations, such as immunocompromised individuals or those with underlying health conditions, there exists a potential risk for systemic infections associated with probiotic supplementation. This heightened susceptibility arises from compromised immune function, which may render individuals more vulnerable to opportunistic infections from ingested probiotic microorganisms. As such, diligent monitoring and caution are warranted when considering probiotic use in these vulnerable populations to mitigate the risk of adverse outcomes. Rigorous clinical evaluations, encompassing randomized controlled trials and long-term safety assessments, are essential for elucidating the safety profile of probiotic strains and identifying any potential adverse effects (Sanders et al., 2010). Furthermore, robust post-market surveillance mechanisms play a critical role in monitoring real-world usage patterns and identifying adverse events associated with probiotic consumption. By implementing comprehensive safety measures, including stringent quality control protocols, standardized manufacturing practices, and ongoing monitoring of adverse events, regulatory authorities can safeguard public health and ensure the safe and responsible use of probiotics as dietary supplements or adjunctive therapies. These proactive measures serve to instill confidence in the safety and efficacy of probiotic products while facilitating informed decision-making by healthcare professionals and consumers alike (Zhu et al., 2023). Prebiotics, revered for their role in fostering gut health by selectively promoting the growth and activity of beneficial gut bacteria, can occasionally elicit discomfort in certain individuals, particularly when consumed in large quantities. The fermentation of prebiotics by gut microbiota yields short-chain fatty acids (SCFAs) and other metabolites, which can lead to symptoms such as bloating, gas, and abdominal pain. These gastrointestinal disturbances arise from the increased production of gas during microbial fermentation, exerting pressure on the intestinal lumen and causing discomfort. While these symptoms are typically mild and transient, they may pose challenges for individuals sensitive to dietary changes or those predisposed to gastrointestinal distress (Davani et al., 2019). Consequently, it is imperative to carefully assess the tolerance and optimal dosage of prebiotic supplements to minimize the risk of adverse effects while maximizing their health benefits. This necessitates personalized approaches tailored to individual preferences, tolerances, and gut microbiota compositions, with gradual dose titration and monitoring of symptomatology to optimize outcomes. Additionally, selecting prebiotic sources with lower fermentability profiles or incorporating them into meals to mitigate rapid fermentation may help alleviate gastrointestinal discomfort. By elucidating the factors influencing prebiotic tolerance and refining dosage recommendations through systematic evaluation, healthcare professionals can empower individuals to reap the benefits of prebiotics while minimizing the risk of adverse reactions, thereby optimizing gut health and overall well-being (Davani et al., 2019).

Postbiotics, comprising non-viable microbial products or metabolic byproducts generated during fermentation, are esteemed for their purported safety profile compared to live microbial supplements. Unlike probiotics, which contain live microorganisms, postbiotics offer the advantage of circumventing concerns related to microbial viability and the potential for pathogenicity, making them inherently safer for consumption. However, the intricate and diverse compositions of postbiotics underscore the need for comprehensive toxicological assessments to ascertain their safety and tolerability (Vinderola et al., 2022). The safety profile of postbiotics is influenced by various factors, including their source, the specific microorganisms involved in their production, and the health status of the host. Postbiotics derived from food-grade microorganisms and produced under controlled fermentation conditions are generally regarded as safe. Nonetheless, rigorous evaluations are necessary to identify any potential harmful effects associated with postbiotic supplementation, particularly in vulnerable populations or individuals with underlying health conditions. By conducting thorough toxicological assessments encompassing acute and chronic exposure studies, genotoxicity assessments, and evaluation of potential allergenicity, researchers can elucidate the safety profile of postbiotics and establish evidence-based guidelines for their use in clinical practice Kango & Nath, 2024). Furthermore, ongoing post-market surveillance and monitoring of adverse events are essential to ensure the continued safety and efficacy of postbiotic products. By integrating these measures, regulatory authorities and healthcare professionals can confidently recommend postbiotic interventions as a safe and effective strategy for promoting gut health and overall well-being. While probiotics, prebiotics, and postbiotics are celebrated for their myriad health benefits, it's imperative to acknowledge the importance of assessing their safety and potential toxicological effects. Despite their generally favorable safety profiles, these bioactive compounds and microbial products can elicit adverse reactions or pose risks, particularly in certain populations or under specific circumstances. Continuous research endeavors and rigorous clinical trials are essential to comprehensively evaluate their impact on human health and elucidate any potential adverse effects. These investigations encompass a broad spectrum of toxicological assessments, including acute and chronic exposure studies, genotoxicity evaluations, assessment of allergic reactions, and monitoring of long-term outcomes(Thorakkattu et al., 2022). Moreover, systematic reviews and meta-analyses play a crucial role in synthesizing existing evidence and identifying emerging trends or safety concerns associated with probiotic, prebiotic, and postbiotic interventions. By leveraging a multidisciplinary approach that integrates expertise from fields such as microbiology, toxicology, nutrition, and clinical medicine, researchers can gain deeper insights into the safety profile of these bioactive compounds and establish evidence-based guidelines for their use in healthcare practice (Asha et al., 202). Furthermore, proactive measures such as post-market surveillance and pharmacovigilance programs are indispensable for monitoring real-world usage patterns and identifying potential safety signals or adverse events associated with probiotic, prebiotic, and postbiotic consumption. Through these concerted efforts, the scientific community can ensure the responsible and informed utilization of probiotics, prebiotics, and postbiotics to optimize health outcomes while mitigating potential risks, thereby advancing our understanding of their role in promoting human health and well-being (Raj et al., 2019).

6.8 FUTURE ROLE OF PROBIOTIC, PREBIOTIC AND POSTBIOTICS TO REDUCE INFLAMMATION

Prebiotics, probiotics, and postbiotics, which are often derived from foods or the human gut microbiota have all been shown via significant study and long-term use to be safe. Unlike to conventional artificial or chemically produced pharmacological therapies, they often show lower risks of drug interactions, making them a safer choice for people taking numerous medicines (Bronzini et al., 2023). Probiotics, prebiotics, and postbiotics provide individualized and tailored therapies for a range of ailments, which has the potential to transform public health and the healthcare system (Naito, 2024). Boost immunity, sustain a healthy gut flora, and treat a variety of conditions ranging from mental health to gastrointestinal illnesses by using the benefits of probiotics and prebiotics (Sánchez et al., 2022). Widespread use of probiotic treatments may result in better patient outcomes, less need for antibiotics, and fewer medical expenses. In order to fully use the gut microbiota for a variety of medical diseases, probiotic, prebiotic, and postbiotic development is being directed towards tailored and targeted therapeutics (Wang et al., 2024). Thanks to developments in next-generation sequencing, metagenomics, and bioinformatics, scientists can now better comprehend the intricate relationships between gut microbiota and host health, which can help create novel therapies (Lim & Nam, n.d.). They are better suited for long-term usage and more amiable towards people who are susceptible than other traditional medications, which might cause tolerance, dependency, or other unpleasant responses. Probiotics work on a variety of intricate methods that vary depending on the bacteria (Meng et al., 2020). Some common processes have been found, however. Probiotics have the ability to modify the makeup of the microbiota in the intestines, improve the operation of the intestinal barrier, fight with microorganisms for micronutrients and binding sites on the intestinal lining, and affect the immune system (Dixit et al., 2021). They are also capable of producing metabolites and antimicrobial compounds that may have a direct or indirect impact on the health of the host. There is currently a limited selection of compounds that have been verified to be prebiotics, with galactans and fructans (like inulin) dominating the market

(Tran et al., 2022). Novel potential prebiotic substances have been developed as a result of the goal to support a larger range of commensal species. Yeast-based substances, many non-carbohydrate substances like polyphenolics, fatty acids, herbs, and other micronutrients, and carbohydrates derived from plants—the source of traditional prebiotics like inulin may also be included. These may also mimic substrates derived from animals (oligosaccharides found in milk, O-linked glycans found in mucins). The primary manner in which that prebiotics function is by giving the good bacteria in the stomach material to digest (Haak et al., 2018). Since human enzymes cannot break them down, they pass through to the colon undigested, where the gut flora ferments them. Acetate, propionate, and butyrate are among the short-chain fatty acids (SCFAs) produced during this fermentation process. Prebiotics aid to maintain a healthy gut microbiota by reducing the amount of harmful bacteria and specifically promoting the development and activity of good gut bacteria like *Lactobacilli* and *Bifidobacteria*. With an increasing emphasis on affordability, size, and sustainability, prebiotics will probably be separated from emerging sources in future generations. One abundant and efficient resource for bioactive compounds that are naturally occurring is the 1.3 billion tons of food debris produced yearly in the food chainii. Prebiotics including pectin from citrus peel and arabinoxylans from distillery and brewing waste, are found in many of the byproducts of processing fruits, vegetables, and grains. Subsequent prebiotic molecules might potentially undergo structural or chemical modifications to alter their usefulness via the use of oxidation, excessive pressure, sonication, acid, and enzyme treatments (Eom et al., 2018). Moreover, novel benefit profiles might be able to be created by combining certain prebiotic compounds in optimal mixes. Prebiotics are also being used with increasing attention to influence various host microbiomes, including those in the female urogenital tract, oral cavity, and skin. For instance, topical administration of prebiotic glucomannan hydrolysates has been shown to alter the skin microbiota and lessen acne. Additionally, more focused prebiotic administration inside the distal colon is being considered for ulcerative colitis and colorectal cancer therapy or prevention. By delivering intact prebiotics to the distal colon and selectively stimulating the genera that metabolize carbohydrates there, a combination of prebiotics with varying chain lengths or particular delivery methods may be able to lessen local proteolysis and the consequent generation of unwanted metabolites. Beyond only promoting microbial growth, prebiotics are expected to play a major role in modulating the intestinal bacterial metabolome to a better configuration (Holleran et al., 2017).

6.9 CLINICAL TRIALS OF THE PROBIOTICS SUGGESTING THEIR THERAPEUTIC POTENTIAL IN GASTROINTESTINAL HEALTH

A major focus of microbiome research is seeking for both individual and community microbial characteristics to predict illness occurrence, development, and response to therapy. This study is made possible by the effective and potent processing of massive data sets. Large cross-sectional studies have linked distinct taxonomic patterns and particular genera and species to host biomarkers, nutritional and lifestyle variables, and health and illness status. These observations indicate a strong interest in focused approaches to individually or within population subgroups modify the microbial makeup of hosts (Ji et al., 2023). Prebiotics have been shown in several animal experiments to specifically promote the development and activity of gut-dwelling beneficial bacteria, enhancing host health. Adolescent rats' anxiety-like behavior was reduced and their gut microbiota was altered when they were fed a prebiotic mixture that included fructo-oligosaccharides (FOS) and galacto-oligosaccharides (GOS) (Liang & Xing, 2023). This finding suggests that prebiotics could be a novel therapy for anxiety disorders. Research findings shown prebiotic GOS and polydextrose (PDX) have the intriguing potential to improve the quality of rats' sleep by mitigating stress-induced sleep disruptions. Prebiotics may have a role in treating stress-related sleep problems, as shown by the GOS/PDX prebiotic diet, which enhanced the fecal bile acid pool and raised the relative abundance of *Parabacteroides distasonis* in stressed rats (Salminen et al., 2021). Prebiotic xylooligosaccharides (XOS) have the potential to alleviate the effects of high-fat diet (HFD)-induced metabolic syndrome (MS) in rats by augmenting the quantity of *Bifidobacteria* and *Lactobacilli* while diminishing the prevalence of *Proteobacteria*. The most prevalent metabolic illness in the world, nonalcoholic fatty liver disease (NAFLD), may also lower insulin resistance in NAFLD patients, causing hyperinsulinemia and raising the risk of bladder cancer in the process. Prebiotics are one of the natural products that 168 doctors are presently examining to help with NAFLD (Gorreja & Walker, n.d.). By controlling the expression of genes related to lipogenic enzymes and lowering fatty acid synthase activity, FOS and inulin decrease de novo fat synthesis in rats given a high-fat diet (HFD) while decreasing liver weight and steatosis. Moreover, they counteract the dysbiosis of the gut microbiota brought on by HFD and reinstate intestinal permeability. In mice, prebiotic GOS and polyphenol supplements may also lessen the buildup of liver fat, lessen systemic inflammation, and enhance insulin sensitivity (Mahdizade Ari et al., 2022). Additionally, prebiotics are good for gastrointestinal disorders. In their investigation on the effectiveness of prebiotic trans-GOS in altering the colonic microbiota and easing IBS symptoms, GOS selectively activated gut *Bifidobacteria* in IBS patients and successfully reduced symptoms of IBD (Registered in ISRCTN, ISRCTN54052375). In patients with diarrhea-predominant IBS, a randomized double-blind, placebo-controlled research (NCT04206410) found that a synbiotic formulation including *Lactobacillus*, *Bifidobacterium* probiotic bacteria, and short-chain fructooligosaccharides (FOS) improves related symptoms like gas and bowel habits. According to these results, prebiotics may be used as therapeutic agents to treat a range of illnesses (Skoufou et al., 2024).

6.10 CONCLUSION

The mechanisms and efficacy of probiotics, prebiotics, and postbiotics in reestablishing gut microbial equilibrium and treating a range of disorders have been extensively studied and clinically supported. Prebiotics work to cure and preserve health by specifically promoting good bacteria and enhancing the gut microbiome. Current issues will be addressed and probiotic-related disciplines will grow thanks to recent advancements in research tools like artificial intelligence and sequencing, as well as in-depth studies of probiotic, prebiotic, and postbiotic processes. One important strategy for preserving gastrointestinal health and avoiding illnesses is the manipulation of the gut microbiota by probiotics, prebiotics, and postbiotics.

Additionally, we have identified a number of concerns that need to be taken into account while researching postbiotics, including the beginning material, inactivation methods, and safety assurance. For research to be dependable and reproducible, these characteristics must be carefully controlled. Probiotics, prebiotics, and postbiotics may be the next generation of representative medications that completely transform the method by which doctors treat and manage illnesses, and they will likely play an increasingly more significant part in therapeutic therapies going forward, according to research and clinical data.

REFERENCES

Arnold, J. W., Roach, J., Fabela, S., Moorfield, E., Ding, S., Blue, E., Dagher, S., Magness, S., Tamayo, R., Bruno-Barcena, J. M., & Azcarate-Peril, M. A. (2021). The pleiotropic effects of prebiotic galacto-oligosaccharides on the aging gut. *Microbiome*, *9*(1), 31. https://doi.org/10.1186/s40168-020-00980-0

Asha, M. Z., & Khalil, S. F. H. (2020). Efficacy and safety of probiotics, prebiotics and synbiotics in the treatment of irritable bowel syndrome: A systematic review and meta-analysis. *Sultan Qaboos University Medical Journal*, *20*(1), e13–e24. https://doi.org/10.18295/squmj.2020.20.01.003

Ashique, S., Mishra, N., Garg, A., Sibuh, B. Z., Taneja, P., Rai, G., Djearamane, S., Wong, L. S., Al-Dayan, N., Roychoudhury, S., & Kesari, K. K. (2023, July). Recent updates on correlation between reactive oxygen species and synbiotics for effective management of ulcerative colitis. Frontiers in Nutrition, *10*, 1126579.

Berkowitz, L., Schultz, B. M., Salazar, G. A., Pardo-Roa, C., Sebastián, V. P., Álvarez-Lobos, M. M., & Bueno, S. M. (2018). Impact of cigarette smoking on the gastrointestinal tract inflammation: Opposing effects in Crohn's disease and ulcerative colitis. *Frontiers in Immunology*, *9*, 74. https://doi.org/10.3389/fimmu.2018.00074

Bronzini, M., Maglione, A., Rosso, R., Matta, M., Masuzzo, F., Rolla, S., & Clerico, M. (2023). Feeding the gut microbiome: Impact on multiple sclerosis. *Frontiers in Immunology*, *14*, 1176016. https://doi.org/10.3389/fimmu.2023.1176016

Bull, M. J., & Plummer, N. T. (2014). Part 1: The human gut microbiome in health and disease. *Integrative Medicine: A Clinician's Journal*, *13*(6), 17–22. www.ncbi.nlm.nih.gov/pmc/articles/PMC4566439/

Cai, L., Gao, P., Wang, Z., Dai, C., Ning, Y., Ilkit, M., Xue, X., Xiao, J., & Chen, C. (2022). Lung and gut microbiomes in pulmonary aspergillosis: Exploring adjunctive therapies to combat the disease. *Frontiers in Immunology*, *13*, 988708. https://doi.org/10.3389/fimmu.2022.988708

Cao, C., Zhu, H., Yao, Y., & Zeng, R. (2022). Gut dysbiosis and kidney diseases. *Frontiers in Medicine*, *9*, 829349. https://doi.org/10.3389/fmed.2022.829349

Chen, Q., Ren, Y., Lu, J., Bartlett, M., Chen, L., Zhang, Y., Guo, X., & Liu, C. (2017). A novel prebiotic blend product prevents irritable bowel syndrome in mice by improving gut microbiota and modulating immune response. *Nutrients*, *9*(12), 1341. https://doi.org/10.3390/nu9121341

Cohen, L. J., Cho, J. H., Gevers, D., & Chu, H. (2019). Genetic factors and the intestinal microbiome guide development of microbe-based therapies for inflammatory bowel diseases. *Gastroenterology*, *156*(8), 2174–2189. https://doi.org/10.1053/j.gastro.2019.03.017

Collado, M. C., Cernada, M., Neu, J., Pérez-Martínez, G., Gormaz, M., & Vento, M. (2015). Factors influencing gastrointestinal tract and microbiota immune interaction in preterm infants. *Pediatric Research*, *77*(6), 726–731. https://doi.org/10.1038/pr.2015.54

Cosnes, J. (2010). Smoking, physical activity, nutrition and lifestyle: Environmental factors and their impact on IBD. *Digestive Diseases (Basel, Switzerland)*, *28*(3), 411–417. https://doi.org/10.1159/000320395

Cremon, C., Barbaro, M. R., Ventura, M., & Barbara, G. (2018). Pre- and probiotic overview. *Current Opinion in Pharmacology*, *43*, 87–92. https://doi.org/10.1016/j.coph.2018.08.010

Davani-Davari, D., Negahdaripour, M., Karimzadeh, I., Seifan, M., Mohkam, M., Masoumi, S. J., Berenjian, A., & Ghasemi, Y. (2019). Prebiotics: Definition, types, sources, mechanisms, and clinical applications. *Foods*, *8*(3), 92. https://doi.org/10.3390/foods8030092

Davis, M. K. (2001). Breastfeeding and chronic disease in childhood and adolescence. *Pediatric Clinics of North America*, *48*(1), 125–141, ix. https://doi.org/10.1016/s0031-3955(05)70289-3

de Simone, C. (2019). The unregulated probiotic market. *Clinical Gastroenterology and Hepatology: The Official Clinical Practice Journal of the American Gastroenterological Association*, *17*(5), 809–817. https://doi.org/10.1016/j.cgh.2018.01.018

Dixit, K., Chaudhari, D., Dhotre, D., Shouche, Y., & Saroj, S. (2021). Restoration of dysbiotic human gut microbiome for homeostasis. *Life Sciences*, *278*, 119622. https://doi.org/10.1016/j.lfs.2021.119622

Domagała-Rodacka, R., Cibor, D., Szczeklik, K., Rodacki, T., Mach, T., & Owczarek, D. (2016). Gastrointestinal tract as a side-effect target of medications. *Przeglad Lekarski*, *73*(9), 652–658.

Eom, T., Kim, Y. S., Choi, C. H., Sadowsky, M. J., & Unno, T. (2018). Current understanding of microbiota- and dietary-therapies for treating inflammatory bowel disease. *Journal of Microbiology (Seoul, Korea)*, *56*(3), 189–198. https://doi.org/10.1007/s12275-018-8049-8

Fakharian, F., Thirugnanam, S., Welsh, D. A., Kim, W.-K., Rappaport, J., Bittinger, K., & Rout, N. (2023). The role of gut dysbiosis in the loss of intestinal immune cell functions and viral pathogenesis. *Microorganisms*, *11*(7), 1849. https://doi.org/10.3390/microorganisms11071849

Feng, Z., Wang, T., Dong, S., Jiang, H., Zhang, J., Raza, H. K., & Lei, G. (2021). Association between gut dysbiosis and chronic kidney disease: A narrative review of the literature. *The Journal of International Medical Research*, *49*(10), 3000605211053276. https://doi.org/10.1177/03000605211053276

Gargaro, M., Manni, G., Scalisi, G., Puccetti, P., & Fallarino, F. (2021). Tryptophan metabolites at the crossroad of immune-cell interaction via the aryl hydrocarbon receptor: Implications for tumor immunotherapy. *International Journal of Molecular Sciences*, *22*(9), 4644. https://doi.org/10.3390/ijms22094644

Gasaly, N., De Vos, P., & Hermoso, M. A. (2021, May). Impact of bacterial metabolites on gut barrier function and host immunity: a focus on bacterial metabolism and its relevance for intestinal inflammation. *Frontiers in Immunology*, *12*, 658354.

Gorreja, F., & Walker, W. A. (n.d.). The potential role of adherence factors in probiotic function in the gastrointestinal tract of adults and pediatrics: A narrative review of experimental and human studies. *Gut Microbes*, *14*(1), 2149214. https://doi.org/10.1080/19490976.2022.2149214

Guarino, M. P. L., Altomare, A., Emerenziani, S., Di Rosa, C., Ribolsi, M., Balestrieri, P., Iovino, P., Rocchi, G., & Cicala, M. (2020). Mechanisms of action of prebiotics and their effects on gastro-intestinal disorders in adults. *Nutrients*, *12*(4), 1037. https://doi.org/10.3390/nu12041037

Gupta, N., Yadav, V. K., Gacem, A., Al-Dossari, M., Yadav, K. K., El-Gawaad, N. S. A., Khedher, N. B., Choudhary, N., Kumar, P., & Cavalu, S. (2022). Deleterious effect of air pollution on human microbial community and bacterial flora: A short review. *International Journal of Environmental Research and Public Health*, *19*(23), 15494. https://doi.org/10.3390/ijerph192315494

Haak, B. W., Prescott, H. C., & Wiersinga, W. J. (2018). Therapeutic potential of the gut microbiota in the prevention and treatment of sepsis. *Frontiers in Immunology*, *9*, 2042. https://doi.org/10.3389/fimmu.2018.02042

Haldar, S., Jadhav, S. R., Gulati, V., Beale, D. J., Balkrishna, A., Varshney, A., Palombo, E. A., Karpe, A. V., & Shah, R. M. (2023). Unravelling the gut-lung axis: Insights into microbiome interactions and traditional indian medicine's perspective on optimal health. *FEMS Microbiology Ecology*, *99*(10), fiad103. https://doi.org/10.1093/femsec/fiad103

Jensen-Kroll, J., Demetrowitsch, T., Clawin-Rädecker, I., Klempt, M., Waschina, S., & Schwarz, K. (2022, Oct). Microbiota independent effects of oligosaccharides on Caco-2 cells-a semi-targeted metabolomics approach using DI-FT-ICR-MS coupled with pathway enrichment analysis. *Frontiers in Molecular Bioscience*, *9*, 968643.

Henry, D. A., Ostapowicz, G., & Robertson, J. (1994). Drugs as a cause of gastrointestinal disease. *Baillière's Clinical Gastroenterology*, *8*(2), 271–300. https://doi.org/10.1016/0950-3528(94)90005-1

Holle, J., Bartolomaeus, H., Löber, U., Behrens, F., Bartolomaeus, T. U. P., Anandakumar, H., Wimmer, M. I., Vu, D. L., Kuhring, M., Brüning, U., Maifeld, A., Geisberger, S., Kempa, S., Schumacher, F., Kleuser, B., Bufler, P., Querfeld, U., Kitschke, S., Engler, D., … & Müller, D. (2022). Inflammation in children with CKD linked to gut dysbiosis and metabolite imbalance. *Journal of the American Society of Nephrology: JASN*, *33*(12), 2259–2275. https://doi.org/10.1681/ASN.2022030378

Holleran, G., Lopetuso, L. R., Ianiro, G., Pecere, S., Pizzoferrato, M., Petito, V., Graziani, C., Mcnamara, D., Gasbarrini, A., & Scaldaferri, F. (2017). Gut microbiota and inflammatory bowel disease: So far so gut! *Minerva Gastroenterologica E Dietologica*, *63*(4), 373–384. https://doi.org/10.23736/S1121-421X.17.02386-8

Hosseinkhani, F., Heinken, A., Thiele, I., Lindenburg, P. W., Harms, A. C., & Hankemeier, T. (2021). The contribution of gut bacterial metabolites in the human immune signaling pathway of non-communicable diseases. *Gut Microbes*, *13*(1), 1–22. https://doi.org/10.1080/19490976.2021.1882927

Huang, C., Yu, Y., Li, Z., Yan, B., Pei, W., & Wu, H. (2022). The preparation technology and application of xylo-oligosaccharide as prebiotics in different fields: A review. *Frontiers in Nutrition*, *9*. https://doi.org/10.3389/fnut.2022.996811

Huang, H., Fang, M., Jostins, L., Umićević Mirkov, M., Boucher, G., Anderson, C. A., Andersen, V., Cleynen, I., Cortes, A., Crins, F., D'Amato, M., Deffontaine, V., Dmitrieva, J., Docampo, E., Elansary, M., Farh, K. K.-H., Franke, A., Gori, A.-S., Goyette, P., … & Barrett, J. C. (2017). Fine-mapping inflammatory bowel disease loci to single-variant resolution. *Nature*, *547*(7662), 173–178. https://doi.org/10.1038/nature22969

Ibrahim, S. A., Yeboah, P. J., Ayivi, R. D., Eddin, A. S., Wijemanna, N. D., Paidari, S., & Bakhshayesh, R. V. (2023). A review and comparative perspective on health benefits of probiotic and fermented foods. *International Journal of Food Science & Technology*, *58*(10), 4948–4964. https://doi.org/10.1111/ijfs.16619

Ji, J., Jin, W., Liu, S., Jiao, Z., & Li, X. (2023). Probiotics, prebiotics, and postbiotics in health and disease. *MedComm*, *4*(6), e420. https://doi.org/10.1002/mco2.420

Kango, N., Jana, U. K., Choukade, R., & Nath, S. (2022). Advances in prebiotic mannooligosaccharides. *Current Opinion in Food Science*, *47*, 100883. https://doi.org/10.1016/j.cofs.2022.100883

Kango, N. & Nath, S. (2024, June). Prebiotics, probiotics and postbiotics: The changing paradigm of functional foods. *Journal of Dietary Supplements, 1*, 1–27.

Khanijow, V., Prakash, P., Emsellem, H. A., Borum, M. L., & Doman, D. B. (2015). Sleep dysfunction and gastrointestinal diseases. *Gastroenterology & Hepatology*, *11*(12), 817–825. www.ncbi.nlm.nih.gov/pmc/articles/PMC4849511/

Kim, B.-Y., & Park, S. (2021). The concepts and applications of postbiotics for the development of health functional food product. *Current Topic in Lactic Acid Bacteria and Probiotics*, *7*, 14–22. https://doi.org/10.35732/ctlabp.2021.7.1.14

Kinnucan, J. A., Rubin, D. T., & Ali, T. (2013). Sleep and inflammatory bowel disease: Exploring the relationship between sleep disturbances and inflammation. *Gastroenterology & Hepatology*, *9*(11), 718–727. www.ncbi.nlm.nih.gov/pmc/articles/PMC3995194/

Konstantinidis, T., Tsigalou, C., Karvelas, A., Stavropoulou, E., Voidarou, C., & Bezirtzoglou, E. (2020). Effects of antibiotics upon the gut microbiome: A review of the literature. *Biomedicines*, *8*(11), 502. https://doi.org/10.3390/biomedicines8110502

Kutty, P. K. (2014). Breastfeeding and risk of parasitic infection–A review. *Asian Pacific Journal of Tropical Biomedicine*, *4*(11), 847–858. https://doi.org/10.12980/APJTB.4.201414B355

Lee, J. G., Lee, S., Jeon, J., Kong, H. G., Cho, H. J., Kim, J. H., Kim, S. Y., Oh, M. J., Lee, D., Seo, N., & Park, K. H. (2022, Jan). Host tp53 mutation induces gut dysbiosis eliciting inflammation through disturbed sialic acid metabolism. *Microbiome*, *10*(1), 3.

Li, H.-Y., Zhou, D.-D., Gan, R.-Y., Huang, S.-Y., Zhao, C.-N., Shang, A., Xu, X.-Y., & Li, H.-B. (2021). Effects and mechanisms of probiotics, prebiotics, synbiotics, and postbiotics on metabolic diseases targeting gut microbiota: A narrative review. *Nutrients*, *13*(9), 3211. https://doi.org/10.3390/nu13093211

Liang, B., & Xing, D. (2023). The current and future perspectives of postbiotics. *Probiotics and Antimicrobial Proteins*, *15*(6), 1626–1643. https://doi.org/10.1007/s12602-023-10045-x

Lim, M. Y., & Nam, Y.-D. (n.d.). Gut microbiome in healthy aging versus those associated with frailty. *Gut Microbes*, *15*(2), 2278225. https://doi.org/10.1080/19490976.2023.2278225

Liu, J., Tan, Y., Cheng, H., Zhang, D., Feng, W., & Peng, C. (2022). Functions of gut microbiota metabolites, current status and future perspectives. *Aging and Disease*, *13*(4), 1106–1126. https://doi.org/10.14336/AD.2022.0104

Liu, Y., Wang, J., & Wu, C. (2022). Modulation of gut microbiota and immune system by probiotics, pre-biotics, and post-biotics. *Frontiers in Nutrition*, *8*, 634897. https://doi.org/10.3389/fnut.2021.634897

Lou, X., Xue, J., Shao, R., Mo, C., Wang, F., & Chen, G. (2023). Postbiotics as potential new therapeutic agents for sepsis. *Burns & Trauma*, *11*, tkad022. https://doi.org/10.1093/burnst/tkad022

Maathuis, A. J. H., van den Heuvel, E. G., Schoterman, M. H. C., & Venema, K. (2012). Galacto-oligosaccharides have prebiotic activity in a dynamic in vitro colon model using a (13)C-labeling technique. *The Journal of Nutrition*, *142*(7), 1205–1212. https://doi.org/10.3945/jn.111.157420

Mahdizade Ari, M., Teymouri, S., Fazlalian, T., Asadollahi, P., Afifirad, R., Sabaghan, M., Valizadeh, F., Ghanavati, R., & Darbandi, A. (2022). The effect of probiotics on gestational diabetes and its complications in pregnant mother and newborn: A systematic review and meta-analysis during 2010–2020. *Journal of Clinical Laboratory Analysis*, *36*(4), e24326. https://doi.org/10.1002/jcla.24326

Manning, T. S., & Gibson, G. R. (2004). Microbial-gut interactions in health and disease. Prebiotics. *Best Practice & Research Clinical Gastroenterology*, *18*(2), 287–298. https://doi.org/10.1016/j.bpg.2003.10.008

Mansour, S. R., Moustafa, M. A. A., Saad, B. M., Hamed, R., & Moustafa, A.-R. A. (2021). Impact of diet on human gut microbiome and disease risk. *New Microbes and New Infections*, *41*, 100845. https://doi.org/10.1016/j.nmni.2021.100845

Markowiak, P., & Śliżewska, K. (2017). Effects of probiotics, prebiotics, and synbiotics on human health. *Nutrients*, *9*(9), 1021. https://doi.org/10.3390/nu9091021

Mavrogeni, M. E., Asadpoor, M., Henricks, P. A., Keshavarzian, A., Folkerts, G., & Braber, S. (2022, Nov). Direct action of non-digestible oligosaccharides against a leaky gut. *Nutrients*, *14*(21), 4699.

Meng, X., Zhang, G., Cao, H., Yu, D., Fang, X., de Vos, W. M., & Wu, H. (2020). Gut dysbacteriosis and intestinal disease: Mechanism and treatment. *Journal of Applied Microbiology*, *129*(4), 787–805. https://doi.org/10.1111/jam.14661

Mercenier, A. (1990, Sep). Molecular genetics of Streptococcus thermophilus. *FEMS Microbiology Reviews*, *7*(1-2), 61–77.

Moradi, M., Kousheh, S. A., Almasi, H., Alizadeh, A., Guimarães, J. T., Yılmaz, N., & Lotfi, A. (2020). Postbiotics produced by lactic acid bacteria: The next frontier in food safety. *Comprehensive Reviews in Food Science and Food Safety*, *19*(6), 3390–3415. https://doi.org/10.1111/1541-4337.12613

Morniroli, D., Vizzari, G., Consales, A., Mosca, F., Giannì, M. L. (2021, Feb). Postbiotic supplementation for children and newborn's health. *Nutrients*, *13*(3), 781.

Mukherjee, S., Joardar, N., Sengupta, S., & Sinha Babu, S. P. (2018). Gut microbes as future therapeutics in treating inflammatory and infectious diseases: Lessons from recent findings. *Journal of Nutritional Biochemistry*, *61*, 111–128. https://doi.org/10.1016/j.jnutbio.2018.07.010

Nagao-Kitamoto, H., Kitamoto, S., Kuffa, P., & Kamada, N. (2016). Pathogenic role of the gut microbiota in gastrointestinal diseases. *Intestinal Research*, *14*(2), 127–138. https://doi.org/10.5217/ir.2016.14.2.127

Naito, Y. (2023). Gut frailty: Its concept and pathogenesis. *Digestion*, *105*(1), 49–57. https://doi.org/10.1159/000534733

Nogal, A., Valdes, A. M., & Menni, C. (2021). The role of short-chain fatty acids in the interplay between gut microbiota and diet in cardio-metabolic health. *Gut Microbes*, *13*(1), 1–24. https://doi.org/10.1080/19490976.2021.1897212

Pandey, Kavita. R., Naik, Suresh. R., & Vakil, Babu. V. (2015). Probiotics, prebiotics and synbiotics–A review. *Journal of Food Science and Technology*, *52*(12), 7577–7587. https://doi.org/10.1007/s13197-015-1921-1

Pérez-Reytor, D., Puebla, C., Karahanian, E., & García, K. (2021). Use of short-chain fatty acids for the recovery of the intestinal epithelial barrier affected by bacterial toxins. *Frontiers in Physiology*, *12*, 650313. https://doi.org/10.3389/fphys.2021.650313

Peshev, D., & Van den Ende, W. (2014). Fructans: Prebiotics and immunomodulators. *Journal of Functional Foods*, *8*, 348–357. https://doi.org/10.1016/j.jff.2014.04.005

Popovic, D., Kulas, J., Tucovic, D., Popov Aleksandrov, A., Malesevic, A., Glamoclija, J., Brdaric, E., Sokovic Bajic, S., Golic, N., Mirkov, I., & Tolinacki, M. (2023). Gut microbial dysbiosis occurring during pulmonary fungal infection in rats is linked to inflammation and depends on healthy microbiota composition. *Microbiology Spectrum*, *11*(5), e0199023. https://doi.org/10.1128/spectrum.01990-23

Prajapati, N., Patel, J., Singh, S., Yadav, V. K., Joshi, C., Patani, A., Prajapati, D., Sahoo, D. K., & Patel, A. (2023). Postbiotic production: Harnessing the power of microbial metabolites for health applications. *Frontiers in Microbiology*, *14*, 1306192. https://doi.org/10.3389/fmicb.2023.1306192

Puccetti, M., Pariano, M., Costantini, C., Giovagnoli, S., Ricci, M. (2022, March). Pharmaceutically active microbial AhR agonists as innovative biodrugs in inflammation. *Pharmaceuticals*, *15*(3), 336.

Raj, N., Fernandes, S., Charyulu, N. R., Dubey, A., G S, R., & Hebbar, S. (2019). Postmarket surveillance: A review on key aspects and measures on the effective functioning in the context of the United Kingdom and Canada. *Therapeutic Advances in Drug Safety*, *10*, 2042098619865413. https://doi.org/10.1177/2042098619865413

Roberfroid, M. (2007). Prebiotics: The concept revisited. *The Journal of Nutrition*, *137*(3 Suppl 2), 830S–837S. https://doi.org/10.1093/jn/137.3.830S

Rodríguez, L., Cervantes, E., & Ortiz, R. (2011). Malnutrition and gastrointestinal and respiratory infections in children: A public health problem. *International Journal of Environmental Research and Public Health*, *8*(4), 1174–1205. https://doi.org/10.3390/ijerph8041174

Saarela, M., Mogensen, G., Fondén, R., Mättö, J., & Mattila-Sandholm, T. (2000). Probiotic bacteria: Safety, functional and technological properties. *Journal of Biotechnology*, *84*(3), 197–215. https://doi.org/10.1016/s0168-1656(00)00375-8

Saito, Y. A. (2011). The role of genetics in IBS. *Gastroenterology Clinics of North America*, *40*(1), 45–67. https://doi.org/10.1016/j.gtc.2010.12.011

Saito, Y. A., Mitra, N., & Mayer, E. A. (2010). Genetic approaches to functional gastrointestinal disorders. *Gastroenterology*, *138*(4). https://doi.org/10.1053/j.gastro.2010.02.037

Salminen, S., Collado, M. C., Endo, A., Hill, C., Lebeer, S., Quigley, E. M. M., Sanders, M. E., Shamir, R., Swann, J. R., Szajewska, H., & Vinderola, G. (2021). The International Scientific Association of Probiotics and Prebiotics (ISAPP) consensus statement on the definition and scope of postbiotics. *Nature Reviews Gastroenterology & Hepatology*, *18*(9), 649–667. https://doi.org/10.1038/s41575-021-00440-6

Salminen, S., von Wright, A., Morelli, L., Marteau, P., Brassart, D., de Vos, W. M., Fondén, R., Saxelin, M., Collins, K., Mogensen, G., Birkeland, S. E., & Mattila-Sandholm, T. (1998). Demonstration of safety of probiotics—A review. *International Journal of Food Microbiology*, *44*(1–2), 93–106. https://doi.org/10.1016/s0168-1605(98)00128-7

Sánchez y Sánchez de la Barquera, B., Martínez Carrillo, B. E., Aguirre Garrido, J. F., Martínez Méndez, R., Benítez Arciniega, A. D., Valdés Ramos, R., & Soto Piña, A. E. (2022). Emerging evidence on the use of probiotics and prebiotics to improve the gut microbiota of older adults with frailty syndrome: A narrative review. *Journal of Nutrition, Health & Aging*, *26*(10), 926–935. https://doi.org/10.1007/s12603-022-1842-4

Sanders, M. E., Akkermans, L. M., Haller, D., Hammerman, C., Heimbach, J., Hörmannsperger, G., Huys, G., Levy, D. D., Lutgendorff, F., Mack, D., Phothirath, P., Solano-Aguilar, G., & Vaughan, E. (2010). Safety assessment of probiotics for human use. *Gut Microbes*, *1*(3), 164–185. https://doi.org/10.4161/gmic.1.3.12127

Skoufou, M., Tsigalou, C., Vradelis, S., & Bezirtzoglou, E. (2024). The networked interaction between probiotics and intestine in health and disease: A promising success story. *Microorganisms*, *12*(1), Article 1. https://doi.org/10.3390/microorganisms12010194

Slavin, J. (2013). Fiber and prebiotics: Mechanisms and health benefits. *Nutrients*, *5*(4), 1417–1435. https://doi.org/10.3390/nu5041417

Snydman, D. R. (2008). The safety of probiotics. *Clinical Infectious Diseases*, *46*(Supplement_2), S104–S111. https://doi.org/10.1086/523331

Strati, F., Pujolassos, M., Burrello, C., Giuffrè, M. R., Lattanzi, G., Caprioli, F., Troisi, J., & Facciotti, F. (2021). Antibiotic-associated dysbiosis affects the ability of the gut microbiota to control intestinal inflammation upon fecal microbiota transplantation in experimental colitis models. *Microbiome*, *9*(1), 39. https://doi.org/10.1186/s40168-020-00991-x

Tedelind, S., Westberg, F., Kjerrulf, M., & Vidal, A. (2007). Anti-inflammatory properties of the short-chain fatty acids acetate and propionate: A study with relevance to inflammatory bowel disease. *World Journal of Gastroenterology: WJG*, *13*(20), 2826–2832. https://doi.org/10.3748/wjg.v13.i20.2826

Thorakkattu, P., Khanashyam, A. C., Shah, K., Babu, K. S., Mundanat, A. S., Deliephan, A., Deokar, G. S., Santivarangkna, C., & Nirmal, N. P. (2022). Postbiotics: Current trends in food and pharmaceutical industry. *Foods (Basel, Switzerland)*, *11*(19), 3094. https://doi.org/10.3390/foods11193094

Tomasik, P., & Tomasik, P. (2020). Probiotics, non-dairy prebiotics and postbiotics in nutrition. *Applied Sciences*, *10*(4), Article 4. https://doi.org/10.3390/app10041470

Tran, C., Cock, I. E., Chen, X., & Feng, Y. (2022, Jan). Antimicrobial Bacillus: metabolites and their mode of action. *Antibiotics*, *11*(1), 88.

Vinderola, G., Sanders, M. E., & Salminen, S. (2022). The concept of postbiotics. *Foods*, *11*(8), 1077. https://doi.org/10.3390/foods11081077

Wang, S., Xiao, Y., Tian, F., Zhao, J., Zhang, H., Zhai, Q., & Chen, W. (2020). Rational use of prebiotics for gut microbiota alterations: Specific bacterial phylotypes and related mechanisms. *Journal of Functional Foods*, *66*, 103838. https://doi.org/10.1016/j.jff.2020.103838

Wang, X.-M., Fan, L., Meng, C.-C., Wang, Y.-J., Deng, L.-E., Yuan, Z., Zhang, J.-P., Li, Y.-Y., & Lv, S.-C. (2024). Gut microbiota influence frailty syndrome in older adults: Mechanisms and therapeutic strategies. *Biogerontology*, *25*(1), 107–129. https://doi.org/10.1007/s10522-023-10082-7

Wittkopf, N., Neurath, M. F., & Becker, C. (2014). Immune-epithelial crosstalk at the intestinal surface. *Journal of Gastroenterology*, *49*(3), 375–387. https://doi.org/10.1007/s00535-013-0929-4

Wu, H.-J., & Wu, E. (2012). The role of gut microbiota in immune homeostasis and autoimmunity. *Gut Microbes*, *3*(1), 4–14. https://doi.org/10.4161/gmic.19320

Yang, D., Xing, Y., Song, X., & Qian, Y. (2020). The impact of lung microbiota dysbiosis on inflammation. *Immunology*, *159*(2), 156–166. https://doi.org/10.1111/imm.13139

Yaqoob, M. U., Abd El-Hack, M. E., Hassan, F., El-Saadony, M. T., Khafaga, A. F., Batiha, G. E., Yehia, N., Elnesr, S. S., Alagawany, M., El-Tarabily, K. A., & Wang, M. (2021, July). The potential mechanistic insights and future implications for the effect of prebiotics on poultry performance, gut microbiome, and intestinal morphology. *Poultry Science*, *100*(7), 101143.

Ye, X., Li, H., Anjum, K., Zhong, X., Miao, S., Zheng, G., Liu, W., & Li, L. (2022). Dual role of indoles derived from intestinal microbiota on human health. *Frontiers in Immunology*, *13*, 903526. https://doi.org/10.3389/fimmu.2022.903526

Yeşilyurt, N., Yılmaz, B., Ağagündüz, D., & Capasso, R. (2021). Involvement of probiotics and postbiotics in the immune system modulation. *Biologics*, *1*(2), Article 2. https://doi.org/10.3390/biologics1020006

Zhu, J., Pitre, T., Ching, C., Zeraatkar, D., & Gruchy, S. (2023, Mar). Safety and efficacy of probiotic supplements as adjunctive therapies in patients with COVID-19: a systematic review and meta-analysis. *PLoS One*, *18*(3), e0278356.

7 Role of Genetic Factor in Gastrointestinal Inflammation

*Priyanka Ray[1], Tamalika Chakraborty[1], and Debojyoti Adak[2]**
[1]Guru Nanak Institute of Pharmaceutical Science & Technology, Sodepur, Kolkata, India
[2]Quest Vitamins Middle East FZE, Dubai, UAE

7.1 INTRODUCTION

Gastrointestinal inflammation is a complex biological response to harm triggered by several stimuli like germs, damaged cells, or irritants. The symbiotic microbiome in the digestive tract is thought to protect the gut by removing harmful substances and aiding in the healing process[1]. Dysbiosis or lack of microbiome can result in heightened type 2 immune response and the development of allergy, infectious, and inflammatory consequences, including parasite illnesses. Therefore, the microbiota plays a crucial role in controlling type 2 immune responses and serves as a vital component in coordinating immune responses at mucosal surfaces. Although the exact process by which microbiota influences type 2 immunity is not well understood, it is recognized as a potent stimulator of proinflammatory T-helper 17 cells and regulatory T-cells (Tregs) in the colon[2]. Inflammation signals help recruit and differentiate cells quickly to remove inflammatory chemicals and restore tissue balance. However, the presence of ongoing inflammatory stimuli or disruption in the processes responsible for healing can result in the development of chronic inflammation. Various stressors can impact the immune system and heighten the vulnerability to infectious disorders, such as gastritis in postinfectious irritable bowel syndrome (IBS)[3]. Conversely, IBS patients are more prone to developing infectious gastroenteritis. Chronic inflammatory bowel illnesses can be caused by a range of viral (such as norovirus), microbial (such as *Campylobacter jejuni, Clostridium, Mycobacterium*), and parasite agents (such as Giardia and helminthes). However, the precise processes by which it operates remain poorly understood and there is currently no existing remedy. Specific nutritional elements like n-3 polyunsaturated fatty acids, antioxidants, probiotics, and prebiotics impact the immune system of the gastrointestinal tract. High-fat diets alter the normal microbiome in the digestive tract and interfere with communication between the gut and the brain, leading to obesity and inflammation[3].

Inflammatory bowel disease (IBD) is characterized by chronic inflammation of the gastrointestinal tract due to an incorrect immune response to gut microbiota, leading to recurrent episodes of inflammation[4]. Two different types of unknown origin intestinal disorders are included under the umbrella of inflammatory bowel disease. Differentiate these disorders by their specific location and extent of impact on the intestinal wall. Ulcerative colitis, often known as UC, is characterized by widespread inflammation of the mucosa of the colon. UC often affects the rectum (also known as proctitis), but it can also expand into the sigmoid (also known as proctosigmoiditis), go beyond the sigmoid (also known as distal ulcerative colitis), or include the entire colon, all the way up to the cecum (also known as pancolitis). Individuals with Crohn's disease are prone to transmural ulcers in any section of the gastrointestinal system, with the terminal ileum and colon being the most often impacted regions. Each disorder is classified based on its severity (mild, moderate, or severe) and location[5]. There are three other phenotypes that can be used to categorize CD: inflammatory, stricturing, and penetrating.

Both Crohn's disease and ulcerative colitis will exhibit themselves in a variety of extraintestinal areas in addition to the gastrointestinal tract. Although it is easy to discriminate between the two disorders in the majority of patients, there are at least ten percent of patients in whom the characteristics are so similar that it is not possible to differentiate between the two disorders at first[6].

Both conditions are characterized by a hereditary propensity; neither can be cured, and both are associated with a significant amount of morbidity[7]. In conclusion, both of these factors raise the likelihood of developing colorectal cancer.

In recent decades, significant progress has been made in comprehending the genetic factors that contribute to IBD. The ability to conduct genome-wide association studies (GWAS) has been made possible by advancements in genetic testing and DNA sequencing[8]. These studies have successfully found new single nucleotide polymorphisms (SNPs).

The discovery of Nucleotide-binding oligomerization domain containing 2 (NOD2) in 2001 was the first identification of a gene that heightens vulnerability to Crohn's disease (CD). This gene encodes a protein that acts as an intracellular receptor for bacterial products in monocytes[9]. It then transmits signals that result in the activation of NFkB. Dendritic cells (DCs) undergo autophagy when NOD2 is activated by muramyl dipeptide. Dendritic cells from individuals with Crohn's disease who have genetic abnormalities in the NOD2 gene show reduced activation of autophagy and have lower levels of bacterial containment within autophagolysosomes[6].

IBDs have been thoroughly examined by genome-wide association studies, leading to the identification

DOI: 10.1201/9781003493143-7

and confirmation of many susceptibility loci. Identifying susceptibility loci has enhanced our comprehension of the condition's origins by providing valuable insights into crucial and disturbed pathways of the gut immune system.

Two more autophagy-related genes, IRGM and ATG16L1, have been found through genetic research. These genes are crucial in immune responses associated to inflammatory bowel disease (IBD)[10,11]. Genetic variants linked to a higher risk of CD emphasize the importance of innate immunity, autophagy, and phagocytosis in the disease's progression. Additional genes, such as IL23R and PTPN2, are linked to autoimmune disease, indicating another factor in the development of Crohn's disease[12]. The latest advancements in the field of genetics have provided insights into the fundamental causes of IBD, shedding light on its etiology. Conventional inflammatory bowel disease (IBD) refers to a collection of polygenic illnesses where numerous susceptibility loci collectively contribute to the overall risk of developing the disease. The latest and largest genetic association study utilized genome-wide association data from more than 75,000 individuals with adolescent and adult-onset inflammatory bowel disease (IBD) as well as control subjects. This analysis revealed 163 regions of the genome that are associated with susceptibility to IBD, which include over 300 possible candidate genes[13]. Out of the total 163 loci, 110 loci were found to increase the risk for both subtypes of inflammatory bowel disease (IBD), whereas 30 loci were specific to Crohn's disease (CD) and 23 loci were specific to ulcerative colitis (UC). A recent study conducted a trans-ethnic investigation involving more than 20,000 individuals of both European and non-European descent. This research revealed 38 new genetic locations associated with inflammatory bowel disease (IBD), indicating a shared genetic risk among different groups[14]. As a result, the total number of known genetic locations associated with IBD has increased to 200.

Combining genotyping data from cohorts of varied ancestry will allow the finding of additional loci linked with inflammatory bowel disease (IBD), as indicated by the evidence of shared IBD risk loci across distinct populations. Nevertheless, each identified locus only accounts for a little proportion of the anticipated heritability in IBD[15].

The increasing amount of susceptibility gene loci discovered in IBD indicates that genetic factors have a substantial impact on the development of the disease[16]. Genetic factors that have been identified explain just a small part of the variation in the disease: 13.1% for CD and 8.2% for UC.

7.2 COMMON GASTROINTESTINAL INFLAMMATORY DISORDERS

Commonly referred to as inflammatory bowel disease (IBD), Crohn's disease (CD) and ulcerative colitis (UC) are two illnesses that fall under this particular category. Both of these conditions are inflammatory disorders of the gastrointestinal tract that are mediated by the immune system[17]. Figure 7.1 represents the various types of gut inflammation along with the associated disease.

7.2.1 Crohn's Disease

Crohn's disease is defined by inflammation that extends along the entire thickness of the intestinal wall, from the inner mucosa to the outer serosa. The illness is characterized by a pattern of relapse and remission. CD can progress from mild to moderate inflammatory diseases to severe penetrating (fistulization) or structuring illness after experiencing many relapses[18,19]. This progression can occur during the course of the disease.

Crohn's disease can affect any part of the gastrointestinal tract[20]. Approximately 33% of individuals had small bowel involvement, mainly in the terminal ileum; 20% have only colon involvement, and over 50% have both colon and small bowel involvement[21]. There is no cure for this condition, and the majority of patients go through periods of remission followed by relapses at unpredictable intervals. This condition results in an extremely low quality of life for the patient.

7.2.1.1 Pathophysiology

The pathogenesis is complex and involves a combination of genetic predisposition, infectious agents, immunological factors, environmental factors, and nutritional factors. The typical transmural inflammation can affect the entire gastrointestinal (GI) tract, spanning from the mouth to the perianal area. However, it most commonly involves the terminal ileum and right colon. The initial lesion begins as an infiltration surrounding an intestinal crypt[22]. This progresses

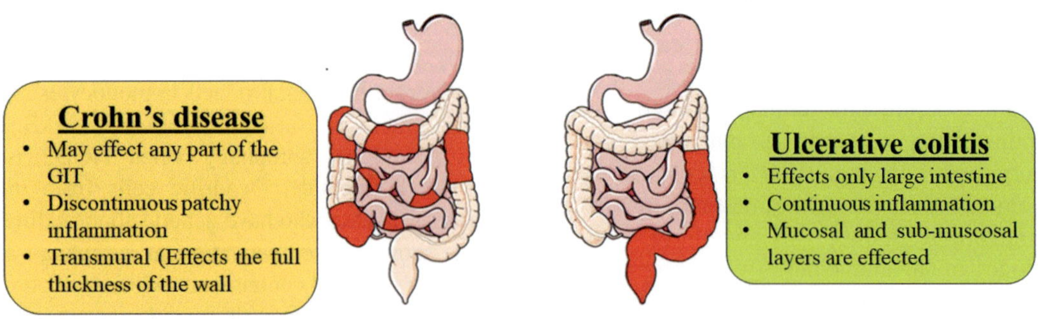

FIGURE 7.1 Diagrammatic illustration of the GIT inflammations with different associated diseases.

to the formation of ulcers, starting in the outermost layer of the mucosa and extending to the underlying layers[23]. During the course of the inflammation, non-caseating granulomas develop, affecting all layers of the intestinal wall. The condition can progress to exhibit the characteristic cobblestone mucosal appearances and skip lesions over the whole length of the gut, while leaving parts with normal mucosa unaffected. Once the inflammation of Crohn's disease subsides, fibrous tissue forms in the previously affected sections of the intestines. Granuloma production is a frequent occurrence in Crohn's disease; however its absence does not rule out the diagnosis. Continual inflammation and scarring result in the blockage of the gut and the development of strictures. Crohn's disease is additionally linked to the presence of enterovesical, enteroenteral, enterocutaneous, and enterovaginal fistulas[24].

7.2.1.2 Treatment Strategy

Medical treatment can be broadly categorized into two classes:

- Oral mesalamine, immunomodulators such thiopurines (mercaptopurines, azathioprine), methotrexate, and steroids are effective treatments for mild to moderate illness.
- The most effective treatment for moderate to severe disease, especially fistulizing disease, is a combination of immunomodulators and biologics (such as infliximab, adalimumab, golimumab, and vedolizumab), or biologics alone. Biologics are genetically modified immunoglobulins designed to target and inhibit specific cytokines or receptors that have a role in the inflammatory process[25]. Each biologic drug targets a distinct molecular location. Monoclonal antibodies targeting tumor necrosis factor (TNF) alpha can inhibit the inflammatory effects of circulating TNF. Anti-integrin drugs are substances that block adhesion molecules by binding to the subunits of the MAdCAM receptors on endothelial cells located in regions of inflammation. They impede the transportation of lymphocytes from the bloodstream to the intestinal wall, therefore preventing the inflammatory reaction specifically aimed at the gut. Some examples of anti-TNF agents are infliximab, adalimumab, and golimumab. Natalizumab and vedolizumab are examples of adhesion molecule inhibitors[26]. Vedolizumab exhibits gut-specificity and possesses a reduced incidence of systemic side effects. Several recently developed therapeutic treatments for inflammatory bowel disease are now being developed and tested[27].
- Surgical interventions are employed to address issues such as intestinal blockages, abscesses, fistulas, or perforated bowels.
- Prior to and throughout the treatment of Crohn's disease, it is strongly advised to seek the guidance of a dietitian and consider nutritional supplementation. Fecal calprotectin can aid in distinguishing between inflammatory bowel disease (IBD) and irritable bowel syndrome (IBS).
- Chromoendoscopy should be employed during colonoscopy as it has the potential to enhance the diagnostic yield for colorectal dysplasia[28].
- Avoid nonsteroidal anti-inflammatory drugs (NSAIDs) as they have the potential to worsen the condition.
- Refrain from engaging in the act of smoking.
- Seek mental health counseling since a significant number of people experience the onset of depression[29].
- Sulfasalazine demonstrates efficacy in treating moderate conditions.
- Ileal release under control Budesonide is effective in inducing remission in people with mild ileocecal illness.
- Refrain from using metronidazole as it lacks efficacy in treating Crohn's disease.
- Antidiarrheals are effective in controlling mild cases of diarrhea.
- Thiopurines are effective in reducing the need for steroids.
- Anti-tumor necrosis factors (TNFs) can be administered to patients who exhibit resistance to steroids[30].

7.2.2 Ulcerative Colitis

Ulcerative colitis is a disorder characterized by inflammation of the colon (Figure 7.2), which causes the colon wall to become fragile and develop shallow erosions, often leading to bleeding. This is the prevailing type of inflammatory bowel disease on a global scale. It typically causes inflammation that is limited to the mucosa and submucosa of the colon[31]. Usually, the disease originates in the rectum and spreads in a consistent manner towards the proximal direction. In the United States, the disease results in around 250,000 healthcare provider visits each year, and the direct medical expenses associated with the disease are anticipated to surpass four billion dollars yearly. Ulcerative colitis is an incurable chronic condition that has a substantial influence on both physical and emotional well-being[32].

7.2.2.1 Pathophysiology

Ulcerative colitis is characterized by abnormalities in the epithelial barrier, immunological response, leukocyte recruitment, and colon microbiota. The colonic mucus and potentially the tight connections of the epithelial barrier are impaired, resulting in heightened absorption of luminal antigens. The mucosa's lamina propria exhibits an elevated quantity of activated and fully developed dendritic cells, which encompass a substantial amount of toll-like receptors (TLR), notably TLR2 and TLR4[33]. In patients with ulcerative colitis, there is an unusual T-helper (Th) cell response, particularly Th2, that causes a cytotoxic reaction against epithelial cells. Additional immune-related variables involved in the pathogenesis of ulcerative colitis are tumor necrosis factor-alpha (TNF-alpha), interleukin 13, and natural killer T-cells. Inflammatory bowel illness is characterized by higher levels of IgM, IgA, and IgG[34]. Patients with ulcerative colitis show a significant increase in IgG1 antibodies. Leukocyte recruitment is affected by two variables. Ulcerative colitis results in elevated production of the chemoattractant CXCL8, causing

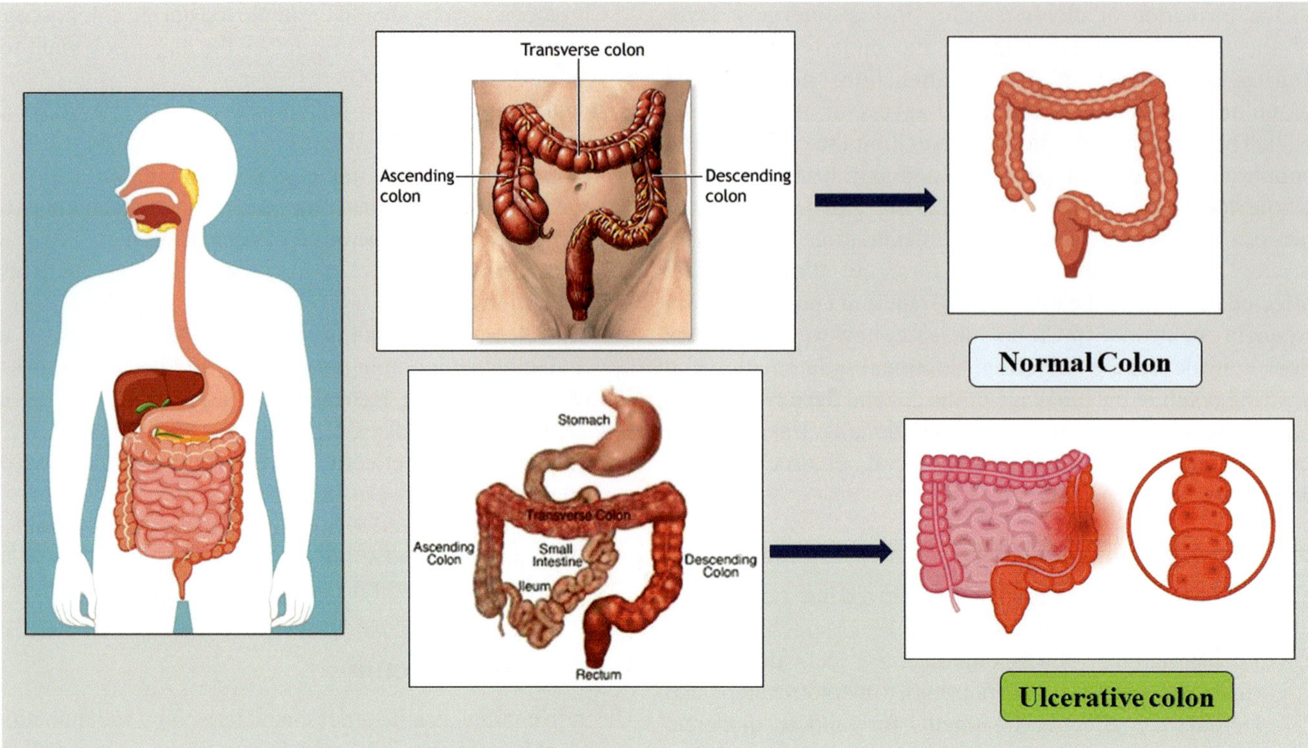

FIGURE 7.2 Figure showing differences between a normal healthy colon from ulcerative colon in Ulcerative colitis.

leukocytes to be drawn from the systemic circulation to the mucosa. Furthermore, there is an elevation in the expression of mucosal cellular adhesion molecule-1 (Mad-CAM1) on the endothelium of mucosal blood vessels. This increase promotes the attachment of white blood cells and their migration from the blood arteries to the mucosal tissue[35].

Research has demonstrated that the intestinal microbiota plays a crucial role in the development and intensity of inflammation and disease characteristics. Ulcerative colitis appears to be caused, at least in part, by a disruption in the balance between the microorganisms in the intestines and the immune system of the individual. This leads to an abnormal reaction to non-pathogenic microorganisms[36].

7.2.2.2 Treatment Strategy

The type of treatment for people with ulcerative colitis relies on how bad the disease is and how widespread it is. Within the first ten years after being diagnosed, the outlook is usually good, and most people achieve remission. Rectal administration of medical treatment, either through suppository or enema, is typically suitable for localized distal disease (proctitis). However, rectal application is commonly employed in conjunction with systemic therapy to specifically target the distal colon and so alleviate tenesmus[37].

The initial therapy consists of sulfasalazine and 5-aminosalicylates, administered either orally or rectally, with an approximately 50% remission rate. For individuals who do not achieve remission within a two-week period, glucocorticoids might be administered orally or rectally as an additional treatment. With the exception of glucocorticoids, all of these drugs can be utilized for the purpose of sustaining remission. Moreover, there is substantiated evidence indicating that probiotics are beneficial in achieving remission. Fecal microbiota transplantation exhibits potential in the management of ulcerative colitis to facilitate the establishment of a healthy gut flora[38].

If individuals do not respond to glucocorticoids, thiopurines or biological medicines can be included in the treatment regimen. Thiopurines, such as azathioprine or 6-mercaptopurine, are drugs that inhibit the immune system. Biological medications encompass anti-TNF-alpha medications, specifically infliximab, adalimumab, and golimumab. Infliximab is the predominant treatment option for ulcerative colitis and is particularly employed in cases of high severity necessitating hospitalization. Vedolizumab, an anti-adhesion molecule inhibitor, is a cutting-edge biological medication that specifically targets and blocks the alpha-4-beta-7 integrin[39].

Because people with ulcerative colitis have less peroxisome proliferator-activated receptor-gamma (PPAR-gamma) in their colonocytes, increasing PPAR-gamma activity may be a possible future treatment. PPAR-gamma functions as a suppressor of NF-KB-mediated inflammation. Novel analogs of 5-aminosalicylic acid (5-ASA) are currently being created with enhanced PPAR-gamma activity. The utilization of current PPAR-gamma agonists is limited due to their cardiotoxic and metabolic toxic effects[40,41].

Because ulcerative colitis only affects the colon, the only way to cure it is to remove the colon. The surgical indications include the ineffectiveness of medical treatment, severe and uncontrollable inflammation of the colon, an abnormally enlarged colon that is toxic, a hole in the colon, unmanageable bleeding, unacceptable adverse effects of drugs, narrowing of the colon, high-grade or several areas of abnormal cell growth that cannot be removed, cancer, or impaired growth in children[42]. The preferred surgical approach is proctocolectomy with ileal pouch-anal anastomosis (IPAA). However, for patients who are not eligible for IPAA, proctocolectomy with ileostomy is a suitable alternative. Regular intervals are indicated for colonoscopy due to the potential danger of colon cancer[43]. Every patient needs ongoing therapy to prevent a recurrence. Oral amino salicylates are the preferred medications, while some individuals may also benefit from azathioprine and 6-mercaptopurine. Patients with ulcerative colitis do not have a prescribed diet, however a significant number of them may develop lactose intolerance. Elemental or parenteral feeding does not play a role in treating Crohn's disease[44].

7.2.3 Celiac Disease

People who are more likely to get celiac disease have an autoimmune disease where their immune system reacts badly to gluten-containing foods. The initial documentation of celiac disease dates back to 1888 when Samuel Gee first characterized it[45]. However, it was not until 1953 that the significance of gluten in the development of this condition was fully understood. In individuals with celiac disease, the consumption of gluten causes damage to the lining of the small intestine, resulting in a reduced ability to absorb nutrients. Celiac disease can be classified as a syndrome due to its diverse range of clinical symptoms and its impact on multiple human systems. Celiac disease exhibits distinctive characteristics when compared to other autoimmune conditions, such as the complete restoration of mucosal damage and the reversibility of its progression and chronic nature, achieved with strict avoidance of gluten. On the other hand, it is now confirmed that undiagnosed celiac disease can have serious effects in both children and adults. In addition to celiac disease and wheat allergy, a new condition called non-celiac gluten sensitivity (NCGS) has been recognized[46]. Unlike the other two conditions, NCGS does not appear to be caused by an immunological response. The cause of NCGS is still not fully understood, however it has been established that it involves a combination of factors. In this article, we provide a comprehensive analysis of the epidemiology, pathophysiology, clinical presentation, diagnostic testing, and therapeutic therapy of celiac disease[47].

7.2.3.1 Treatment Strategy

Permanent Gluten-Free Diet

Celiac disease is treated with a lifelong gluten-free diet. Clinical improvement occurs in weeks and mucosal damage improves in 1–2 years. Celiac disease patients may have brush border lactase insufficiency due to surface epithelial cell destruction, hence milk and dairy products should be avoided in the beginning. All patients should take a gluten-free multivitamin because GFDs often cause vitamin B deficiency. Pediatric celiac disease requires early identification and treatment since certain effects are irreversible: Growth retardation, malformed teeth, osteoporosis. Several observational studies suggest that prolonged breastfeeding and delaying and gradually introducing gluten in the first year may lower childhood celiac disease risk[48]. Even goods for celiac disease diets may include trace levels of gluten proteins. Food labeling laws vary by country, adding to uncertainty. Take gluten-free diet and refractory celiac disease into account. GFD fails in a minority of refractory celiac disease (RCD) patients despite dietary restriction. Its higher mortality rate than RCD type 1 is due to severe malnutrition and a higher probability of overt lymphoma. In most individuals, corticosteroids and immunosuppressants such azathioprine or cyclosporin can relieve clinical symptoms temporarily. But viable therapies for both forms of RCD have yet to be developed. These medications may increase the risk of T-cell lymphoma; thus they should be used cautiously in RCD type 2 patients at risk of this consequence. Cladribine and Pentostatine, anti-T-cell nucleoside analogues, have been used successfully at chemotherapy, and stem cell transplantation has been suggested.

Other drug-based therapies are being studied, but patients say GFD is good and well-tolerated[49,50].

Degrading Gluten Enzymes

Enzyme supplement therapy with bacterial prolyl-endopeptidasis from diverse microorganisms may speed gluten digestion and remove T-cell epitopes[97]. Proline-specific endopeptidases cleave gluten peptides. Two medication candidates—ALV003 and AN-PEP—are in clinical testing. Two phase 1 clinical trials found that pre-treating gluten with ALV003 removes the peripheral blood T-cell response in celiac disease patients, suggesting that gluten-specific enzymes may treat the condition. Current clinical phase IIa testing shows a significant reduction in gluten-related T-cell responses compared to placebo but no reduction in gluten-induced symptoms[51].

AN-PEP efficiently destroys gluten peptides at stomach pH. This enzyme may be acceptable for oral supplementation, although more research is needed.

Modified grains can be developed by selecting breeding early wheat species or by mutating or silencing immunostimulatory sequences with siRNA[52].

Zonulin inhibitor larazotide (AT-1001) prevents gluten penetration through intestinal epithelium and repairs intestinal barrier abnormalities[53]. Animal models have been used. AT-1001 is the best-studied celiac disease drug, having undergone phase II clinical trials. AT-1001 reduced autoantibody reaction, proinflammatory production, and urine nitrate excretion compared to placebo. Rho/Rho kinase block intestinal permeability increases with ROCK activity. ROCK controls tight junction shape and function and axon growth. The medicine could investigate if ROCK inhibition reverses gluten-dependent intestinal permeability in these patients[54].

Immunotherapy

When celiac disease was first seen in a person with acute leukemia who had an allogenic bone marrow donation, it showed the role of T-lymphocytes[55]. Many immunologic treatment techniques are being developed since then due to recognition. For severe autoimmune illnesses, cytokine treatments that amplify regulatory or suppress inflammatory cytokines are spreading. Immunomodulators are being used for celiac disease resistant to diet and refractory celiac disease. In intestinal epithelium of human IL-15 transgenic mice, IL-15 blocking antibodies promote intra-epithelial lymphocyte death. A celiac disease investigation is pending, as these antibodies have only been tested on rheumatoid arthritis patients[56]. Gluten-dependent T-cell activation was slowed down in celiac disease gut mucosa by human recombinant IL-10 alone. However, a test study on people with refractory celiac disease showed that this monoclonal antibody did not work as a drug. Anti-IFN-γ antibodies showed good tolerance in people with inflammatory bowel disease, but more study is needed for people with celiac disease. Similar problems arise when anti-TNF-α antibodies are used, which are commonly prescribed for inflammatory bowel disease but not often seen in cases of celiac disease that won't go away. Chemokines and their receptors recruit gut lymphocytes to cause celiac disease, making them a potential immunotherapy target. T-lymphocytes that live in the small intestine have the protein CCR9, which binds to CCL25 released by intestinal epithelial cells, and integrin $\alpha 4^2 7$, which binds to MAdCAM-1.Thus, both substances are required for T-cell migration to the intestinal mucosa and may be therapeutic targets. Recently, Natalizumab, a monoclonal antibody against integrin α4, has shown efficacy in Crohn's disease, suggesting potential efficacy in celiac disease. Additionally, CCX282-B, a CCR9 inhibitor, and LDP-02, both studied in Crohn's patients, are being investigated for celiac disease. Because helpful immunosuppressive regulatory T-cells are similarly blocked, limiting lymphocyte homing to the colon may have long-term risks. Since its demonstration, gluten tolerance in celiac disease patients has become a promising immunomodulatory approach. Oral injection of a genetically engineered *Lactococcus lactis* bacteria that secretes deamidated DQ8-restricted gliadin epitope has been explored to induce gluten tolerance. In previously sensitized transgenic mice, Foxp3+ regulatory T-lymphocytes were induced and local and systemic T-cell responses to the gliadin peptide were significantly suppressed. This method still needs human trials, however nonspecific immunomodulation clinical trials have tested the function of parasite infection in gluten immune response inhibition. In fact, people in wealthy countries are more likely to get autoimmune diseases now that they don't have any intestinal parasites[57]. A phase I clinical study was started using hookworm Necator Americanus larvae to change T-cells toward a suppressive T-regulatory response. However, this infection did not have a big effect on gluten-induced enteropathy. A vaccine based on gluten peptides identified by HLA-DQ2 in an immunodominant way has been tested in clinical trials, with phase I just completed. Vaccination is popular, especially for increasing patient compliance, which could benefit from a single-dose delivery rather than daily treatment. The possibility of immune system activation from vaccination therapy and associated negative effects needs further study. Most of the alternative therapy approaches need extensive long-term clinical trials to answer questions about their clinical efficacy, safety, and price[58].

7.3 GENETIC BASIS OF GASTROINTESTINAL INFLAMMATION

7.3.1 Role of Genetic Factors

BOX 2 demonstrates that a significant fraction of mice with various genetic abnormalities experience the occurrence of mucosal inflammation without any external cause. The first idea says that the same signs of gut inflammation could be caused by different genetic problems. If there is a major genetic defect, the second principle says that the host's genetic background shows how likely they are to get intestinal inflammation. For example, some types of mice that were bred to not have interleukin-10 (IL-10) are more apt to get colitis than others[59]. The variations in disease susceptibility and resistance among different strains of mice provide a chance to discover mouse genes that control the expression of genes crucial in disease. This has the potential to facilitate the discovery of genes that are crucial for the onset of diseases in people.

7.3.2 The Role of Genetic Variables in the Development of Inflammatory Bowel Disease

Thorough studies including the spread of diseases and analysis of family connections have definitively proven that genetic variables have a substantial impact on determining vulnerability to IBD. The most compelling evidence is obtained from a study carried out with twins[60]. The concordance rate for Crohn's disease has been seen to reach up to 58% in identical twins, but the concordance rate in dizygotic twins does not show any significant difference compared to that in all siblings. The documented rates of concurrence for monozygotic and dizygotic twins with ulcerative colitis range from 6% to 17% and from 0% to 5%, respectively. The statistics strongly indicate that inheriting vulnerability to IBD, namely Crohn's disease, is quite likely. IBD is not inherited in a straightforward Mendelian pattern but has a complex genetic basis including multiple contributing genes. The intricate nature of genetics, which encompasses genetic variability, interactions between genes, and interactions between genes and the environment, has long hindered the discovery of precise genetic disorders. Recently, advancements have been achieved in this field of study. We examine a recently found gene that increases the likelihood of developing Crohn's disease in a specific subset of patients, together with recent advancements in detecting genetic elements in animal models of inflammatory bowel disease (IBD)[61].

7.3.3 Detection of NOD_2 Gene Mutations in Individuals With Crohn's Disease

Identifying mutations for monogenic disorders has become relatively simple, however finding genes linked to complex genetic disorders is still challenging. An effective method to identify genes associated with susceptibility to inflammatory bowel disease involves examining the genomes of families with a high incidence of sick individuals using polymorphic microsatellite markers. These whole-genome screens operate on the assumption that a genetic marker near a disease gene is less likely to separate during meiosis and will be passed down together with the disease-causing gene. Relatives who are impacted by the condition, such as siblings who also have the condition, will possess a greater number of alleles than what would be statistically expected for genetic markers that are located near a gene or genes responsible for disease susceptibility. Studying large groups of sick siblings enables the identification of wide genomic areas containing illness genes. Hugot and colleagues utilized this methodology to investigate families with individuals diagnosed with Crohn's disease. IBD1, a susceptibility locus, was identified in the pericentromeric region of a chromosome in this analysis and other independent studies. Further investigation of this region uncovered a strong association with a single gene, NOD2, also known as caspase-recruitment domain protein 15 (CARD15). This gene encodes an endogenous molecule belonging to the NOD family, which is theorized to play a role in bacterial detection. The examination of the gene in persons diagnosed with Crohn's disease uncovered the existence of a cytosine insertion at location 3020 in exon 11. This insertion causes the formation of a stop codon and leads to the production of a shortened NOD_2 protein. Additional polymorphisms were discovered in individuals diagnosed with Crohn's disease, specifically two prevalent missense mutations—2722G C [Gly908Arg] and 2104C T [Arg702Trp][62] conducted parallel investigations using the candidate-gene technique to identify genes inside the IBD1 region. The authors were aware that NOD_2 plays a role in identifying pathogens within cells and is closely related to the IBD1 locus. Through the analysis of the genetic composition of a cohort of individuals with Crohn's disease and a comparison group, the researchers discovered compelling indications of the previously mentioned 3020-frameshift mutation in this specific gene. ATG16L1, a crucial element in the autophagic reaction to invading pathogens, has been linked to a higher likelihood of developing Crohn's disease. The prevalent and extensively researched genetic variation of ATG16L1 (rs2241880; resulting in a T300A conversion) is strongly linked to an increased chance of developing Crohn's disease[63]. The rs2241880 variation is essential for clearing pathogens, causing an imbalance in cytokine production, and is associated with biological processes, such endoplasmic reticulum stress/unfolded protein response as per Figure 7.3.

7.3.4 Association Between NOD_2 Mutations and Susceptibility to Inflammatory Bowel Disease (IBD)

Studies conducted by Hugot et al. and Ogura et al. show that NOD2 mutations are specific to individuals with Crohn's disease and are absent in those with ulcerative colitis. Based on Hugot et al.'s data, the relative risk of developing Crohn's disease is 3 for simple heterozygotes and 38 for homozygotes. Furthermore, compound heterozygotes, which refers to

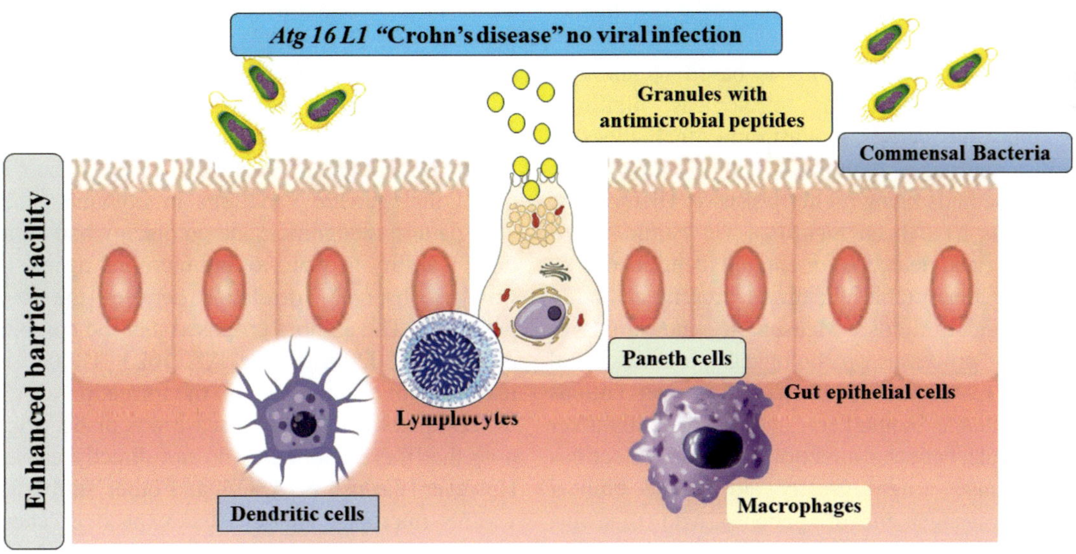

FIGURE 7.3 ATG16L1, a crucial element in the autophagic reaction to invading pathogens, has been linked to a higher likelihood of developing Crohn's disease. The prevalent and extensively researched genetic variation of ATG16L1 (rs2241880; resulting in a T300A conversion) is strongly linked to an increased chance of developing Crohn's disease. The rs2241880 variation is essential for clearing pathogens, causing an imbalance in cytokine production, and is associated with biological processes such endoplasmic reticulum stress/ unfolded protein response.

individuals with one mutation on each chromosome, exhibit the same characteristics as homozygotes, with a relative risk of 44. Hampe et al. conducted a study on a significant number of individuals and discovered that the presence of 3020 insertions in the NOD2 gene is prevalent in both familial and sporadic instances of Crohn's disease. Moreover, they demonstrated that no healthy individuals exhibited homozygosity for this mutation. The observations indicate that a solitary recessive mutation in NOD2 could be sufficient to induce the disease. 8–17% of Caucasian patients with Crohn's disease have been verified to possess two copies of the major risk alleles of NOD2[64]. Individuals in good health can have NOD2 mutations on both chromosomes without showing any illness symptoms. Furthermore, none of the three mutations in the NOD2 gene were found in a group of 483 Japanese patients who had been diagnosed with Crohn's disease. The latter provides convincing proof of genetic diversity among patients from various ethnic backgrounds.

7.4 GENETIC BIOMARKERS AND DIAGNOSTICS

Ulcerative colitis (UC) and Crohn's disease (CD) are long-lasting conditions characterized by inflammation in the bowels. In ulcerative colitis (UC), the colon is affected specifically, whereas in Crohn's disease (CD), any part of the gastrointestinal system, spanning from the mouth to the anus, may be affected[65]. Thus yet, no specific microbe implicated in the development of inflammatory bowel disease has been discovered. The bacterial pathogens commonly responsible for diarrhea include *Campylobacter*, *Salmonella*, *Shigella*, and *Escherichia coli*. Diarrhea is caused by the following viral pathogens: rotavirus, norovirus, herpes simplex virus, and viral hepatitis[66]. Gastritis, which affects the lining of the stomach, can be categorized into two primary forms—acute and chronic—according to the clinical manifestations that are observed[64].

The diagnosis of digestive issues is based on clinical, radiographic, endoscopic, and serological findings. Determining the root cause of gastrointestinal illnesses poses difficulties and incurs high costs, with the majority of commonly used methods being highly sensitive. Historically, bacterial pathogens have been identified via culture-based techniques, nucleic acid detection, or antigen identification[67].

Biomarkers are measurable chemicals present in bodily tissues or biological fluids. Measurable indicators of normal biological processes, pathological processes, and pharmacological responses to therapeutic interventions constitute a dependable biomarker. An excellent biomarker that can differentiate between infectious and non-infectious gastrointestinal causes might greatly aid in the clinical assessment and management of patients with digestive illnesses. In addition, an optimal biomarker has the ability to rapidly distinguish between viral, bacterial, and parasitic origins[68].

Biomarkers have a significant and occasionally distinctive function in the diagnosis and treatment of cancer patients[69]. Biomarkers can be used as a screening tool for early diagnosis of cancer or pre-cancerous disorders in persons who do not show any symptoms. The use of symptomatic patients can aid in distinguishing between cancer and benign illnesses in the process of differential diagnosis. Biomarkers may be utilized to assess the efficacy of surgery in eliminating tumors and preventing disease relapse in cancer patients[70]. Biomarkers are a valuable tool for assessing the efficacy of medication therapy in patients. Genetic biomarkers such as micro-RNAs and long non-coding RNAs have proven to be beneficial in identifying digestive cancers[71].

7.4.1 Serologic Biomarkers in Inflammatory Bowel Disease (IBD)

Before the recent advancement of IgA-SEQ technologies, it was widely recognized that the immune responses of patients with inflammatory bowel disease (IBD) are altered in response to the antigens of the luminal commensal microbiota. Immunological responses that specifically target a particular oligomannosidic epitope situated on the cell wall of yeast were identified in response to the presence of a dietary antigen, such as celiac disease. The presence of *Saccharomyces cerevisiae* was elevated in individuals with CD[72]. Following this, 20% of asymptomatic first-degree relatives of individuals with Crohn's disease exhibited ASCA, suggesting that these antibodies may function as a subclinical indicator for CD within familial contexts. During the early investigations, the authors raised questions about whether ASCA, a marker for CD, was influenced by genetic factors, environmental variables, or a combination of both, as the inheritance patterns of CD were not well understood at that time. Subsequent genetic findings have provided researchers with the chance to investigate the correlation between genotypes and serotypes linked with IBD. In order to construct a coherent narrative, it is necessary to establish a correlation between the interleukin-23 signaling pathway (IGLUM, STAT3, NOD2, TLR5, IRGM, ATG16L1) and the synthesis of antimicrobial antibodies in individuals afflicted with Crohn's disease (CD), via the genes accountable for bacterial sensing and autophagy. A study conducted on a group of more than 600 patients with Crohn's disease (CD) found that genetic variations in immune genes associated with pattern recognition and autophagy had an impact on the body's ability to react to antimicrobial agents in CD. There was no observed correlation between the IL-23 signaling pathways. The authors proposed that the combined influence of NOD_2 3020insC and ATG16L1 T300A indicates the involvement of autophagy in the formation of ASCA[73]. While the comparison to celiac disease was made, it is important to note that these antibodies do not directly cause the disease. However, the use of ASCA and other antibodies that target various microbial antigens has expanded. These antibodies now serve as markers for the state and behavior of the disease, and their ability to predict the development of the disease before clinical symptoms appear is becoming increasingly recognized. Microbial antigens that have been referenced encompass a variety of substances, including CBir1 flagellin, *E. coli* outer-membrane porin C (anti-OmpC), *Pseudomonas fluorescens*-related protein (anti-I2), a protein found in bacterial flagella, and various glycan epitopes or bacterial flagella. Additionally, antigen combinations consisting of cell lysates or membrane-associated antigens from diverse *Bacteroides* spp., *Klebsiella pneumoniae*, *Enterococcus faecalis*, and *Candida albicans* are also included. Although the immune responses induced by microbes are well established and comprehended, the precise target of the perinuclear antinuclear cytoplasmic antibody (pANCA) is still unknown. pANCA identifies a recurring structural domain present in mycobacteria that cross-reacts with a DNA-binding domain of histone H165, according to Kohavy. The antigen responsible for IBD pANCA is different from the pANCA associated with small vessel systemic vasculitides. It is most typically linked to a phenotype that resembles ulcerative colitis (UC). There is no genetic susceptibility condition associated with ASCA. In order to function as dependable pre-clinical predictors, prognosticators, or diagnostic indicators, seromarkers must demonstrate stability within a given patient. The dilemma at hand is whether it is the mere existence of something that needs to be consistently stable, or if it is the level of sero-response that must be maintained at the highest standard. Reider et al discovered a notable variation in ASCA levels, but did not see any difference between negative and positive status[72]. There is a lack of significant research examining the correlation between disease activity and the levels of the serologic response. Positive or negative ASCA status in a particular patient is predominately determined by the genetic susceptibility of an individual to inflammatory bowel disease (IBD), according to the proposed hypothesis. Furthermore, the degree of intestinal inflammation is more strongly correlated with the magnitude or intensity of the serologic response. This phenomenon may elucidate the enduring nature of the antibody status observed in certain studies over a period of time, as well as the correlation between elevated ASCA levels and disease prognosis. The genetic association with serologies may elucidate the presence of ASCA, namely, in individuals with non-inflammatory bowel disease (IBD) immunological disorders. The findings revealed that patients with early-stage celiac disease exhibited seroreactivity to many microbial antigens, indicating a common intestinal target shared with CD. Notably, it was observed that ASCA antibodies are dependent on gluten. The researchers inferred that microbial targets may have a role in the initial stages of celiac disease.

7.4.2 Diagnosis and Phenotype

The identification of IBD and subsequent diagnostic assessment can generally be uncomplicated when the clinical presentation is clear and unequivocal. Nevertheless, a diagnostic dilemma develops in individuals who exhibit overlapping, nonspecific, and slow-developing symptoms that are indicative of both organic and non-organic illnesses. When faced with uncertainty in diagnosis, clinicians frequently have to use invasive diagnostic tests, such as cross-sectional imaging and colonoscopy with biopsies, to rule out inflammatory bowel disease (IBD). Patients who are suspected of having IBD often undergo thorough diagnostic examinations, only to be finally diagnosed with a functional bowel condition. On the other hand, the identification of IBD, especially CD, can be overlooked or postponed because the symptoms in the intestines and other parts of the body are not precise enough when the condition first appears. In light of these complex clinical obstacles, there is an ongoing quest to identify a precise non-invasive diagnostic indicator that can assist doctors in promptly identifying IBD and distinguishing it from similar conditions. An optimal non-invasive diagnostic test should possess a high degree of sensitivity and specificity. Furthermore, it should be of comparable quality to the gold standard. Currently, there is no test that has been created for this purpose. However, improvements in testing methods and the inclusion of new indicators have enhanced the qualities of existing tests. The diagnostic utility of these biomarkers, namely ASCA and pANCA, in individuals with and without inflammatory bowel disease (IBD), has been the subject of numerous studies[74]. The majority of studies consistently find that serological markers have a high level of specificity for inflammatory bowel disease (IBD), but a poor level of sensitivity, which limits their usefulness as diagnostic assays. A prospective study was conducted on children with non-alarm symptoms who were getting a comprehensive diagnostic evaluation to determine the presence or absence of inflammatory bowel disease (IBD), while simultaneously undergoing serologic testing. The distinction between IBD and non-IBD was determined using the most reliable method available and without knowledge of the results of serological investigations. The results of the research indicated that the overall accuracy rate of serologies was 84%. Antibodies that are specific to mannobioside (AMCA), laminaribioside (ALCA), and chitobioside (ACCA) bond to glycans in a selective manner. These antibodies are significantly more prevalent and abundant in individuals with CD in comparison to those with UC and healthy individuals. Analogous to ASCA, these antibodies have been identified as being associated with NOD_2 mutation[75].

Anti-zymogen glycoprotein (GP2) antibodies and Crohn's disease are gaining increasing attention due to their potential utility in differentiating this condition from ulcerative colitis, non-inflammatory bowel disease, and Crohn's disease. GP2 is also present on the microfold (M) cells of the follicle-associated epithelium (FAE) in intestinal Peyer's patches, according to the available evidence. Evidence indicates that inflammation in the ileum is necessary for the formation of anti-GP2 antibodies in Crohn's disease. Furthermore, it suggests that the intestine, rather than pancreatic juice, is the source of the antigen that triggers the production of anti-GP2 antibodies. Furthermore, there was no discernible association between ASCA and anti-GP2 titers[76]. A novel assay for concurrent analysis of IgA and IgG antibodies against MZGP2 was recently devised; it achieved a specificity of 96% and a sensitivity of 31%. A comprehensive analysis was conducted on a total of 832 serum samples. This included 617 consecutive patients with inflammatory bowel disease (IBD), consisting of 323 patients with Crohn's disease (CD) and 294 patients with ulcerative colitis (UC), who were being monitored at a specialized medical center. Additionally, there were 112 samples from individuals with pathological conditions and 103 samples from healthy individuals serving as controls. The presence of either ASCA (IgA or IgG) or anti-MZGP2 (IgA or IgG) antibodies demonstrated a sensitivity of 75% (with a range of 70% to 80%) and a specificity of 84% (with a range of 79% to 89%). Similar to ASCA, the presence of anti-GP2 antibodies may have a greater influence on the location and behavior of the disease[77]. The issue at hand is whether a malfunctioning intestinal barrier, which leads to the inability to tolerate microbial antigens, may also be applied to the loss of immunological tolerance to food antigens. Several research have been conducted to establish the correlation between food and the development of inflammatory bowel disease (IBD). Prior studies have shown that individuals with inflammatory bowel disease (IBD) have a higher occurrence of antibodies against food derivatives, common dietary allergens, and gliadin compared to individuals without IBD. Frehn and colleagues demonstrated that anti-dietary antibodies (extracts of wheat and milk, pure ovalbumin) did not exhibit any overall changes in patients with inflammatory bowel disease (IBD). This result diminishes the significance of testing for antibodies against food antigens as a diagnostic tool or as an indicator of disease development[78]. While UC and CD may have similarities in terms of epidemiology, immunology, therapy, and clinical characteristics, they are currently recognized as separate subtypes of inflammatory bowel disease (IBD). IBD unspecified (IBDU) occurs in roughly 10–15% of persons with inflammatory bowel disease (IBD). Indeterminate colitis (IC) is currently used to describe patients whose diagnosis remains uncertain even after thorough evaluation of the surgically removed specimens. Nevertheless, there is a significant amount of discrepancy in the literature about the definition of IBDU, primarily due to imprecise clinical criteria and limited retrospective investigations. It is crucial to highlight that surgical interventions frequently depend on an accurate diagnosis.

• **A Pre-Clinical Marker for Predicting Disease**

IBD often results in increased intestinal permeability, which is caused by several alterations in the physiology of the epithelial cells in response to changes in inflammation and the microbial environment of the mucosal lining, even during periods of remission. The seminal study conducted by Hollander and colleagues showed that individuals with a genetic predisposition to inflammatory bowel disease (IBD) have heightened intestinal permeability to ingested polyethylene glycol-400. This finding was supported by the evidence from at-risk family members of IBD patients[79]. This observation has been corroborated by other researchers [80] and expanded to include individuals without health issues who carry disease-related versions of the NOD_2 gene. This likely indicates that the impaired regulation of gut bacteria is caused by these fewer functioning forms of the microbial-detecting protein. One hundred fourteen given the widespread use of non-invasive permeability assays that specifically target the colon, it is necessary to reevaluate the inclusion of intestinal permeability as a trait in future studies that aim to assess persons at risk for inflammatory bowel disease (IBD). Several data indicate that unaffected relatives of individuals with CD exhibit higher levels of ASCA compared to spouses and healthy individuals. ASCA has demonstrated significant levels in twins who have the same diagnosis of CD, but not in twins who have different diagnoses, indicating the presence of an environmental factor[81]. Family members who are not afflicted by NOD_2 have been discovered to exhibit elevated levels of serologies in comparison to patients who do not have NOD_2, indicating a connection between the innate and adaptive immune responses[82].

Unaffected family members were also assessed for the presence of anti-OmpC in addition to ASCA. Curiously, the unaffected family members of CD patients[83] had heightened expression. The hereditary aspect of these serologies is widely recognized. It is yet uncertain whether these markers can accurately predict the occurrence of illness. Van Schaik FD and colleagues conducted a study to investigate if serologies might be used to predict a diagnosis of inflammatory bowel disease (IBD). The study participants were adults who were recruited in the European Prospective Investigation into Cancer and Nutrition (EPIC) study[84].

7.4.3 Implication for Personalized Medicine

Personalized medicine, or precision medicine, seeks to customize medical therapy based on the unique characteristics of each patient, such as their genetic composition, lifestyle, and environmental influences. Personalized medicine has major significance in the context of gastrointestinal inflammation, namely in inflammatory bowel diseases (IBD) such as Crohn's disease and ulcerative colitis. Here are numerous significant consequences:

1. **Predictive Biomarkers:** Predictive biomarkers play a crucial role in personalized medicine for gastrointestinal inflammation. They are used to identify and anticipate disease development as well as predict treatment effects. For instance, biomarkers such as C-reactive protein (CRP) and fecal calprotectin can serve as indicators of the existence and intensity of inflammation in the gastrointestinal tract. Employing these indicators might inform treatment choices, such as modifying medication dosage or shifting to alternative therapies.
2. **Genetic Profiling:** Recent advancements in genomic sequencing have enabled the identification of many genetic variations linked to gastrointestinal inflammation. Personalized medicine can employ genetic testing to uncover precise genetic markers that make individuals more susceptible to particular types of gastrointestinal inflammation. This data can be utilized to forecast the occurrence, intensity, and possible therapeutic reactions of diseases.
3. **Customized Treatment Plans:** Instead of using a universal strategy, personalized medicine enables

healthcare providers to create tailored treatment plans that take into account a patient's distinct qualities. When deciding on the most appropriate treatment approach, various factors are considered, including genetic profile, illness subtype, intensity of symptoms, and any other existing medical conditions. This may involve the process of choosing the most efficacious drugs, making dietary adjustments, or implementing lifestyle treatments tailored to each individual patient.

4. **Pharmacogenomics:** Pharmacogenomics is the study of how a person's genetic composition affects their reaction to drugs. Pharmacogenomic testing can be used to identify patients with gastrointestinal inflammation who are more likely to have positive responses to specific medicines, as well as those who may have adverse events or resistance to therapy. By providing this information, practitioners can select pharmaceuticals that are highly probable to be efficacious and secure for individual patients, so reducing the need for trial-and-error methods and enhancing treatment results.
5. **Precision Nutrition:** Nutrition is essential for effectively managing gastrointestinal inflammation. Personalized medicine acknowledges the fact that individuals with IBD have different dietary triggers and tolerances. Customized nutritional therapies, taking into account an individual's unique food sensitivities, microbiota makeup, and metabolic profile, can effectively ease symptoms, mitigate inflammation, and enhance overall gastrointestinal well-being.
6. **Targeted Treatment and Therapy:** Personalized medicine enables the advancement of focused medicines that directly target the fundamental biological pathways responsible for gastrointestinal inflammation. Biologic drugs, such as monoclonal antibodies that specifically target cytokines like tumor necrosis factor-alpha (TNF-α) or interleukins, are a type of therapy that has greatly transformed the treatment of inflammatory bowel disease (IBD). Through targeted inhibition of crucial inflammatory pathways, these medicines have the ability to initiate and sustain remission in patients with treatment-resistant conditions.

7.5 GENE-ENVIRONMENT INTERACTION

7.5.1 Risk Factors Associated With the Environment

The rising number of IBD-associated risk alleles highlights the importance of environmental exposures in the development of the illness in genetically susceptible individuals (Figure 7.4). The gut microbiome environment is defined by decreased species diversity and changes in metabolic pathways. Previous reviews have fully covered the significant function of the gut flora in IBD, and it will not be described in depth here. The external environment impacts an individual's gut microbiota and immunological functions throughout their lives. Below, the existing research about modifiable environmental factors in IBD is examined. Exhibiting decreased species diversity and changes in metabolic pathways

7.5.2 Early Environmental Exposures

The method of delivery has been proposed to impact the likelihood of acquiring inflammatory bowel disease (IBD). The gut microbiome of newborns delivered by cesarean section differs significantly from that of infants born through vaginal birth, indicating that the method of delivery may impact the likelihood of developing IBD[85.] A Danish cohort study found that cesarean birth significantly increased the likelihood of getting IBD.

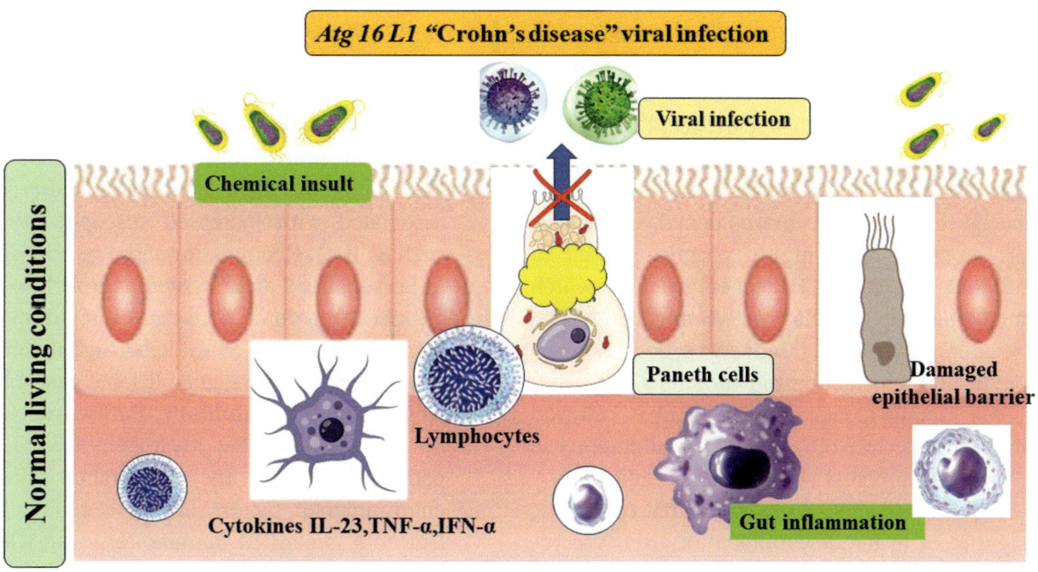

FIGURE 7.4 Various factors influencing IBS and IBD.

Inflammatory Bowel Disease. A recent systematic review did not confirm the connection. IBD has been linked to breastfeeding. Non-breastfed newborns had higher levels of harmful bacteria, indicating that nursing provides health benefits by passing on antibacterial substances such lactoferrin, lysozyme, and maternal antibodies to the infant. IBD research indicates that breastfeeding offers protection against illness progression in a manner that depends on the duration of breastfeeding[86]. The association's intensity varies among ethnicities, being most pronounced in Asian populations and when breastfeeding lasts more than 12 months[87].

7.5.3 Epigenetics in Inflammatory Disease

Inflammatory bowel illnesses may result from the interplay between the patient and several environmental factors, including the genome, immune system, intestinal microbiota, and specific elements like nursing, diet, smoking, and medications. Epigenetics refers to heritable changes in gene expression that occur without affecting the DNA sequence during cell division. Chromatin modifications can impact gene expression. Various cells in the body exhibit distinct functions and varying levels of gene expression, despite all containing the same genetic code. Epigenetics refers to the systems and processes that lead to variations in gene activity among different cells. Key epigenetic mechanisms are DNA methylation, histone modification, RNA interference, and nucleosome placement. Epigenetic mechanisms, especially DNA methylation, play a crucial role in the interplay between the environment and the genome. It can potentially be reversed and inherited over multiple rounds of cell division. Differences in DNA methylation are a well-known factor in human disease and are thought to be a key factor in the development of complicated disorders (Figure 7.4). Various recognized disorders of imprinting include Beckwith–Wiedemann syndrome, Temple syndrome, Wang–Ogata syndrome, Silver–Russell syndrome, Angelman syndrome, and Prader–Willi syndrome. Imprinted genes are believed to have a significant impact on embryonic development, with their precise expression being crucial for regular cellular functions and human behavior. The task is to pinpoint reliable epigenetic changes that have a causal relevance, considering that DNA epigenetic modifications vary among tissues, happen at various stages of development in the same tissue, and are influenced by ongoing environmental influences. Epigenetics has emerged as a highly promising idea in various fields of biological study[88]. Recent epigenetic research has demonstrated that the interplay between genetics and environmental factors significantly influences how diseases manifest phenotypically, elucidating the variations in disease manifestation even among identical twins. IBD is believed to result from an intricate interplay of genetic alterations in several genes and environmental conditions, as indicated by substantial evidence. Increasing data suggests that epigenetic variables are significant in the development of IBD. Various clinical uses of epigenetics in diagnostics and therapies are now being explored. Epigenetics is used in diagnostics to utilize biomarkers for confirming diagnosis, categorizing disease progression and response to chemotherapy, and forecasting cancer development. DNA methylation is the most researched epigenetic alteration, and its connection to IBD development has been well-established during the past decade. Multiple investigations indicate notable differences in DNA methylation levels between healthy and inflamed tissues in people with Crohn's disease (CD) and ulcerative colitis (UC). Thus, there is evidence linking hypermethylation of several gene promoters to people with inflammatory bowel disease (IBD). Various biological pathways' genes have been examined, although there is currently no established database of methylation genes in IBD[89]. DNA methylation patterns as per Figure 7.4 are highly effective for detecting disease with great sensitivity. DNA methylation methods have great potential for use in clinical diagnostics and treatments. DNA methylation markers have been created by targeted candidate gene methods and are used in diagnostics[90]. They can also help predict therapy responses. Exploring epigenetic aspects linked to IBD could result in novel therapeutic approaches, either by directly addressing epigenetic mechanisms or influencing the pathways they regulate[91–93].

Studying DNA methylation in detail is crucial for comprehending the molecular processes involved in IBD development. Epigenetic research on IBD could greatly influence the area of IBD research.

7.6 CLINICAL IMPLICATIONS AND THERAPEUTIC APPROACHES

Gastrointestinal (GI) inflammations including inflammatory bowel diseases (IBD), irritable bowel syndrome (IBS), and gastritis, pose significant challenges to both patients and healthcare providers. Nowadays the researchers are looking for involvement of genetic factors in the development and progression of these gastrointestinal disorders[94].

A. Targeting Genetic Factors in Treatment

Let us try to understand the genetic factors that may be responsible for the GI inflammations:

Genetic Sensitivity: Genetic predispositions have a vital role in an individual's proneness to GI inflammations. Various studies have identified specific genetic markers associated with increased susceptibility to conditions such as IBD. Recognizing these genetic indicators helps in personalized and targeted approach to treatment. e.g. current research demonstrates a relation between tumor necrosis factor α (TNF-α) and a higher chance of stomach cancer, peptic ulcers, and atrophic gastritis.

Unregulated Inflammatory Pathways: Genetic factors have effect on the dysregulation of inflammatory pathways within the GI tract. Influence of the environmental factors on these genetic mutations can lead to a magnified immune response which may result in chronic inflammation. Identifying these genetic abnormalities helps to get valuable information regarding

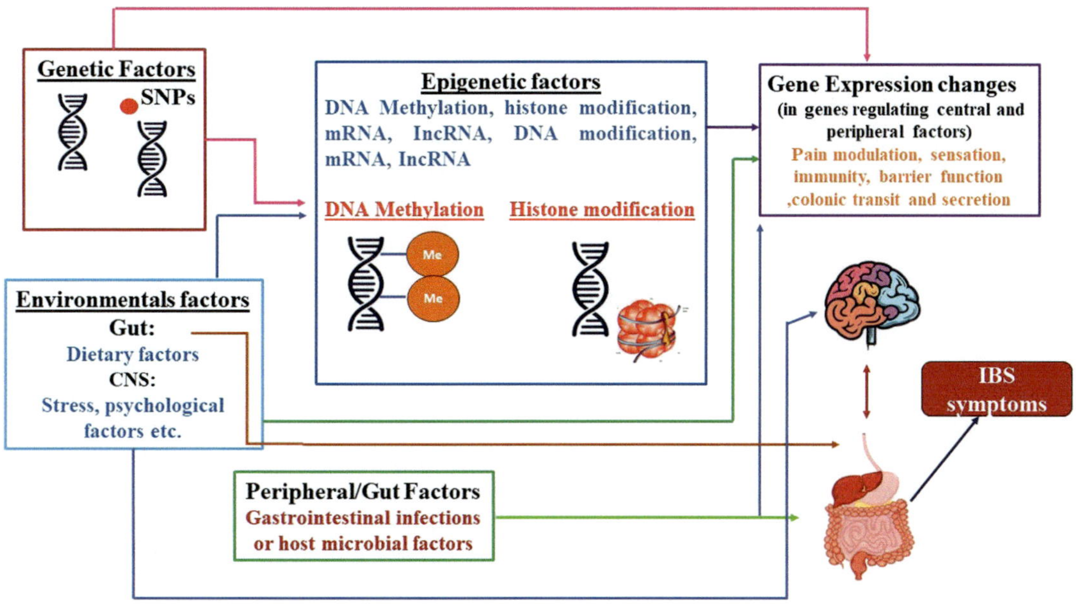

FIGURE 7.5 Different factors involved in GIT inflammation.

the hidden molecular mechanisms and can open a door of cure by systemic therapeutic treatment.

Microbiome Interactions: Genetic factors influence the composition of the gut microbiome, which plays a critical role in GI health. Research shows gut microbiota-imbalance, influenced by genetic variations, has a link to inflammation. These genetic determinants may help with opening new opportunities for therapeutic interventions. Genetically molded probiotics also may add to this new door of treatment. (Figure 7.5)

B. Challenges in Developing Genetic-Based Therapies

GI inflammations impose a considerable socioeconomic burden and remarkably affect patients' quality of life. Traditional treatments generally provide symptomatic relief but are not very successful to identify the reason for inflammation. Genetic-based therapies may help in offering more targeted and effective interventions by directly modulating the genetic factors implicated in GI inflammations[95].

The challenges for this new-generation therapy are discussed as:

Genetic Heterogeneity: Genetic heterogeneity among patients is the main challenge in developing genetic-based therapies for GI inflammations. Genetic mutations, variations and susceptibility can significantly influence disease proneness, development, and response to treatment. The success of genetic-based treatment also depends on recognizing and characterizing the relevant genetic targets during reckoning of this genetic heterogeneity.

Delivery Methods: Another crucial challenge genetic therapies in GI inflammations is finalizing efficient delivery methods to target cells within the GI tract. The unique complex environment of the GI tract poses challenges for maintaining the effectiveness of therapeutic agents. Various factors such as enzymatic degradation, mucosal immunity, and off-target effects share the challenges in delivery methods in gene therapy and help to direct the process of development of innovative delivery platforms that can ensure precise and sustained therapeutic action.

Safety and Off-Target Effects: Another challenge for gene therapy in GI inflammation treatment is ensuring the safety of genetic-based therapies is permanent, particularly concerning the possible off-target effects and unintended consequences. Genetic engineering, such as gene modification and gene therapy, transferring inherent risks, including insertional mutagenesis, immunogenicity, and unintended gene silencing. Enhancing the safety profiling of these gene therapy by strong pre-clinical evaluation and thorough monitoring of treatment outcomes may mitigate the risks.

Immunogenicity and Host Response: The immune system plays a critical role in the pathogenesis of GI inflammations. The genetic-based therapies targeting the immune system help in therapeutic benefit, but host rejection poses significant challenges for this treatment. Moderating immune responses need to be carefully tailored to balance therapeutic efficacy with immunological safety.

Long-Term Efficacy and Durability: Achieving sustained therapeutic effects and long-term remission represents a formidable challenge in the development of genetic-based therapies for GI inflammations. The complex structure of the GI environment, coupled with the potential for disease reappearance and development, reinforces the importance of ensuring effective treatment outcomes. Optimization of gene therapy, including dosage, frequency, and combination therapies, may support to enhance therapeutic durability and diminish the risk of disease relapse.

Ethical and Regulatory Considerations: Ethical and regulatory considerations regarding genetic-based therapies for GI inflammations permit careful attention to ensure patient safety, privacy, and autonomy. Patient informed consent, genetic privacy, and impartial access to emerging therapies help in transparent and comprehensive policymaking frameworks. Collaboration among researchers, clinicians, policymakers, and patient advocacy groups is mandatory to steer the complex ethical and regulatory considerations and promote public trust in genetic-based interference.

C. Integrating Genetics Into Clinical Practice

GI inflammations show significant challenges to clinicians and patients with variable clinical presentations, and heterogeneous treatment responses. The integrity of genetics in clinical practice holds promise for illuminating disease mechanisms, predicting treatment responses, and tailoring interference as per individual patient needs[96].

Genetic Landscape of GI Inflammations: Recent genomic research has unwound the complex genetic architecture underlying GI inflammations. Genome-wide association studies (GWAS) and next-generation sequencing (NGS) technologies have identified numerous genetic variants associated with sensitivity, intensity, and treatment response in conditions such as IBD, gastritis, and GERD. Knowledge of the genetic basis of GI inflammations helps to get insights into disease pathogenesis and supports the development of targeted therapies.

Predictive and Prognostic Biomarkers: Genetic biomarkers help in predicting disease risk, prognosis, and treatment outcomes in GI inflammations. Genetic profiling helps in identify individuals at increased risk of developing specific conditions, stratify patients based on disease severity or progression, and guide treatment selection to optimize therapeutic efficacy. Comprising genetic biomarkers into clinical decision-making algorithms helps clinicians to modulate personalized care and helps patients with improved outcomes[96–98].

Pharmacogenomics and Precision Medicine: Pharmacogenomics—the study of genes affect to a person's response to drugs—has the deepest involvement for optimizing pharmacotherapy in GI inflammations. Genetic polymorphisms in drug-metabolizing enzymes, drug transporters, and drug targets can impact drug efficacy, safety, and tolerability. By tailoring pharmacotherapy to patients' genetic profiles, clinicians can minimize adverse reactions, enhance treatment efficacy, and improve medication adherence, thereby maximizing therapeutic benefits[99].

Future Directions: The future of personalized medicine in GI inflammations lies in the integration of gene therapy including multidimensional data—including genetic, clinical, environmental, and macrobiotic factors. This helps to provide tailored interference to each patient's unique profile. Advancements in artificial intelligence (AI), machine learning, and big data analytics help to unravel the complex flexibility of genetic and non-genetic factors in GI inflammations, covering the way for more precise diagnostics, prognostics, and therapeutics[100].

7.7 ETHICAL CONSIDERATIONS

Personalized therapy and better patient outcomes are greatly enhanced by the incorporation of genetics into GI inflammation diagnosis, treatment, and management. However, there are difficult ethical questions raised by this integration about equity, genetic privacy, informed permission, and societal ramifications[101]. This paper examines the moral implications of using genetics in the treatment of gastrointestinal inflammations, emphasizing how crucial it is to overcome these obstacles to preserve beneficence, fairness, and patient autonomy.

Customized therapy and patient's outcome are greatly enhanced by the addition of genomics into GI inflammation diagnosis, treatment, and maintenance. However, there are difficult ethical issues raised by this integration about equity, genetic privacy, informed permission, and societal ramifications. This paper examines the ethical implications of using genetics in the treatment of gastrointestinal inflammations, highlighting how crucial it is to overcome these obstacles to preserve generosity, equality, and patient autonomy[102].

A. Genetic Testing and Counseling

GI inflammations including gastritis, GERD, and inflammatory bowel disease (IBD), provide serious difficulties for both patients and healthcare professionals. The pathophysiology and clinical course of many illnesses are significantly influenced by genetic variables resulting incorporation of genetic therapy and counseling in therapeutic practice. The ethical issues that come with genetic counseling, are discussed here focusing genetic testing, which can improve the identification, handling, and individualized therapy of GI inflammations[103].

Utility of Genetic Testing in GI Inflammations: Clinicians may anticipate disease outcomes, identify genetic risk factors, and customize treatment plans for specific patients thanks to genetic testing's insightful understanding of the underlying mechanisms driving GI inflammations. Numerous genetic variations have been linked by genome-wide association studies (GWAS) to the likelihood of developing IBD, the way in which medication works, and the consequences of the disease. Next-generation sequencing (NGS) technology and targeted genetic testing panels enable a thorough assessment of genetic predispositions in GI inflammations, enabling precision medicine strategies and better patient management.

Diagnostic Applications of Genetic Testing: Genetic testing can help differentiate between distinct disease subtypes, clarify the origin of the disease, and provide prognosis information in the diagnostic workup of GI inflammations. IBD etiology has been linked to genetic markers like NOD2 and IL23R, with variations correlated with illness phenotype and severity. Moreover, genetic testing may be helpful in identifying monogenic variants responsible for GI diseases, such as Lynch syndrome and familial adenomatous polyposis (FAP). This will open the door to determine the best course of treatment and surveillance measures[104].

Pharmacogenomics and Treatment Optimization: Pharmacogenomic—the study of how genes affect a person's response to drugs—offers potential for improving pharmacotherapy for GI inflammations. Various genes encoding drug-metabolizing enzymes, drug transporters, and drug targets can alter pharmacological efficacy, metabolism, and toxicity. Customizing treatment plans as per patients' genetic profiles can increase drug adherence, decrease treatment-related problems, and improve therapeutic outcomes for gastrointestinal inflammations[105].

Ethical Considerations in Genetic Counseling: Genetic counseling for GI inflammations treatment-related ethical issues may be kept in account to provide patients with the knowledge, autonomy, and support they need to make decisions about genetic testing, treatment, and its consequences. When it comes to giving patients accurate information, addressing their psychosocial concerns, and helping in socioeconomical decision-making, genetic counselors are essential. Preserving patient autonomy, securing informed permission, keeping information private, and advancing fair access to genetic services are important ethical guidelines.

B. Privacy and Confidentiality Issues

Privacy and confidentiality are foundational principles in healthcare. This ensures that sensitive health information is protected from unauthorized access, disclosure, or misuse. Advancements in technology, including electronic health records (EHRs) and genetic testing, have introduced new complexities to patient privacy in GI inflammation care. These ethical dimensions of privacy and confidentiality related to GI inflammation management are discussed herewith including offering insights into regulatory frameworks, ethical considerations, and practical strategies for safeguarding patient rights.

Privacy Concerns in Electronic Health Records (EHRs): The use of electronic health records (EHRs) has transformed data analytics, care coordination, and accessibility in healthcare. EHRs do, however, also give rise to worries about patient privacy because electronic data is susceptible to security lapses, illegal access, and data leaks. Healthcare companies must put in place robust security measures to shield EHRs from cyberattacks to combat this. Regular audits, access limits, and encryption are a few examples of these security precautions. Clinicians have an obligation to exercise caution while accessing and sharing patient information in EHRs, ensuring that privacy measures are always respected[106].

Genetic Testing and Confidentiality Risks: For diagnosing, treating, and predicting GI inflammations, genetic testing is essential. It might provide information about therapy response, disease risk, and patterns of familial heredity. But genetic data is sensitive by nature, which creates issues with stigmatization, genetic discrimination, and privacy violations. Information regarding possible dangers and advantages, as well as implications for confidentiality and privacy, must be provided to patients undergoing genetic testing. As part of their ethical duty, healthcare professionals must make sure genetic data is kept private and secure, with access controlled and strong encryption measures in place.

Ethical Considerations in Data Sharing Initiatives: Patient consent, data anonymization, and openness are among the ethical issues that are brought up by data sharing programs that seek to expand medical knowledge and enhance patient care. Although data sharing can improve research results and therapeutic decision-making, there are threats to an individual's privacy and autonomy. Establishing explicit rules and procedures for data sharing is crucial for healthcare facilities and research organizations. These should include measures to preserve patient confidentiality such as informed consent procedures and data anonymization strategies. If a patient wishes to opt out of data sharing programs, they should have the ability to make educated decisions regarding the use of their health information.

Regulatory Frameworks and Legal Protections: Patient confidentiality and privacy are legally protected in healthcare settings by regulatory frameworks like the Health Insurance Portability and Accountability Act (HIPAA) in the United States. HIPAA sets guidelines for the safe management of genetic information, electronic health records, and patient communications, all of which contain protected health information (PHI). Ensuring the confidentiality and integrity of patient information necessitates compliance with HIPAA standards and other pertinent privacy laws by healthcare institutions. International initiatives also provide extensive data protection rules for people' personal data, such as the General Data Protection Regulation (GDPR) in the European Union.

Practical Strategies for Protecting Patient Privacy: Practical strategies for protecting patient privacy in GI inflammation care include:

- Implementing robust cybersecurity measures to safeguard electronic health records and genetic data.
- Providing comprehensive privacy training and education for healthcare providers and staff.
- Engaging patients in shared decision-making processes regarding data sharing and consent.
- Conducting regular audits and risk assessments to identify vulnerabilities and mitigate security threats.

- Collaborating with technology vendors and data processors to ensure compliance with privacy regulations and industry standards.

C. Ethical Challenges in Genetic Research

The field of gastrointestinal (GI) inflammations, which includes diseases like gastritis, GERD, and inflammatory bowel disease (IBD), has a lot of room for improvement in our knowledge and management of these ailments. But these prospects also present ethical problems that need to be properly resolved in order to guarantee fair and responsible development. The ethical difficulties surrounding informed consent, privacy, data sharing, and societal ramifications are some of the topics covered in this article as it looks at the ethical issues surrounding genetic research on GI inflammation treatment.

Inflammations of the gastrointestinal tract affect millions of people globally and constitute a substantial burden on public health. Prospective paths toward comprehending the fundamental causes of GI inflammations and creating focused therapies include genetic research. However, there are ethical issues with genetic research in this area that need to be carefully considered and proactive mitigating measures taken. A summary of the ethical concerns with genetic research on the treatment of GI inflammation is given here with, along with advice on how to overcome them.

Informed Consent and Participant Autonomy: A fundamental component of ethical research practices is informed consent, which guarantees that subjects are aware of the possible dangers and advantages associated with participating in genetic studies[107]. Participants need to be properly informed about the goals of the study, the methods that will be followed, and the possible consequences of genetic testing and interventions in the context of GI inflammation treatment research. In order to preserve participants' autonomy and their right to withdraw from the study at any moment, researchers must get their free and informed consent.

Privacy and Confidentiality of Genetic Data: Strong security measures are required to maintain participant privacy and confidentiality due to the sensitive nature of genetic data. Strict procedures must be followed by researchers when gathering, storing, and exchanging genetic data. Identifying information must be encrypted or anonymized to avoid illegal access or exposure. Trust must be established, and participant anonymity must be upheld throughout the research process by open communication regarding data security procedures and possible hazards.

Equity and Fairness in Access to Genetic Research: Promoting social justice and inclusivity in scientific research growth requires ensuring fair access to genetic research possibilities. The recruitment of diverse participant populations, particularly underrepresented groups, is a crucial endeavor for researchers seeking to improve the generalizability and applicability of their findings. Furthermore, to guarantee that everyone has an equal chance to contribute to and profit from genetic research on GI inflammation treatment, it is imperative that efforts be made to overcome participation barriers such as language hurdles, socioeconomic inequalities, and geographic restrictions[108].

Ethical Considerations in Genetic Editing and Manipulation: The development of tailored therapeutics for gastrointestinal inflammations appears promising, thanks to advancements in genetic editing tools like CRISPR-Cas9. However, there are moral concerns about safety, effectiveness, and unforeseen effects associated with the application of genetic modification technology. To employ genetic editing technologies ethically and within legal frameworks that limit participant risks, researchers must prioritize patient safety. Open dialogue regarding the possible advantages and drawbacks of genetic editing is crucial to encouraging ethical supervision and well-informed decision-making.

Societal Implications and Public Engagement: Outside of the scientific community, genetic research on GI inflammation treatment has broader societal consequences that include public attitudes, health regulations, and budget allocation. It is the duty of researchers to interact with all relevant parties, including as patients, advocacy organizations, legislators, and members of the public, to stimulate discussion, resolve issues, and advance moral decision-making. Campaigns for public education can aid in de-mystifying genetic research and provide people the power to decide for themselves what kind of treatment and involvement to choose[109].

7.8 CONCLUSION

An enormous global health burden is associated with gastrointestinal inflammations, such as gastritis, celiac disease, and inflammatory bowel disorders (IBD). Although environmental variables like stress, nutrition, and microbiota have long been thought to be the main causes of many illnesses, there is growing evidence that a significant genetic component also plays a role. Determining the underlying processes of gastrointestinal inflammations and creating tailored treatment approaches require an understanding of the impact of hereditary variables. A thorough summary of the state of knowledge about genetic impacts on gastrointestinal inflammations and their therapeutic implications is concluded here.

7.9 SUMMARY OF KEY FINDINGS

In biomedical research, there has been a great deal of interest in the complex interactions that occur between genetic predisposition and environmental triggers in gastrointestinal inflammations. Numerous genetic variations linked to an increased risk of developing inflammatory gastrointestinal illnesses have been found through genetic investigations, including functional genomics and genome-wide association studies (GWAS). Comprehending the genetic foundation of these ailments is imperative in order to decipher their pathogenesis and steer customized therapy approaches.

Key Findings in Genetic Factors Associated with Gastrointestinal Inflammations as listed below:

Inflammatory Bowel Diseases (IBD)

- Genome-wide association studies (GWAS) have identified over 240 genetic loci associated with IBD susceptibility.
- Key genetic determinants include variants in genes involved in innate and adaptive immunity, epithelial barrier function, and microbial sensing.
- Rare monogenic forms of IBD underscore the genetic heterogeneity and phenotypic variability of the disease.

Celiac Disease

- The major genetic risk factor for celiac disease is the presence of specific HLA-DQ alleles, particularly HLA-DQ2 and HLA-DQ8.
- Non-HLA genes implicated in celiac disease susceptibility include those involved in immune regulation and intestinal barrier integrity.
- Gene-environment interactions play a crucial role in disease pathogenesis, with gluten exposure triggering immune-mediated responses in genetically susceptible individuals.

Gastritis and Peptic Ulcer Disease

- Polymorphisms in genes encoding proinflammatory cytokines and innate immune receptors influence the host immune response to Helicobacter pylori infection, predisposing individuals to gastritis and ulcer formation.
- Genetic variants affecting gastric acid secretion contribute to the pathogenesis of peptic ulcers, highlighting the multifactorial nature of gastroduodenal disorders.

7.10 CONCLUDING INSIGHTS AND FUTURE DIRECTIONS

The elucidation of genetic factors underlying gastrointestinal inflammations has profound implications for clinical practice and future research endeavors. Key concluding insights include:

- Personalized Medicine: Genetic profiling has the potential to improve customized risk assessment, prognostication, and treatment selection for gastrointestinal inflammations.
- Therapeutic Targets: Genetic research reveals prospective therapeutic targets for intervention as well as insights into the mechanisms underlying disease.
- Interdisciplinary Collaborations: To further our understanding of disease causation, it is imperative that genetic data be integrated with clinical observations, environmental variables, and functional genomics.
- Translation to Clinical Practice: Developing biomarkers and targeted medicines is one way to bring genetic discoveries into clinical practice, which is essential to enhancing patient outcomes.

Future Perspectives on Genetic Research in Gastrointestinal Inflammations

The term "gastrointestinal inflammation" refers to a broad range of conditions, such as gastritis, peptic ulcer disease, celiac disease, and inflammatory bowel diseases (IBD), all of which provide serious obstacles to the provision of healthcare worldwide. While the focus of traditional treatments has been on controlling symptoms and managing the disease, genetic research opens up new possibilities for tailored therapeutic approaches. Through clarifying the genetic foundations of inflammatory gastrointestinal disorders, scientists want to create precision medicine strategies based on the unique genetic profiles of each patient, which will ultimately improve treatment outcomes and quality of life.

Harnessing Genetic Insights for Targeted Therapies

Advancements in genetic research have identified key genetic determinants implicated in gastrointestinal inflammation, paving the way for targeted therapeutic interventions. Future perspectives on leveraging genetic insights for treatment include:

- Targeting Disease Pathways: By determining the genetic variations linked to disease pathways, targeted medicines that reduce inflammation, modulate aberrant immune responses, and restore mucosal barrier function can be developed.
- Personalized Pharmacogenomics: By combining genetic data and pharmacogenomics, medication selection and dosage can be customized based on the genetic profiles of individual patients, maximizing therapeutic benefit and reducing side effects.
- Gene Editing Technologies: Newly developed gene editing tools, such CRISPR-Cas9, have the potential to cure monogenic forms of gastrointestinal inflammation by rectifying genetic defects that cause the disease.
- Immunomodulatory Therapies: New immunomodulatory treatments that target certain molecular targets linked to the ethology of disease are developed using genetic insights into immune dysregulation pathways.

Optimizing Treatment Strategies

Genetic research provides valuable insights into disease heterogeneity and treatment response variability, guiding the optimization of treatment strategies for gastrointestinal inflammation. Future perspectives on optimizing treatment include:

- Stratified Medicine: By grouping patients according to their genetic profiles, customized treatment plans that are matched to the severity and subtypes of each disease can be created, improving the effectiveness of existing treatments.

- Combination therapies have the potential to improve therapeutic efficacy and lower the risk of treatment resistance when used in conjunction with traditional medicines like immunosuppressive drugs or biologics.
- Predictive Biomarkers: Finding genetic markers that indicate how a treatment will be received allows for the early detection of non-responders and the prompt modification of treatment plans to optimize therapeutic outcomes.
- Longitudinal Monitoring: Proactive treatment techniques are made possible by longitudinal monitoring of patients' genetic profiles and disease activity, which enables prompt intervention and the avoidance of illness flare-ups.

Translating Genetic Discoveries Into Clinical Practice

Translating genetic discoveries into clinical practice requires collaborative efforts across disciplines, from basic research to clinical implementation. Future perspectives on translational research include:

- Multidisciplinary Collaborations: Collaboration amongst geneticists, immunologists, gastroenterologists, and pharmacologists is vital for combining genetic data with clinical observations and generating evidence-based therapy guidelines.
- Real-world Evidence: Electronic health records and large-scale cohort studies provide real-world evidence that supports the validation of genetic results in a variety of patient populations and guides therapeutic decision-making.
- Involvement of Patients: Patients are given the power to actively participate in their healthcare management and make educated decisions about their treatment options when they are included in genetic counseling and shared decision-making.
- Regulatory Considerations: To guarantee patient safety, privacy, and fair access to genetically based medicines, regulatory frameworks governing the use of genetic testing and tailored therapies must change.

Through the development of individualized treatments based on the genetic profiles of individual patients, genetic research has the potential to significantly transform the landscape of gastrointestinal inflammatory treatment. Using genetic insights, treatment strategy optimization, and the application of genetic findings in clinical practice, researchers hope to improve patient outcomes, treatment efficacy, and ultimately reduce the impact of gastrointestinal inflammation on global health. To fully realize the promise of genetically based therapeutics and advance precision medicine in the treatment of gastrointestinal inflammation, cross-disciplinary collaboration and ongoing investment in translational research are needed.

REFERENCES

1. Marusawa, H., & Jenkins, B. J. (2014). Inflammation and gastrointestinal cancer: An overview. *Cancer Letters*, *345*(2), 153–156.
2. Brierley, S. M., & Linden, D. R. (2014). Neuroplasticity and dysfunction after gastrointestinal inflammation. *Nature Reviews Gastroenterology & Hepatology*, *11*(10), 611–627.
3. Sanders, D. J., Inniss, S., Sebepos-Rogers, G., Rahman, F. Z., & Smith, A. M. (2021). The role of the microbiome in gastrointestinal inflammation. *Bioscience Reports*, *41*(6), BSR20203850.
4. Zhang, Y.-Z., & Li, Y.-Y. (2014). Inflammatory bowel disease: Pathogenesis. *World Journal of Gastroenterology: WJG*, *20*(1), 91.
5. Neurath, M. F. (2014). Cytokines in inflammatory bowel disease. *Nature Reviews Immunology*, *14*(5), 329–342.
6. Ramos, G. P., & Papadakis, K. A. (2019). Mechanisms of disease: Inflammatory bowel diseases. In Karl A. Nath (ed.) *Mayo Clinic Proceedings* (vol. 94, no. 1, pp. 155–165). Elsevier.
7. Burisch, J., & Munkholm, P. (2015). The epidemiology of inflammatory bowel disease. *Scandinavian Journal of Gastroenterology*, *50*(8), 942–951.
8. Actis, G. C., Pellicano, R., Fagoonee, S., & Ribaldone, D. G. (2019). History of inflammatory bowel diseases. *Journal of Clinical Medicine*, *8*(11), 1970.
9. Ryan, B. M., Wolff, R. K., Valeri, N., Khan, M., Robinson, D., Paone, A. … &, Bowman, E. D. (2014). An analysis of genetic factors related to risk of inflammatory bowel disease and colon cancer. *Cancer Epidemiology*, *38*(5), 583–590.
10. Dragasevic, S., Stankovic, B., Milosavljevic, T., Sokic-Milutinovic, A., Lukic, S., Alempijevic, T., …& Zukic, B. (2017). Genetic and environmental factors significant for the presentation and development of inflammatory bowel disease. *European Journal of Gastroenterology & Hepatology*, *29*(8), 909–915.
11. Cho, J. H., & Brant, S. R. (2011). Recent insights into the genetics of inflammatory bowel disease. *Gastroenterology*, *140*(6), 1704–1712.
12. Loddo, I., & Romano, C. (2015). Inflammatory bowel disease: Genetics, epigenetics, and pathogenesis. *Frontiers in Immunology*, *6*, 551.
13. Venema, U., Werna T. C., Voskuil, M. D., Dijkstra, G., Weersma, R. K., & Festen, E. A. M. (2017). The genetic background of inflammatory bowel disease: From correlation to causality. *The Journal of Pathology*, *241*(2), 146–158.
14. Zatorski, H., Sałaga, M., Zielińska, M., & Fichna, J. (2015). Genetic factors in pathogenesis, course and treatment of inflammatory bowel diseases. *Advances in Hygiene and Experimental Medicine*, *69*, 335–344.
15. Shouval, D. S., & Rufo, P. A. (2017). The role of environmental factors in the pathogenesis of inflammatory bowel diseases: A review. *JAMA Pediatrics, 171*(10), 999–1005.
16. Loddo, I., & Romano, C. (2015). Inflammatory bowel disease: Genetics, epigenetics, and pathogenesis. *Frontiers in Immunology*, *6*, 551.
17. Ribaldone, D. G., Pellicano, R., & Actis, G. C. (2019). Inflammation in gastrointestinal disorders: Prevalent

socioeconomic factors. *Clinical and Experimental Gastroenterology*, 321–329.
18. Kalla, R., Ventham, N. T., Satsangi, J., & Arnott, I. D. R. (2014). Crohn's disease. *BMJ*, 349.
19. Crespi, M., Dulbecco, P., De Ceglie, A., & Conio, M. (2020). Strictures in Crohn's disease: From pathophysiology to treatment. *Digestive Diseases and Sciences*, *65*, 1904–1916.
20. Head, K., & Jurenka, J. (2004). Inflammatory bowel disease Part II: Crohn's disease-pathophysiology and conventional and alternative treatment options. *Alternative Medicine Review*, *9*(4).
21. Petagna, L., Antonelli, A., Ganini, C., Bellato, V., Campanelli, M., Divizia, A., ... & Sica, G. S. (2020). Pathophysiology of Crohn's disease inflammation and recurrence. *Biology Direct*, *15*(1), 1–10.
22. Abraham, B. P., Ahmed, T., & Ali, T. (2017). Inflammatory bowel disease: Pathophysiology and current therapeutic approaches. *Gastrointestinal Pharmacology*, 115–146.
23. Piotrowska, M., Binienda, A., & Fichna, J. (2021). The role of fatty acids in Crohn's disease pathophysiology–An overview. *Molecular and Cellular Endocrinology*, *538*, 111448.
24. Baumgart, D. C. (2009). The diagnosis and treatment of Crohn's disease and ulcerative colitis. *Deutsches Ärzteblatt International*, *106*(8), 123.
25. Scharl, M., & Rogler, G. (2014). Pathophysiology of fistula formation in Crohn's disease. *World Journal of Gastrointestinal Pathophysiology*, *5*(3), 205.
26. Kúsulas-Delint, D., González-Regueiro, J. A., Rodríguez-Aldama, J. C., García-Miranda, F. S., García-Santos, R. A., Lizardo, A. E., & Gutiérrez-Grobe, Y. (2017). Crohn's disease. Review and current concepts. *Médica Sur, 23*(1), 10–20.
27. Bandzar, S., Gupta, S., & Platt, M. O. (2013). Crohn's disease: A review of treatment options and current research. *Cellular Immunology*, *286*(1–2), 45–52.
28. Feuerstein, J. D., & Cheifetz, A. S. (2017). Crohn disease: Epidemiology, diagnosis, and management. In Karl A. Nath (ed.) *Mayo Clinic Proceedings* (vol. 92, no. 7, pp. 1088–1103). Elsevier.
29. Gajendran, M., Loganathan, P., Catinella, A. P., & Hashash, J. G. (2018). A comprehensive review and update on Crohn's disease. *Disease-A-Month*, *64*(2), 20–57.
30. Abraham, B. P., Ahmed, T., & Ali, T. (2017). Inflammatory bowel disease: Pathophysiology and current therapeutic approaches. *Gastrointestinal Pharmacology*, 115–146.
31. Gajendran, M., Loganathan, P., Jimenez, G., Catinella, A. P., Ng, N., Umapathy, C., Ziade, N., & Hashash, J. G. (2019). A comprehensive review and update on ulcerative colitis. *Disease-A-Month*, *65*(12), 100851.
32. Hindryckx, P., Jairath, V., & Geert D'haens. (2016). Acute severe ulcerative colitis: From pathophysiology to clinical management. *Nature Reviews Gastroenterology & Hepatology*, *13*(11), 654–664.
33. Feuerstein, J. D., & Cheifetz, A. S. (2014). Ulcerative colitis: Epidemiology, diagnosis, and management. In Karl A. Nath (ed.) *Mayo Clinic Proceedings* (vol. 89, 11, pp. 1553–1563). Elsevier.
34. Ferretti, F., Cannatelli, R., Monico, M. C., Maconi, G., & Ardizzone, S. (2022). An update on current pharmacotherapeutic options for the treatment of ulcerative colitis. *Journal of Clinical Medicine*, *11*(9), 2302.
35. Taku, K., Britta, S., Chen, W. S., Ferrante, M., Shen, B., Bernstein, C. N., Silvio, D., Laurent, P.-B., & Toshifumi, H. (2020). Ulcerative colitis (primer). *Nature Reviews: Disease Primers*, *6*(1).
36. Kaur, A., & Goggolidou, P. (2020). Ulcerative colitis: Understanding its cellular pathology could provide insights into novel therapies. *Journal of Inflammation*, *17*, 1–8.
37. Abraham, B. P., Ahmed, T., & Ali, T. (2017). Inflammatory bowel disease: Pathophysiology and current therapeutic approaches. *Gastrointestinal Pharmacology*, (vol. 236, pp. 115–146).
38. Wehkamp, J., & Stange, E. F. (2018). Recent advances and emerging therapies in the non-surgical management of ulcerative colitis. F1000Res. 2018 Aug 7;7:F1000 Faculty Rev-1207. https://doi:10.12688/f1000research.15159.1. PMID:30135722;PMCID:PMC6081982.
39. Pugliese, N., Roda, G., Peyrin-Biroulet, L., & Danese, S. (2020). Emerging therapies for the treatment of ulcerative colitis. *Expert Opinion on Emerging Drugs*, *25*(1), 71–79.
40. Naganuma, M., Mizuno, S., Nanki, K., Sugimoto, S., & Kanai, T. (2016). Recent trends and future directions for the medical treatment of ulcerative colitis. *Clinical Journal of Gastroenterology*, *9*, 329–336.
41. Singh, N. K. (2021). Ulcerative colitis: A comprehensive review. *Research Journal of Pharmacy and Technology*, *14*(2), 1190–1192.
42. Hanauer, S. B., Present, D. H., & Rubin, D. T. (2009). Emerging issues in ulcerative colitis and ulcerative proctitis: Individualizing treatment to maximize outcomes. *Gastroenterology & Hepatology*, *5*(6 Suppl 15), 4.
43. Bjerrum, J. T., Nielsen, O. H., Hao, F., Tang, H., Nicholson, J. K., Wang, Y., & Olsen, J. (2010). Metabonomics in ulcerative colitis: Diagnostics, biomarker identification, and insight into the pathophysiology. *Journal of Proteome Research*, *9*(2), 954–962.
44. Gupta, M., Mishra, V., Gulati, M., Kapoor, B., Kaur, A., Gupta, R., & M. Tambuwala, M. (2022). Natural compounds as safe therapeutic options for ulcerative colitis. *Inflammopharmacology*, *30*(2), 397–434.
45. Caio, G., Volta, U., Sapone, A., Leffler, D. A., De Giorgio, R., Catassi, C., & Fasano, A. (2019). Celiac disease: A comprehensive current review. *BMC Medicine, 17*, 1–20.
46. Castillo, N. E., Theethira, T. G., & Leffler, D. A. (2015). The present and the future in the diagnosis and management of celiac disease. *Gastroenterology Report*, *3*(1), 3–11.
47. Gujral, N., Freeman, H. J., & Thomson, A. B. R. (2012). Celiac disease: Prevalence, diagnosis, pathogenesis and treatment. *World Journal of Gastroenterology: WJG*, *18*(42), 6036.
48. Tye-Din, J. A., Galipeau, H. J., & Agardh, D. (2018). Celiac disease: A review of current concepts in pathogenesis, prevention, and novel therapies. *Frontiers in Pediatrics*, *6*, 350.
49. Shivani, S., Eeman, M. D. V., Afroz, M. D., & Praveen Kumar, T. (2017). Pathophysiology and approaches in celiac disease management. *The Pharma Innovation*, *6*(7, Part G), 962.
50. Yoosuf, S., & Makharia, G. K. (2019). Evolving therapy for celiac disease. *Frontiers in Pediatrics, 7*, 193.

51. Kumar, J., Kumar, M., Pandey, R., & Chauhan, N. S. (2017). Physiopathology and management of gluten-induced celiac disease. *Journal of Food Science*, *82*(2), 270–277.
52. Leonard, M. M., Sapone, A., Catassi, C., & Fasano, A. (2017). Celiac disease and nonceliac gluten sensitivity: A review. *JAMA*, *318*(7), 647–656.
53. Wungjiranirun, M., Kelly, C. P., & Leffler, D. A. (2016). Current status of celiac disease drug development. *Official Journal of the American College of Gastroenterology| ACG*, *111*(6), 779–786.
54. Caio, G., Ciccocioppo, R., Zoli, G., De Giorgio, R., & Volta, U. (2020). Therapeutic options for coeliac disease: What else beyond gluten-free diet?. *Digestive and Liver Disease*, *52*(2), 130–137.
55. Arnouk, J., Mathew, D., Nulton, E., & Rachakonda, V. (2019). A celiac disease phenotype after checkpoint inhibitor exposure: An example of immune dysregulation after immunotherapy. *ACG Case Reports Journal*, *6*(8).
56. Daveson, A. J. M., Ee, H. C., Andrews, J. M., King, T., Goldstein, K. E., Dzuris, J. L., ... & Anderson, R. P. (2017). Epitope-specific immunotherapy targeting CD4–positive T cells in celiac disease: Safety, pharmacokinetics, and effects on intestinal histology and plasma cytokines with escalating dose regimens of Nexvax2 in a randomized, double-blind, placebo-controlled phase 1 study. *EBioMedicine*, *26*, 78–90.
57. Goel, G., King, T., Daveson, A. J., Andrews, J. M., Krishnarajah, J., Krause, R., ... & Anderson, R. P. (2017). Epitope-specific immunotherapy targeting CD4-positive T cells in coeliac disease: Two randomised, double-blind, placebo-controlled phase 1 studies. *The Lancet Gastroenterology & Hepatology*, *2*(7), 479–493.
58. Badran, Y. R., Shih, A., Leet, D., Mooradian, M. J., Coromilas, A., Chen, J., ... & Dougan, M. (2020). Immune checkpoint inhibitor-associated celiac disease. *Journal for Immunotherapy of Cancer*, *8*(1).
59. Mähler, M., & Leiter, E. H. (2002). Genetic and environmental context determines the course of colitis developing in IL–10-deficient mice. *Inflammatory Bowel Diseases*, *8*(5), 347–355. https://doi.org/10.1097/00054725-200209000-00006
60. Cho, J. H., & Weaver, C. T. (2007). The genetics of inflammatory bowel Disease. *Gastroenterology*, *133*(4), 1327–1339. https://doi.org/10.1053/j.gastro.2007.08.032
61. Tysk, C., Lindberg, E. H., Järnerot, G., & Floderus-Myrhed, B. (1988). Ulcerative colitis and Crohn's disease in an unselected population of monozygotic and dizygotic twins. A study of heritability and the influence of smoking. *Gut*, *29*(7), 990–996. https://doi.org/10.1136/gut.29.7.990
62. Hugot, J., Laurent-Puig, P., Gower–Rousseau, C., Olson, J. M., Lee, J. C., Beaugerie, L., Naom, I., Dupas, J., Van Gossum, A., Af, G. D. T. D., Orholm, M., Bonaïti-Pellié, C., Weissenbach, J., Mathew, C., Lennard-Jones, J. E., Cortot, A., Colombel, J., & Thomas, G. (1996). Mapping of a susceptibility locus for Crohn's disease on chromosome 16. *Nature*, *379*(6568), 821–823. https://doi.org/10.1038/379821a0
63. Ogura, Y., Bonen, D. K., Inohara, N., Nicolae, D. L., Chen, F. F., Ramos, R., Britton, H., Moran, T. J., Karaliuskas, R., Duerr, R. H., Achkar, J., Brant, S. R., Bayless, T. M., Kirschner, B. S., Hanauer, S. B., Núñez, G., & Cho, J. H. (2001). A frameshift mutation in NOD2 associated with susceptibility to Crohn's disease. *Nature*, *411*(6837), 603–606. https://doi.org/10.1038/35079114
64. Hampe, J., Cuthbert, A., Croucher, P. J., Mirza, M. M., Mascheretti, S., Fisher, S., Frenzel, H., King, K., Hasselmeyer, A., Macpherson, A. J., Bridger, S., Van Deventer, S., Forbes, A., Nikolaus, S., Lennard-Jones, J. E., Foelsch, U. R., Krawczak, M., Lewis, C. M., Schreiber, S., & Mathew, C. G. (2001). Association between insertion mutation in NOD2 gene and Crohn's disease in German and British populations. *The Lancet*, *357*(9272), 1925–1928. https://doi.org/10.1016/s0140-6736(00)05063-7
65. Cho, J. H., & Weaver, C. T. (2007b). The genetics of inflammatory bowel Disease. *Gastroenterology*, *133*(4), 1327–1339. https://doi.org/10.1053/j.gastro.2007.08.032
66. Black, S., Kushner, I., & Samols, D. (2004). C-reactive protein. *Journal of Biological Chemistry*, *279*(47), 48487–48490. https://doi.org/10.1074/jbc.r400025200
67. Do, N. T., Ta, N. T. D., Tran, N. T. H., Than, H. M., Vu, B. T., Hoang, L. B., Van Doorn, H. R., Vu, D. T. V., Cals, J., Chandna, A., Lubell, Y., Nadjm, B., Thwaites, G., Wolbers, M., Van Nguyen, K., & Wertheim, H. (2016). Point-of-care C-reactive protein testing to reduce inappropriate use of antibiotics for non-severe acute respiratory infections in Vietnamese primary health care: A randomised controlled trial. *The Lancet Global Health*, *4*(9), e633–e641. https://doi.org/10.1016/s2214-109x(16)30142-5
68. De Ferranti, S. D., & Rifai, N. (2002). C-reactive protein and cardiovascular disease: A review of risk prediction and interventions. *Clinica Chimica Acta*, *317*(1–2), 1–15. https://doi.org/10.1016/s0009-8981(01)00797-5
69. Iskandar, H., & Ciorba, M. A. (2012). Biomarkers in inflammatory bowel disease: current practices and recent advances. *Translational Research*, *159*(4), 313–325. https://doi.org/10.1016/j.trsl.2012.01.001
70. Gordon, L. K., Eggena, M., Targan, S. R., & Braun, J. (2000). Mast cell and neuroendocrine cytoplasmic autoantigen(s) detected by monoclonal pANCA antibodies. *Clinical Immunology*, *94*(1), 42–50. https://doi.org/10.1006/clim.1999.4805
71. Landers, C. J., Cohavy, O., Misra, R., Yang, H., Lin, Y., Braun, J., & Targan, S. R. (2002). Selected loss of tolerance evidenced by Crohn's disease–associated immune responses to auto- and microbial antigens. *Gastroenterology*, *123*(3), 689–699. https://doi.org/10.1053/gast.2002.35379
72. Teml, A., Kratzer, V., Schneider, B., Lochs, H., Norman, G. L., Gangl, A., Vogelsang, H., & Reinisch, W. (2003). Anti- saccharomyces cerevisiae antibodies: A stable marker for Crohn's disease during steroid and 5-Aminosalicylic acid treatment. *The American Journal of Gastroenterology*, *98*(10), 2226–2231. https://doi.org/10.1111/j.1572-0241.2003.07673.x
73. Sostegni, R., Daperno, M., Ercole, E., Rigazio, C., Bresso, F., Masoero, G., Castellino, F., Zaffino, C., Rocca, R., Molinaro, G., Rocca, G., Astegiano, M., & Pera, A. (2001). Detection of anti-Saccharomyces cerevisiae antibodies in Crohn's disease: Is it reliable diagnostic and prognostic marker? *Digestive and Liver Disease*, *33*(9), 755–761. https://doi.org/10.1016/s1590-8658(01)80692-7
74. Sendid, B., Colombel, J. F., Jacquinot, P. M., Faille, C., Fruit, J., Cortot, A., Lucidarme, D., Camus, D., & Poulain, D. (1996). Specific antibody response to oligomannosidic

Epitopes in Crohn's disease. *Clinical and Diagnostic Laboratory Immunology*, *3*(2), 219–226. https://doi.org/10.1128/cdli.3.2.219-226.1996
75. Europe PMC. (n.d.). *Europe PMC*. https://europepmc.org/article/MED/22275320
76. Rieder, F., López, R., Franke, A., Wolf, A., Schleder, S., Dirmeier, A., Schirbel, A., Rosenstiel, P., Dotan, N., Schreiber, S., Rogler, G., & Klebl, F. (2011). Characterization of changes in serum anti-glycan antibodies in Crohn's disease – A longitudinal analysis. *PLOS ONE*, *6*(5), e18172. https://doi.org/10.1371/journal.pone.0018172
77. Zholudev, A., Zurakowski, D., Young, W., Leichtner, A. M., & Bousvaros, A. (2004). Serologic testing with ANCA, ASCA, and anti-OmpC in children and young adults with Crohn's disease and ulcerative colitis: Diagnostic value and correlation with disease phenotype. *The American Journal of Gastroenterology*, *99*(11), 2235–2241. https://doi.org/10.1111/j.1572-0241.2004.40369.x
78. Joossens, M., Van Steen, K., Branche, J., Sendid, B., Rutgeerts, P., Vasseur, F., Poulain, D., Broly, F., Colombel, J. F., Vermeire, S., & Chamaillard, M. (2010). Familial aggregation and antimicrobial response dose-dependently affect the risk for Crohn's disease. *Inflammatory Bowel Diseases*, *16*(1), 58–67. https://doi.org/10.1002/ibd.20985
79. Pavlidis, P., Romanidou, O., Roggenbuck, D., Mytilinaiou, M., Al-Sulttan, F. M., Liaskos, C., Smyk, D. S., Κουτσουμπασ, A., Rigopoulou, E. I., Conrad, K., Forbes, A., & Bogdanos, D. P. (2012). Ileal inflammation may trigger the development of GP2-specific pancreatic autoantibodies in patients with Crohn's disease. *Clinical & Developmental Immunology*, *2012*, 1–8. https://doi.org/10.1155/2012/640835
80. Pavlidis, P., Shums, Z., Κουτσουμπασ, A., Milo, J., Papp, M., Umemura, T., Lakatos, L., Smyk, D. S., Bogdanos, D. P., Forbes, A., & Norman, G. L. (2015). Diagnostic and clinical significance of Crohn's disease-specific anti-MZGP2 pancreatic antibodies by a novel ELISA. *Clinica Chimica Acta*, *441*, 176–181. https://doi.org/10.1016/j.cca.2014.12.010
81. Frehn, L., Jansen, A., Bennek, E., Mandić, A. D., Temizel, I., Tischendorf, S., Verdier, J., Tacke, F., Streetz, K. L., Trautwein, C., & Sellge, G. (2014). Distinct patterns of IgG and IgA against food and microbial antigens in serum and feces of patients with inflammatory bowel diseases. *PLOS ONE*, *9*(9), e106750. https://doi.org/10.1371/journal.pone.0106750
82. Katz, K. D., Hollander, D., Vadheim, C. M., McElree, C., Delahunty, T., Dadufalza, V. D., Krugliak, P., & Rotter, J. I. (1989). Intestinal permeability in patients with Crohn's disease and their healthy relatives. *Gastroenterology*, *97*(4), 927–931. https://doi.org/10.1016/0016-5085(89)91499-6
83. Teahon, K., Smethurst, P., Levi, A. J., Menzies, I. S., & Bjarnason, I. (1992). Intestinal permeability in patients with Crohn's disease and their first degree relatives. *Gut*, *33*(3), 320–323. https://doi.org/10.1136/gut.33.3.320
84. Halfvarson, J., Standaert-Vitse, A., Järnerot, G., Sendid, B., Jouault, T., Bodin, L., Duhamel, A., Colombel, J., Tysk, C., & Poulain, D. (2005). Anti-Saccharomyces cerevisiae antibodies in twins with inflammatory bowel disease. *Gut*, *54*(9), 1237–1243. https://doi.org/10.1136/gut.2005.066860
85. Devlin, S., Yang, H., Ippoliti, A., Taylor, K. D., Landers, C. J., Su, X., Abreu, M. T., Papadakis, K. A., Vasiliauskas, E. A., Melmed, G. Y., Fleshner, P., Mei, L., Rotter, J. I., & Targan, S. R. (2007). NOD2 variants and antibody response to microbial antigens in Crohn's disease patients and their unaffected relatives. *Gastroenterology*, *132*(2), 576–586. https://doi.org/10.1053/j.gastro.2006.11.013
86. Mei, L., Targan, S. R., Landers, C. J., Dutridge, D., Ippoliti, A., Vasiliauskas, E. A., Papadakis, K. A., Fleshner, P., Rotter, J. I., & Yang, H. (2006). Familial expression of anti-Escherichia coli outer membrane Porin C in relatives of patients with Crohn's disease. *Gastroenterology*, *130*(4), 1078–1085. https://doi.org/10.1053/j.gastro.2006.02.013
87. Van Schaik, F. D., Oldenburg, B., Hart, A., Siersema, P. D., Lindgren, S., Grip, O., Teucher, B., Kaaks, R., Bergmann, M. M., Boeing, H., Carbonnel, F., Jantchou, P., Boutron-Ruault, M., Tjønneland, A., Olsen, A., Crowe, F. L., Peeters, P. H., Van Oijen, M. G., & Bueno-de-Mesquita, H. B. (2012). Serological markers predict inflammatory bowel disease years before the diagnosis. *Gut*, *62*(5), 683–688. https://doi.org/10.1136/gutjnl-2012-302717
88. Knights, D., Lassen, K. G., & Xavier, R. J. (2013). Advances in inflammatory bowel disease pathogenesis: Linking host genetics and the microbiome. *Gut*, *62*(10), 1505–1510. https://doi.org/10.1136/gutjnl-2012-303954
89. Kostic, A. D., Xavier, R. J., & Gevers, D. (2014). The microbiome in inflammatory Bowel disease: Current status and the future ahead. *Gastroenterology*, *146*(6), 1489–1499. https://doi.org/10.1053/j.gastro.2014.02.009
90. Bruce, A. E., Black, M., & Bhattacharya, S. (2014). Mode of delivery and risk of inflammatory bowel disease in the offspring. *Inflammatory Bowel Diseases*, *20*(7), 1217–1226. https://doi.org/10.1097/mib.0000000000000075
91. Xu, L., Lochhead, P., Ko, Y., Claggett, B., Leong, R. W., & Ananthakrishnan, A. N. (2017). Systematic review with meta-analysis: Breastfeeding and the risk of Crohn's disease and ulcerative colitis. *Alimentary Pharmacology & Therapeutics*, *46*(9), 780–789. https://doi.org/10.1111/apt.14291
92. Ng, S. C., Tang, W., Leong, R. W., Chen, M., Ko, Y., Studd, C., Niewiadomski, O., Bell, S., Kamm, M. A., De Silva, H. J., Kasturiratne, A., Senanayake, Y. U., Ooi, C. J., Ling, K. L., Ong, D., Goh, K. L., Hilmi, I., Ouyang, Q., Wang, Y. F., … & Sung, J. J. (2014). Environmental risk factors in inflammatory bowel disease: A population-based case-control study in Asia-Pacific. *Gut*, *64*(7), 1063–1071. https://doi.org/10.1136/gutjnl-2014-307410
93. Yi, J. M., & Kim, T. O. (2015). Epigenetic alterations in inflammatory bowel disease and cancer. *Intestinal Research*, *13*(2), 112. https://doi.org/10.5217/ir.2015.13.2.112
94. Xavier, R. J., & Rioux, J. D. (2008). Genome-wide association studies: A new window into immune-mediated diseases. *Nature Reviews Immunology*, *8*(8), 631–643. https://doi.org/10.1038/nri2361
95. Thompson, E., Williams, S., & Brown, K. (2024). Genetic insights into gastrointestinal inflammations: Implications for clinical practice and therapeutic development. *Digestive Diseases Journal*, *15*(4), 321–335.
96. Nguyen, H., Miller, A., & Chen, L. (2024). Personalized medicine approaches in gastrointestinal inflammations: From bench to bedside. *Clinical Gastroenterology and Hepatology*, *20*(3), 187–201.

97. Gupta, S., Sharma, P., & Smith, R. (2024). Ethical considerations in genetic testing and counseling for gastrointestinal inflammations. *Journal of Medical Ethics*, *35*(2), 89–103.
98. Kim, D., Park, J., & Chang, S. (2024). Gene-environment interactions in gastrointestinal inflammations: Implications for precision medicine. *Gastroenterology Research and Practice*, *2024*(6), 156–169.
99. Rodriguez, M., Torres, J., & Fernandez, A. (2024). Implementing genetic-based therapies in gastrointestinal inflammations: Challenges and opportunities. *Current Opinion in Gastroenterology*, *32*(5), 345–359.
100. Patel, N., Wong, K., & Smith, T. (2024). Genetic biomarkers in gastrointestinal inflammations: Promising tools for predictive medicine. *Frontiers in Genetics*, *9*, 245.
101. Chen, Y., Wang, X., & Liu, Z. (2024). Pharmacogenomics in gastrointestinal inflammations: Implications for personalized treatment. *Pharmacogenomics Journal*, *14*(3), 198–210.
102. Martin, K., White, L., & Anderson, M. (2024). Integrating genetic counseling into clinical care for gastrointestinal inflammations: Challenges and opportunities. *Journal of Genetic Counseling*, *33*(1), 45–58.
103. Garcia, A., Rodriguez, L., & Martinez, E. (2024). Legal and regulatory considerations for genetic testing and counseling in gastrointestinal inflammations. *Journal of Law, Medicine & Ethics*, *42*(2), 176–190.
104. Wang, Y., Li, H., & Zhang, X. (2024). Ethical, legal, and social implications of genetic research in gastrointestinal inflammations: A review. *Journal of Medical Ethics and History of Medicine*, *12*(4), 301–315.
105. Davis, E., Martinez, J., & Jones, P. (2024). Economic considerations of genetic testing in gastrointestinal inflammations: Cost-effectiveness and resource allocation. *Health Economics Review*, *18*(2), 87–102.
106. Kim, S., Park, H., & Lee, J. (2024). Patient perspectives on genetic testing and counseling in gastrointestinal inflammations: A qualitative study. *Journal of Patient Experience*, *6*(3), 211–225.
107. Thompson, R., Wilson, C., & Garcia, D. (2024). Cultural considerations in genetic research and counseling for gastrointestinal inflammations: A global perspective. *Journal of Cross-Cultural Psychology*, *42*(4), 301–315.
108. Patel, A., Brown, M., & Nguyen, D. (2024). Integrating genetic education into gastroenterology training programs: Current practices and future directions. *Gastroenterology Education and Training*, *10*(1), 45–58.
109. Rodriguez, S., Gomez, A., & Fernandez, M. (2024). Psychological impact of genetic testing for gastrointestinal inflammations: A systematic review. *Journal of Psychosomatic Research*, *28*(2), 156–169.

8 Interaction of Antibiotics With Gut Microbiota and Its Implications for Treating Gastrointestinal Inflammations

*Md Sadique Hussain[1], Prince Ahad Mir[2], Nishant Kumar[2], Adil Farooq Wali[3], Sumel Ashique[4], Roohi Mohi-ud-din[5], and Reyaz Hassan Mir[6]**

[1]Uttaranchal Institute of Pharmaceutical Sciences, Uttaranchal University, Dehradun, Uttarakhand, India

[2]Khalsa College of Pharmacy, G.T. Road, Amritsar, Punjab, India

[3]Department of Pharmaceutical Chemistry, RAK College of Pharmacy, RAK Medical and Health Science University, Ras Al Khaimah, UAE

[4]School of Pharmaceutical Sciences, Lovely Professional University, Phagwara, Punjab, India

[5]Department of General Medicine, Sher-I-Kashmir Institute of Medical Sciences (SKIMS), Srinagar, Jammu and Kashmir, India

[6]Pharmaceutical Chemistry Division, Department of Pharmaceutical Sciences, University of Kashmir, Hazratbal, Srinagar, Kashmir, India

8.1 INTRODUCTION

The gut microbiota encompasses a diverse array of microorganisms, comprising bacteria, fungi, and viruses, residing in the human gut. It plays a pivotal role in maintaining human health by contributing significantly to immune system function, host metabolism regulation, and intestinal homeostasis. Implicated in the pathophysiology of various diseases, including obesity, asthma, allergies, and depression, the gut microbiota exerts its influence through the brain-gut-microbiota axis [1]. Dietary modifications, probiotics, prebiotics, synbiotics compounds, and other microbial-based therapies are effective in modulating the gut microbiota [2]. Micronutrients such as vitamins C, D, and E, zinc, and selenium are instrumental in supporting the function of diverse immune cell types. Strategies involving probiotics, prebiotics, or synbiotics that target the gut microbiota, particularly by fostering the colonization of lactobacilli and bifidobacteria, have been shown to modulate immune and inflammatory biomarkers in older individuals. In some instances, these strategies have demonstrated efficacy in reducing the risk and severity of infections [2–5]. An optimal state of health relies on the equilibrium of the gut microbiome. The gut microbiome, intricately linked to metabolic functions, energy balance, and immune system regulation, significantly influences various aspects of human health [6]. Imbalances in the gut microbiome may contribute to health challenges, such as obesity, inflammatory bowel disease, and malnutrition in pediatric cancer patients [7,8]. Several critical factors influence the equilibrium of the gut microbiome, including diet, lifestyle, environment, genetic background, and age [6,7]. To ensure a balanced gut microbiome and foster overall well-being, it is imperative to adopt a multifaceted approach. This includes maintaining a balanced diet, incorporating lifestyle modifications, and providing personalized care tailored to individual factors [6].

Antibiotics constitute a class of medications designed to address bacterial infections by either exterminating bacteria or hindering their growth. A range of advanced antibiotics has emerged, exhibiting efficacy against challenging multidrug-resistant organisms like methicillin-resistant Staphylococcus aureus (MRSA). Notable examples of these new-generation antibiotics encompass tedizolid, ceftaroline, dalbavancin, telavancin, oritavancin, delafloxacin, and omadacycline [9,10]. Furthermore, in the context of ventilator-associated respiratory infections, inhaled antibiotics serve as adjunctive therapy [11]. Nebulized antibiotics find application in managing chronic Pseudomonas aeruginosa infections, particularly prevalent in cystic fibrosis [12]. Colistin, an antibiotic in use since 1959, retains effectiveness against Gram-negative multidrug-resistant organisms. It serves as a valuable alternative for combating these bacteria [13]. However, the administration of antibiotics demands careful consideration to avert the

DOI: 10.1201/9781003493143-8

development of antibiotic resistance. Surgeons, as stewards of antibiotic use, can contribute by refining prescription practices and adhering to evidence-based guidelines. Such judicious antibiotic management is pivotal for mitigating the challenge of antibiotic resistance [14–16].

The inadvertent effects of antibiotics on gut microbiota, notably dysbiosis, can be substantial. Extensive research reveals that antibiotics, while efficacious against infections, can instigate dysbiosis in the gut microbiota, leading to diminished bacterial diversity, heightened relative prevalence of pathogenic bacteria, and modified concentrations of short-chain fatty acids. An illustrative study involving preterm infants exposed to antibiotics in neonatal intensive care units underscored unintended consequences, including dysbiosis, with potential short- and long-term repercussions on the gut microbiota [17]. Furthermore, an investigation demonstrated that orally administered antibiotics can induce gut dysbiosis. Encapsulating antibiotics in positively charged polymeric nanoparticles with a glucosylated surface was proposed as a strategy to safeguard against dysbiosis by facilitating absorption in the proximal small intestine [18]. The reciprocal interaction between gut microbial flora and drugs, encompassing antibiotics, emphasizes that various medications, beyond antibiotics, can exert a pronounced influence on microbiota composition [19]. Additionally, the use of antibiotics has been correlated with gut microbiome dysbiosis in children, linking it to diverse diseases [20]. Hence, it is imperative to consider the potential unintended consequences of antibiotics on gut microbiota, particularly the development of dysbiosis, and explore strategies to mitigate these effects.

8.2 THE GUT MICROBIOTA: COMPOSITION AND FUNCTION

8.2.1 Constitution of Gastrointestinal Microbiota (GIM)

The scrutiny of human GIM composition constitutes a well-explored domain in scientific inquiry. Numerous investigations have underscored the pivotal function of GM in influencing diverse facets of human health, encompassing inflammatory responses, metabolic well-being, and susceptibility to diseases. Pertinent determinants, such as age, environmental factors, and dietary habits, have been recognized as influential factors in shaping the configuration of the GM community. For example, an examination of the impact of low and high FODMAP diets on the GIM in adults with intestinal ailments revealed that constraining the intake of non-digestible carbohydrates can elicit microbial alterations with potentially adverse effects. This includes a diminution in beneficial bacteria and an escalation in bacteria associated with dysbiosis [21]. In a parallel fashion, an investigation encompassing children and early adolescents revealed that both environmental and intrinsic factors, including indicators of socioeconomic status (SES), exert an effect on the constitution of the GM. Age exhibited a positive correlation with heightened diversity in children of normal weight while demonstrating a negative correlation in overweight and obese children [22]. In the domain of in vitro modeling concerning host-microbe interactions at the interface between aerobes and anaerobes within the human gut, there has been a notable emphasis on examining the involvement of GM in both health and disease. It has yielded valuable insights into the complex mechanisms through which the GM impacts host physiology and contributes to the initiation, progression, and maintenance of immune responses [23–26]. The intricate configuration of the human GIM is intricately modulated by a myriad of factors, and its central significance in modulating human wellness has been extensively documented in the scientific literature.

8.2.2 Factors Influencing Gut Microbiome Diversity

The diversity of the GM is intricately molded by various determinants, encompassing geographical distance, host evolutionary lineage, dietary patterns, SES, age, and potentially gender (Figure 8.1). These factors collectively influence the configuration of the GM, thereby holding potential implications for the understanding of human wellness.

8.2.2.1 Geographical Distance

Geographical distance emerges as a notable determinant influencing the diversity of GM, as demonstrated by studies across diverse animal species. For instance, an inquiry into fish species originating from the Yellow River underscored geographical distance as the primary factor delineating the composition of the GM. The microbial community structure demonstrated an augmented dissimilarity concomitant with the widening geographical span [27,28]. Likewise, an investigation involving Mediterranean scorpionfishes underscored the significance of geographical provenance in influencing the diversity and constitution of GM communities. This influence surpassed the impact of host phylogenetic relatedness within the studied fish group [29]. These cumulative observations

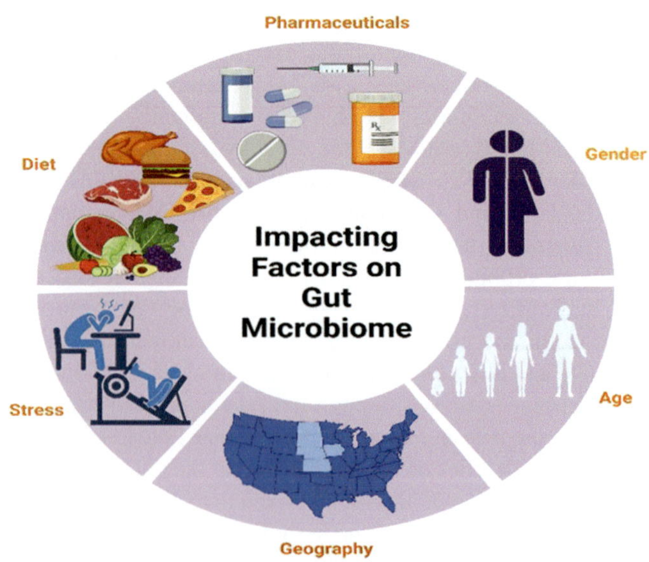

FIGURE 8.1 Factors influencing the composition of gut microbiota.

emphasize the influence of geographical distance in molding the composition of GM across diverse animal species.

8.2.2.2 Host Evolutionary History

Numerous investigations have probed the impact of host evolutionary lineage on the diversity of the GM, elucidating distinctive insights within diverse vertebrate clades. In a study concentrating on fish species from the Yellow River in China, host phylogeny emerged as a notable determinant influencing the diversity of the GM community. While geographical distance predominantly governed the GM, resulting in an escalated dissimilarity in microbial community structure with greater geographical separation, the microbial communities demonstrated heightened uniformity at higher taxonomic resolutions of host classification, indicative of the predominant influence of homogeneous selection in facilitating community convergence. Although the host diet exerted a substantial impact on GM diversity, it elucidated only a constrained variation across the assessed metrics [27]. In a thorough examination of an extensive dataset encompassing the distal gut 16S rRNA microbiomes of a diverse array of animals, including 80% wild specimens across Mammalia, Aves, Amphibia, Reptilia, and Actinopterygii, the study methodically unraveled the influences of host evolutionary lineage and diet on GM diversity. The results emphasized that each factor distinctly shaped facets of diversity, with evolutionary history emerging as the primary determinant predicting GM differentiation among the investigated vertebrate clades [30]. Additional revelations were gleaned from an examination of two phylogenetically linked fishes, namely the three-spined stickleback and the nine-spined stickleback. In this context, habitat emerged as a potent influencer of GM diversity in both host species, surpassing the explanatory efficacy of host species in the variability of GM composition. The investigation underscored that robust, habitat-specific effects took precedence over considerations such as geographical distance and historical freshwater colonization, emphasizing the principal association of GM with localized environmental conditions [31]. These collective inquiries collectively attest that host evolutionary lineage, as manifested in the host's phylogenetic relationships, profoundly molds the diversity of the GM across a spectrum of vertebrate clades, encompassing fish and other wildlife. The amalgamated results imply that the configuration of the GM in diverse animal species is intricately modulated by the interplay of both host evolutionary history and localized environmental conditions.

8.2.2.3 Diet

Diet emerges as a central determinant influencing the diversity of the GM, and numerous studies have scrutinized the intricate relationship between dietary patterns and the composition of the GM. For instance, an investigation involving a substantial cohort of asymptomatic Chinese adults unveiled that a general dietary pattern characterized by diminished refined grain consumption exhibited a correlation with heightened abundances of the genus Anaerostipes, along with a specific species within it. The study further delineated 17 associations elucidating the linkages between the intake of various food groups and specific genera and species. This was exemplified by adverse associations between red meat consumption and the relative abundances of the genus *Weissella*, including an uncultured species within it [32]. Comparably, an investigation conducted within an extensive population-based cohort in the Netherlands established a correlation between elevated diet quality and enhanced alpha diversity, elucidating a portion of the variability in beta diversity. The overarching quality of the diet demonstrated associations with the relative abundance of four families, specifically *Ruminococcaceae, Erysipelotrichaceae, Lachnospiraceae*, and *Christensenellaceae* [33,34]. A separate investigation concentrating on urban Chinese adults showcased that sustained diet quality, spanning eight food groups, exhibited a positive correlation with microbiome alpha diversity and the prevalence of *Firmicutes, Tenericutes,* and *Actinobacteria*, along with specific genera/species within these phyla. Notably, this included *Coprococcus, Bifidobacterium* (specifically *Bifidobacterium adolescentis*), *Faecalibacterium* (specifically *Faecalibacterium prausnitzii*), and the order RF39 [30,35,36]. These collective findings underscore the influence of dietary patterns in shaping the composition of the GM, suggesting potential implications for human health and disease.

8.2.2.4 Socioeconomic Status

Several investigations have explored the correlation between SES and the diversity of the GM, yielding noteworthy insights. One study detected a link between the SES of neighborhoods and the diversity of the GM. Notably, it revealed that lower diversity in GM and the presence of multidrug-resistant microorganisms were associated with neighborhoods characterized by lower SES. This research underscores the conceivable influence of the socioeconomic environment as a determinant affecting the diversity of the GM [37]. An additional investigation explored the relationship between family SES and host genetics in conjunction with the composition of the microbiome in infants and children. The findings suggested that family SES played a substantial role in predicting a latent variable reflecting the abundance of six genera identified through whole-genome shotgun sequencing. While SES did not predict alpha diversity, it did exert an influence on beta diversity, underscoring the impact of malleable environmental factors, including family SES, on the initial composition of the GM [38]. In a study conducted on a population of school-aged children in Latin America, associations between 14 clinical and environmental covariates and inter-individual variability in GM composition were discerned. Significantly, indicators of SES were identified as influential factors affecting the prevalence of the most abundant taxa and the distribution of enterotypes. Furthermore, age exhibited a positive correlation with heightened diversity in children of normal weight but manifested a negative correlation in overweight and obese children, thereby emphasizing the role of SES in molding the composition of GM in children [22]. In aggregate, these results imply an association between SES, encompassing both neighborhood and family levels, and the diversity of the

GM. This highlights the impact of social and environmental determinants on the intricate configuration of the GM.

8.2.2.5 Age

Age emerges as a crucial determinant significantly shaping the diversity of the GM, and its influence on both the composition and functionality of the GM is subject to a multitude of influencing factors. These alterations, occurring across the lifespan, are intricately shaped by a complex interplay of variables, including but not limited to diet, genetics, gender, SES, disease state, pregnancy, geography, and environmental exposures [39]. In a population-based study concentrating on school-aged children in Latin America, age exhibited a positive correlation with heightened diversity in children of normal weight, while displaying a negative correlation in those who were overweight and obese [22]. Likewise, investigations on six-year-old children underscored the impact of early life determinants, such as exclusive breastfeeding, antibiotic utilization between ages 1–2 years, and the mode of delivery, on the alpha diversity of the GM [40]. Acknowledging the growing importance of early-life GM in human health, a study emphasized their pivotal role in diverse aspects, encompassing the digestion of food substances for nutrient release, regulation of pathogens, modulation of the immune system, and impact on systems such as the liver, brain, and endocrine system [41]. Findings from a murine investigation reaffirmed the influence of age on microbiome composition, with gender also exerting a discernible effect, while geographical location demonstrated a relatively modest impact on diversity. Remarkably, shifts in serum cytokine expression, specifically IL-10 and IL-6, were noted, revealing distinct regulatory patterns in the context of both gender and aging [42]. Fundamentally, age exerts a substantive influence on modulating the diversity of the GM, as the configuration of the GM undergoes alterations driven by a multifaceted interplay of diverse factors, spanning individual behavioral patterns to environmental exposures.

8.2.2.6 Gender

Gender emerges as a noteworthy factor capable of influencing GM diversity, as highlighted by various studies. An exhaustive review article investigating the correlation between diet and GM diversity identified gender as one of the individual parameters manifesting robust associations with microbiome diversity [43]. Contrasting findings emerged from a study conducted on Chinese forest musk deer, where gender did not significantly impact bacterial diversity within the GM [44]. In contrast, an examination centered on hyperlipidemic outbred mice unveiled noteworthy gender-related distinctions. Female mice exhibited elevated microbial alpha diversity in comparison to their male counterparts, and beta diversity analysis revealed substantial dissimilarities in the overarching microbial community structure between the two genders [45]. Investigating irritable bowel syndrome (IBS), another study found that females were more susceptible to IBS and exhibited more severe gut dysbiosis compared to males [46]. These findings collectively underscore the potential influence of gender on GM diversity. However, it is crucial to note that the extent of this influence may vary based on factors such as the animal species under investigation and other contextual considerations.

8.2.3 Role of Gut Microbiota in Digestion and Nutrient Absorption

The gut microbiota assumes a pivotal role in digestion and the absorption of nutrients, contributing to the breakdown of complex carbohydrates, vitamin synthesis, and the development of the intestinal epithelium. Additionally, it participates in the metabolism of dietary compounds and the regulation of energy homeostasis. Recent research underscores the impact of the gut microbiota on nutrient absorption in the small intestine and energy extraction from dietary fiber in the large intestine, thereby influencing overall energy balance and metabolism [47,48]. Moreover, investigations reveal that alterations in the composition of the gut microbiota, induced by weight loss therapy or surgical and dietary interventions, can exert effects on nutrient absorption and energy metabolism [49]. The intricate interplay among gut microbiota, digestion, and immune activities has also been explored, elucidating a direct relationship between changes in the gut bacterial community and the activities of digestive and immune enzymes. These interactions subsequently influence the host's physiological adaptation to stressors such as starvation [50].

8.2.4 Interactions Between Gut Microbiota and the Immune System

The gut microbiota engages with the immune system through diverse mechanisms, playing a crucial role in preserving intestinal homeostasis and shaping immune responses. The dynamic interplay between the gut microbiota and the host's innate and adaptive immune systems is indispensable for restraining inflammation and regulating immune function. Metabolic products produced by the gut microbiota facilitate communication between gut epithelial and immune cells. Impaired interactions can lead to dysregulated immune responses, inflammation, and an increased susceptibility to infections [51]. Additionally, the gut microbiota contributes to the development of memory cells and the diversity of the human gut microbiota. Dysbiosis in the gut microbiota can negatively impact microbial composition, disrupt the epithelial barrier, and heighten susceptibility to various diseases [51,52]. Ongoing research indicates that the gut microbiota plays a role in influencing responses to immunotherapy. Investigations are actively exploring therapeutic strategies involving the manipulation of the gut microbiota to enhance the efficacy of immune checkpoint inhibitors [53]. The intricate interplay between the gut microbiota and the immune system constitutes a complex and pivotal aspect of human health and the development of diseases [51,52].

8.3 ANTIBIOTICS AND THEIR MECHANISMS OF ACTION

Antibiotics exert their impact via a diverse range of modes of operation. Many β-lactam antibiotics function by blocking the formation of bacterial cell walls [54]. The process of bacterial cell wall production involves the partial construction of wall components inside the cell, transportation of these structures over the cell membrane to the developing wall, assembly of the components into the wall, and ultimately the interconnection of the strands of wall material [55,56]. Antibiotics that hinder the production of the cell wall have a distinct impact on one or more stages. The outcome is a modification in the cellular membrane and morphology of the organism, ultimately leading to the demise of the bacteria [57].

Aminoglycosides like clindamycin, erythromycin and chloramphenicol are examples of antibiotics that hinder the process of protein production in bacteria [58]. The core mechanism of protein synthesis is comparable across bacteria and mammalian cells, however the specific proteins synthesized vary. Selective toxicity is achieved by antibiotics that use these disparities to specifically attach to or hinder the activity of bacterial proteins, hence impeding the production of fresh proteins and new bacterial cells [58,59].

Polymyxin E and polymyxin B (colistin) are antibiotics that attach to lipid compounds in the bacterial cell wall thus disrupting its role as a specific barrier. As a consequence, vital macromolecules within the cytoplasm escape, leading to the collapse of the cell [60]. Due to the presence of comparable or identical lipids in a variety of cells, notably human cells, various antibiotics exhibit a specific degree of toxicity [61].

Certain antibiotics, like sulfonamides, act as potential blockers of folic acid (folate) production, which is a crucial first stage in the production of nucleic acids. Sulfonamides can hinder the manufacture of folic acid due to their resemblance to a specific precursor chemical (PABA) that undergoes conversion by an enzyme into folic acid [62]. The structural resemblance of these molecules leads to a rivalry between sulfonamide and PABA for the enzyme that converts the precursor into folic acid. The reversibility of this process may be achieved by eliminating the chemical, leading to the prevention rather than the mortality of the bacteria [63]. Rifampin, an antibiotic, disrupts the process of RNA production in pathogens by attaching itself to a component of the microbial enzyme that is crucial in replicating RNA. Due to the higher selectivity of rifampin for a microbial enzyme compared to the human enzyme, medicinal levels of the drug do not damage cells in humans [64]. Frailty is a condition characterized by a reduction in reserve and physiological activity within many organ systems, which occurs with age and limits an individual's capacity to withstand mild shocks [65]. Frailty, a regrettable but unavoidable aspect of becoming older, exposes persons to a higher likelihood of experiencing negative health consequences. Muscular skeleton fragility is characterized by sarcopenia, which refers to the reduction in muscle mass and function that occurs with aging [66,67]. The decline in muscular function in older adults renders them susceptible to many catastrophic occurrences such as fractures, falls, incapacity, and death [67]. Physical frailty and Sarcopenia arise from several pathophysiologic causes, such as insufficient food intake and physical exercise, infection, immunity decline, anabolic reaction resistance, and oxidative stress [68]. The gut microbiota has a significant impact on activities like infection and immunity. Multiple research has clearly shown a correlation between modified bacterial diversity and diminished physical activity in elderly patients. Specific research found that the bacteria makeup of aged people living in society who met the demands for physical frailty was comparable to that of patients in nursing homes [69]. This suggests that an imbalance in bacterial populations may contribute to the deterioration in physical activity associated with aging. Adding to the evidence of the intestinal microbiome's involvement in age-associated decline, Buigues et al. showed that prebiotics featuring inulin and fructooligosaccharides had a positive impact on two indicators of frailty, fatigue and grip endurance (which is also an official measure for sarcopenia) [70].

The preclinical research has provided more evidence for the connection between gut dysbiosis and age-associated physical frailty and reduction in function. This results in the theory that there is a relationship between the gut and muscles, where the microbiome affects overall physique, muscle tone, and physical fitness [71]. According to this concept, changes in the gut microbiota and gut metagenome affect important cellular functions, including inflammation (e.g., higher permeability of the gut epithelium), accessibility to nutrients, and digestion of lipids. These changes may lead to the reduction of muscular activity associated with aging. While human data are scarce, some research conducted on animal models provides substantial evidence in favor of this idea [72]. Backhed et al. conducted ground-breaking research showing that germ-free mice, when exposed to fecal specimens from conventionally-raised mice, have a major reduction in lean body composition, including muscular skeleton mass [73]. Recent investigation compared pathogen-free and germ-free mice, revealing that the germ-free animals had skeletal muscle collapse, reduced levels of insulin-like growth factor 1, and decreased levels of genes associated with skeletal muscle growth and mitochondrial function. Collectively, these studies indicate that the impact of microorganisms on skeletal muscle is probably contingent on the specific taxonomic group [74]. The bacterial species that have been shown to promote muscle mass growth and function comprises *Lactobacillus casei*, *Bifidobacterium longum*, *Eubacterium rectale*, *Clostridium coccoides*, and *Barnesiella intestinihominis*. The suggested method by which these organisms improve skeletal muscle physiology is by generating short-chain fatty acids (SCFAs) such as acetate, propionate, and butyrate [75]. Siddarth et al. have shown that as individuals age, there is a loss in lean body mass and gastrocnemius muscle mass. This decline is directly tied to the levels of vitamin B12 in the bloodstream and lipid metabolism. These findings suggest that these processes have a role in the age-related decline of muscle mass [76]. On the other hand, the increase in pro-inflammatory species, namely those belonging to the Gram-negative bacteria family called Enterobacteriaceae, is linked to a decline in the strength and structure of skeletal muscles [77]. The potential ways in which these bacteria affect skeletal muscle include increased permeability of the epithelial layer due to inflammation, circulation of bacterial toxins such as lipopolysaccharide (LPS) and indoxyl sulfate, and the subsequent development of insulin resistance and skeletal muscle atrophy [78,79].

The emergence and accessibility of culture-independent methodologies for assessing microbial diversity, namely 16S ribosomal RNA (rRNA) sequencing, has led to a significant increase in scientific investigations on the human gut microbiota in recent years [80]. This study has yielded a ground-breaking comprehension of the formation, structure, and long-term consistency of the gut microbiome, while also pinpointing changes linked to several disease processes [81]. Although microbial diversity has been studied concerning skeletal muscle mass and function in animal models and young adult humans, there is a lack of published research on microbiome changes linked with physical frailty and muscular deterioration in older individuals [82]. Furthermore, as far as we know, there are no existing papers that assess the gut microbiota in clinical sarcopenia. Through the use of 16S rRNA analysis, we have detected significant variations in microbial diversity between older persons who are in good condition and those who suffer from sarcopenia [83]. A new investigation discovered a decrease in microbial diversity in sarcopenia patients, as well as a drop in certain species that produce butyrate. This provides a new understanding of how changes in the microbial composition may impact the development of the illness [84].

8.3.1 Classes of Antibiotics Commonly Used in Gastrointestinal Infections

Bacterial diseases may lead to gastrointestinal problems, and antibiotics are commonly used to treat these illnesses [85]. The choice of antibiotics depends on the specific organism causing the disease, its susceptibility to the antibiotic, and other factors. The selection of an antibiotic is determined by several criteria, including the specific kind of organism, local resistance trends, and the extent of the illness [86]. The improper or unselective utilization of antibiotics may encourage the development of drug resistance. Therefore, it is important to exercise prudence and seek the direction of medical experts while using these drugs. Furthermore, the rise of antibiotic-resistant bacteria highlights the need for continuous investigation and surveillance in the realm of infectious illnesses [87]. Many GIT illnesses are caused by viruses or parasites, and antibiotics may not be efficient against these pathogens. For such instances, drugs that combat viruses or parasites may be more suitable. The issue of drug resistance

TABLE 8.1
Antibiotics Used in the Treatment of Gastrointestinal Infections

Class	Examples	MOA	Indications	References
Fluoroquinolones	Ciprofloxacin, Levofloxacin	Inhibition of DNA gyrase, preventing DNA replication and repair	Broad spectrum of gastrointestinal infections, including traveler's diarrhea and certain bacterial infections in the stomach and intestines	[91]
Macrolides	Azithromycin, Clarithromycin, Fidaxomicin	Inhibition of bacterial protein synthesis	Effective against some gastrointestinal pathogens, including *Campylobacter* species and certain types of atypical pneumonia	[92]
Cephalosporins	Ceftriaxone, Cefixime, Cefpodoxime	Inhibition of bacterial cell wall synthesis	Used for a variety of bacterial infections, including some gastrointestinal infections caused by enteric pathogens	[93]
Penicillins	Amoxicillin, Ampicillin	Inhibition of bacterial cell wall synthesis	Used for certain gastrointestinal infections caused by susceptible bacteria	[94]
Tetracyclines	Doxycycline, Minocycline	Inhibition of bacterial protein synthesis	Effective against a range of bacteria and often used for treating certain gastrointestinal infections, such as those caused by *Vibrio cholerae*	[95]
Nitroimidazoles	Metronidazole, Tinidazole	Disruption of bacterial DNA and inhibition of nucleic acid synthesis	Commonly used for anaerobic bacterial infections in the gastrointestinal tract, including infections caused by *Clostridium difficile*	[96]
Aminoglycosides	Gentamicin, Tobramycin	Inhibition of bacterial protein synthesis	Less commonly used for gastrointestinal infections due to their potential side effects, but they may be employed in specific cases	[97]
Rifamycins	Rifampin, Rifaximin	Inhibition of bacterial RNA synthesis	Rifaximin, in particular, is often used for gastrointestinal infections, including traveler's diarrhea and hepatic encephalopathy. It is not significantly absorbed systemically, making it suitable for treating infections in the gastrointestinal tract	[98]

is increasing, and healthcare providers must exercise caution when administering antibiotics to minimize the likelihood of resistance emergence [88,89].

Before initiating any medical treatment, patients need to seek guidance from healthcare specialists to get a reliable diagnosis and receive suitable antibiotic medication. Engaging in self-prescription or using leftover medicines is strictly prohibited owing to the possibility of incorrect treatment, potential risks, and the onset of antibiotic resistance [90]. Table 8.1 provides the list of antibiotics commonly prescribed in the treatment of gastrointestinal infections.

8.3.2 Broad-Spectrum vs. Narrow-Spectrum Antibiotics

Broad-spectrum antibiotics refer to a category of antibiotics that may effectively target a diverse array of bacteria, including both Gram-positive and Gram-negative types. These antibiotics are formulated to selectively attack a wide range of bacterial species, rendering them advantageous in scenarios where the identity of the bacteria causing an illness is uncertain or when there is a combination of several bacterial kinds involved in the infection [99]. Broad-spectrum antibiotics can combat a wide range of germs, including those that are often linked to different kinds of diseases. Their adaptability renders them beneficial in the primary treatment of serious or life-threatening infections when the precise etiology is unknown [100]. Empirical therapy is initiating antibiotic treatment before the identification of a particular infection by laboratory testing. Empirical treatment frequently involves the selection of broad-spectrum antibiotics due to their ability to give protection against a wide range of possible pathogens [101]. Nevertheless, the use of broad-spectrum antibiotics is linked to the probable emergence of antibiotic resistance. The excessive or incorrect use of these antibiotics might lead to the formation of resistant strains, hence diminishing the efficacy of these medications over time [102]. In addition to targeting harmful bacteria, they also have an impact on the body's natural microbial ecology. This disturbance may result in adverse consequences and provide a favorable environment for the proliferation of opportunistic infections [103].

Although Narrow-spectrum antibiotics are a kind of antibiotics that selectively target a certain subset of microorganisms. Unlike broad-spectrum antibiotics, which have efficacy against a wide array of bacteria, including both Gram-positive and Gram-negative types, narrow-spectrum antibiotics have a more targeted and specific action [100]. These substances may be engineered to selectively attack either Gram-positive or Gram-negative bacteria, or certain species within those categories. Specialized antibiotics, known as narrow-spectrum antibiotics, are specifically formulated to target a restricted range of microorganisms [104]. Antimicrobial agents may selectively focus on certain bacterial structures or activities, hence exhibiting efficacy against a more limited spectrum of infections. These antibiotics are frequently prescribed upon identification of the specific bacteria responsible for an infection through laboratory tests [105]. This enables healthcare practitioners to customize the therapy based on the specific attributes of the diagnosed infection. Narrow-spectrum antibiotics have a more targeted effect on the microbial balance in the body as compared to broad-spectrum antibiotics [106]. Minimizing changes to the natural microbial population may be favorable by reducing the influence on regular flora. Due to their ability to specifically target a certain type of bacteria, narrow-spectrum antibiotics have a reduced risk of generating resistance as compared to broad-spectrum antibiotics [104]. Nevertheless, resistance may develop if these antibiotics are overused or handled incorrectly. The selection of narrow-spectrum antibiotics is often determined by the specific kind and site of the illness. For instance, a specific antibiotic that targets respiratory microorganisms may be chosen to treat a respiratory tract infection [107].

8.3.3 Impact of Antibiotics on Non-Target Microorganisms, Including Gut Microbiota

Antibiotics may have a substantial influence on non-target microorganisms, namely the advantageous bacteria that constitute the natural flora of the human body, especially the gut microbiota [108]. The gut microbiota is essential for sustaining health since it aids in digestion, absorbs nutrients, regulates the immune system, and protects against harmful pathogens. Antibiotics are specifically formulated to selectively eliminate or hinder the proliferation of certain bacteria responsible for causing illnesses [109]. Nevertheless, they fail to differentiate between detrimental and beneficial microbes. Consequently, the use of antibiotics might cause an imbalance in the natural microbial equilibrium in different bodily systems, such as the gastrointestinal tract [110]. Antibiotics can substantially modify the makeup of the gut microbiota. The presence of antibiotics may result in a reduction in the population and variety of certain bacterial species while promoting the proliferation of other bacteria that possess higher resistance to the antibiotic. Diminished microbial diversity in the gastrointestinal tract has been linked to a range of health problems [111]. The use of antibiotics, particularly those with a wide range of effectiveness, may lead to a reduction in the variety of microorganisms present in the gut microbiota, which may have an impact on its overall ability to adapt and operate optimally [112]. Disturbance of the indigenous microbiota might provide a favorable milieu for the proliferation of opportunistic pathogens, resulting in subsequent illnesses. For instance, the excessive proliferation of *Clostridium difficile* (*C. difficile*) bacteria in the large intestine may happen as a result of antibiotic therapy, leading to antibiotic-associated diarrhea and, in extreme instances, colitis [113]. The gut microbiota has a vital function in instructing and regulating the immune system. The alteration in the gut microbiota caused by antibiotics might affect the functioning of the immune system, possibly compromising the body's capacity to effectively react to infections and other challenges [114]. Several studies indicate that changes in the gut microbiota resulting from early or recurrent antibiotic use may have enduring impacts on metabolism and overall well-being [115]. This is a subject that is still being studied, and the exact outcomes may differ based on variables such as the precise antibiotics used, length of therapy, and individual variations. Although antibiotic therapy may cause substantial changes to the gut microbiota, it often can restore itself gradually. Nevertheless, the rate and magnitude of recuperation might differ across people, and in some instances, the microbiota may not completely revert to its initial condition [116].

To minimize the effects of antibiotics on unintended bacteria, healthcare practitioners strive to use antibiotics prudently, taking into account aspects like as the kind and severity of the illness, the particular antibiotic used, and the possibility of unintended harm to the microbiota [117]. Additional approaches, such as the use of probiotic supplements or the transplantation of fecal microbiota (FMT), may be taken into account to aid in the recovery of a balanced gut microbiota after antibiotic therapy [118].

8.4 ANTIBIOTICS AND GUT MICROBIOTA DYSBIOSIS

8.4.1 Definition and Causes of Gut Microbiota Dysbiosis

Dysbiosis is frequently referred to as a state of "imbalance" in the microbial population of the digestive tract, which is linked to the presence of illness [119]. This discrepancy may arise from the acquisition or depletion of community members or modifications in the overall prevalence of microorganisms. This hypothesis largely emphasizes alterations in the taxonomic composition of the microorganisms and the roles attributed to individual members or specific subgroups within the population [120]. The characterization of dysbiosis has traditionally depended on taxonomic inventories of gut microorganisms obtained from various people at certain periods via the use of 16S or 18S rRNA sequencing [121]. Recent endeavors have aimed to compile the microbial genes present in the gut by the use of shotgun metagenomics, rather than focusing on categorizing the microbial species [122,123].

Numerous studies have shown substantial disparities in the taxonomic makeup of gut microbial communities between individuals with inflammatory bowel disease (IBD) and those who are healthy [124]. The distinctions have been identified in the bacterial, fungal, and viral constituents of the community [125]. Nevertheless, the precise variations in microbial taxonomy vary significantly across different studies, and a conclusive community structure related to the illness has not yet been established [126]. This phenomenon might be attributed to the substantial disparity in the makeup of gut microbial communities across individuals, and/or to technical discrepancies throughout the conducted investigations. Thus, the precise definition and diagnostic criteria for dysbiosis and a "dysbiotic community" remain uncertain [123,127]. The relationship between dysbiosis and illness is not well understood since it is uncertain if dysbiosis is a cause, a consequence, or both a cause and effect of the disease. This uncertainty arises because inflammation may independently alter the makeup of the gut's microbial population [128]. Dysbiosis may arise from host-specific variables, including genetic background, health state (infections, inflammation), and lifestyle behaviors. However, it is mostly influenced by external variables like nutrition (high sugar, low fiber), xenobiotics (antibiotics, medications, food additives), and cleanliness [129,130]. Alterations in the composition of gut bacteria and fungi may be swiftly accomplished by modifications in macronutrients. These alterations have notable physiological ramifications [131]. For instance, diets abundant in simple sugars disturb the digestive barrier, provoke inflammation within the gut, and detrimentally impact host metabolism [132]. Nevertheless, in the majority of instances, the harmful consequences are dependent on the interplay between food and microbiota, as they do not manifest in the absence of gut microbiota, specifically in a germ-free condition. Moreover, the replacement of gut microbiota frequently leads to the transmission of the disease phenotype [133].

The impact of food ingredients on the gut microbiota has been underestimated for a long time. However, recent research from various groups, including ours, has revealed that certain types of human intestinal microbiota are highly susceptible to additives [134]. Furthermore, it has been observed that exposure to commonly used food preservatives leads to excessive development of proteobacteria. Additional classifications of chemicals have also shown detrimental impacts on human health [135]. As an example, food emulsifiers have a direct impact on the makeup of the bacteria in the human stomach and may cause discomfort in the intestines. Sugar substitutes, like Splenda, contribute to an imbalance of proteobacteria and worsen inflammation in the intestines of SAMP1/YitFc (SAMP) mice [136]. Ironically, non-caloric artificial sweeteners were first developed with the main purpose of preventing metabolic syndrome. However, they inadvertently cause an imbalance in the gut microbiota and contribute to the development of glucose intolerance [137,138]. This occurs in a way that is reliant on the presence of bacteria, ultimately resulting in the undesired metabolic consequences that these sweeteners were originally designed to protect against [139]. Host-related variables that modify the quantity and makeup of the gut microbiota consist of substances produced by the gastric glands and liver that kill bacteria, namely gastric acid and bile, as well as antimicrobial compounds like defensins, lysozymes, and antibacterial lectins (Reg3γ) produced by Paneth cells, or SIgA produced by plasma cells [140,141].

The repercussions of disrupting the gut microbiota ecology may be categorized into two main effects: disruption of the gut barrier and imbalance of the host immune and metabolic systems [142]. Acetaldehyde, which is generated by the microbiota from ethanol that comes from outside the body or is created inside the body, might potentially weaken the integrity of the intestinal wall. This can occur by direct mucolytic activity and other processes [143]. Microbiota-derived chemicals may affect the host immune system by inflammasome signaling, Toll-like receptor (TLR) signaling, or NOD-like receptor (NLR) signaling. Another method involves a change in the equilibrium between regulatory and pro-inflammatory immune cells [144,145]. Changes in bile acid composition, synthesis of short-chain fatty acids (SCFAs) from dietary fiber, conversion of choline to trimethylamine (TMA), and other factors influence the effects on the human metabolic system, namely glucose and lipid metabolism [131,146].

8.4.2 Studies on the Short-Term and Long-Term Effects of Antibiotics on Gut Microbiota

The awareness towards the variation of the gut microbiota caused by antibiotics has grown in recent years. Several studies have shown a clear correlation between the use of antibiotics and changes in the composition of the gut microbiota [147–149]. Furthermore, it has been shown that antibiotics may cause alterations in the gut microbiota, which can in turn disturb metabolic balance and have an impact on the proper functioning of the immune system or immunological status [150,151].

The negative effects of broad-spectrum antibiotics are often overlooked due to their extensive usage in the prevention and cure of diseases in both humans and animals [114]. Nevertheless, the prolonged and high dosage administration of antibiotics that target a wide range of bacteria may disrupt the balance of microorganisms in the intestines. This disruption can diminish the capacity of normal bacteria to colonize, while simultaneously increasing the population of potentially harmful bacteria. In extreme circumstances, it can even result in endogenous infection [147,152]. The level of disruption is intricately linked to the spectrum, dose, and method of delivery, as well as the pharmacokinetic and pharmacodynamic characteristics [153]. Fifteen β-lactams, such as penicillin and cephalosporins, are extensively used in the treatment of infectious disorders. They function by interfering with the creation of cell walls and peptidoglycan, ultimately resulting in the demise of the bacterial cells. Piperacillin/tazobactam (TZP) and ceftriaxone (CRO) are frequently used broad-spectrum antimicrobials in clinical practice [154]. They have very comparable pharmacokinetic and pharmacodynamic profiles, as well as broad spectrums of efficacy against therapeutically significant infections [155]. Gamage and colleagues discovered that the treatment of either of two broad-spectrum antibiotics, TZP or CRO, led to significant alterations in the fecal microbiota of ICU patients [156]. Tigecycline administered intravenously is a significant antibiotic belonging to the glycylcycline class. It was specifically developed to overcome the typical mechanisms of resistance seen in tetracycline antibiotics. On the other hand, Levofloxacin is a fluoroquinolone antibiotic that effectively treats a broad spectrum of infections caused by both Gram-positive and Gram-negative bacteria and has a high level of efficacy as well as safety [157]. Nevertheless, the precise relationship and mechanism linking these wide-ranging antibiotics to the makeup of the gut microbiota are still unknown, perhaps leading to an imbalance in the body's immune system [158].

8.4.3 Alterations in Microbial Diversity and Abundance

The human gut microbiome is a rich complicated collection of bacteria that inhabit the whole length of the digestive system and undergo continuous changes throughout time. The gut microbiome experiences the most significant variations in composition throughout infancy and old age, which are the two ends of the age spectrum [159]. Several studies have shown significant variations in the makeup of gut microbiota between older persons and younger adults [160,161]. Furthermore, the gut microbiome of elderly individuals exhibits substantial variation across and among individuals, while there is a considerable level of consistency over time in this particular population [162]. The microbiome's dynamic nature is believed to be the underlying factor in the physiological and pathological changes that occur with aging. For instance, the disruption of older individuals' ability to regulate their immune system (known as immunosenescence or inflammation) and the subsequent rise in pro-inflammatory cytokines such as tumor necrosis factor (TNF)-α, interleukin (IL)-6, and IL-8, are associated with changes in the composition of gut bacteria [163,164]

Frailty is a condition characterized by a reduction in reserve and physiological function across various organ systems due to aging. This deterioration reduces an individual's capacity to handle mild shocks. Frailty, a regrettable but unavoidable aspect of becoming older, exposes persons to a higher likelihood of experiencing negative health consequences [165,166]. Skeletal muscle fragility is characterized by sarcopenia, which refers to the reduction in muscle mass and function that occurs with aging. The decline in muscular function in older adults renders them susceptible to several detrimental occurrences such as falls, fractures, incapacity, and mortality [167]. Sarcopenia and physical frailty arise from several pathophysiologic causes, such as insufficient nutrition and physical exercise, inflammation, immunosenescence, anabolic resistance, and oxidative stress [168–170]. The gut microbiota has a significant impact on processes such as inflammation and the immune system. Multiple research has indeed shown a correlation between modified microbial diversity and diminished physical performance in elderly patients [171–173]. Specific research found that the microbial composition of older individuals living in the community who met the criteria for physical frailty was comparable to that of patients in nursing homes. This suggests that an imbalance in microbial communities may contribute to the deterioration in physical performance associated with aging. Buigues et al. provided additional evidence supporting the involvement of the gut microbiome in age-related deterioration. They showed that a prebiotic supplement containing inulin and fructooligosaccharides had a positive impact on two indicators of frailty, exhaustion and handgrip strength (which is also used to assess sarcopenia) [70].

The preclinical research has provided more evidence for the connection between gut dysbiosis and age-related physical frailty and reduction in performance [174]. This has led to the theory that there is a relationship between the gut and muscles, where the microbiome affects overall lean mass, skeletal muscle mass, and physical functioning. According to this model, changes in the gut microbiota and gut metagenome affect important biological processes, such as the inflammatory environment (e.g., increased permeability of the gut epithelial barrier), availability of nutrients, and metabolism of lipids [175,176]. These changes can contribute to the decline of muscle function associated with aging. While human data are scarce, many research conducted on animal models provides substantial evidence to support this notion. In their groundbreaking study, Backhed et al. showed that germ-free mice, when colonized with fecal samples from conventionally-raised mice, experience a significant decrease in lean body composition, including skeletal muscle mass [177]. A recent study compared pathogen-free and germ-free mice and found that the germ-free animals showed skeletal muscle atrophy, decreased expression of insulin-like growth factor 1, and lower expression of genes related to skeletal muscle growth and mitochondrial function [74]. Collectively, these studies indicate that the impact of microorganisms on skeletal muscle is probably contingent on the specific taxonomic group. Some particular types of bacteria, such as *Barnesiella intestinihominis, Eubacterium rectale*, *Lactobacillus casei*, *Clostridium coccoides*, and *Bifidobacterium longum*, [178–180] have been shown to promote the development and functioning of skeletal muscles. The suggested method by which these species improve skeletal muscle physiology is by generating short-chain fatty acids (SCFAs) such as acetate, propionate, and butyrate [181]. Siddharth and colleagues have shown that when individuals age, there is a decrease in lean body mass and gastrocnemius muscle mass, which is directly connected to the levels of vitamin B12 and lipid metabolism in the bloodstream. This suggests that these processes play a part in the reduction of muscle mass associated with aging. Conversely, the increase in pro-inflammatory types of bacteria, namely those belonging to the Gram-negative bacteria family called Enterobacteriaceae, is linked to a decline in the strength and health of skeletal muscles [76]. The potential ways in which these bacteria affect skeletal muscle include increased permeability of the epithelial layer due to inflammation, circulation of bacterial toxins such as lipopolysaccharide (LPS) and indoxyl sulfate, and the subsequent development of insulin resistance and skeletal muscle atrophy [182].

The emergence and accessibility of culture-independent methods for assessing microbial diversity, namely 16S ribosomal RNA (rRNA) sequencing, has led to a significant increase in studies on the human gut microbiota in recent years [183]. This study has yielded a remarkable comprehension of the formation, structure, and long-term consistency of the gut microbiome, while also pinpointing changes linked to several pathological processes [184,185]. Although microbial diversity has been studied concerning skeletal muscle mass and function in animal models and young adult humans, there is a scarcity of published research investigating changes in the microbiome linked with physical frailty and muscular deterioration in older individuals [186]. Furthermore, as far as we know, there are no existing papers that assess the gut microbiota in clinical sarcopenia. Through the use of 16S rRNA analysis, we have detected significant variations in microbial diversity between older persons who are in good condition and those who suffer from sarcopenia [187]. A new investigation discovered a decrease in microbial diversity in sarcopenia patients, as well as a drop in specific species that produce butyrate. This provides a new understanding of how changes in the microbial community may impact the development of the illness [175,187].

8.4.4 Consequences of Gut Microbiota Dysbiosis for Overall Health

Gut microbiota dysbiosis, characterized by an aberration or disturbance in the structure and operation of the gut microbiota, may have extensive ramifications for general well-being. The gut microbiota is essential for regulating several physiological activities, and disturbances in this intricate equilibrium have been associated with a variety of health conditions [188]. Imbalance in the gut microbiota, known as dysbiosis, may have a role in the development of digestive illnesses such as irritable bowel syndrome (IBS), inflammatory bowel diseases (IBD) such as Crohn's disease and ulcerative colitis, and functional gastrointestinal disorders [189]. It has a crucial function in training and regulating the immune system. It has the potential to disrupt the immune system, which may contribute to the development of autoimmune illnesses, allergies, and an increased vulnerability to infections. Changes in the makeup of the gut microbiota have been linked to metabolic illnesses such as obesity, insulin resistance, and type 2 diabetes [190]. The microbiota has an effect on energy metabolism, food absorption, and the control of inflammation, all of which may have an impact on metabolic health. The gut-brain axis, a two-way communication channel connecting the gut and the central nervous system, is impacted by the gut bacteria [191]. The presence of dysbiosis has been associated with mental health diseases such as anxiety, depression, and neurological illnesses such as Parkinson's disease. An imbalance in the gut microbial population, known as dysbiosis, may lead to chronic inflammation, which is a significant component in the development of several chronic illnesses such as cardiovascular disease, certain malignancies, and neurological disorders [192]. The gut bacteria also contribute to the process of absorbing nutrients and extracting energy from the meal. Dysbiosis may impact various physiological systems, possibly resulting in deficits in essential nutrients and disruptions in energy equilibrium [193]. Dysbiosis often occurs as a result of antibiotic usage, since it may disturb the natural equilibrium of the gut microbiota. Extended or frequent use of antibiotics might heighten the likelihood of experiencing antibiotic-associated complications, such as *Clostridium difficile* infection and other opportunistic infections [194]. The gut microbiota plays a role in maintaining a robust intestinal barrier. Dysbiosis may undermine the integrity of the intestinal barrier, resulting in heightened permeability, sometimes referred to as "leaky gut". This might potentially facilitate the entry of noxious chemicals into the circulatory system, inducing systemic inflammation [72]. The gut microbiota has a role in the breakdown and processing of certain dietary substances, such as fiber and specific polyphenols. Dysbiosis may influence various metabolic pathways, which might disrupt the generation of beneficial metabolites that possess anti-inflammatory and antioxidant characteristics [195]. An inclusive and harmonious gut microbiota aids in preventing the establishment of detrimental infections. Dysbiosis may compromise this response, leaving people more vulnerable to infections [196]. The effects of an imbalance in gut microbiota might differ significantly across people, and the connection between this imbalance and certain health results is now being actively investigated. Nevertheless, other lifestyle variables such as dietary choices, antibiotic use, stress levels, and genetic predisposition may all have an impact on the makeup of the gut microbiota [197]. Implementing strategies aimed at fostering a healthy gut microbiota, such as adhering to a well-balanced diet abundant in fiber, prebiotics, and probiotics, might potentially bolster general well-being and serve as a preventive or mitigating measure against the adverse effects of dysbiosis [198].

8.5 IMPLICATIONS FOR GASTROINTESTINAL INFLAMMATIONS

8.5.1 The Link Between Gut Microbiota and Gastrointestinal Inflammations

The relationship between gastrointestinal inflammations and gut microbiota is a complicated and dynamic relationship that has received significant attention in recent years [199]. The microbiota of the GI tract is a complex community of microorganisms, including viruses, bacteria, archaea and fungi, that live in the digestive system. These microbes play a crucial role in sustaining an equilibrium in the GI tract thus having a substantial influence on many aspects of the host's health [200]. A disruption or dysbiosis of the gut microbiota might result in irritation inside the digestive tract. Changes in the relative abundance of particular microbial species may lead to the growth of pathogenic bacteria or a decrease in beneficial bacteria, disrupting the delicate balance required for a healthy gastrointestinal tract [152,201]. The gut microbiota has a role in maintaining the structural integrity of the intestinal mucosal barrier. The breakdown of the protective mucosal barrier may lead to the infiltration of bacteria and microbial chemicals into the intestinal tissues, triggering an immune response and inducing inflammation [202].

The gut microbiota collaborates with the immune system of the host, influencing the development and functioning of immune cells. Interference with this link might lead to an incorrect immune response, leading to persistent inflammation in the GI tract [203]. Particular beneficial bacteria in the gut microbiota produce short-chain fatty acids (SCFAs) via the process of digesting dietary fibers, that show anti-inflammatory properties and contribute to maintaining gut health by improving the functioning of the intestinal tract and regulating immune responses [204]. The presence of modified gut microbiota has been associated with inflammatory bowel diseases (IBD) such as Crohn's disease and ulcerative colitis. Studies suggest that individuals suffering from inflammatory bowel disease (IBD) exhibit reduced diversity and a unique composition of microbiota in comparison to those in a healthy state [205]. One's food has a significant impact on the gut microbiome. Consuming diets high in saturated fats and low in fiber may promote the growth of pro-inflammatory bacteria, leading to gut inflammation [206]. The investigation of the intricate connection between gut microbiota and gastrointestinal inflammation is a rapidly progressing area of inquiry. Exploring the modification of the gut microbiota with medicines like probiotics, prebiotics, and fecal microbiota transplantation (FMT) is being examined as a potential therapeutic approach for treating gastrointestinal inflammatory illnesses [118]. However, more research is necessary to fully understand the underlying mechanisms and develop targeted strategies for maintaining healthy gut microbiota and reducing or controlling gastrointestinal inflammations.

8.5.2 Impact of Antibiotics on the Development and Progression of Inflammatory Bowel Diseases (IBD)

Ongoing research is being conducted to investigate the influence of antibiotics on the onset as well as the progression of inflammatory bowel illnesses (IBD), such as Crohn's disease and ulcerative colitis [207]. Although antibiotics are frequently prescribed for the treatment of bacterial infections, their impact on the gut microbiota, which is a varied collection of microorganisms resident in the gastrointestinal system, might have consequences for inflammatory bowel disease (IBD) [114]. Antibiotics have the potential to disturb the equilibrium and variety of the gut microbiota. The use of broad-spectrum antibiotics may result in the reduction of helpful bacteria and an increase in the proliferation of potentially dangerous bacteria [208]. Alterations in the microbial makeup may lead to dysbiosis, a condition that is linked to inflammatory bowel disease (IBD). Compelling data indicates that being exposed to antibiotics at a young age may be associated with a higher likelihood of getting inflammatory bowel disease (IBD) [209]. The use of antibiotics throughout infancy or adolescence, a period when the intestinal microbiota is still maturing, has been linked to an increased susceptibility to inflammatory bowel disease (IBD) in later stages of life [210]. This correlation might be linked to the influence of antibiotics on the development and resilience of the gut microbiota. Antibiotic usage may affect the development of inflammatory bowel disease (IBD) in persons who already have the condition [211]. Several studies have shown that the repetition or prolongation of antibiotic treatments may potentially worsen the progression of diseases. This is believed to be caused by the disturbance of the intricate equilibrium in the gut microbiota and the possible facilitation of an inflammatory milieu [212]. However, antibiotics are sometimes used as a treatment strategy for addressing specific issues of IBD, such as the excessive development of bacteria in the small intestine. Rifaximin, an antibiotic, has been used to treat some gastrointestinal problems, such as small intestinal bacterial overgrowth (SIBO) in patients with inflammatory bowel disease (IBD) [213]. The effect of antibiotics on inflammatory bowel disease (IBD) might differ among people. This variability is influenced by factors like the particular antibiotic used, the length of the therapy, and changes in the makeup of gut microbiota across individuals [214]. Certain individuals with inflammatory bowel disease (IBD) may see amelioration of symptoms with antibiotic therapy, however, others may not exhibit a response or may even encounter exacerbation of symptoms [215]. Healthcare practitioners exercise caution when prescribing antibiotics to persons with IBD due to the possible influence on the gut flora. The prescription of antibiotics is usually determined by the particular clinical scenario and the assessment of the possible advantages compared to the hazards involved [216]. It is crucial to acknowledge that research in this field is still developing, and the use of antibiotics in the context of IBD is a multifaceted and intricate subject [217]. It is important to carefully evaluate the possible influence of antibiotics on the gut microbiota and IBD for each person, and choices should be taken in collaboration with healthcare specialists, taking into account the particular circumstances of each patient. Furthermore, continuing research in the area of IBD is focused on reducing needless antibiotic usage and investigating alternative treatment approaches [124,205].

8.5.3 Antibiotic-Induced Changes in the Gut Microbiome as a Potential Trigger for Gastrointestinal Inflammation

The gut microbiota may be altered by antibiotics, which can initiate gastrointestinal inflammation. The gut microbiome, an intricate assemblage of bacteria inhabiting the GI tract, has a vital function in preserving gut health and regulating the immune system [218]. Although antibiotics are successful in curing bacterial illnesses, they may inadvertently impact the makeup and variety of the gut microbiota, which may result in an imbalance and inflammatory reactions [219]. The use of antibiotics, particularly broad-spectrum ones, might disturb the equilibrium of the gut microbiota by eradicating both pathogenic and beneficial microorganisms. This disturbance may lead to dysbiosis, which is marked by a change in the microbial makeup and a possible increase in harmful bacteria [208,220]. The use of antibiotics has been linked to a reduction in the variety of microorganisms present in the gastrointestinal tract. Decreased variety is often associated with a microbiota that is less robust and less stable, rendering the gut more vulnerable to disruptions and subsequent inflammation [221]. The use of antibiotics may lead to the eradication of helpful bacteria, which in turn creates a favorable environment for the proliferation of potentially hazardous microorganisms [222]. The excessive development of cells may result in an elevated synthesis of toxins and inflammatory chemicals, hence leading to inflammation in the gastrointestinal tract. The gut microbiota contributes to the preservation of the intestinal barrier's integrity and the regulation of immune responses [223]. The microbiome may be altered by antibiotics, which can weaken its protective capabilities. This can lead to the movement of bacteria and microbial products over the intestinal barrier, causing an immunological response and inflammation [224]. Modifications in the makeup of the gut microbiota may influence the metabolic activities of microorganisms, resulting in changes in the synthesis of metabolites, such as short-chain fatty acids (SCFAs). Short-chain fatty acids (SCFAs) possess anti-inflammatory characteristics and have a role in preserving the well-being of the gastrointestinal tract [225]. Interference in the formation of short-chain fatty acids (SCFAs) may lead to the development of an inflammatory environment in the gastrointestinal tract. Studies have shown a correlation between the use of antibiotics and a heightened susceptibility to inflammatory bowel illnesses (IBD), like Crohn's disease and ulcerative colitis [226]. Prolonged or recurrent use of antibiotics, especially during crucial stages of development, may be associated with an increased likelihood of chronic gastrointestinal inflammation. Specific antibiotics, such as clindamycin and fluoroquinolones, have been linked to the occurrence of antibiotic-induced colitis, which is often caused by the bacteria *Clostridium difficile*. This disorder, known as antibiotic-associated colitis, may lead to significant inflammation of the colon and is a well-recognized consequence of antibiotic treatment [227]. The gut microbiota has a vital function in instructing and regulating the immune system. The gut microbiota may be disrupted by antibiotics, which can disturb the delicate balance and possibly result in immune system dysregulation and a higher vulnerability to inflammatory reactions [116]. Distinct antibiotics have diverse impacts on the gut microbiome. Certain antibiotics have a higher degree of selectivity towards certain bacterial species, whilst others have a broader impact, affecting a wide array of microorganisms [228]. The choice of antibiotic may impact the kind and magnitude of alterations in the microbiome. The timing and duration of antibiotic administration are critical variables. Exposure to antibiotics in the early stages of life, particularly during crucial times of microbiome formation, may have a more significant effect on the health of the gut [229]. Extended or recurrent administrations of antibiotics may potentially lead to persistent disturbances in the microbiome. The capacity and aptitude of the gut microbiota to recuperate after antibiotic administration exhibit variability across people. Although the microbiome can largely restore itself, some enduring changes may remain, thereby impacting long-term vulnerability to gastrointestinal inflammation [230]. Various individual variables, including genetic predisposition, general health state, and pre-existing illnesses, might impact an individual's response to changes in the gut microbiota caused by antibiotics. Certain people may exhibit more resilience, whilst others may have heightened susceptibility to the onset of gastrointestinal inflammation [231].

8.5.4 Strategies to Mitigate the Negative Effects of Antibiotics on Gut Health During Inflammation Treatment

The human gastrointestinal tract contains a complex assemblage of diverse bacteria species and many viruses, commonly known as the gut microbiome [232]. Collectively, they enhance our ability to use nutrients more effectively and impede the colonization of harmful microorganisms in our gastrointestinal tract. Nevertheless, using medicines to combat a bacterial illness has the potential hazard of harming the gut flora [233].

Multiple antibiotics impede the proliferation of diverse harmful microorganisms. "The wide range of activities exhibited by this treatment is beneficial for combating infections, but it also raises the likelihood of the gut microbiota being affected," [234].

Dysbiosis, a term widely used to describe an imbalance in the makeup of our microbiota, may occur as a result of antibiotics treatment if particular gut bacteria are disproportionately affected [235]. Diarrhea is a frequently occurring temporary outcome, although allergic diseases like asthma or food allergies and obesity have potential persistent consequences. The knowledge that antibiotics also exhibit activity against gut bacteria has been established for a considerable duration [236]. However, the impact of antibiotics on the extensive array of microbes present in the human gut has not been thoroughly investigated systematically, mostly owing to technological obstacles [232].

Probiotics are live microorganisms that provide health benefits when administered in adequate amounts [237]. They may help restore and maintain a balanced and harmonious population of bacteria in the intestines. Choosing probiotics that have been scientifically studied and shown to have a positive effect on gastrointestinal health, especially while taking antibiotics, may be beneficial [238]. Prebiotics are substances, often dietary fibers, that function as nutrients for beneficial bacteria in the gastrointestinal system. Including foods that have a high prebiotic content in one's diet may encourage the growth and multiplication of these bacteria [239]. Introducing a time gap between taking antibiotics and prebiotics may be beneficial to avoid any potential negative impacts of antibiotics on bacteria that are responsive to prebiotics. Synbiotics are a combination of probiotics and prebiotics [240]. Following the completion of antibiotic treatment, including probiotics or other supplements that promote gut health may aid in the recovery and restoration of the gut microbiota. A diet rich in dietary fiber promotes the growth of beneficial bacteria and the production of short-chain fatty acids (SCFAs), which have anti-inflammatory properties. Incorporating fermented foods like yogurt, kefir, sauerkraut, and kimchi into one's diet might introduce beneficial bacteria [241].

Recent research suggests that the timing of antibiotic administration may influence their impact on the gut flora. Administering antibiotics at the time of day when bacteria are most actively growing, known as the diurnal peak, might help lessen the negative effects on the microbiota caused by antibiotics [242]. Various nutrients, such as glutamine, zinc, and vitamin D, have been studied to determine their capacity to enhance gut health and preserve its integrity. Integrating these vital nutrients into a comprehensive treatment plan can mitigate the detrimental effects of antibiotics on gastrointestinal well-being [243]. Chronic stress may adversely impact the digestive system's well-being and exacerbate the body's inflammatory reaction. Participating in stress-reduction techniques such as mindfulness, meditation, and regular physical activity may enhance gut health during antibiotic treatment [244]. Ensuring an adequate and optimal quantity of sleep is essential for maintaining immune function and overall well-being, including gut health. Adequate sleep may improve the body's ability to recover from the effects of antibiotics and decrease inflammation [245].

Extended monitoring may be necessary for those who have pre-existing gastrointestinal issues or other risk factors. Various natural and herbal treatments have been studied to assess their efficacy in enhancing gastrointestinal health and reducing inflammation. However, the information about the efficacy of these therapies in mitigating the negative effects of antibiotics on gut health is still evolving [246]. It is advisable to consult healthcare specialists before including these treatments in treatment regimens.

8.6 THERAPEUTIC APPROACHES

8.6.1 Probiotics and Prebiotics in Mitigating Antibiotic-Induced Dysbiosis

Antibiotics have a vital role in the treatment of bacterial infections, but they may also disturb the equilibrium of beneficial bacteria in the gastrointestinal tract, resulting in antibiotic-induced dysbiosis [247]. Probiotics and prebiotics are two approaches that have been investigated to alleviate the consequences of antibiotic-induced dysbiosis [248].

Probiotics are living bacteria that provide health advantages to the host when given in sufficient quantities. They are often known as "beneficial" or "amiable" microorganisms [249]. The rationale for using probiotics during or after antibiotic therapy is to replenish and reinstate the equilibrium of microorganisms in the gastrointestinal tract. Typical probiotic strains include species such as *Lactobacillus* and *Bifidobacterium* [85]. It is advisable to consume probiotics concurrently or after antibiotic therapy while ensuring that there is sufficient time between doses to prevent any negative impact on the effectiveness of the antibiotics. It is crucial to choose a probiotic that has been shown effective in reducing antibiotic-induced dysbiosis since different strains of probiotics might have different impacts [250]. Utilizing a mixture of many strains may provide a more thorough strategy for replenishing gut microbial diversity. Probiotics may be obtained in several formats, including as capsules, powders, and fermented meals. The selection of shape might vary according to personal tastes and the particular strains required [251].

Prebiotics are indigestible substances that stimulate the proliferation and functionality of advantageous bacteria in the gastrointestinal tract [252]. They operate as a nourishment for these bacteria, facilitating their growth and development. Prevalent prebiotics consist of inulin, fructooligosaccharides [253], and galactooligosaccharides [225] [254].

Prebiotics are present in certain foods, including garlic, onions, bananas, and whole grains. Incorporating these items into one's diet may promote the proliferation of advantageous gut flora. Additionally, there are prebiotic supplements that may be used to conveniently increase prebiotic consumption [255]. Synbiotics, which include the combination of prebiotics and probiotics, may provide synergistic advantages by supplying both the essential bacteria and their nutrient source [256]. Gradually introducing prebiotics helps mitigate the likelihood of experiencing gastrointestinal discomfort, particularly in persons who may have sensitivity to certain types of fibers [257]. It is crucial to acknowledge that while probiotics and prebiotics have the potential to alleviate antibiotic-induced dysbiosis, the outcomes might differ across individuals. Furthermore, current research is actively investigating the most suitable strains, doses, and combinations to provide the most efficient control of antibiotic-induced dysbiosis [247,248,258].

8.6.2 Fecal Microbiota Transplantation (FMT) as a Potential Intervention

Fecal microbiota transplantation (FMT) has become a recognized therapy for chronic infections caused by *Clostridium difficile* (formerly known as *Clostridioides difficile*) (CDI), with success rates ranging from 80% to 90% [259]. Fecal Microbiota Transplantation (FMT) involves transferring bacterial material from a healthy person into the gastrointestinal (GI) tract of a person who is sick [260]. The precise mechanism by which the perceived impact occurs is not fully understood. However, it seems that the restoration of a disrupted microbiota, also known as dysbiosis, which is necessary for the development of *Clostridioides difficile* infection (CDI), is responsible for this effect [261]. Dysbiosis may also contribute to the development of several other illnesses, including metabolic syndrome, hepatic encephalopathy (HE), Irritable bowel syndrome (IBS), inflammatory bowel disease (IBD) and multiple sclerosis (MS) [262]. By focusing on the imbalanced microbiota, one might potentially alter the progression of these illnesses using various methods such as dietary treatments, antibiotics, prebiotics, probiotics and fecal microbiota transplantation (FMT). Fecal microbiota transplantation (FMT) seems to be the most potent yet unregulated approach for manipulating the microbiota [263]. While FMT may seem safe and straightforward, it should be used cautiously due to the uncertain or unacknowledged long-term consequences. Fecal microbiota transplantation (FMT) is now approved exclusively for the treatment of recurrent *Clostridioides difficile* infection (rCDI) [264]. Nevertheless, several clinical studies are now examining the impacts of FMT on various illnesses, indicating that FMT might potentially emerge as a therapeutic strategy for a certain group of patients with IBD, HE, and other disorders soon [265].

8.6.3 Developing Targeted Antibiotics to Minimize Collateral Damage to the Gut Microbiota

Current research focuses on the creation of specific antibiotics that have a lesser influence on the gut microbiota, to minimize the negative effects and unintended consequences often associated with broad-spectrum antibiotics [114]. Pinpointing distinct bacterial species crucial for preserving gastrointestinal well-being might inform the creation of medications that preserve these advantageous microorganisms. Developing antibiotics that have a limited range of effectiveness, specifically targeting certain diseases while not impacting a wide variety of bacteria, may reduce disturbance to the gut microbiota [266]. One must possess an intricate comprehension of the particular bacteria responsible for the illness and their vulnerability to various medications to address this issue. Applying precision medicine techniques, such as personalized genomes and microbiome testing, may assist in customizing antibiotic therapies for specific patients [267]. This enables the use of antibiotics that selectively target the bacteria responsible for the illness while preserving helpful microorganisms. To lessen the effect on the gut flora, it is beneficial to develop combination treatments that include targeted antibiotics together with supporting medicines like probiotics or prebiotics [240]. The combination of antibiotics with chemicals that specifically promote the development of beneficial bacteria may help restore the equilibrium of microbial populations. The process of prodrug development entails the creation of antibiotics in inert forms that may be selectively transformed into active medications at the specific location of the illness. This may amplify the localized effects and decrease systemic exposure, hence minimizing the influence on the gut flora [268]. Creating antibiotics with targeted delivery systems may guarantee that the medication largely affects the infection site, minimizing its exposure to the wider systemic circulation and gut [269]. One potential avenue to consider is the investigation of nanoparticle-based drug delivery systems or formulations that may release antibiotics in a regulated way. Developing medications that selectively target crucial microbial processes or exploit distinct weaknesses of infections might effectively reduce the development of antibiotic resistance [270,271]. The long-term efficacy of targeted antibiotics may be enhanced by combining them with measures to decrease resistance processes. It is crucial to use animal models that closely resemble the human gut microbiota and undertake well-designed clinical studies to assess the effectiveness and safety of targeted antibiotics [272]. Extended-duration studies may evaluate the long-term effects on the gut microbiota. It is crucial to acknowledge that the development of targeted antibiotics is a complex undertaking, and achieving a balance between effectiveness and minimizing unintended harm requires a multidisciplinary approach. Effective development and implementation of these targeted therapies need collaboration among microbiologists, pharmacologists, doctors, and other specialists [273].

8.6.4 Personalized Medicine Approaches for Optimizing Antibiotic Use Based on Individual Microbiome Profiles

Personalized medicine strategies for optimizing antibiotic use, by considering individual microbiome profiles, include customizing antibiotic prescriptions to align with the distinct attributes of a patient's gut flora [274]. The objective of this strategy is to optimize the effectiveness of therapy while minimizing any negative impact on the beneficial microorganisms in the gut [275]. Analyzing the composition and functional capabilities of an individual's gut microbiota using sophisticated sequencing technology. Determining the precise bacterial species and their respective quantities, while comprehending the total range of microbial variety [276]. Before commencing antibiotic treatment, it is advisable to do microbiome profiling to create a foundational comprehension of the patient's gut flora. Having this fundamental knowledge may assist in choosing the appropriate antibiotics and determining the correct dosage to minimize any negative impact on the current microbial ecology. Choosing antibiotics depends on the particular bacterium responsible for the illness and its susceptibility profile [277]. By using a focused strategy, the use of wide-ranging antibiotics is minimized, hence mitigating the effects on non-pathogenic microorganisms. Combining microbiome data with knowledge of antibiotic resistance genes to forecast the probability of resistance emergence [278]. Selecting antibiotics with reduced anticipated resistance based on the individual's microbiome composition. Developing antibiotic treatment strategies that include the patient's microbiome makeup, including considerations of microbial diversity, the prevalence of crucial species, and possible susceptibilities [266]. Modifying the dose and length of treatment according to the individual's capacity to restore a balanced microbiome after antibiotic exposure [114]. Suggesting the use of probiotics or prebiotics tailored to the patient's microbiome profile to promote the re-establishment of a balanced microbial community. These supplements may be selected to particularly target deficiencies or imbalances discovered in the individual's gut flora [279]. Consistently evaluate the influence of antibiotics on the gut microbiota during and after therapy by conducting recurrent analysis of the microbiome. Modifying treatment strategies as necessary to facilitate the restoration of the microbiome and avoid persistent dysbiosis [195]. Providing patients with information on the significance of microbiome well-being and the possible consequences of antibiotics on the gut's bacterial composition. Promoting compliance with prescription antibiotics and post-treatment guidelines for the restoration of the microbiota [280]. By incorporating microbiome data with other clinical information, such as a patient's medical history, one may make more knowledgeable choices about the use of antibiotics [281]. Collaboration among healthcare practitioners, microbiologists, and bioinformaticians is necessary to implement personalized medicine strategies for optimizing antibiotics. It also requires keeping up-to-date with developments in microbiome research and using this information in clinical practice to enhance patient outcomes while minimizing the unexpected negative effects of antibiotic use [282,283].

8.7 CONCLUSION

The gut microbiota is a complicated network of metabolically interconnected bacteria, as recent study has demonstrated. In addition to assisting with digestion, a symbiotic gut microbiota also stimulates and regulates the immune system and inhibits the growth of infections, among other essential roles. Since antibiotics alter the gut flora, their rising usage is a serious cause for concern. To tackle this obstacle, additional investigation is required to enhance our comprehension of the gut microbiota and tactics to undo these disturbances, such as the utilization of probiotics. The use of antibiotics nowadays calls for cautious stewardship that minimizes side effects and emphasizes the use of antibiotic substitutes. In order to achieve this, we must create, develop, and implement novel antibiotic substitutes from the bench to the bedside, along with effective techniques for maintaining and repairing the microbial population following drug-related disruptions.

REFERENCES

1. Quigley, E.M.M. Microbiota-brain-gut axis and neurodegenerative diseases. *Current Neurology and Neuroscience Reports* **2017**, *17*, 94, doi:10.1007/s11910-017-0802-6
2. Sheykhsaran, E.; Abbasi, A.; Ebrahimzadeh Leylabadlo, H.; Sadeghi, J.; Mehri, S.; Naeimi Mazraeh, F.; Feizi, H.; Bannazadeh Baghi, H. Gut microbiota and obesity: An overview of microbiota to microbial-based therapies. *Postgraduate Medical Journal* **2023**, *99*, 384–402, doi:10.1136/postgradmedj-2021-141311
3. Ursell, L.K.; Metcalf, J.L.; Parfrey, L.W.; Knight, R. Defining the human microbiome. *Nutrition Reviews* **2012**, *70 Suppl 1*, S38–S44, doi:10.1111/j.1753-4887.2012.00493.x
4. Yao, W.; Gong, Y.; Li, L.; Hu, X.; You, L. The effects of dietary fibers from rice bran and wheat bran on gut microbiota: An overview. *Food Chemistry: X* **2022**, *13*, 100252, doi:10.1016/j.fochx.2022.100252
5. Calder, P.C.; Ortega, E.F.; Meydani, S.N.; Adkins, Y.; Stephensen, C.B.; Thompson, B.; Zwickey, H. Nutrition, immunosenescence, and infectious disease: An overview of the scientific evidence on micronutrients and on modulation of the gut microbiota. *Advances in Nutrition (Bethesda, Md.)* **2022**, *13*, S1–s26, doi:10.1093/advances/nmac052
6. LePine, T. 09B. It takes guts to be healthy: Assessing the gut microbiome to maintain optimal health. *Global Advances in Health and Medicine* **2013**, *2*, gahmj. 2013.
7. Zsálig, D.; Berta, A.; Tóth, V.; Szabó, Z.; Simon, K.; Figler, M.; Pusztafalvi, H.; Polyák, É. A review of the relationship between gut microbiome and obesity. *Applied Sciences* **2023**, *13*, 610.
8. Zhou, S.; Martin, M.; Powell, C.; Sutton, K.S.; George, B.; Olson, T.; Konstantinidis, K.T.; Bruner, D.W.; Bai, J. How to maintain a healthy gut microbiome in children with cancer? Gut microbiome association with diet in children

with solid tumors postchemotherapy. *Omics: A Journal of Integrative Biology* **2022**, *26*, 236–245, doi:10.1089/omi.2022.0002
9. Koulenti, D.; Xu, E.; Mok, I.Y.S.; Song, A.; Karageorgopoulos, D.E.; Armaganidis, A.; Lipman, J.; Tsiodras, S. Novel antibiotics for multidrug-resistant gram-positive microorganisms. *Microorganisms* **2019**, *7*, 270. doi:10.3390/microorganisms7080270
10. Escolà-Vergé, L.; Los-Arcos, I.; Almirante, B. New antibiotics for the treatment of infections by multidrug-resistant microorganisms. *Medicina Clinica* **2020**, *154*, 351–357, doi:10.1016/j.medcli.2019.11.002
11. Sweeney, D.A.; Kalil, A.C. Why don't we have more inhaled antibiotics to treat ventilator-associated pneumonia? *Clinical Microbiology and Infection: The Official Publication of the European Society of Clinical Microbiology and Infectious Diseases* **2019**, *25*, 1195–1199, doi:10.1016/j.cmi.2019.04.018
12. Elborn, J.S.; Vataire, A.L.; Fukushima, A.; Aballea, S.; Khemiri, A.; Moore, C.; Medic, G.; Hemels, M.E. Comparison of inhaled antibiotics for the treatment of chronic Pseudomonas aeruginosa lung infection in patients with cystic fibrosis: Systematic literature review and network meta-analysis. *Clinical Therapeutics* **2016**, *38*, 2204–2226, doi:10.1016/j.clinthera.2016.08.014
13. Loho, T.; Dharmayanti, A. Colistin: An antibiotic and its role in multiresistant Gram-negative infections. *Acta Medica Indonesiana* **2015**, *47*, 157–168.
14. English, B.K.; Gaur, A.H. The use and abuse of antibiotics and the development of antibiotic resistance. *Advances in Experimental Medicine and Biology* **2010**, *659*, 73–82, doi:10.1007/978-1-4419-0981-7_6
15. Lee, C.R.; Cho, I.H.; Jeong, B.C.; Lee, S.H. Strategies to minimize antibiotic resistance. *International Journal of Environmental Research and Public Health* **2013**, *10*, 4274–4305, doi:10.3390/ijerph10094274
16. Cunha, C.B.; Opal, S.M. Antibiotic stewardship: Strategies to minimize antibiotic resistance while maximizing antibiotic effectiveness. *The Medical Clinics of North America* **2018**, *102*, 831–843, doi:10.1016/j.mcna.2018.04.006
17. Mulinge, M.M.; Mwanza, S.S.; Kabahweza, H.M.; Wamalwa, D.C.; Nduati, R.W. The impact of neonatal intensive care unit antibiotics on gut bacterial microbiota of preterm infants: A systematic review. *Frontiers in Microbiomes* **2023**, *2*, 1180565.
18. Zhang, G.; Wang, Q.; Tao, W.; Jiang, W.; Elinav, E.; Wang, Y.; Zhu, S. Glucosylated nanoparticles for the oral delivery of antibiotics to the proximal small intestine protect mice from gut dysbiosis. *Nature Biomedical Engineering* **2022**, *6*, 867–881, doi:10.1038/s41551-022-00903-4
19. Conte, L.; Caldararo, C.; Lupo, R. The bidirectional interaction between our gut flora and drugs. *JDREAM. Journal of InterDisciplinary REsearch Applied to Medicine* **2021**, *5*, 7–10.
20. McDonnell, L.; Gilkes, A.; Ashworth, M.; Rowland, V.; Harries, T.H.; Armstrong, D.; White, P. Association between antibiotics and gut microbiome dysbiosis in children: Systematic review and meta-analysis. *Gut Microbes* **2021**, *13*, 1–18, doi:10.1080/19490976.2020.1870402
21. Vandeputte, D.; Joossens, M. Effects of low and high FODMAP diets on human gastrointestinal microbiota composition in adults with intestinal diseases: A systematic review. *Microorganisms* **2020**, *8*, 1638. doi:10.3390/microorganisms8111638
22. Moran-Ramos, S.; Lopez-Contreras, B.E.; Villarruel-Vazquez, R.; Ocampo-Medina, E.; Macias-Kauffer, L.; Martinez-Medina, J.N.; Villamil-Ramirez, H.; León-Mimila, P.; Del Rio-Navarro, B.E.; Ibarra-Gonzalez, I.; et al. Environmental and intrinsic factors shaping gut microbiota composition and diversity and its relation to metabolic health in children and early adolescents: A population-based study. *Gut Microbes* **2020**, *11*, 900–917, doi:10.1080/19490976.2020.1712985
23. Wipperman, M.F.; Bhattarai, S.K.; Vorkas, C.K.; Maringati, V.S.; Taur, Y.; Mathurin, L.; McAulay, K.; Vilbrun, S.C.; Francois, D.; Bean, J.; et al. Gastrointestinal microbiota composition predicts peripheral inflammatory state during treatment of human tuberculosis. *Nature Communications* **2021**, *12*, 1141, doi:10.1038/s41467-021-21475-y
24. von Martels, J.Z.H.; Sadaghian Sadabad, M.; Bourgonje, A.R.; Blokzijl, T.; Dijkstra, G.; Faber, K.N.; Harmsen, H.J.M. The role of gut microbiota in health and disease: In vitro modeling of host-microbe interactions at the aerobe-anaerobe interphase of the human gut. *Anaerobe* **2017**, *44*, 3–12, doi:10.1016/j.anaerobe.2017.01.001
25. Lepage, P. The human gut microbiota: Interactions with the host and dysfunctions. *Revue des maladies respiratoires* **2017**, *34*, 1085–1090, doi:10.1016/j.rmr.2016.11.003
26. Li, J.; Li, D.; Chen, Y.; Chen, W.; Xu, J.; Gao, L. Gut microbiota and aging: Traditional Chinese medicine and modern medicine. *Clinical Interventions in Aging* **2023**, *18*, 963–986, doi:10.2147/cia.S414714
27. Pan, B.; Han, X.; Yu, K.; Sun, H.; Mu, R.; Lian, C.A. Geographical distance, host evolutionary history and diet drive gut microbiome diversity of fish across the Yellow River. *Molecular Ecology* **2023**, *32*, 1183–1196, doi:10.1111/mec.16812
28. Ge, Y.; Jing, Z.; Diao, Q.; He, J.Z.; Liu, Y.J. Host species and geography differentiate honeybee gut bacterial communities by changing the relative contribution of community assembly processes. *mBio* **2021**, *12*, e0075121, doi:10.1128/mBio.00751-21
29. Lilli, G.; Sirot, C.; Campbell, H.; Brophy, D.; Graham, C.; George, I. Geographic origin and host's phylogeny are predictors of the gut mucosal microbiota diversity and composition in Mediterranean scorpionfishes (Scorpaena sp.). *Frontiers in Marine Science* **2023**, 10, 1286706.
30. Youngblut, N.D.; Reischer, G.H.; Walters, W.; Schuster, N.; Walzer, C.; Stalder, G.; Ley, R.E.; Farnleitner, A.H. Host diet and evolutionary history explain different aspects of gut microbiome diversity among vertebrate clades. *Nature Communications* **2019**, *10*, 2200, doi:10.1038/s41467-019-10191-3
31. Shankregowda, A.M.; Siriyappagouder, P.; Kuizenga, M.; Bal, T.M.P.; Abdelhafiz, Y.; Eizaguirre, C.; Fernandes, J.M.O.; Kiron, V.; Raeymaekers, J.A.M. Host habitat rather than evolutionary history explains gut microbiome diversity in sympatric stickleback species. *Frontiers in Microbiology* **2023**, *14*, 1232358, doi:10.3389/fmicb.2023.1232358
32. Zhang, Y.; Chen, H.; Lu, M.; Cai, J.; Lu, B.; Luo, C.; Dai, M. Habitual diet pattern associations with gut microbiome diversity and composition: Results from a Chinese adult cohort. *Nutrients* **2022**, *14*, doi:10.3390/nu14132639

33. Miao, Z.; Du, W.; Xiao, C.; Su, C.; Gou, W.; Shen, L.; Zhang, J.; Fu, Y.; Jiang, Z.; Wang, Z.; et al. Gut microbiota signatures of long-term and short-term plant-based dietary pattern and cardiometabolic health: A prospective cohort study. *BMC Medicine* **2022**, *20*, 204, doi:10.1186/s12916-022-02402-4
34. Chen, Z.; Radjabzadeh, D.; Ikram, A.; Uitterlinden, A.; Kraaij, R.; Voortman, T. Dietary patterns and gut microbiome composition in a large population-based cohort. *Current Developments in Nutrition* **2020**, *4*, 514–514.
35. Yu, D.; Nguyen, S.M.; Yang, Y.; Xu, W.; Cai, H.; Wu, J.; Cai, Q.; Long, J.; Zheng, W.; Shu, X.O. Long-term diet quality is associated with gut microbiome diversity and composition among urban Chinese adults. *The American Journal of Clinical Nutrition* **2021**, *113*, 684–694, doi:10.1093/ajcn/nqaa350
36. Hussain, M.S.; Sharma, A.; Kumar, R. Prebiotics and probiotics: A focused review of applications in respiratory disorders. *Carpathian Journal of Food Science and Technology* **2023**, 15, 183–207.
37. Zuniga-Chaves, I.; Eggers, S.; Kates, A.E.; Safdar, N.; Suen, G.; Malecki, K.M.C. Neighborhood socioeconomic status is associated with low diversity gut microbiomes and multi-drug resistant microorganism colonization. *NPJ Biofilms and Microbiomes* **2023**, *9*, 61, doi:10.1038/s41522-023-00430-3
38. Lewis, C.R.; Bonham, K.S.; McCann, S.H.; Volpe, A.R.; D'Sa, V.; Naymik, M.; De Both, M.D.; Huentelman, M.J.; Lemery-Chalfant, K.; Highlander, S.K.; et al. Family SES is associated with the gut microbiome in infants and children. *Microorganisms* **2021**, *9*, 1608.doi:10.3390/microorganisms9081608
39. Lephart, E.D.; Naftolin, F. Estrogen action and gut microbiome metabolism in dermal health. *Dermatology and Therapy* **2022**, *12*, 1535–1550, doi:10.1007/s13555-022-00759-1
40. Gschwendtner, S.; Kang, H.; Thiering, E.; Kublik, S.; Fösel, B.; Schulz, H.; Krauss-Etschmann, S.; Heinrich, J.; Schöler, A.; Schloter, M.; et al. Early life determinants induce sustainable changes in the gut microbiome of six-year-old children. *Scientific Reports* **2019**, *9*, 12675, doi:10.1038/s41598-019-49160-7
41. Sarkar, A.; Yoo, J.Y.; Valeria Ozorio Dutra, S.; Morgan, K.H.; Groer, M. The association between early-life gut microbiota and long-term health and diseases. *Journal of Clinical Medicine* **2021**, *10*, doi:10.3390/jcm10030459
42. Webster, S.E.; Vos, D.; Rothstein, T.L.; Holodick, N.E. Modulation of microbiome diversity and cytokine expression is influenced in a sex-dependent manner during aging. *Frontiers in Microbiomes* **2022**, *1*, doi:10.3389/frmbi.2022.994464
43. Daniel, H. Diet and gut microbiome and the "Chicken or egg" problem. *Frontiers in Nutrition* **2021**, *8*, 828630, doi:10.3389/fnut.2021.828630
44. Zhao, G.; Ma, T.; Tang, W.; Li, D.; Mishra, S.K.; Xu, Z.; Wang, Q.; Jie, H. Gut microbiome of Chinese forest musk deer examined across gender and age. *BioMed Research International* **2019**, *2019*, 9291216, doi:10.1155/2019/9291216
45. Kim, M.; Huda, M.N.; Que, E.; Gertz, E.; Bennett, B. Sexual dimorphism of atherosclerosis by gut microbiome in a hyperlipidemic diversity outbred F1 mouse population. *Current Developments in Nutrition* **2020**, *4*, 1569–1569.
46. Su, Q.; Tun, H.M.; Liu, Q.; Yeoh, Y.K.; Mak, J.W.Y.; Chan, F.K.; Ng, S.C. Gut microbiome signatures reflect different subtypes of irritable bowel syndrome. *Gut Microbes* **2023**, *15*, 2157697, doi:10.1080/19490976.2022.2157697
47. Zhang, H.S.; Li, C.; Li, Y.T.; Jin, Y.P.; Liu, W.; Chen, H.T. Research progress on the regulation of mammalian energy metabolism by the circadian clock system and gut microbiota. *Sheng li xue bao: [Acta physiologica Sinica]* **2022**, *74*, 443–460.
48. Sundaram, S.; Borthakur, A. Altered intestinal epithelial nutrient transport: An underappreciated factor in obesity modulated by diet and microbiota. *The Biochemical Journal* **2021**, *478*, 975–995, doi:10.1042/bcj20200902
49. Damms-Machado, A.; Mitra, S.; Schollenberger, A.E.; Kramer, K.M.; Meile, T.; Königsrainer, A.; Huson, D.H.; Bischoff, S.C. Effects of surgical and dietary weight loss therapy for obesity on gut microbiota composition and nutrient absorption. *BioMed Research International* **2015**, *2015*, 806248, doi:10.1155/2015/806248
50. Dai, W.F.; Zhang, J.J.; Qiu, Q.F.; Chen, J.; Yang, W.; Ni, S.; Xiong, J.B. Starvation stress affects the interplay among shrimp gut microbiota, digestion and immune activities. *Fish & Shellfish Immunology* **2018**, *80*, 191–199, doi:10.1016/j.fsi.2018.05.040
51. Yoo, J.Y.; Groer, M.; Dutra, S.V.O.; Sarkar, A.; McSkimming, D.I. Gut microbiota and immune system interactions. *Microorganisms* **2020**, *8*, 1587. doi:10.3390/microorganisms8101587
52. Saini, A.; Dalal, P.; Sharma, D. Deciphering the interdependent labyrinth between gut microbiota and the immune system. *Letters in Applied Microbiology* **2022**, *75*, 1122–1135, doi:10.1111/lam.13775
53. Lu, Y.; Yuan, X.; Wang, M.; He, Z.; Li, H.; Wang, J.; Li, Q. Gut microbiota influence immunotherapy responses: Mechanisms and therapeutic strategies. *Journal of Hematology & Oncology* **2022**, *15*, 47, doi:10.1186/s13045-022-01273-9
54. Zango, U.U.; Ibrahim, M.; Shawai, S.A.A.; Shamsuddin, I.M. A review on β-lactam antibiotic drug resistance. *MOJ Drug Design Development Therapy* **2019**, *3*, 52–58.
55. Rose, J.K.; Catalá, C.; Gonzalez-Carranza, Z.H.; Roberts, J.A. Cell wall disassembly. *Annual Plant Reviews Online* **2018**, 264–324.
56. Ghuysen, J.-M. Biosynthesis and assembly of bacterial cell walls. In The sysnthesis, assembly and turnover of cell surface components; North-Holland Publishing Company: **1977**.
57. Kohanski, M.A.; Dwyer, D.J.; Collins, J.J. How antibiotics kill bacteria: From targets to networks. *Nature Reviews Microbiology* **2010**, *8*, 423–435.
58. McCoy, L.S.; Xie, Y.; Tor, Y. Antibiotics that target protein synthesis. *Wiley Interdisciplinary Reviews: RNA* **2011**, *2*, 209–232.
59. Lengyel, P.; Söll, D. Mechanism of protein biosynthesis. *Bacteriological Reviews* **1969**, *33*, 264–301.
60. Ayoub Moubareck, C. Polymyxins and bacterial membranes: A review of antibacterial activity and mechanisms of resistance. *Membranes* **2020**, *10*, 181.
61. Yılmaz, Ç.; Özcengiz, G. Antibiotics: Pharmacokinetics, toxicity, resistance and multidrug efflux pumps. *Biochemical Pharmacology* **2017**, *133*, 43–62.

62. Gilbert, P.; Allison, D.; Lambert, P. Antibiotics that act on nucleic acids and protein biosynthesis. In *Molecular medical microbiology*; Elsevier: 2002; pp. 599–608.
63. Swarbrick, J.; Iliades, P.; Simpson, J.S.; Macreadie, I. Folate biosynthesis–reappraisal of old and novel targets in the search for new antimicrobials. *The Open Enzyme Inhibition Journal* **2008**, *1*, 12–33.
64. Adams, R.A.; Leon, G.; Miller, N.M.; Reyes, S.P.; Thantrong, C.H.; Thokkadam, A.M.; Lemma, A.S.; Sivaloganathan, D.M.; Wan, X.; Brynildsen, M.P. Rifamycin antibiotics and the mechanisms of their failure. *The Journal of Antibiotics* **2021**, *74*, 786–798.
65. Franceschi, C.; Garagnani, P.; Morsiani, C.; Conte, M.; Santoro, A.; Grignolio, A.; Monti, D.; Capri, M.; Salvioli, S. The continuum of aging and age-related diseases: Common mechanisms but different rates. *Frontiers in Medicine* **2018**, *5*, 61.
66. Higgs, P.; Gilleard, C. Frailty, abjection and the 'othering' of the fourth age. *Health Sociology Review* **2014**, *23*, 10–19.
67. Larsson, L.; Degens, H.; Li, M.; Salviati, L.; Lee, Y.I.; Thompson, W.; Kirkland, J.L.; Sandri, M. Sarcopenia: Aging-related loss of muscle mass and function. *Physiological Reviews* **2019**, *99*, 427–511.
68. Ali, A.M.; Kunugi, H. Physical frailty/sarcopenia as a key predisposing factor to coronavirus disease 2019 (COVID-19) and its complications in older adults. *BioMed* **2021**, *1*, 11–40.
69. Ticinesi, A.; Nouvenne, A.; Cerundolo, N.; Catania, P.; Prati, B.; Tana, C.; Meschi, T. Gut microbiota, muscle mass and function in aging: A focus on physical frailty and sarcopenia. *Nutrients* **2019**, *11*, 1633.
70. Buigues, C.; Fernández-Garrido, J.; Pruimboom, L.; Hoogland, A.J.; Navarro-Martínez, R.; Martínez-Martínez, M.; Verdejo, Y.; Mascarós, M.C.; Peris, C.; Cauli, O. Effect of a prebiotic formulation on frailty syndrome: A randomized, double-blind clinical trial. *International Journal of Molecular Sciences* **2016**, *17*, 932.
71. D'Amico, F.; Barone, M.; Brigidi, P.; Turroni, S. Gut microbiota in relation to frailty and clinical outcomes. *Current Opinion in Clinical Nutrition and Metabolic Care* **2023**, *26*, 219–225.
72. Di Vincenzo, F.; Del Gaudio, A.; Petito, V.; Lopetuso, L.R.; Scaldaferri, F. Gut microbiota, intestinal permeability, and systemic inflammation: A narrative review. *Internal and Emergency Medicine* **2023**, *19*, 275–293.
73. Bäckhed, F.; Ding, H.; Wang, T.; Hooper, L.V.; Koh, G.Y.; Nagy, A.; Semenkovich, C.F.; Gordon, J.I. The gut microbiota as an environmental factor that regulates fat storage. *Proceedings of the National Academy of Sciences* **2004**, *101*, 15718–15723.
74. Lahiri, S.; Kim, H.; Garcia-Perez, I.; Reza, M.M.; Martin, K.A.; Kundu, P.; Cox, L.M.; Selkrig, J.; Posma, J.M.; Zhang, H. The gut microbiota influences skeletal muscle mass and function in mice. *Science Translational Medicine* **2019**, *11*, eaan5662.
75. Arip, M.; Yap, V.L.; Rajagopal, M.; Selvaraja, M.; Dharmendra, K.; Chinnapan, S.J.F.I.P. Evidence-based management of uterine fibroids with botanical drugs-A review. **2022**, *13*, 878407.
76. Siddharth, J.; Chakrabarti, A.; Pannerec, A.; Karaz, S.; Morin-Rivron, D.; Masoodi, M.; Feige, J.N.; Parkinson, S.J. Aging and sarcopenia associate with specific interactions between gut microbes, serum biomarkers and host physiology in rats. *Aging (Albany NY)* **2017**, 9 , 1698.
77. Strasser, B.; Wolters, M.; Weyh, C.; Krüger, K.; Ticinesi, A. The effects of lifestyle and diet on gut microbiota composition, inflammation and muscle performance in our aging society. *Nutrients* **2021**, *13*, 2045.
78. Boopathi, S.; Priya, P.S.; Haridevamuthu, B.; Nayak, S.R.R.; Chandrasekar, M.; Arockiaraj, J.; Jia, A.-Q. Expanding germ-organ theory: Understanding non-communicable diseases through enterobacterial translocation. *Pharmacological Research* **2023**, *194*, 106856.
79. Mamic, P.; Chaikijurajai, T.; Tang, W.W. Gut microbiome-A potential mediator of pathogenesis in heart failure and its comorbidities: State-of-the-art review. *Journal of Molecular and Cellular Cardiology* **2021**, *152*, 105–117.
80. Cocolin, L.; Alessandria, V.; Dolci, P.; Gorra, R.; Rantsiou, K. Culture independent methods to assess the diversity and dynamics of microbiota during food fermentation. *International Journal of Food Microbiology* **2013**, *167*, 29–43.
81. Hill, D.R.; Spence, J.R. Gastrointestinal organoids: Understanding the molecular basis of the host–microbe interface. *Cellular and Molecular Gastroenterology and Hepatology* **2017**, *3*, 138–149.
82. Ticinesi, A.; Lauretani, F.; Milani, C.; Nouvenne, A.; Tana, C.; Del Rio, D.; Maggio, M.; Ventura, M.; Meschi, T. Aging gut microbiota at the cross-road between nutrition, physical frailty, and sarcopenia: Is there a gut–muscle axis? *Nutrients* **2017**, *9*, 1303.
83. Peng, J.; Gong, H.; Lyu, X.; Liu, Y.; Li, S.; Tan, S.; Dong, L.; Zhang, X. Characteristics of the fecal microbiome and metabolome in older patients with heart failure and sarcopenia. *Frontiers in Cellular and Infection Microbiology* **2023**, *13*, 1127041.
84. Liu, C.; Cheung, W.H.; Li, J.; Chow, S.K.H.; Yu, J.; Wong, S.H.; Ip, M.; Sung, J.J.Y.; Wong, R.M.Y. Understanding the gut microbiota and sarcopenia: A systematic review. *Journal of Cachexia, Sarcopenia and Muscle* **2021**, *12*, 1393–1407.
85. Quigley, E.M.; Quera, R. Small intestinal bacterial overgrowth: Roles of antibiotics, prebiotics, and probiotics. *Gastroenterology* **2006**, *130*, S78–S90.
86. Levy, S.B. Antibiotic resistance: An ecological imbalance. In Derek J. Chadwick, Jamie Goode (eds.) *Proceedings of the Ciba Foundation Symposium 207-Antibiotic Resistance: Origins, Evolution, Selection and Spread: Antibiotic Resistance: Origins, Evolution, Selection and Spread: Ciba Foundation Symposium 207*, 2007; pp. 1–14.
87. Majumder, M.A.A.; Rahman, S.; Cohall, D.; Bharatha, A.; Singh, K.; Haque, M.; Gittens-St Hilaire, M. Antimicrobial stewardship: Fighting antimicrobial resistance and protecting global public health. *Infection and Drug Resistance* **2020**, *13*, 4713–4738.
88. Calo-Mata, P.; Ageitos, J.M.; Böhme, K.; Barros-Velázquez, J. Intestinal microbiota: First barrier against gut-affecting pathogens. In Tomas G. Villa, Miguel Vinas (eds.) *New Weapons to Control Bacterial Growth* **2016**, 281–314.
89. Mukherjee, S.; Joardar, N.; Sengupta, S.; Babu, S.P.S. Gut microbes as future therapeutics in treating inflammatory

and infectious diseases: Lessons from recent findings. *The Journal of Nutritional Biochemistry* **2018**, *61*, 111–128.
90. Leekha, S.; Terrell, C.L.; Edson, R.S. General principles of antimicrobial therapy. *Mayo Clinic Proceedings* **2011**, 86(2), 156–167.
91. Nepal, D.; Balasubramanian, S.; Simonian, A.L.; Davis, V.A. Strong antimicrobial coatings: single-walled carbon nanotubes armored with biopolymers. *Nano Letters* **2008**, 8(7), 1896–901..
92. Lewis, J.S.; Bush, K. Antibacterial agents. In James H. Jorgensen, Karen C. Carroll, Guido Funke, Michael A. Pfaller, Marie Louise Landry, Sandra S. Richter, David W. Warnock (eds.) *Manual of Clinical Microbiology* **2015**, 11th edition; 1169–1211.
93. Calderón, C.B.; Sabundayo, B.P. Antimicrobial classifications. *Antimicrobial Susceptibility Testing Protocols* **2007**, *7*, 60–88.
94. Raynor, B.D. Penicillin and ampicillin. *Primary Care Update for OB/GYNS* **1997**, *4*, 147–152.
95. Chopra, I.; Roberts, M. Tetracycline antibiotics: Mode of action, applications, molecular biology, and epidemiology of bacterial resistance. *Microbiology and Molecular Biology Reviews* **2001**, *65*, 232–260.
96. Ceruelos, A.H.; Romero-Quezada, L.C.; Ledezma, J.R.; Contreras, L.L. Therapeutic uses of metronidazole and its side effects: an update. *European Review for Medical and Pharmacological Sciences* **2019**, 23(1), 397–401..
97. Poulikakos, P.; Falagas, M.E. Aminoglycoside therapy in infectious diseases. *Expert Opinion on Pharmacotherapy* **2013**, *14*, 1585–1597.
98. Cottreau, J.; Baker, S.F.; DuPont, H.L.; Garey, K.W. Rifaximin: A nonsystemic rifamycin antibiotic for gastrointestinal infections. *Expert Review of Anti-Infective Therapy* **2010**, *8*, 747–760.
99. Richter, M.F.; Hergenrother, P.J. The challenge of converting Gram-positive-only compounds into broad-spectrum antibiotics. *Annals of the New York Academy of Sciences* **2019**, *1435*, 18–38.
100. Melander, R.J.; Zurawski, D.V.; Melander, C. Narrow-spectrum antibacterial agents. *Medchemcomm* **2018**, *9*, 12–21.
101. Álvarez-Lerma, F.; Alvarez, B.; Luque, P.; Ruiz, F.; Dominguez-Roldan, J.-M.; Quintana, E.; Sanz-Rodriguez, C.; Group, A.S. Empiric broad-spectrum antibiotic therapy of nosocomial pneumonia in the intensive care unit: A prospective observational study. *Critical Care* **2006**, *10*, 1–11.
102. Cantón, R.; Horcajada, J.P.; Oliver, A.; Garbajosa, P.R.; Vila, J. Inappropriate use of antibiotics in hospitals: The complex relationship between antibiotic use and antimicrobial resistance. *Enfermedades infecciosas y microbiologia clinica* **2013**, *31*, 3–11.
103. Flandroy, L.; Poutahidis, T.; Berg, G.; Clarke, G.; Dao, M.-C.; Decaestecker, E.; Furman, E.; Haahtela, T.; Massart, S.; Plovier, H. The impact of human activities and lifestyles on the interlinked microbiota and health of humans and of ecosystems. *Science of the Total Environment* **2018**, *627*, 1018–1038.
104. Cook, M.A.; Wright, G.D. The past, present, and future of antibiotics. *Science Translational Medicine* **2022**, *14*, eabo7793.
105. Bessa, L.J.; Fazii, P.; Di Giulio, M.; Cellini, L. Bacterial isolates from infected wounds and their antibiotic susceptibility pattern: Some remarks about wound infection. *International Wound Journal* **2015**, *12*, 47–52.
106. Ruppé, E.; Burdet, C.; Grall, N.; De Lastours, V.; Lescure, F.-X.; Andremont, A.; Armand-Lefèvre, L. Impact of antibiotics on the intestinal microbiota needs to be re-defined to optimize antibiotic usage. *Clinical Microbiology and Infection* **2018**, *24*, 3–5.
107. Niederman, M.S. Principles of appropriate antibiotic use. *International Journal of Antimicrobial Agents* **2005**, *26*, S170–S175.
108. Van Zyl, K.N.; Matukane, S.R.; Hamman, B.L.; Whitelaw, A.C.; Newton-Foot, M. Effect of antibiotics on the human microbiome: A systematic review. *International Journal of Antimicrobial Agents* **2022**, *59*, 106502.
109. Etebu, E.; Arikekpar, I. Antibiotics: Classification and mechanisms of action with emphasis on molecular perspectives. *International Journal of Applied Microbiology and Biotechnology Research* **2016**, *4*, 90–101.
110. Sullivan, Å.; Edlund, C.; Nord, C.E. Effect of antimicrobial agents on the ecological balance of human microflora. *The Lancet Infectious Diseases* **2001**, *1*, 101–114.
111. Sommer, F.; Anderson, J.M.; Bharti, R.; Raes, J.; Rosenstiel, P. The resilience of the intestinal microbiota influences health and disease. *Nature Reviews Microbiology* **2017**, *15*, 630–638.
112. Ianiro, G.; Tilg, H.; Gasbarrini, A. Antibiotics as deep modulators of gut microbiota: Between good and evil. *Gut* **2016**, *65*(11):1906-1915.
113. Borriello, S.P. *Antibiotic associated diarrhoea and colitis: The role of Clostridium difficile in gastrointestinal disorders*; Springer Science & Business Media: 1984; Vol. 5.
114. Ribeiro, C.F.A.; Silveira, G.G.d.O.S.; Candido, E.d.S.; Cardoso, M.H.; Espinola Carvalho, C.M.; Franco, O.L. Effects of antibiotic treatment on gut microbiota and how to overcome its negative impacts on human health. *ACS Infectious Diseases* **2020**, *6*, 2544–2559.
115. Bejaoui, S.; Poulsen, M. The impact of early life antibiotic use on atopic and metabolic disorders: Meta-analyses of recent insights. *Evolution, Medicine, and Public Health* **2020**, *2020*, 279–289.
116. Duan, H.; Yu, L.; Tian, F.; Zhai, Q.; Fan, L.; Chen, W. Antibiotic-induced gut dysbiosis and barrier disruption and the potential protective strategies. *Critical Reviews in Food Science and Nutrition* **2022**, *62*, 1427–1452.
117. Gyssens, I.C. Antibiotic policy. *International Journal of Antimicrobial Agents* **2011**, *38*, 11–20.
118. Cammarota, G.; Ianiro, G.; Bibbò, S.; Gasbarrini, A. Gut microbiota modulation: Probiotics, antibiotics or fecal microbiota transplantation? *Internal and Emergency Medicine* **2014**, *9*, 365–373.
119. Toor, D.; Wasson, M.K.; Kumar, P.; Karthikeyan, G.; Kaushik, N.K.; Goel, C.; Singh, S.; Kumar, A.; Prakash, H. Dysbiosis disrupts gut immune homeostasis and promotes gastric diseases. *International Journal of Molecular Sciences* **2019**, *20*, 2432.
120. Tamboli, C.P.; Neut, C.; Desreumaux, P.; Colombel, J.F. Dysbiosis in inflammatory bowel disease. *Gut* **2004**, *53*, 1.
121. Frank, D.N.; St. Amand, A.L.; Feldman, R.A.; Boedeker, E.C.; Harpaz, N.; Pace, N.R. Molecular-phylogenetic characterization of microbial community imbalances in human inflammatory bowel diseases. *Proceedings of the National Academy of Sciences* **2007**, *104*, 13780–13785.

122. Wang, W.-L.; Xu, S.-Y.; Ren, Z.-G.; Tao, L.; Jiang, J.-W.; Zheng, S.-S. Application of metagenomics in the human gut microbiome. *World Journal of Gastroenterology: WJG* **2015**, *21*, 803.
123. Morgan, X.C.; Tickle, T.L.; Sokol, H.; Gevers, D.; Devaney, K.L.; Ward, D.V.; Reyes, J.A.; Shah, S.A.; LeLeiko, N.; Snapper, S.B. Dysfunction of the intestinal microbiome in inflammatory bowel disease and treatment. *Genome Biology* **2012**, *13*, 1–18.
124. Miyoshi, J.; Chang, E.B. The gut microbiota and inflammatory bowel diseases. *Translational Research* **2017**, *179*, 38–48.
125. Ott, S.J.; Kühbacher, T.; Musfeldt, M.; Rosenstiel, P.; Hellmig, S.; Rehman, A.; Drews, O.; Weichert, W.; Timmis, K.N.; Schreiber, S. Fungi and inflammatory bowel diseases: Alterations of composition and diversity. *Scandinavian Journal of Gastroenterology* **2008**, *43*, 831–841.
126. Louca, S.; Jacques, S.M.; Pires, A.P.; Leal, J.S.; Srivastava, D.S.; Parfrey, L.W.; Farjalla, V.F.; Doebeli, M. High taxonomic variability despite stable functional structure across microbial communities. *Nature Ecology & Evolution* **2016**, *1*, 0015.
127. Singh, V.; Proctor, S.D.; Willing, B.P. Koch's postulates, microbial dysbiosis and inflammatory bowel disease. *Clinical Microbiology and Infection* **2016**, *22*, 594–599.
128. Dalal, S.R.; Chang, E.B. The microbial basis of inflammatory bowel diseases. *The Journal of Clinical Investigation* **2014**, *124*, 4190–4196.
129. Popkes, M.; Valenzano, D.R. Microbiota–host interactions shape ageing dynamics. *Philosophical Transactions of the Royal Society B* **2020**, *375*, 20190596.
130. Altajar, S.; Moss, A. Inflammatory bowel disease environmental risk factors: Diet and gut microbiota. *Current Gastroenterology Reports* **2020**, *22*, 1–7.
131. Vernocchi, P.; Del Chierico, F.; Putignani, L. Gut microbiota metabolism and interaction with food components. *International Journal of Molecular Sciences* **2020**, *21*, 3688.
132. Garcia, K.; Ferreira, G.; Reis, F.; Viana, S. Impact of dietary sugars on gut microbiota and metabolic health. *Diabetology* **2022**, *3*, 549–560.
133. Illiano, P.; Brambilla, R.; Parolini, C. The mutual interplay of gut microbiota, diet and human disease. *The FEBS Journal* **2020**, *287*, 833–855.
134. Naimi, S.; Viennois, E.; Gewirtz, A.T.; Chassaing, B. Direct impact of commonly used dietary emulsifiers on human gut microbiota. *Microbiome* **2021**, *9*, 1–19.
135. Cao, Y.; Liu, H.; Qin, N.; Ren, X.; Zhu, B.; Xia, X. Impact of food additives on the composition and function of gut microbiota: A review. *Trends in Food Science & Technology* **2020**, *99*, 295–310.
136. Laudisi, F.; Stolfi, C.; Monteleone, G. Impact of food additives on gut homeostasis. *Nutrients* **2019**, *11*, 2334.
137. Samreen, H.; Dhaneshwar, S. Artificial sweeteners: Perceptions and realities. *Current Diabetes Reviews* **2023**, *19*, 131–144.
138. Colella, M.; Charitos, I.A.; Ballini, A.; Cafiero, C.; Topi, S.; Palmirotta, R.; Santacroce, L. Microbiota revolution: How gut microbes regulate our lives. *World Journal of Gastroenterology* **2023**, *29*, 4368.
139. Carocho, M.; Morales, P.; Ferreira, I.C. Sweeteners as food additives in the XXI century: A review of what is known, and what is to come. *Food and Chemical Toxicology* **2017**, *107*, 302–317.
140. Srivastava, N.; Ibrahim, S.A.; Chattopadhyay, J.; Arbab, M.H. *The gut microbiota in health and disease*; John Wiley & Sons: 2023.
141. Chairatana, P.; Nolan, E.M. Defensins, lectins, mucins, and secretory immunoglobulin A: Microbe-binding biomolecules that contribute to mucosal immunity in the human gut. *Critical Reviews in Biochemistry and Molecular Biology* **2017**, *52*, 45–56.
142. Gomaa, E.Z. Human gut microbiota/microbiome in health and diseases: A review. *Antonie Van Leeuwenhoek* 2020, 113, 2019–2040.
143. Zhang, D.; Liu, Z.; Bai, F. Roles of Gut Microbiota in Alcoholic Liver Disease. *International Journal of General Medicine* **2023**, 3735–3746.
144. Won, S.-M.; Park, E.; Jeong, J.-J.; Ganesan, R.; Gupta, H.; Gebru, Y.A.; Sharma, S.; Kim, D.-J.; Suk, K.-T. The gut microbiota-derived immune response in chronic liver disease. *International Journal of Molecular Sciences* **2021**, *22*, 8309.
145. Arifuzzaman, M.; Collins, N.; Guo, C.-J.; Artis, D. Nutritional regulation of microbiota-derived metabolites: Implications for immunity and inflammation. *Immunity* **2024**, *57*, 14–27.
146. Agus, A.; Clément, K.; Sokol, H. Gut microbiota-derived metabolites as central regulators in metabolic disorders. *Gut* **2020**, *70*(6), 1174–1182.
147. Greenwood, C.; Morrow, A.L.; Lagomarcino, A.J.; Altaye, M.; Taft, D.H.; Yu, Z.; Newburg, D.S.; Ward, D.V.; Schibler, K.R. Early empiric antibiotic use in preterm infants is associated with lower bacterial diversity and higher relative abundance of Enterobacter. *The Journal of Pediatrics* **2014**, *165*, 23–29.
148. Ferrer, M.; Méndez-García, C.; Rojo, D.; Barbas, C.; Moya, A. Antibiotic use and microbiome function. *Biochemical Pharmacology* **2017**, *134*, 114–126.
149. Palleja, A.; Mikkelsen, K.H.; Forslund, S.K.; Kashani, A.; Allin, K.H.; Nielsen, T.; Hansen, T.H.; Liang, S.; Feng, Q.; Zhang, C. Recovery of gut microbiota of healthy adults following antibiotic exposure. *Nature Microbiology* **2018**, *3*, 1255–1265.
150. Sun, L.; Zhang, X.; Zhang, Y.; Zheng, K.; Xiang, Q.; Chen, N.; Chen, Z.; Zhang, N.; Zhu, J.; He, Q. Antibiotic-induced disruption of gut microbiota alters local metabolomes and immune responses. *Frontiers in Cellular and Infection Microbiology* **2019**, *9*, 99.
151. Zhang, N.; Liu, J.; Chen, Z.; Chen, N.; Gu, F.; He, Q. Integrated analysis of the alterations in gut microbiota and metabolites of mice induced after long-term intervention with different antibiotics. *Frontiers in Microbiology* **2022**, *13*, 832915.
152. Pickard, J.M.; Zeng, M.Y.; Caruso, R.; Núñez, G. Gut microbiota: Role in pathogen colonization, immune responses, and inflammatory disease. *Immunological Reviews* **2017**, *279*, 70–89.
153. Bhalodi, A.A.; van Engelen, T.S.; Virk, H.S.; Wiersinga, W.J. Impact of antimicrobial therapy on the gut microbiome. *Journal of Antimicrobial Chemotherapy* **2019**, *74*, i6–i15.

154. Magill, S.S.; Edwards, J.R.; Beldavs, Z.G.; Dumyati, G.; Janelle, S.J.; Kainer, M.A.; Lynfield, R.; Nadle, J.; Neuhauser, M.M.; Ray, S.M. Prevalence of antimicrobial use in US acute care hospitals, May-September 2011. *JAMA* **2014**, *312*, 1438–1446.
155. Michelle Ng, K.; Aranda-Díaz, A.; Tropini, C.; Frankel, M.; Van Treuren, W.; O'Loughlin, C. Recovery of the gut microbiota after antibiotics depends on host diet, community context, and environmental reservoirs. *Cell Host Microbe* **2019**, *26*, 650–665.
156. Gamage, H.K.; Venturini, C.; Tetu, S.G.; Kabir, M.; Nayyar, V.; Ginn, A.N.; Roychoudhry, B.; Thomas, L.; Brown, M.; Holmes, A. Third generation cephalosporins and piperacillin/tazobactam have distinct impacts on the microbiota of critically ill patients. *Scientific Reports* **2021**, *11*, 7252.
157. Jump, R.L.; Kraft, D.; Hurless, K.; Polinkovsky, A.; Donskey, C.J. Impact of tigecycline versus other antibiotics on the fecal metabolome and on colonization resistance to Clostridium difficile in mice. *Pathogens & Immunity* **2017**, *2*, 1.
158. Bist, P.; Choudhary, S. Impact of heavy metal toxicity on the gut microbiota and its relationship with metabolites and future probiotics strategy: A review. *Biological Trace Element Research* **2022**, *200*, 5328–5350.
159. Odamaki, T.; Kato, K.; Sugahara, H.; Hashikura, N.; Takahashi, S.; Xiao, J.-Z.; Abe, F.; Osawa, R. Age-related changes in gut microbiota composition from newborn to centenarian: A cross-sectional study. *BMC Microbiology* **2016**, *16*, 1–12.
160. Woodmansey, E.J.; McMurdo, M.E.; Macfarlane, G.T.; Macfarlane, S. Comparison of compositions and metabolic activities of fecal microbiotas in young adults and in antibiotic-treated and non-antibiotic-treated elderly subjects. *Applied and Environmental Microbiology* **2004**, *70*, 6113–6122.
161. He, F.; Ouwehand, A.C.; Isolauri, E.; Hosoda, M.; Benno, Y.; Salminen, S. Differences in composition and mucosal adhesion of bifidobacteria isolated from healthy adults and healthy seniors. *Current Microbiology* **2001**, *43*, 351–354.
162. Claesson, M.J.; Cusack, S.; O'Sullivan, O.; Greene-Diniz, R.; de Weerd, H.; Flannery, E.; Marchesi, J.R.; Falush, D.; Dinan, T.; Fitzgerald, G. Composition, variability, and temporal stability of the intestinal microbiota of the elderly. *Proceedings of the National Academy of Sciences* **2011**, *108*, 4586–4591.
163. Franceschi, C.; Garagnani, P.; Parini, P.; Giuliani, C.; Santoro, A. Inflammaging: A new immune–metabolic viewpoint for age-related diseases. *Nature Reviews Endocrinology* **2018**, *14*, 576–590.
164. Biagi, E.; Nylund, L.; Candela, M.; Ostan, R.; Bucci, L.; Pini, E.; Nikkıla, J.; Monti, D.; Satokari, R.; Franceschi, C. Through ageing, and beyond: Gut microbiota and inflammatory status in seniors and centenarians. *PloS One* **2010**, *5*, e10667.
165. Xue, Q.-L. The frailty syndrome: Definition and natural history. *Clinics in Geriatric Medicine* **2011**, *27*, 1–15.
166. Kojima, G.; Liljas, A.E.; Iliffe, S. Frailty syndrome: Implications and challenges for health care policy. *Risk management and Healthcare Policy* **2019**, *12*, 23–30.
167. Cruz-Jentoft, A.J.; Bahat, G.; Bauer, J.; Boirie, Y.; Bruyère, O.; Cederholm, T.; Cooper, C.; Landi, F.; Rolland, Y.; Sayer, A.A. Sarcopenia: Revised European consensus on definition and diagnosis. *Age and Ageing* **2019**, *48*, 16–31.
168. Dalle, S.; Rossmeislova, L.; Koppo, K. The role of inflammation in age-related sarcopenia. *Frontiers in Physiology* **2017**, *8*, 1045.
169. Haran, P.H.; Rivas, D.A.; Fielding, R.A. Role and potential mechanisms of anabolic resistance in sarcopenia. *Journal of Cachexia, Sarcopenia and Muscle* **2012**, *3*, 157–162.
170. Wilson, D.; Jackson, T.; Sapey, E.; Lord, J.M. Frailty and sarcopenia: The potential role of an aged immune system. *Ageing Research Reviews* **2017**, *36*, 1–10.
171. Jeffery, I.B.; Lynch, D.B.; O'Toole, P.W. Composition and temporal stability of the gut microbiota in older persons. *The ISME journal* **2016**, *10*, 170–182.
172. Maffei, V.J.; Kim, S.; Blanchard IV, E.; Luo, M.; Jazwinski, S.M.; Taylor, C.M.; Welsh, D.A. Biological aging and the human gut microbiota. *Journals of Gerontology Series A: Biomedical Sciences and Medical Sciences* **2017**, *72*, 1474–1482.
173. Jackson, M.A.; Jeffery, I.B.; Beaumont, M.; Bell, J.T.; Clark, A.G.; Ley, R.E.; O'Toole, P.W.; Spector, T.D.; Steves, C.J. Signatures of early frailty in the gut microbiota. *Genome Medicine* **2016**, *8*, 1–11.
174. Ticinesi, A.; Tana, C.; Nouvenne, A.; Prati, B.; Lauretani, F.; Meschi, T. Gut microbiota, cognitive frailty and dementia in older individuals: A systematic review. *Clinical Interventions in Aging* **2018**, *13*, 1497–1511.
175. Picca, A.; Fanelli, F.; Calvani, R.; Mulè, G.; Pesce, V.; Sisto, A.; Pantanelli, C.; Bernabei, R.; Landi, F.; Marzetti, E. Gut dysbiosis and muscle aging: Searching for novel targets against sarcopenia. *Mediators of Inflammation* **2018**, *2018*, 7026198.
176. Allam-Ndoul, B.; Castonguay-Paradis, S.; Veilleux, A. Gut microbiota and intestinal trans-epithelial permeability. *International Journal of Molecular Sciences* **2020**, *21*, 6402.
177. Backhed, F.; Ding, H.; Wang, T.; Hooper, L.; Koh, G. Nagy A.; Semenkovich, C.F.; Gordon, J.I. The gut microbiota as an environmental factor that regulates fat storage. *Proceedings of the National Academy of Sciences of the United States of America* **2004**, *101*, 15718–15723.
178. Ni, Y.; Yang, X.; Zheng, L.; Wang, Z.; Wu, L.; Jiang, J.; Yang, T.; Ma, L.; Fu, Z. Lactobacillus and Bifidobacterium improves physiological function and cognitive ability in aged mice by the regulation of gut microbiota. *Molecular Nutrition & Food Research* **2019**, *63*, 1900603.
179. Huang, W.-C.; Chen, Y.-H.; Chuang, H.-L.; Chiu, C.-C.; Huang, C.-C. Investigation of the effects of microbiota on exercise physiological adaption, performance, and energy utilization using a gnotobiotic animal model. *Frontiers in Microbiology* **2019**, *10*, 1906.
180. Fielding, R.A.; Reeves, A.R.; Jasuja, R.; Liu, C.; Barrett, B.B.; Lustgarten, M.S. Muscle strength is increased in mice that are colonized with microbiota from high-functioning older adults. *Experimental Gerontology* **2019**, *127*, 110722.
181. Lustgarten, M.S. The role of the gut microbiome on skeletal muscle mass and physical function: 2019 update. *Frontiers in Physiology* **2019**, *10*, 1435.

182. Grosicki, G.J.; Fielding, R.A.; Lustgarten, M.S. Gut microbiota contribute to age-related changes in skeletal muscle size, composition, and function: Biological basis for a gut-muscle axis. *Calcified Tissue International* **2018**, *102*, 433–442.
183. Jovel, J.; Patterson, J.; Wang, W.; Hotte, N.; O'Keefe, S.; Mitchel, T.; Perry, T.; Kao, D.; Mason, A.L.; Madsen, K.L. Characterization of the gut microbiome using 16S or shotgun metagenomics. *Frontiers in Microbiology* **2016**, *7*, 459.
184. Marchesi, J.R.; Adams, D.H.; Fava, F.; Hermes, G.D.; Hirschfield, G.M.; Hold, G.; Quraishi, M.N.; Kinross, J.; Smidt, H.; Tuohy, K.M. The gut microbiota and host health: A new clinical frontier. *Gut* **2016**, *65*, 330–339.
185. Cho, I.; Blaser, M.J. The human microbiome: At the interface of health and disease. *Nature Reviews Genetics* **2012**, *13*, 260–270.
186. Steves, C.J.; Bird, S.; Williams, F.M.; Spector, T.D. The microbiome and musculoskeletal conditions of aging: A review of evidence for impact and potential therapeutics. *Journal of Bone and Mineral Research* **2016**, *31*, 261–269.
187. Kang, L.; Li, P.; Wang, D.; Wang, T.; Hao, D.; Qu, X. Alterations in intestinal microbiota diversity, composition, and function in patients with sarcopenia. *Scientific Reports* **2021**, *11*, 4628.
188. Potrykus, M.; Czaja-Stolc, S.; Stankiewicz, M.; Kaska, Ł.; Małgorzewicz, S. Intestinal microbiota as a contributor to chronic inflammation and its potential modifications. *Nutrients* **2021**, *13*, 3839.
189. Passos, M.d.C.F.; Moraes-Filho, J.P. Intestinal microbiota in digestive diseases. *Arquivos de gastroenterologia* **2017**, *54*, 255–262.
190. Waldner, H. The role of innate immune responses in autoimmune disease development. *Autoimmunity Reviews* **2009**, *8*, 400–404.
191. Janssen, A.W.; Kersten, S. The role of the gut microbiota in metabolic health. *The FASEB Journal* **2015**, *29*, 3111–3123.
192. Felice, V.D.; Quigley, E.M.; Sullivan, A.M.; O'Keeffe, G.W.; O'Mahony, S.M. Microbiota-gut-brain signalling in Parkinson's disease: Implications for non-motor symptoms. *Parkinsonism & Related Disorders* **2016**, *27*, 1–8.
193. Weiss, G.A.; Hennet, T. Mechanisms and consequences of intestinal dysbiosis. *Cellular and Molecular Life Sciences* **2017**, *74*, 2959–2977.
194. Khanna, S.; Pardi, D.S. Clinical implications of antibiotic impact on gastrointestinal microbiota and Clostridium difficile infection. *Expert Review of Gastroenterology & Hepatology* **2016**, *10*, 1145–1152.
195. Belizário, J.E.; Faintuch, J.; Garay-Malpartida, M. Gut microbiome dysbiosis and immunometabolism: New frontiers for treatment of metabolic diseases. *Mediators of Inflammation* **2018**, *2018,* 203783.
196. Walker, A.W.; Lawley, T.D. Therapeutic modulation of intestinal dysbiosis. *Pharmacological Research* **2013**, *69*, 75–86.
197. Matute, S.P.; Iyavoo, S. Exploring the gut microbiota: Lifestyle choices, disease associations, and personal genomics. *Frontiers in Nutrition* **2023**, *10,* 1225120.
198. Thanabalan, A. *Investigations of the impact of dietary n-3 polyunsaturated fatty acids on performance, metabolism, and immunocompetence responses in broiler breeders and their progeny*: University of Guelph; 2023.
199. Meng, C.; Bai, C.; Brown, T.D.; Hood, L.E.; Tian, Q. Human gut microbiota and gastrointestinal cancer. *Genomics, Proteomics & Bioinformatics* **2018**, *16*, 33–49.
200. Neish, A.S. Microbes in gastrointestinal health and disease. *Gastroenterology* **2009**, *136*, 65–80.
201. Sekirov, I.; Russell, S.L.; Antunes, L.C.M.; Finlay, B.B. Gut microbiota in health and disease. *Physiological Reviews* **2010**, 859–904.
202. Salim, S.a.Y.; Söderholm, J.D. Importance of disrupted intestinal barrier in inflammatory bowel diseases. *Inflammatory Bowel Diseases* **2011**, *17*, 362–381.
203. Geuking, M.B.; Köller, Y.; Rupp, S.; McCoy, K.D. The interplay between the gut microbiota and the immune system. *Gut Microbes* **2014**, *5*, 411–418.
204. Vinelli, V.; Biscotti, P.; Martini, D.; Del Bo', C.; Marino, M.; Meroño, T.; Nikoloudaki, O.; Calabrese, F.M.; Turroni, S.; Taverniti, V. Effects of dietary fibers on short-chain fatty acids and gut microbiota composition in healthy adults: A systematic review. *Nutrients* **2022**, *14*, 2559.
205. Khan, I.; Ullah, N.; Zha, L.; Bai, Y.; Khan, A.; Zhao, T.; Che, T.; Zhang, C. Alteration of gut microbiota in inflammatory bowel disease (IBD): Cause or consequence? IBD treatment targeting the gut microbiome. *Pathogens* **2019**, *8*, 126.
206. Tomasello, G.; Mazzola, M.; Leone, A.; Sinagra, E.; Zummo, G.; Farina, F.; Damiani, P.; Cappello, F.; Gerges Geagea, A.; Jurjus, A. Nutrition, oxidative stress and intestinal dysbiosis: Influence of diet on gut microbiota in inflammatory bowel diseases. *Biomed Pap Med Fac Univ Palacky Olomouc Czech Repub* **2016**, *160*, 461–466.
207. Shaw, S.Y.; Blanchard, J.F.; Bernstein, C.N. Association between the use of antibiotics and new diagnoses of Crohn's disease and ulcerative colitis. *Official Journal of the American College of Gastroenterology| ACG* **2011**, *106*, 2133–2142.
208. Pilmis, B.; Le Monnier, A.; Zahar, J.-R. Gut microbiota, antibiotic therapy and antimicrobial resistance: A narrative review. *Microorganisms* **2020**, *8*, 269.
209. Li, J.; Butcher, J.; Mack, D.; Stintzi, A. Functional impacts of the intestinal microbiome in the pathogenesis of inflammatory bowel disease. *Inflammatory Bowel Diseases* **2015**, *21*, 139–153.
210. Rosen, M.J.; Dhawan, A.; Saeed, S.A. Inflammatory bowel disease in children and adolescents. *JAMA Pediatrics* **2015**, *169*, 1053–1060.
211. Hviid, A.; Svanström, H.; Frisch, M. Antibiotic use and inflammatory bowel diseases in childhood. *Gut* **2010**, *60*(1), 49–54.
212. Sulaiman, J.E.; Lam, H. Evolution of bacterial tolerance under antibiotic treatment and its implications on the development of resistance. *Frontiers in Microbiology* **2021**, *12*, 617412.
213. Peralta, S.; Cottone, C.; Doveri, T.; Almasio, P.L.; Craxi, A. Small intestine bacterial overgrowth and irritable bowel syndrome-related symptoms: Experience with Rifaximin. *World Journal of Gastroenterology: WJG* **2009**, *15*, 2628.
214. Elvers, K.T.; Wilson, V.J.; Hammond, A.; Duncan, L.; Huntley, A.L.; Hay, A.D.; Van Der Werf, E.T. Antibiotic-induced changes in the human gut microbiota for the most

commonly prescribed antibiotics in primary care in the UK: A systematic review. *BMJ Open* **2020**, *10*, e035677.
215. Nitzan, O.; Elias, M.; Peretz, A.; Saliba, W. Role of antibiotics for treatment of inflammatory bowel disease. *World Journal of Gastroenterology* **2016**, *22*, 1078.
216. Costelloe, C.; Metcalfe, C.; Lovering, A.; Mant, D.; Hay, A.D. Effect of antibiotic prescribing in primary care on antimicrobial resistance in individual patients: Systematic review and meta-analysis. *BMJ* **2010**, 340.
217. Kostic, A.D.; Xavier, R.J.; Gevers, D. The microbiome in inflammatory bowel disease: Current status and the future ahead. *Gastroenterology* **2014**, *146*, 1489–1499.
218. Sanders, D.J.; Inniss, S.; Sebepos-Rogers, G.; Rahman, F.Z.; Smith, A.M. The role of the microbiome in gastrointestinal inflammation. *Bioscience Reports* **2021**, *41*, BSR20203850.
219. Bhardwaj, M.; Yadav, M. Microbiome therapeutics: A boon to modern therapeutics. In *Microbiome therapeutics*; Elsevier: 2023; pp. 1–40.
220. Zhang, S.; Chen, D.-C. Facing a new challenge: The adverse effects of antibiotics on gut microbiota and host immunity. *Chinese Medical Journal* **2019**, *132*, 1135–1138.
221. Chassaing, B.; Gewirtz, A.T. Gut microbiota, low-grade inflammation, and metabolic syndrome. *Toxicologic Pathology* **2014**, *42*, 49–53.
222. Serwecińska, L. Antimicrobials and antibiotic-resistant bacteria: A risk to the environment and to public health. *Water* **2020**, *12*, 3313.
223. Takiishi, T.; Fenero, C.I.M.; Câmara, N.O.S. Intestinal barrier and gut microbiota: Shaping our immune responses throughout life. *Tissue barriers* **2017**, *5*, e1373208.
224. Yu, L.C.-H.; Wang, J.-T.; Wei, S.-C.; Ni, Y.-H. Host-microbial interactions and regulation of intestinal epithelial barrier function: From physiology to pathology. *World Journal of Gastrointestinal Pathophysiology* **2012**, *3*, 27.
225. Portincasa, P.; Bonfrate, L.; Vacca, M.; De Angelis, M.; Farella, I.; Lanza, E.; Khalil, M.; Wang, D.Q.-H.; Sperandio, M.; Di Ciaula, A. Gut microbiota and short chain fatty acids: Implications in glucose homeostasis. *International Journal of Molecular Sciences* **2022**, *23*, 1105.
226. Parada Venegas, D.; De la Fuente, M.K.; Landskron, G.; González, M.J.; Quera, R.; Dijkstra, G.; Harmsen, H.J.; Faber, K.N.; Hermoso, M.A. Short chain fatty acids (SCFAs)-mediated gut epithelial and immune regulation and its relevance for inflammatory bowel diseases. *Frontiers in Immunology* **2019**, 277.
227. Hookman, P.; Barkin, J.S. Clostridium difficile associated infection, diarrhea and colitis. *World Journal of Gastroenterology: WJG* **2009**, *15*, 1554.
228. Grenni, P.; Ancona, V.; Caracciolo, A.B. Ecological effects of antibiotics on natural ecosystems: A review. *Microchemical Journal* **2018**, *136*, 25–39.
229. Butel, M.-J.; Waligora-Dupriet, A.-J.; Wydau-Dematteis, S. The developing gut microbiota and its consequences for health. *Journal of Developmental Origins of Health and Disease* **2018**, *9*, 590–597.
230. Lozupone, C.A.; Stombaugh, J.I.; Gordon, J.I.; Jansson, J.K.; Knight, R. Diversity, stability and resilience of the human gut microbiota. *Nature* **2012**, *489*, 220–230.
231. Miyauchi, E.; Shimokawa, C.; Steimle, A.; Desai, M.S.; Ohno, H. The impact of the gut microbiome on extra-intestinal autoimmune diseases. *Nature Reviews Immunology* **2023**, *23*, 9–23.
232. Lagier, J.-C.; Million, M.; Hugon, P.; Armougom, F.; Raoult, D. Human gut microbiota: Repertoire and variations. *Frontiers in Cellular and Infection Microbiology* **2012**, *2*, 136.
233. Lawley, T.D.; Walker, A.W. Intestinal colonization resistance. *Immunology* **2013**, *138*, 1–11.
234. Durack, J.; Lynch, S.V. The gut microbiome: Relationships with disease and opportunities for therapy. *Journal of Experimental Medicine* **2019**, *216*, 20–40.
235. Winter, S.E.; Bäumler, A.J. Gut dysbiosis: Ecological causes and causative effects on human disease. *Proceedings of the National Academy of Sciences* **2023**, *120*, e2316579120.
236. Sicherer, S.H. Clinical aspects of gastrointestinal food allergy in childhood. *Pediatrics* **2003**, *111*, 1609–1616.
237. Ducarmon, Q.; Zwittink, R.; Hornung, B.; Van Schaik, W.; Young, V.; Kuijper, E. Gut microbiota and colonization resistance against bacterial enteric infection. *Microbiology and Molecular Biology Reviews* **2019**, *83*, doi:10.1128/mmbr.00007-00019
238. Dicks, L.; Botes, M. Probiotic lactic acid bacteria in the gastro-intestinal tract: Health benefits, safety and mode of action. *Beneficial Microbes* **2010**, *1*, 11–29.
239. Blaut, M. Relationship of prebiotics and food to intestinal microflora. *European Journal of Nutrition* **2002**, *41*, i11–i16.
240. Sekhon, B.S.; Jairath, S. Prebiotics, probiotics and synbiotics: An overview. *Journal of Pharmaceutical Education & Research* **2010**, *1*, P13.
241. Tan, J.; McKenzie, C.; Potamitis, M.; Thorburn, A.N.; Mackay, C.R.; Macia, L. The role of short-chain fatty acids in health and disease. *Advances in Immunology* **2014**, *121*, 91–119.
242. Mikkelsen, K.H.; Frost, M.; Bahl, M.I.; Licht, T.R.; Jensen, U.S.; Rosenberg, J.; Pedersen, O.; Hansen, T.; Rehfeld, J.F.; Holst, J.J. Effect of antibiotics on gut microbiota, gut hormones and glucose metabolism. *PloS One* **2015**, *10*, e0142352.
243. Adedokun, S.A.; Olojede, O.C. Optimizing gastrointestinal integrity in poultry: The role of nutrients and feed additives. *Frontiers in Veterinary Science* **2019**, *5*, 348.
244. Oligschlaeger, Y.; Yadati, T.; Houben, T.; Condello Oliván, C.M.; Shiri-Sverdlov, R. Inflammatory bowel disease: A stressed "gut/feeling". *Cells* **2019**, *8*, 659.
245. Petkoska, A.T.; Trajkovska-Broach, A. Mediterranean way of living as an optimal lifestyle and a dietary pattern for healthy gut and strong immunity. *EC Nutrition* **2021**, *16*, 141–171.
246. Krupavaram, B.; Babu, A.K.; Pillay, S.M.; Siang, T.C.; Menon, B.V.; Paneerselvam, G.S.; Ming, L.C.; Al-Worafi, Y.M.; Chooi, W.H. A review of herbal treatment for functional gastrointestinal disorders and infection. *Progress In Microbes & Molecular Biology* **2023**, 6.
247. Rosa, C.P.; Brancaglion, G.A.; Miyauchi-Tavares, T.M.; Corsetti, P.P.; de Almeida, L.A. Antibiotic-induced dysbiosis effects on the murine gastrointestinal tract and their systemic repercussions. *Life Sciences* **2018**, *207*, 480–491.
248. Kesavelu, D.; Jog, P. Current understanding of antibiotic-associated dysbiosis and approaches for its management. *Therapeutic Advances in Infectious Disease* **2023**, *10*, 20499361231154443.

249. Singh, N. Chapter-7 functional foods: Probiotics and prebiotics. *Food Technology and Nutrition* **2020**, *83*, 145.
250. Goodman, C.; Keating, G.; Georgousopoulou, E.; Hespe, C.; Levett, K. Probiotics for the prevention of antibiotic-associated diarrhoea: A systematic review and meta-analysis. *BMJ Open* **2021**, *11*, e043054.
251. Wang, G.; Chen, Y.; Xia, Y.; Song, X.; Ai, L. Characteristics of probiotic preparations and their applications. *Foods* **2022**, *11*, 2472.
252. Peng, M.; Tabashsum, Z.; Anderson, M.; Truong, A.; Houser, A.K.; Padilla, J.; Akmel, A.; Bhatti, J.; Rahaman, S.O.; Biswas, D. Effectiveness of probiotics, prebiotics, and prebiotic-like components in common functional foods. *Comprehensive Reviews in Food Science and Food Safety* **2020**, *19*, 1908–1933.
253. Shawer, H.; Norman, K.; Cheng, C.W.; Foster, R.; Beech, D.J.; Bailey, M.A.J.F.i.C.; Biology, D. ORAI1 Ca2+ channel as a therapeutic target in pathological vascular remodelling. **2021**, *9*, 653812.
254. Orsi, C.; Zambrini, A.V. Probiotics and prebiotics. In Francesco Conto (ed.) *Advances in Dairy Products* **2017**, 37–57.
255. Madrid, E. The benefits of prebiotics. *Benefits* **2020**, *20*.
256. Ouwehand, A.; Tiihonen, K.; Mäkivuokko, H.; Rautonen, N. Synbiotics: Combining the benefits of pre-and probiotics. In *Functional dairy products*; Elsevier: 2007; pp. 195–213.
257. Eswaran, S.; Muir, J.; Chey, W.D. Fiber and functional gastrointestinal disorders. *Official Journal of the American College of Gastroenterology| ACG* **2013**, *108*, 718–727.
258. de Nies, L.; Kobras, C.M.; Stracy, M. Antibiotic-induced collateral damage to the microbiota and associated infections. *Nature Reviews Microbiology* **2023**, *21*, 789–804.
259. Voth, E.; Khanna, S. Fecal microbiota transplantation for treatment of patients with recurrent Clostridioides difficile infection. *Expert Review of Anti-Infective Therapy* **2020**, *18*, 669–676.
260. Borody, T.J.; Khoruts, A. Fecal microbiota transplantation and emerging applications. *Nature Reviews Gastroenterology & Hepatology* **2012**, *9*, 88–96.
261. Kachrimanidou, M.; Tsintarakis, E. Insights into the role of human gut microbiota in Clostridioides difficile infection. *Microorganisms* **2020**, *8*, 200.
262. Banfi, D.; Moro, E.; Bosi, A.; Bistoletti, M.; Cerantola, S.; Crema, F.; Maggi, F.; Giron, M.C.; Giaroni, C.; Baj, A. Impact of microbial metabolites on microbiota–gut–brain axis in inflammatory bowel disease. *International Journal of Molecular Sciences* **2021**, *22*, 1623.
263. Qu, Z.; Tian, P.; Yang, B.; Zhao, J.; Wang, G.; Chen, W. Fecal microbiota transplantation for diseases: Therapeutic potential, methodology, risk management in clinical practice. *Life Sciences* **2022**, *304*, 120719.
264. de Stefano, M.C.; Mazzanti, B.; Vespasiano, F.; Lombardini, L.; Cardillo, M. The regulatory approach for faecal microbiota transplantation as treatment for Clostridioides difficile infection in Italy. *Antibiotics* **2022**, *11*, 480.
265. Berry, P.; Khanna, S. Recurrent clostridioides difficile infection: Current clinical management and microbiome-based therapies. *BioDrugs* **2023**, *37*, 757–773.
266. Langdon, A.; Crook, N.; Dantas, G. The effects of antibiotics on the microbiome throughout development and alternative approaches for therapeutic modulation. *Genome Medicine* **2016**, *8*, 1–16.
267. Rizkallah, M.R.; Saad, R.; Aziz, R.K. The Human Microbiome Project, personalized medicine and the birth of pharmacomicrobiomics. *Current Pharmacogenomics and Personalized Medicine (Formerly Current Pharmacogenomics)* **2010**, *8*, 182–193.
268. Husain, A.; Monga, J.; Narwal, S.; Singh, G.; Rashid, M.; Afzal, O.; Alatawi, A.; Almadani, N.M. Prodrug rewards in medicinal chemistry: An advance and challenges approach for drug designing. *Chemistry & Biodiversity* **2023**, *20*, e202301169.
269. Nazli, A.; He, D.L.; Liao, D.; Khan, M.Z.I.; Huang, C.; He, Y. Strategies and progresses for enhancing targeted antibiotic delivery. *Advanced Drug Delivery Reviews* **2022**, *189*, 114502.
270. Karavolos, M.; Holban, A. Nanosized drug delivery systems in gastrointestinal targeting: Interactions with microbiota. *Pharmaceuticals* **2016**, *9*, 62.
271. Relman, D.A.; Lipsitch, M. Microbiome as a tool and a target in the effort to address antimicrobial resistance. *Proceedings of the National Academy of Sciences* **2018**, *115*, 12902–12910.
272. Gough, E.K. The impact of mass drug administration of antibiotics on the gut microbiota of target populations. *Infectious Diseases of Poverty* **2022**, *11*, 76.
273. Muteeb, G.; Rehman, M.T.; Shahwan, M.; Aatif, M. Origin of antibiotics and antibiotic resistance, and their impacts on drug development: A narrative review. *Pharmaceuticals* **2023**, *16*, 1615.
274. Sharpton, S.; Schnabl, B.; Knight, R.; Loomba, R. Current concepts, opportunities, and challenges of gut microbiome-based personalized medicine in nonalcoholic fatty liver disease. *Cell Metabolism* **2021**, *33*, 21–32.
275. Zmora, N. *Microbiome-based personalized interventions*: The Weizmann Institute of Science (Israel); 2019.
276. Zoetendal, E.G.; Rajilić-Stojanović, M.; de Vos, W.M. High-throughput diversity and functionality analysis of the gastrointestinal tract microbiota. *Gut* **2008**, *57*, 1605–1615.
277. Fishbein, S.R.; Mahmud, B.; Dantas, G. Antibiotic perturbations to the gut microbiome. *Nature Reviews Microbiology* **2023**, *21*, 772–788.
278. Ellabaan, M.M.; Munck, C.; Porse, A.; Imamovic, L.; Sommer, M.O. Forecasting the dissemination of antibiotic resistance genes across bacterial genomes. *Nature Communications* **2021**, *12*, 2435.
279. Sehrawat, N.; Yadav, M.; Singh, M.; Kumar, V.; Sharma, V.R.; Sharma, A.K. Probiotics in microbiome ecological balance providing a therapeutic window against cancer. *Seminars in Cancer Biology* **2021**, 70, 24–36.
280. Aziz, Q.; Doré, J.; Emmanuel, A.; Guarner, F.; Quigley, E. Gut microbiota and gastrointestinal health: Current concepts and future directions. *Neurogastroenterology & Motility* **2013**, *25*, 4–15.
281. Tamma, P.D.; Miller, M.A.; Cosgrove, S.E. Rethinking how antibiotics are prescribed: Incorporating the 4 moments of antibiotic decision making into clinical practice. *JAMA* **2019**, *321*, 139–140.
282. Young, V.B. The role of the microbiome in human health and disease: An introduction for clinicians. *BMJ* **2017**, *356*, j831.
283. Tamburini, S.; Shen, N.; Wu, H.C.; Clemente, J.C. The microbiome in early life: Implications for health outcomes. *Nature Medicine* **2016**, *22*, 713–722.

9 Role of the Immune System in Gastrointestinal Inflammation

*Tahreen Taj[1], Mithun Bhowmick[2], Pratibha Bhowmick[2], Sumel Ashique[3], Neeraj Mishra[3], and Farzad Taghizadeh-Hesary[5,6]**

[1]Department of Pharmacology, Yenepoya Pharmacy College and Research Centre, Yenepoya (Deemed to be) University, Mangalore, India

[2]Department of Pharmaceutical Sciences, Bengal College of Pharmaceutical Sciences & Research, Durgapur, West Bengal, India

[3]Department of Pharmaceutical Technology, Bharat Technology, Uluberia, West Bengal, India

[4]Amity Institute of Pharmacy, Amity University Gwalior, Madhya Pradesh, India

[5]ENT and Head and Neck Research Center and Department, The Five Senses Health Institute, School of Medicine, Iran University of Medical Sciences, Tehran, Iran

[6]Clinical Oncology Department, Iran University of Medical Sciences, Tehran, Iran

9.1 INTRODUCTION TO GIT INFLAMMATION

Gastrointestinal inflammation is mainly caused when the body's immune system responds to the pathogens within the digestive tract and demonstrates broad spectrum of conditions which includes diverse range of abnormalities that affects digestive lining, with own causes and effects. The common gastrointestinal disorders include inflammatory bowel disease (IBD), gastritis, peptic ulcers, and irritable bowel syndrome (IBD) has two types known as, Crohn's disease and ulcerative colitis, which are highly chronic inflammatory conditions, characterized by immune responses against the gut microbiota which results in mucosal inflammation [1]. Gastritis is a condition defined by the inflammation in the stomach lining. It has two types depending upon the severity of the gastritis, acute gastritis results from the exposure of pathogens like *Helicobacter pylori* and prolonged usage of non-steroidal anti-inflammatory drugs (NSAIDs). Chronic gastritis occurs due to persistent infection, leading to mucosal damage [2]. Another gastrointestinal disorder is irritable bowel syndrome (IBS) caused mainly by dysregulation of gut motility causing the disruptions in the gut microbiota [3]. The importance of gut homeostasis aids in determining the overall health and well- being. Maintaining gut homeostasis ensures nutrient uptake and efficient support to overall metabolic process [4]. The gut homeostasis is critical for preserving the integrity of the intestinal barrier. The barrier acts as a protective shield, actively preventing the entry of pathogens [5]. In addition to its role in digestion and barrier function, the gut is known to be the hub of immune interactions. A balanced gut is essential for overall immune function and preventing excessive immune responses, when it is balanced it trains the immune system to withhold harmless antigens and respond to harmful ones. Dysregulation of immune responses in the gut usually induces autoimmune conditions and various other gastrointestinal disorders [6]. The gut microbiota is composed of diverse microorganisms, which play an essential role in digestion, metabolism and immune function. Maintaining a balanced microbiota composition through gut homeostasis is vital for overall health, alteration in the composition causes dysbiosis and significantly contributes to gastrointestinal disorders [7]. Gastrointestinal inflammation is a multifactorial process which involves a combination of immune dysregulation, genetic susceptibility and environmental influences. The comprehensive understanding of the combinational nature of gastrointestinal inflammation is crucial for developing effective diagnostic and therapeutic strategies. The findings on the relationship involved between immune dysregulation and the major other factors can strive towards improved management and prevention of these conditions, ultimately enhancing the gastrointestinal well-being of the individuals across the world.

9.2 PATHOGENESIS OF GIT INFLAMMATION

9.2.1 Role of the Intestinal Barrier

The gastrointestinal tract is responsible for digestion, nutrient absorption and plays an essential role in maintaining the overall health. It is made of a cellular lining which functions as a protective barrier in between the external environment and internal tissues. This barrier is composed of innate immune cells and epithelial cells (IECs), serves as an important role in preventing the infiltration of harmful substances [8].

DOI: 10.1201/9781003493143-9

The defects in this barrier can give rise to gastrointestinal issues such as abnormal intestinal permeability observed in patients with Crohn's disease. The abnormal intestinal permeability indicates the significance of the epithelial barrier in predisposition to inflammatory bowel diseases [9]. The junction protein E- cadherin plays an important role in maintaining the integrity of this barrier, which is down regulated in Crohn's disease patients. E-cadherin as its role in the formation of tight junctions within intestinal epithelium, the reduction of E- cadherin indicates the disruption of the tight junctions leading to a compromised intestinal barrier.

9.2.2 Dysregulation of the Immune System

The presence of dysregulation in the immune system contributes to the gastrointestinal inflammation, which causes defects in the intestinal epithelial barrier and mucosal immune system. There are various cytokines such as TNF-α, TGF-β, and IFN-γ which modulate IBD, and it is known that targeting the cytokines and chemokines can conclude strategies for curing IBD. The increased infiltration of immune cells in the mucous membrane (macrophages, natural killer cells, dendritic cells, and mast cells) contributes efectively to gastrointestinal inflammation, resulting in higher levels of pro-inflammatory factors TNF-α, IFN-γ, IL-1B and IL-23/Th17-pathway cytokines [8]. The balance of effectors and regulatory populations of T cells is essential for initiating the immune response in the GALT, and the presence of insufficient regulatory cell activity results in abnormal inflammation [10] and [8].

9.2.3 Balance of Th1 & Th2 Responses

Certain T cell sub types (Th1, Th2, or Th17) have an essential role in the pathogenesis of gastrointestinal inflammation [11] and [12]. The lamina propria is an important site of immune cell activation in the gut and specific abnormalities in immune cell activation have been reported in the patients with IBD. Intestinal macrophages in IBD patients exhibit high levels of CD14 content and increased activity in the NF-κB transcription pathway, which leads to the recruitment of peripheral macrophages. Similar studies have noted elevated counts and cytotoxic activity of natural killer cells, indicating the abnormal immune cell activation in the lamina propria [13].

9.2.4 NLRP6 Inflammasome in Gut Defense

NLRP6 is a protein that forms an inflammasome in the intestinal mucosa, playing a role in the regulation of the composition of gut microbiome for proper maintenance of intestinal homeostasis [14]. The NLRP6 inflammasome controls the secretion of mucus by sentinel goblet cells in the intestine, providing essential protection against pathogens [15]. This inflammasome regulates the secretion of interlukin-18(IL-18) by intestinal epithelial cells and the expression profiles of antimicrobial peptides(AMPs) [16].The balance between Th1 and Th2 responses is critical for maintaining immune homeostasis in the gut. The imbalance in the overexpression of interleukin-23(IL-23) within the intestinal mucosa, compromises the guts protective barrier and disrupts the local immune regulation by further inhibition to produce interleukin-10(IL-10) and regulatory T (Treg) cells [17]. IL-10, is an important cytokine involved in promoting Th2 responses and plays an important role in adaptive immunity. In patients with gastrointestinal inflammation specifically, IBD patients IL-10 levels are significantly less and the production is itself reduced. This dysregulation in IL-10 contributes to the pathogenesis of IBD [18].

9.2.5 Genetic & Environmental Triggers

The development of gastrointestinal inflammation is mainly because of both genetic and environmental factors. The genetic triggers are often more prone to gastrointestinal disorders [19]. Environmental factors also highly contribute to gastrointestinal inflammation. Foods such as high-fat diets, low fiber diets, and consumption of food additives are one of the major reasons of disease progression [20]. Stress is another factor that triggers the inflammation [21]. The gut microbiota and intestinal barrier integrity can be altered by the use of antibiotics [22].

9.3 FACTORS RESPONSIBLE FOR GIT INFLAMMATION

There are various factors that contribute to inflammation of the gastrointestinal tract. The factors like dietary choices, stress levels, and the balance of microbial communities together play important roles in influencing the inflammation. Proper understanding of these factors helps to manage and prevent gastrointestinal inflammation. The factors are as follows-

9.3.1 Dietary Factors and GIT Inflammation

Excessive consumption of the dietary monosaccharaides has been linked to the development of inflammatory bowel disease particularly in patients with Crohn's disease [23]. The consumption of chocolate and cola type beverages has been associated with a higher incidence of IBD [24]. The intake of fats and proteins has been explored in relation to the occurrence of inflammatory bowel disease (IBD). The intake of animal protein is associated with slight elevation in the risk of onset of IBD and the consumption of high-fat diet rich in cholesterol and animal fats has increased risk for the disease [23]. There are studies that highlight the potential adverse effect of trans fatty acids and linoleic acid on the risk of developing ulcerative colitis [25]. The type of fatty acids consumed, particularly omega-6 and monosaturated fatty acids play role in influencing the risk of UC [26] and [27].The fiber consumption is mentioned to provide protective effect on IBD development content of 24.3 g/d and high-fiber diet prevents in developing IBD but there were no protective effects observed in patients with UC [23]. The deficiency of vitamin D is common in patients with inflammatory bowel disease and some researchers claim it as a dietary factor that may increase the risk of IBD. The fruits with anti-inflammatory and anti-oxidative activity, particularly citrus fruits can manage the inflammation associated with IBD [24] and [28].

9.3.2 Stress and Psychological Impact on GIT Inflammation

There are studies showcasing a clear connection between IBD and stress. Stress can cause mood swings in IBD patients leading to worsening of the symptoms like abdominal pain. Stress can alter the established tolerance and augment the immune responses in chronic intestinal inflammation. There is a study that examines the impact of psychological stress on inflammatory bowel disease and its implications for neuroimmunomodulation [29]. Stress can alter the established tolerance and augment the immune responses in chronic intestinal inflammation. A group of researchers conducted a study that investigated how stress and antibiotics impact both luminal and wall adhered microbiota along with effects on visceral sensory system in murine models [30]. Increased permeability in the intestine caused as a result of stress allows the microbiota to travel throughout the gut epithelial barrier and trigger mucosal immune response. The psychological stress can impact intestinal disorders and the gut microbiota. For instance, in experiments with mice, exposure to stressful situations similar to post-traumatic stress in humans led to changes in the gut microbiota. This suggests a link between mental health, gut health and how stress can affect the balance of microorganisms in the digestive tract and disorder in humans led to changes in the gut microbiota [31]. When individuals experience psychological stresses in overcrowded aspects, it can influence the composition of bacteria in the gut. Certain types of bacteria tend to increase in abundance in response to stress, these include *Staphylococcus, Streptococcus, Corynebacterium,* and *Enterobacteriaceae.* The presence of *Peptococcaceae, Bacteroidaceae,* and *Clostridium* also increased while the levels of *Lactobacillus* were reduced in comparison [32]. The reduction of *Lactobacillus* can highly increase the intestinal inflammation. When stress reduces the amount of *Lactobacillus* bacteria in the gut, it can lead to inflammation in the intestine, but administration of *L. reuteri* helped to prevent inflammation in the intestine. They are stressed and then exposed to a gut infection. Studies concluded that, when stressed mice were exposed to harmful gut bacteria, their gut inflammation worsened. This was increasing with prolonged stress. Tests using germ-free mice showed that those colonized with bacteria from stressed mice had more severe inflammation. The presence of *Bifidobacterium* was absent, which usually helps in reducing inflammation caused by the harmful bacteria [33]. In silico studies stated that the stress-induced changes in the gut microbiota led to a decrease in the production of fatty acids known as SCFAs, including butyrate. The reduction in the butyrate helps prevent inflammation; it is also reported that stress has lowered the levels of mucin- 2 and lysozyme, which are protective substances in the gut. This imbalance in the gut protection may lead to decrease in butyrate producing bacteria like *Lachnospiraceae.* The transfer of gut microbiota from stressed mice were transferred to other mice, it triggered a type of intestinal inflammation known as trigger colitis, a type of intestinal inflammation [34]. All these conditions together make the gut system more prone to inflammation.

9.3.3 Microbiota Imbalances and GIT Inflammation

Healthy gut compromises of *Firmicutes* and *Bacteroidetes,* which aids in maintaining a balanced microbiota [35] and [36]. The imbalance in the gut microbiota can disrupt the equilibrium, leading to the induction of disease [37]. Dysbiosis contributes to the presence of harmful bacteria and this imbalance can result in the release of enterotoxins by these bacteria. These toxins contribute to higher intestinal permeability and production of immunosuppressive proteins. The immune dysfunction, in turn, may lead to damage the intestinal epithelial cells and cascade of events forms the connection between microbiota imbalances and the multifaceted mechanisms involved in inducing inflammation within the gastrointestinal tract. The presence of *Enterobacteriaceae* is commonly found in gastrointestinal problems like Inflammatory bowel disease (IBD). They thrive in the inflamed gut environment, making the intestine susceptible to inflammation and disease development [36] and [38]. The enteric bacteria and usual gut microbe can violate the intestinal barrier by triggering the inflammation in a susceptible host. This clearly states that any bacteria under certain conditions can notably have the potential to turn into a harmful pathogen [39]. *Faecalibacterium prausnitzii* plays a vital role in the inflammatory bowel disease, as it is reduced in IBD conditions. This bacterium aids in producing the metabolite butyrate, which prevents the IBD development. *Faecalibacterium prausnitzii* converts acetate, also produced by *Ruminococcus albus* (usually reduced under IBD conditions), into butyrate [40] and [41]. Butyrate has the ability to regenerate colonocytes and maintains the intestinal integrity. It promotes the development of regulatory T cell (T reg), which contributes to intestinal mucosal tolerance and maintains equilibrium of Th 17 and Treg cells to prevent intestinal inflammation [42] and [43]. In addition, clinical studies have explored the use of probiotics, specifically *Bifidobacterium* and *Lactobacillus*. These probiotics have demonstrated the ability to enhance the secretion of anti-inflammatory factors, restoration of the intestinal barrier, and protection from harmful microbes. Gut microbiota is an essential determinant in the onset of gastrointestinal tract inflammation linked due to inflammatory bowel disease.

9.4 Immune Cells Involvement in GIT Inflammation

9.4.1 Innate Immunity

In the context of gastrointestinal inflammation, the innate immune system serves as the frontline defense against pathogens within the gut lumen [44]. Though, it lacks the immunological memory but plays a crucial role in immediate protection against foreign [45]. Key cellular components contributing to this immediate defense include enterocytes which provides overall integrity of the mucosal lining [46], Paneath cells plays a vital role by providing response against bacterial (AMPs), goblet cells secretes mucus which leads to the formation of protective barrier that prevents the direct contact between luminal pathogens and the intestinal epithelium, crucial for maintaining the inflammation and preserving homeostasis [47]. Dendritic cells are considered as antigen presenting cell, initiating adaptive immune response by activating T cells [48]. Macrophages phagocytize pathogens and release inflammatory molecules in response to TLR activation and modulate the inflammatory environment [49]. Neutrophils are recruited to infection sites for pathogen elimination, while mast cells respond to pathogens, regulate inflammation and contribute to mucosal defense [50]. Innate immunity serves as the body's initial defense mechanism against the invasion of pathogens in the gastrointestinal tract, by proper maintenance of gut homeostasis and regulation of inflammatory responses [51]. Macrophages are important immune cells in the gut and key keep safeguarding the gut lining. They are equipped with pattern recognition receptors (PRRs) that allow the detection of microbial products and initiate immune responses. Macrophages can produce pro-inflammatory cytokines, which are essential mediators in the gut mucosa and also exhibit anti-inflammatory functions, contributing to the resolution of inflammation [52]. The classes of sentinel guardians are dendritic cells, known as specialized antigen presenting cells that play major role in initiating adaptive responses in the gut responses in gut mucosa. Dendritic cells maintain immune tolerance to dietary antigens and microorganisms while showcasing effective immune responses against pathogens. DCs presents antigenic peptides to T lymphocytes, thereby activating and directing T cell responses. They secrete various cytokines and chemokine that modulate the activity of immune cells, promoting immune regulation to the current context [53]. The forefront defense is usually exhibited by neutrophils are rapidly recruited to sites of inflammation across the gastrointestinal tract in response to chemokine signals released by activated immune cells. These are highly phagocytic cells play a critical role in clearance of pathogens through release of antimicrobial factors. Neutrophils consist of antimicrobial peptides and reactive oxygen species which contribute to microbial killing. Figure 9.1 depicts the role of an innate immune system in intestinal inflammation [54].

9.4.2 Adaptive Immune Response

The adaptive immune system present across the gastrointestinal tract involves primarily, T lymphocytes (T cells) and B lymphocytes (B cell), which aids in gut homeostasis. [55]. T cells aid in regulation of immune responses, cytotoxic T cells

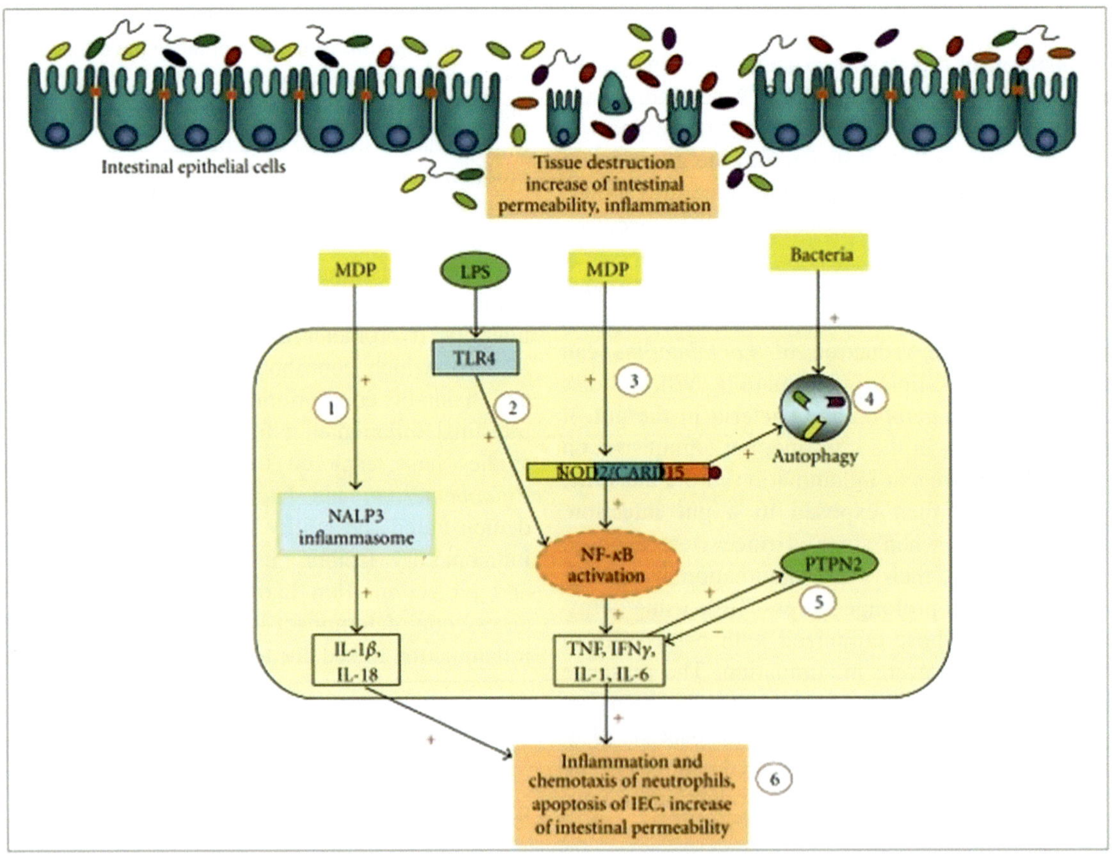

FIGURE 9.1 The innate immune system during intestinal inflammation. (Adapted from Scharl M et al., 2010; © 2010 Michael Scharl and Gerhard Rogler. This is an open access article distributed under the Creative Commons Attribution License.)

eliminate abnormal cells, regulatory cells maintain the tolerance to the commensal microbes in the gastrointestinal tract, ensuring a balanced immune environment [56]. B cells contribute to antigen presentation and immune regulation, they are essential for mucosal immunity in the gut through antibody production particularly immunoglobulin A (IgA). They neutralize pathogens and shape the gut microbiota composition [6]. Mucosa- associated lymphoid tissue is composed of structures like Peyer's patches, mesenteric lymph nodes and lymphoid follicles in the gastrointestinal mucosa. These points initiate immune cell interaction and initiate the antibody production, essential for gut health [57]. The connection between Peyer's patches (PPs) and Crohn's Disease lies in the abnormal T cell mediated immune responses to the gut microbiota. [58]. The involvement of Peyer's patches (PPs) in Crohn's disease (CD) highlights their essential role in maintaining gut immune balance [59]. The genetic susceptibility to Crohn's disease involves genes linked to bacterial recognition and innate immunity, like NOD2, ATG16L1 and IRGM [60] and [61]. MALT ensures proper coordination between innate and adaptive immune cells in the gastrointestinal mucosa. They aid in initiation and activation of T and B lymphocytes and ensure effective immune responses [62]. The interaction between the commensal bacteria with immune cells through pattern recognition receptors (PRRs), such as Toll like receptors (TLRs) and NOD-like receptors (NLRs) triggers immune responses. Microbiota- derived metabolites actively influence immune cell function in the gut [63].

9.5 FACTORS RESPONSIBLE FOR IMMUNITY INHIBITION DONE

There are certain factors that can inhibit the immune system, making it weak and diminishing its efficiency. One such prominent factor is gastrointestinal inflammation. When inflammation occurs in the gastrointestinal tract it causes the disruption of the integrity of the mucosal barrier, which results in the immune dysfunction, the dysregulated immune response in turn effectively contributes to the progression of various gastrointestinal disorders [64]. Antimicrobial peptides working alongside the innate immune response, act as the body's initial defense mechanism at mucosal surface. The proper balance of AMP expression in gut mucosa is disrupted by inflammatory various conditions such as ulcerative colitis (UC), Crohn's disease (CD) [65]. The reduction in expression of AMPs, such as defensins and cathelicidins, in the gut mucosa of inflammatory bowel disease patients compromises the guts' ability to defend. Conversely, unregulated inflammation can lead to excessive production of AMPs [66] This overexpression may result in tissue damage, for instance cathelicidin LL-37, when overexpressed, has been associated with increased intestinal inflammation in IBD [67]. The strategies aimed in modulating AMP production may help restore mucosal barrier function. The chronic inflammation of the gut tract frequently contributes to the excessive production of pro- inflammatory cytokines and mediators. The cytokines like tumor necrosis factor- alpha (TNF-α), interleukin-6 (IL-6), and interleukin-1β (IL-1β) are highly produced in inflamed gut tissues and initiates persistent tissue damage [68]. In gastrointestinal inflammation there is an activation of immunosuppressive cells, such as T cells (Tregs) and myeloid- derived suppressor cells (MDSCs). The release of the immunosuppressive cytokines further contributes to the immune inhibition within the inflamed gut environment [69]. Chronic gastrointestinal inflammation often results in loss of immune tolerance against the commensal bacteria present in the gut. This dysregulated immune response contributes to the immunity inhibition [70]. The gastrointestinal inflammation leads to immune dysfunction and high progression of gastrointestinal disorders. It involves dysregulation of antimicrobial peptides and excess production of pro-inflammatory cytokines and in addition to these, there is a bulk release of immunosuppressive cells and loss of immune tolerance, Several environmental factors can modulate the immune system ranging from exposure to abiotic chemicals, genetic programming, and developmental defects due to disrupted metabolism mainly nutritional and toxicological, and inappropriate decision making by the immune system further inhibiting immune function.

9.6 SEVERAL SIGNALING PATHWAYS INVOLVED IN GIT INFLAMMATION

The process of moving information or signals from the outward circumstances into the inward of the cell and vice versa is referred to as the signaling route, in this process the involvement of molecular variations as well as intracellular interactions are noted. Similarly, the gastrointestinal tract's inflammation is generated by a number of signaling pathways. Such as, NF-κB (Nuclear Factor-κB) Pathway, JAK-STAT (Janus Kinase-Signal Transducer and Activator of Transcription) Pathway, Wnt Signaling, TGF-β (Transforming Growth Factor-β) Pathway [71]. Table 9.1 depicts several signaling pathways associated with GIT inflammation.

9.6.1 NF-κB (Nuclear Factor-κB) Pathway

The Nuclear Factor-κB Pathway, commonly known as the NF-κB signaling pathway, possesses several tasks in the body, including modulating inflammation and maintaining innate immunity. In addition to inflammation cancer has been connected to NF-κB. With the help of two principles such as classical also known as canonical pathway and the alternative pathway, the NF-κB will get activated influenced through multiple yet distinct stimuli. Furthermore, by triggering the action of multiple variables such as IL-2, growth factor, CD40L and granulocyte monocyte colony stimulator factor [72]. Due to the presence of infections caused by *Helicobacter pylori* leads to chronic inflammation, which causes persistent, long-term peripheral gastritis followed by intestine metaplasia, then cancer in sequence [73]. Vyas D et al examined the reaction of the intestinal epithelium to damage and, consequently, the impact of NF-κB on the gastrointestinal tract [74]. By triggering downstream proteins with carcinogenic properties, NF-κB stimulates the formation of gastrointestinal tract

TABLE 9.1
Overview of Various Signaling Pathways Involved in Gastrointestinal Inflammation

Sl. No	Name of the Pathway	Property	Mechanism	Role in GIT Inflammation	Outcome
1	NF-κB Pathway	Induction of pro-inflammatory genes	Activation by pro-inflammatory stimuli	Regulation of cytokine production, inflammation, and cell survival	Central in inflammatory response
2	JAK-STAT Pathway	Transcriptional regulation of inflammation-related genes	Cytokine receptor activation	Regulates cell survival, proliferation, and inflammation	Key in cytokine signaling
3	Wnt/β-Catenin Pathway	Differentiation and Regulation of cell proliferation	Wnt ligand binding activates pathway	Involved in intestinal homeostasis, cell proliferation, and inflammation	Implicated in tissue response
4	TGF-β Pathway	Regulating cellular processes	Activation of downstream Smad proteins	To inflammation, fibrosis, and tumorigenesis	Carcinogenesis
5	TLR Signaling	Recognition of pathogen-associated molecular patterns	Activation of immune response	Mediates cytokine signaling, regulation of immune responses, and inflammation	Initiates innate immune response
6	MAPK Pathway	Regulation of cell proliferation, apoptosis	Activation by growth factors, cytokines	Regulation of inflammatory gene expression, cell proliferation, and apoptosis	Promotes inflammation
7	PI3K-Akt pathway	Binding of extracellular ligands	Activation of growth factor 1 (IGF-1), epidermal growth factor (EGF)	Regulates cell growth, survival, proliferation, and inflammation	Angiogenesis and cell cycle progression

malignancies, such as esophageal, gastric, particularly intestinal type, and colonic cancers [75].

9.6.2 JAK-STAT (Janus Kinase-Signal Transducer and Activator of Transcription) Pathway

Janus Kinase-Signal Transducer and Activator of Transcription Pathway commonly known as JAK-STAT Pathway. According to the research, Lamb A et al., reported that a JAK inhibitor was proven to be useful in the treatment of ulcerative colitis. The JAK/STAT signaling pathways are pivotal in numerous essential physiological activities, such as cell development, differentiation, growth, and regulatory immunological functions. They have also been connected to the pathophysiology of a number of disorders, including inflammatory bowel disease. Lei H et al. state that while the focus of JAK inhibitors affects immune cells; it also directly impacts the pathways of epithelial cells that regulate tight junction proteins. This suggests that the JAK-STAT Pathway might be a promising beneficial possibility in the management of inflammatory bowel disease [76].

9.6.3 Wnt Signaling

The large family of glycoproteins known as wnt proteins is highly conserved and has a role in development, regeneration, and tissue homeostasis [77]. Wnt signaling has developed as a potential objective in inflammatory bowel disease. According to Taheri N et al., it is usually regarded as a crucial component in the regulation of cell destiny, polarization of cells, and organ development. It is also connected to an inflammatory pathway in IBD, which may represent an opportunity for therapy [78]. Wnt signaling is dysregulated in esophagitis and gastritis brought on by *H. pylori,* indicating its involvement in inflammation. Wnt/β-catenin signaling plays critical roles in wound healing in inflammatory bowel disease [79].

9.6.4 TGF-β (Transforming Growth Factor-β) Pathway

TGF-βs are cytokines with a number of functions that help to control various essential cellular processes such as cell division, migration, cell growth and death, synthesis of extracellular matrix [80]. The regulatory function of TGF-β can be disrupted by dysregulation of the system, where intestinal homeostasis and GI tract carcinogenesis linked to inflammation and both are prevented by TGF-β signaling which can effect in the subsequent development of cancer and chronic inflammation [81].

9.7 GUT MICROBIOTA ASSOCIATED WITH GIT INFLAMMATION

The gut microbiome balances the microorganisms in the gut, those microorganisms which are essential for maintaining health as well as influencing the disease to get developed. By changes in nutrition, the complex community gut microbiota is constantly formed. It is primarily composed of bacteria, such as *Firmicutes* and Bacteroidetes, furthermore it also includes fungi, parasites, viruses, and archaea [82]. The environmental factors as well as the genetic factors, this both can get disrupt the microbiota and its production of metabolites, furthermore other factors include sleep patterns, stress, physical activity, stressors like antibiotics, diet, and environmental conditions, which causes the inflammation in the intestine as well as inflammatory bowel diseases, due to the above factors mentioned will affect the immune system and gut barrier integrity [83]. Intestinal health depends on the microbial population, and the dysbiosis can lead to disturbances similar to those observed in inflammatory bowel disease [84]. Toxins that are injurious to the body can enter the circulation when there is an imbalance known as dysbiosis brought on by disruption of the gut barrier and changes in the makeup and function of the gut microbiome [85]. This inflammation of bowel and intestine is exacerbated by imbalances that promote pathogenic microorganisms such as Proteobacteria. The gut homeostasis can be restored and also anti-inflammatory factors can be promoted by using the probiotics such as *Bacteroides*, *Lactobacillus*, *Firmicutes,* and *Bifidobacterium* to treat IBD in which gut microbiota plays a major role. Figure 9.2 depicts factors interlinked to GIT inflammation [80].

9.8 ALTERATIONS IN BARRIER FUNCTION AND IMPLICATIONS FOR IMMUNE ACTIVATION

Changes in the function of the intestinal barrier have an essential function in gastrointestinal inflammation, influencing the development of disease and immunological activation [87]. The largest human barrier, the intestinal mucosa serves as a barrier against food and microbial antigens. This barrier is one of the most important interfaces among gut bacteria as well as the human immune system. It is composed of layers of mucus which consist of antimicrobial peptides, secretory immunoglobulin A, acts as the initial contact with the gut microbiota and mucins, besides tight junctions with epithelial monolayer encompassing different cells and each cell has a distinct immune, endocrine, or protective role. Those each various cells such as enterocytes as well as specialized cells like Paneth, Goblet, and enterochromaffin cells, are present among others [88]. This maintains the integrity of the intestinal barrier, which prevents foreign substances from entering the bloodstream. This layer also forms a connection between the body's internal tissues, the tissue beneath it, and the region wherein the immune system combat infections. If the intestinal permeability increases then there is a possibility

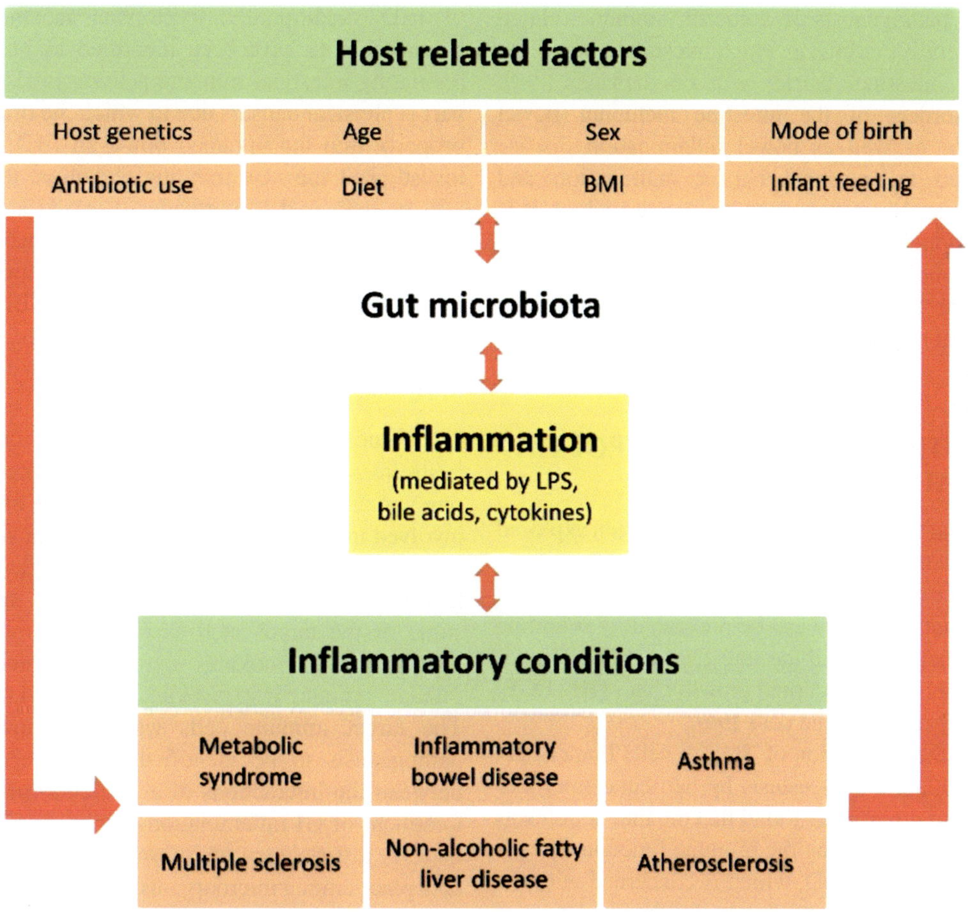

FIGURE 9.2 Interrelationships between the gut microbiota, inflammation and inflammatory conditions. (Adapted from Al Bander et al., 2020. © 2020 by the authors. Licensee MDPI, Basel, Switzerland. This article is an open access article distributed under the terms and conditions of the Creative Commons Attribution (CC BY) license; http://creativecommons.org/licenses/by/4.0/.)

of foreign substances invading the bloodstream through intestinal epithelial junctions and later on start affecting the different organs and systems in the body. The constituents of the intestinal microbiota and the homeostasis of the intestinal barrier, ultimately contribute to increased intestinal permeability [89]. According to Eleftheriotis G et al., conducted a study where gut dysbiosis occurred due to SARS-CoV-2 a viral infections by which mucosal immune cell function impaired and intestinal immunological balance and the barrier got disrupted, it may be not only due to viral infection but also due to cellular molecules changes, from autoimmune diseases immune responses got dysregulated or injury occurred in mucosa [90]. Under the effect of signals of commensal bacteria and metabolites, this dysbiosis strengthens the alterations of immunology and affects epithelial integrity of the intestine. Furthermore, the inflammatory responses start triggering due to the increased passage of antigen through epithelium leads to disruption of intestinal barrier [91]. Several aspects are critical to the pathophysiology of inflammatory bowel disease (IBD), including as altered mucus composition, decreased tight junctions, integrity that exacerbates inflammatory responses, and disruption of processes like autophagy and innate immunity [92]. Various diseases are associated due to the intestinal barrier disruption [93]. In the gastrointestinal tract, mast cells are widely distributed and are supposed to be related to the pathogenesis of a chronic immune-related disease that is ulcerative colitis, in which intestinal microbiota gets altered and intestinal barrier will be damaged [94]. Furthermore, disorders of the intestine including bowel irritable syndrome as well as bowel inflammation disease are interconnected to intestinal barrier malfunction and leaky gut syndrome. Moreover, intestinal barrier which it is weakened and implicated in various diseases other than the gastrointestinal conditions, including chronic liver disease, heart diseases, type 1 diabetes, celiac disease and obesity and a syndrome of viral sepsis [95].

9.9 TARGETING IMMUNE CELLS AS THERAPEUTIC STRATEGY TO CONTROL GIT INFLAMMATION

Many strategies that target the immune cell have been explored to manage GIT inflammation. As per the findings of Zhou et al., the intestinal tumor necrosis factor leads to cellular death and persistent inflammation, which can be prevented by shielding the intestinal epithelium. As in the intestine, a growth factor identified heparin-binding epidermal growth factor (HB-EGF) produces relatively rare immune cells [96].

By restraining the production of TGF- β cells Ihara et al., investigated the immune homeostasis by particular microbes in the intestine, also researchers found that the interaction was found to be complicated among the immune function and gut microbiota [97]. In patient IBD, which is concerned with the inflammation of intestine is seen due to the dysregulation of signaling of TGF- β. Controlling inflammation is crucial for controlling IBD and achieving the best possible outcomes for illnesses [98]. The intestinal epithelium's destruction can be prevented by certain other immune cells, among which are Group 3 innate lymphoid cells (ILC3s), which work independently of interleukin-22 and TNF-induced cell death. TNF inhibitors function by neutralizing the immune system's TNF protein, which can prevent inflammatory processes. According to Monteleone et al., intestine immunity will be boosted by TGF-β-mediated suppression of inflammatory responses in addition to the formation of immunological tolerance [99]. It's interesting to observe that another growth factor which is largely derived from human ILC3s named Heparin-binding EGF (HB-EGF) decreases the inflammation in the gut by suppressing the production of cytokines and chemokine. TGF-β also suppresses the cytokines, which are mediators of inflammation and also it promotes the immune tolerance, and by means of this TGF-β signaling pathway, the immunological reactions of the mucosa are controlled. Also it is very helpful in the treatment of IBD, which is taken into consideration as it is a potential agent. Any disturbance to the TGF-β signaling system will lead to development of inflammation in the intestine [100]. Zhang et al., emphasized in the maintaining intestinal epithelial integrity and mucosal immunity, also investigated the various metabolites mechanism and its role, some of the metabolites including serotonin, and d-tryptophan, indole and its derivatives, which shows the promising therapeutic practice in the management of IBD. Additionally, tryptophan metabolites formed by the microbiota have been identified as potential targets for regulating intestinal immune activity [101]. The challenging part is physical barriers due to which the drug may not deliver properly into the inflamed bowel in IBD. So Li et al., has investigated the new trending technique that is drug which will be delivered to particular diseased area in the form of nanoparticles, as it has the special physicochemical properties, enhanced permeation, as well as retention capabilities of nanoparticles in inflammatory tissues in IBD therapy [102].

9.10 FUTURE PROSPECTIVE

The immune system remained a crucial factor in gastrointestinal inflammation, and research on it has been ongoing. Understanding the pathogenesis of different diseases involved in GI inflammation and different signaling pathways of the various diseases which lead for the identification of novel therapeutic targets for gastrointestinal tract inflammation, some of the target such as particular immune cell surface receptors or chemokines, cytokines, which can be used to regulate immune responses and subside GI tract inflammation. The target immune cells which are involved in disease pathogenesis to be identified and understand the relation between the interactions of a particular immune cell for the treatment of GI inflammation-related diseases. Subsequently there is a deficiency of information on the interactions that take place among microorganisms, comprising those involving separate bacteria, variations in microbial activities, alterations to microbial metabolism, and the human microbiome. Even though the substantial advancements made in understanding the numerous facets of GIT inflammation, factors such as immune response, genetics, and microbial dysbiosis may still be involved. The various techniques which are targeted at the composition of the gut microbiota modification or beneficial microbial activities enhancement may have an impact on the functions of immune cell and support the GI homeostasis maintenance. Additionally, the nanoparticles will cross the tough barriers and deliver the drug to the particular area and resolve the GIT related issues. The initial beginning and gradually started expanding the incidence of diseases related to the GI inflammation has brought into the focus, both of which have increased the complexity of healthcare systems. Consequently, greater emphasis should be provided on the complex interactions which show a major part in the pathophysiology of GI inflammation [102].

9.11 CONCLUSION

Gastrointestinal inflammation, which is believed to be the body's immunological response to gut microbes, leads to intestinal epithelial barrier disruption and mucosal immune system dysregulation, which further causes damage of tissues and results in chronic inflammation. Because of which many GIT related diseases have emerged. Moreover a variety of factors such as dietary, stress, environmental, imbalance of microbiota leads to the development and progression of gastrointestinal inflammation. Furthermore, a variety of mediators, including chemokines and cytokines, have the capacity to specifically target and inhibit certain inflammatory cytokines and chemokines, which is useful in the treatment of GIT related diseases. It is important to know the relationship between gut microorganisms as well as the human immune system since the gut microbiota is crucial to maintaining immunological homeostasis. By understanding the interactions and pathogenesis of different signaling pathways and immune cells, there has been various treatments that came to existence. Although there are numerous treatments for the GI associated diseases, there are few that are incurable such as IBD etc., which requires thorough investigation.

REFERENCES

1. Xavier, R. J., & Podolsky, D. K. (2007). Unravelling the pathogenesis of inflammatory bowel disease. *Nature*, 448(7152), 427–434. https://doi.org/10.1038/nature06005
2. Malfertheiner, P., & Schulz, C. (2020). Peptic ulcer: Chapter closed?. *Digestive Diseases*, 38(2), 112–116. https://doi.org/10.1159/000505367
3. Ford, A. C., Sperber, A. D., Corsetti, M., & Camilleri, M. (2020). Functional gastrointestinal disorders 2 irritable bowel syndrome. *Lancet*, 396(10263), 1675–1688. https://doi.org/10.1016/s0140-6736(20)31548-8
4. Saulnier, D. M., Riehle, K., Mistretta, T. A., Diaz, M. A., Mandal, D., Raza, S., ... & Versalovic, J. (2011). Gastrointestinal microbiome signatures of pediatric patients with irritable bowel syndrome. *Gastroenterology*, 141(5), 1782–1791. https://doi.org/10.1053/j.gastro.2011.06.072
5. Cani, P. D. (2017). Gut microbiota—at the intersection of everything?. *Nature Reviews Gastroenterology & Hepatology*, 14(6), 321–322. https://doi.org/10.1038/nrgastro.2017.54
6. Belkaid, Y., & Hand, T. W. (2014). Role of the microbiota in immunity and inflammation. *Cell*, 157(1), 121–141. https://doi.org/10.1016/j.cell.2014.03.011
7. Shreiner, A. B., Kao, J. Y., & Young, V. B. (2015). The gut microbiome in health and in disease. *Current Opinion in Gastroenterology*, 31(1), 69–75. https://doi.org/10.1097/MOG.0000000000000139
8. Ramos, G. P., & Papadakis, K. A. (2019, January). Mechanisms of disease: inflammatory bowel diseases. In Guilherme Piovezani Ramos, MD, and Konstantinos A. Papadakis, MD (eds.) *Mayo Clinic Proceedings* (Vol. 94, No. 1, pp. 155–165). Elsevier. https://doi.org/10.1016/j.mayocp.2018.09.013
9. Zeissig, S., Bürgel, N., Günzel, D., Richter, J., Mankertz, J., Wahnschaffe, U., ... & Schulzke, J. D. (2007). Changes in expression and distribution of claudin 2, 5 and 8 lead to discontinuous tight junctions and barrier dysfunction in active Crohn's disease. *Gut*, 56(1), 61–72. https://doi.org/10.1136/gut.2006.104182
10. Ince, M. N., & Elliott, D. E. (2007). Immunologic and molecular mechanisms in inflammatory bowel disease. *Surgical Clinics of North America*, 87(3), 681–696. https://doi.org/10.1016/j.suc.2007.03.005
11. Lopetuso, L. R., Felice, C., Pugliese, D., Scaldaferri, F., Gasbarrini, A., & Armuzzi, A. (2014). Innate immune response and gut microbiota in inflammatory bowel disease. *CellR4*, 2, e1212. https://doi.org/10.32113/cellr4_20239_3411
12. Zheng, D., Liwinski, T., & Elinav, E. (2020). Interaction between microbiota and immunity in health and disease. *Cell Research*, 30(6), 492–506. https://doi.org/10.1038/s41422-020-0332-7
13. Aldars-García, L., Marin, A. C., Chaparro, M., & Gisbert, J. P. (2021). The interplay between immune system and microbiota in inflammatory bowel disease: A narrative review. *International Journal of Molecular Sciences*, 22(6), 3076. https://doi.org/10.3390/ijms22063076
14. Anand, P. K., Malireddi, R. S., Lukens, J. R., Vogel, P., Bertin, J., Lamkanfi, M., & Kanneganti, T. D. (2012). NLRP6 negatively regulates innate immunity and host defence against bacterial pathogens. *Nature*, 488(7411), 389–393. https://doi.org/10.1038/nature11250
15. Wlodarska, M., Thaiss, C. A., Nowarski, R., Henao-Mejia, J., Zhang, J. P., Brown, E. M., & Flavell, R. A. (2014). NLRP6 inflammasome orchestrates the colonic host-microbial interface by regulating goblet cell mucus secretion. *Cell*, 156(5), 1045–1059. https://doi.org/10.1016/j.cell.2014.01.026
16. Weaver, C. T., & Hatton, R. D. (2009). Interplay between the TH17 and TReg cell lineages: A (co-) evolutionary perspective. *Nature Reviews Immunology*, 9(12), 883–889. https://doi.org/10.1038/nri2660
17. Hueber, W., Sands, B. E., Lewitzky, S., Vandemeulebroecke, M., Reinisch, W., Higgins, P. D., & Secukinumab in Crohn's Disease Study Group. (2012). Secukinumab, a human anti-IL-17A monoclonal antibody, for moderate to severe Crohn's disease: Unexpected results of a randomised, double-blind placebo-controlled trial. *Gut*, 61(12), 1693–1700. https://doi.org/10.1136/gutjnl-2011-301668

18. Zeng, C., Shi, X., Zhang, B., Liu, H., Zhang, L., Ding, W., & Zhao, Y. (2012). The imbalance of Th17/Th1/ Tregs in patients with type 2 diabetes: Relationship with metabolic factors and complications. *Journal of Molecular Medicine*, 90, 175–186. https://doi.org/ 10.1007/s00109-011-0816-5
19. Khor, B., Gardet, A., & Xavier, R. J. (2011). Genetics and pathogenesis of inflammatory bowel disease. *Nature*, 474(7351), 307–317. https://doi.org/10.1038/nature10209
20. Scaldaferri, F., Gerardi, V., Lopetuso, L. R., Del Zompo, F., Mangiola, F., Boškoski, I., ... & Gasbarrini, A. (2013). Gut microbial flora, prebiotics, and probiotics in IBD: Their current usage and utility. *BioMed Research International*, 2013, 435268. https://doi.org/10.1155/2013/435268
21. Mikocka-Walus, A., Pittet, V., Rossel, J. B., von Känel, R., Anderegg, C., Bauerfeind, P., & Thorens, J. (2016). Symptoms of depression and anxiety are independently associated with clinical recurrence of inflammatory bowel disease. *Clinical Gastroenterology and Hepatology*, 14(6), 829–835. https://doi.org/10.1016/j.cgh.2015.12.045
22. Wallace, J. L. (2008). Prostaglandins, NSAIDs, and gastric mucosal protection: Why doesn't the stomach digest itself? *Physiological Reviews*, 88(4), 1547–1565. https://doi.org/ 10.1152/physrev.00004.2008
23. Reif, S., Klein, I., Lubin, F., Farbstein, M., Hallak, A., & Gilat, T. (1997). Pre-illness dietary factors in inflammatory bowel disease. *Gut*, 40(6), 754–760. https://doi.org/ 10.1136/gut.40.6.754
24. Russel, M. G., Engels, L. G., Muris, J. W., Limonard, C. B., Volovics, A., Brummer, R. J. M., & Stockbrugger, R. W. (1998). 'Modern life'in the epidemiology of inflammatory bowel disease: A case-control study with special emphasis on nutritional factors. *European Journal of Gastroenterology & Hepatology*, 10(3), 243–250.
25. Ananthakrishnan, A. N., Khalili, H., Konijeti, G. G., Higuchi, L. M., de Silva, P., Fuchs, C. S., & Chan, A. T. (2014). Long-term intake of dietary fat and risk of ulcerative colitis and Crohn's disease. *Gut*, 63(5), 776–784. https://doi.org/10.1136/gutjnl-2013-305304
26. Hart, A. R. (2009). Linoleic acid, a dietary N-6 polyunsaturated fatty acid, and the aetiology of ulcerative colitis-A European prospective cohort study. *Gut*, 58, 1577. https://doi.org/10.1136/gut.2008.169078
27. de Silva, P. S., Luben, R., Shrestha, S. S., Khaw, K. T., & Hart, A. R. (2014). Dietary arachidonic and oleic acid intake in ulcerative colitis etiology: A prospective cohort study using 7-day food diaries. *European Journal of Gastroenterology & Hepatology*, 26(1), 11–18. https://doi. org/10.1097/MEG.0b013e328365c372
28. Zhang, Y. Z., & Li, Y. Y. (2014). Inflammatory bowel disease: Pathogenesis. *World Journal of Gastroenterology: WJG*, 20(1), 91. https://doi.org/ 10.3748%2Fwjg.v20.i1.91
29. Arck, P., & Paus, R. (2007). From the brain-skin connection: The neuroendocrine-immune misalliance of stress and itch. *Neuroimmunomodulation*, 13(5–6), 347–356. https://doi.org/10.1159/000104863
30. Aguilera, M., Vergara, P., & Martínez, V. (2013). Stress and antibiotics alter luminal and wall-adhered microbiota and enhance the local expression of visceral sensory-related systems in mice. *Neurogastroenterology & Motility*, 25(8), e515–e529. https://doi.org/10.1111/nmo.12154
31. Gautam, A., Kumar, R., Chakraborty, N., Muhie, S., Hoke, A., Hammamieh, R., & Jett, M. (2018). Altered fecal microbiota composition in all male aggressor-exposed rodent model simulating features of post-traumatic stress disorder. *Journal of Neuroscience Research*, 96(7), 1311–1323. https://doi.org/10.1002/jnr.24229
32. Mitsuoka, T. (2014). Establishment of intestinal bacteriology. *Bioscience of Microbiota, Food and Health*, 33(3), 99–116. https://doi.org/10.12938/bmfh.33.99
33. Galley, J. D., Parry, N. M., Ahmer, B. M., Fox, J. G., & Bailey, M. T. (2017). The commensal microbiota exacerbate infectious colitis in stressor-exposed mice. *Brain, Behavior, and Immunity*, 60, 44–50. https://doi.org/ 10.1016/j.bbi.2016.09.010
34. Gao, X., Cao, Q., Cheng, Y., Zhao, D., Wang, Z., Yang, H., ... & Yang, Y. (2018). Chronic stress promotes colitis by disturbing the gut microbiota and triggering immune system response. *Proceedings of the National Academy of Sciences*, 115(13), E2960–E2969. https://doi.org/10.1073/ pnas.1720696115
35. Becker, C., Neurath, M. F., & Wirtz, S. (2015). The intestinal microbiota in inflammatory bowel disease. *ILAR Journal*, 56(2), 192–204. https://doi.org/10.1093/ilar/ ilv030
36. Zhang, Y. J., Li, S., Gan, R. Y., Zhou, T., Xu, D. P., & Li, H. B. (2015). Impacts of gut bacteria on human health and diseases. *International Journal of Molecular Sciences*, 16(4), 7493–7519. https://doi.org/10.3390/ijms16047493
37. Morgan, X. C., Tickle, T. L., Sokol, H., Gevers, D., Devaney, K. L., Ward, D. V., ... & Huttenhower, C. (2012). Dysfunction of the intestinal microbiome in inflammatory bowel disease and treatment. *Genome Biology*, 13, 1–18. https://doi.org/10.1186/gb-2012-13-9-r79
38. Balfour Sartor, R. (1997). Enteric microflora in IBD: Pathogens or commensals?. *Inflammatory Bowel Diseases*, 3(3), 230–235. https://doi.org/10.1097/00054 725-199709000-00008
39. Cao, Y., Shen, J., & Ran, Z. H. (2014). Association between Faecalibacterium prausnitzii reduction and inflammatory bowel disease: A meta-analysis and systematic review of the literature. *Gastroenterology Research and Practice*, 2014, 872725. https://doi.org/10.1155/2014/872725
40. Louis, P., & Flint, H. J. (2009). Diversity, metabolism and microbial ecology of butyrate-producing bacteria from the human large intestine. *FEMS Microbiology Letters*, 294(1), 1–8. https://doi.org/10.1111/j.1574-6968.2009.01514.x
41. Zhou, L., Zhang, M., Wang, Y., Dorfman, R. G., Liu, H., Yu, T., & Yu, C. (2018). Faecalibacterium prausnitzii produces butyrate to maintain Th17/Treg balance and to ameliorate colorectal colitis by inhibiting histone deacetylase 1. *Inflammatory Bowel Diseases*, 24(9), 1926–1940. https:// doi.org/10.1093/ibd/izy182
42. Furusawa, Y., Obata, Y., Fukuda, S., Endo, T. A., Nakato, G., Takahashi, D., ... & Ohno, H. (2013). Commensal microbe-derived butyrate induces the differentiation of colonic regulatory T cells. *Nature*, 504(7480), 446–450. https://doi.org/10.1038/nature12721
43. Isobe, J., Maeda, S., Obata, Y., Iizuka, K., Nakamura, Y., Fujimura, Y., ... & Hase, K. (2020). Commensal-bacteria-derived butyrate promotes the T-cell-independent IgA response in the colon. *International Immunology*, 32(4), 243–258. https://doi.org/10.1093/intimm/dxz078

44. Janeway Jr, C. A., & Medzhitov, R. (2002). Innate immune recognition. *Annual Review of Immunology*, 20(1), 197–216. https://doi.org/10.1146/annurev.immunol.20.083001.084359
45. Turner, J. R. (2009). Intestinal mucosal barrier function in health and disease. *Nature Reviews Immunology*, 9(11), 799–809. https://doi.org/10.1038/nri2653
46. Merad, M., Sathe, P., Helft, J., Miller, J., & Mortha, A. (2013). The dendritic cell lineage: Ontogeny and function of dendritic cells and their subsets in the steady state and the inflamed setting. *Annual Review of Immunology*, 31, 563–604. https://doi.org/10.1146/annurev-immunol-020711-074950
47. Wynn, T. A., Chawla, A., & Pollard, J. W. (2013). Macrophage biology in development, homeostasis and disease. *Nature*, 496(7446), 445–455. https://doi.org/10.1038/nature12034
48. Abraham, S. N., & St. John, A. L. (2010). Mast cell-orchestrated immunity to pathogens. *Nature Reviews Immunology*, 10(6), 440–452. https://doi.org/10.1038/nri2782
49. Rakoff-Nahoum, S., & Medzhitov, R. (2007). Regulation of spontaneous intestinal tumorigenesis through the adaptor protein MyD88. *Science*, 317(5834), 124–127. https://doi.org/10.1126/science.1140488
50. Bain, C. C., & Mowat, A. M. (2014). Macrophages in intestinal homeostasis and inflammation. *Immunological Reviews*, 260(1), 102–117. https://doi.org/10.1111/imr.12192
51. Rescigno, M., & Di Sabatino, A. (2009). Dendritic cells in intestinal homeostasis and disease. *The Journal of Clinical Investigation*, 119(9), 2441–2450. https://doi.org/10.1172/JCI39134
52. Kolaczkowska, E., & Kubes, P. (2013). Neutrophil recruitment and function in health and inflammation. *Nature Reviews Immunology*, 13(3), 159–175. https://doi.org/10.1038/nri3399
53. Mowat, A. M., & Agace, W. W. (2014). Regional specialization within the intestinal immune system. *Nature Reviews Immunology*, 14(10), 667–685. https://doi.org/10.1038/nri3738
54. Scharl, M., & Rogler, G. (2010). Microbial sensing by the intestinal epithelium in the pathogenesis of inflammatory bowel disease. *International Journal of Inflammation*, 2010, 671258. https://doi.org/10.4061/2010/671258
55. Macpherson, A. J., & Uhr, T. (2004). Induction of protective IgA by intestinal dendritic cells carrying commensal bacteria. *Science*, 303(5664), 1662–1665. https://doi.org/10.1126/science.1091334
56. Uhlig, H. H., & Powrie, F. (2018). Translating immunology into therapeutic concepts for inflammatory bowel disease. *Annual Review of Immunology*, 36, 755–781. https://doi.org/10.1146/annurev-immunol-042617-053055
57. Loh, L., Ivarsson, M. A., Michaëlsson, J., Sandberg, J. K., & Nixon, D. F. (2014). Invariant natural killer T cells developing in the human fetus accumulate and mature in the small intestine. *Mucosal Immunology*, 7(5), 1233–1243. https://doi.org/10.1038/mi.2014.13
58. Frank, D. N., St. Amand, A. L., Feldman, R. A., Boedeker, E. C., Harpaz, N., & Pace, N. R. (2007). Molecular-phylogenetic characterization of microbial community imbalances in human inflammatory bowel diseases. *Proceedings of the National Academy of Sciences*, 104(34), 13780–13785. https://doi.org/10.1073/pnas.0706625104
59. Scanlan, P. D., Shanahan, F., O'Mahony, C., & Marchesi, J. R. (2006). Culture-independent analyses of temporal variation of the dominant fecal microbiota and targeted bacterial subgroups in Crohn's disease. *Journal of Clinical Microbiology*, 44(11), 3980–3988. https://doi.org/10.1128/jcm.00312-06
60. Xue, Y., He, J., Xiao, C., Guo, Y., Fu, T., Liu, J., ... & Li, Z. (2018). The mouse autonomic nervous system modulates inflammation and epithelial renewal after corneal abrasion through the activation of distinct local macrophages. *Mucosal Immunology*, 11(5), 1496–1511. https://doi.org/10.1038/s41385-018-0031-6
61. Randolph, G. J., Ivanov, S., Zinselmeyer, B. H., & Scallan, J. P. (2017). The lymphatic system: Integral roles in immunity. *Annual Review of Immunology*, 35, 31–52. https://doi.org/10.1146/annurev-immunol-041015-055354
62. Turner, J. R. (2009). Intestinal mucosal barrier function in health and disease. *Nature Reviews Immunology*, 9(11), 799–809. https://doi.org/10.1038/nri2653
63. Salzman, N. H. (2010). Paneth cell defensins and the regulation of the microbiome: Detente at mucosal surfaces. *Gut Microbes*, 1(6), 401–406. https://doi.org/10.4161/gmic.1.6.14076
64. Bevins, C. L., & Salzman, N. H. (2011). Paneth cells, antimicrobial peptides and maintenance of intestinal homeostasis. *Nature Reviews Microbiology*, 9(5), 356–368. https://doi.org/10.1038/nrmicro2546
65. Schauber, J., Svanholm, C., Termen, S., Iffland, K., Menzel, T., Scheppach, W., ... & Gudmundsson, G. (2003). Expression of the cathelicidin LL-37 is modulated by short chain fatty acids in colonocytes: Relevance of signalling pathways. *Gut*, 52(5), 735–741. https://doi.org/10.1136/gut.52.5.735
66. Neurath, M. F. (2014). Cytokines in inflammatory bowel disease. *Nature Reviews Immunology*, 14(5), 329–342. https://doi.org/10.1038/nri3661
67. Gabrilovich, D. I., & Nagaraj, S. (2009). Myeloid-derived suppressor cells as regulators of the immune system. *Nature Reviews Immunology*, 9(3), 162–174. https://doi.org/10.1038/nri2506
68. Maloy, K. J., & Powrie, F. (2011). Intestinal homeostasis and its breakdown in inflammatory bowel disease. *Nature*, 474(7351), 298–306. https://doi.org/10.1038/nature10208
69. Moparthi, L., & Koch, S. (2019). Wnt signaling in intestinal inflammation. *Differentiation*, 108, 24–32. https://doi.org/10.1016/j.diff.2019.01.002
70. Gambhir, S., Vyas, D., Hollis, M., Aekka, A., & Vyas, A. (2015). Nuclear factor kappa B role in inflammation associated gastrointestinal malignancies. *World Journal of Gastroenterology. WJG*, 21(11), 3174. https://doi.org/10.3748%2Fwjg.v21.i11.3174
71. Lamb, A., Yang, X. D., Tsang, Y. H. N., Li, J. D., Higashi, H., Hatakeyama, M., ... & Chen, L. F. (2009). Helicobacter pylori CagA activates NF-κB by targeting TAK1 for TRAF6-mediated Lys 63 ubiquitination. *EMBO Reports*, 10(11), 1242–1249. https://doi.org/10.1038/embor.2009.210
72. Vyas, D., Robertson, C. M., Stromberg, P. E., Martin, J. R., Dunne, W. M., Houchen, C. W., & Coopersmith, C. M. (2007). Epithelial apoptosis in mechanistically

distinct methods of injury in the murine small intestine. *Histology and Histopathology*, 22(6), 623. https://doi.org/10.14670%2Fhh-22.623

73. Brown, K. D., Claudio, E., & Siebenlist, U. (2008). The roles of the classical and alternative nuclear factor-kappaB pathways: Potential implications for autoimmunity and rheumatoid arthritis. *Arthritis Research & Therapy*, 10, 1–14. https://doi.org/10.1186/ar2457
74. Lei, H., Crawford, M. S. S., & McCole, D. F. (2021). JAK-STAT pathway regulation of intestinal permeability: Pathogenic roles and therapeutic opportunities in inflammatory bowel disease. *Pharmaceuticals*, 14(9), 840. https://doi.org/10.3390/ph14090840
75. Jridi, I., Canté-Barrett, K., Pike-Overzet, K., & Staal, F. J. (2021). Inflammation and Wnt signaling: Target for immunomodulatory therapy?. *Frontiers in Cell and Developmental Biology*, 8, 615131. https://doi.org/10.3389/fcell.2020.615131
76. Taheri, N., Choi, E. L., Nguyen, V. T. T., Chandra, A., & Hayashi, Y. (2023). Wnt signaling in the gastrointestinal tract in health and disease. *Physiologia*, 3(1), 86–97. https://doi.org/10.3390/physiologia3010007
77. Frei, S. M., Hemsley, C., Pesch, T., Lang, S., Weber, A., Jehle, E., ... & Scharl, M. (2013). The role for dickkopf-homolog-1 in the pathogenesis of Crohn's disease-associated fistulae. *PLoS One*, 8(11), e78882. https://doi.org/10.1371/journal.pone.0078882
78. Hatamzade Esfahani, N., & Day, A. S. (2023). The role of TGF-β, activin and follistatin in inflammatory bowel disease. *Gastrointestinal Disorders*, 5(2), 167–186. https://doi.org/10.3390/gidisord5020015
79. Hong, S., Lee, H. J., Kim, S. J., & Hahm, K. B. (2010). Connection between inflammation and carcinogenesis in gastrointestinal tract: Focus on TGF-β signaling. *World Journal of Gastroenterology: WJG*, 16(17), 2080. https://doi.org/10.3748%2Fwjg.v16.i17.2080
80. Al Bander, Z., Nitert, M. D., Mousa, A., & Naderpoor, N. (2020). The gut microbiota and inflammation: An overview. *International Journal of Environmental Research and Public Health*, 17(20), 7618. https://doi.org/10.3390/ijerph17207618
81. Zheng, D., Liwinski, T., & Elinav, E. (2020). Interaction between microbiota and immunity in health and disease. *Cell Research*, 30(6), 492–506. https://doi.org/10.1038/s41422-020-0332-7
82. Lobionda, S., Sittipo, P., Kwon, H. Y., & Lee, Y. K. (2019). The role of gut microbiota in intestinal inflammation with respect to diet and extrinsic stressors. *Microorganisms*, 7(8), 271. https://doi.org/10.3390/microorganisms7080271
83. Lew, S. Q., & Radhakrishnan, J. (2020). Chronic kidney disease and gastrointestinal disorders. *Chronic Renal Disease*, 1(2), 521–539. https://doi.org/10.1016/B978-0-12-815876-0.00033-4
84. Ghosh, T. S., Shanahan, F., & O'Toole, P. W. (2022). The gut microbiome as a modulator of healthy ageing. *Nature Reviews Gastroenterology & Hepatology*, 19(9), 565–584. https://doi.org/10.1038/s41575-022-00605-x
85. Stolfi, C., Maresca, C., Monteleone, G., & Laudisi, F. (2022). Implication of intestinal barrier dysfunction in gut dysbiosis and diseases. *Biomedicines*, 10(2), 289. https://doi.org/10.3390/biomedicines10020289
86. Al Bander, Z., Nitert, M. D., Mousa, A., & Naderpoor, N. (2020). The gut microbiota and inflammation: An overview. *International Journal of Environmental Research and Public Health*, 17(20), 7618. https://doi.org/10.3390/ijerph17207618
87. Aleman, R. S., Moncada, M., & Aryana, K. J. (2023). Leaky gut and the ingredients that help treat it: A review. *Molecules*, 28(2), 619. https://doi.org/10.3390/molecules28020619
88. Eleftheriotis, G., Tsounis, E. P., Aggeletopoulou, I., Triantos, C., & Assimakopoulos, S. F. (2023). Alterations in gut immunological barrier in SARS-CoV-2 infection and their prognostic potential. *Frontiers in Immunology*, 14, 1129190. https://doi.org/10.3389/fimmu.2023.1129190
89. Vindigni, S. M., Zisman, T. L., Suskind, D. L., & Damman, C. J. (2016). The intestinal microbiome, barrier function, and immune system in inflammatory bowel disease: A tripartite pathophysiological circuit with implications for new therapeutic directions. *Therapeutic Advances in Gastroenterology*, 9(4), 606–625. https://doi.org/10.1177/1756283X16644242
90. Vancamelbeke, M., & Vermeire, S. (2017). The intestinal barrier: A fundamental role in health and disease. *Expert Review of Gastroenterology & Hepatology*, 11(9), 821–834. https://doi.org/10.1080/17474124.2017.1343143
91. Haroun, E., Kumar, P. A., Saba, L., Kassab, J., Ghimire, K., Dutta, D., & Lim, S. H. (2023). Intestinal barrier functions in hematologic and oncologic diseases. *Journal of Translational Medicine*, 21(1), 233. https://doi.org/10.1186/s12967-023-04091-w
92. Shao, M., Liu, J., & Luo, H. (2023). Colitis aggravated by Mrgprb2 knockout is associated with altered immune response, intestinal barrier function and gut microbiota. *Experimental Physiology*, 108(1), 63–75. https://doi.org/10.1113/EP090635
93. Yu, L. C. H., Wang, J. T., Wei, S. C., & Ni, Y. H. (2012). Host-microbial interactions and regulation of intestinal epithelial barrier function: From physiology to pathology. *World Journal of Gastrointestinal Pathophysiology*, 3(1), 27. https://doi.org/10.4291%2Fwjgp.v3.i1.27
94. Zhou, L., Zhou, W., Joseph, A. M., Chu, C., Putzel, G. G., Fang, B., ... & Sonnenberg, G. F. (2022). Group 3 innate lymphoid cells produce the growth factor HB-EGF to protect the intestine from TNF-mediated inflammation. *Nature Immunology*, 23(2), 251–261. https://doi.org/10.1038/s41590-021-01110-0
95. Ihara, S., Hirata, Y., & Koike, K. (2017). TGF-β in inflammatory bowel disease: A key regulator of immune cells, epithelium, and the intestinal microbiota. *Journal of Gastroenterology*, 52, 777–787. https://doi.org/10.1007/s00535-017-1350-1
96. McInnes, I. B., & Gravallese, E. M. (2021). Immune-mediated inflammatory disease therapeutics: Past, present and future. *Nature Reviews Immunology*, 21(10), 680–686. https://doi.org/10.1038/s41577-021-00603-1
97. Monteleone, G., Kumberova, A., Croft, N. M., McKenzie, C., Steer, H. W., & MacDonald, T. T. (2001). Blocking Smad7 restores TGF-β1 signaling in chronic inflammatory bowel disease. *The Journal of Clinical Investigation*, 108(4), 601–609. https://doi.org/10.1172/JCI12821

98. Song, D., Lai, L., & Ran, Z. (2020). Metabolic regulation of group 3 innate lymphoid cells and their role in inflammatory bowel disease. *Frontiers in Immunology*, 11, 580467. https://doi.org/10.3389/fimmu.2020.580467
99. Zhang, J., Zhu, S., Ma, N., Johnston, L. J., Wu, C., & Ma, X. (2021). Metabolites of microbiota response to tryptophan and intestinal mucosal immunity: A therapeutic target to control intestinal inflammation. *Medicinal Research Reviews*, 41(2), 1061–1088. https://doi.org/10.1002/med.21752
100. Li, D. F., Yang, M. F., Xu, H. M., Zhu, M. Z., Zhang, Y., Tian, C. M., ... & Wang, L. S. (2022). Nanoparticles for oral delivery: Targeted therapy for inflammatory bowel disease. *Journal of Materials Chemistry B*, 10(31), 5853–5872. https://doi.org/10.1039/D2TB01190E
101. Kostic, A. D., Xavier, R. J., & Gevers, D. (2014). The microbiome in inflammatory bowel disease: Current status and the future ahead. *Gastroenterology*, 146(6), 1489–1499. https://doi.org/10.1053/j.gastro.2014.02.009
102. Ahlawat, S., Kumar, P., Mohan, H., Goyal, S., & Sharma, K. K. (2021). Inflammatory bowel disease: Tri-directional relationship between microbiota, immune system and intestinal epithelium. *Critical Reviews in Microbiology*, 47(2), 254–273. https://doi.org/10.1080/1040841X.2021.1876631

10 Advancements in Diagnostic Techniques and Technology for Gastroenterology

Sunny Rathee[1,2]*, Vishal Pandey*[1]*, and Sanjay K. Jain*[1]*
[1]Pharmaceutics Research Projects Laboratory, Department of Pharmaceutical Sciences, Dr. Harisingh Gour Vishwavidyalaya, Sagar, Madhya Pradesh, India
[2]College of Pharmaceutical Sciences (COPS), School of Health Sciences, Dayananda Sagar University, Devarakaggalahalli, Harohalli, Kanakapura Road, Ramanagara Dt., Bengaluru, Karnataka, India

10.1 INTRODUCTION

Within the realm of gastroenterology, the pursuit of effective treatment strategies for gastrointestinal inflammation has ushered in an era marked by the intricate interplay of diagnostic precision, technological breakthroughs, and the profound influence of gut microbiota. This chapter delves into the evolving landscape, emphasizing the therapeutic potential of gut microbiota-based targeting (Chauhan & Kumar, 2023). Commencing with an insightful overview of diagnostic methodologies, traditional methods like endoscopy and imaging techniques converge with cutting-edge approaches, including blood tests and state-of-the-art microbiome analyses. This foundational understanding sets the stage for a meticulous examination of the intricate relationship between gut microbiota and gastrointestinal inflammation. The chapter unfolds with a detailed exploration of dysbiosis the disruption of the delicate balance within the gut microbiota and its implications for targeted therapies. Central to this discussion is the pivotal role played by the gut microbiota in modulating the inflammatory milieu within the gastrointestinal tract. Venturing further, the narrative transitions into a comprehensive analysis of advancements in targeted treatment approaches (Sanders et al., 2021). Probiotics, prebiotics, fecal microbiota transplantation (FMT), and microbiota-modulating pharmaceuticals present a panoramic view of the arsenal available to clinicians and researchers in the pursuit of precision medicine for gastrointestinal inflammation. A particularly illuminating exploration of personalized medicine unfolds, showcasing its promise in tailoring therapeutic interventions to the unique profiles of individual patients (Parigi et al., 2023).

The technological revolution in gastroenterology takes center stage as we shift focus to innovations in monitoring and assessment. Wearable devices, artificial intelligence, and remote monitoring emerge as game-changers, augmenting our ability to comprehensively understand and manage gastrointestinal health. This section underscores the transformative power of technology in shaping the future of gastroenterological care (Kuo & Chiu, 2021). However, the path to progress is not without its challenges. The chapter candidly addresses the current limitations of therapies, navigates regulatory considerations, and boldly envisions future technological trends that promise to reshape the landscape of gastroenterology. Through insightful case studies and clinical applications, tangible examples of successful gut microbiota-based treatments are presented, enriched by the voices of patients and ongoing research endeavors. Approaching the conclusion of this comprehensive chapter, key findings are distilled, and their implications for the future of gastrointestinal inflammation treatment are underscored. The chapter stands as a valuable resource for clinicians, researchers, and healthcare professionals navigating the rapidly evolving landscape of gut microbiota-based strategies, a landscape that holds immense promise for the effective management of gastrointestinal inflammation (Jones & Neish, 2021).

10.2 GUT MICROBIOTA AND GASTROINTESTINAL INFLAMMATION

The gut microbiota, often referred to as the gut microbiome, is a vast and intricate community of microorganisms that reside in the gastrointestinal tract, exerting a profound influence on host health and various physiological processes. Comprising bacteria, viruses, fungi, and other microbes, this dynamic ecosystem plays a crucial role in maintaining a delicate balance that contributes to digestive function, immune system regulation, and metabolic processes (Al Bander et al., 2020).

The gut microbiota is incredibly diverse, harboring an estimated 100 trillion microorganisms, representing thousands of different species. While bacteria are the most predominant members, archaea, viruses, fungi, and other microorganisms contribute to the overall complexity of the community. The

DOI: 10.1201/9781003493143-10

composition of the gut microbiota varies widely among individuals, influenced by factors such as genetics, diet, lifestyle, and environmental exposures (Dupont et al., 2020).

10.2.1 Functions of the Gut Microbiota

10.2.1.1 Digestion and Nutrient Metabolism

The gut microbiota plays a pivotal role in the breakdown of complex carbohydrates, fibers, and other dietary components that are resistant to digestion by human enzymes. Microbes in the colon ferment these substrates, producing short-chain fatty acids (SCFAs) as byproducts. SCFAs not only serve as an energy source for the host but also contribute to the maintenance of the intestinal epithelial barrier (Rowland et al., 2018).

10.2.1.2 Immune System Modulation

The gut microbiota interacts closely with the host immune system, contributing to the development and education of immune cells. Commensal bacteria aid in distinguishing between harmless substances and potential threats, promoting immune tolerance to prevent inappropriate inflammatory responses. Dysregulation of this balance can lead to inflammatory conditions within the gastrointestinal tract (Yoo et al., 2020).

10.2.1.3 Protection Against Pathogens

The gut microbiota serves as a protective barrier against invading pathogens by occupying ecological niches and producing antimicrobial substances. Commensal bacteria compete for resources with potential pathogens, preventing their colonization and overgrowth (Iacob et al., 2019).

10.2.1.4 Metabolism and Energy Homeostasis

Emerging research suggests a connection between the gut microbiota and metabolic processes, influencing host metabolism and energy balance. Imbalances in the microbiota composition have been associated with conditions such as obesity and metabolic syndrome (Amabebe et al., 2020).

10.2.1.5 Neurological and Behavioral Effects

The gut-brain axis, a bidirectional communication system between the gut and the central nervous system, is influenced by the gut microbiota. Microbial metabolites and signaling molecules can impact brain function and behavior, with potential implications for mental health (Morais et al., 2021).

10.2.2 Factors Influencing Gut Microbiota Composition

10.2.2.1 Diet

Dietary patterns, particularly the intake of fiber, prebiotics, and fermented foods, significantly influence the composition and diversity of the gut microbiota (Hasan & Yang, 2019).

10.2.2.2 Drugs

Antibiotics can alter the gut microbiota by selectively targeting both pathogenic and commensal bacteria, leading to disruptions in microbial balance (Ribeiro et al., 2020).

10.2.2.3 Age

The gut microbiota evolves throughout life, with distinct microbial compositions in infancy, adulthood, and old age (Rinninella et al., 2019).

10.2.2.4 Geography and Lifestyle

Environmental factors, including geography, sanitation, and lifestyle choices, contribute to variations in gut microbiota composition among populations (Parizadeh & Arrieta, 2023).

10.2.2.5 Dysbiosis and Health Implications

Dysbiosis refers to an imbalance or disruption in the composition and function of the gut microbiota. This condition has been associated with various health conditions, including inflammatory bowel disease (IBD), irritable bowel syndrome (IBS), metabolic disorders, and autoimmune diseases. Understanding dysbiosis is crucial for developing targeted interventions aimed at restoring a healthy microbial balance (Brüssow, 2020).

10.2.3 Therapeutic Strategies

10.2.3.1 Probiotics

Live microorganisms, typically beneficial bacteria, are administered to confer health benefits by restoring or maintaining a balanced gut microbiota (Zhang et al., 2019).

10.2.3.2 Prebiotics

Substances that promote the growth and activity of beneficial microorganisms in the gut are typically non-digestible fibers and compounds (Rezende et al., 2021).

10.2.3.3 Fecal Microbiota Transplantation (FMT)

The transfer of fecal material from a healthy donor to a recipient aims to restore diverse and functional gut microbiota (Bibbò et al., 2020).

10.2.3.4 Microbiota-Modulating Pharmaceuticals

Developing pharmaceutical agents that selectively target dysbiotic microbial populations while preserving beneficial microbes.

The gut microbiota is a dynamic and intricate ecosystem with far-reaching implications for host health. Its influence extends beyond digestion to impact immune function, metabolism, and even mental well-being. Ongoing research continues to unveil the complexities of this symbiotic relationship, offering insights that may lead to innovative therapeutic approaches and personalized interventions for a range of health conditions. As our understanding of the gut microbiota deepens, it opens avenues for harnessing its potential to promote health and prevent disease (Alshehri et al., 2021).

Gastrointestinal inflammation represents a broad category of disorders characterized by inflammation within the digestive tract, encompassing a spectrum of conditions that range in severity and clinical manifestations. The gastrointestinal (GI) tract, comprising the stomach, small intestine, and large intestine, is vulnerable to inflammation due to its continuous exposure to various environmental factors, dietary antigens, and the intricate interplay between the gut microbiota and the host immune system (Bhushan et al., 2019).

I) Inflammatory Bowel Disease (IBD)

IBD comprises chronic inflammatory conditions, primarily Crohn's disease and ulcerative colitis, where the immune system mistakenly targets the digestive tract. Crohn's disease can affect any part of the GI tract and is characterized by deep-seated inflammation, while ulcerative colitis predominantly involves the colon and rectum, causing inflammation in the mucosal layer. Both conditions result in debilitating symptoms such as abdominal pain, diarrhea, and weight loss, significantly impacting the quality of life for those affected (Ribaldone et al., 2019).

II) Irritable Bowel Syndrome (IBS)

In contrast to IBD, irritable bowel syndrome (IBS) is a functional GI disorder characterized by abdominal pain, bloating, and alterations in bowel habits without evidence of inflammation or structural abnormalities. While IBS does not involve inflammation in the classical sense, low-grade inflammation and alterations in the gut microbiota have been observed in some individuals with IBS, highlighting the complexity of these conditions (Leylabadlo et al., 2022).

III) Celiac Disease

Celiac disease is an autoimmune disorder triggered by the ingestion of gluten, a protein found in wheat, barley, and rye. In individuals with celiac disease, the immune system responds to gluten by attacking the small intestine, leading to inflammation and damage to the villi, small finger-like projections that aid in nutrient absorption. This can result in malabsorption of nutrients and a range of symptoms, including diarrhea, weight loss, and fatigue (Sharma et al., 2020).

IV) Gastroenteritis

Acute gastrointestinal inflammation often referred to as gastroenteritis, can be caused by infections, viruses, or ingesting contaminated food or water. Inflammation in the stomach and intestines leads to symptoms such as nausea, vomiting, abdominal cramps, and diarrhea. While gastroenteritis is typically short-lived, it can be severe and may require medical attention, especially in vulnerable populations such as young children and the elderly (Negrut et al., 2020). Vraious host-related factors influencing gut microbiota composition and various inflammatory conditions are represented in Figure 10.1.

The underlying causes of gastrointestinal inflammation are multifactorial and may involve genetic predispositions, environmental factors, dysregulation of the immune system, and alterations in the gut microbiota. The intricate crosstalk between these factors contributes to the initiation and perpetuation of inflammation within the GI tract. Table 10.1 describes the role of pathogenic gut microbiota in gastrointestinal diseases.

Management of gastrointestinal inflammation often involves a multifaceted approach, including dietary modifications, pharmacological interventions, and, in some cases, surgical procedures. Recent advancements in therapeutic strategies have explored the role of gut microbiota-based targeting, aiming to restore the balance of the microbiome and modulate the immune response.

Gastrointestinal inflammation represents a diverse array of conditions that impact millions of individuals worldwide. The complex interplay between genetic, environmental, and microbial factors underscores the challenges in understanding and effectively treating these disorders. As our understanding of the gut microbiota and its role in gastrointestinal health deepens, novel therapeutic approaches hold promise for more targeted and personalized interventions in the management of these inflammatory conditions.

The human gastrointestinal (GI) tract harbors a diverse and dynamic ecosystem known as the gut microbiota, comprising trillions of microorganisms. This intricate community, dominated by bacteria but also including viruses, fungi, and other microorganisms, plays a pivotal role in maintaining homeostasis within the GI tract. The interplay between the gut microbiota and the host's immune system is a finely tuned relationship, with implications for the development and resolution of gastrointestinal inflammation (Srivastava et al., 2021).

A) Microbial Diversity and Composition

The gut microbiota is characterized by its remarkable diversity, consisting of numerous bacterial species, each contributing uniquely to the overall ecosystem. The composition of the gut microbiota is influenced by a myriad of factors, including genetics, diet, age, and environmental exposures. This microbial diversity is crucial for the proper functioning of the digestive system and the overall health of the host (Lozupone et al., 2012).

B) Protective Role Against Pathogens

One primary function of the gut microbiota is to act as a barrier against invading pathogens. Commensal bacteria occupy ecological niches, competing for resources and preventing the colonization of harmful microorganisms. Additionally, these beneficial microbes produce antimicrobial substances that help regulate the growth of potential pathogens, thereby contributing to the overall defense mechanisms of the GI tract (Pickard et al., 2017).

C) Immune System Modulation

The gut microbiota actively interacts with the host's immune system, playing a fundamental role in the development and education of immune cells. Commensal bacteria help train the

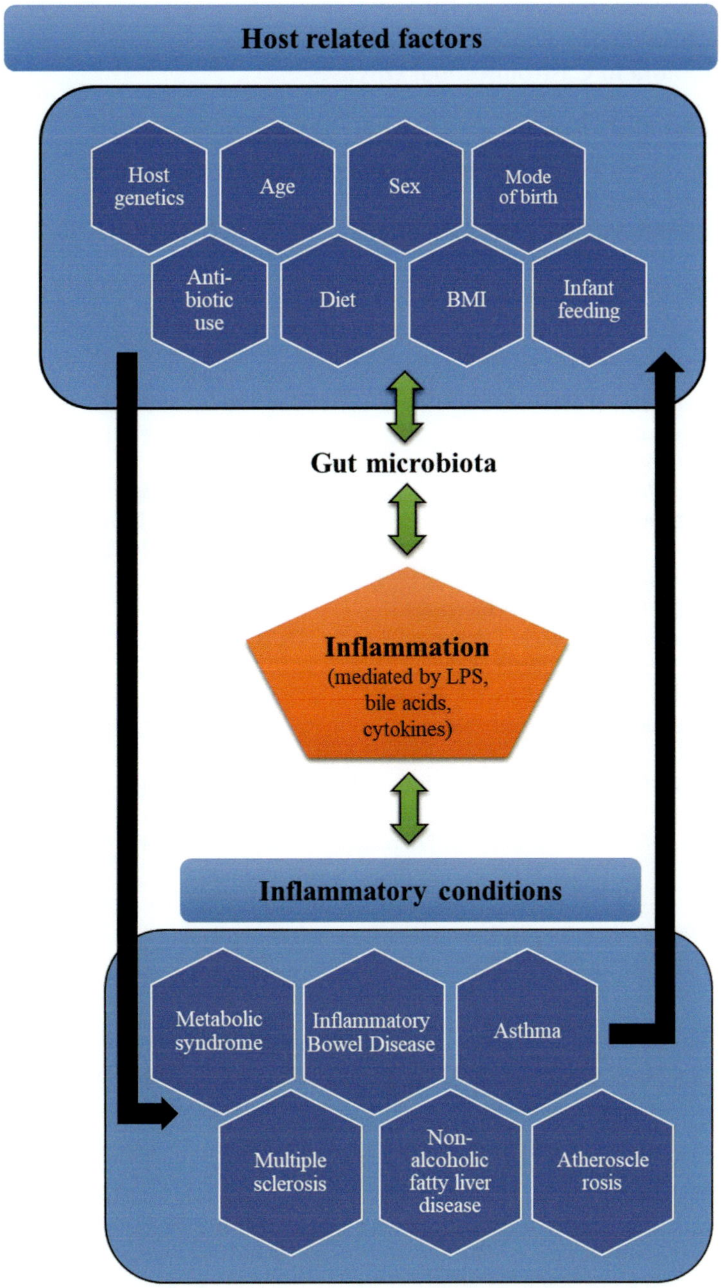

FIGURE 10.1 Host-related factors influencing gut microbiota composition and various inflammatory conditions.

immune system to recognize and tolerate harmless substances, promoting immune tolerance. This delicate balance is essential for preventing inappropriate immune responses and maintaining a state of equilibrium within the GI tract (Zheng et al., 2020).

D) Influence on the Intestinal Epithelial Barrier

The gut microbiota contributes to the integrity of the intestinal epithelial barrier, a crucial line of defense that separates the microbiota from the host's tissues. A healthy gut microbiota promotes the production of mucus and other protective factors, reinforcing the barrier function and preventing the translocation of microbes and microbial products into the bloodstream. Dysbiosis, or an imbalance in the gut microbiota, can compromise this barrier, leading to increased permeability and the initiation of inflammatory responses (Gieryńska et al., 2022).

E) Metabolic Functions and Short-Chain Fatty Acids (SCFAs)

Microbes in the gut participate in the metabolism of dietary components, producing metabolites such as short-chain fatty acids (SCFAs) through the fermentation of fibers. SCFAs, particularly acetate, propionate, and butyrate, serve as an energy source for the host and contribute to the maintenance of the intestinal epithelial barrier. These metabolites also have

TABLE 10.1
Role of Pathogenic Gut Microbiota in Gastrointestinal Diseases

S. No.	Risk Factor	Microbial Change	Possible Mechanisms	References
		Inflammatory Bowel Disease		
1	Genetics (HLA-B27)	*Bacteroides fragilis* (monocolonization)	Unknown	(Rath et al., 1999)
2	Genetics (IL-10, IL-2 deficient)	*Escherichia coli* or *Enterococcus faecalis* (monocolonization)	IL-12, IFN-γ Increases	(Kim et al., 2007)
3	Genetics (Nlrp6 deficient)	*Prevotellaceae* Increases, TM7 Increases	IL-18 Decreases, CCL5 Increases, and innate and adaptive immune cell recruitment	(Henao-Mejia et al., 2012; Lucke et al., 2006)
4	Smoking	*Anaerostipes* Decreases	Butyrate Decreases	(Morgan et al., 2012)
5	Diet (high fat derived from milk)	Firmicutes Decreases, *Bilophilawadsworthia* Increases	Immune system (Th1) disruption	(Devkota et al., 2012; DeVoss & Diehl, 2014)
6	Diet (high protein)	*Desulfovibrio spp.* Increases, *Desufuromonas spp.* Increases	Genotoxic Increases, DNA damage Increases, inflammation Increases	(Muyzer & Stams, 2008; Rowan et al., 2010)
7	Diet (High fat, high beef)	*Eryspielotrichaceae* Increases, *Bacteroides fragilis* Increases,	Unknown	(Hentges et al., 1977; Zhang et al., 2012)
8	Antibiotics	*Clostridium scindens* Decreases, *Clostridium difficile* Increases	DCA Decreases	(Buffie et al., 2015)
9	Antibiotics (ciprofloxacin, metronidazole)	*Dorea* Decreases, *Butyricicoccus* Decreases, *Coriobacteriaceae* Decreases	Organic acid (e.g., formic acid, butyrate)	(Musa et al., 2010; Rupnik et al., 2009)
10	Unknown	*Pks+ Escherichia coli* Increases	Colibactin Increases, DNA damage Increases	(Arthur et al., 2012)
11	Unknown	*Faecalibacteriumprausnitzii* Decreases	Anti-inflammatory effect Decreases	(Dudík et al., 2022; Sokol et al., 2008)
		Irritable Bowel Syndrome		
1	Infection	*Salmonella* Increases, *Campylobacter* Increases, *Shigella* Increases	Unknown	(Ericsson et al., 2008)
2	Unknown	*Methanobrevibactersmithii* Increases	Methane-driven gut dysfunction	(Kim et al., 2012)
		Colorectal Cancer		
1	Aging	*Faecalibacteriumprausnitzii* Decreases, *Roseburia intestinalis* Decreases	Butyrate, anti-inflammatory property Decreases	(Biagi et al., 2010; Marchesi et al., 2011)
2	Alcohol	*Erysipelotrichaceae* Increases	Unknown	(Nistal et al., 2015; Yan et al., 2011)
3	Diet (high beef)	*Bacteroides fragilis* (ETBF) Increases	Wnt/ β-catenin signaling Increases IL-17-driven STAT3 Increases	(Hentges et al., 1977; Rhee et al., 2009; Ulger Toprak et al., 2006)
4	Diet (high fat, fiber)	*Butyrate-producing bacteria* Decreases, *hydrogen sulfide-producing bacteria* Increases	SCFAs Decreases (e.g., acetate, butyrate, propionate), secondary bile acids Increases (e.g., LCA, DCA)	(Canani et al., 2011; Gill & Rowland, 2002; Ou et al., 2013)
5	Unknown	*Fusobacterium nucleatum* Increases	E-cadherin/β-catenin signaling Increases	(Qin et al., 2010; Rubinstein et al., 2013)
6	Unknown	*pks+Escherichia coli* Increases	Colibactin Increases, DNA damage Increases	(Arthur et al., 2012)
		Celiac Disease		
1	Antibiotic (Vancomycin)	*Bacteroides* Decreases, *Parabacteroides* Decreases, *Escherichia* Increases, Helicobacter Increases	Unknown	(Galipeau et al., 2015)
2	Genetic (FUT2 mutation)	*Bifidobacterium spp.* Decreases, *Candida albicans* Increases	Unknown	(Hurd & Domino, 2004; Nieuwenhuizen et al., 2003)
3	Breast feeding (by mother with celiac disease)	*Bifidobacterium spp.* Decreases	Unknown	(Fälth-Magnusson et al., 1996; Olivares et al., 2015; Peters et al., 2001)

anti-inflammatory properties, influencing the balance between pro-inflammatory and anti-inflammatory signals within the gut (Morrison & Preston, 2016).

F) Implications for Gastrointestinal Inflammation

Dysregulation of the delicate balance within the gut microbiota can lead to gastrointestinal inflammation. In conditions such as inflammatory bowel disease (IBD) and irritable bowel syndrome (IBS), alterations in the gut microbiota composition have been observed. Dysbiosis may contribute to increased immune activation, disruptions in the intestinal epithelial barrier, and the perpetuation of chronic inflammation within the GI tract.

Understanding the role of the gut microbiota in gastrointestinal inflammation has paved the way for innovative therapeutic approaches. Probiotics, which are live microorganisms with potential health benefits, aim to restore a healthy microbial balance. Prebiotics, non-digestible fibers that promote the growth of beneficial bacteria, provide nutritional support for the gut microbiota. Fecal microbiota transplantation (FMT) involves transferring fecal material from a healthy donor to a recipient, aiming to restore a diverse and functional gut microbiota (Yang & Jobin, 2014).

The intricate interplay between the gut microbiota and gastrointestinal inflammation highlights the pivotal role of this microbial community in maintaining digestive health. The ongoing research in this field offers insights into the mechanisms by which dysbiosis contributes to inflammation and provides a foundation for developing targeted interventions aimed at restoring a balanced and functional gut microbiota. As the symbiotic relationship between the gut microbiota and the host continues to unfold, it opens new avenues for therapeutic strategies that harness the potential of the microbiome in preventing and managing gastrointestinal inflammation (Fujimura et al., 2010). Various environmental and immunological factors associated with gastrointestinal diseases are represented in Figure 10.2.

10.3 DIAGNOSTIC TECHNIQUES IN GASTROENTEROLOGY

Gastroenterology, a branch of medicine focused on the study and management of disorders affecting the digestive system, relies on a diverse array of diagnostic techniques to unravel the complexities within the gastrointestinal (GI) tract. These techniques range from traditional methods to cutting-edge technologies, providing clinicians with a comprehensive toolkit for accurate diagnosis and targeted treatment strategies (Buie et al., 2010).

10.3.1 Endoscopy

Endoscopy is a cornerstone in gastroenterological diagnostics, allowing direct visualization of the GI tract. Techniques such as upper endoscopy (esophagogastroduodenoscopy or EGD) and colonoscopy enable examination of the esophagus, stomach, duodenum, and colon, facilitating the detection of abnormalities such as ulcers, polyps, and inflammation.

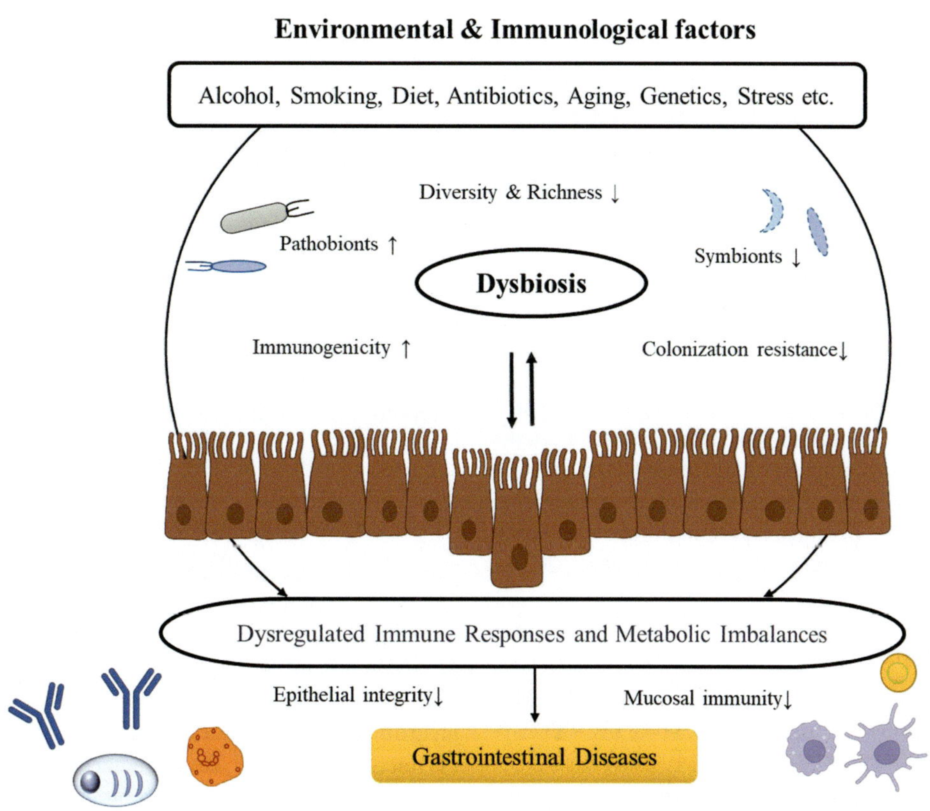

FIGURE 10.2 Various environmental and immunological factors associated with gastrointestinal diseases.

Advances in endoscopic technology, including high-definition imaging and narrow-band imaging, enhance the visualization of subtle lesions and improve diagnostic precision (Klenske et al., 2018).

10.3.2 Imaging Techniques

10.3.2.1 CT Scan and MRI

Computed tomography (CT) scans and magnetic resonance imaging (MRI) provide detailed cross-sectional images of the abdomen, aiding in the diagnosis of conditions such as inflammatory bowel disease (IBD), tumors, and structural abnormalities (Mackalski & Bernstein, 2006).

10.3.2.2 Barium Studies

Barium contrast studies, such as barium swallow and barium enema, involve the use of a contrast medium to enhance X-ray images of the upper and lower GI tract, respectively. These studies are valuable for detecting anatomical abnormalities, strictures, and motility disorders (Périard, 2003).

10.3.3 Blood Tests

Blood tests play a crucial role in assessing various aspects of gastrointestinal health. Markers such as C-reactive protein (CRP) and erythrocyte sedimentation rate (ESR) help evaluate inflammation. Liver function tests assess hepatic health, while specific blood markers aid in the diagnosis of conditions such as celiac disease and pancreatic disorders (Ansar et al., 2016).

10.3.4 Stool Tests

Stool tests provide valuable information about GI function and detect abnormalities. The fecal occult blood tests screen for gastrointestinal bleeding, while fecal calprotectin levels help assess inflammation, particularly in the context of IBD. Stool cultures aid in identifying infectious agents causing gastrointestinal infections (Kim et al., 2014).

10.3.5 Breath Tests

Breath tests are employed to diagnose certain gastrointestinal conditions, notably Helicobacter pylori infection. Patients ingest a substrate, and the detection of specific gases in exhaled breath indicates the presence of certain bacterial or metabolic processes, aiding in the diagnosis of conditions like small intestinal bacterial overgrowth (SIBO) (Sabbagh et al., 2019).

10.3.6 Capsule Endoscopy

Capsule endoscopy involves the ingestion of a small, pill-sized camera that captures images as it traverses the digestive tract. This technique is particularly useful for visualizing the small intestine, which is challenging to reach with traditional endoscopy or imaging methods (Mackiewicz, 2011).

10.3.7 Manometry

Esophageal and anorectal manometry assess the muscular function of the GI tract. Esophageal manometry evaluates the motility of the esophagus, aiding in the diagnosis of conditions such as gastroesophageal reflux disease (GERD). Anorectal manometry assesses the function of the anal sphincters and pelvic floor muscles, assisting in the evaluation of disorders like fecal incontinence (Fox et al., 2018).

10.3.8 Biopsy and Histopathology

Biopsies obtained during endoscopic procedures or surgeries are crucial for histopathological evaluation. Histopathology provides insights into the nature of lesions, confirming or ruling out conditions such as inflammatory disorders, malignancies, or infections (Group et al., 2010).

10.3.9 Genetic Testing

Genetic testing, particularly in conditions like hereditary colorectal cancer syndromes, aids in identifying individuals at increased risk. Understanding genetic predispositions enables personalized screening and management strategies.

The diagnostic landscape in gastroenterology is diverse and continually evolving. The integration of traditional methods with advanced technologies allows for a comprehensive understanding of GI conditions, facilitating precise diagnoses and tailored treatment plans. The ongoing refinement of diagnostic techniques in gastroenterology underscores the commitment to improving patient outcomes through accurate and timely assessments of digestive health (Stoll & Kupfer, 2019).

10.4 ADVANCEMENTS IN TARGETED TREATMENT APPROACHES

The landscape of gastrointestinal (GI) disorders has witnessed notable advancements in targeted treatment approaches, aiming to address specific molecular, microbial, or immunological aspects underlying these conditions. These innovative strategies represent a shift towards personalized and precision medicine, offering more effective and tailored interventions for individuals with GI disorders (Schupack et al., 2022).

10.4.1 Probiotics and Prebiotics

The advent of targeted treatment approaches in gastroenterology has been prominently marked by the utilization of probiotics and prebiotics. Probiotics, consisting of live beneficial microorganisms, and prebiotics, non-digestible compounds fostering their growth, represent a paradigm shift in addressing gastrointestinal disorders. Probiotics, such as certain strains of Lactobacillus and Bifidobacterium, have shown promise in modulating the gut microbiota, restoring balance, and exerting anti-inflammatory effects. This therapeutic avenue is particularly relevant in conditions like irritable bowel syndrome (IBS), where disruptions in the gut microbiota are implicated. Concurrently, prebiotics, including fibers and oligosaccharides, serve as substrates for beneficial bacteria, promoting their proliferation and enhancing overall gut health. The synergy between probiotics and prebiotics aims to optimize the microbial environment and alleviate symptoms associated with GI disorders (Akutko & Stawarski, 2021; Salminen et al., 2021).

Shah et al. (2020) elucidated the potential of dietary fibers (DFs) as formulations with significant implications for human health, particularly in combating life-threatening chronic diseases such as cardiovascular disease (CVD), diabetes mellitus, obesity, and cancer. Their review underscores the burgeoning interest among scientists worldwide in investigating the intricate relationship between DFs and gastrointestinal (GIT) microbiota. Drawing from various epidemiological and clinical trials, the authors highlighted how DFs modulate the composition and metabolic activities of microbial communities within the human GIT, thus playing pivotal roles in sustaining health and well-being. Moreover, they pointed out an emerging body of literature demonstrating the immunomodulatory effects of DFs, particularly their role as prebiotics in stimulating innate, cellular, and humoral immune responses. This immunomodulation is intricately linked with the modulation of probiotics, beneficial microorganisms that reside in the gut. The scope and approach of their review centered on emphasizing the significance of DFs as prebiotics in human health, particularly in their influence on gut microbiota. To fully realize the promising health benefits offered by these prebiotics, the authors advocated for the development of novel formulations aimed at enhancing and scaling up their efficacy. In conclusion, the insights provided by them offer a valuable platform for the development of novel functional foods, addressing the escalating demand for dietary solutions that promote health and well-being in modern times (Shah et al., 2020).

Liu et al. (2024) conducted a comprehensive review focusing on advancements in hydrogel-based drug delivery systems for the treatment of inflammatory bowel disease (IBD). They addressed the pressing need for novel therapeutic approaches to overcome the limitations of current IBD drug therapies, including significant side effects, low efficacy, and poor patient compliance. Hydrogels, characterized by their three-dimensional networks of hydrophilic polymers capable of swelling and retaining water, have garnered attention for their potential in IBD treatment due to their biocompatibility, tunability, and responsiveness to various stimuli. In their review, they first identified three key pathophysiological alterations in IBD that necessitate addressing: damage to the intestinal mucosal barrier, dysbiosis of intestinal flora, and activation of inflammatory signaling pathways leading to disequilibrium within the intestines. They then delved into the processes involved in preparing hydrogel drug delivery systems, covering aspects such as the selection of hydrogel materials, types of drugs to be loaded, methods of drug loading, drug release mechanisms, and crucial points in the preparation process. Furthermore, the authors highlighted the progress and impact of hydrogel-based drug delivery systems in IBD treatment, emphasizing their role in regulating physical barrier immune responses, promoting mucosal repair, and improving gut microbiota. They concluded by analyzing the challenges faced by hydrogel-based drug delivery systems in clinical applications for IBD treatment and proposed potential solutions from their perspective. Overall, the review provides valuable insights into the advancements, challenges, and potential solutions in the development and application of hydrogel-based drug delivery systems for treating IBD (Liu et al., 2024).

10.4.2 Fecal Microbiota Transplantation (FMT)

Fecal microbiota transplantation has emerged as a groundbreaking intervention for gastrointestinal disorders, revolutionizing the treatment landscape. FMT involves the transfer of fecal material from a healthy donor to a recipient, intending to restore a diverse and functional gut microbiota. Initially proven highly effective in the management of recurrent Clostridioides difficile infection, FMT has since sparked interest in its potential applications across various gastrointestinal conditions. Ongoing research investigates its efficacy in inflammatory bowel disease (IBD), irritable bowel syndrome (IBS), and other disorders characterized by dysbiosis. FMT represents a pioneering approach that directly addresses the root cause of certain GI conditions by restoring microbial balance. FMT's success in treating recurrent Clostridioides difficile infection has paved the way for exploring its application in broader gastrointestinal contexts. In IBD, where the immune system mistakenly attacks the gut, FMT offers a unique strategy by modulating the gut microbiota to promote anti-inflammatory responses. Preliminary studies show promising results, indicating that FMT could induce clinical remission and potentially transform the management of IBD. Similarly, in the realm of irritable bowel syndrome (IBS), characterized by abdominal pain and altered bowel habits, FMT's ability to address dysbiosis holds the potential for alleviating symptoms and improving the overall gut environment (Verhoef et al., 2023). As researchers delve deeper into the intricate relationships between the gut microbiota and various gastrointestinal disorders, FMT stands out as a targeted intervention that addresses the root cause, offering a personalized and effective approach to reshape the treatment landscape for these complex conditions. Further investigations and clinical trials are imperative to unlock the full potential of FMT and establish its role as a mainstream therapeutic option in the evolving field of gastroenterology. Table 10.2. illustrates the impact of fecal microbiota transplantation on gastrointestinal disease. Fecal microbiota transplantation (FMT) has demonstrated effectiveness in treating certain gastrointestinal disorders, but it is essential to consider potential adverse effects associated with the procedure. One concern is the risk of infection transmission, as, despite donor screening for known pathogens, there remains a possibility of transmitting undetected or emerging infections. Allergic

TABLE 10.2
The Impact of Fecal Microbiota Transplantation on Gastrointestinal Disease

S. No.	Disease	Species	Dosage and Duration	Action	References
1	Ulcerative colitis	Human; mice	Pooled analysis of two prospective clinical trials for moderate to severe UC from November 2012 to July 2017 200 μL fecal suspension by enema, on days 1,3,5 and 7.	Diarrhea, haematochezia, partial mayo score, EQ-5D index Decreases TNF-α, IL-1β, MPO activity, Increases IL-10	(Ding et al., 2019) (Zhou et al., 2019)
2	Clostridium difficile infection	Human	500 ml of a solution consisting of >75 g of donor feces mixed with isotonic saline is infused through a nasoduodenal tube at a rate of 20 ml min^{-1}	Normalization of the disturbed microbiota: *Bacteroidetes; Clostridium clusters IV and XIVa;* *Fecalibacteriumprausnitzii; Butyrivibriocrossotus; Enterococcus, Lactobacillus, Veillonella.* *C. difficile; E. aerogenes, Caudovirales*	(Borody et al., 2019; Fuentes et al., 2014)
3	D-Lactic acidosis	Human	2 tbsp of the father's stool was mixed with 75 ml of normal saline and slowly infused via the patient's gastrostomy tube.	Decreases D-Lactate Increases Weight gain Change in the smell of the patient's feces and breath	(Davidovics et al., 2017)
4	Irritable bowel syndrome	Human	200 mL fecal suspension by a colonoscope	Lower IBS-SSS scores Improvement in the GSRS scores Improvement in the HAMA and HAMD cores Improvement in the BS-QOL scores	(Johnsen et al., 2018) (Huang et al., 2019)
5	Hepatic Encephalopathy	Rat; Human	200 μL/d Via intestinal intubation 90 mL instilled by enema	Decreases TLR4, TLR9, TNF-α, IL-β and IL-6 in liver, Increases Claudin-1, Claudin-6 and Occludin Increases Lactobacillaceae, Bifidobacteriaceae, Lachnospiraceae and Ruminococcaeceae Decreases Hospitalizations improved cognitive tests	(Wang et al., 2017) (Bajaj et al., 2017)
6	Primary sclerosing cholangitis	Human	200 mL by an endoscope to the second part of the duodenum once weekly for 4 weeks	Decreases total bilirubin, direct bilirubin, aspartate transaminase, alanine transaminase, alkaline phosphatase, gamma-glutamyl Transferase, total bile acids, cholic acid, deoxycholic acid, chenodeoxycholic acid, ALP, AST Decreases Proteobacteria Increases Bacteroidetes and Firmicutes	(Allegretti et al., 2019; Philips et al., 2018)

reactions to components in the donor fecal material, leading to symptoms such as hives or itching, are also reported. Immediate gastrointestinal symptoms, including abdominal discomfort and diarrhea, can occur following FMT. Long-term consequences are not fully understood, and alterations in the recipient's gut microbiota may have unforeseen effects on health. Overgrowth of certain bacterial species, especially in the context of introducing a large microbial load, could lead to imbalances and associated symptoms. Individuals with compromised immune systems may face immunological risks, and the psychological and ethical considerations surrounding the transfer of fecal material are also noteworthy. Additionally, the lack of standardized protocols for donor screening and FMT procedures contributes to variability in safety and efficacy (Gaonkar, 2021). Continuous research and monitoring are crucial to address these concerns, ensuring the refinement of FMT protocols and the comprehensive understanding of potential adverse effects associated with this evolving therapeutic approach. Table 10.3 represents various adverse effects of fecal microbiota transplantation.

Liu et al. (2024) investigated the potential of fecal microbiota transplantation (FMT) in alleviating intestinal inflammatory diarrhea, particularly in the context of oxidative stress and pyroptosis, through the reduction of gut microbiota-derived lipopolysaccharides (LPS). They highlighted infancy as a critical period in gut microbiota maturation, wherein

TABLE 10.3
Various Adverse Effects of Fecal Microbiota Transplantation

S. No.	Adverse Effects of Fecal Microbiota Transplantation	References
1	Mechanical damage (Physical injury, Intestinal perforation)	(Baxter & Colville, 2016; Mattila et al., 2012)
2	Induction of chronic disease (Diabetes, Parkinson's, Cancer, metabolic syndrome)	(Alang & Kelly, 2015; Yu et al., 2018)
3	Contamination of pathogenic microorganisms (multi-drug resistant organisms, fungi, parasites, viruses, etc.)	(Baxter & Colville, 2016; DeFilipp et al., 2019; Quera et al., 2014)
4	Complications arising from sedative medication during endoscopy or colonoscopy	(Dailey et al., 2019; Vindigni & Surawicz, 2017)
5	Abdominal discomfort, abdominal cramping, constipation, flatulence, bloating, belching, abdominal swelling, nausea, vomiting, diarrhea, blood in stool, dysbiosis, temperature reading	(Baxter & Colville, 2016; Dailey et al., 2019)
6	Fatalities (mainly occur due to the exacerbation or remission of chronic diseases, leading to weakened immunity, infections by drug-resistant bacteria, sepsis, or pneumonia)	(Hefazi et al., 2017; Quraishi et al., 2017; Wang et al., 2019)

disturbances can lead to long-term effects on host physiology, including intestinal injury and diarrhea. Despite ethical challenges in human infant studies, their research utilized piglets as a model to elucidate mechanisms underlying infant diarrhea and the regulatory effects of FMT on gut microbiota balance. Their study revealed elevated LPS levels in piglets with diarrhea, attributed to colon microbiota dysbiosis induced by early weaning. Additionally, they found that LPS upregulated NLRP3 levels via TLR4 activation and induction of reactive oxygen species (ROS), leading to pyroptosis, intestinal barrier disruption, bacterial translocation, and subsequent inflammation, ultimately resulting in diarrhea in piglets. Through microbiota regulation, FMT was shown to modulate β-PBD-2 secretion in the colon by increasing butyric acid levels. This modulation mitigated gut microbiota dysbiosis, reduced LPS levels, attenuated oxidative stress, and pyroptosis, suppressed the inflammatory response, preserved intestinal barrier integrity, and ultimately alleviated diarrhea in piglets with colitis. These findings offer a novel perspective on the pathogenesis, pathophysiology, prevention, and treatment of diarrhea diseases, emphasizing the critical interaction between FMT and gut microbiota as a promising strategy for managing diarrhea and intestinal disorders in both infants and farm animals (Liu et al., 2024).

10.4.3 Microbiota-Modulating Pharmaceuticals

Advancements in pharmaceutical research have given rise to a new class of drugs designed to selectively modulate the gut microbiota. These microbiota-modulating pharmaceuticals aim to restore a healthy microbial balance, alleviate inflammation, and regulate immune responses. The development of antibiotics with narrow-spectrum activity, microbial metabolites, and other compounds holds promise in reshaping the gut microbiome composition. By specifically targeting dysbiotic microbial populations, these drugs offer a more precise and targeted approach to gastrointestinal disorders. The ongoing refinement of microbiota-modulating pharmaceuticals underscores the potential for innovative treatments that go beyond conventional approaches (Zhang et al., 2018).

Ziemons et al. (2024) delve into the intricate role of the gut microbiota in human cancer. The review highlights the emerging understanding of how microbial communities residing in the gastrointestinal tract can influence carcinogenesis, tumor progression, and treatment outcomes in various types of cancer. The authors elucidate how dysbiosis, or imbalance in the gut microbiota composition, can contribute to cancer development through multiple mechanisms including chronic inflammation, alteration of immune responses, production of carcinogenic metabolites, and modulation of host cell signaling pathways. Furthermore, they discussed the potential of the gut microbiota as a biomarker for cancer risk assessment and prognosis, as well as its role in modulating responses to cancer therapies such as chemotherapy, immunotherapy, and radiotherapy. By comprehensively synthesizing recent research findings, this review underscores the importance of understanding the complex interplay between gut microbiota and human cancer for the development of novel preventive and therapeutic strategies (Ziemons, 2024).

10.4.4 Personalized Medicine in Gastrointestinal Inflammation

The era of personalized medicine has dawned upon gastroenterology, bringing forth tailored treatment strategies for individuals with gastrointestinal inflammation. Personalized medicine in this context involves an intricate understanding of individual characteristics, encompassing genetic, microbial, and clinical factors. Genetic profiling enables the identification of specific mutations or variations associated with increased susceptibility or responsiveness to certain therapies. Integrating an individual's gut microbiota composition into treatment decisions allows for targeted interventions, and optimizing therapeutic outcomes. This personalized approach is particularly crucial in conditions like inflammatory bowel disease (IBD), where heterogeneity among patients necessitates precise and individualized therapeutic strategies. The integration of personalized medicine into gastroenterological care holds the promise of enhancing treatment efficacy, minimizing adverse effects, and ultimately improving the quality of life for patients grappling with gastrointestinal inflammation (Wang & Camilleri, 2019).

Kocabas et al. (2024) present a study showcasing the application of three-dimensional (3D) printing technology in the compounding of personalized rectal preparations for pediatric patients with inflammatory bowel diseases. They address the challenges associated with conventional medicines in pediatric populations, such as difficulties in administration and dosing, by leveraging 3D printing to tailor medication to individual needs. The authors investigate the material requirements for printing prednisolone phosphate-loaded suppositories with tunable doses and rapid drug release. They explore different formulations containing prednisolone sodium phosphate (PSP) and various excipients, such as hydroxypropyl cellulose (HPC) and mannitol, to optimize drug delivery for the treatment of gastrointestinal inflammation. Their findings demonstrate that incorporating higher weight fractions of mannitol as a pore former enhances the release rate of PSP, with rapid drug release achieved within 30 minutes for formulations containing 48% w/w mannitol. Additionally, they showed that 3D-printed suppositories with varying doses of PSP (ranging from 6 to 30 mg) can be compounded without significantly altering dissolution kinetics, ensuring consistent therapeutic efficacy across different doses. The study highlights the potential of 3D printing technology for automated compounding of personalized medication, offering flexibility in dose adjustment through changes in infill density or dimensions. By enabling precise control over drug release profiles, 3D-printed suppositories hold promise for optimizing treatment outcomes and enhancing patient adherence in the management of gastrointestinal inflammation (Kocabas et al., 2024).

10.5 TECHNOLOGICAL INNOVATIONS IN MONITORING AND ASSESSMENT

In the ever-advancing landscape of gastroenterology, the integration of cutting-edge technologies has revolutionized the way we monitor and assess gastrointestinal health. This section illuminates the transformative impact of technological innovations, highlighting key advancements that enhance our understanding and management of gastrointestinal conditions.

Wearable devices have emerged as pivotal tools in real-time monitoring, providing continuous data streams that offer insights into various physiological parameters. These devices, ranging from smartwatches to ingestible sensors, enable the non-invasive collection of data related to heart rate, activity levels, and even specific gastrointestinal markers. The accessibility and convenience of wearable technology empower both patients and healthcare providers to track and analyze vital information, fostering a more comprehensive understanding of gastrointestinal well-being (Tian et al., 2014).

Artificial intelligence (AI) plays a prominent role in data analysis, offering unparalleled capabilities in pattern recognition, data interpretation, and predictive modeling. AI algorithms can sift through vast datasets, identifying subtle patterns or anomalies that may escape human observation. This assists in the early detection of potential issues, enabling timely intervention and personalized treatment strategies. Moreover, AI-driven decision support systems contribute to more accurate diagnoses and prognoses, ultimately improving patient outcomes (Ahmed et al., 2020).

Remote monitoring platforms further expand the reach of healthcare beyond traditional clinical settings. Telemedicine and remote patient monitoring enable healthcare providers to remotely assess patients' gastrointestinal health, collect relevant data, and make informed decisions without the need for physical presence. This not only enhances patient convenience but also facilitates proactive management and early intervention, particularly for individuals with chronic gastrointestinal conditions (Field & Grigsby, 2002).

Capsule endoscopy represents a groundbreaking advancement in imaging technologies. This minimally invasive technique involves swallowing a small, pill-sized camera that captures high-resolution images as it travels through the gastrointestinal tract. Capsule endoscopy provides a detailed visualization of the small intestine, an area challenging to reach with traditional endoscopic procedures. This technology aids in the diagnosis of conditions such as Crohn's disease, obscure gastrointestinal bleeding, and small intestinal tumors (Schembre, 2021).

Furthermore, molecular imaging techniques, such as positron emission tomography (PET) and magnetic resonance imaging (MRI), have evolved to provide functional insights into gastrointestinal physiology. These techniques enable the visualization of metabolic and molecular processes, aiding in the detection and characterization of inflammatory lesions, tumors, and other abnormalities (Maccioni et al., 2023).

The integration of big data analytics facilitates the aggregation and analysis of vast datasets from diverse sources, including electronic health records, genetic information, and patient-reported outcomes. This holistic approach allows for a more comprehensive understanding of the factors influencing gastrointestinal health, paving the way for personalized and precision medicine strategies.

In conclusion, technological innovations in monitoring and assessment have propelled gastroenterology into a new era of precision healthcare. Wearable devices, artificial intelligence, remote monitoring, capsule endoscopy, and advanced imaging techniques collectively contribute to a more nuanced and patient-centric approach to gastrointestinal health. These innovations not only enhance diagnostic accuracy and treatment efficacy but also empower individuals to actively participate in the management of their gastrointestinal well-being. As the field continues to evolve, the integration of these technologies promises to redefine the standard of care and foster continuous improvements in patient outcomes.

10.6 CHALLENGES AND FUTURE DIRECTIONS

As we navigate the ever-evolving landscape of gastrointestinal inflammation treatment, addressing the challenges that currently impede progress is crucial while simultaneously exploring potential future directions to advance the field. One formidable challenge lies in the heterogeneity of gastrointestinal conditions, such as inflammatory bowel disease (IBD), irritable bowel syndrome (IBS), and celiac disease. The diversity in presentation and characteristics poses a significant hurdle for developing universal treatment strategies, necessitating the intricate tailoring of interventions to specific subtypes within these conditions (Rauniyar et al., 2023).

Regulatory considerations constitute another challenge, especially for novel therapies centered around gut microbiota-based strategies. The continuously evolving regulatory landscape demands a delicate balance between ensuring patient safety, validating treatment efficacy, and expediting the translation of innovative therapies from bench to bedside. Moreover, the long-term efficacy and safety of gut microbiota-based interventions remain under scrutiny. While some interventions show promise in the short term, understanding their durability and potential adverse events over extended periods is critical for establishing reliability and sustainability (Sharpton et al., 2021).

Implementing personalized and targeted treatment approaches faces the challenge of patient compliance and adherence. Factors such as the complexity of treatment regimens, potential side effects, and the extended duration of some interventions may impact patients' ability to adhere to prescribed therapies, ultimately influencing the overall success of treatments. The inherent variability of the gut microbiome among individuals adds complexity to microbiota-based treatments, posing challenges in identifying consistent patterns and establishing robust correlations between specific microbial compositions and treatment responses.

Looking toward future directions, the advent of precision medicine in gastroenterology holds immense potential. Integrating patient-specific data, including genetic information, microbiome profiles, and clinical parameters, could pave the way for tailored therapeutic interventions. Biomarker discovery and validation may further aid in identifying individuals more likely to respond positively to specific treatments. Advancements in microbiome research, encompassing metagenomics, metatranscriptomics, and metabolomics, are expected to unravel the complexities of microbial communities in the gut. This deeper understanding may lead to the identification of key microbial signatures associated with health and disease, fostering the development of more targeted and effective interventions (Maddux et al., 2018).

The integration of digital health technologies, including mobile applications, wearable devices, and telehealth platforms, will play a pivotal role in monitoring patients, collecting real-time data, and facilitating remote consultations. This not only enhances patient engagement but also allows for continuous monitoring and adjustment of treatment plans. The future may witness the emergence of combination therapies that leverage synergies between different gut microbiota-based interventions. Combinatorial approaches, such as integrating probiotics with personalized dietary interventions or microbiota-modulating pharmaceuticals, have the potential to enhance treatment efficacy and address the multifactorial nature of gastrointestinal inflammation (Fong et al., 2020).

Continued advancements in machine learning and predictive modeling hold promise for developing algorithms that can analyze complex datasets and predict treatment responses. Integrating these tools into clinical practice may assist healthcare providers in making more informed decisions, ultimately improving the overall management of gastrointestinal conditions. In navigating these challenges and embracing future directions, the field of gut microbiota-based targeting for gastrointestinal inflammation treatment is poised for transformative growth. Interdisciplinary collaboration, ongoing research endeavors, and a commitment to patient-centered care will be paramount in shaping the trajectory of this dynamic field (Javaid et al., 2022).

10.7 CASE STUDIES AND CLINICAL APPLICATIONS

The application of gut microbiota-based targeting strategies in the treatment of gastrointestinal inflammation is underscored by compelling case studies and clinical applications, offering tangible evidence of the effectiveness of these innovative approaches (Gu et al., 2019).

Case Study 1: Probiotics in Inflammatory Bowel Disease (IBD)

A notable case involves a patient diagnosed with Crohn's disease, a form of IBD, who experienced recurrent flares despite conventional treatments. The introduction of a probiotic regimen, consisting of specific strains with anti-inflammatory properties, resulted in a significant reduction in symptom severity and a prolonged period of remission. Fecal microbiota analysis revealed a favorable shift in microbial composition, emphasizing the potential of probiotics in modulating the gut microbiome and alleviating inflammation (Mishra et al., 2022).

Case Study 2: Fecal Microbiota Transplantation (FMT) in Ulcerative Colitis (UC)

In a clinical trial focusing on patients with refractory ulcerative colitis, FMT emerged as a promising intervention. After a series of fecal microbiota transplants from healthy donors, a substantial number of patients exhibited clinical improvement, evidenced by reduced disease activity scores and improved quality of life. The success of FMT in inducing remission suggests its potential as a transformative therapy for individuals with UC resistant to traditional treatments (Saleh et al., 2023).

Case Study 3: Microbiota-Modulating Pharmaceuticals in Irritable Bowel Syndrome (IBS)

A multicenter trial investigated the efficacy of a microbiota-modulating pharmaceutical in patients with irritable bowel syndrome (IBS). The pharmaceutical, designed to selectively target dysbiotic microbial populations, demonstrated a notable reduction in IBS symptoms. The study showcased the potential of microbiota-modulating pharmaceuticals in addressing the complex microbial imbalances associated with IBS, offering a targeted and personalized therapeutic approach (Chong et al., 2019; Marasco et al., 2022).

Case Study 4: Personalized Medicine in Gastrointestinal Inflammation

A personalized medicine approach was applied to a cohort of patients with varying forms of gastrointestinal inflammation. Integrating genetic information, microbiome profiles, and clinical data, tailored treatment plans were devised. Patients experienced improved treatment responses and reduced adverse effects compared to traditional one-size-fits-all approaches. This case study highlights the potential of precision medicine in optimizing therapeutic outcomes by addressing individual variations in disease presentation and treatment response (Bastaki et al., 2023).

These case studies collectively illustrate the diverse applications of gut microbiota-based targeting in the clinical realm. From probiotics and FMT to microbiota-modulating pharmaceuticals and personalized medicine, these interventions offer promising results in mitigating gastrointestinal inflammation. Importantly, these clinical applications signify a paradigm shift towards more personalized, effective, and targeted treatments, paving the way for a new era in gastroenterology. Continued research and the accumulation of additional clinical evidence will further refine these approaches, solidifying their role in the mainstream management of gastrointestinal inflammation (Strati et al., 2022).

10.8 CONCLUSION

The exploration of gut microbiota-based targeting strategies for the treatment of gastrointestinal inflammation has unveiled a transformative landscape at the intersection of diagnostics, technology, and innovative therapies. In navigating this terrain, a synthesis of challenges, advancements, and promising future directions emerges, shaping the overarching narrative of this comprehensive chapter.

Challenges such as the heterogeneity of gastrointestinal conditions, regulatory intricacies, and ensuring long-term efficacy and patient compliance underscore the complexity of implementing gut microbiota-based interventions. The inherent variability of the gut microbiome further adds layers of intricacy to treatment strategies, necessitating a nuanced approach to achieve consistent and reproducible outcomes.

However, the promising future directions present a beacon of hope for overcoming these challenges. Precision medicine, with its emphasis on tailoring interventions based on individual characteristics, offers a paradigm shift in gastroenterological care. Biomarker discovery, advancements in microbiome research, and the integration of digital health technologies collectively contribute to a more refined and personalized approach to treatment.

The case studies and clinical applications underscore the tangible impact of gut microbiota-based interventions. Probiotics, fecal microbiota transplantation, microbiota-modulating pharmaceuticals, and personalized medicine showcase successful outcomes, offering evidence of the transformative potential of these strategies in real-world clinical scenarios. These case studies provide a roadmap for future research and implementation, emphasizing the relevance of these interventions across various gastrointestinal conditions.

In conclusion, the journey through this chapter illuminates a dynamic landscape where the fusion of diagnostic precision, technological innovation, and gut microbiota-based therapies holds immense promise for the effective management of gastrointestinal inflammation. The challenges outlined serve as guideposts for further refinement, while the future directions beckon towards a more personalized and patient-centric era in gastroenterology. As interdisciplinary collaboration, ongoing research endeavors, and a commitment to patient-centered care continue to drive the field forward, gut microbiota-based targeting strategies are poised to redefine the standard of care for gastrointestinal inflammation. This chapter stands as a testament to the evolving nature of gastroenterology, where the symbiotic relationship between science and clinical application propels us toward a future of enhanced precision and improved patient outcomes.

REFERENCES

Ahmed, Z., Mohamed, K., Zeeshan, S., & Dong, X. (2020). Artificial intelligence with multi-functional machine learning platform development for better healthcare and precision medicine. *Database*, *2020*, baaa010.

Akutko, K., & Stawarski, A. (2021). Probiotics, prebiotics and synbiotics in inflammatory bowel diseases. *Journal of Clinical Medicine*, *10*(11), 2466.

Al Bander, Z., Nitert, M. D., Mousa, A., & Naderpoor, N. (2020). The gut microbiota and inflammation: An overview. *International Journal of Environmental Research and Public Health*, *17*(20), 7618.

Alang, N., & Kelly, C. R. (2015). Weight gain after fecal microbiota transplantation. In *Open Forum Infectious Diseases*. (Vol. 2, No. 1, p. ofv004). Oxford University Press.

Allegretti, J. R., Kassam, Z., Carrellas, M., Mullish, B. H., Marchesi, J. R., Pechlivanis, A., Smith, M., Gerardin, Y., Timberlake, S., & Pratt, D. S. (2019). Fecal microbiota transplantation in patients with primary sclerosing cholangitis: Apilot clinical trial. *Official Journal of the American College of Gastroenterology| ACG*, *114*(7), 1071–1079.

Alshehri, D., Saadah, O., Mosli, M., Edris, S., Alhindi, R., & Bahieldin, A. (2021). Dysbiosis of gut microbiota in inflammatory bowel disease: Current therapies and potential for microbiota-modulating therapeutic approaches. *Bosnian Journal of Basic Medical Sciences*, *21*(3), 270.

Amabebe, E., Robert, F. O., Agbalalah, T., & Orubu, E. S. (2020). Microbial dysbiosis-induced obesity: Role of gut microbiota in homoeostasis of energy metabolism. *British Journal of Nutrition*, *123*(10), 1127–1137.

Ansar, W., Ghosh, S., Ansar, W., & Ghosh, S. (2016). Inflammation and inflammatory diseases, markers, and mediators: Role of CRP in some inflammatory diseases. *Biology of C Reactive Protein in Health and Disease*, *4*, 67–107.

Arthur, J. C., Perez-Chanona, E., Mühlbauer, M., Tomkovich, S., Uronis, J. M., Fan, T.-J., Campbell, B. J., Abujamel, T., Dogan, B., & Rogers, A. B. (2012). Intestinal inflammation targets cancer-inducing activity of the microbiota. *Science*, *338*(6103), 120–123.

Bajaj, J. S., Kassam, Z., Fagan, A., Gavis, E. A., Liu, E., Cox, I. J., Kheradman, R., Heuman, D., Wang, J., & Gurry, T. (2017). Fecal microbiota transplant from a rational stool donor improves hepatic encephalopathy: A randomized clinical trial. *Hepatology*, *66*(6), 1727–1738.

Bastaki, K., Umlai, U.-K. I., & Jithesh, P. V. (2023). Personalized medicine. In Mahbuba Rahman (ed.) *Metabolomics* (pp. 1–32). Elsevier.

Baxter, M., &Colville, A. (2016). Adverse events in faecal microbiota transplant: Areview of the literature. *Journal of Hospital Infection*, *92*(2), 117–127.

Bhushan, B., Singh, B. P., Kumari, M., Mishra, V., Saini, K., & Singh, D. (2019). Microbes, their metabolites, and effector molecules: A pharmacological perspective for host-microbiota interaction. *Pharmaceuticals from Microbes: Impact on Drug Discovery*, *28*, 155–206.

Biagi, E., Nylund, L., Candela, M., Ostan, R., Bucci, L., Pini, E., Nikkïla, J., Monti, D., Satokari, R., & Franceschi, C. (2010). Through ageing, and beyond: Gut microbiota and inflammatory status in seniors and centenarians. *PloSOne*, *5*(5), e10667.

Bibbò, S., Settanni, C. R., Porcari, S., Bocchino, E., Ianiro, G., Cammarota, G., & Gasbarrini, A. (2020). Fecal microbiota transplantation: Screening and selection to choose the optimal donor. *Journal of Clinical Medicine*, *9*(6), 1757.

Borody, T. J., Eslick, G. D., & Clancy, R. L. (2019). Fecal microbiota transplantation as a new therapy: From Clostridioides difficile infection to inflammatory bowel disease, irritable bowel syndrome, and colon cancer. *Current Opinion in Pharmacology*, *49*, 43–51.

Brüssow, H. (2020). Problems with the concept of gut microbiota dysbiosis. *Microbial Biotechnology*, *13*(2), 423–434.

Buffie, C. G., Bucci, V., Stein, R. R., McKenney, P. T., Ling, L., Gobourne, A., No, D., Liu, H., Kinnebrew, M., & Viale, A. (2015). Precision microbiome reconstitution restores bile acid mediated resistance to Clostridium difficile. *Nature*, *517*(7533), 205–208.

Buie, T., Campbell, D. B., Fuchs III, G. J., Furuta, G. T., Levy, J., VandeWater, J., Whitaker, A. H., Atkins, D., Bauman, M. L., & Beaudet, A. L. (2010). Evaluation, diagnosis, and treatment of gastrointestinal disorders in individuals with ASDs: A consensus report. *Pediatrics*, *125*(Supplement_1), S1–S18.

Canani, R. B., Di Costanzo, M., Leone, L., Pedata, M., Meli, R., & Calignano, A. (2011). Potential beneficial effects of butyrate in intestinal and extraintestinal diseases. *World Journal of Gastroenterology: WJG*, *17*(12), 1519.

Chauhan, N. S., &Kumar, S. (2023). *Microbiome Therapeutics: Personalized Therapy Beyond Conventional Approaches*. Elsevier.

Chong, P. P., Chin, V. K., Looi, C. Y., Wong, W. F., Madhavan, P., & Yong, V. C. (2019). The microbiome and irritable bowel syndrome–a review on the pathophysiology, current research and future therapy. *Frontiers in Microbiology*, *10*, 424646.

Dailey, F. E., Turse, E. P., Daglilar, E., & Tahan, V. (2019). The dirty aspects of fecal microbiota transplantation: A review of its adverse effects and complications. *Current Opinion in Pharmacology*, *49*, 29–33.

Davidovics, Z. H., Vance, K., Etienne, N., & Hyams, J. S. (2017). Fecal transplantation successfully treats recurrent D-Lactic acidosis in a child with short bowel syndrome. *Journal of Parenteral and Enteral Nutrition*, *41*(5), 896–897.

DeFilipp, Z., Bloom, P. P., Torres Soto, M., Mansour, M. K., Sater, M. R., Huntley, M. H., Turbett, S., Chung, R. T., Chen, Y.-B., & Hohmann, E. L. (2019). Drug-resistant E. coli bacteremia transmitted by fecal microbiota transplant. *New England Journal of Medicine*, *381*(21), 2043–2050.

Devkota, S., Wang, Y., Musch, M. W., Leone, V., Fehlner-Peach, H., Nadimpalli, A., Antonopoulos, D. A., Jabri, B., & Chang, E. B. (2012). Dietary-fat-induced taurocholic acid promotes pathobiont expansion and colitis in Il10−/− mice. *Nature*, *487*(7405), 104–108.

DeVoss, J., & Diehl, L. (2014). Murine models of inflammatory bowel disease (IBD) challenges of modeling human disease. *Toxicologic Pathology*, *42*(1), 99–110.

Ding, X., Li, Q., Li, P., Zhang, T., Cui, B., Ji, G., Lu, X., & Zhang, F. (2019). Long-term safety and efficacy of fecal microbiota transplant in active ulcerative colitis. *Drug Safety*, *42*, 869–880.

Dudík, B., KiňováSepová, H., Greifová, G., Bilka, F., & Bilková, A. (2022). Probiotikáďalšejgenerácie: Prehľad najperspek tívnejšíchkandidátov. *Epidemiol MikrobiolImunol*, *71*(1), 48–56.

Dupont, H. L., Jiang, Z.-D., Dupont, A. W., & Utay, N. S. (2020). The intestinal microbiome in human health and disease. *Transactions of the American Clinical and Climatological Association*, *131*, 178.

Ericsson, C. D., Hatz, C., & DuPont, A. W. (2008). Postinfectious irritable bowel syndrome. *Clinical Infectious Diseases*, *46*(4), 594–599.

Fälth-Magnusson, K., Franzen, L., Jansson, G., Laurin, P., & Stenhammai, L. (1996). Infant feeding history shows distinct differences between Swedish celiac and reference children. *Pediatric Allergy and Immunology*, *7*(1), 1–5.

Field, M. J., & Grigsby, J. (2002). Telemedicine and remote patient monitoring. *JAMA*, *288*(4), 423–425.

Fong, W., Li, Q., &Yu, J. (2020). Gut microbiota modulation: A novel strategy for prevention and treatment of colorectal cancer. *Oncogene*, *39*(26), 4925–4943.

Fox, M. R., Kahrilas, P. J., Roman, S., Gyawali, C. P., Scott, S. M., Rao, S. S., Keller, J., Camilleri, M., Motility, I. W. G. f. D. o. G., & Function. (2018). Clinical measurement of gastrointestinal motility and function: Who, when and which test? *Nature Reviews Gastroenterology & Hepatology*, *15*(9), 568–579.

Fuentes, S., Van Nood, E., Tims, S., Heikamp-de Jong, I., Ter Braak, C. J., Keller, J. J., Zoetendal, E. G., & De Vos, W. M. (2014). Reset of a critically disturbed microbial ecosystem: Faecal transplant in recurrent Clostridium difficile infection. *The ISME Journal*, *8*(8), 1621–1633.

Fujimura, K. E., Slusher, N. A., Cabana, M. D., & Lynch, S. V. (2010). Role of the gut microbiota in defining human health. *Expert Review of Anti-Infective Therapy*, *8*(4), 435–454.

Galipeau, H. J., McCarville, J. L., Huebener, S., Litwin, O., Meisel, M., Jabri, B., Sanz, Y., Murray, J. A., Jordana, M., & Alaedini, A. (2015). Intestinal microbiota modulates gluten-induced immunopathology in humanized mice. *The American Journal of Pathology*, *185*(11), 2969–2982.

Gaonkar, P. (2021). Safety and potential risks with fecal microbiota transplantation. In Stanislow P. Stawicki and Michael S. Firstenberg (eds.) *Contemporary Topics in Patient Safety-Volume 1*. IntechOpen.

Gieryńska, M., Szulc-Dąbrowska, L., Struzik, J., Mielcarska, M. B., & Gregorczyk-Zboroch, K. P. (2022). Integrity of the intestinal barrier: The involvement of epithelial cells and microbiota—A mutual relationship. *Animals*, *12*(2), 145.

Gill, C., &Rowland, I. (2002). Diet and cancer: Assessing the risk. *British Journal of Nutrition*, *88*(S1), s73–s87.

Group, W. I. G. S., Washabau, R., Day, M., Willard, M., Hall, E., Jergens, A., Mansell, J., Minami, T., & Bilzer, T. (2010). Endoscopic, biopsy, and histopathologic guidelines for the evaluation of gastrointestinal inflammation in companion animals. *Journal of Veterinary Internal Medicine*, *24*(1), 10–26.

Gu, Y., Zhou, G., Qin, X., Huang, S., Wang, B., & Cao, H. (2019). The potential role of gut mycobiome in irritable bowel syndrome. *Frontiers in Microbiology*, *10*, 1894.

Hasan, N., & Yang, H. (2019). Factors affecting the composition of the gut microbiota, and its modulation. *PeerJ*, *7*, e7502.

Hefazi, M., Patnaik, M. M., Hogan, W. J., Litzow, M. R., Pardi, D. S., & Khanna, S. (2017). Safety and efficacy of fecal microbiota transplant for recurrent Clostridium difficile infection in patients with cancer treated with cytotoxic chemotherapy: A single-institution retrospective case series. *Mayo Clinic Proceedings*, *92*(11), 1617–1624.

Henao-Mejia, J., Elinav, E., Jin, C., Hao, L., Mehal, W. Z., Strowig, T., Thaiss, C. A., Kau, A. L., Eisenbarth, S. C., & Jurczak, M. J. (2012). Inflammasome-mediated dysbiosis regulates progression of NAFLD and obesity. *Nature*, *482*(7384), 179–185.

Hentges, D. J., Maier, B. R., Burton, G. C., Flynn, M. A., & Tsutakawa, R. K. (1977). Effect of a high-beef diet on the fecal bacterial flora of humans. *Cancer Research*, *37*(2), 568–571.

Huang, H. L., Chen, H. T., Luo, Q. L., Xu, H. M., He, J., Li, Y. Q., Zhou, Y. L., Yao, F., Nie, Y. Q., & Zhou, Y. J. (2019). Relief of irritable bowel syndrome by fecal microbiota transplantation is associated with changes in diversity and composition of the gut microbiota. *Journal of Digestive Diseases*, *20*(8), 401–408.

Hurd, E. A., & Domino, S. E. (2004). Increased susceptibility of secretor factor gene Fut2-null mice to experimental vaginal candidiasis. *Infection and Immunity*, *72*(7), 4279–4281.

Iacob, S., Iacob, D. G., & Luminos, L. M. (2019). Intestinal microbiota as a host defense mechanism to infectious threats. *Frontiers in Microbiology*, *9*, 3328.

Javaid, A., Shahab, O., Adorno, W., Fernandes, P., May, E., & Syed, S. (2022). Machine learning predictive outcomes modeling in inflammatory bowel diseases. *Inflammatory Bowel Diseases*, *28*(6), 819–829.

Johnsen, P. H., Hilpüsch, F., Cavanagh, J. P., Leikanger, I. S., Kolstad, C., Valle, P. C., & Goll, R. (2018). Faecal microbiota transplantation versus placebo for moderate-to-severe irritable bowel syndrome: Adouble-blind, randomised, placebo-controlled, parallel-group, single-centre trial. *The Lancet Gastroenterology &Hepatology*, *3*(1), 17–24.

Jones, R. M., & Neish, A. S. (2021). Gut microbiota in intestinal and liver disease. *Annual Review of Pathology: Mechanisms of Disease*, *16*, 251–275.

Kim, B. S. M., Li, B. T., Engel, A., Samra, J. S., Clarke, S., Norton, I. D., & Li, A. E. (2014). Diagnosis of gastrointestinal bleeding: A practical guide for clinicians. *World Journal of Gastrointestinal Pathophysiology*, *5*(4), 467.

Kim, G., Deepinder, F., Morales, W., Hwang, L., Weitsman, S., Chang, C., Gunsalus, R., & Pimentel, M. (2012). Methanobrevibacter smithii is the predominant methanogen in patients with constipation-predominant IBS and methane on breath. *Digestive Diseases and Sciences*, *57*, 3213–3218.

Kim, S. C., Tonkonogy, S. L., Karrasch, T., Jobin, C., & Sartor, B. R. (2007). Dual-association of gnotobiotic IL-10−/− mice with 2 nonpathogenic commensal bacteria induces aggressive pancolitis. *Inflammatory Bowel Diseases*, *13*(12), 1457–1466.

Klenske, E., Neurath, M. F., Atreya, R., & Rath, T. (2018). Molecular imaging in gastroenterology: A route for personalized endoscopy. *Digestive and Liver Disease*, *50*(9), 878–885.

Kocabas, L., Ayyoubi, S., Tajqurishi, M., Quodbach, J., Vermonden, T., & Kok, R. (2024). 3D-printed prednisolone phosphate suppositories with tunable dose and rapid release for the treatment of inflammatory bowel disease. *International Journal of Pharmaceutics*, *649*, 123639.

Kuo, C. Y., & Chiu, H. M. (2021). Application of artificial intelligence in gastroenterology: Potential role in clinical practice. *Journal of Gastroenterology and Hepatology*, *36*(2), 267–272.

Leylabadlo, H. E., Heravi, F. S., Soltani, E., Abbasi, A., Kafil, H. S., Parsaei, M., Sanaie, S., Ahmadian, Z., & Ghotaslou, R. (2022). The role of gut microbiota in the treatment of irritable bowel syndrome. *Reviews and Research in Medical Microbiology*,*33*(1), e89–e104.

Liu, Y., Huang, J., Li, S., Li, Z., Chen, C., Qu, G., Chen, K., Teng, Y., Ma, R., & Wu, X. (2024). Advancements in hydrogel-based drug delivery systems for the treatment of inflammatory bowel disease: A review. Biomaterials Science, *12*, 837–862.

Lozupone, C. A., Stombaugh, J. I., Gordon, J. I., Jansson, J. K., & Knight, R. (2012). Diversity, stability and resilience of the human gut microbiota. *Nature*, *489*(7415), 220–230.

Lucke, K., Miehlke, S., Jacobs, E., & Schuppler, M. (2006). Prevalence of Bacteroides and Prevotella spp. in ulcerative colitis. *Journal of Medical Microbiology*, *55*(5), 617–624.

Ma, H., Mueed, A., Ma, Y., Ibrahim, M., Su, L., & Wang Q. (2024, Nov). Fecal microbiota transplantation activity of *Floccularia luteovirens* polysaccharides and their protective effect on cyclophosphamide-induced immunosuppression and intestinal injury in mice. *Foods, 13*(23), 3881.

Maccioni, F., Busato, L., Valenti, A., Cardaccio, S., Longhi, A., & Catalano, C. (2023). Magnetic resonance imaging of the gastrointestinal tract: Current role, recent advancements and future prospectives. *Diagnostics*, *13*(14), 2410.

Mackalski, B., & Bernstein, C. (2006). New diagnostic imaging tools for inflammatory bowel disease. *Gut*, *55*(5), 733–741.

Mackiewicz, M. (2011). Capsule endoscopy-State of the technology and computer vision tools after the first decade. *New Techniques in Gastrointestinal Endoscopy*,*1*.

Maddux, M. H., Ricks, S., Bass, J. A., Daniel, J. F., Carpenter, E., & Radford, K. (2018). Practice survey: Adherence monitoring and intervention in pediatric gastroenterology and hepatology. *Therapeutics and Clinical Risk Management*, 1227–1234.

Marasco, G., Cremon, C., Barbaro, M. R., Stanghellini, V., & Barbara, G. (2022). Gut microbiota signatures and modulation in irritable bowel syndrome. *Microbiome Research Reports*, *1*(2), 1–12.

Marchesi, J. R., Dutilh, B. E., Hall, N., Peters, W. H., Roelofs, R., Boleij, A., & Tjalsma, H. (2011). Towards the human colorectal cancer microbiome. *PloSOne*, *6*(5), e20447.

Mattila, E., Uusitalo–Seppälä, R., Wuorela, M., Lehtola, L., Nurmi, H., Ristikankare, M., Moilanen, V., Salminen, K., Seppälä, M., & Mattila, P. S. (2012). Fecal transplantation, through colonoscopy, is effective therapy for recurrent Clostridium difficile infection. *Gastroenterology*, *142*(3), 490–496.

Mishra, J., Stubbs, M., Kuang, L., Vara, N., Kumar, P., & Kumar, N. (2022). Inflammatory bowel disease therapeutics: A focus on probiotic engineering. *Mediators of Inflammation*, 2022(1), 9621668.

Morais, L. H., Schreiber IV, H. L., & Mazmanian, S. K. (2021). The gut microbiota–brain axis in behaviour and brain disorders. *Nature Reviews Microbiology*, *19*(4), 241–255.

Morgan, X. C., Tickle, T. L., Sokol, H., Gevers, D., Devaney, K. L., Ward, D. V., Reyes, J. A., Shah, S. A., LeLeiko, N., & Snapper, S. B. (2012). Dysfunction of the intestinal microbiome in inflammatory bowel disease and treatment. *Genome Biology*, *13*(9), 1–18.

Morrison, D. J., & Preston, T. (2016). Formation of short chain fatty acids by the gut microbiota and their impact on human metabolism. *Gut Microbes*, *7*(3), 189–200.

Musa, S., Thomson, S., Cowan, M., & Rahman, T. (2010). Clostridium difficile infection and inflammatory bowel disease. *Scandinavian Journal of Gastroenterology*, *45*(3), 261–272.

Muyzer, G., & Stams, A. J. (2008). The ecology and biotechnology of sulphate-reducing bacteria. *Nature Reviews Microbiology*, *6*(6), 441–454.

Negrut, N., Khan, S. A., Bungau, S., Zaha, D. C., Anca, C. A., Bratu, O., Diaconu, C. C., & Ionita-Radu, F. (2020). Diagnostic challenges in gastrointestinal infections. *Romanian Journal of Military Medicine*, *123*, 83–90.

Nieuwenhuizen, W., Pieters, R., Knippels, L., Jansen, M., & Koppelman, S. (2003). Is Candida albicans a trigger in the onset of coeliac disease? *The Lancet*, *361*(9375), 2152–2154.

Nistal, E., Fernández-Fernández, N., Vivas, S., & Olcoz, J. L. (2015). Factors determining colorectal cancer: The role of the intestinal microbiota. *Frontiers in Oncology*, *5*, 220.

Olivares, M., Albrecht, S., De Palma, G., Ferrer, M. D., Castillejo, G., Schols, H. A., & Sanz, Y. (2015). Human milk composition differs in healthy mothers and mothers with celiac disease. *European Journal of Nutrition*, *54*, 119–128.

Ou, J., Carbonero, F., Zoetendal, E. G., DeLany, J. P., Wang, M., Newton, K., Gaskins, H. R., & O'Keefe, S. J. (2013). Diet, microbiota, and microbial metabolites in colon cancer risk in rural Africans and African Americans. *American Journal of Clinical Nutrition*, *98*(1), 111–120.

Parigi, T. L., Vieujean, S., Paridaens, K., Dalgaard, K., Peyrin-Biroulet, L., & Danese, S. (2023). Efficacy, safety, and concerns on microbiota modulation, antibiotics, probiotics, and fecal microbial transplant for inflammatory bowel disease and other gastrointestinal conditions: Results from an international survey. *Microorganisms*, *11*(11), 2806.

Parizadeh, M., & Arrieta, M.-C. (2023). The global human gut microbiome: Genes, lifestyles, and diet. *Trends in Molecular Medicine*, *12*.

Périard, M. A. (2003). Adverse effects and complications related to the use of barium sulphate contrast media for radiological examinations of the gastrointestinal tract:-A literature review. *Canadian Journal of Medical Radiation Technology*, *34*(3), 3–9.

Peters, U., Schneeweiss, S., Trautwein, E. A., & Erbersdobler, H. F. (2001). A case-control study of the effect of infant feeding on celiac disease. *Annals of Nutrition and Metabolism*, *45*(4), 135–142.

Philips, C. A., Augustine, P., & Phadke, N. (2018). Healthy donor fecal microbiota transplantation for recurrent bacterial cholangitis in primary sclerosing cholangitis–a single case report. *Journal of Clinical and Translational Hepatology*, *6*(4), 438.

Pickard, J. M., Zeng, M. Y., Caruso, R., & Núñez, G. (2017). Gut microbiota: Role in pathogen colonization, immune responses, and inflammatory disease. *Immunological Reviews*, *279*(1), 70–89.

Qin, J., Li, R., Raes, J., Arumugam, M., Burgdorf, K. S., Manichanh, C., Nielsen, T., Pons, N., Levenez, F., & Yamada, T. (2010). A human gut microbial gene catalogue established by metagenomic sequencing. *Nature*, *464*(7285), 59–65.

Quera, R., Espinoza, R., Estay, C., & Rivera, D. (2014). Bacteremia as an adverse event of fecal microbiota transplantation in a patient with Crohn's disease and recurrent Clostridium difficile infection. *Journal of Crohn's and Colitis*, *8*(3), 252–253.

Quraishi, M. N., Widlak, M., Bhala, N. a., Moore, D., Price, M., Sharma, N., & Iqbal, T. (2017). Systematic review with meta-analysis: The efficacy of faecal microbiota transplantation for the treatment of recurrent and refractory Clostridium difficile infection. *Alimentary Pharmacology &Therapeutics*, *46*(5), 479–493.

Rath, H. C., Wilson, K. H., & Sartor, R. B. (1999). Differential induction of colitis and gastritis in HLA-B27 transgenic rats selectively colonized with Bacteroides vulgatus or Escherichia coli. *Infection and Immunity*, *67*(6), 2969–2974.

Rauniyar, A., Hagos, D. H., Jha, D., Håkegård, J. E., Bagci, U., Rawat, D. B., & Vlassov, V. (2023). Federated learning for medical applications: A taxonomy, current trends, challenges, and future research directions. *IEEE Internet of Things Journal*, 11, 7374–7398.

Rezende, E. S. V., Lima, G. C., & Naves, M. M. V. (2021). Dietary fibers as beneficial microbiota modulators: A proposed classification by prebiotic categories. *Nutrition*, *89*, 111217.

Rhee, K.-J., Wu, S., Wu, X., Huso, D. L., Karim, B., Franco, A. A., Rabizadeh, S., Golub, J. E., Mathews, L. E., & Shin, J. (2009). Induction of persistent colitis by a human commensal, enterotoxigenic Bacteroides fragilis, in wild-type C57BL/6 mice. *Infection and Immunity*, *77*(4), 1708–1718.

Ribaldone, D. G., Pellicano, R., & Actis, G. C. (2019). The gut and the inflammatory bowel diseases inside-out: The extra-intestinal manifestations. *Minerva Gastroenterologicae Dietologica*, *4*, 309–318.

Ribeiro, C. F. A., Silveira, G. G. D. O. S., Candido, E. D. S., Cardoso, M. H., Espinola Carvalho, C. M., & Franco, O. L. (2020). Effects of antibiotic treatment on gut microbiota and how to overcome its negative impacts on human health. *ACS Infectious Diseases*, *6*(10), 2544–2559.

Rinninella, E., Raoul, P., Cintoni, M., Franceschi, F., Miggiano, G. A. D., Gasbarrini, A., & Mele, M. C. (2019). What is the healthy gut microbiota composition? A changing ecosystem across age, environment, diet, and diseases. *Microorganisms*, *7*(1), 14.

Rowan, F., Docherty, N. G., Murphy, M., Murphy, B., Coffey, J. C., & O'Connell, P. R. (2010). Desulfovibrio bacterial species are increased in ulcerative colitis. *Diseases of the Colon & Rectum*, *53*(11), 1530–1536.

Rowland, I., Gibson, G., Heinken, A., Scott, K., Swann, J., Thiele, I., & Tuohy, K. (2018). Gut microbiota functions: Metabolism of nutrients and other food components. *European Journal of Nutrition*, *57*, 1–24.

Rubinstein, M. R., Wang, X., Liu, W., Hao, Y., Cai, G., & Han, Y. W. (2013). Fusobacterium nucleatum promotes colorectal carcinogenesis by modulating E-cadherin/β-catenin signaling via its FadA adhesin. *Cell Host &Microbe*, *14*(2), 195–206.

Rupnik, M., Wilcox, M. H., & Gerding, D. N. (2009). Clostridium difficile infection: New developments in epidemiology and pathogenesis. *Nature Reviews Microbiology*, *7*(7), 526–536.

Sabbagh, P., Mohammadnia-Afrouzi, M., Javanian, M., Babazadeh, A., Koppolu, V., Vasigala, V. R., Nouri, H. R., & Ebrahimpour, S. (2019). Diagnostic methods for Helicobacter pylori infection: Ideals, options, and limitations. *European Journal of Clinical Microbiology & Infectious Diseases*, *38*, 55–66.

Saleh, A., Parsa, S., Garza, M., Quigley, E. M., & Abraham, B. P. (2023). The role of fecal microbiota transplantation in the induction of remission in ulcerative colitis. *Digestive Diseases*, *41*(4), 656–665.

Salminen, S., Collado, M. C., Endo, A., Hill, C., Lebeer, S., Quigley, E. M., Sanders, M. E., Shamir, R., Swann, J. R., & Szajewska, H. (2021). The International Scientific Association of Probiotics and Prebiotics (ISAPP) consensus statement on the definition and scope of postbiotics. *Nature Reviews Gastroenterology &Hepatology*, *18*(9), 649–667.

Sanders, D. J., Inniss, S., Sebepos-Rogers, G., Rahman, F. Z., & Smith, A. M. (2021). The role of the microbiome in gastrointestinal inflammation. *Bioscience Reports*, *41*(6), BSR20203850.

Schembre, D. B. (2021). Innovations in gastrointestinal endoscopy: Challenges in advancing the frontier. *Gastroenterology & Hepatology*, *17*(10), 485.

Schupack, D. A., Mars, R. A., Voelker, D. H., Abeykoon, J. P., & Kashyap, P. C. (2022). The promise of the gut microbiome as part of individualized treatment strategies. *Nature Reviews Gastroenterology &Hepatology*, *19*(1), 7–25.

Shah, B. R., Li, B., Al Sabbah, H., Xu, W., & Mráz, J. (2020). Effects of prebiotic dietary fibers and probiotics on human health: With special focus on recent advancement in their encapsulated formulations. *Trends in Food Science &Technology*, *102*, 178–192.

Sharma, N., Bhatia, S., Chunduri, V., Kaur, S., Sharma, S., Kapoor, P., Kumari, A., & Garg, M. (2020). Pathogenesis of celiac disease and other gluten related disorders in wheat and strategies for mitigating them. *Frontiers in Nutrition*, *7*, 6.

Sharpton, S., Schnabl, B., Knight, R., & Loomba, R. (2021). Current concepts, opportunities, and challenges of gut microbiome-based personalized medicine in nonalcoholic fatty liver disease. *Cell Metabolism*, *33*(1), 21–32.

Sokol, H., Pigneur, B., Watterlot, L., Lakhdari, O., Bermúdez-Humarán, L. G., Gratadoux, J.-J., Blugeon, S., Bridonneau, C., Furet, J.-P., & Corthier, G. (2008). Faecalibacteriumprausnitzii is an anti-inflammatory commensal bacterium identified by gut microbiota analysis of Crohn disease patients. *Proceedings of the National Academy of Sciences*, *105*(43), 16731–16736.

Srivastava, A., Prabhakar, M. R., Mohanty, A., & Meena, S. S. (2021). Influence of gut microbiome on the human physiology. *Systems Microbiology and Biomanufacturing*, *2*, 1–15.

Stoll, J., & Kupfer, S. S. (2019). Risk assessment and genetic testing for inherited gastrointestinal syndromes. *Gastroenterology & Hepatology*, *15*(9), 462.

Strati, F., Lattanzi, G., Amoroso, C., & Facciotti, F. (2022). Microbiota-targeted therapies in inflammation resolution. *Seminars in Immunology*, *59*, 101599.

Tian, K., Sautter, P., Fisher, D., Fischbach, S., Luna-Nevarez, C., Boberg, K., Kroger, J., & Vann, R. (2014). Transforming health care: Empowering therapeutic communities through technology-enhanced narratives. *Journal of Consumer Research*, *41*(2), 237–260.

Ulger Toprak, N., Yagci, A., Gulluoglu, B., Akin, M., Demirkalem, P., Celenk, T., & Soyletir, G. (2006). A possible role of Bacteroides fragilis enterotoxin in the aetiology of colorectal cancer. *Clinical Microbiology and Infection*, *12*(8), 782–786.

Verhoef, J. I., Klont, E., van Overveld, F. J., & Rijkers, G. T. (2023). The long and winding road of faecal microbiota transplants to targeted intervention for improvement of immune checkpoint inhibition therapy. *Expert Review of Anticancer Therapy*, *23*(11), 1179–1191.

Vindigni, S. M., & Surawicz, C. M. (2017). Fecal microbiota transplantation. *Gastroenterology Clinics*, *46*(1), 171–185.

Wang, J.-W., Kuo, C.-H., Kuo, F.-C., Wang, Y.-K., Hsu, W.-H., Yu, F.-J., Hu, H.-M., Hsu, P.-I., Wang, J.-Y., & Wu, D.-C. (2019). Fecal microbiota transplantation: Review and update. *Journal of the Formosan Medical Association*, *118*, S23–S31.

Wang, W.-W., Zhang, Y., Huang, X.-B., You, N., Zheng, L., & Li, J. (2017). Fecal microbiota transplantation prevents hepatic encephalopathy in rats with carbon tetrachloride-induced acute hepatic dysfunction. *World Journal of Gastroenterology*, *23*(38), 6983.

Wang, X. J., & Camilleri, M. (2019). Personalized medicine in functional gastrointestinal disorders: Understanding pathogenesis to increase diagnostic and treatment efficacy. *World Journal of Gastroenterology*, *25*(10), 1185.

Yan, A. W., E. Fouts, D., Brandl, J., Stärkel, P., Torralba, M., Schott, E., Tsukamoto, H., E. Nelson, K., A. Brenner, D., & Schnabl, B. (2011). Enteric dysbiosis associated with a mouse model of alcoholic liver disease. *Hepatology*, *53*(1), 96–105.

Yang, Y., & Jobin, C. (2014). Microbial imbalance and intestinal pathologies: Connections and contributions. *Disease Models &Mechanisms*, *7*(10), 1131–1142.

Yoo, J. Y., Groer, M., Dutra, S. V. O., Sarkar, A., & McSkimming, D. I. (2020). Gut microbiota and immune system interactions. *Microorganisms*, *8*(10), 1587.

Yu, L. C.-H., Wei, S.-C., &Ni, Y.-H. (2018). Impact of microbiota in colorectal carcinogenesis: Lessons from experimental models. *Intestinal Research*, *16*(3), 346–357.

Zhang, C.-X., Wang, H.-Y., & Chen, T.-X. (2019). Interactions between intestinal microflora/probiotics and the immune system. *BioMed Research International*, *2019*(1), 6764919.

Zhang, C., Zhang, M., Pang, X., Zhao, Y., Wang, L., & Zhao, L. (2012). Structural resilience of the gut microbiota in adult mice under high-fat dietary perturbations. *The ISME Journal*, *6*(10), 1848–1857.

Zhang, J., Zhang, J., & Wang, R. (2018). Gut microbiota modulates drug pharmacokinetics. *Drug Metabolism Reviews*, *50*(3), 357–368.

Zheng, D., Liwinski, T., & Elinav, E. (2020). Interaction between microbiota and immunity in health and disease. *Cell Research*, *30*(6), 492–506.

Zhou, J., Zhou, Z., Ji, P., Ma, M., Guo, J., & Jiang, S. (2019). Effect of fecal microbiota transplantation on experimental colitis in mice. *Experimental and Therapeutic Medicine*, *17*(4), 2581–2586.

Ziemons, J. (2024). The role of the gut microbiota in human cancer: The power of an equilibrium. Thesis of cris. maastrichtuniversity.

11 Gut Microbiota Based Strategies for the Management of Gastrointestinal Inflammations

Sowmiya S.[1,2]*, Jasmine R.*[1]*, Selvakumar B.N.*[3]*, Deena Priscilla H.*[1]*, and Rameshkumar Santhanam*[4*]

[1]PG and Research Department of Biotechnology, Bishop Heber College (Autonomous), Affiliated to Bharathidasan University, Tiruchirappalli, TamilNadu, India

[2]PG and Research Department of Biotechnology, Vivekanandha College of Arts and Sciences for Women (Autonomous), Affiliated to Periyar University, Elayampalayam, Tiruchengode, Namakkal, TamilNadu, India

[3]Department of Microbiology, Educare Institute of Dental Sciences, Malappuram, Kerala, India

[4*]Faculty of Science and Marine Environment, Universiti Malaysia Terengganu, Kuala Nerus, Terengganu, Malaysia

11.1 INTRODUCTION

The human gut microbiota develops in a synergistic relationship with its host and has a major role playing in many functions such immunological response control, nutrition, digestion, and biological functioning, all of which support physiological equilibrium (Bajinka et al., 2020). Numerous investigations have been done over the past decade to assess the connection that exists between the host and gut microbiota, primarily the goal of explaining the mechanisms behind disease genesis. The microbiome of the digestive tract is an ecosystem of over 1,000 bacteria, each of which hosts at least 160 different species that cohabit in the human gastrointestinal system and influence a variety of physiological processes (Gomaa, 2020). Recent study has shown the microbiome as a viable nutritional target, with diverse dietary patterns possibly affecting the microbiome in ways that affect metabolism. Recent investigation has indicated that the gut microbiome is a viable dietary target, with different eating habits possibly affecting the gut in ways affecting digestion and immune system. Microbiome-based treatments aim at reestablishing gut microbial balance, also known as eubiosis (Yoo et al., 2020). With the development of genome sequencing, dysbiosis, or alterations in the composition and function of the microbiota in the gut, has emerged as a significant pathological factor in IBD. Dysbiosis in IBD patients can impair any significant pathogenic processes, notably impaired epithelial barrier integrity, incorrect microbial identification, antigen presentation and autophagy, and dysfunctional responses by T-cells that result in aberrant adaptive immune responses (Santana et al., 2022).

Numerous studies have established correlations between intestinal dysbiosis and persistent minimal inflammation and metabolic problems, eventually resulting in a condition called metabolic syndrome, weight gain, and diabetes. Modifications in the diversity of the intestinal microbiome are linked to gut infections. The most common forms of inflammatory bowel disease (IBD) appear to be linked with IBS, CD, and UC, which are immediately influenced by numerous variables such as the environment, genetics, and immunogenic factors (Crudele et al., 2023). Initially, Americans were the most severely impacted by IBD however, it gradually evolved into a disease of worldwide scope, impacting people in all regions regardless of ethnic origin or racial background. Data reveal that IBD arises from abnormalities in the microbiota of the intestinal tract and exerts a part in illness development and prognosis (Kochar, 2020).

Some of the microbes involved in the pathogenesis of IBD include *Mycobacterium avium* subspecies *Para tuberculosis*, *Escherichia coli*, *Shigella*, *Enterococcus*, *Bacteroides*, *Lactobacillus spp.*, *Clostridium*, *Faecalibacterium*, and *Streptococcus spp*. On a most intriguing finding, it came to light that the use of antibiotics could lead to microbial dysbiosis and increase the risk of developing IBD. It also ensures the suppression of mucosal inflammation in IBD. Some present-day treatment strategies include the use of prebiotics, probiotics, symbiotic and fecal microbial transplant (FMT). It has been reported that dietary habits influence the way microbes behave in relation to immunity and metabolic processes. Additionally, changes in the local native species could lead to gastrointestinal cancer (Liu et al., 2021). The

DOI: 10.1201/9781003493143-11

results of the case study indicated that systemic antibiotic therapy was linked to an aggravated risk of new-onset IBD. Therefore, researchers and clinicians are on the lookout for novel therapeutics to attenuate or reverse microbial dysbiosis, which could prove effective in the treatment of IBD. The individual's microbiome is a generic term that describes the genetic information of microbes, mostly bacteria, including Archaea, viruses, fungi, and Protista. It has been suggested that a balanced microbiota leads to a normal healthy lifestyle, whereas an impaired balance (dysbiosis) could disrupt homeostasis, thereby paving the way for the onset of disease (Li et al., 2022).

11.2 MICROBIOME AND INTESTINAL HOMEOSTASIS

The human microbiome can be distinguished based on those microorganisms that belong to the skin and mucosa of the mouth cavities, gastrointestinal system, respiration, and the genitourinary tract. However, the most diverse array of microbiome seems to occupy the gut, especially the colon. The digestive system, or GI tract, extends through the oral cavity to the anus and serves an essential function in the absorption, digestion and assimilation of nutrients, also elimination of waste from the body (Dekaboruah et al., 2020). The intestinal mucosa serves as a host to a large population of resident microbes and dietary antigens that play diverse roles such as aiding in nutrient uptake and eliciting immune responses to pathogenic infections.

The host and microorganisms in their intestines share a mutually beneficial relationship. The microbes seem to communicate with the host by means of microbe-associated molecular patterns (MAMPs), which include peptides and (LPSs), peptidoglycan (PGN) and flagellum, have exceptionally well-preserved structural elements that are recognized by the host (Zheng et al., 2020). Gut microbiome associated variations develop from early life for instance differences exist between breastfed and formula-fed infants. It is observed that breastfed infants possess *Lactobacilli* whereas *Enterococci, Bacteroides, Clostridia* and *Streptococci* occupy the gut of formula-fed infants (Davis et al., 2022). Binding of pattern-recognition receptors (PRRs) with MAMPs takes place on the surface of epithelial cells lining the intestine and other immune cells. PRRs belong to an innate immune receptor family comprising the transmembrane and cytoplasmic space similar to toll-like receptors and NOD-like receptors. NOD refers to nucleotide-binding oligomerization domain (Li & Wu, 2021).

A PPRs trigger activates a subset of signaling cascades intracellularly leading to the development of early immune responses resulting in the inflammation of mucosa and the upregulation of dual immune response (innate and adaptive). It is found that pathogen and pathobionts activated PRRs give rise to mucosal inflammation but commensal microbes help to maintain intestinal homeostasis and reduce inflammation by their activities. An intestinal commensal *Bacteroides fragilis* exhibit PSA (polysaccharide A) that is recognized by TLR1/TLR2 duo in conjunction with a C-type Lectin PRR Dectin-1, which activates the signaling cascade. The P13K (phosphoinositide 3-kinase) cascade is further induced to transcribe the anti-inflammatory genes via cAMP-CREB (i.e.) 3'5'-cyclic adenosine monophosphate (cAMP) response element binding protein. NOD2 activation by Muramyl-dipeptide (MDP), a peptidoglycan (PGN) fragment, stimulates gastrointestinal leucine-rich repeat (Lgr5)+ embryonic cell survival and rejuvenation of intestinal epithelial cells (Van Hensbergen & Hu, 2024).

11.3 DISEASE AND DYSBIOSIS

The human body provides a nutrient-rich habitat for intestinal bacteria, which, in turn, benefits the host by producing short-chain fatty acids and important vitamins. Symbiosis refers to the collaborative relationship that exists between the host and intestinal microbes. Short-chain fatty acids (SCFAs) also promote various signaling pathways, modulating epithelial immune responses and microbial symbiosis. Recent developments in sequencing technology for the next generation enabled culture independent investigation of the gut microbiota, revealing that a modified balance of gut microbiota elements, rather than specific pathogens, has a part in the pathogenesis of several diseases. This condition is known as dysbiosis and is a disturbance in the balance among the microbes which are present in the gut (Colella et al., 2023). The two major factors responsible for dysbiosis are environmental changes and host genetic susceptibility (Figure 11.1).

A dysbiotic state arising from altered abundance of microbial species and their metabolic derivatives leads to the loss of integrity of the intestinal barrier and subsequent host immune responses. Dysbiosis leads to IBD, MS (multiple sclerosis), RA means rheumatoid arthirits, and various metabolic syndromes (Chu et al., 2023).

11.3.1 IBD

Patients with IBD are known to harbor a reduced microbiota size, functional stability, and diversity compared to their healthy counterparts. IBD includes CD and UC. The microbiome portrays a reduction in Firmicutes the major group like *Clostridium sp.,* especially *F. prasunitzii,* and increasing *Proteobacteria* and *Bacteroidetes* like *D. desulfuricans, Escherichia coli* (Tan et al., 2023).

According to numerical estimates, people with this disorder possess 25% fewer microbes than a healthy individual. The major change in the gut microbiota has been found to be responsible for simultaneous action of inhibition or enhancing the inflammatory actions. They also contain small quantities of SCFA-producing bacteria, responsible for promoting good gut health via upregulation of Tregs and certain anti-inflammatory cytokines. Additionally, an increase in sulfate-reducing bacteria leads to the overproduction of hydrogen sulfide that could damage intestinal epithelial cells, thereby inducing mucosal inflammation. Also, mucosa-associated *E. coli* richness paves the way for increased gut permeability and inflammation. Along with the symptoms, a reduction in

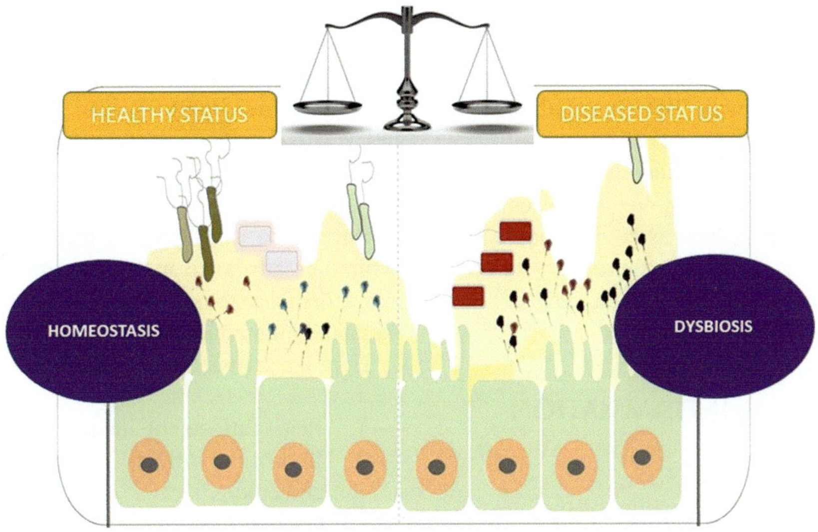

FIGURE 11.1 Balance of gut microbiota leads to dysbiosis.

tryptophan serum levels has been inversely proportional to elevated IL-22 levels and an increase in the diseased state (Pellegrini et al., 2023).

11.3.2 Coeliac Disease

Recent evidence suggests that an imbalance is a key factor in causing celiac disease, involving an immune system malfunction with loss of gastrointestinal stability leading to inflammation that affects the mucous membrane. No continuous microbiological pattern has been found in celiac patients; nonetheless, investigations reveal that a disparity particularly raising in the *Bacteroidetes* to normal *Lactobacilli* that possess a protective anti-inflammatory activity. The exact mechanism of coeliac disease still remains unclear (Mędza & Szlagatys-Sidorkiewicz, 2023).

11.3.3 Rheumatoid Arthritis (RA)

RA is characterized by a systemic autoimmune disorder that leads to the destruction of joints. Affected individuals were known to exhibit reduced microbial gut diversity and dysbiosis of gut bacteria with an increase in *Prevotella*, *Lactobacilli,* and *Collinsella* (Coradduzza et al., 2023).

11.3.4 Multiple Sclerosis (MS)

This disease is a type of neurodegenerative autoimmune illness that largely impacts the brain and the central nervous system (CNS). Patients afflicted with MS exhibit a reduction in species such as *Bacteroides* and *Prevotella,* while there is an increase in *Faecalibacteria* and *Pseudomonas* (Chu et al., 2018).

11.3.5 Metabolic Syndrome

This includes some parameters such as obesity, hyperglycemia and dyslipidemia that could lead to several conditions, including cardiovascular diseases (CVD). Only in recent years, it has come to light that gut dysbiosis could be a risk factor for metabolic syndrome. Dysbiosis is the disruption of gut microbiota. Microbial diversity of the gut ensures the proper maintenance of human health along with the balance of beneficial microorganisms; when this integrity between diversity and balance is lost, it leads to dysbiosis. Metabolic disorders associated with gut microbial dysbiosis result in consuming excessive amounts of fat and can lead to decreased intestinal cell activity and an increase in gut barrier (Mayden, 2023).

11.4 SYMPTOMS AND CAUSES OF GASTROINTESTINAL INFLAMMATION

The bacterial diversity and structural composition of the gutbiota vary remarkably from individual to individual, relying on intrinsic and environmental factors. Notable environmental factors responsible for the composition of the microbiome include diet and medication, which seem to dominate host genetics. Apart from these, age, geography, and birthing practices also contribute to the selection of microbiomes (Sanders et al., 2021).

Dysbiosis is known to affect host health in a variety of ways, with increased intestinal inflammation among the few symptoms. Dysbiosis exerts an important part in the development of gastrointestinal diseases.

Chronic stress seems to negatively impact our guts by increasing inflammation and susceptibility to pathogens. IBD arises from a weakened immune system. The symptoms might differ depends on the stages of inflammation and symptoms include diarrhea, weight loss, stomach pain, blood-tainted stool, loss of appetite. The exact cause of IBD is unknown. Earlier studies suggested diet and stress as causes of IBD, but later researchers felt they only exacerbated the diseased condition. Immune system malfunction has been found to be the most feasible cause of IBD. Gene mutations have also been linked to IBD. Cigarette smoking would be attributed mainly one of the controllable risk parameters impacting CD. Nonsteroidal anti-inflammatory medications influence the increased risk of being affected with IBD (Sun et al., 2019).

The incidence of IBD has greatly increased annually, with patients being exposed to a 2-3-fold increased lifetime risk of colorectal cancer. The pathophysiological mechanisms pertaining to IBD involve genes such as NOD2 (nucleotide-binding oligomerization-2), ATG16L1 (autophagy-related), and immunity-related guanosine (Park & Yeen, 2019).

11.5 DISEASES ASSOCIATED WITH GASTROINTESTINAL INFLAMMATION

11.5.1 Crohn's Disease

This disease condition arises from swelling of the intestinal tract, mainly involving deeper layers, commonly impacting the small intestine. Complications include bowel obstruction, malnutrition due to difficulty in absorbing essential nutrients, leading to anemia and vitamin B12 deficiency, fistulas, and anal fissures. It arises from genetic predisposition as well as influences from environmental factors. CD mainly known as recurrent, systemic inflammatory illness in the gut that causes extra intestinal symptoms and immunological disorders. CD symptoms include abdominal pain, fever, bowel blockage, and diarrhea with blood, mucus, or both (Seyedian et al., 2019).

Apart from genetics, environmental factors contributing to the increased incidence of intestinal inflammation include rapid industrialization, changes in lifestyle focusing on career and higher education, nuclear families, improved hygiene and sanitation, quality of water with improved features, pollution hazards, sedentary lifestyle, consumption of inappropriate nutrient-lacking foodstuffs, alcohol, and tobacco (Clemente-Suárez et al., 2023). CD seems to occur after infectious gastroenteritis. There is an ideology suggesting that CD arises when there is an impaired interaction of commensals residing in the intestine with the immunological system of the host. Research has shown that a drop in *Faecalibacterium prausnitzii* (Firmicute) is responsible for the post-operative recurrence of Crohn's disease. In CD, tight junctions possess permeability defects due to NOD2 mutation. Paneth cells, part of the mucosal barrier responsible for defense function and regulation of commensal microbiota, dysregulation is known to induce inflammation. The specific genes linked with CD are NoD2, ATG16L1 genes, and 23R gene (Linares et al., 2021).

In order to diagnose a disease we rely on a variety of techniques that are either used individually or in combination to predict the disease state. In the case of IBD, techniques such as radiography, endoscopy and pathological studies are employed. Abdomen computer tomography (CT) enterography are the diagnostic procedures used for the recognition of small-bowel disorder CD. Endoscopy scores are valued to be gold standards in the prediction of CD and are commonly employed in research studies to identify the efficiency of medicines that has the ability to sustain and bring about mucosal repair (Li et al., 2023).

Crohn's Disease Activity Index (CDAI), short inflammatory bowel disease questionnaire (SIBDQ), Harvey-Bradshaw Index (HBI) and score of Lehman or the Lehman score are the notable and preferred scoring system that could indicate the activity or nature of disease clinically. Crohn's disease is recognized as a chronic recurring disease. Some of the symptoms attributed to CD are intestinal ulcers, discomfort and pain in the stomach. It is also known to be an autoimmune-mediated inflammation condition (Swenson et al., 2023).

Here listed major types of Crohn's disease:

1. Ileitis: affects the ileum.
2. Ileocolitis: affects the end of the small intestine and colon.
3. Gastroduodenal Crohn's disease: affects the stomach and the duodenum.
4. Jejunoileitis: forms patches of inflammation in the jejunum.
5. Crohn's colitis: affects only the colon.
6. Defined by irritation and ulcers (sores) on the inner membrane of the alimentary tract (colon) and rectum.

11.5.2 Ulcerative Colitis

Ulcerative colitis (UC) defined by irritation, ulcers (sores) then inflammation on the inner membrane of the alimentary tract (colon) and the rectum. Complications of UC include toxic megacolon, which results from rapid widening and swelling of the colon. Toxic megacolon can lead to a perforated colon, a condition that occurs due to holes in the colon. Individuals affected by ulcerative colitis have a reduced life expectancy and are at high risk of colectomy and colorectal cancer. UC tends to cause irritation, inflammation, and ulcers in the inner membrane of the gastro-intestinal tract. There is no cure for UC. Factors influencing the risk of developing UC include age (between 15 and 30; above 60), ethnicity, family history, and genetics (Franjić, 2023).

Here we mentioned different types of UC depends on the occurrence of the body parts:

- Ulcerative proctitis: the mildest form occurring in the anus. Rectal bleeding seems to be the only indicator of the disease.
- Proctosigmoiditis: also referred to as sigmoid colon. Occurs near the rectum and end of the colon. Symptoms include belly cramps, pain, and bloody diarrhea.
- Left-sided colitis: cramping of the left side of the belly, bloody diarrhea, and unexplained weight loss. Inflammation arises from the rectum upwards towards the left side of the colon.
- Pancolitis: occurs in the colon. It causes bloody diarrhea, pain, cramps, fatigue, etc.
- Acute severe UC: a rare condition usually accompanied by severe pain, excessive diarrhea, bleeding, fever (Li et al., 2023).

One of the major symptoms of UC is bloody diarrhea along with some pus in stools. Other symptoms include belly cramps, fatigue, weight loss, fever, dehydration, sores, eye pain, anemia, skin sores, inability to control stools, joint pain, and pain with bowel movements.

The measures used to diagnose UC include blood tests, analysis of stool samples, flexible sigmoidoscopy, colonoscopy, and X-rays (Cankurtaran et al., 2023). The preferred treatment options for UC include a balanced diet, medicines containing antibiotics, aminosalicylates, corticosteroids, immunomodulators, JAK inhibitors, Loperamide, and surgery to remove the colon. UC affects people whose genetic predisposition is triggered by environmental exposure, defects in the epithelial barrier, weakened microbiota, and a dysregulated immune response. UC was first described by Samuel Wilks in 1859 and found to be more common than CD across the globe. Heocolonoscopy with biopsy is considered one of the definitive diagnoses of UC. Histopathology helps detect the severity of the disease and identify dysplasia or cancer (Da et al., 2023).

11.6 MEASURES TO TREAT GASTROINTESTINAL INFLAMMATION

Conventional treatments are entirely focused on reducing symptoms by means of antibiotics, steroids, immunomodulators, biological agents, and surgical resection. The medications for IBD are aminosalicylates, immunomodulators and oral small molecules (Imbrizi et al., 2023) (Figure 11.2).

11.6.1 Antibiotics and Anti-inflammatory Drugs

11.6.1.1 Aminosalicylates

Aminosalicylates comprise sulfasalazine (SASP), then another type of drugs mainly (5-ASA) 5-aminosalicylic acid. SASP is made up of 5-ASA and sulpha-pyridine (SP) formed by a diazo bonding conventionally employed in the treatment of IBD for almost 80 years. In the case of SASP, SP is a carrier while 5'-ASA would be the major active part. The prescribed mode of action is said to interfere with the metabolism of arachidonic acid, scavenging reactive oxygen species (ROS) with durable effects on white blood cells, and production of cytokines (Eronen, 2023). In ulcerative colitis (UC), 5-ASA induces via the aryl gasoline receptor mechanism. This pathway activates T-cells that regulate (Tregs) in the intestines, which also activates transforming growth factor receptors (TGF-β). More efficiency of ASAP with 5-ASA preparations in the case of UC are almost similar but comparing the cost-effectiveness of the drugs, SASP is widely preferred in terms of clinical applications rather than 5-ASA preparations (Jang & Jeong, 2023).

11.6.1.2 Corticosteroids/Anti-Inflammatory Drugs (CS)

Oral corticosteroids have been practiced for the treatment of IBD since the 1950s. The mechanism of action of corticosteroids is as follows: CS interacts with CS receptors present in the cytoplasm and activates it. The activated CS receptor enters the genome and interacts via specific pro-inflammatory factors like nuclear factor-kappa β then activator protein-1 that engages a co-activator complex, instance, histone acetylation enzymes to suppress the transcription of certain inflammatory genes. Also, activated CS receptors attach themselves to suitable response elements present in the promoter segment of genes that regulate inflammation within the nucleus capable of regulating the expression of anti-inflammatory genes. CS is a preferred treatment option for individuals with UC, people are not responding to mesalazine during a couple of weeks of treatment, as well as for people with intermediate to severe Crohn's disease having severe lesions (Zhdanava et al., 2024). Systemic oral CSs can cause a number of adverse effects, including infections that are opportunistic, hyperglycemia, hypertension, ophthalmic symptoms, venous thromboembolism (VTE), and osteoporosis. The subsequent generations' oral CSs, such as budesonide, are being employed because they have a greater effectiveness and safety profile than classic CSs. Targeted administration of steroids to the region of inflammation helps to reduce any systemic side effects (Roda et al., 2023).

11.6.2 Pain Relievers

11.6.2.1 Immunomodulators

Patients suffering from IBD are administered medications popular by the term immunomodulators. These agents work by modulating immune response and reduce inflammation. Some of the drugs widely applied in this category include thiopurines (TP), methotrexate (MTX), calcineurin inhibitors and Janus Kinase (JAK) inhibitors.

i. Thiopurines (TP)

Azathiopurine (AZA), 6-mercapto purine (MP), and 6-thioguanine (TG), belong to the class of TP. The mechanism of action of this drug corresponds to inhibition of proliferation and activation of T-cells that play a pivotal role in the inflammatory cascade. Among these, AZA is said to be the most favorable and effective in the treatment of CD and ulcerative colitis. This drug helps to inhibit the inflammatory response thereby promoting reduced hospitalization (Mantzaris et al., 2017).

ii. Methotrexate (MTX)

MTX also works by suppressing the immune system and thereby reducing inflammation. It mainly blocks the enzyme dihydrofolate reductase that prevents cell proliferation due to interference in the synthesis of DNA.

iii. Calcineurin Inhibitors

Tacrolimus and cyclolspinne causes disturbances in the calcineurin pathway that is involved in the process of activation of T-cells. In this manner, the drugs tend to lower the immune response and bring about a reduction in inflammation. These drugs are used to treat acute severe flares of IBD especially in patients who do not respond to corticosteroid treatment.

iv. Janus Kinase (JAK) Inhibitors

Tofacitinib and upadacitinib work by interfering with Janus Kinase, a signal transducer and activator of JAK-STAT signaling pathway. This pathway plays a crucial role in the activation of immune cells resulting in inflammation. As these drugs inhibit JAK, they control the immune response which helps to prevent inflammation in IBD patients who are moderate to severe in their diseased condition and have failed to comply with conventional therapy.

11.6.2.2 Biological Agents

These agents target particular components of the immune system that are responsible for inflammation. Certain pro-inflammatory cytokines like TNF-α and IL-12/23 have been predicted to possess a prominent role in the pathogenesis of IBD (Liu et al., 2023).

i. Anti-TNF Therapy

TNF-α, a cytokine involved in inflammatory processes, is often overexpressed in IBD, leading to chronic inflammation and tissue damage. Monoclonal antibodies like Infliximab (IFX) and adalimumab (ADA) work by inhibiting the inflammatory response and tissue damage caused by TNF-α. IFX therapy is typically administered to patients who do not respond to corticosteroids and other immunomodulators and studies have shown promising results in both UC and CD patients (Jang et al., 2021).

ii. Anti-Il-12/23 Therapy

IL-12 and IL-23 are cytokines implicated in intestinal inflammation and are produced by antigen-presenting cells. Ustekinumab, a fully humanized Ig G1 monoclonal antibody, prevents the binding of these interleukins to receptors on T-cells and NK cells, thereby inhibiting intestinal inflammation. This therapy has been approved for the treatment of moderate to severe CD and UC, and research suggests that IL-12/23 antagonists are a valuable therapeutic option for patients with IBD (Tang et al., 2012).

11.6.3 Surgical Treatment

Though biologists seem to case the process of treatment of IBD, surgery still remains a necessary option for IBD patients. The need for surgery in UC patients was mostly related to complicated, massive bleeding perforation of the intestine and carcinogenesis. Patients may also need a colectomy, which involves the removal of the colon. In this procedure, a small opening is made in front of the abdominal wall, and the tip of the end of the small intestine is led into this opening forming a stoma that helps to eliminate feces from the body. A pouch is fitted to collect this fecal matter. Patients can lead a normal, active life. If the surgeon removes the deceased section and reconnects the healthy areas, there is no need for a stoma. Raising the possibility of persistence of illness following the surgery is inevitable (Cai et al., 2021).

11.6.4 Nature Treatment/Diet

There is no cure for CD. Some of the natural approaches to treat the symptom include:

- **Diet**: avoidance of high fiber foods, dairy, sugar, high fat, and too spicy foodstuffs. Intake of plenty of water, small meals at well-spaced intervals, boil or steam food rather than frying.
- **Behavioral therapy**: quitting smoking, practicing relaxation techniques like meditation, yoga, exercise.
- **Use of supplements**: curcumin, probiotics could prove effective in the treatment of IBD.
- **Essential oils**: certain essential oils like peppermint could also help in reducing inflammation (De Mattos et al., 2015).

11.7 THE MAJOR ROLE OF GUT MICROBIOME IN HEALTH ISSUES

GM exerts an essential role in physiology of the host, providing metabolic, trophic, along with defensive functions. Technologies with next-generation sequencing technologies have made a significant contribution to our present knowledge of the GM impacts on the injury, inflammation, and ulcers in the intestines. Furthermore, the gut microbiota appears to have a significant impact on host physiology, including mood, inspiration, and improved cognition. Multi-omics analysis conducted as part of the Human Mucosa Project found that individuals with IBD disorders and biologically disrupted bacterial community equilibrium in the gut microflora throughout the disease (Guinane et al., 2013). Animal research conducted on germ-free mice demonstrated the importance within the intestinal microbiome in the progression of the disease. Studies with animals utilizing germ-free mice demonstrated the importance of the GM in the development of IBD, since inflammation did not occur under germ-free settings, but did arise following colony formation of the GI tract loaded with beneficial microbes.

Colonialism of the intestine in standard research animals with bacteria found in those suffering from UC can exacerbate gastrointestinal inflammation including UC signs. Furthermore, findings demonstrated an obvious rise in anaerobes that are facultative at the expense of obligatory anaerobes, as well as genetic changes in bacterial expression and compounds as a result of mucosal inflammation (Zhang et al., 2022).

The growth of some species to the detriment of others may result in a shift in metabolic activity with the distinct characteristics of produced metabolites from microbes like SCFA and tryptophan as seen in patients with IBD. The preservation of intestinal eubiosis, which is defined as major microbial diversity with microbial population as well as harmonious microbial metabolism, is critical in the pathological process of UC. Microbial equilibrium becomes disrupted with the consequence by synergetic effect that exists between environmental factors and the host's genetic

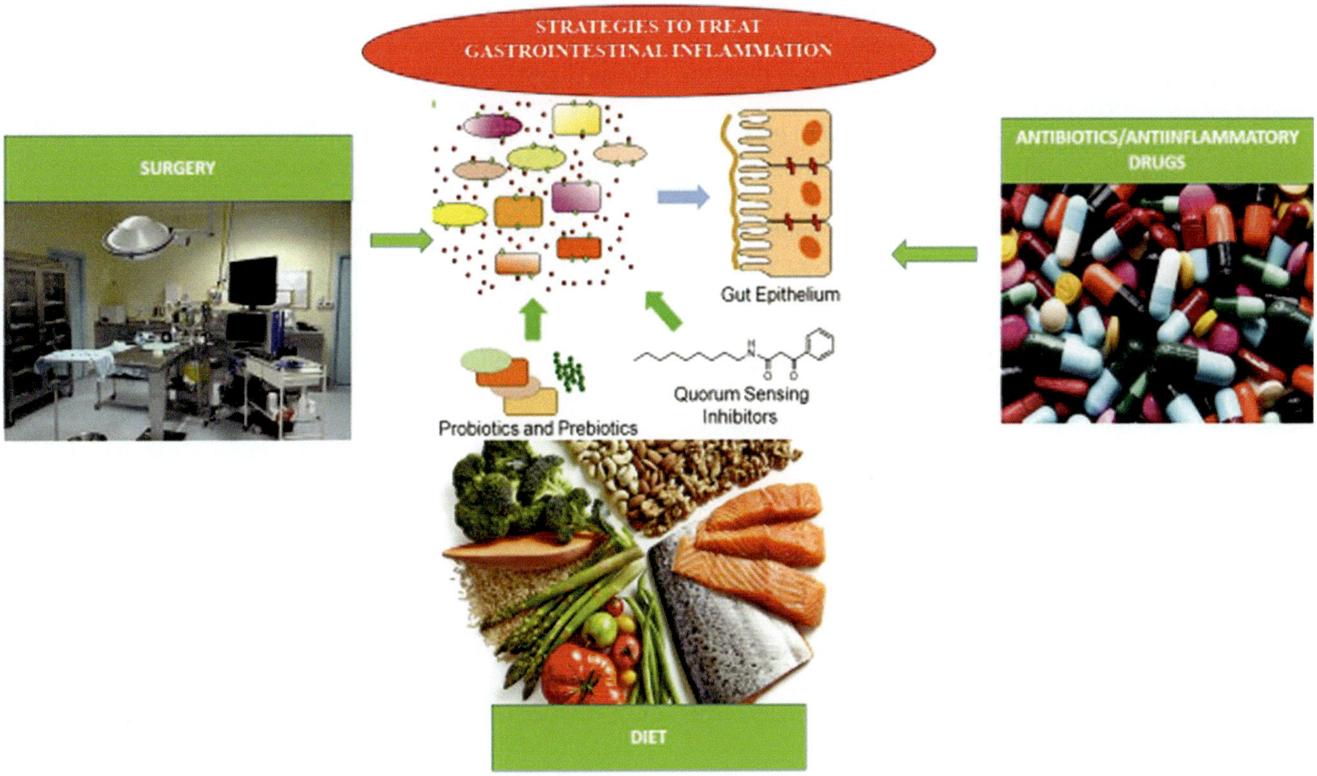

FIGURE 11.2 Strategies to treat gastro intestinal inflammation.

predisposition, triggering a pro-inflammation and affects the equilibrium state at GT leads to dysbiosis (Nogal et al., 2021).

11.8 INFLAMMATORY ACTION OF DISTURBED GUT MICROBIOTA

BD has grown into a worldwide condition, especially rising prevalence across emerging countries with societies that recently have grown more westernized in nature. IBD occurrence and prevalence have been rising globally, with approximately a rate of 0. of the European population affected. One of the most prevalent types of IBD are UC and CD. UC is characterized by persistent external aggravation of the skin and mucous membranes that spreads through the upper section of the rectum to the more proximate region of the bowel. Inflammation in UC causes only mucosal lesions; the harm never reaches the muscle membrane (Ng et al., 2017).

Since the pathophysiology of UC is unknown, factors related to the environment, together with individual genetics, an imbalanced immune response, and microbial dysbiosis, play essential roles in its development. Genetic background is an important factor of the etiology depending on the intestinal bowel disorders. However, mostly transmissible compounds appeared slightly lower in UC and higher in CD. According to this finding presented here, it is reasonable to conclude that external factors have a significant role in causing inflammation in the intestines, particularly via their effect on microbial ecology. The diversity of microbial life in the gut differs dramatically between healthy persons and those suffering from UC, indicating which the microbial community (GM) performs an essential function on the development diseases like UC (Popov et al., 2019). Nonetheless, it is unclear whether the gut imbalance in micro associated with unbalanced immune system reactions is something else or the result of UC. Considering the abundance of frequently available medicines, many individuals are unresponsive to them or have subsequent failure during therapy. As a result, novel medicines and other tactics aimed at microbial dysbiosis are required for UC management.

Probiotics have been used to control the condition by modifying the gut microbiota, improving the barrier's strength and function, and maintaining an appropriate immune response (Stofilova et al., 2022). However, traditional probiotics frequently fail to deliver the predicted human wellness therefore NGP, FMT has been added as advantages that are able to compassionate the human health. Recent research has focused on postbiotic treatments which target an imbalance in UC. The current state of probiotic-based therapy is reviewed in this article along with the potential applications for NGP and Postbiotics, Fecal Microbiota Transplantation, and most effective traditional probiotic strains that are currently available.

11.9 MICROBIOTA AND INTESTINAL HOMEOSTASIS

Mostly microbial community present in the GI involves in the human wellness to its host, especially by maintaining symbiosis. Furthermor, changes in microbial populations in the gut have recently been shown to cause immunological imbalance, which can lead to immune-mediated illnesses (autoimmune disorders).

Antigen-presenting cells (APCs) Intestinal APCs have developed alongside microbiota, and one of their most essential features is their ability to protect the host from infection simultaneously retaining adaptive immunity for the typical microbiota of the gut. Dendritic cells, also referred to as DCs, originating place would be Peyer's patches, located in lymphocytes embedded in the intestinal wall, generate increased IL-10 compared to splenic DCs activated with comparable conditions. (Bibbò et al., 2017). Gut phagocytes, like DCs, live in very close contact to the microbiota of the intestine and have a unique phenotype termed as "Inflammation Energy," which refers to intestinal macrophages' non-inflammatory profile that occurs when they encounter microbiological stimuli in homeostasis conditions. As an instance, intestinal macrophage does not produce such as Toll-like receptor (TLR) ligands, that mimic genetically linked to microorganisms. The cytoplasmic receptor-nucleotide oligomerization domain 1 (NOD1) identified peptidoglycan from the gut microbiome, increasing bone marrow neutrophils killing activity. This study effectively revealed how gut bacteria might modulate systemic immune responses (Dietrich, 2021).

Conventional natural killer cells, or NK cells for short, are naturally occurring lymphocytes which generate IFNγ or performing to identify and eliminate infected target cells.

Mast cells account for between two and three per of the propria lamination (LP) layers in the digestive system. Mast cells are responsible for controlling blood circulation and coagulating blood, swollen muscle peristalsis, permeability, and electrolytes transport through intestinal epithelial cells (IECs).

The gastrointestinal epithelium, which consists of the solitary membrane of IECs, acts as the initial layer of physical barrier between the commensals which inhabit in the lumen of the intestine and underneath sterile tissues. Besides from their biomechanical defensive function, IECs, which cannot be defined as innate immune cells, perform a number of immunological regulatory functions, which includes as the release of peptides that inhibit microbes, inflammatory substances, and chemokines into treated mice with GF and broad-spectrum antibiotics, IECs grew more slowly and expressed less antimicrobial genes (Portales-Cervantes et al., 2019).

MC (Mast cells) account for 3% of the lamina propria (LP) which is present in the gut. These cells regulate blood circulation and coagulate blood, muscle smoothness peristalsis, permeability, including electrochemical transport through (IECs).

The intestinal epithelium, which is made up of a single layer of IECs, serves as the principal physical barrier between the commensals that live in the GI tract and sterile tissues. Besides having a mechanical barrier functioning, which are not normally categorized as strong immunized cells play a variety of immune regulatory tasks, including the release of antimicrobial peptides, cytokines, and chemokines. IECs proliferated at a slower rate and expressed less antimicrobial genes (Krystel-Whittemore et al., 2016). These findings imply that the gut microbiota can influence IECs. The microbial community present in the GI tract serves significant human wellness and benefits to their host immune system especially via maintaining immune homeostasis. Furthermore, this would been demonstrated by the changes in the microbial population in the gut can cause immunological malfunction, culminating in autoimmune diseases.

Antigen-presenting cells (APCs) Intestinal APCs have developed alongside microbiota, and one of their most essential features is their ability to protect the host from infection simultaneously preserving immunological tolerance to a typical gut microbiome. Dendritic cells (DCs) from Peyer's patches, which are lymphatic nodules transplanted in the intestinal wall, produce higher interleukin-10 (IL-10) than splenic DCs activated beneath comparable circumstances to regulate antimicrobial factor expression. The findings suggest that antimicrobial factor expression is regulated by the gut microbiota, which can impact IEC immune regulatory functions.

Gut-associated B lymphocytes are typically found in Peyer's patches, and the bulk of them release immunoglobulin A. Every day, approximately 0.8 milligrams of IgA each centimeter of gut is released, considerably outnumbering the entire creation of all other Ig subclasses. The quantity and cell proliferation of patches identified as Peyer's decreased substantially in GF mice, resulting in lower IgA levels and fewer plasma cells in their intestines. Thus, GM performs a significant role in epithelial IgA production in higher doses at GF rats (CFU) and then beneficial active microbes had to be administered to elicit a high secretory IgA level (Yoo et al., 2020).

Intestinal IgA production fails to result in a memory reacting, hence indicates that gastrointestinal IgA is recognized by a majority of the microbiome. Historically, GF the mice spleens have lower and shorter the embryonic centers, which are crucial for B cell differentiation and affinity maturation. As a result, serum natural IgG levels were dramatically reduced in GF animals, whereas serum natural IgM levels were unchanged. Unexpectedly, IgE antibodies, an allergy-related immunoglobulin isotype, demonstrated to be elevated internally in the GI tract and globally in the bloodstream of GF mice. Few researchers reported that Th2- predisposed characteristics, increased spontaneous antibody production and a Th2 humoral immunity (Shi et al., 2017).

11.10 MUCOSAL IMMUNE REGULATION BY GUT MICROBIOTA

The gut mucosa provides a tangible, biochemical processes, physical and immunological barrier that helps in promoting a symbiotic relationship with controlled microbial population and reduced contact with the host. It is essential to maintain these barriers in a healthy way so as to prevent invasion by pathogens that could lead to elevated levels of immune response and inflammation of the mucous. It also serves as an effective defense against pathogens that compete for nutrients and production of antimicrobial products.

This was previously hypothesized that the intestinal microbiota has a significant impact in the the maintenance of host mucosal barriers that aid in the prevention of colonization by opportunistic pathogens, reducing inflammation of the mucous membrane thereby preserving homeostasis of the gut (Sanders et al., 2021) (Figure 11.3).

The gastrointestinal mucosal immune response is crucial to maintenance of intensity of intestinal barrier along with inflammation. Mostly the major responsible elements seem to be the microbiota. That aids childhood integrity, aging and host immune modulation are crucial to prevent swelling or inflammation of the mucus is under control and the host possesses the intrusive ability to distinguish between beneficial commensals and pathogenic groups. Microbial colonization is said to influence the progression of initial stage B cell lineage in case of mucosal membrane. Intestinal microbiota is necessary to form an immune regulatory pathway that offers protection against excessive production level of IgE and seems that will be directly connected to hypersensitivity of immune system, mucosal membrane inflammation and allergic reactions (Zheng et al., 2021).

11.11 MICROBIOTA DERIVED METABOLITES AS THERAPEUTICS

To accomplish this, innovative treatment possibilities based on gut microbiota manipulation have piqued researchers' curiosity. This chapter presents recent findings on the relationship between dysbiosis and IBD pathogenesis, focuses on the significance of the intestinal microbiota in IBD therapy, and discusses innovative and presently accessible methods of therapy for influencing the makeup and operation of gut microbial community in IBD patients (Figure 11.4).

Probiotics, which postbiotics, FMT, and exogenous polys are all applicable and developing microbiome treatment techniques that show promise. Fortunately, the therapeutic potential of the aforementioned techniques is still being investigated, and additional research is needed to determine their role in current IBD therapy algorithms (Alshehri et al., 2021).

Alternative therapies for manipulating and restoring the equilibrium in the variety and abundance of the gut microbiome are being investigated. Probiotics could bring beneficial activities into the gastrointestinal system or boost the workings of the present-day microbial populations. NGP have the ability to alter the chemical composition also enhances the role of bacterial populations by competing for resources, producing development precursors or inhibiting agents and modulating gut immunity.

Gastrointestinal microorganisms utilize the nutrients and convert the absorbed nutrients taken by the host, like vitamins, amino acids, and dietary fibers. Some of the by-products of the beneficial microbes enhances the metabolic transformations, such as SCFAs, histamines, and other amino-acid-derived metabolites like serotonin or GABA, might be physiologically active in both healthy and diseased states (Ma et al., 2023). The metabolism of these chemicals could additionally be capable of altering microbial diversity. Nutritional carbohydrates that are not digestible may metabolize in the lumen of the intestines, producing SCFAs such as acetate, propionate, and butyrate. Through metabolism active SCFAs, which participate in many biological activities, serve as mitochondrial sources of energy that nourish human colonic epithelial cells. Furthermore,

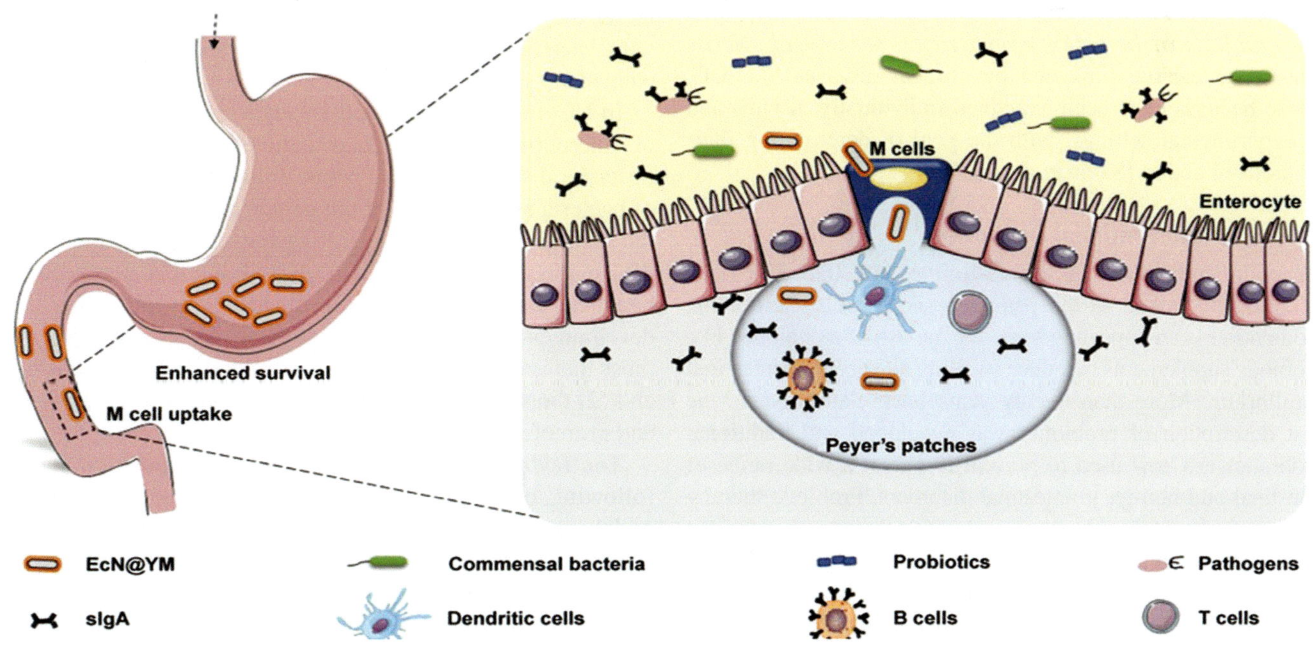

FIGURE 11.3 Mucosal immunity – mediated gut microbiota homeostasis.

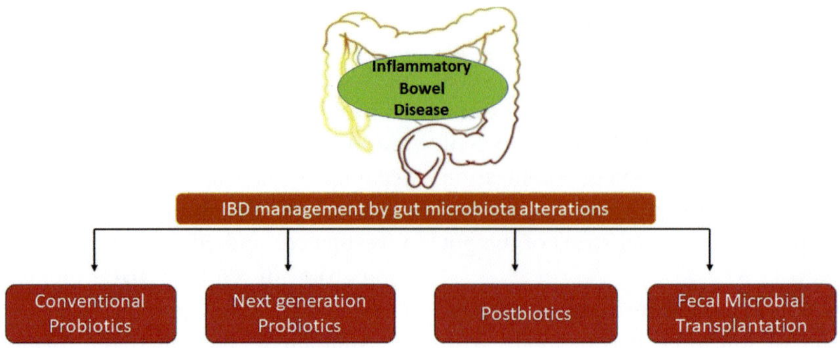

FIGURE 11.4 IBD management by gut microbiota alterations.

digestion of prebiotic polysaccharides such as inulin and fructose proliferate the growth of commensal bacterial (mostly *Bifidobacterium spp.* and *Lactobacillus spp.*) in the intestinal tract (Koppel et al., 2017).

11.11.1 Next Generation Probiotics

Probiotic exploration has grown to include bacteria that are not included under the auspices of traditional probiotic species, sometimes known as "next-generation probiotics" (NGP). This is defined as newly discovered functional microorganisms having beneficial qualities, and the phrase refers to freshly isolated bacteria, primarily hypoxic ones. In the majority of instances, NGP are members of ubiquitous microbiota adhering to various genera, with the exception of *Lactobacillus spp.* and *Bifidobacterium spp.,* discovered and identified through examination of both healthy and sick animals/people. This would be produced from the microbes of the gut representing novel approaches to prevention and therapy for the treatment of a variety of disorders. There are numerous NGP prospects, including *Christensenella minuta, A. muciniphila, Clostridium butyricum, F. prausnitzii,* non-pathogenic microbes like *Bacteroides fragilis* als*Anaerobutyricum soehngenii,* whose prevalence can be impacted by clinical diseases like UC. These bacteria have been examined individually in vitro and in experimental settings, with the goal of determining their methods of action (Savin et al., 2019).

11.11.1.1 Conventional Probiotics

Since the microbiota of the gastrointestinal tract has been scientifically linked to the pathogenesis of UC, microbiome modulation is an appealing topic for the use of techniques like probiotic supplements to reduce swelling and promote intestinal equilibrium. More than twenty years have elapsed since the first description of probiotics was developed and traditional probiotics are now used to prevent and treat a wide range of intestinal and non-gastrointestinal disorders. Probiotic therapy is extensively used in the treatment of illnesses to alter the gut microbiota, enhance intestinal barrier functioning, and preserve an appropriate immune response (Stofilova et al., 2022). *Lactic acid bacteria, Bifidobacteria,* other bacteria like *Streptococcus thermophilus* or *Escherichia coli* and certain yeasts (*Saccharomyces boulardii*) are the most commonly utilized microbes in traditional probiotics. Most standard probiotics were well-characterized and *labeled* as GRAS—*generally recognized as safe.* Nonetheless, traditional probiotics may not always give the anticipated health advantages to patients. Probiotics' biological consequences are strain- and dose-dependent; thus, the efficacy or rejection of one strain cannot be generalized to others. In addition, the probiotic benefit may differ based on the stage and kind of the disease, and also on the immune system of the host (Nyanzi et al., 2021).

11.11.1.2 Probiotics' Mechanisms of Action—UC

The exact mechanism that regulates the activity of probiotics delivered in many clinical research to those having UC is occasionally studied. Recent research is more concerned with tracking the relieving impact of probiotic during the clinical symptoms of the disorder. Preclinical research employing numerous animal models of colitis has validated the particular processes behind the therapeutic effects of specific probiotics, which are still being explored. In addition, it has been demonstrated that the process of action varies depending on the strain, dose and degree of colitis. Probiotics primarily target the gut mucosa, where they attach and communicate with bacteria, cells of the epithelium, and immune system elements (Plaza-Diaz et al., 2019).

In UC, the intestinal mucosa exaggerates to an abundance of microorganisms and their antibodies, resulting in gut obstacles disruption, raised permeability, overbearing transfer of bacteria and exposure of the person receiving treatment to the luminal content, and a greatly exaggerated reaction about the mucosal immune system. Based on this perspective, beneficial bacteria bacteria's mechanisms for mitigating UC development can be divided into two distinct groups: (1) those that impact path biogenesis immigration and expansion, and (2) those which regulate over-activated immune systems and promote barrier function (Shi et al., 2017).

The favorable changes in microbial composition found following probiotic use in UC animal models have been widely reported in terms of microbial alterations, probiotics employ both competitive exclusion and inhibitory chemical synthesis. Probiotics compete with bacteria that are pathogenic for adhesion sites and nutrition probiotics additionally produce substances that include lactic acid and SCFA, lowering the pH of the gut and making it hostile to infections. Further, beneficial bacteria create a wide range of chemicals with specific antimicrobial properties, including hydrogen peroxide and species-specific bacteriocins (Tegegne et al., 2022).

Several strains of *Lactobacilli* and *Bifidobacteria* would been demonstrated to dramatically inhibit the adherence and penetration of adherent-invasive *E. coli* LF82 in human cells of the intestines. In biofilm tests, these microbial strains decreased the adherence ratio of harmful microbes to the abiotic factors surface, indicating that they may impact the level of expression of *E. coli* LF82 adhering characteristics instead of employing a competitive method for host cell receptor locations on epithelial cells (Vlasova et al., 2016) (Stofilova et al., 2022).

11.11.2 Postbiotics

The term "Postbiotics" refers to metabolites generated during anaerobic digestion of indigestible nutrients from food, or even digested ones (e.g., complex carbs, protein molecules, and lipids), as well as those produced by the metabolism of bacteria in the gastrointestinal tract. The term endogenous substances formed by bacteria-host interaction engaged in both favorable and unpleasant behaviors to the living thing are referred to as "Postbiotics" (Thorakkattu et al., 2022).

One of the greatest obstacles in nutrition investigation is understanding the biological processes that explain how the compounds generated by the microbiota of the digestive tract (postbiotics) generate beneficial responses, hinder diseases and operate variously in response to drastic eating habits and changes in the environment. Such understanding paves the way for the creation of novel therapies towards against IBD (Conlon et al., 2014).

Scientists recently discovered numerous types of postbiotic compounds derived from external and intra cellular probiotic bacteria. Such substances include peptidoglycan-derived muropeptides, a substance called ex (EPS), teichoic acids, and exterior sticking molecules that are like filaments, pili, or flagella which make up cell wall elements, released proteins/peptides, bacterial antibiotics like acidophilin, reuterin and bifidin, cell-free supernatant, organic compounds such as lactate and acetic acid, nutrients, fatty acids with short chains like butyric acid and propionate, the neurotransmitters bio surfactants and etc (Team et al., 2020). As a result of their developed chemical compositions and long storage stability, postbiotics as are known to provide health benefits by demonstrating local impacts on certain gut epithelial tissues, as well as affecting various other tissues and organs. They attempt to imitate the therapeutic impact of probiotics, eliminating the risk of administering live bacteria to altered intestinal barriers or impaired immune systems, while additionally being more durable and possessing an extended shelf life (Liang et al., 2023).

11.11.2.1 Short-Chain Fatty Acids

Recent research has demonstrated that acetic acid, propionic acid, and butyric acid can suppress the multiplication of specific bacteria under certain pH circumstances, minimizing the improper functioning of the human digestive tract that results from overwhelming expansion of bacteria. It is essential to remember that gastrointestinal cells in the human organism generally choose the compound butyrate as their initial source of energy, so the compound butyrate may encourage the growth of healthy colon cells while improving the function of the intestinal barrier to some extent, yet butyrate may additionally cause apoptosis, which is the death of transformed cells (Tian et al., 2020).

Previous research has demonstrated that propionic acid and butyric acid can limit the generation and function of regulatory T-cells by decreasing Histone Deacetylase (HDAC) induction. This also implies that SCFA plays a key role in establishing a relationship between the microbiome and the immune system. However, whether SCFA indirectly induces tolerance to the host-associated microbiota via signaling or directly serves as a signal to lower inflammatory response has not been well investigated (Liu et al., 2021).

11.11.2.2 Exopolysaccharides

Probiotics like *Lactobacillus rhamnosus*, *L. fermentum*, *L. amylovorus*, *Leuconostoc* sp., etc. are capable of producing EPS. An enormous variety of polysaccharide structures, categorized as homo-polysaccharides or hetero-polysaccharides, have been observed. These structures are made up of one, two, or more monosaccharides and can be either unbranched or branched.

Experimental study has shown that *lactic acid bacteria*-EPS possesses numerous functional properties, including an anti-oxidative impact by eliminating a wide range of radicals that are free, an immune-modulating effect, the microbiota of the gut regulation, and cholesterol-lowering activity by supplying to free cholesterol. EPS can help prevent and treat the metabolic disorders (Lee et al., 2021).

11.11.2.3 Enzymes

Lactic acid bacteria may neutralize harmful free radicals using their own unique natural antioxidant enzyme systems, including the enzyme catalase (CAT), NADH-oxidase, glutathione peroxidase and peroxide dismutase (SOD), which dismutate free radical changes into O_2 and peroxide from hydrogen (H_2O_2), which plays a significant part in preventing ROS and oxidative stresses. *Lactobacillus casei strain* BL23 is capable of producing CAT or SOD, stimulating initial loss of weight recovery in mice with Crohn's disease and also enhancing enzymatic activities then decreasing intestinal inflammation. *L. acidophilus strain 900* was more efficient in preventing the process of inflammation (Feng et al., 2020).

11.11.3 Fecal Microbiota Transplantation

Fecal microbiota transplantation (FMT) was successfully identified as a life-saving therapy for recurring *Clostridioides* (previously *Clostridium*) *difficile* infection. The documented contribution of the microbiota of the gut in the etiopathology of IBD, as well as the presence of dysbiosis, has prompted further clinical study into the application of FMT as a combination UC and CD treatment. The initial attempts at FMT in UC were focused at establishing equilibrium by treating microbial dysbiosis and providing the microbial community from an ideal donor to a patient, as well as generating intestinal equilibrium linked with inflammation elimination in order to establish recovery. Previous investigations using creatures as models of colitis shown that FMT had promising effects on both microbiota repair and inflammatory reduction, as well as improved overall appearance.

The most recent research studies on FMT have indicated that it may play a role in the management of mild to moderate UC. Tian et al. found that FMT provides prospective beneficial effects for the therapy of UC since it increased the quantity of intestinal microbiota and decreased grades for diarrhea, discomfort in the stomach, and skin ulcers in UC patients (Mahmoudi & Hossainpour, 2023).

Probiotics intentionally introduce intestinal bacteria that are thought to benefit the surrounding microenvironment and may be linked to IBD symptom alleviation. FMT introduces microorganisms found in higher proportions in healthy people, albeit more directly than probiotics. FMT has been linked to higher rates of clinical remission in IBD.

However, variation in FMT efficacy may be affected by IBD category, FMT, which stands donor selection, and administration techniques (Gianotti et al., 2017).

11.12 BIOTHERAPEUTICS MECHANISM OF GUT MICROBIOTA IN TARGETING GASTROINTESTINAL INFLAMMATION

Microbial-targeted therapies, such as antibiotics, pre-/probiotics, and fecal microbiota transplant (FMT), are based on the idea that restoring a healthy gut microbiome will reduce inflammation in the intestinal tract. Antibiotics immediately suppress or destroy particular intestinal microbes. Antibiotics may help induce clinical improvement and cure recalcitrant pouchitis, and they may also be linked to hypersensitivity to anti-TNF biologics in IBD. Prebiotics are substances that are the molecular metabolic components of beneficial gut bacteria (Kim et al., 2021) (Figure 11.5).

Despite there are several medicines that can treat ulcerative colitis (UC), many individuals do not respond to these therapies or have secondary failure during treatment. Therefore, the creation of new medicines or alternative tactics with fewer side effects is unavoidable. Interventions addressing an imbalance of the intestinal microbiota were attempted in the treatment of UC due to the undeniable significance of gut microbiota in the etiology. Comprehensive molecular investigations of the intestinal microbiomes demonstrated clear dysbiosis in UC patients, as evidenced by a lower diversification of symbiotic microbiota. The consumption of traditional probiotic varieties is a frequent method in treating illnesses to alter the microbiome of the gut, enhance the strength of the intestinal barrier and functioning, and preserve an appropriate immune response (Štofilová et al., 2022). However, traditional probiotics do occasionally give the anticipated health advantages to patients. Their advantages vary substantially based on the kind and stage associated with the disease, as well as the kind of strain and dosage of probiotics used. The way they operate of action is also strain-related. In recent years, more possibilities for feasible next-generation probiotics were identified. This might offer novel techniques in restoring microbiome equilibrium and UC treatment.

Inflammatory bowel disease (IBD) has emerged as an international epidemic, with a rising prevalence in highly industrialized nations whose cultures are currently growing more western. IBD rates and incidences have risen worldwide, with around 0.2% of the European population suffering from the condition. According to this it is reasonable to conclude that environmental influences have a significant role in causing inflammation in the intestines, particularly through their influence on microbiota composition. The diversity of the microbes in the intestines differs dramatically between healthy persons and UC patients, indicating that the microbiota plays an important role in the etiology of UC. Nonetheless, it is unclear whether the gut dysbiosis associated with unbalanced immune responses is the cause or the result of UC (Park et al., 2021).

The purpose of this chapter is to offer an updated overview of the present probiotic-based treatments, emphasize operational traditional probiotic varieties, and define the future opportunities of contemporary beneficial bacteria and postbiotic supplemental intake and transplantation of fecal microbiota in the treatment of UC.

11.13 FUTURE PERSPECTIVES

The prime key factor regulating the communication between the gut microbiota and the host is known to be aided by the action of short-chain fatty acids (SCFA). However, the inexplicable role of microbes in regulating immune metabolism is yet to be analyzed on a wider perspective. The main agent responsible for the complexity arises from the microbial end products as they tend to vary based on the nature of the environment and the type of cell. The application of functional metagenomics to help in the identification of bioactive molecules linked to IBD is a technology that holds great promise for the future of medicine.

Multiomics technology, comprising genomics, transcriptomics, proteomics, immunomics, and microbiomics, helps to predict the onset of disease with a certain degree of accuracy. Additionally, 16S rRNA sequencing could pave the way for analyzing the gut microbiota and offer personalized treatment suiting an individual's genome.

In the future, the combined use of intestinal microbiology, gastroenterology epidemiology with rapid analysis of gut microbiota, microbial metabolites, molecular signaling studies, and the application of genetic engineering seem to be the novel scope of therapeutics that could facilitate the diagnosis and treatment of IBD.

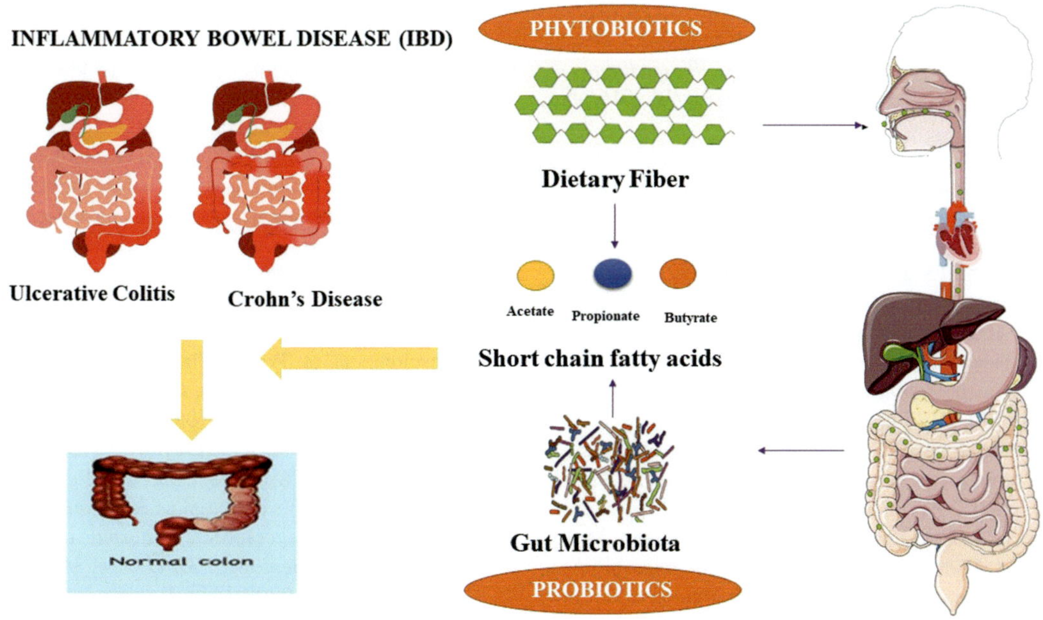

FIGURE 11.5 Biotherpeutic mechanism of gut microbiota in targeting gastrointestinal inflammation.

11.14 CONCLUSION

The treatment of intestinal disorders is time-consuming and involves a complex set of processes owing to the complicated pathogenesis accompanying IBD. Present-day treatment scenarios come with a subset of limitations and decreased effectiveness with an escalating incidence of IBD. To cope with this drawback, Microbe-based therapies like probiotics, prebiotics, and postbiotics seem to open up new avenues and better prospects for the effective treatment of IBD incidence.

Multidisciplinary approaches might help us to navigate through the shortcomings by providing a deeper and more comprehensive understanding of the signaling mechanism with a detailed idea of the molecular metabolism and pathways involved in forming the regulatory network between the metabolites, microbiota, and the host system.

To enhance the specificity and selectivity of therapeutic medications targeting intestinal microbiota relies on the screening of microflora to identify the individual differences and offer personalized microbiota based therapies to treat the patients. Though we have come a long way in the pursuit of getting to know the pathogenesis of IBD and providing the necessary support to treat IBD, we are yet to scale up higher so as to develop effective treatment strategies still remains a distant stream.

Understanding of the relationship between microbiota and host could simplify the process of identifying potential targets providing solutions to numerous untold queries pertaining to treatment schedules, age, suitable precise strategy, employability of applicable optimal strain, personalized medication, and explicit use of live or dead organisms to treat IBD. IBD relies on several genetic environmental and phenotypic characteristics when it comes to disease outcome, cure, and treatment. Hence, the future therapy is entitled to the initiation of personalized medicine to treat intestinal disorders after precise analysis of gut microbiota and the immune profile of the diseased individuals. Ultimately, we rely on safe, economic, and efficient therapeutic approaches for the effective management of IBD.

REFERENCES

Alshehri, D., Saadah, O., Mosli, M., Edris, S., Alhindi, R., & Bahieldin, A. (2021). Dysbiosis of gut microbiota in inflammatory bowel disease: Current therapies and potential for microbiota-modulating therapeutic approaches. *Bosnian Journal of Basic Medical Sciences*, 21(3), 270–283. https://doi.org/10.17305/bjbms.2020.5016

Bajinka, O., Darboe, A., Tan, Y., Abdelhalim, K. A., & Cham, L. B. (2020). Gut microbiota and the human gut physiological changes. *Annals of Microbiology*, 70(1), 1–9. https://doi.org/10.1186/s13213-020-01608-2

Bibbò, S., Dore, M. P., Pes, G. M., Delitala, G., & Delitala, A. P. (2017). Is there a role for gut microbiota in type 1 diabetes pathogenesis?. *Annals of Medicine*, 49(1), 11–22. https://doi.org/10.1080/07853890.2016.1222449

Cai, Z., Wang, S., & Li, J. (2021). Treatment of inflammatory bowel disease: A comprehensive review. *Frontiers in Medicine*, 8, 765474. https://doi.org/10.3389/fmed.2021.765474

Cankurtaran, R. E., Polat, Y. H., Aydemir, N. G., Umay, E., & Yurekli, O. T. (2023). Reliability and usefulness of ChatGPT for inflammatory bowel diseases: An analysis for patients and healthcare professionals. *Cureus*, 15(10). https://doi.org/10.7759/cureus.46736

Chu, F., Shi, M., Lang, Y., Shen, D., Jin, T., Zhu, J., & Cui, L. (2018). Gut microbiota in multiple sclerosis and experimental autoimmune encephalomyelitis: Current applications and future perspectives. *Mediators of Inflammation*, 2018, 8168717. https://doi.org/10.1155/2018/8168717

Chu, J., Feng, S., Guo, C., Xue, B., He, K., & Li, L. (2023). Immunological mechanisms of inflammatory diseases caused by gut microbiota dysbiosis: A review. *Biomedicine & Pharmacotherapy*, 164, 114985. https://doi.org/10.1016/j.biopha.2023.114985

Clemente-Suárez, V. J., Beltrán-Velasco, A. I., Redondo-Flórez, L., Martín-Rodríguez, A., & Tornero-Aguilera, J. F. (2023). Global impacts of western diet and its effects on metabolism and health: A narrative review. *Nutrients*, 15(12), 2749. https://doi.org/10.3390/nu15122749

Colella, M., Charitos, I. A., Ballini, A., Cafiero, C., Topi, S., Palmirotta, R., & Santacroce, L. (2023). Microbiota revolution: How gut microbes regulate our lives. *World Journal of Gastroenterology*, 29(28), 4368. https://doi.org/10.3748/wjg.v29.i28.4368

Conlon, M. A., & Bird, A. R. (2014). The impact of diet and lifestyle on gut microbiota and human health. *Nutrients*, 7(1), 17–44. https://doi.org/10.3390/nu7010017

Coradduzza, D., Bo, M., Congiargiu, A., Azara, E., De Miglio, M. R., Erre, G. L., & Carru, C. (2023). Decoding the microbiome's influence on rheumatoid arthritis. *Microorganisms*, 11(9), 2170. https://doi.org/10.3390/microorganisms11092170

Crudele, L., Gadaleta, R. M., Cariello, M., & Moschetta, A. (2023). Gut microbiota in the pathogenesis and therapeutic approaches of diabetes. *EBioMedicine*, 97. https://doi.org/10.1016/j.ebiom.2023.104821

Dai, Q., Dai, C., & Han, J. (2023). The method of combining internal and external treatment of ulcerative colitis in Chinese medicine. *Chinese Medicine and Natural Products*, 3(4), e158–e164. https://doi.org/10.1055/s-0043-1777310

Davis, E. C., Castagna, V. P., Sela, D. A., Hillard, M. A., Lindberg, S., Mantis, N. J., Seppo, A. E., & Järvinen, K. M. (2022). Gut microbiome and breast-feeding: Implications for early immune development. *Journal of Allergy and Clinical Immunology*, 150(3), 523–534. https://doi.org/10.1016/j.jaci.2022.07.014

De Mattos, B. R., Garcia, M. P., Nogueira, J. B., Paiatto, L. N., Albuquerque, C. G., Souza, C. L., ... & Simioni, P. U. (2015). Inflammatory bowel disease: An overview of immune mechanisms and biological treatments. *Mediators Inflammation*, 2015(493012), 493012. https://doi.org/10.1155/2015/493012

Dekaboruah, E., Suryavanshi, M. V., Chettri, D., & Verma, A. K. (2020). Human microbiome: An academic update on human body site specific surveillance and its possible role. *Archives of Microbiology*, 202(8), 2147–2167. https://doi.org/10.1007/s00203-020-01931-x

Dietrich, C. (2021). The role of the gut-brain axis in at-genetic-risk and clinical Alzheimer's disease (Doctoral dissertation, University of East Anglia). https://ueaeprints.uea.ac.uk/id/eprint/84221

Eronen, H. (2023). Old and novel therapies in inflammatory bowel disease. https://urn.fi/URN:ISBN:978-952-03-2818-4

Feng, T., & Wang, J. (2020). Oxidative stress tolerance and antioxidant capacity of lactic acid bacteria as probiotic: A systematic review. *Gut Microbes*, 12(1), 1801944. https://doi.org/10.1080/19490976.2020.1801944

Franjić, S. (2023). Briefly about ulcerative colitis. *Journal of Emergency Medicine OA*, 1(1), 1–5. www.opastpublishers.com/open-access-articles/briefly-about-ulcerative-colitis.pdf

Gianotti, R. J., & Moss, A. C. (2017). Fecal microbiota transplantation: From clostridium difficile to inflammatory bowel disease. *Gastroenterology & Hepatology*, 13(4), 209–213. https://pubmed.ncbi.nlm.nih.gov/28546791/

Gomaa, E. Z. (2020). Human gut microbiota/microbiome in health and diseases: A review. *Antonie Van Leeuwenhoek*, 113(12), 2019–2040. https://doi.org/10.1007/s10482-020-01474-7

Guinane, C. M., & Cotter, P. D. (2013). Role of the gut microbiota in health and chronic gastrointestinal disease: Understanding a hidden metabolic organ. *Therapeutic Advances in Gastroenterology*, 6(4), 295–308. https://doi.org/10.1177/1756283X13482996

Hayden M. R. (2023). Overview and new insights into the metabolic syndrome: Risk factors and emerging variables in the development of Type 2 diabetes and cerebrocardiovascular disease. *Medicina (Kaunas, Lithuania)*, 59(3), 561. https://doi.org/10.3390/medicina5903056

Hussain, M. S., Gupta, G., Goyal, A., Thapa, R., Almalki, W. H., Kazmi, I., ... & Dua, K. (2023). From nature to therapy: Luteolin's potential as an immune system modulator in inflammatory disorders. *Journal of Biochemical and Molecular Toxicology*, 37(11), e23482. https://doi.org/10.1007/s10482-020-01474-7

Imbrizi, M., Magro, F., & Coy, C. S. R. (2023). Pharmacological therapy in inflammatory bowel diseases: A narrative review of the past 90 years. *Pharmaceuticals*, 16(9), 1272. https://doi.org/10.3390/ph16091272

Jang, D. I., Lee, A. H., Shin, H. Y., Song, H. R., Park, J. H., Kang, T. B., Lee, S. R., & Yang, S. H. (2021). The role of tumor necrosis factor alpha (TNF-α) in autoimmune disease and current TNF-α inhibitors in therapeutics. *International Journal of Molecular Sciences*, 22(5), 2719. https://doi.org/10.3390/ijms2205271

Jang, J., & Jeong, S. (2023). Inflammatory bowel disease: Pathophysiology, treatment, and disease modeling. *BioChip Journal*, 17(4), 403–430. https://doi.org/10.1007/s13206-023-00118-y

Kim, J. H., Kim, K., & Kim, W. (2021). Gut microbiota restoration through fecal microbiota transplantation: A new atopic dermatitis therapy. *Experimental & Molecular Medicine*, 53(5), 907–916. https://doi.org/10.1038/s12276-021-00627-6

Kochar, B. (2020). Inflammatory bowel diseases in South Asian patients: Underappreciated and understudied. *Inflammatory Bowel Diseases*, 26(12), 1943–1944. https://doi.org/10.1093/ibd/izaa040

Koppel, N., Maini Rekdal, V., & Balskus, E. P. (2017). Chemical transformation of xenobiotics by the human gut microbiota. *Science*, 356(6344), eaag2770. https://doi.org/10.1126/science.aag2770

Krystel-Whittemore, M., Dileepan, K. N., & Wood, J. G. (2016). Mast cell: A multi-functional master cell. *Frontiers in Immunology*, 6, 620. https://doi.org/10.3389/fimmu.2015.00620

Lee, J., Park, S., Oh, N., Park, J., Kwon, M., Seo, J., & Roh, S. (2021). Oral intake of Lactobacillus plantarum L-14 extract alleviates TLR2-and AMPK-mediated obesity-associated disorders in high-fat-diet-induced obese C57BL/6J mice. *Cell Proliferation*, 54(6), e13039. https://doi.org/10.1111/cpr.13039

Li, C., Gong, L., Jiang, Y., Huo, X., Huang, L., Lei, H., ... & Deng, Y. (2023). Sanguisorba officinalis ethyl acetate extract attenuates ulcerative colitis through inhibiting PI3K-AKT/NF-κB/STAT3 pathway uncovered by single-cell RNA sequencing. *Phytomedicine*, 120, 155052. https://doi.org/10.1016/j.phymed.2023.155052

Li, D., & Wu, M. (2021). Pattern recognition receptors in health and diseases. *Signal Transduction and Targeted Therapy*, 6(1), 291. https://doi.org/10.1038/s41392-021-00687-0

Li, M., Yang, L., Mu, C., Sun, Y., Gu, Y., Chen, D., Liu, T., & Cao, H. (2022). Gut microbial metabolome in inflammatory bowel disease: From association to therapeutic perspectives. *Computational and Structural Biotechnology Journal*, 20, 2402–2414. https://doi.org/10.1016/j.csbj.2022.03.038

Li, R., Ye, S., Zhou, C., Liu, F., & Li, X. (2023). A systematic review and meta-analysis of magnetic resonance and computed tomography enterography in the diagnosis of small intestinal tumors. *PeerJ*, 11, e16687. https://doi.org/10.7717/peerj.16687

Liang, B., & Xing, D. (2023). The current and future perspectives of postbiotics. *Probiotics and Antimicrobial Proteins*, 15(6), 1626–1643. https://doi.org/10.1007/s12602-023-10045-x

Linares, R., Francés, R., Gutiérrez, A., & Juanola, O. (2021). Bacterial translocation as inflammatory driver in Crohn's disease. *Frontiers in Cell and Developmental Biology*, 9, 703310. https://doi.org/10.3389/fcell.2021.703310

Liu, J., Di, B., & Xu, L. L. (2023). Recent advances in the treatment of IBD: Targets, mechanisms and related therapies. *Cytokine & Growth Factor Reviews,* 71, 1–12. https://doi.org/10.1016/j.cytogfr.2023.07.001

Liu, L., Li, Q., Yang, Y., & Guo, A. (2021). Biological function of short-chain fatty acids and its regulation on intestinal health of poultry. *Frontiers in Veterinary Science*, 8, 736739. https://doi.org/10.3389/fvets.2021.736739

Liu, S., Zhao, W., Lan, P., & Mou, X. (2021). The microbiome in inflammatory bowel diseases: From pathogenesis to therapy. *Protein & Cell*, 12(5), 331–345. https://doi.org/10.1007/s13238-020-00745-3

Ma, T., Shen, X., Shi, X., Sakandar, H. A., Quan, K., Li, Y., ... & Sun, Z. (2023). Targeting gut microbiota and metabolism as the major probiotic mechanism-An evidence-based review. *Trends in Food Science & Technology*, 138, 178–198. https://doi.org/10.1016/j.tifs.2023.06.013

Mahmoudi, H., & Hossainpour, H. (2023). Application and development of fecal microbiota transplantation in the treatment of gastrointestinal and metabolic diseases: A review. *Saudi Journal of Gastroenterology: Official Journal of the Saudi Gastroenterology Association*, 29(1), 3–11. https://doi.org/10.4103/sjg.sjg_131_22

Mantzaris G. J. (2017). Thiopurines and methotrexate use in IBD patients in a biologic era. *Current Treatment Options in Gastroenterology*, 15(1), 84–104. https://doi.org/10.1007/s11938-017-0128-0

Mędza, A., & Szlagatys-Sidorkiewicz, A. (2023). Nutritional status and metabolism in celiac disease: Narrative review. *Journal of Clinical Medicine*, 12(15), 5107. https://doi.org/10.3390/jcm12155107

Ng, S. C., Shi, H. Y., Hamidi, N., Underwood, F. E., Tang, W., Benchimol, E. I., Panaccione, R., Ghosh, S., Wu, J. C. Y., Chan, F. K. L., Sung, J. J. Y., & Kaplan, G. G. (2017). Worldwide incidence and prevalence of inflammatory bowel disease in the 21st century: A systematic review of population-based studies. *Lancet (London, England)*, 390(10114), 2769–2778. https://doi.org/10.1016/S0140-6736(17)32448-0

Nogal, A., Valdes, A. M., & Menni, C. (2021). The role of short-chain fatty acids in the interplay between gut microbiota and diet in cardio-metabolic health. *Gut Microbes*, 13(1), 1–24. https://doi.org/10.1080/19490976.2021.1897212

Nyanzi, R., Jooste, P. J., & Buys, E. M. (2021). Invited review: Probiotic yogurt quality criteria, regulatory framework, clinical evidence, and analytical aspects. *Journal of Dairy Science*, 104(1), 1–19. https://doi.org/10.3168/jds.2020-19116

Park, J., & Cheon, J. H. (2021). Incidence and prevalence of inflammatory bowel disease across Asia. *Yonsei Medicine Journal*, 62(2), 99–108. https://doi.org/10.3349/ymj.2021.62.2.99

Park, S. C., & Jeen, Y. T. (2019). Genetic studies of inflammatory bowel disease-focusing on Asian patients. *Cells*, 8(5), 404. https://doi.org/10.3390/cells8050404

Pellegrini, C., Fornai, M., D'Antongiovanni, V., Antonioli, L., Bernardini, N., & Derkinderen, P. (2023). The intestinal barrier in disorders of the central nervous system. *The Lancet Gastroenterology & Hepatology,* 8(1), 66–80. https://doi.org/10.1016/S2468-1253(22)00241-2

Plaza-Diaz, J., Ruiz-Ojeda, F. J., Gil-Campos, M., & Gil, A. (2019). Mechanisms of action of probiotics. *Advances in Nutrition (Bethesda, Md.)*, 10(suppl_1), S49–S66. https://doi.org/10.1093/advances/nmy063

Popov, J., Caputi, V., Nandeesha, N., Rodriguez, D. A., & Pai, N. (2021). Microbiota-immune interactions in ulcerative colitis and coaitis associated cancer and emerging microbiota-based therapies. *International Journal of Molecular Sciences*, 22(21), 11365. https://doi.org/10.3390/ijms222111365

Portales-Cervantes, L., Dawod, B., & Marshall, J. S. (2019). Mast cells and natural killer cells-A potentially critical interaction. *Viruses*, 11(6), 514. https://doi.org/10.3390/v11060514

Sanders, D. J., Inniss, S., Sebepos-Rogers, G., Rahman, F. Z., & Smith, A. M. (2021). The role of the microbiome in gastrointestinal inflammation. *Bioscience Reports*, 41(6), BSR20203850. https://doi.org/10.1042/bsr20203850

Santana, P. T., Rosas, S. L. B., Ribeiro, B. E., Marinho, Y., & de Souza, H. S. P. (2022). Dysbiosis in inflammatory bowel disease: Pathogenic role and potential therapeutic targets. *International Journal of Molecular Sciences*, 23(7), 3464. https://doi.org/10.3390/ijms23073464

Savin, K. W., Zawadzki, J., Auldist, M. J., Wang, J., Ram, D., Rochfort, S., & Cocks, B. G. (2019). Faecalibacterium diversity in dairy cow milk. *PLoS One*, 14(8), e0221055. https://doi.org/10.1371/journal.pone.0221055

Seyedian, S. S., Nokhostin, F., & Malamir, M. D. (2019). A review of the diagnosis, prevention, and treatment methods of inflammatory bowel disease. *Journal of Medicine and Life*, 12(2), 113–122. https://doi.org/10.25122/jml-2018-0075

Shi, N., Li, N., Duan, X., & Niu, H. (2017). Interaction between the gut microbiome and mucosal immune system. *Military Medical Research*, 4, 1–7. https://doi.org/10.1186/s40779-017-0122-9

Štofilová, J., Kvaková, M., Kamlárová, A., Hijová, E., Bertková, I., & Guľašová, Z. (2022). Probiotic-based intervention in the treatment of ulcerative colitis: Conventional and new approaches. *Biomedicines*, 10(9), 2236. https://doi.org/10.3390/biomedicines10092236

Sun, Y., Li, L., Xie, R., Wang, B., Jiang, K., & Cao, H. (2019). Stress triggers flare of inflammatory bowel disease in children and adults. *Frontiers in Pediatrics*, 7, 432. https://doi.org/10.3389/fped.2019.00432

Swenson, A., Womack, K., Quinn, M., & Curhan, G. (2023). Methods to derive a crohn's disease activity index from electronic health records data. *Inflammatory Bowel Diseases*, 29(Supplement_1), S20–S21. https://doi.org/10.1093/ibd/izac247.039

Tan, J., Taitz, J., Nanan, R., Grau, G., & Macia, L. (2023). Dysbiotic gut microbiota-derived metabolites and their role in non-communicable diseases. *International Journal of Molecular Sciences*, 24(20), 15256. https://doi.org/10.3390/ijms242015256

Tang, C., Chen, S., Qian, H., & Huang, W. (2012). Interleukin-23: As a drug target for autoimmune inflammatory diseases. *Immunology*, 135(2), 112–124. https://doi.org/10.1111/j.1365-2567.2011.03522.x

Teame, T., Wang, A., Xie, M., Zhang, Z., Yang, Y., Ding, Q., ... & Zhou, Z. (2020). Paraprobiotics and postbiotics of probiotic Lactobacilli, their positive effects on the host and action mechanisms: A review. *Frontiers in Nutrition*, 7, 570344. https://doi.org/10.3389/fnut.2020.570344

Tegegne, B. A., & Kebede, B. (2022). Probiotics, their prophylactic and therapeutic applications in human health development: A review of the literature. *Heliyon*, 8(6), e09725. https://doi.org/10.1016/j.heliyon.2022.e09725

Thorakkattu, P., Khanashyam, A. C., Shah, K., Babu, K. S., Mundanat, A. S., Deliephan, A., Deokar, G. S., Santivarangkna, C., & Nirmal, N. P. (2022). Postbiotics: Current trends in food and pharmaceutical industry. *Foods (Basel, Switzerland)*, 11(19), 3094. https://doi.org/10.3390/foods11193094

Tian, Z., Zhuang, X., Luo, M., Yin, W., & Xiong, L. (2020). The propionic acid and butyric acid in serum but not in feces are increased in patients with diarrhea-predominant irritable bowel syndrome. *BMC Gastroenterology*, 20(1), 73. https://doi.org/10.1186/s12876-020-01212-3

Van Hensbergen, V. P., & Hu, X. (2024). Pattern recognition receptors and the innate immune network. In Jane Doe (ed.) *Molecular Medical Microbiology* (pp. 407–441). Academic Press. https://doi.org/10.1016/B978-0-12-818619-0.00131-3

Vlasova, A. N., Kandasamy, S., Chattha, K. S., Rajashekara, G., & Saif, L. J. (2016). Comparison of probiotic lactobacilli and bifidobacteria effects, immune responses and rotavirus vaccines and infection in different host species. *Veterinary Immunology and Immunopathology*, 172, 72–84. https://doi.org/10.1016/j.vetimm.2016.01.003

Yoo, J. Y., Groer, M., Dutra, S. V. O., Sarkar, A., & McSkimming, D. I. (2020). Gut microbiota and immune system interactions. *Microorganisms*, 8(10), 1587. https://doi.org/10.3390/microorganisms8101587

Zhang, Y., Si, X., Yang, L., Wang, H., Sun, Y., & Liu, N. (2022). Association between intestinal microbiota and inflammatory bowel disease. *Animal Models and Experimental Medicine*, 5(4), 311–322. https://doi.org/10.1002/ame2.12255

Zhdanava, M., Zhao, R., Manceur, A. M., Ding, Z., Boudreau, J., Kachroo, S., ... & Pilon, D. (2024). Burden of chronic corticosteroid use among patients with ulcerative colitis initiated on targeted treatment or conventional therapy in the United States. *Journal of Managed Care & Specialty Pharmacy*, 30(2), 141–152. https://doi.org/10.18553/jmcp.2024.30.2.141

Zheng, D., Liwinski, T., & Elinav, E. (2020). Interaction between microbiota and immunity in health and disease. *Cell Research*, 30, 492–506. https://doi.org/10.1038/s41422-020-0332-7

12 Nanoparticles-Based Approaches for the Management of Gastrointestinal Inflammations

*Vasu Peddinti[1], Biswajit Rout[1], Shyam Sudhakar Gomte[1], Tejas Girish Agnihotri[1], and Aakanchha Jain[1]**

[1]Department of Pharmaceutics, National Institute of Pharmaceutical Education and Research (NIPER)-Ahmedabad, Palaj, Gandhinagar, Gujarat, India

12.1 INTRODUCTION

Nanoparticles (NPs) are promising nanocarriers for treating and preventing chronic diseases, cancer, surgery, and personalized medicine due to their specificity through synthetic design (Jayant & Nair, 2016; Kaushik et al., 2016). NPs are small particles with a size range of 1–1000 nanometers. NPs are useful in the treatment of diseases due to their distinctive characteristics such as high surface area, small size, and surface modification (Janrao et al., 2023). Compared to organic materials, they are often less susceptible to biological regulation, and many NPs can be controlled externally. This benefit makes it possible to target drugs more precisely, bind receptors more sensitively for more sophisticated imaging, and detect disease biomarkers with greater sensitivity (Jong & Paul, 2008). Inflammatory bowel disease (IBD) is a chronic, idiopathic inflammatory disorder that affects the entire gastrointestinal tract (GIT) and has been linked to an increased risk of colon cancer (Kaser et al., 2010). IBD is classified into two main clinically distinct forms, viz., ulcerative colitis (UC) and Crohn's disease (CD) (Abraham & Cho, 2009a). While UC is limited to the large intestine and can extend proximally from the rectum to possibly include the whole colon in a continuous pattern with extensive superficial mucosal ulceration, CD typically affects the ileum and the colon, though it can affect any part of the GI tract irregularly and the inflammation is often transmural (Abraham & Cho, 2009a). High morbidity and reduced quality of life are linked to both CD and UC (Moradkhani et al., 2013). IBD is becoming increasingly prevalent globally, placing a significant strain on public healthcare systems (Kaplan, 2015, Molodecky et al., 2012). Presently, 1.4 million people in the United States suffer from IBD, and the disease is projected to have a direct annual cost of over $6 billion (Kappelman et al., 2008).

IBD is caused by a variety of factors, including microbial exposure, environmental factors, immunoregulatory failure, and genetic predisposition (Chassaing & Darfeuille–Michaud, 2011; Cho, 2008). There have also been reports of the involvement of pathogenic infections and dysbiosis, which refers to changes in the makeup or growth of the microbiota (Chassaing & Darfeuille–Michaud, 2011). Medications used to treat IBD often include corticosteroids, 5-aminosalicylic acid, and anti-inflammatory medications. Immunosuppressive medications include azathioprine and 6-mercaptopurine. The therapeutic options for IBD have been greatly improved with the development of anti-tumor necrosis factor (TNF)- a monoclonal antibody as a biological therapy. Unfortunately, there is no known treatment for IBD, and many patients find that the treatments they are currently receiving are ineffective in managing their symptoms.

Due to the loss of cellular integrity brought on by the activation of pro-inflammatory cytokines, the intestinal tissue permeability is enhanced in the inflammatory colon at the level of both the endothelium and the epithelium, where tight junctions (TJs) between cells are dilated (Barua & Mitragotri, 2014). The "leaky" gut allows for the passive accumulation of NPs at inflammatory sites due to increased permeability (Tolstanova et al., 2012). This is likely accompanied by increased NP absorption by immune cells (Lamprecht, 2010). NPs can enter the target tissue deeply because of their small size, which may be advantageous for treating CD, which is characterized by transmural inflammation (Schmidt et al., 2013). Further, NPs provide the possibility of altering the stability, immunogenicity, and solubility of drugs. In addition, surface modification of NPs enables precise and regulated drug release to optimize therapeutic concentration at the site of inflammation for an extended duration while reducing systemic adverse effects (Schmidt et al., 2013).

This chapter first discusses epidemiology, pathophysiology, and general factors for targeted drug delivery in IBD. Following that, we covered different routes of administration used in IBD therapy, as well as the various mechanisms used in drug delivery systems. We also provided an overview of nanoparticle-based IBD diagnostic techniques. Further, we have discussed several treatment strategies based on nanoparticles. We also provided information on patents and clinical studies regarding IBD and the use of nanocarriers. Lastly, we have also contemplated the challenges and potential of using nanoparticle-based strategies to treat IBD.

DOI: 10.1201/9781003493143-12

12.2 VARIOUS ROUTES OF ADMINISTRATION IN THE TREATMENT OF IBD

12.2.1 Oral Administration

Oral delivery is the most convenient route for drug administration. This route is non-invasive, self-administrable, safe, and easy to discontinue at any time. Despite its popularity, drug delivery through the oral route for the treatment of IBD faces multiple challenges in the form of low gastric pH, abundant metabolic enzymes throughout GIT, uneven gastrointestinal transit time, and a complex microenvironment of inflammation sites. To counter these challenges, several formulation strategies have been developed (Yang & Merlin, 2019). One such strategy is to use pH-sensitive polymer like Eudragit to protect drug molecules from gastric pH and deliver them to the colonic region. Deshmukh and colleagues developed sulfasalazine-loaded pectin microparticles coated with Eudragit S 100 for the management of IBD. The prepared microparticles showed selective release of sulfasalazine at the alkaline pH of the colon (7.4), protecting the drug from the acidic pH of the stomach (1.2). The selective release of the drug at the intended site also contributes to reducing dose frequency and side effects associated with sulfasalazine (Deshmukh et al., 2020). A similar result was observed by Naeem et al. when they tested cyclosporin A loaded poly (lactic-co-glycolic acid) PLGA NPs coated with Eudragit FS30D in dextran sodium sulfate (DSS)-induced colitis in mice. The Eudragit coating prevents the burst release of cyclosporin in the stomach and promotes the selective release of the drug in the ileum and colon region (Naeem et al., 2018). Comparable outcomes can be seen in the works of Gao et al. (Gao et al., 2021), Abdollahy et al. (Abdollahy et al., 2024), El-Maghawry et al. (El-Maghawry et al., 2020), and many others. Using the specificity of colon-specific microorganisms and their metabolic enzymes offers a further tactic to get around the problems with oral delivery, in addition to Eudragit coating. These methods can be combined to create a thorough and practical solution for better drug administration. In one such study, a team of scientists developed Eudragit (pH-responsive) and an azo group containing polyurethane (enzyme-responsive) polymeric NPs for targeted delivery of curcumin-6 (as a model drug) to the colon. The *in vitro* studies revealed that these NPs can exhibit triggered drug release in the presence of the enzyme azo-reductase produced by the colonic microflora (obtained from a rat model of colitis). Additionally, they demonstrated increased drug release in an alkaline pH environment of 7.4 (Naeem et al., 2014). Finally, reactive oxygen species (ROS) present in the inflammatory microenvironment of IBD can be used as a triggering agent to release medications at the inflamed site. For example, Wang and coworkers developed a ROS-responsive oral nanocarrier system for the simultaneous delivery of dexamethasone acetate and LY294002 (a potent protein inhibitor) to the inflamed colon. As expected, the release of drug molecules from nanocarriers was triggered by excess ROS present in the microenvironment of the inflamed colon. This approach represents an innovative strategy to improve oral route efficacy (Wang et al., 2024).

12.2.2 Rectal Administration

As discussed earlier, IBD includes UC and CD and these two pathological conditions predominantly occur in the ileum and colon region. Therefore, targeting these regions makes sense for an effective treatment of IBD (Abraham & Cho, 2009b). Though there are many oral formulations have been developed that can target the colon, they still need to cover a long path of GIT to reach their target. Therefore, as an alternative to the oral route and to target the colonic region of GIT, the rectal route for drug administration can be used (Hua, 2019, 2020). The rectal route offers the shortest path and a more direct route to target the colonic region of GIT (Watts & Lllum, 1997). The rectal route of drug administration in the form of enemas, rectal foams, or suppositories presents a form of topical drug delivery system that can be handy to build up drug concentration in the colon. For individuals with mild to moderate distal colitis, rectal administration of the drug is advised as the initial line of treatment. The rectal route is thought to be practical and successful for both inducing and maintaining the effects of drugs (Klotz & Schwab, 2005). In comparison to oral and systemic drug delivery, rectal route drug delivery offers two major advantages in IBD treatment. One, it can avoid pharmacokinetic problems associated with oral drugs; and second, it offers less systemic exposure and a more local effect of the drug in comparison to the systemic route (Li et al., 2020). However, regardless of the acknowledged benefits rectal administration can offer, the administration of nano medications through the rectal route is still underutilized. One such study was done by Fan and colleagues. In this study, they developed an inflammation-targeting polyphenol-poloxamer nanoparticle entrapping infliximab, an anti-inflammatory monoclonal antibody, and were tagged with radioactive 89Zr for tracking and positron emission tomography (PET) imaging. The prepared NPs were administered through the rectal route. From PET imaging, it was found that rectal administration of NPs had a remarkable colon-targeting ability, which led to less systemic exposure to infliximab in comparison to intravenous (IV) administration. The study also showcased that rectal administration of infliximab through NPs had a similar anti-inflammatory effect as intrarenal infliximab injection. These findings support the benefits of the rectal route that we discussed earlier (Fan et al., 2023).

12.2.3 Systemic Administration

Systemic administration of medication through the intravenously or subcutaneously (SC) route is less preferred than oral medications because it is painful, requires specialized personnel to administer, or calls for patient education in the case of SC injections (Trinchard-Lugan et al., 2001). Generally, the systemic route is used for the administration of proteins and peptides class of drugs (Parker & Li, 2021).

Currently, antibodies against tumor necrosis factor-alpha (TNF-α), antibodies against integrins, and corticosteroids are given through a systemic route for the treatment of IBD (Mowat et al., 2011). Systemic drug administration can showcase several advantages like increased bioavailability, rapid onset of action, and reduced first-pass metabolism (Keller et al., 2022). However, high bioavailability can lead to increased incidents of side effects associated with the drugs. For instance, anti-TNF-α antibodies may depress the immune system non-specifically when given intravenously. Therefore, the systemic route of drug administration is preferable and helpful only when inflammatory bowel disease appears outside of the intestine. On the other hand, the oral and/or rectal route of drug delivery is more favorable if the IBD is restricted to the GIT only (Mowat et al., 2011).

12.3 DIFFERENT MECHANISMS EMPLOYED IN THE DRUG DELIVERY SYSTEMS FOR IBD TREATMENT

In IBD, targeted administration of drugs may increase local drug concentrations in inflammatory tissue, enhance medication effectiveness, cut down on dosage frequency, and lessen adverse drug reactions. There has been a growing interest in NP systems that target inflammation in IBD. NPs that target the site of inflammation in IBD are generally administered using two basic approaches, viz., IV administration from the endothelium and oral or rectal administration from the epithelium. When NPs are found in the human body, the mononuclear phagocytic system (MPS) of the host usually detects them and triggers an immune reaction. The interaction between NPs and the MPS is influenced by their size, shape, and surface chemistry. This interaction in turn has a significant impact on the destiny of NPs following administration, including NP internalization, biodistribution, and host clearance. Phagocyte internalization of NPs is frequently viewed as a "problem" because, once internalized, NPs typically undergo fast clearance. This reduces the effectiveness of NP targeting and may have toxicological effects on the liver, kidney, and spleen, among other clearance organs (Gustafson et al., 2015). Consequently, NP systems have been purposefully created to be "bio-invisible" to the host to provide a "stealth effect" (such as PEGylation (Agnihotri et al., 2023; Owens & Peppas, 2006)), prevent MPS recognition, increase NP targeting efficacy, and extend NP circulation duration. This use of the "stealth effect" is still applicable to the intravenous delivery of NPs intended to target inflammation in IBD, as these NPs are anticipated to gather at the site of inflammation following bloodstream circulation (where evading MPS detection is required). However, NPs might be particularly engineered to take advantage of this host recognition mechanism for oral or rectal delivery of NPs targeting inflammation in IBD. This would increase the NPs' targeting efficiency by optimizing NPs absorption by the infiltrating immune cells. The following section discusses various approaches used by existing NP systems to optimize NP absorption by the invading immune cells, hence increasing the targeting effectiveness of NPs and accomplishing inflammation-targeted drug delivery in IBD.

12.3.1 Size-Mediated Targeting

The most common approach for NPs to target the inflamed colon is size-mediated targeting since IBD (Tolstanova et al., 2012) increases both vascular permeability and epithelial permeability. Lamprecht et al. compared fluorescent polystyrene particles at sizes of 10 μm, 1 μm, and 100 nm (in diameter) and observed a size-dependent accumulation of NPs in the inflamed gut (A. Lamprecht et al., 2001). Among the particles studied, the 100 nm particles showed the strongest adhesion to inflamed colons compared to healthy colons when given orally to rats with colitis induced by trinitrobenzenesulfonic acid (TNBS) (A. L. F. Lamprecht et al., 2001). In a follow-up study, Lamprecht et al. showed that, in a TNBS-induced colitis model in rats, rolipram-loaded PLGA NPs had a longer anti-inflammatory effect and less systemic absorption (470 nm and 330 nm, respectively) when compared to rolipram solution (Ali et al., 2016). Subsequently, *in vivo*, fluorescence imaging demonstrated the localization of 200 nm budesonide-loaded PLGA NPs at the colon's inflammation in mice with colitis induced by oxazolone (Beloqui et al., 2013). Budesonide's effectiveness in treating dextran sodium sulfate (DSS)-induced colitis in mice was studied using nanostructured lipid carriers (NLCs). When compared to DSS-induced colitic mice treated with budesonide solution, the budesonide-loaded NLCs protected the mucosal architecture of the colon, decreased neutrophil infiltration, and lowered colonic levels of interleukin (IL)-1β and TNF-α. The targeting of NPs made using multiple approaches to the inflamed mouse colon was compared by Coco et al. These NPs included PLGA NPs (~180 nm), peptidic ligand grafted PLGA NPs (~210 nm), mannosylated PLGA NPs (~210 nm), trimethylchitosan NPs (~340 nm), and Eudragit® S100 NPs (~620 nm) (Coco et al., 2013). In the inflammatory colon, the mannosylated PLGA NPs that could selectively engage with the mannose receptors on macrophages exhibited the greatest accumulation (two to three times greater than other formulations). NPs have the potential to penetrate deeper into inflamed tissue compared to microparticles that stay at the mucosal surface. In recent research, the targeting impact of fluorescently tagged PLGA particles, measuring 250 nm and 3 μm, respectively, was compared to the mucosa of the colon in both healthy individuals and patients with IBD (Schmidt et al., 2013). In comparison to the 250 nm NPs, which were only detected in traces at the mucosal surface, there was a noticeably greater concentration of 3 μm particles in the ulcerous lesions. Subsequent analysis of biopsies obtained from both the patients and the controls revealed that most of the 250 nm NPs entered the tissue and moved to the "serosal" compartment. The difference in NP accumulation between rodent colitis models and human patients with IBD indicates a significant species difference in size-mediated targeting, even though it was explained by NPs moving to a deeper tissue compartment in human biopsy studies. This highlights the need for more research to fill the gap between data obtained from animal models and human patients, including the use of large animal models. Additionally, in animal models of IBD, assessments of the therapeutic effectiveness of drugs delivered by these particles of varying sizes need to be conducted.

12.3.2 Charge-Mediated Targeting

Colonic mucosal inflammation is associated with mucus layer depletion and *in situ* accumulation of positively charged proteins, which leads to the accumulation of positive charges at the injured epithelium and serves as a molecular target and anchor for drug carriers whose surface charges are negative. For instance, Jubeh et al. observed that negatively charged drug carriers preferentially bound to the inflamed colon. They studied the differentiative binding of neutral, cationic, and anionic liposomes to inflamed colons compared with healthy colons using colon explants isolated from rats with dinitrobenzene sulfonic acid (DNBS)-induced colitis (Jubeh et al., 2004). The findings highlighted that cationic liposomes bound to the healthy colonic mucosa were three times higher than neutral or anionic liposomes, whereas anionic liposomes adhered to the inflammatory colonic mucosa two times higher than neutral and cationic liposomes. Further, Researchers used anionic liposomes to provide antioxidants for treating DNBS-induced colitis in rats (Jubeh et al., 2006). The study found that the administration of antioxidants via anionic liposomes was more efficient to free antioxidants in the management of colitis. This was attributed to the liposomes' preferred adherence to the inflammatory colon, which led to an extended residence time and improved antioxidant uptake by the inflamed epithelium. Zhang et al. developed a hydrogel microfiber system for local administration of drugs in IBD using ascorbyl palmitate, a substance that is "generally recognized as safe" (GRAS) (Zhang et al., 2015). Negatively charged hydrogel microfibers were shown to adhere more preferentially to inflamed mucosa in tissue samples from UC patients as well as two murine colitis models: DSS-induced colitis and T-bet −/−Rag2−/− ulcerative colitis. When administered by enema, hydrogel microfibers loaded with dexamethasone exhibited greater therapeutic efficacy and reduced systemic drug exposure in comparison to free dexamethasone solution.

12.3.3 Ligand-Mediated Targeting

Specific ligands or receptors that are overexpressed on cell surfaces during the inflammatory process provide molecular targets for drug carriers to be anchored through particular interactions. One example of ligand-receptor interaction comes from leukocyte-endothelial biochemistry, which selectively recruits leukocytes to a site of inflammation (Kim et al., 2012). In response to inflammation, inflammatory cytokines stimulate endothelial cells, causing them to express more adhesion molecules on their luminal surfaces, including P-selectin, E-selectin, vascular cell adhesion molecule (VCAM)-1, and intercellular adhesion molecule (ICAM)-1. On leukocyte surfaces, ligands involved in rolling and facilitating firm leukocyte adhesion are expressed at high levels: these include carbohydrate glycoconjugates like P-selectin glycoprotein ligand-1 (PSGL-1), which primarily binds to P-selectin to start leukocyte binding to vascular endothelium and also binds to L-selectin to mediate rolling of neutrophils on adherent neutrophils (Spertini et al., 1996) and integrins that bind strongly to the adhesion molecules overexpressed on inflammatory endothelial cells, such as $\alpha_4\beta_7$, $\alpha_4\beta_1$, $\alpha_L\beta_2$, and $\alpha_M\beta_2$ (Kim et al., 2012). For example, Sakhalkar et al. developed PEGylated PLA particles coupled with recombinant PSGL-1 to imitate the leukocyte-endothelial biochemistry and construct tailored systems to the site of inflammation. These particles showed considerable rolling adherence to the inflamed endothelium *in vivo* (Sakhalkar et al., 2003). In another study, Sialyl Lewis^X (sLe^x), a carbohydrate that binds to both P- and E-selectins, was used to simulate leukocyte adhesion to selectins by adhering to the surface of PLGA spheres (Eniola et al., 2002). None of the previously mentioned particles had any medication loaded for use as a treatment in IBD. However, Peer et al. formulated liposome-based, β_7 integrin-targeted NPs (β_7 I-tsNPs) entrapped with siRNAs to target particular subsets of leukocytes and treat DSS-induced murine colitis by silencing the cell cycle-regulating molecule Cyclin D1 (CyD1), which is disproportionately increased in IBD-affected immune and epithelial cells 150 (Peer et al., 2008). CyD1-siRNA given by β_7 I-tsNP resulted in a significant decrease in damage to intestinal tissue, strong inhibition of leukocyte infiltration into the colon, and a reversal of the decline in hemoglobin and body weight loss.

In addition to adhesion molecules that are overexpressed on endothelial cells, other ligands and/or receptors that are elevated on immune cells and/or colonic epithelial cells in IBD have been used to direct NPs to the inflamed intestine through ligand-receptor interactions. In inflammatory conditions, colonic epithelial cells, and immune cells that infiltrate the body express more glycoprotein CD98. To target these cells, single-chain CD98 antibody-linked chitosan/polyethyleneimine (PEI) NPs (B. Xiao, Laroui, et al., 2014) and poly (lactic acid) PLA NPs carrying CD98 Fab (B. Xiao, Yang, et al., 2014) were developed respectively and used to manage colitis in mice models. In NP-based therapies for IBD, targeting macrophages via overexpressed surface receptors/ligands has also been extensively studied. For example, mannose receptors have been targeted by mannosylated poly (amido amine)-based NPs (Xiao et al., 2013), and galactosylated chitosan NPs have been shown to target macrophage galactose/N-acetyl-galactosamine-specific lectins (MGL) in IBD (Zhang et al., 2013; Zuo et al., 2010).

12.3.4 Microbiota-Mediated Targeting

In IBD, the host-bacterial interaction can be harmful, causing pathological inflammation rather than beneficially regulating physiological inflammation. Prebiotics promote the growth of probiotics and other helpful microorganisms, whereas probiotics might introduce suitable microbes (Guha et al., 2023; Preidis & Versalovic, 2009). To regain intestinal homeostasis, the host-gut microbiota relationship can be altered from one that is harmful to one that is favorable. Probiotics that have been genetically modified are meant to thrive in a certain niche in the colon and then release proteins that are biologically active to cause the intended pharmacological effect. For example, *Lactococcus lactis* has been utilized to treat colitis in mice by delivering anti-inflammatory cytokine IL-10 (Kaye et al., 1994), immunomodulatory Yersinia LcrV protein (Foligne et al., 2007), and cytoprotective trefoil factors (Vandenbroucke et al., 2004). In addition to *Lactococcus lactis*, a wide range of modified bacteria strains have been utilized to transport different compounds to treat colitis in animal models. These strains include *Streptococcus gordonii* (Ricci et al., 2003), *Lactobacillus casei* BL23 (Watterlot et al., 2010), and *Lactococcus plantarum* (Han et al., 2006). However, studies led researchers to include controllers in the delivery systems as a safety precaution due to worries about the quantity of proteins generated and released *in vivo*. For instance, the xylanase promoter was added to the xylan operon of the *Bacteroides ovatus* in order to use the modified strain of the bacteria to treat DSS-induced colitis in mice by delivering transforming growth factor (TGF)-β under the influence of the plant polysaccharide xylan (Hamady et al., 2011). As an alternative, *Lactococcus lactis* was employed in a stress-inducible regulated expression system to control the synthesis and delivery of IL-10 to treat DNBS-induced murine colitis (Benbouziane et al., 2013).

Proteins on bacterial surfaces have also been used in novel ways to transport drugs by acting as docking sites for NPs. Plasmid DNA-loaded streptavidin-coated polystyrene NPs that adhere to the surfaces of biotinylated bacteria, often known as "microbots," have proven the combined use of NPs and bacteria to deliver medications *in vivo* (Akin et al., 2007). The *Listeria monocytogenes* bacteria in these systems have a length of approximately 1 μm and were coated with NPs with a size range of 40 to 200 nm on their surfaces. These "microbots" have been introduced into mice, and their success in expressing the loaded genes *in vivo* can be seen by their ability to illuminate various organs such as the kidney, liver, intestine, spleen, etc.

12.3.5 Degradation-Mediated Targeting

Delivery systems can be locally broken down by certain molecules or enzymes found in the extracellular matrix (ECM) or microbiota of an inflamed colon, which enables degradation-mediated targeting of the site of inflammation. There is strong evidence linking excessive ROS production to persisting intestinal inflammation, and that the higher quantities of ROS in the mucosa are specific to the site of inflammation and are associated with the advancement of the disease (Lih-Brody et al., 1996; Zhu & Li, 2012). Wilson et al. formulated thioketal NPs that release TNF-α siRNA locally at the site of inflammation when exposed to ROS, aiding in the treatment of colitis in mice (Wilson et al., 2010). NPs that were broken down when exposed to hydrogen peroxide were also synthesized by de Gracia et al. using polyesters with boronic ester-triggering groups to deliver drugs only to inflamed tissue (de Gracia Lux et al., 2012). It is known that the inflamed extracellular matrix (ECM) upregulates proteolytic enzymes such as matrix metalloproteinases (MMPs) (Biancheri et al., 2015). It has been reported that MMP3 and MMP12 in the inflammatory mucosa of IBD patients may break down anti-TNF-α antibodies and lead to a lack of response to such treatment. Synthesized MMP-sensitive delivery systems (Lutolf et al., 2003) offer the potential to both treat inflammation in IBD and safeguard biologics. Additionally, an increasing amount of evidence suggests that certain ECM components, including hyaluronan, may have a significant impact on IBD disease phenotypes (Järveläinen et al., 2009). This requires further research into NP systems that target these ECM components to manage IBD.

Bacteria in the colon release several reductive and hydrolytic enzymes, resulting in enzyme-based fermentation systems. For instance, a great number of polysaccharidases of microbial origin found in the human colon (Karrout et al., 2015; Sinha & Kumria, 2004) have prompted in-depth research on polysaccharide-based delivery methods, which include the use of amylose, guar gum, chitosan, and pectin (N. Shah et al., 2011). Probiotic extracts were administered to the inflammatory colon via chitosan-coated PLGA NPs to manage TNBS-induced colitis in rats (Saadatzadeh et al., 2012). Lately, real-time "snapshots" of the complex gut bacterial community have been made available for comparative research through the use of terminal restriction fragment length polymorphism (TRFLP) methods based on small subunit ribosomal RNA (16s rRNA) (Samanta et al., 2012). With the advancement of technology, it is now possible to identify the bacteria linked to colitis and use their secreted enzymes to specifically break down NP systems. This allows the inflammatory dysbiotic state to release drugs, potentially leading to new treatment methods for IBD.

12.4 NANOPARTICLES-BASED DIAGNOSIS OF IBD

Utilizing advanced methodologies in the detection of IBD can markedly decelerate the progression of this disease. Within this realm, a promising avenue for exploration lies in nanotechnology, particularly the strategic utilization of iron oxide nanoparticles. Among other nanoparticles, iron oxide nanoparticles have a great attraction in biomedical applications due to their low toxicity, colloidal stability, and surface engineering capability in the biological system. They have both magnetic behavior and semiconductor properties leading to multifunctional biomedical applications (Agnihotri et al., 2023; Gomte et al., 2024a ; Wu et al., 2016). These nanoparticles have garnered significant attention due to their unique attributes like superparamagnetic behavior, high surface-to-volume ratio, and efficient separation techniques (Dadfar et al., 2019; Prakash et al., 2024). For instance, Wu and the team explored the dual role of single photon emission computed tomography/computed tomography (SPECT/CT) imaging and magnetic resonance imaging (MRI) with Indium

/iron oxide nanoparticles labeled macrophages in evaluating IBD in experimental mice models. *Ex vivo* analysis using mass spectrometry demonstrated significant uptake of iron oxide nanoparticles by cells. Further SPECT/CT imaging revealed that approximately 3% of the injected dose remained in the intestine after 24 hrs, which was confirmed by MRI and histological examination. Cells labeled with Indium 111 were detected in all tissues associated with the reticuloendothelial system or mononuclear phagocyte system 24 hours post-injection (Wu et al., 2014). There have also been several reports buttressing the utility of imaging techniques in the assessment and imaging of IBD (Barani et al., 2021; Klöppel et al., 1989; Naha et al., 2020; Rieber et al., 2000). The scientists are exploring the amalgamation of nanotechnology with imaging techniques at cellular and molecular level which would assist in better diagnosis and evaluation of IBD (Liu et al., 2015).

In recent decades, there has been rapid development in the field of nanosensors, which are increasingly contributing to biological research, particularly in the context of IBD. Nanostructures have gained traction in the diagnosis of IBD in recent years (Barani et al., 2021; Ruckh & Clark, 2014). Karban et al. reported on chemical sensors utilizing configurable molecularly modified gold nanoparticles for distinguishing between irritable bowel syndrome (IBS) and IBD. The findings demonstrated an 81% accuracy in discriminating between IBD and IBS, and a 75% accuracy in distinguishing between UC and CD (Karban et al., 2016). Gholami et al. explored surface-enhanced Raman spectroscopy quenching nanosensors to identify TNF-α in blood. With the help of this sensor, estimation of TNF-α was possible down to 173 pg/L. Enzyme-linked immunosorbent assay (ELISA) technique was used for comparison and 93% agreement was found between the two (Gholami et al., 2020). In another experiment by Ma et al. ionic liquid-molybdenum disulfide/gold nanoparticles and iridium nanoparticles-loaded graphene were prepared as a diagnostic tool for the estimation of C-reactive protein (CRP). CRP is considered to be an indicator of early heart disease. The developed nanosensor was found to have a limit of detection of 3.3 pg/mL and was found to be linear from 0.01 to 100 ng/mL. The added advantages like the high surface area of gold nanoparticles, and the electrocatalytic properties of iridium nanoparticles made it a superior platform for the detection of CRP in serum (Ma et al., 2020). In an interesting study conducted by Lv et al., a quantum dot labelled immunosorbent assay was developed for the detection of CRP with a range from 1.56 to 400 ng/mL. The limit of detection was found to be 0.46 ng/mL while the limit of quantification was found to be 1.53 ng/mL. In comparison with the ELISA technique, the total reaction time for this developed method was found to be 50 min only eliciting a sensitive, accurate, and precise method for the detection of CRP (Lv et al., 2017).

12.5 NPS-BASED THERAPEUTIC APPROACHES IN GI INFLAMMATIONS (IBD)

Nanotechnology represents an innovative treatment approach harnessing NPs or nanocarriers to precisely deliver drugs to specific sites within the body, enhancing therapeutic efficacy for various medical applications. In recent times, NPs have been promising nanocarriers for targeted colonic drug delivery because of their smaller size, shape, high surface area, and easy surface modification. In addition, it protects the active therapeutic agent against harsh acidic environment i.e., acidic pH and enzymatic degradation. The NPs also provide stability to various therapeutic agents (Barani et al., 2021). Various types of nanoparticulate systems for targeted colonic drug delivery are discussed (Table 12.1) (Figure 12.1).

12.5.1 Metallic NPs

Metal-based NPs, including gold and silver NPs, are significant types of NPs commonly utilized in various applications due to their distinctive properties and versatile functionalities (Agnihotri et al., 2023). These types of NPs possess unique features like catalytic, optical, physical, and electronic attributes along with exceptional versatility and biocompatibility. In comparison to lipid NPs or polymer-based NPs, metal-based NPs have higher stability under abdominal fluid pressure. These NPs can be easily functionalized and loaded with therapeutic agents. In current times, metallic NPs are employed in diverse fields like therapeutic, diagnostic, and drug delivery. Specifically, metal-based NPs attained significant interest in the colon-targeted drug delivery for the management of IBD. Gold NPs (AuNPs) are found to be promising candidates for the treatment of IBD, due to their inherent anti-inflammatory and antimicrobial attributes. For instance, bare AuNPs could inhibit the secretion of inflammatory cytokines and diminish colonic inflammation in the colitis model as compared to the control untreated model (Abdelmegid et al., 2019). Sulfasalazine-loaded AuNPs can increase colon-targeted delivery thereby enhancing therapeutic efficacy compared with plain sulfasalazine administration in colitis models (Bhol & Schechter, 2007; Li et al., 2022). Hussein et al. developed AuNPs for the management of trinitrobenzene sulfonic acid (TNBS)-induced colitis in rats. It was observed from the studies that exposure of AuNPs for 72h. enhanced the regeneration of damaged colonic tissue, reduced the inflammatory cytokines levels, and decreased the colonic myeloperoxidase and malondialdehyde activities. Moreover, there was a promising enhancement in the antioxidant defense system (Hussein & Saleh, 2018). In another study, Li et al. developed ceria-loaded AuNPs for increased antioxidant activity in the management of IBD. The prepared ceria/AuNPs exhibit distinctly increased enzymatic activities compared to commercial ceria NPs. This was possibly due to the efficient exposure of catalytic sites, increased content of Ce (III), and enhanced reduction of Ce (IV) to Ce (III). The ceria-loaded AuNPs coated with hyaluronic acid helped to promote accumulation in inflamed colonic tissues through oral administration. The prepared NPs showed promising biocompatibility and suitable therapeutic options for the management of IBD (Li et al., 2023).

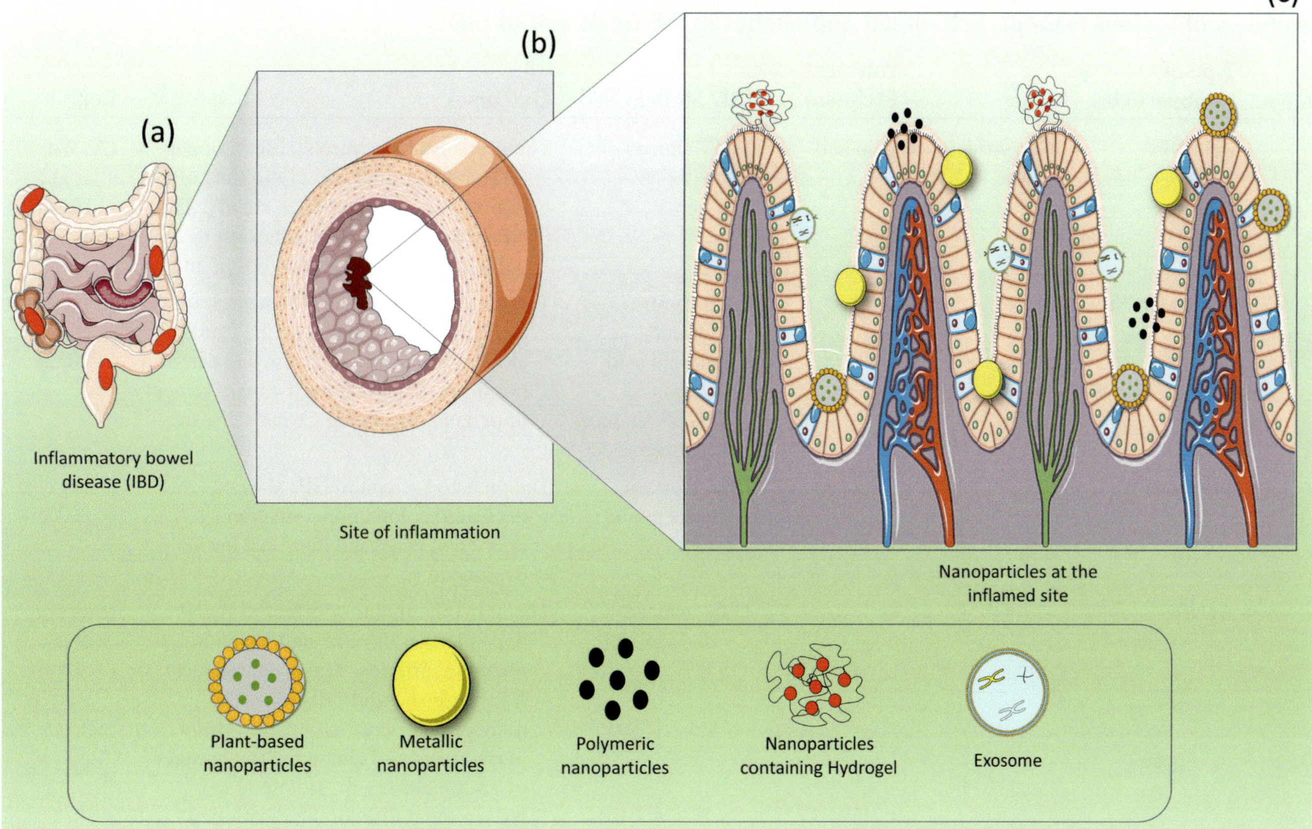

FIGURE 12.1 The use of various nanoparticles in treating inflammatory bowel disease (IBD) where (**a**) gastrointestinal tract (GIT) inflammation in IBD, (**b**) cross-section of an inflamed colon site, and (**c**) microscopic view showing nanoparticles at the site of inflammation.

TABLE 12.1
Overview of Various Nanoparticle-Based Approaches in the Treatment of IBD

Sr. No.	Type of Nanoparticles	Drug	Polymer/ Excipient	IBD Model Used	Outcomes	Ref.
1	PLGA NPs	Garcinol	PLGA, Eudragit S100	-	The NPs were biocompatible and showed pH-dependent release due to ES00 coating. The NPs exhibited a reduction in inflammation in the colonic tissues.	(Jacob et al., 2021)
2	PCL NPs	Budesonide and Glycyrrhizic acid	PCL	-	The prepared NPs have a particle size of 230nm with smooth and spherical morphology. The NPs helped to reduce the colonic mast cellular infiltration and enhanced the histological properties of the colonic site. The dual drug-loaded NPs help to enhance the various inflammatory markers such as COX-2, iNOS, IL1-β, TNF-α, etc.	(Ahmad et al., 2021)
3	Chitosan NPs	Rifaximin	Chitosan	-	The prepared NPs were found to be in the nano-size range. The maximum drug release was found in the colonic region as compared to the upper GIT region. In addition, NPs improved the solubility of rifaximin.	(Kumar & Newton, 2017)

(Continued)

TABLE 12.1 (Continued)
Overview of Various Nanoparticle-Based Approaches in the Treatment of IBD

Sr. No.	Type of Nanoparticles	Drug	Polymer/ Excipient	IBD Model Used	Outcomes	Ref.
4	Zein NPs	Glycyrrhizic acid	Zein and pectin	DSS-induced colitis	The prepared NPs were stable in upper GIT and degraded slowly in the colonic region as pectin degrades slowly. The NPs showed excellent adhesion to the colon using DSS-induced colitis model.	(A. Wu et al., 2021)
5	Iron oxide NPs	-	-	2,4,6-trinitro-benzene-sulfonic acid solution (TNBS)-induced proctitis model	The prepared iron oxide NPs were safe and helped to totally close the fistula along with the whole length and can be a promising approach for the management of perianal fistulizing Crohn's disease.	(Cazelles et al., 2022)
6	Selenium NPs	-	Chitosan	DSS-induced colitis model	The prepared selenium NPs showed decreased colonic inflammation injury and enhanced intestinal barrier dysfunction.	(Peng et al., 2022)
7	AgNPs	-	-	DSS-induced colitis model	The anti-inflammatory effect of AgNPs depended on the size and shape of the nanoparticles. The prepared NPs were shown to decrease the number of *E. coli, and Clostridium perfringens* and enhanced the number of *Lactobacillus sp.* It helped to reduce the colitis in the developed experimental colitis model.	(Siczek et al., 2017)
8	PLGA NPs loaded into microparticles		Eudragit FS30D	-	The prepared NPs sustained release at colonic pH. The prepared NPs into microparticles were shown to improve histopathological analysis, and reduced neutrophils, macrophage infiltration, and myeloperoxidase activity.	(Naeem et al., 2020)
9	PLGA NPs	Curcumin	EudragitS100 and PLGA	DSS-induced colitis model	The prepared NPs help to permeate across CaCo2 cells as compared to curcumin suspension. The NPs significantly decreased TNF-α and neutrophil secretion and improved colonic tissue damage.	(Beloqui et al., 2014)
10	Chitosan NPs	Quercetin	Chitosan	DSS-induced colitis model	The prepared NPs effectively reduced the disease activity index, and decreased the expression of COX-2, and iNOS in colonic tissues. The NPs helped to reduce pro-inflammatory cytokines and can be a suitable alternative for the management of IBD.	(Khater et al., 2022)

12.5.2 Polymeric NPs

Polymer-based drug delivery systems have demonstrated significant biomedical applications and clinical efficacy across various diseases. Their straightforward synthesis methods, self-assembling structures, ability to modulate chemical compositions, excellent biocompatibility, and biodegradability contribute to their versatility and success in diverse medical applications. Polymers are classified as natural and synthetic polymers based on their origin (Janrao et al., 2023; Pu et al., 2023). Natural polymers include chitosan, alginate, and hyaluronic acid (HA) whereas synthetic polymers include poly (caprolactone) (PCL), polyvinyl pyrrolidine (PVP), poly (ethylene) glycol (PEG), etc (Gomte et al., 2024b). For instance, Mahami et al. developed berberine and 5-aminosalicylic acid (5-ASA) loaded chitosan NPs cross-linked by hydroxy propyl methylcellulose phthalate (HPMCP) for the management of ulcerative colitis (UC). The prepared formulation showed promising drug release in the simulated intestinal fluid and did not exhibit drug release in the simulated gastric fluid. Further, the histopathological studies of the colon showed improved therapeutic effect of prepared NPs. The prepared NPs can be one of the promising solutions for the effective management of ulcerative colitis (Mahami et al., 2023). Maleki et al. prepared *Achillea wilhelmsii*-loaded chitosan-Eudragit NPs for effective management of UC. The prepared NPs have a particle size of 305±34 nm with 88.6±7.3% entrapment efficiency. The NPs showed reduced release in acidic pH and enhanced release at the colonic region. The loaded NPs diminished induced UC in rats, and decreased inflammation, ulcer severity, and UC-related symptoms. The histopathological investigation showed reduced colonic inflammation at the colonic site (Maleki et al., 2023). In another study, Xiao et al. prepared cyclosporine-loaded PEG-PCL micelles, which were further loaded into hydrogel for the treatment of UC. The prepared NPs showed enhanced drug loading with a quick gelation rate after administration into the body. The developed hydrogel showed longer retention in the inflamed colon which was confirmed from *in vitro* and *in vivo* distribution experiments. The developed formulation showed a reduction in colitis symptoms in TNBS-induced colitis mice by controlling the expression levels of pro-inflammatory cytokines (TNF-α, IL-1, iNOS2, and COX-2) and anti-inflammatory cytokines (Nrf-2, Il-10, HO-1, and NQO1). The prepared hydrogel could be a promising strategy for increasing the NP concentration in the colonic site and promoting effective treatment for UC (Xiao et al., 2023).

12.5.3 Plant-Based NPs

The plant-based compounds like flavonoids, alkaloids, volatile oils, polyphenols, polysaccharides, and tannins have excellent biological and therapeutic advantages in nano-drug delivery for the management of different diseases (Dubey et al., 2018; Jadhav et al., 2023; Paradia et al., 2023). Recently, there has been a resurgence of interest in leveraging nanotechnology by incorporating plant-based elements. This innovative approach has paved the way for the development of nanomedicines tailored to address IBD (Kumari et al., 2019; Watkins et al., 2015). Some natural products derived from plant sources such as curcumin, resveratrol, thymoquinone, berberine, ginger, and piceatannol have proven excellent potential in the management of IBD (Singh et al., 2016). For instance, curcumin and embelin have antioxidant and anti-inflammatory properties, and NPs loaded with them showed promising results in the management of UC in the colitis mice model. Further, *in vivo* investigation showed that prepared NPs enter the colonic region against DSS-induced colitis model (Sreepriya & Bali, 2006; Tekeli et al., 2019). In another study, ginger-loaded NPs were shown to improve the reduction in the levels of TNF-α, IL-6, and IL-1β and promoted levels of anti-inflammatory cytokines like IL-1β, and IL-10. The experiments indicate that ginger-based NPs can be effectively prepared and employed for the management of IBD and colitis-related cancer (Tian et al., 2020). Thymoquinone possesses various immunomodulating and anti-inflammatory activities targeting TNF-α, IL-1β, and IL-10 signaling. It helps in the reduction of the levels of myeloperoxidase and malondialdehyde activity with gradual enhancement in the level of glutathione that showed improvement in colitis-related tissue damage (Lei et al., 2012). Phytomolecules obtained from plants can be used as a promising option for the management of IBD. For instance, zein, a food-derived protein obtained from *Zea mays* L. is useful in the management of IBD. It possesses an inherent hydrophobic nature that could help to load hydrophobic molecules inside it. Generally, zein has been used as a coating material to prevent degradation of therapeutic agents from gastric acid and discharge them into the intestinal region. It has been employed in the preparation of various nano-drug delivery systems such as nanoparticles, micelles, hydrogel, and microspheres because of their biodegradability, mucoadhesiveness, and reduced cost (Barani et al., 2021; Iravani & Varma, 2019; Shah et al., 2020; Zhang et al., 2015) Resveratrol, a natural polyphenol has found to be promising in the management of IBD. For example, resveratrol loaded into β-lactoglobulin nanospheres was prepared for the management of IBD. The developed nanospheres showed enhanced stability and solubility of resveratrol due to the complexation of β-lactoglobulin. The resveratrol solubility was enhanced by 1.7 times with β-lactoglobulin with 10% loading efficiency of resveratrol. The oral administration of resveratrol results in distinctively enhanced body weight percentage and disease activity index of UC mice. It enhances the expression of IL-10, enhances the number of goblet cells, promotes an anti-inflammatory effect, and reduces the destruction of the colonic epithelium (Li et al., 2023; Pujara et al., 2020).

12.5.4 Hydrogel

Hydrogels can be described as a network of cross-linked polymeric chains that swell when they come in contact with water (Ahmed, 2015; Battu et al., 2021). In the last few years, hydrogels have gained significant attention due to their versatile applications in the fields of drug delivery, wound dressing, tissue engineering, and many others (Caló & Khutoryanskiy, 2015). They can be used as a carrier system for drug molecules or as an external compartment to contain drug-loaded nanoparticles. In the latter case, hydrogels can be considered as a hybrid drug delivery system whose primary function is to protect the nanoparticles from adverse environmental conditions and release them at the targeted site (Zhang et al., 2017). Wei et al. developed a hybrid hydrogel system made up of chitosan/alginate and containing hyaluronic acid-modified L-arginine CO_2 nanoparticles loaded with pterostilbene. Pterostilbene is a derivative of resveratrol with antioxidant and anti-inflammatory activities. The oral administration of hydrogels in DSS-induced colitis mice displayed an excellent protective effect against a harsh gastrointestinal environment and colon-targeting ability. The presence of hydrogel enabled the NPs to deliver pterostilbene at the inflammation site, which leads to the downregulation of pro-inflammatory cytokines and reduced intestinal permeability, resulting in a significant improvement in the symptoms of UC (Wei et al., 2022). Laroui et al. also conducted a similar study in which they dispersed anti-inflammatory tripeptide Lys-Pro-Val KPV-loaded nanoparticles in an alginate-chitosan hydrogel. The potency of hydrogels was tested in DSS-induced colitis mice. The result showcased the ability of the hydrogel to overcome physiological barriers and deliver KPV-loaded NPs to the targeted colonocytes. This proves the potential application of hydrogel as a protective and colon-targeting hybrid drug carrier in the treatment of IBD (Laroui et al., 2010).

6-shogaol is a phytoconstituent present in dried ginger and has many therapeutic benefits. For instance, it showcases anti-inflammatory, antioxidant, and anti-carcinogenic effects. Zhang and colleagues tried to explore the effect of 6-shogaol against UC due to its therapeutic benefits. For this purpose, they prepared folate-conjugated PLGA/PLA NPs containing 6-shogaol [NPs-PEG-FA/6-shogaol] by using a single-step surface-functionalizing technique. For oral administration, the formulated NPs were then encapsulated in chitosan/alginate hydrogel. The hydrogel was then given to mice with DSS-induced colitis orally. The ELISA result showcased a marked reduction of fecal lipocalin-2, a widely accepted biomarker for intestinal inflammation, indicating the effect anti-inflammatory effect of 6-shogaol. This result was further confirmed by *in vivo* bioluminescence study. At the same time, in the histocompatibility study, mice treated with oral hydrogel showed a marked decrease in inflammatory cells and a decreased mRNA expression level of pro-inflammatory cells in the real-time polymerase chain reaction (RT-PCR) test. RTPCR also showcased the potential of 6-shogaol to promote the expression of nuclear factor Nrf-2 (nuclear factor erythroid 2–related factor 2), which is a potent anti-inflammatory factor. Overall, the hydrogel-encapsulated NPs-PEG-FA/6-shogaol showed effective symptom relief and accelerated wound healing in UC mice (Zhang et al., 2018).

The ROS in local tissues is one important factor responsible for inflammation. Thus, a potential avenue to manage UC is to get rid of ROS in the inflammatory colon. To reduce ROS in UC, Liu and colleagues developed platinum nanoparticles (Pt-NPs) that could eliminate excess ROS and protect cells against oxidative stress. These NPs were prepared by mixing chloroplatinic acid hexahydrate with polyvinylpyrrolidone via the reduction method. Finally, the NPs were dispersed in a pH-responsive oral hydrogel (chitosan or alginate), which could release entrapped NPs at the alkaline pH of the colon, thereby achieving colon-specific delivery. Once at the target site, these NPs showcased excellent enzymatic activities similar to those of catalase (CAT) and superoxide dismutase (SOD), as well as a potent ability to scavenge hydroxyl radicals. All of which enable the Pt-NPs to reduce ROS effectively. Overall, the specific oral hydrogel helps with the site-specific and pH-responsive release of Pt-NPs (Liu et al., 2023). In a similar study, a group of scientists developed a pH-responsive sodium alginate (SA) hydrogel system containing both curcumin and emodin nanoemulsions for control release. The formulated hydrogel could target colon macrophages and release drug molecules in a controlled manner at the inflammation site. This leads to a significant reduction in TNF-α, IL-6, and ROS expression present in the colon inflammatory microenvironment. The presence of nanoemulsion in the inflammation site also improved IL-10 expression and restored intestinal mucosal tight junction protein expression. Overall, the favorable outcomes showcased by the hydrogel medication helped rapid improvement in the pathophysiology of IBD (Lei et al., 2023).

12.5.5 Exosomes

Exosomes are nano-sized cell-derived vesicles secreted by several cells with sizes ranging from 30–150nm (Patel et al., 2023). In recent years, exosomes have grabbed much attention in the scientific community due to their potential role in the exosomal pathway involved in the pathophysiology of IBD. Several cells can release exosomes. These exosomes can perform different functions in IBD according to their vesicular components (RNA and proteins) (Wani et al., 2020; Zhang et al., 2019).

Later, scientists started using exosomes as a therapeutic agent against IBD and many have succeeded in utilizing exosomes against IBD. In one such study, Cai et al. explored the potential application of exosomes derived from TGF-β1 gene-modified bone marrow-derived dendritic cells (TGF-β1-EXO) against IBD. The extracted exosomes were tested in mice with DSS-induced IBD. Once inside the body, these exosomes showed a clear increase in Foxp3 expressing CD4+ T cells. These specific T cells help in maintaining intestinal homeostasis. Also, these exosomes showcased a marked decrease in Th17 in lymphocytes from mesentery lymph nodes of the inflammatory site. Th17 immune cells mostly produce IL-17, which plays an important role in the development of IBD. The overall results showcase the potential immune suppressive role and the protective effect of exosomes in IBD (Cai et al., 2012).

More recently, it was found that ferroptosis contributes to the chronic aberrant inflammation associated with IBD by causing intestinal epithelial cell death and epithelial erosion. Therefore, limiting ferroptosis activity could be beneficial in IBD patients (Ocansey et al., 2023). Considering this, Wei and coworkers prepared exosomes derived from human umbilical cord mesenchymal stem cells (hucMSC-Ex), anticipating their inhibitory effect on ferroptosis. They successfully isolated and collected the exosomes from cultured human umbilical cord cells through a series of centrifugation steps. The isolated exosomes were found to have a size range between 30 and 300 nm. Identification and characterization of the exosomes were conducted using transmission electron microscopy (TEM) and western blotting techniques (Figure 12.2). Small RNA sequencing revealed a high expression of miR-129-5p in hucMSC-Ex. The presence of miR-129-5p facilitated hucMSC-Ex in suppressing the protein expression of ACSL4, a crucial isoenzyme involved in ferroptosis. These consistent findings

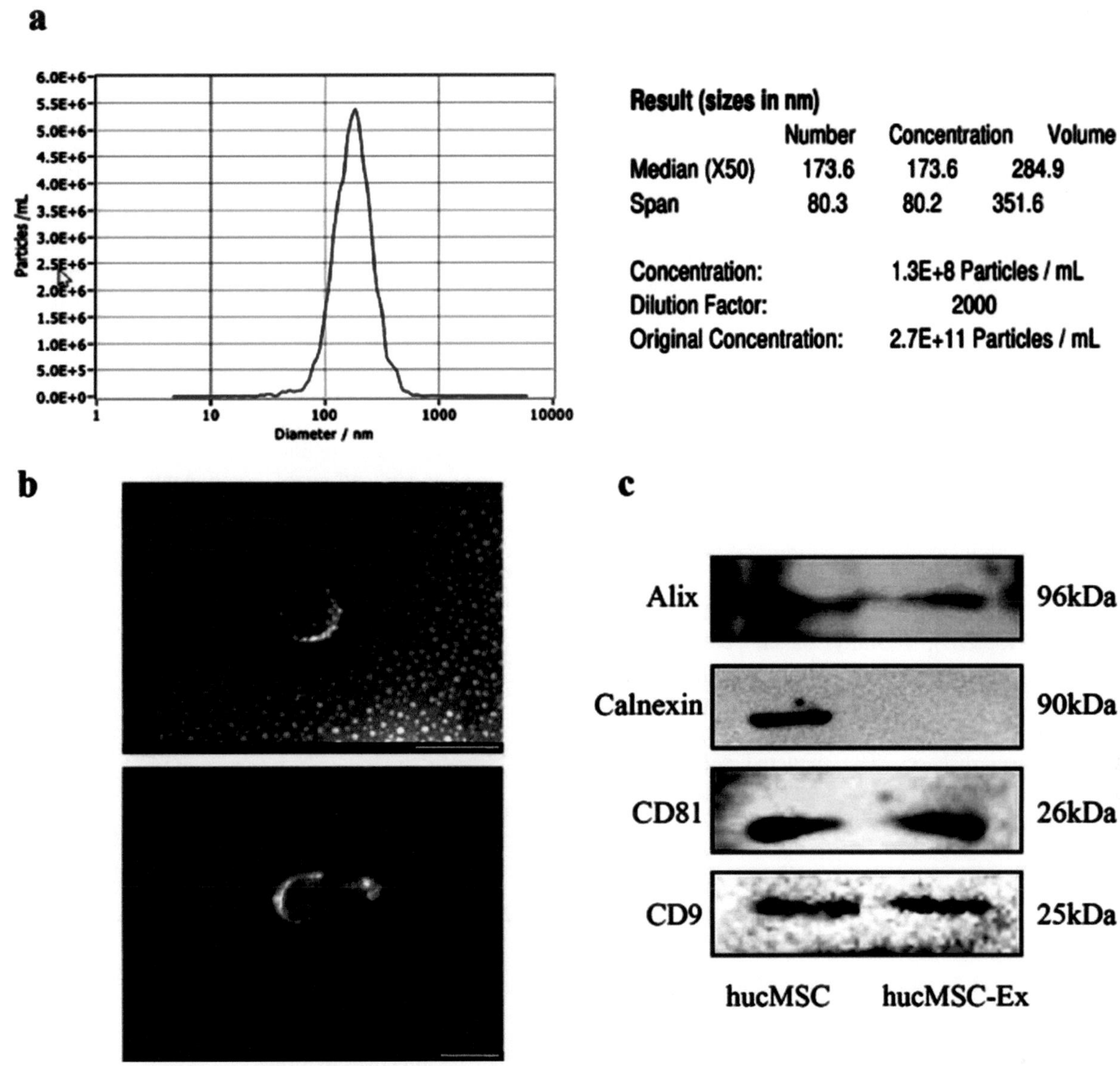

FIGURE 12.2 Identification of hucMSC-Ex. (**a**) the detection of the size and number of hucMSC-Ex by using a nanoparticle tracking analyzer (NanoSight), (**b**) identification of hucMSC-Ex by transmission electron microscopy (TEM) (the scale of the top picture is 200 nm and the scale of the bottom picture is 100 nm), (**c**) Identification of hucMSC-Ex surface markers by western blot. (Reprinted with permission from Wei et al., 2023.)

underscore the potential application of hucMSC-Ex in the prevention and treatment of IBD (Wei et al., 2023).

In a similar study, Cai et al also isolated hucMSC-derived exosomes and demonstrated their protective effect in IBD. These exosomes exhibited a high expression of micro-RNA miR-378a-5p, capable of inhibiting the activation of NLRP3 (NOD-like receptor family, pyrin domain-containing 3) inflammasomes. These inflammasomes play a role in various inflammatory disorders, including IBD. The inhibition of inflammasomes by exosomes resulted in the suppression of IL-18, IL-1β secretion, and Caspase-1 cleavage, all of which are essential for cell pyroptosis (colon mucosal cells). Therefore, hucMSC-derived exosomes offer protection against IBD by preventing cell pyroptosis (Cai et al., 2021).

12.6 CLINICAL TRIALS

There are very few clinical trials that associate IBD with nanocarriers. Table 12.2 represents details of those clinical trials.

12.7 PATENTS REGARDING IBD AND NANOCARRIERS

There have been a few patents that have been filed by researchers all over the world. In the following section, a brief overview of those patents is illustrated. The patent entitled, "Pharmaceutical composition of nanoparticles" (US8192718B1) talks about nanoparticles consisting of positively charged chitosan, negatively charged substrate, and at least one bioactive agent for treating diseases of TJs of animal subjects. The claim suggested that a negatively charged substrate is polyglutamic acid (PGA) consisting of γ-PGA, α-PGA, salts of PGA, derivatives of PGA or combination and nanoparticles are coated enteric coating polymer while a bioactive agent is selected from an anti-epileptic drug, an anti-inflammatory drug, meningitis antagonist, or antioxidant drug. Another patent entitled, "Immune-modifying nanoparticles for the treatment of inflammatory diseases" (US202400008301A1) essentially revolved around the administration of negatively charged particles like polystyrene, PLGA, or diamond in ameliorating inflammatory diseases. This invention also provides a method of removing pro-inflammatory mediators from an inflammatory environment and administration of the prepared formulation to subjects suffering from inflammatory diseases. The patent entitled, "Particles for the treatment of inflammatory bowel disease" (WO2014071207A1), which was filed in 2013, is concerned with a particle-based delivery system and treatment methods for UC or CD. The method consists of oral administration of a drug delivery vehicle consisting of a therapeutic agent while a drug delivery vehicle consists of particles encapsulated in PLGA polymer. Another patent (CN103315959B), "Orally taken colon-targeted preparation for treatment of inflammatory bowel diseases and preparation method thereof" is related to Eudragit S100 wrapped 5-aminosalicylic acid nanoparticles for treatment of inflammatory bowel diseases. The prepared colon-targeted nanoparticles have characteristics like small particle size, narrow distribution, and low solvent residue.

12.8 CHALLENGES

In recent years, the investigation into NPs and their potential use against IBD has surged intensely. The promising results of NPs are mostly due to their ability to accumulate in damaged and inflamed tissue, thereby delivering therapeutic agents precisely at the required site. Despite the favorable results obtained from the laboratory studies, there is still no marketed formulation of NPs available for IBD. This may be attributed to the several challenges associated with NPs (Barani et al., 2021). One such challenge faced by NPs is their complexity. NPs preparation generally involves several components. A thorough understanding of the physiochemical characteristics of these components is essential to generating NPs with the desired properties. This process often makes the NPs preparation tedious and unsuccessful. Therefore, the proper understanding of components is a must for the development of a successful nanoparticle (Rajput, 2015).

Another challenge faced by NPs is the lack of a well-defined and robust preparation technique, which impacts their reproducibility. The development of a successful NP mostly depends on multiple factors like drug-excipient compatibility, excipient ratio, solvent concentration, preparation conditions like mixing speed, reaction time, temperature, and many more. Change in any one of them can lead to reproducibility problems. This is a major drawback as keeping all the parameters in an optimized state is difficult if not impossible (Desai, 2012; Herdiana et al., 2022).

Additionally, NPs face challenges like lack of uniform and repeated pharmacological effects. The pharmacological

TABLE 12.2
Overview of Clinical Trials Pertaining to Gastrointestinal Disorders and Nanocarriers

Sr.No.	Title of Study	Type of Study	Number of Subjects	Objectives of Study	Clinical Trial ID
1	Anemia in IBD (2016–2018)	Interventional, treatment, open-label, with oral liposomal iron as a drug (28 mg per day over 8 weeks).	200	Primary: Hemoglobin level improvement Secondary: Improvement of quality of life and fatigue, activation of inflammation, improvement of anxiety and depression.	NCT02760940
2	Intraoperative TAP block with bupivacaine/ dexamethasone against liposomal bupivacaine (Exparel) (2018–2019).	Interventional, phase 4, treatment, randomized, parallel, assignment, open-label.	102	Primary: Cumulative 48-hour post-surgical opiate use. Secondary: Post-up pain score, bowel motility, postoperative length of stay, postoperative complications.	NCT03723447
3	Plant exosomes +/- curcumin to abrogate symptoms of IBD (2018–2022).	Treatment, randomized, parallel assignment, open-label.	4	Primary: Change in inflammation on colonoscopy Secondary: Change in subjective symptoms.	NCT04879810
4	Safety of injection of placental mesenchymal stem cell-derived exosomes for treatment of resistant perianal fistula in Crohn's patients (2022–ongoing).	Treatment, randomized, parallel assignment, double-masked.	80	Primary: Safety of injected exosomes, clinical efficacy of injected exosomes and inflammatory markers for 3 months.	NCT05499156

effect of NPs widely depends on factors like shape, size, and surface modifications. These factors control the nanoparticle's distribution, clearance, and drug release inside the body. Therefore, it is necessary to keep these factors constant to get the desired and uniform pharmacological effects. Any change in these parameters could alter the therapeutic outcome (Petschauer et al., 2015). Again, it is difficult to translate results obtained from preclinical animal studies to clinical studies. It is often seen that the therapeutic effects of an NP observed in small lab animals do not correlate with human subjects. This unevenness is associated with almost all types of formulation but it is quite prominent in the case of nanoparticles. This lack of correlation between preclinical and clinical data is often responsible for the lack of marketed nanoformulations (Bi et al., 2023; Metselaar & Lammers, 2020).

Environmental concern is also associated with NPs. Due to their small size, there is a fair chance that NPs could escape to the environment during their preparation, handling, or use. Once entering into the environment, NPs could undergo physical, chemical, and biological transformations and could impact the ecosystem in several ways. Therefore there is a concern in the form of ecotoxicity will remain associated with NPs (Gomte et al., 2024; Rambaran & Schirhagl, 2022; Sharma et al., 2021). Besides toxicity, NPs also present several health hazards. For instance, due to their small size, there is a fair chance that dry NPs could escape outside as aerosol and when inhaled could impact the respiratory system (Song et al., 2009). Also, some NPs can penetrate skin barriers; therefore dermal exposure to these NPs could be hazardous to health (Nel et al., 2006). Therefore, preparing and using NPs without causing any ecological or health-related problems is a real challenge. Finally, NPs also face regulatory challenges due to their multi-component composition.

In contrast to conventional medication, which is mostly simple formulations with one or two active pharmaceutical ingredients (API) and few excipients, NPs represent a complex formulation consisting of multiple components many of which have their pharmacological impact. This complexity of nanoformulation makes their regulatory aspect more difficult and complex. This regulatory aspect is one of the reasons why there is a scarcity of nonmedication in the market (Sainz et al., 2015). In the end, NP-based approaches represent a promising way to treat many serious health challenges, including IBD, however, their clinical application is still limited due to the challenges discussed above. These challenges need to be addressed as soon as possible to develop the future of NPs in the field of pharmaceutics.

12.9 CONCLUSION AND FUTURE PERSPECTIVES

NPs offer a promising avenue for managing GI inflammations due to their unique characteristics such as high surface area, tunable size, and surface chemistry. Through various studies, it has been demonstrated that NPs can effectively target inflamed sites within the GIT, delivering therapeutic agents with increased specificity and efficacy. Additionally, the ability of NPs to protect the payloads from degradation and facilitate controlled release of therapeutic agents further enhances their utility in GI inflammation treatment. Furthermore, the uniqueness of nanoparticle formulations allows for the incorporation of diverse therapeutic agents, including anti-inflammatory drugs, antioxidants, and immunomodulators, thereby enabling tailored treatment strategies for different types and severities of GI inflammations. Moreover, the potential for surface modification of NPs with various targeting ligands facilitates specific binding to receptors overexpressed on inflamed tissues, thereby minimizing off-target effects and maximizing therapeutic outcomes.

Looking ahead, the field of NP-based approaches for GI inflammation treatment holds several exciting prospects. Firstly, recent advancements in nanotechnology and materials science will likely lead to the development of novel nanoparticle formulations with improved biocompatibility, stability, and targeting capabilities. This could involve the integration of stimuli-responsive materials that enable site-specific drug release in response to pH, enzymes, or inflammation-associated biomarkers, thereby enhancing therapeutic efficacy while minimizing systemic side effects. Additionally, the emergence of personalized medicine approaches could revolutionize the treatment of GI inflammations by tailoring nanoparticle formulations and drug combinations to individual patient characteristics, including genetic makeup, microbiome composition, and disease phenotype. Furthermore, the integration of advanced imaging modalities, such as nanoparticle-based contrast agents for molecular imaging and theranostics, could enable real-time monitoring of disease progression and treatment response, facilitating early intervention and personalized adjustments to therapy regimens.

Overall, nanoparticle-based approaches hold significant promise for revolutionizing the management of GI inflammations, offering targeted and personalized treatment strategies that could improve patient outcomes and quality of life. However, continued research efforts are needed to overcome remaining challenges and translate these innovations from the laboratory to clinical practice effectively.

ACKNOWLEDGMENTS

The authors are thankful to the Director, NIPER-A for providing necessary literature searching facilities. AJ is thankful to the Science and Engineering Board, Department of Science and Technology, Government of India (Grant No. SRG/2023/000096).

CONFLICT OF INTEREST

All the authors report no Conflict of Interest.

ABBREVIATIONS

API, active pharmaceutical ingredient; CD, Crohn's disease; CRP, C-reactive protein; DSS, dextran sodium sulfate; DNBS, dinitrobenzene sulfonic acid; ECM, extracellular matrix; ELISA, enzyme-linked immunosorbent assay; GIT, gastrointestinal tract; GRAS, generally recognized as safe; HA, hyaluronic acid; HPMCP, hydroxy propyl methylcellulose phthalate; IBD, inflammatory bowel disease; IBS, irritable bowel syndrome; ICAM, intercellular adhesion molecule; IL, interleukin; IV, intravenous; MMPs, matrix metalloproteinases; MPS, mononuclear phagocytic system; MRI, magnetic resonance imaging; NPs, Nanoparticles; NLCs, nanostructured lipid carriers; PCL, poly (caprolactone); PEI, polyethyleneimine; PEG, poly (ethylene) glycol; PET, positron emission tomography; PGA, polyglutamic acid; PLA, poly (lactic acid); PLGA poly (lactic-co-glycolic acid); PSGL, P-selectin glycoprotein ligand; PVP, polyvinyl pyrrolidine; ROS, reactive oxygen species; SC, subcutaneous; SPECT, single photon emission computed tomography; TJs tight junctions; TNBS, trinitrobenzene sulfonic acid; TNF, tumor necrosis factor; TRFLP, terminal restriction fragment length polymorphism; UC, ulcerative colitis; VCAM, vascular cell adhesion molecule.

REFERENCES

Abdelmegid, A. M., Abdo, F. K., Ahmed, F. E., & Kattaia, A. A. A. (2019). Therapeutic effect of gold nanoparticles on DSS-induced ulcerative colitis in mice with reference to interleukin-17 expression. *Scientific Reports*, 9(1), 10176. https://doi.org/10.1038/s41598-019-46671-1

Abdollahy, A., Salehi, M., Mahami, S., Bernkop-Schnürch, A., Vahedi, H., Gharravi, A. M., & Mehrabi, M. (2024). Therapeutic effect of 5-ASA and hesperidin-loaded chitosan/Eudragit® S100 nanoparticles as a pH-sensitive carrier for local targeted drug delivery in a rat model of ulcerative colitis. *International Journal of Pharmaceutics*, 652, 123838. https://doi.org/https://doi.org/10.1016/j.ijpharm.2024.123838

Abraham, C., & Cho, J. H. (2009a). Inflamatory bowel disease mehanism of disease. *The New England Journal of Medicine*, 361, 2066–2078. https://doi.org/10.1056/NEJMra0804647

Abraham, C., & Cho, J. H. (2009b). Inflammatory bowel disease. *The New England Journal of Medicine*, 361(21), 2066–2078. https://doi.org/10.1056/NEJMra0804647

Agnihotri, T. G., Alexander, A., Agrawal, M., Dubey, S. K., & Jain, A. (2023). In vitro-in vivo correlation in nanocarriers: From protein corona to therapeutic implications. *Journal of Controlled Release*, 354, 794–809. https://doi.org/10.1016/j.jconrel.2023.01.063

Ahmad, A., Ansari, M. M., Kumar, A., Bishnoi, M., Raza, S. S., & Khan, R. (2021). Aminocellulose – grafted polycaprolactone-coated core–shell nanoparticles alleviate the severity of ulcerative colitis: A novel adjuvant therapeutic approach. *Biomaterials Science*, 9(17), 5868–5883. https://doi.org/10.1039/D1BM00877C

Ahmed, E. M. (2015). Hydrogel: Preparation, characterization, and applications: A review. *Journal of Advanced Research*, 6(2), 105–121. https://doi.org/10.1016/j.jare.2013.07.006

Akin, D., Sturgis, J., Ragheb, K., Sherman, D., Burkholder, K., Robinson, J. P., Bhunia, A. K., Mohammed, S., & Bashir, R. (2007). Bacteria-mediated delivery of nanoparticles and cargo into cells. *Nature Nanotechnology*, 2(7), 441–449. https://doi.org/10.1038/nnano.2007.149

Ali, H., Weigmann, B., Collnot, E.-M., Khan, S. A., Windbergs, M., & Lehr, C.-M. (2016). Budesonide loaded PLGA nanoparticles for targeting the inflamed intestinal mucosa--Pharmaceutical characterization and fluorescence imaging. *Pharmaceutical Research*, 33(5), 1085–1092. https://doi.org/10.1007/s11095-015-1852-6

Barani, M., Rahdar, A., Sargazi, S., Amiri, M. S., Sharma, P. K., & Bhalla, N. (2021). Nanotechnology for inflammatory bowel disease management: Detection, imaging and treatment. *Sensing and Bio-Sensing Research*, 32, 100417. https://doi.org/10.1016/j.sbsr.2021.100417

Barua, S., & Mitragotri, S. (2014). Challenges associated with penetration of nanoparticles across cell and tissue barriers: A review of current status and future prospects. *Nano Today*, 9(2), 223–243. https://doi.org/10.1016/j.nantod.2014.04.008

Battu, G., Gadiraju, S., Pepakayala, H., Pidaparthy, L. M., Devi, A., Vasu, P., & Thulluru, A. (2021). Swellable hydrogels for controlled drug delivery: An updated review. *International Journal of Pharmaceutical Technology Letters*, 1(1), 1–8.

Beloqui, A., Coco, R., Alhouayek, M., Solinís, M. Á., Rodríguez-Gascón, A., Muccioli, G. G., & Préat, V. (2013). Budesonide-loaded nanostructured lipid carriers reduce inflammation in murine DSS-induced colitis. *International Journal of Pharmaceutics*, 454(2), 775–783.

Beloqui, A., Coco, R., Memvanga, P. B., Ucakar, B., des Rieux, A., & Préat, V. (2014). pH-sensitive nanoparticles for colonic delivery of curcumin in inflammatory bowel disease. *International Journal of Pharmaceutics*, 473(1–2), 203–212. https://doi.org/10.1016/j.ijpharm.2014.07.009

Benbouziane, B., Ribelles, P., Aubry, C., Martin, R., Kharrat, P., Riazi, A., Langella, P., & Bermúdez-Humarán, L. G. (2013). Development of a Stress-Inducible Controlled Expression (SICE) system in Lactococcus lactis for the production and delivery of therapeutic molecules at mucosal surfaces. *Journal of Biotechnology*, 168(2), 120–129. https://doi.org/10.1016/j.jbiotec.2013.04.019

Bhol, K. C., & Schechter, P. J. (2007). Effects of nanocrystalline silver (NPI 32101) in a rat model of ulcerative colitis. *Digestive Diseases and Sciences*, 52(10), 2732–2742. https://doi.org/10.1007/s10620-006-9738-4

Bi, J., Mo, C., Li, S., Huang, M., Lin, Y., Yuan, P., Liu, Z., Jia, B., & Xu, S. (2023). Immunotoxicity of metal and metal oxide nanoparticles: From toxic mechanisms to metabolism and outcomes. *Biomaterials Science*, 11(12), 4151–4183. https://doi.org/10.1039/D3BM00271C

Biancheri, P., Brezski, R. J., Di Sabatino, A., Greenplate, A. R., Soring, K. L., Corazza, G. R., Kok, K. B., Rovedatti, L., Vossenkämper, A., Ahmad, N., Snoek, S. A., Vermeire, S., Rutgeerts, P., Jordan, R. E., & MacDonald, T. T. (2015). Proteolytic cleavage and loss of function of biologic agents that neutralize tumor necrosis factor in the mucosa of patients with inflammatory bowel disease. *Gastroenterology*, 149(6), 1564–1574.e3. https://doi.org/10.1053/j.gastro.2015.07.002

Cai, X., Zhang, Z.-Y., Yuan, J.-T., Ocansey, D. K. W., Tu, Q., Zhang, X., Qian, H., Xu, W.-R., Qiu, W., & Mao, F. (2021). hucMSC-derived exosomes attenuate colitis by regulating macrophage pyroptosis via the miR-378a-5p/NLRP3 axis. *Stem Cell Research & Therapy*, 12(1), 416. https://doi.org/10.1186/s13287-021-02492-6

Cai, Z., Zhang, W., Yang, F., Yu, L., Yu, Z., Pan, J., Wang, L., Cao, X., & Wang, J. (2012). Immunosuppressive exosomes from TGF-β1 gene-modified dendritic cells attenuate Th17-mediated inflammatory autoimmune disease by inducing regulatory T cells. *Cell Research*, 22(3), 607–610. https://doi.org/10.1038/cr.2011.196

Caló, E., & Khutoryanskiy, V. V. (2015). Biomedical applications of hydrogels: A review of patents and commercial products. *European Polymer Journal*, 65, 252–267. https://doi.org/https://doi.org/10.1016/j.eurpolymj.2014.11.024

Cazelles, A., Collard, M. K., Lalatonne, Y., Doblas, S., Zappa, M., Labiad, C., Cazals-Hatem, D., Maggiori, L., Treton, X., & Panis, Y. (2022). A preclinical validation of iron oxide nanoparticles for treatment of perianal fistulizing Crohn's disease. *International Journal of Molecular Sciences*, 23(15), 8324. https://doi.org/10.3390/ijms23158324

Chassaing, B., & Darfeuille–Michaud, A. (2011). The commensal microbiota and enteropathogens in the pathogenesis of inflammatory bowel diseases. *Gastroenterology*, 140(6), 1720–1728. https://doi.org/10.1053/j.gastro.2011.01.054

Cho, J. H. (2008). The genetics and immunopathogenesis of inflammatory bowel disease. *Nature Reviews Immunology*, 8(6), 458–466. https://doi.org/10.1038/nri2340

Coco, R., Plapied, L., Pourcelle, V., Jérôme, C., Brayden, D. J., Schneider, Y.-J., & Préat, V. (2013). Drug delivery to inflamed colon by nanoparticles: Comparison of different strategies. *International Journal of Pharmaceutics*, 440(1), 3–12. https://doi.org/10.1016/j.ijpharm.2012.07.017

Dadfar, S. M., Roemhild, K., Drude, N. I., von Stillfried, S., Knüchel, R., Kiessling, F., & Lammers, T. (2019). Iron oxide nanoparticles: Diagnostic, therapeutic and theranostic applications. *Advanced Drug Delivery Reviews*, 138, 302–325. https://doi.org/10.1016/j.addr.2019.01.005

de Gracia Lux, C., Joshi-Barr, S., Nguyen, T., Mahmoud, E., Schopf, E., Fomina, N., & Almutairi, A. (2012). Biocompatible polymeric nanoparticles degrade and release cargo in response to biologically relevant levels of hydrogen peroxide. *Journal of the American Chemical Society*, 134(38), 15758–15764.

Desai, N. (2012). Challenges in development of nanoparticle-based therapeutics. *AAPS Journal*, 14(2), 282–295. https://doi.org/10.1208/s12248-012-9339-4

Deshmukh, R., Harwansh, R. K., Paul, S. Das, & Shukla, R. (2020). Controlled release of sulfasalazine loaded amidated pectin microparticles through Eudragit S 100 coated capsule for management of inflammatory bowel disease. *Journal of Drug Delivery Science and Technology*, 55, 101495. https://doi.org/10.1016/j.jddst.2019.101495

Dubey, S. K., Singhvi, G., Krishna, K. V., Agnihotri, T., Saha, R. N., & Gupta, G. (2018). Herbal medicines in neurodegenerative disorders: An evolutionary approach through novel drug delivery system. *Journal of Environmental Pathology, Toxicology and Oncology*, 37(3), 199–208. https://doi.org/10.1615/JEnvironPatholToxicolOncol.2018027246

El-Maghawry, E., Tadros, M. I., Elkheshen, S. A., & Abd-Elbary, A. (2020). Eudragit(®)-S100 coated PLGA nanoparticles for colon targeting of etoricoxib: Optimization and pharmacokinetic assessments in healthy human volunteers. *International Journal of Nanomedicine*, 15, 3965–3980. https://doi.org/10.2147/IJN.S244124

Eniola, A. O., Rodgers, S. D., & Hammer, D. A. (2002). Characterization of biodegradable drug delivery vehicles with the adhesive properties of leukocytes. *Biomaterials*, 23(10), 2167–2177.

Fan, Y., Wang, X., Yan, G., Gao, H., & Yang, M. (2023). Rectal delivery of (89)Zr-labeled infliximab-loaded nanoparticles enables PET imaging-guided localized therapy of inflammatory bowel disease. *Journal of Materials Chemistry. B*, 11(47), 11228–11234. https://doi.org/10.1039/d3tb02128a

Foligne, B., Dessein, R., Marceau, M., Poiret, S., Chamaillard, M., Pot, B., Simonet, M., & Daniel, C. (2007). Prevention and treatment of colitis with lactococcus lactis secreting the immunomodulatory yersinia LcrV protein. *Gastroenterology*, 133(3), 862–874. https://doi.org/10.1053/j.gastro.2007.06.018

Gao, C., Yu, S., Zhang, X., Dang, Y., Han, D.-D., Liu, X., Han, J., & Hui, M. (2021). Dual functional Eudragit(®) S100/L30D-55 and PLGA colon-targeted nanoparticles of iridoid glycoside for improved treatment of induced ulcerative colitis. *International Journal of Nanomedicine*, 16, 1405–1422. https://doi.org/10.2147/IJN.S291090

Gholami, M. D., Sonar, P., Ayoko, G. A., & Izake, E. L. (2020). A highly sensitive SERS quenching nanosensor for the determination of tumor necrosis factor alpha in blood. *Sensors and Actuators, B: Chemical*, 310(1), 127867. https://doi.org/10.1016/j.snb.2020.127867

Gomte, S. S., Agnihotri, T. G., Khopade, S., & Jain, A. (2024a). Exploring the potential of pH-sensitive polymers in targeted drug delivery. *Journal of Biomaterials Science, Polymer Edition*, 35(2), 228–268. https://doi.org/10.1080/09205063.2023.2279792

Gomte, S. S., Jadhav, P. V., Jothi Prasath V. R, N., Agnihotri, T. G., & Jain, A. (2024b). From lab to ecosystem: Understanding the ecological footprints of engineered nanoparticles. *Journal of Environmental Science and Health, Part C*, 42(1), 33–73. https://doi.org/10.1080/26896583.2023.2289767

Guha, L., Agnihotri, T. G., Jain, A., & Kumar, H. (2023). Gut microbiota and traumatic central nervous system injuries: Insights into pathophysiology and therapeutic approaches. *Life Sciences*, 334, 122193. https://doi.org/10.1016/j.lfs.2023.122193

Gustafson, H. H., Holt-Casper, D., Grainger, D. W., & Ghandehari, H. (2015). Nanoparticle uptake: The phagocyte problem. *Nano Today*, 10(4), 487–510.

Hamady, Z. Z. R., Scott, N., Farrar, M. D., Wadhwa, M., Dilger, P., Whitehead, T. R., Thorpe, R., Holland, K. T., Lodge, J. P. A., & Carding, S. R. (2011). Treatment of colitis with a commensal gut bacterium engineered to secrete human tgf-β1 under the control of dietary xylan 1. *Inflammatory Bowel Diseases*, 17(9), 1925–1935. https://doi.org/10.1002/ibd.21565

Han, W., Mercenier, A., Ait-belgnaoui, A., Pavan, S., Lamine, F., Swam, I. I. Van, Kleerebezem, M., Salvador-cartier, C., Hisbergues, M., Bueno, L., Theodorou, V., & Fioramonti, J. (2006). Improvement of an experimental colitis in rats by lactic acid bacteria producing SOD.pdf. *Inflammatory Bowel Disease*, 12(11), 1044 1052.

Herdiana, Y., Wathoni, N., Shamsuddin, S., & Muchtaridi, M. (2022). Scale-up polymeric-based nanoparticles drug delivery systems: Development and challenges. *OpenNano*, 7, 100048. https://doi.org/10.1016/j.onano.2022.100048

Hua, S. (2019). Physiological and pharmaceutical considerations for rectal drug formulations. *Frontiers in Pharmacology*, 10, 1196. https://doi.org/10.3389/fphar.2019.01196

Hua, S. (2020). Advances in oral drug delivery for regional targeting in the gastrointestinal tract-influence of physiological, pathophysiological and pharmaceutical factors. *Frontiers in Pharmacology*, 11, 524. https://doi.org/10.3389/fphar.2020.00524

Hussein, R. M., & Saleh, H. (2018). Promising therapeutic effect of gold nanoparticles against dinitrobenzene sulfonic acid-induced colitis in rats. *Nanomedicine*, 13(14), 1657–1679. https://doi.org/10.2217/nnm-2018-0009

Iravani, S., & Varma, R. S. (2019). Plant-derived edible nanoparticles and miRNAs: Emerging frontier for therapeutics and targeted drug-delivery. *ACS Sustainable Chemistry & Engineering*, 7(9), 8055–8069. https://doi.org/10.1021/acssuschemeng.9b00954

Jacob, E. M., Borah, A., Pillai, S. C., & Kumar, D. S. (2021). Garcinol encapsulated pH-sensitive biodegradable nanoparticles: A novel therapeutic strategy for the treatment of inflammatory bowel disease. *Polymers*, 13(6), 862. https://doi.org/10.3390/polym13060862

Jadhav, P. V., Prasath, N. J., Gajbhiye, S., Rane, U., Agnihotri, T. G., Gomte, S. S., & Jain, A. (2023). Empowering the battle: Bioenhancers as allies against cancer drug resistance. *Current Pharmaceutical Biotechnology*. https://doi.org/10.2174/0113892010192038231107051715

Janrao, C., Khopade, S., Bavaskar, A., Gomte, S. S., Agnihotri, T. G., & Jain, A. (2023). Recent advances of polymer based nanosystems in cancer management. *Journal of Biomaterials Science, Polymer Edition*, 34(9), 1274–1335. https://doi.org/10.1080/09205063.2022.2161780

Järveläinen, H., Sainio, A., Koulu, M., Wight, T. N., & Penttinen, R. (2009). Extracellular matrix molecules: Potential targets in pharmacotherapy. *Pharmacological Reviews*, 61(2), 198–223. https://doi.org/10.1124/pr.109.001289

Jayant, R., & Nair, M. (2016). Nanotechnology for the treatment of NeuroAIDS. *Journao of Nanomedicine Research*, 3(1), 47. https://doi.org/10.15406/jnmr.2016.03.00047

Jong, W. H. De, & Paul, J. B. (2008). Drug delivery and nanoparticles: Applications and hazards. *International Journal of Nanomedicine*, 3(2), 133–149. https://doi.org/10.2147/ijn.s596

Jubeh, T. T., Barenholz, Y., & Rubinstein, A. (2004). Differential adhesion of normal and inflamed rat colonic mucosa by charged liposomes. *Pharmaceutical Research*, 21(3), 447–453. https://doi.org/10.1023/B:PHAM.0000019298.29561.cd

Jubeh, T. T., Nadler-Milbauer, M., Barenholz, Y., & Rubinstein, A. (2006). Local treatment of experimental colitis in the rat by negatively charged liposomes of catalase, TMN and SOD. *Journal of Drug Targeting*, 14(3), 155–163. https://doi.org/10.1080/10611860600648429

Kaplan, G. G. (2015). The global burden of IBD: From 2015 to 2025. *Nature Reviews Gastroenterology & Hepatology*, 12(12), 720–727. https://doi.org/10.1038/nrgastro.2015.150

Kappelman, M. D., Rifas–Shiman, S. L., Porter, C. Q., Ollendorf, D. A., Sandler, R. S., Galanko, J. A., & Finkelstein, J. A. (2008). Direct health care costs of Crohn's disease and ulcerative colitis in US children and adults. *Gastroenterology*, 135(6), 1907–1913. https://doi.org/10.1053/j.gastro.2008.09.012

Karban, A., Nakhleh, M. K., Cancilla, J. C., Vishinkin, R., Rainis, T., Koifman, E., Jeries, R., Ivgi, H., Torrecilla, J. S., & Haick, H. (2016). Programmed nanoparticles for tailoring the detection of inflammatory bowel diseases and irritable bowel syndrome disease via breathprint. *Advanced Healthcare Materials*, 5(18), 2339–2344. https://doi.org/10.1002/adhm.201600588

Karrout, Y., Dubuquoy, L., Piveteau, C., Siepmann, F., Moussa, E., Wils, D., Beghyn, T., Neut, C., Flament, M.-P., & Guerin-Deremaux, L. (2015). In vivo efficacy of microbiota-sensitive coatings for colon targeting: A promising tool for IBD therapy. *Journal of Controlled Release*, 197, 121–130. https://doi.org/10.1016/j.jconrel.2014.11.006

Kaser, A., Zeissig, S., & Blumberg, R. S. (2010). Inflammatory bowel disease. *Annual Review of Immunology*, 28, 573–621. https://doi.org/10.1146/annurev-immunol-030409-101225

Kaushik, A., Jayant, R. D., & Nair, M. (2016). Advancements in nano-enabled therapeutics for neuroHIV management. *International Journal of Nanomedicine*, 11, 4317–4325. https://doi.org/10.2147/IJN.S109943

Kaye, J., Porcelli, S., Jones, B., janeway Jr, C. A., Steidler, L., Hans, W., Schotte, L., Neirynck, S., Obermeier, F., Falk, W., Fiers, W., & Remaut, E. (1994). 11. C. A janeway jr. *Proceedings of the National Academy of Sciences of the United States of America*, 76, 104.

Keller, L.-A., Merkel, O., & Popp, A. (2022). Intranasal drug delivery: Opportunities and toxicologic challenges during drug development. *Drug Delivery and Translational Research*, 12(4), 735–757. https://doi.org/10.1007/s13346-020-00891-5

Khater, S. I., Lotfy, M. M., Alandiyjany, M. N., Alqahtani, L. S., Zaglool, A. W., Althobaiti, F., Ismail, T. A., Soliman, M. M., Saad, S., & Ibrahim, D. (2022). Therapeutic potential of quercetin loaded nanoparticles: Novel insights in alleviating colitis in an experimental DSS induced colitis model. *Biomedicines*, 10(7), 1654. https://doi.org/10.3390/biomedicines10071654

Kim, E., Schueller, O., & Sweetnam, P. M. (2012). Targeting the leukocyte activation cascade: Getting to the site of inflammation using microfabricated assays. *Lab on a Chip*, 12(12), 2255–2264. https://doi.org/10.1039/C2LC21078A

Klöppel, R., Neumann, T., & Schmidt, F. (1989). Diagnostic imaging of pancreatic cancer. *Zeitschrift Für Die Gesamte Innere Medizin Und Ihre Grenzgebiete*, 44(19), 580–583.

Klotz, U., & Schwab, M. (2005). Topical delivery of therapeutic agents in the treatment of inflammatory bowel disease. *Advanced Drug Delivery Reviews*, 57(2), 267–279. https://doi.org/10.1016/j.addr.2004.08.007

Kumar, J., & Newton, A. M. J. (2017). Rifaximin-chitosan nanoparticles for inflammatory bowel disease (IBD). *Recent Patents on Inflammation & Allergy Drug Discovery*, 11(1), 41–52. https://doi.org/10.2174/1872213X10666161230111226

Kumari, P., Luqman, S., & Meena, A. (2019). Application of the combinatorial approaches of medicinal and aromatic plants with nanotechnology and its impacts on healthcare. *DARU Journal of Pharmaceutical Sciences*, 27(1), 475–489. https://doi.org/10.1007/s40199-019-00271-6

Lamprecht, A. (2010). Selective nanoparticle adhesion can enhance colitis therapy. *Nature Reviews Gastroenterology & Hepatology*, 7(6), 311–312. https://doi.org/10.1038/nrgastro.2010.66

Lamprecht, A. L. F., Ubrich, N., Yamamoto, H., Schäfer, U., Takeuchi, H., Maincent, P., Kawashima, Y., & Lehr, C.-M. (2001). Biodegradable nanoparticles for targeted drug delivery in treatment of inflammatory bowel disease. *Journal of Pharmacology and Experimental Therapeutics*, 299(2), 775–781.

Lamprecht, A., Schäfer, U., & Lehr, C.-M. (2001). Size-dependent bioadhesion of micro-and nanoparticulate carriers to the inflamed colonic mucosa. *Pharmaceutical Research*, 18(6), 788–793. https://doi.org/10.1023/a:1011032328064

Laroui, H., Dalmasso, G., Nguyen, H. T. T., Yan, Y., Sitaraman, S. V, & Merlin, D. (2010). Drug-loaded nanoparticles targeted to the colon with polysaccharide hydrogel reduce colitis in a mouse model. *Gastroenterology*, 138(3), 842–843. https://doi.org/10.1053/j.gastro.2009.11.003

Lei, F., Zeng, F., Yu, X., Deng, Y., Zhang, Z., Xu, M., Ding, N., Tian, J., & Li, C. (2023). Oral hydrogel nanoemulsion co-delivery system treats inflammatory bowel disease via anti-inflammatory and promoting intestinal mucosa repair. *Journal of Nanobiotechnology*, 21(1), 275. https://doi.org/10.1186/s12951-023-02045-4

Lei, X., Liu, M., Yang, Z., Ji, M., Guo, X., & Dong, W. (2012). Thymoquinone prevents and ameliorates dextran sulfate sodium-induced colitis in mice. *Digestive Diseases and Sciences*, 57(9), 2296–2303. https://doi.org/10.1007/s10620-012-2156-x

Li, D., Yang, M., Xu, H., Zhu, M., Zhang, Y., Tian, C., Nie, Y., Wang, J., Liang, Y., & Yao, J. (2022). Nanoparticles for oral delivery: Targeted therapy for inflammatory bowel disease. *Journal of Materials Chemistry B*, 10(31), 5853–5872. https://doi.org/10.1039/D2TB01190E

Li, M., Liu, J., Shi, L., Zhou, C., Zou, M., Fu, D., Yuan, Y., Yao, C., Zhang, L., & Qin, S. (2023). Gold nanoparticles-embedded ceria with enhanced antioxidant activities for treating inflammatory bowel disease. *Bioactive Materials*, 25, 95–106. https://doi.org/10.1016/j.bioactmat.2023.01.015

Li, N., Wang, M., Lyu, Z., Shan, K., Chen, Z., Chen, B., Chen, Y., Hu, X., Dou, B., & Zhang, J. (2023). Medicinal plant-based drug delivery system for inflammatory bowel disease. *Frontiers in Pharmacology*, 14, 1158945. https://doi.org/10.3389/fphar.2023.1158945

Li, X., Lu, C., Yang, Y., Yu, C., & Rao, Y. (2020). Site-specific targeted drug delivery systems for the treatment of inflammatory bowel disease. *Biomedicine & Pharmacotherapy*, 129, 110486. https://doi.org/10.1016/j.biopha.2020.110486

Lih-Brody, L., Powell, S. R., Collier, K. P., Reddy, G. M., Cerchia, R., Kahn, E., Weissman, G. S., Katz, S., Floyd, R. A., & McKinley, M. J. (1996). Increased oxidative stress and decreased antioxidant defenses in mucosa of inflammatory bowel disease. *Digestive Diseases and Sciences*, 41(10), 2078–2086. https://doi.org/10.1007/BF02093613

Liu, H., Zhang, Y., Zhang, M., Yu, Z., & Zhang, M. (2023). Oral administration of platinum nanoparticles with SOD/CAT cascade catalytic activity to alleviate ulcerative colitis. *Journal of Functional Biomaterials*, 14(11), 548. https://doi.org/10.3390/jfb14110548

Liu, Z., Liu, J., Wang, R., Du, Y., Ren, J., & Qu, X. (2015). An efficient nano-based theranostic system for multi-modal imaging-guided photothermal sterilization in gastrointestinal tract. *Biomaterials*, 56, 206–218. https://doi.org/10.1016/j.biomaterials.2015.04.005

Lutolf, M. P., Lauer-Fields, J. L., Schmoekel, H. G., Metters, A. T., Weber, F. E., Fields, G. B., & Hubbell, J. A. (2003). Synthetic matrix metalloproteinase-sensitive hydrogels for the conduction of tissue regeneration: Engineering cell-invasion characteristics. *Proceedings of the National Academy of Sciences*, 100(9), 5413–5418. https://doi.org/10.1073/pnas.0737381100

Lv, Y., Wu, R., Feng, K., Li, J., Mao, Q., Yuan, H., Shen, H., Chai, X., & Li, L. S. (2017). Highly sensitive and accurate detection of C-reactive protein by CdSe/ZnS quantum dot-based fluorescence-linked immunosorbent assay. *Journal of Nanobiotechnology*, 15(1), 1–10. https://doi.org/10.1186/s12951-017-0267-4

Ma, Y., Yang, J., Yang, T., Deng, Y., Gu, M., Wang, M., Hu, R., & Yang, Y. (2020). Electrochemical detection of C-reactive protein using functionalized iridium nanoparticles/graphene oxide as a tag. *RSC Advances*, 10(16), 9723–9729. https://doi.org/10.1039/c9ra10386d

Mahami, S., Salehi, M., Mehrabi, M., Vahedi, H., Hassani, M. S., Bitaraf, F. S., & Omri, A. (2023). pH-sensitive HPMCP-chitosan nanoparticles containing 5-aminosalicylic acid and berberine for oral colon delivery in a rat model of ulcerative colitis. *International Journal of Biological Macromolecules*, 244, 125332. https://doi.org/10.1016/j.ijbiomac.2023.125332

Maleki, H., Doostan, M., Farzaei, M. H., Seifi, P., Miraghaee, S., & Doostan, M. (2023). Achillea wilhelmsii–incorporated Chitosan@ Eudragit nanoparticles intended for enhanced ulcerative colitis treatment. *AAPS PharmSciTech*, 24(5), 112. https://doi.org/10.1208/s12249-023-02568-x

Metselaar, J. M., & Lammers, T. (2020). Challenges in nanomedicine clinical translation. *Drug Delivery and Translational Research*, 10(3), 721–725. https://doi.org/10.1007/s13346-020-00740-5

Molodecky, N. A., Soon, S., Rabi, D. M., Ghali, W. A., Ferris, M., Chernoff, G., Benchimol, E. I., Panaccione, R., Ghosh, S., & Barkema, H. W. (2012). Increasing incidence and prevalence of the inflammatory bowel diseases with time, based on systematic review. *Gastroenterology*, 142(1), 46–54. https://doi.org/10.1053/j.gastro.2011.10.001

Moradkhani, A., Beckman, L. J., & Tabibian, J. H. (2013). Health-related quality of life in inflammatory bowel disease: Psychosocial, clinical, socioeconomic, and demographic predictors. *Journal of Crohn's & Colitis*, 7(6), 467–473. https://doi.org/10.1016/j.crohns.2012.07.012

Mowat, C., Cole, A., Windsor, A., Ahmad, T., Arnott, I., Driscoll, R., Mitton, S., Orchard, T., Rutter, M., Younge, L., Lees, C., Ho, G.-T., Satsangi, J., & Bloom, S. (2011). Guidelines for the management of inflammatory bowel disease in adults. *Gut*, 60(5), 571–607. https://doi.org/10.1136/gut.2010.224154

Naeem, M., Bae, J., Oshi, M. A., Kim, M.-S., Moon, H. R., Lee, B. L., Im, E., Jung, Y., & Yoo, J.-W. (2018). Colon-targeted delivery of cyclosporine A using dual-functional Eudragit® FS30D/PLGA nanoparticles ameliorates murine experimental colitis. *International Journal of Nanomedicine*, 13, 1225–1240. https://doi.org/10.2147/IJN.S157566

Naeem, M., Kim, W., Cao, J., Jung, Y., & Yoo, J.-W. (2014). Enzyme/pH dual sensitive polymeric nanoparticles for targeted drug delivery to the inflamed colon. *Colloids and Surfaces. B, Biointerfaces*, 123, 271–278. https://doi.org/10.1016/j.colsurfb.2014.09.026

Naeem, M., Lee, J., Oshi, M. A., Cao, J., Hlaing, S. P., Im, E., Jung, Y., & Yoo, J. W. (2020). Colitis-targeted hybrid nanoparticles-in-microparticles system for the treatment of ulcerative colitis. *Acta Biomaterialia*, 116, 368–382. https://doi.org/10.1016/j.actbio.2020.09.017

Naha, P. C., Hsu, J. C., Kim, J., Shah, S., Bouché, M., Si-Mohamed, S., Rosario-Berrios, D. N., Douek, P., Hajfathalian, M., Yasini, P., Singh, S., Rosen, M. A., Morgan, M. A., & Cormode, D. P. (2020). Dextran-coated cerium oxide nanoparticles: A computed tomography contrast agent for imaging the gastrointestinal tract and inflammatory bowel disease. *ACS Nano*, 14(8), 10187–10197. https://doi.org/10.1021/acsnano.0c03457

Nel, A., Xia, T., Mädler, L., & Li, N. (2006). Toxic potential of materials at the nanolevel. *Science (New York, N.Y.)*, 311(5761), 622–627. https://doi.org/10.1126/science.1114397

Ocansey, D. K. W., Yuan, J., Wei, Z., Mao, F., & Zhang, Z. (2023). Role of ferroptosis in the pathogenesis and as a therapeutic target of inflammatory bowel disease (Review). *International Journal of Molecular Medicine*, 51(6), 1–16. https://doi.org/10.3892/ijmm.2023.5256

Owens, D. E., & Peppas, N. A. (2006). Opsonization, biodistribution, and pharmacokinetics of polymeric nanoparticles. *International Journal of Pharmaceutics*, 307(1), 93–102. https://doi.org/10.1016/j.ijpharm.2005.10.010

Paradia, P. K., Bhavale, R., Agnihotri, T., & Jain, A. (2023). A review on edible vaccines and biopharmaceutical products from plants. *Current Pharmaceutical Biotechnology*, 24(4), 495–509. https://doi.org/10.2174/138920102366622080 3151039

Parker, M., & Li, Z. (2021). *Chapter 22 – Biotechnology and drugs*. (A. B. T.-R. Twenty-third E. Adejare (ed.); pp. 397–415). Academic Press. https://doi.org/10.1016/B978-0-12-820 007-0.00022-2

Patel, G., Girish, T., Manoj, A., Tanuja, G., Shyam, S., & Gomte, S. (2023). Exosomes: A potential diagnostic and treatment modality in the quest for counteracting cancer. *Cellular Oncology*, 46(5), 1159–1179. https://doi.org/10.1007/s13 402-023-00810-z

Peer, D., Park, E. J., Morishita, Y., Carman, C. V, & Shimaoka, M. (2008). Systemic leukocyte-directed siRNA delivery revealing cyclin D1 as an anti-inflammatory target. *Science*, 319(5863), 627–630. https://doi.org/10.1126/science.1149859

Peng, S.-J., Ye, D.-T., Zheng, J., Xue, Y.-R., Lin, L., Zhao, Y.-D., Miao, W.-H., Song, Y., Wen, Z.-S., & Zheng, B. (2022). Synthesis, characterization of low molecular weight chitosan selenium nanoparticles and its effect on dss-induced ulcerative colitis in mice. *International Journal of Molecular Sciences*, 23(24), 15527. https://doi.org/10.3390/ijms232415527

Petschauer, J. S., Madden, A. J., Kirschbrown, W. P., Song, G., & Zamboni, W. C. (2015). The effects of nanoparticle drug loading on the pharmacokinetics of anticancer agents. *Nanomedicine (London, England)*, 10(3), 447–463. https://doi.org/10.2217/nnm.14.179

Prakash, M., Chandraprabha, M. N., Hari Krishna, R., Satish, H., & Girish Kumar, S. (2024). Iron oxide nanoparticles for inflammatory bowel disease: Recent advances in diagnosis and targeted drug therapy. *Applied Surface Science Advances*, 19, 100540. https://doi.org/10.1016/j.apsadv.2023.100540

Preidis, G. A., & Versalovic, J. (2009). Targeting the human microbiome with antibiotics, probiotics, and prebiotics: Gastroenterology enters the metagenomics era. *Gastroenterology*, 136(6), 2015–2031. https://doi.org/10.1053/j.gastro.2009.01.072

Pu, Y., Fan, X., Zhang, Z., Guo, Z., Pan, Q., Gao, W., Luo, K., & He, B. (2023). Harnessing polymer-derived drug delivery systems for combating inflammatory bowel disease. *Journal of Controlled Release*, 354, 1–18. https://doi.org/10.1016/j.jconrel.2022.12.044

Pujara, N., Wong, K. Y., Qu, Z., Wang, R., Moniruzzaman, M., Rewatkar, P., Kumeria, T., Ross, B. P., McGuckin, M., & Popat, A. (2020). Oral delivery of β-lactoglobulin-nanosphere-encapsulated resveratrol alleviates inflammation in winnie mice with spontaneous ulcerative colitis. *Molecular Pharmaceutics*, 18(2), 627–640. https://doi.org/10.1021/acs.molpharmaceut.0c00048

Rajput, N. (2015). Methods of preparation of nanoparticles-A review. *International Journal of Advances in Engineering & Technology*, 7(6), 1806.

Rambaran, T., & Schirhagl, R. (2022). Nanotechnology from lab to industry–A look at current trends. *Nanoscale Advances*, 4(18), 3664–3675. https://doi.org/10.1039/D2NA00439A

Ricci, S., Macchia, G., Ruggiero, P., Maggi, T., Bossù, P., Xu, L., Medaglini, D., Tagliabue, A., Hammarström, L., Pozzi, G., & Boraschi, D. (2003). In vivo mucosal delivery of bioactive human interleukin I receptor antagonist produced by Streptococcus gordonii. *BMC Biotechnology*, 3, 1–11. https://doi.org/10.1186/1472-6750-3-15

Rieber, A., Wruk, D., Potthast, S., Nüssle, K., Reinshagen, M., Adler, G., & Brambs, H. J. (2000). Diagnostic imaging in Crohn's disease: Comparison of magnetic resonance imaging and convertional imaging methods. *International Journal of Colorectal Disease*, 15(3), 176–181. https://doi.org/10.1007/s003840000219

Ruckh, T. T., & Clark, H. A. (2014). Implantable nanosensors: Toward continuous physiologic monitoring. *Analytical Chemistry*, 86(3), 1314–1323. https://doi.org/10.1021/ac402688k

Saadatzadeh, A., Atyabi, F., Fazeli, M. R., Dinarvand, R., Jamalifar, H., Abdolghaffari, A. H., Mahdaviani, P., Mahbod, M., Baeeri, M., & Baghaei, A. (2012). Biochemical and pathological evidences on the benefit of a new biodegradable nanoparticles of probiotic extract in murine colitis. *Fundamental & Clinical Pharmacology*, 26(5), 589–598. https://doi.org/10.1111/j.1472-8206.2011.00966.x

Sainz, V., Conniot, J., Matos, A. I., Peres, C., Zupančič, E., Moura, L., Silva, L. C., Florindo, H. F., & Gaspar, R. S. (2015). Regulatory aspects on nanomedicines. *Biochemical and Biophysical Research Communications*, 468(3), 504–510. https://doi.org/10.1016/j.bbrc.2015.08.023

Sakhalkar, H. S., Dalal, M. K., Salem, A. K., Ansari, R., Fu, J., Kiani, M. F., Kurjiaka, D. T., Hanes, J., Shakesheff, K. M., & Goetz, D. J. (2003). Leukocyte-inspired biodegradable particles that selectively and avidly adhere to inflamed endothelium in vitro and in vivo. *Proceedings of the National Academy of Sciences*, 100(26), 15895–15900. https://doi.org/10.2174/13892001279 8356916

Samanta, A. K., Torok, V. A., Percy, N. J., Abimosleh, S. M., & Howarth, G. S. (2012). Microbial fingerprinting detects unique bacterial communities in the faecal microbiota of rats with experimentally-induced colitis. *Journal of Microbiology*, 50, 218–225. https://doi.org/10.1007/s12 275-012-1362-8

Schmidt, C., Lautenschlaeger, C., Collnot, E.-M., Schumann, M., Bojarski, C., Schulzke, J.-D., Lehr, C.-M., & Stallmach, A. (2013). Nano-and microscaled particles for drug targeting to inflamed intestinal mucosa—A first in vivo study in human patients. *Journal of Controlled Release*, 165(2), 139–145. https://doi.org/10.1016/j.jconrel.2012.10.019

Shah, B. M., Palakurthi, S. S., Khare, T., Khare, S., & Palakurthi, S. (2020). Natural proteins and polysaccharides in the development of micro/nano delivery systems for the treatment of inflammatory bowel disease. *International Journal of Biological Macromolecules*, 165, 722–737. https://doi.org/10.1016/j.ijbiomac.2020.09.214

Shah, N., Shah, T., & Amin, A. (2011). Polysaccharides: A targeting strategy for colonic drug delivery. *Expert Opinion on Drug Delivery*, 8(6), 779–796. https://doi.org/10.1517/17425 247.2011.574121

Sharma, S., Parveen, R., & Chatterji, B. P. (2021). Toxicology of nanoparticles in drug delivery. *Current Pathobiology Reports*, 9(4), 133–144. https://doi.org/10.1007/s40139-021-00227-z

Siczek, K., Zatorski, H., Chmielowiec-Korzeniowska, A., Pulit-Prociak, J., Śmiech, M., Kordek, R., Tymczyna, L., Banach, M., & Fichna, J. (2017). Synthesis and evaluation of anti-inflammatory properties of silver nanoparticle suspensions in experimental colitis in mice. *Chemical Biology & Drug Design*, 89(4), 538–547. https://doi.org/10.1111/cbdd.12876

Singh, P., Kim, Y.-J., Zhang, D., & Yang, D.-C. (2016). Biological synthesis of nanoparticles from plants and microorganisms. *Trends in Biotechnology*, 34(7), 588–599. https://doi.org/10.1016/j.tibtech.2016.02.006

Sinha, V. R., & Kumria, R. (2004). Polysaccharide matrices for microbially triggered drug delivery to the colon. *Drug Development and Industrial Pharmacy*, 30(2), 143–150. https://doi.org/10.1081/ddc-120028709

Song, Y., Li, X., & Du, X. (2009). Exposure to nanoparticles is related to pleural effusion, pulmonary fibrosis and granuloma. *European Respiratory Journal*, 34(3), 559–567. https://doi.org/10.1183/09031936.00178308

Spertini, O., Cordey, A.-S., Monai, N., Giuffrè, L., & Schapira, M. (1996). P-selectin glycoprotein ligand 1 is a ligand for L-selectin on neutrophils, monocytes, and CD34+ hematopoietic progenitor cells. *Journal of Cell Biology*, 135(2), 523–531. https://doi.org/10.1083/jcb.135.2.523

Sreepriya, M., & Bali, G. (2006). Effects of administration of Embelin and Curcumin on lipid peroxidation, hepatic glutathione antioxidant defense and hematopoietic system during N-nitrosodiethylamine/Phenobarbital-induced hepatocarcinogenesis in Wistar rats. *Molecular and Cellular Biochemistry*, 284(1–2), 49–55.

Tejas Girish Agnihotri, Shyam Sudhakar Gomte, A. J. (2022). Emerging theranostics to combat cancer: A perspective on metal-based nanomaterials. Drug *Development and Industrial Pharmacy*, 48(11), 585–601. https://doi.org/10.1080/03639045.2022.2153862

Tekeli, İ. O., Ateşşahin, A., Sakin, F., Aslan, A., Çeribaşı, S., & Yipel, M. (2019). Protective effects of conventional and colon-targeted lycopene and linalool on ulcerative colitis induced by acetic acid in rats. *Inflammopharmacology*, 27, 313–322. https://doi.org/10.1007/s10787-018-0485-x

Tian, R., Liu, X., Luo, Y., Jiang, S., Liu, H., You, F., Zheng, C., & Wu, J. (2020). Apoptosis exerts a vital role in the treatment of colitis-associated cancer by herbal medicine. *Frontiers in Pharmacology*, 11, 438. https://doi.org/10.3389/fphar.2020.00438

Tolstanova, G., Deng, X., French, S. W., Lungo, W., Paunovic, B., Khomenko, T., Ahluwalia, A., Kaplan, T., Dacosta-Iyer, M., & Tarnawski, A. (2012). Early endothelial damage and increased colonic vascular permeability in the development of experimental ulcerative colitis in rats and mice. *Laboratory Investigation*, 92(1), 9–21. https://doi.org/10.1038/labinvest.2011.122

Trinchard-Lugan, I., Ho-Nguyen, Q., Bilham, W. M., Buraglio, M., Ythier, A., & Munafo, A. (2001). Safety, pharmacokinetics and pharmacodynamics of recombinant human tumour necrosis factor-binding protein-1 (Onercept) injected by intravenous, intramuscular and subcutaneous routes into healthy volunteers. *European Cytokine Network*, 12(3), 391–398.

Vandenbroucke, K., Hans, W., Van Huysse, J., Neirynck, S., Demetter, P., Remaut, E., Rottiers, P., & Steidler, L. (2004). Active delivery of trefoil factors by genetically modified Lactococcus lactis prevents and heals acute colitis in mice. *Gastroenterology*, 127(2), 502–513. https://doi.org/10.1053/j.gastro.2004.05.020

Wang, D., Jiang, Q., Shen, R., Peng, L., Zhou, W., Meng, T., Hu, F., Wang, J., & Yuan, H. (2024). ROS-responsive nanoparticles targeting inflamed colon for synergistic therapy of inflammatory bowel disease via barrier repair and anti-inflammation. *Nano Research*, 5409–5423. https://doi.org/10.1007/s12274-024-6435-6

Wani, S., Man Law, I. K., & Pothoulakis, C. (2020). Role and mechanisms of exosomal miRNAs in IBD pathophysiology. *American Journal of Physiology-Gastrointestinal and Liver Physiology*, 319(6), G646–G654. https://doi.org10.1152/ajpgi.00295.2020

Watkins, R., Wu, L., Zhang, C., Davis, R. M., & Xu, B. (2015). Natural product-based nanomedicine: Recent advances and issues. *International Journal of Nanomedicine*, 10, 6055–6074. https://doi.org10.2147/IJN.S92162

Watterlot, L., Rochat, T., Sokol, H., Cherbuy, C., Bouloufa, I., Lefèvre, F., Gratadoux, J. J., Honvo-Hueto, E., Chilmonczyk, S., Blugeon, S., Corthier, G., Langella, P., & Bermúdez-Humarán, L. G. (2010). Intragastric administration of a superoxide dismutase-producing recombinant Lactobacillus casei BL23 strain attenuates DSS colitis in mice. *International Journal of Food Microbiology*, 144(1), 35–41. https://doi.org/10.1016/j.ijfoodmicro.2010.03.037

Watts, P. J., & Lllum, L. (1997). Colonic drug delivery. *Drug Development and Industrial Pharmacy*, 23(9), 893–913. https://doi.org10.3109/03639049709148695

Wei, W., Zhang, Y., Li, R., Cao, Y., Yan, X., Ma, Y., Zhang, Y., Yang, M., & Zhang, M. (2022). Oral delivery of pterostilbene by L-arginine-mediated "nano-bomb" carrier for the treatment of ulcerative colitis. *International Journal of Nanomedicine*, 17, 603–616. https://doi.org/10.2147/IJN.S347506

Wei, Z., Hang, S., Wiredu Ocansey, D. K., Zhang, Z., Wang, B., Zhang, X., & Mao, F. (2023). Human umbilical cord mesenchymal stem cells derived exosome shuttling mir-129-5p attenuates inflammatory bowel disease by inhibiting ferroptosis. *Journal of Nanobiotechnology*, 21(1), 188. https://doi.org/10.1186/s12951-023-01951-x

Wilson, D. S., Dalmasso, G., Wang, L., Sitaraman, S. V, Merlin, D., & Murthy, N. (2010). Orally delivered thioketal nanoparticles loaded with TNF-α–siRNA target inflammation and inhibit gene expression in the intestines. *Nature Materials*, 9(11), 923–928. https://doi.org10.1038/nmat2859

Wu, A., Chen, C., Lu, J., Sun, J., Xiao, M., Yue, X., Zhou, P., Zhao, S., Zhong, G., & Huang, C. (2021). Preparation of oral core–shell zein nanoparticles to improve the bioavailability of glycyrrhizic acid for the treatment of ulcerative colitis. *Biomacromolecules*, 23(1), 210–225. https://doi.org10.1021/acs.biomac.1c01233

Wu, W., Jiang, C. Z., & Roy, V. A. L. (2016). Designed synthesis and surface engineering strategies of magnetic iron oxide nanoparticles for biomedical applications. *Nanoscale*, 8(47), 19421–19474. https://doi.org/10.1039/c6nr07542h

Wu, Y., Briley-Saebo, K., Xie, J., Zhang, R., Wang, Z., He, C., Tang, C. Y., & Tao, X. (2014). Inflammatory bowel disease: MR- and SPECT/CT-based macrophage imaging for monitoring and evaluating disease activity in experimental mouse model-pilot study. *Radiology*, 271(2), 400–407. https://doi.org/10.1148/radiol.13122254

Xiao, B., Laroui, H., Ayyadurai, S., Viennois, E., Charania, M. A., Zhang, Y., & Merlin, D. (2013). Mannosylated bioreducible nanoparticle-mediated macrophage-specific TNF-α RNA interference for IBD therapy. *Biomaterials*, 34(30), 7471–7482. https://doi.org/10.1016/j.biomaterials.2013.06.008

Xiao, B., Laroui, H., Viennois, E., Ayyadurai, S., Charania, M. A., Zhang, Y., Zhang, Z., Baker, M. T., Zhang, B., Gewirtz, A. T., & Merlin, D. (2014). Nanoparticles with surface antibody against CD98 and carrying CD98 small interfering RNA reduce colitis in mice. *Gastroenterology*, 146(5), 1219–1289. https://doi.org/10.1053/j.gastro.2014.01.056

Xiao, B., Yang, Y., Viennois, E., Zhang, Y., Ayyadurai, S., Baker, M. T., Laroui, H., & Merlin, D. (2014). Glycoprotein CD98 as a receptor for colitis-targeted delivery of nanoparticles. *Journal of Materials Chemistry B*, 2(11), 1499–1508. https://doi.org/10.1039/C3TB21564D

Xiao, C., Li, G., Li, X., Wang, D., Wu, Y., Sun, M., Huang, J., & Si, L. (2023). A topical thermosensitive hydrogel system with cyclosporine A PEG-PCL micelles alleviates ulcerative colitis induced by TNBS in mice. *Drug Delivery and Translational Research*, 13(10), 2447–2462. https://doi.org10.1007/s13346-023-01317-8

Yang, C., & Merlin, D. (2019). Nanoparticle-mediated drug delivery systems for the treatment Of IBD: Current perspectives. *International Journal of Nanomedicine*, 14, 8875–8889. https://doi.org/10.2147/IJN.S210315

Zhang, H., Wang, L., Li, C., Yu, Y., Yi, Y., Wang, J., & Chen, D. (2019). Exosome-induced regulation in inflammatory bowel disease. *Frontiers in Immunology*, 10, 416673. https://doi.org/10.3389/fimmu.2019.01464

Zhang, J., Tang, C., & Yin, C. (2013). Galactosylated trimethyl chitosan–cysteine nanoparticles loaded with Map4k4 siRNA for targeting activated macrophages. *Biomaterials*, 34(14), 3667–3677. https://doi.org/10.1016/j.biomaterials.2013.01.079

Zhang, M., Xu, C., Liu, D., Han, M. K., Wang, L., & Merlin, D. (2018). Oral delivery of nanoparticles loaded with ginger active compound, 6-shogaol, attenuates ulcerative colitis and promotes wound healing in a murine model of ulcerative colitis. *Journal of Crohn's & Colitis*, 12(2), 217–229. https://doi.org/10.1093/ecco-jcc/jjx115

Zhang, S., Ermann, J., Succi, M. D., Zhou, A., Hamilton, M. J., Cao, B., Korzenik, J. R., Glickman, J. N., Vemula, P. K., Glimcher, L. H., Traverso, G., Langer, R., & Karp, J. M. (2015). An inflammation-targeting hydrogel for local drug delivery in inflammatory bowel disease. *Science Translational Medicine*, 7(300), 300ra128. https://doi.org/10.1126/scitranslmed.aaa5657

Zhang, S., Langer, R., & Traverso, G. (2017). Nanoparticulate drug delivery systems targeting inflammation for treatment of inflammatory bowel disease. *Nano Today*, 16, 82–96. https://doi.org/10.1016/j.nantod.2017.08.006

Zhang, Y., Cui, L., Che, X., Zhang, H., Shi, N., Li, C., Chen, Y., & Kong, W. (2015). Zein-based films and their usage for controlled delivery: Origin, classes and current landscape. *Journal of Controlled Release*, 206, 206–219. https://doi.org/10.1016/j.jconrel.2015.03.030

Zhu, H., & Li, Y. R. (2012). Oxidative stress and redox signaling mechanisms of inflammatory bowel disease: Updated experimental and clinical evidence. *Experimental Biology and Medicine*, 237(5), 474–480. https://doi.org/10.1258/ebm.2011.011358

Zuo, L., Huang, Z., Dong, L., Xu, L., Zhu, Y., Zeng, K., Zhang, C., Chen, J., & Zhang, J. (2010). Targeting delivery of anti-TNFα oligonucleotide into activated colonic macrophages protects against experimental colitis. *Gut*, 59(4), 470–479. https://doi.org/10.1136/gut.2009.184556

13 Experimental Animal Models and *In-Vitro* Models for Gastrointestinal Inflammation

Phool Chandra[1], Zeeshan Ali[2], Nishat Fatma[3], and Neetu Sachan[4]*

[1]Teerthanker Mahaveer College of Pharmacy, Teerthanker Mahaveer University, Moradabad-244001, Uttar Pradesh, India

[2]Sahara Degree College, Rampur, Utter Pradesh, India

[3]MIT College of Pharmacy, Moradabad, Uttar Pradesh, India

[4]Maharana Pratap College of Pharmacy, Mandhana, Kanpur, Uttar Pradesh, India

13.1 INTRODUCTION

Inflammatory bowel disease is defined as "chronic intestinal inflammation produced by host direct contact in a genetically vulnerable individual" (IBD). A group of autoimmune diseases known as "inflammatory bowel disease" (IBD) cause inflammation of both the small and large intestines as a result of the body's immune system attacking the digestive system. Classifying diseases according to the organs they affect and the resulting clinical symptoms is standard procedure. People who suffer from intestinal inflammatory disorders find it challenging to effectively digest food due to a wide range of severe medical conditions. This multifactorial disease includes inflammation of the small and large bowels along with other organs, some of which are involved in the intricate connection of digestive systems (1). Gastrointestinal tract inflammation typically upsets the delicate balance inside the digestive system, which affects critical activities like immune system reactions, nutrition absorption, as well as ultimately physiological stability. Understanding this phenomenon thoroughly is very important for academics, people, and healthcare professionals, as it encompasses a wide spectrum of disorders, from terrible and chronic to acute and self-limiting (2). Besides controlling food intake and digestion, the intestinal mucosa is an important organ system that acts as the body's initial defense against pathogens and external objects (3). Animal research is one of the fundamental bases of many academic fields, and it has long been used in science. The ethical considerations surrounding animal models are crucial. Even if the desired ailment is treatable, researchers are not allowed to create problems in humans for research purposes. It could cause permanent psychological and physical harm and is wholly immoral. Thus, for a variety of objectives, including the development of disease and the identification of therapeutic targets, researchers employed animal models. The genetic resemblances between humans and certain animal species, particularly mammals, account for their widespread use. To lessen the suffering and distress of employed animals, there are, nevertheless, guidelines that must be adhered to when dealing with animals (4). A number of recommendations have been created for this reason, such as setting up facilities that adequately supply food and water, care, as well as physical wellbeing; substituting predictive models and advanced techniques for animal models; minimizing the suffering and distress inflicted upon the animals; and minimizing the number of animals utilized in the research. According to earlier Greek writings from the second and fourth centuries, Aristotle and Erasistratus most likely conducted the first animal investigations on living creatures (5). Advances in physician management techniques, personalized medicine methods, and novel drugs that specifically target inflammatory pathways are propelling transformations in the field of digestive healthcare. Nonetheless, certain obstacles persist, such as the requirement for more focused and efficient therapies, enhanced diagnostic instruments, and a more profound comprehension of the elements impacting individual variations in treatment reaction (6). In terms of digestive tract disorders, the research of gastrointestinal inflammation is broad and challenging. With a thorough approach, the researchers want to offer the foundation for a deeper investigation into the nuances of gastrointestinal inflammation's etiology, diagnosis, and changing treatment paradigms. Different leukocytes and cytokines cause inflammation in consequence to an infection, trauma, as well as damage (7). It is hard to deny the importance of animal research in the advancement of science, particularly in the field of medicine. Several criteria should be followed when selecting animals for use in research that aim to create particular symptoms and diseases. They should, first and foremost, resemble humans in phenotype. Second, diseases should arise from the same processes in humans and all those animals. Third, the drugs or non-pharmacological therapies used to treat that illness in humans should not affect the animals adversely. The criteria have been upgraded in recent research, and new aspects have been introduced, such as the cost-effectiveness and biological mechanism similarities (8). Intestinal inflammation diagnosis has advanced significantly

DOI: 10.1201/9781003493143-13

as a result of technological advancements in endoscopy, medical testing, and laboratory testing. A range of diagnostic techniques are employed by medical professionals to precisely select the nature, location, and intensity of inflammation, in contrast to physical investigations and patient histories. Diagnostic aids that support diagnosis and direct treatment decisions include serological indicators and imaging tests like endoscopic techniques (9). The initial protection the body has against an infection, wound, or inflammatory responses is the acute response (APR). After a trigger, APR is a complex series of sudden systemic reactions. Researchers can now more precisely customize treatment plans to the particulars of each person's inflammatory profile because to advancements in molecular and technological studies. There are many different approaches to treating gastrointestinal inflammation, including diet, medication, lifestyle modifications, and, in extreme circumstances, surgery (10,11).

13.2 ANIMAL MODELS FOR GASTROINTESTINAL INFLAMMATION

Animal health and welfare, both human and non-human, are significantly and negatively impacted by acute and chronic inflammatory bowel diseases. It is essential to comprehend the basic causes of inflammatory diseases in order to create efficient preventive and treatment plans. Due to the utilization of animal models, researchers now have a significantly better understanding of gastrointestinal inflammation and the intricate mechanisms behind a wide spectrum of digestive illnesses. The hallmark of inflammatory bowel diseases, such Crohn's disease as well as ulcerative colitis, is glandular inflammation of the gastrointestinal system (12,13). The gastrointestinal ailment known as inflammatory bowel disease (IBD) is persistent and recurrent. Inflammatory bowel diseases include Crohn's disease and ulcerative colitis. Experts have been studying the disease for many years, trying to determine its pathophysiology and cause. To properly identify a therapy and ultimately understand the cause, researchers have developed a number of *ex vivo* systems, *in-vitro* cell lines, and *in-vivo* animal models. Yet, because they are unable to accurately replicate the illness, *in-vivo* animal models are not considered to be useful research tools for inflammatory bowel disease. Moreover, *in-vitro* cell lines and *ex vivo* systems can be used to accurately replicate the condition because species differences are an important area of concern (14). The basis for biological as well as medical research is *in-vitro models*. There are now both *in-vitro* and *ex vivo* model available for simulating gut physiology (15). Rats are more commonly employed as animal models than some other animals because of their greater size and more human-like physiological characteristics. To study the immune system, the operation of epithelial barriers, and the part the microbiota performs in the progression of disease, rats can be housed in laboratory settings that mimic the gastrointestinal inflammation which mice encounter showed in Figure 13.1 (16). Zebrafish and drosophila are two other animals that are utilized as animal models for gastrointestinal inflammation in addition to mammals (fruit flies). These models are useful for cost-effectiveness, high-throughput screening, and genetic editing. They may not fully capture the complexity of the human digestive system, but they allow researchers to investigate fundamental biological processes and identify shared pathways linked to inflammation (17,18). However, exorbitant costs, few resources, and ethical issues prevent apes from being widely used in gastrointestinal inflammatory research. Furthermore, larger animals including pigs and rabbits more closely resemble the anatomy and physiology of humans, exposing characteristics that smaller models would not be able to convey (19). As demonstrated in Table 13.1, animal models are essential to expanding our knowledge of gastrointestinal inflammation. Scholars can investigate many facets of the intricate pathophysiology linked to inflammatory bowel disorders and associated ailments, since each model has distinct benefits and constraints.

The information gathered from these investigations advances our understanding of the complex interactions between genetic or environmental, as well as microbiological aspects in digestive health and diseases in addition to aiding in the generation of targeted medicines (20).

13.3 PATHOPHYSIOLOGY OF EXPERIMENTAL ANIMAL MODELS

Cells cannot be viewed as distinct entities due to the way the human body is organized. The physiological mechanism that permits these elements to endure and carry out their duties in those conditions is known as homeostasis. To simulate some features of gastrointestinal (GI) inflammatory illnesses in humans, like inflammatory bowel disease (IBD), animal models have been developed (29,30). Colitis in animals, especially mice, is a regularly used experimental model for studying GI inflammation caused by dextran sulfate sodium (DSS). This method involves giving the animals potable water that contains the sulfated polysaccharide DSS. The colonic epithelial barrier bursts due to DSS, exposing immune cells to luminal antigens as well as increasing their permeability (31). This initiates an inflammatory reaction that is characterized by immune cells penetrating the intestinal mucosa, especially neutrophils and macrophages (32). Another commonly used model is the colitis model produced in rats by trinitrobenzene sulfonic acid (TNBS). This approach involves infusing TNBS into the colon, which triggers the body to create hapten-protein mimics that trigger an immunological response. This is similar to actual Crohn's disease in that it involves alteration of the intestinal architecture, the secretion of pro-inflammatory chemicals, or even the activation of inflammatory cells. Researching the genetic foundation of GI inflammation requires the use of genetically modified animals, especially mice (33). The role of immune modulation, cytokine signaling, and mucosal barriers function can be studied in relation to the onset and evolution of GI inflammation using knockout or modified mice that either upregulates or lack particular genes in these domains. A series of actions are involved in the pathogenesis of experimental GI inflammation. First,

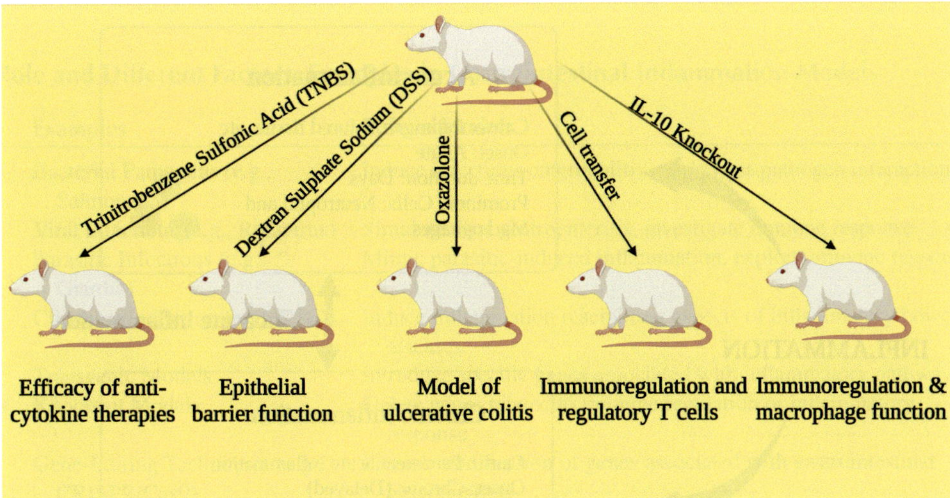

FIGURE 13.1 Experimental models for inflammation with relation groups and their conditions.

TABLE 13.1
Details of Experimental Animal Models

Animal Model	Common Strains/Breeds	Specific Models/Induction Methods	References
Mouse Model	C57BL/6, BALB/c	IL-10 knockout mice, TNF-α overexpressing mice, chemical induction	(21)
Rat Model	Sprague-Dawley, Wistar	TNBS-induced colitis rats, DSS-treated rats, various chemical methods	(22)
Rabbit Model	Various breeds	Acetic acid-induced colitis in rabbits.	(23,24)
Pig Model	Yorkshire, Landrace	DSS-induced colitis in pigs, TNBS models	(25)
Non-Human Primate Model	Rhesus macaque, cynomolgus macaque	Induced colitis in macaques for specific studies	(26)
Zebrafish Model	AB, Tu, TL	Genetic mutants, chemical induction	(27)
Drosophila (Fruit Fly) Model	Oregon-R, Canton-S	Genetically modified flies to study conserved inflammatory pathways	(28)

the integrity of the mucosal barrier is altered, which makes it possible for luminal antigens to interact with the immune system underneath (34). Pro-inflammatory cytokines including tumor necrosis factor-alpha (TNF-α), or interleukin-1β (IL-1β), as well as interleukin-6 are released as a result, inducing an immunological response (IL-6). All such cytokines are essential for bringing immune cells towards the region of inflammation and allowing it to continue. Reactive oxygen species or even proteolytic enzymes are released by activated immune cells, especially neutrophils and macrophages, which penetrate the mucosa and cause tissue destruction. Furthermore, imbalances in regulatory T cells (Tregs) or effector T cells, as well as endothelial dysfunction of T-cell responses, intensify inflammation (34,35). The general structure and functioning of the intestine are impacted by tissue regeneration, fibrosis, and angiogenesis that result from inflammatory reaction in the GI tract. Diarrhea, stomach pain, and losing weight are among the clinical symptoms of GI inflammatory illnesses that are caused by a downregulation immune response and persistent inflammation. An invaluable resource for researching the intricate pathogenesis of GI inflammation is experimental animal models. By allowing researchers to examine particular facets of the immune system response, hereditary variables, and molecular pathways connected to GI inflammation, these models eventually prepare the way for the creation of cutting-edge treatment approaches for inflammatory bowel disorders in humans (36,37).

13.4 ROLE OF PATHOGENIC/GENETICALLY MODIFIED FACTORS IN EXPERIMENTAL ANIMAL MODELS

The creation of safe and effective vaccinations and treatments depends on having a thorough awareness of the benefits and drawbacks of the animal models that scientists use. In some circumstances, the animal model is essential for comprehending the pathophysiology of the disease prior the creation of vaccines or therapies can even be investigated, such as for new or developing diseases for which research data are unavailable. In order to fully understand the

research how various eukaryotic species as well as bacteria interact with one another on a cell-to-cell basis. Due to this technology, researchers can now duplicate complex tissue topologies with greater accuracy. These innovative techniques allow for a detailed investigation of the functions of Toll-like receptors (TLRs) and other sequences recognition receptors in the detection of microbiological components. Co-culture models provide a flexible platform for studying the molecular components of gastrointestinal inflammation. Utilizing this technology, new treatments for ailments such as inflammatory bowel disease can be developed and feasible therapy programs identified (85).

13.6 DEVELOPMENT OF NOVEL THERAPEUTICS

"Novel therapeutics" are therapies that are completely exclusive to the veterinary medicine field. This could be because the treatment is truly original and has never been applied in a medical context, or it could be a reference to treatments that are rather well-known in the scientific world and might even be brand-new to the field of human medicine. Novel pharmaceuticals are often region products that close gaps in the medical community, enhance patient care, or even promote public health in other contexts. The term "inflammatory gastrointestinal" (GI) disorders refers to a broad category of illnesses, ranging from common ailments like gastroenteritis and inflammatory bowel disease (IBD) to far more serious problems like ulcerative colitis and Crohn's disease. Globally, GI inflammatory illnesses are becoming increasingly common, which has prompted the need for additional study and the creation of novel treatment strategies. Because researchers have been concentrating on cutting-edge methods to reduce symptoms, encourage mucosal healing, and address the underlying causes of inflammation, the field of GI therapies has significantly improved in recent years (86).

13.6.1 Biologics Targeting Inflammatory Pathways

Highly regulated in the body, pro-inflammatory cytokines are powerful mediators of many biological processes. Such cytokines can cause and initiate a wide range of pathologies, including cancer and autoimmune diseases, when their levels are unchecked and prolonged. Targeting TNF-α pathways in gastrointestinal inflammation has been demonstrated by biologics (87). Biomarkers can be either recombinant substances that bind to receptors to inhibit binding or antibodies that either deactivate the indications themselves or block the receptors of inflammatory indicators. The introduction of biologics that particularly target inflammatory pathways associated with gastrointestinal (GI) inflammation has had a substantial impact on the development of innovative treatments. Targeting TNF-α pathways in gastrointestinal inflammation has been demonstrated by biologics. TNF-α is a cytokine that plays a vital function in inflammatory pathways and is involved in ulcerative colitis and Crohn's disease (88). Focusing on the Janus kinase-signal transducer and regulator of transcription (JAK-STAT) pathway is an alternative approach to the development of biologics. Small molecules provide tofacitinib, an inhibitor of the JAK pathway, as one of the oral medicines as an option to injectable biologics. This oral route of delivery gives patients with inflammatory bowel disease additional convenient medication options. Consequently, throughout the past few decades, a great deal of research has been done on medicines that control the activity of inflammatory cytokines, either by adding anti-inflammatory recombinant cytokines or by neutralizing them using blocking antibodies (89).

13.6.2 Small Molecules and Oral Therapies

The ability of tiny molecules to enter cells and obstruct intracellular signaling pathways associated with inflammation is one of their most significant advantages. Small compounds differ from some biologics in that they largely rely on specialized mechanisms for their activity rather than primarily depending on cell surface receptors. The use of small molecule medications as treatments has a few unique benefits. A low molecular weight organic compound is referred to as a small molecule. For example, the majority are oral and capable of penetrating cell membranes to get to intracellular destinations. The discovery of oral medications and small molecules has advanced the hunt for new therapies for gastrointestinal (GI) inflammation. Historically, injectable biologics have received the majority of attention; however, studies on small molecules provide an alternate, practical oral delivery option (90). Tofacitinib is one of these Janus kinase inhibitors that has become well-known for treating inflammatory bowel disease (IBD) effectively and offering an oral medication for managing the condition. Small molecule inhibitors of Janus kinases, such as tofacitinib, have been shown to reduce inflammation of the gastrointestinal tract and alter the immune system's response. Tiny molecule development intends to create drugs that suit the preferences and lifestyle of people with GI inflammation in response to a broader trend of treatment alternatives becoming more varied (91). Therefore, smaller molecules might provide a more direct and targeted intervention, decreasing the likelihood of off-target effects and enhancing overall therapeutic efficacy (92).

13.6.3 Microbiome Modulation

To maintain host homeostasis, the gut microbiota performs a variety of essential functions, including the breakdown of complex carbohydrates, the synthesis of vitamins, the maintenance of the integrity of the mucosal lining, including antibiotic defense. The microbial population known as the gut microbiome, found in the digestive tract, is crucial for immune system as well as gut health preservation. Therapeutic interventions that alter the microbiome are attractive because dysbiosis, or an imbalance in the microbiome's makeup, has been connected to a range of gastrointestinal inflammatory illnesses (11,93). Restoring the microbiome through fecal microbiota transplantation is an important method (FMT).

FMT involves transplanting a healthy person's feces to a sick patient in an effort to replace the patient's stomach with a balancing microbial ecology. Positive outcomes from treating illnesses like Clostridium difficile infection indicate that modifying the microbiome can lower inflammation and repair the gut homeostasis (94). These medications target and modify the particular bacteria populations linked to inflammatory diseases. Understanding the intricate relationships between the host and the microbiota will help scientists discover key microbial players and develop treatments that promote a sustainable and anti-inflammatory microbial environment inside the gut (95).

13.6.4 Precision Medicine and Genetic Approaches

Precision medicine takes into account the differences in behavioral or socio-environmental, including genetic traits within subpopulations to provide individualized treatment recommendations. Genomics and precision medicine are transforming the discovery of new treatments for gastrointestinal (GI) inflammation. Comprehending the hereditary variables that impact a person's vulnerability to gastrointestinal illnesses has enabled tailored treatment plans. By customizing care based on a patient's genetic profile, personalized medicine enables more focused and efficient interventions (96). Researchers are working hard to find genetic markers linked to medication reactions, which will allow physicians connect patients with the best treatments. Using this method to optimize care for patients with GI inflammatory diseases represents a significant advancement since it not only avoids adverse effects but also improves therapeutic outcomes. Personalized medicine and genetic techniques are expected to become more important in the creation of individualized and patient-specific therapies as research on the genetic foundations of these ailments progresses (97).

13.6.5 Nutraceuticals and Dietary Interventions

These days, there is a lot of attention in nutraceuticals because of their possible benefits in terms of nutrition, safety, and treatment. Dietary modifications and nutraceuticals are becoming recognized as essential elements in the creation of cutting-edge treatments for gastrointestinal (GI) inflammation. These interventions, which go beyond conventional pharmacological methods, taking advantage of the potential health advantages of bioactive substances obtained from food. For GI diseases, omega-3 fatty acids, for instance, have anti-inflammatory qualities and are being investigated as supplementary therapy. Nutraceuticals provide an alternative to traditional therapies by attempting to reduce inflammation and promote gut health (98). The emphasis on nutrition's role in managing GI inflammation emphasizes the need of a comprehensive strategy, where dietary interventions can play a critical role in boosting the general wellness of patients coping with these challenging conditions and improving patient results. Nutraceuticals and dietary treatments show promise as widely obtainable and natural solutions in the complex field of GI therapies as research in this area continues (99).

13.6.6 Emerging Technologies in Drug Delivery

Modern generation drug delivery methods include a number of drug carriers, including aptamers, magnetic nanoparticle-based delivery, liposomes or polymers, quantum dots, peptides, and metals. A new stage of research and development for novel treatments for gastrointestinal (GI) inflammation has begun as a result of advancements in pharmaceutical delivery technologies. Among these breakthroughs, drug delivery techniques based on nanoparticles have garnered a lot of attention. These methods use microscopic particles to precisely enclose and deliver therapeutic substances. These nanoparticles provided customized drug release across the gastrointestinal system in the situation of GI inflammation, addressing the challenge of limited therapeutic delivery (100). One important advantage of drug delivery by nanoparticles is its ability to navigate the complex gastrointestinal tract. These nanoparticles can be designed to tolerate the stomach's acidic environment and distribute their payloads at particular intestinal locations where inflammation is particularly prominent (101). The systemic drug dissemination adverse effects may be lessened by this focused method, which maximizes the therapeutic efficiency of drugs while minimizing exposures to unaffected regions. It is also possible to incorporate many kinds of therapeutic agents, such as biologics, anti-inflammatory medications, and sometimes even nucleic acid-based therapies, into nanoparticle-based systems because of their adaptability. Researchers can investigate complex treatment plans due to this flexibility, which combines many therapy modalities on a particular drug delivery platforms (102).

13.6.7 Clinical Trials and Biomarker Identification

Treatment selection is optimized by using predictive biomarkers to create (binary) classifiers that classify patients as either suitable or poor responders for a certain treatment. The identification of biomarkers and clinical trials are critical steps in the creation of new treatments for inflammatory gastrointestinal disorders. The safety and effectiveness of novel medicines must be evaluated through rigorous and quite well clinical trials in order to provide critical evidence for their possible integration into clinical practice, as demonstrated in Table 13.3 (103). To assess the treatment effect on gastrointestinal disorders, these trials frequently use strong endpoints and a variety of patient populations. Moreover, biomarker identification is now a crucial step in the research and development process. Measurable indications of biological activities or treatment responds, or biomarkers, enable with patient classification, therapy response prediction, and disease progression tracking. Reliable biomarkers can improve patient selection for targeted medicines, refine treatment approaches, and set alternative endpoints for clinical studies in the area of gastrointestinal inflammation (104). Development of therapies

TABLE 13.3
Clinical Trials and Biomarkers for Therapeutic Approaches

S. No.	Aspect	Examples/Description	References
1.	**Biologics Targeting Inflammatory Pathways**	Infliximab, Adalimumab (targeting TNF-α); Ustekinumab (targeting IL-23)	(106)
2.	**Small Molecules and Oral Therapies**	Tofacitinib (JAK inhibitor for IBD); Vedolizumab (selective gut lymphocyte trafficking)	(107)
3.	**Microbiome Modulation**	Fecal Microbiota Transplantation (FMT); Microbiome-targeted drugs; Probiotics	(108)
4.	**Precision Medicine and Genetic Approaches**	Tailoring treatments based on genetic profiles; Identifying genetic markers (e.g., pharmacogenomics)	(109)
5.	**Nutraceuticals and Dietary Interventions**	Omega-3 Fatty Acids; Dietary Fiber; Exclusive Enteral Nutrition (EEN) for IBD	(110)
6.	**Emerging Technologies in Drug Delivery**	Nanoparticle-based drug delivery systems; Microparticles; Hydrogels	(111)
7.	**Clinical Trials and Biomarker Identification**	Rigorous clinical trials; Identification of biomarkers for patient stratification and treatment monitoring	(112)
8.	**Artificial Intelligence in Treatment Decisions**	Integration of AI and machine learning for analyzing patient data and treatment decision-making	(113)
9.	**Cost-Effective Therapies and Access**	Development of cost-effective therapies to improve accessibility; Biosimilars	(114)
10.	**Patient-Centered Care**	Emphasis on patient preferences and quality of life; Shared decision-making in treatment plans	(115)

for gastrointestinal inflammatory disorders is conducted with precision and evidence-based methodology mainly to the continuous cooperation of researchers, physicians, and pharmaceutical corporations in clinical trial initiatives and the progress made in biomarker discoveries (105).

13.7 TRANSLATIONAL POTENTIAL AND CLINICAL RELEVANCE

Translated research aims to produce results that directly enhance human health and are extremely practical and meaningful. The scientific and medical communities are paying more and more concern to gastrointestinal inflammation, a complex and multifaceted phenomenon, due to its clinical relevance and translational potential. Due to a multitude of factors, such as infections, autoimmune illnesses, and environmental stressors, this intricate process sets off the immune system's response, which results in inflammation within the digestive tract (116). To develop targeted treatments and improve patient outcomes, it is critical to understand the underlying reasons of gastrointestinal inflammation (117). Inflammatory cells and chemokines, like cytokines, play a key role in this mechanism since they set off a series of events that eventually destroy tissue. Recent developments in the fields of molecular biology, particularly genetics, have shed light on the complex mechanisms underlying inflammatory gastrointestinal disorders and opened up new pathways for the discovery of possible therapeutic targets (118). Different people experience gastrointestinal inflammation in various ways, which makes it difficult for physicians to diagnose patients (119). The clinical variability involved with gastrointestinal inflammation is exemplified by disorders like Crohn's disease including ulcerative colitis, as well as inflammatory bowel disease (IBD), which can cause modest discomfort or serious problems. Initiating suitable treatment techniques and limiting long-term consequences require a timely and precise diagnosis (120). Visualizing gastrointestinal inflammation has been made much easier by advancements in medical imaging technology like endoscopy, computed tomography (CT), as well as magnetic resonance imaging (MRI). These instruments support illness severity evaluation, observation, and diagnosis. Furthermore, the identification of trustworthy biomarkers linked to gastrointestinal inflammation offers hope for early disease diagnosis and progression tracking, providing a non-invasive method of evaluating therapy response (121). Personalized therapy approaches are necessary due to the variability of gastrointestinal inflammation. Traditionally, the basis of symptom management has been the use of immunosuppressants and anti-inflammatory medications. On the other hand, personalized treatments based on unique genetic, biochemical, and immune profiles are now possible due to the advent of personalized medicine (122). Anti-TNF medications and other targeted biologic therapy have shown promise in particular patient populations, indicating the promise of customized medicine in the treatment of gastrointestinal inflammation. Studies on the gut microbiota's function in gastrointestinal inflammation is constantly changing. Intestinal homeostasis depends on the precise balancing between the microbiota and the host immune response (123). A disequilibrium in the community of gut microbes known as dysbiosis has been linked to the etiology of inflammatory diseases. Furthermore, a crucial role in controlling inflammation has been identified for the gut-brain axis, a two-way communication mechanism that connects the central nervous system with the gastrointestinal tract. Gaining insight into these intricate relationships creates new opportunities for treatment approaches aimed at the gut-brain axis and microbiome. Despite notable advancements, obstacles continue to arise in comprehending the intricacies of inflammatory gastrointestinal disorders and converting scientific discoveries into practical applications (124). The development of globally successful medicines is still hampered by the variability of patient groups, the unpredictable nature of illness progression, and the absence of conclusive biomarkers. In order to improve our overview of gastrointestinal inflammation and opening the opportunity to more targeted therapeutic approaches, future research efforts should concentrate on clarifying the interactions between genetics, environmental, or microbial variables. With its clinical significance and translational promise, gastrointestinal inflammation is at the forefront of medical investigation and medical technology (125). The route from laboratory to clinic demonstrates the multifaceted nature of scientific investigation, ranging from comprehending molecular subtleties to putting tailored therapy ideas into practice. The ultimate goal is to use our growing understanding of the intricacies of gastrointestinal inflammation to develop clinical therapies that increase patient

TABLE 13.4
Experimental Models Limitations

Limitation	Description	Example	References
Species Differences	The intrinsic differences between animal species & humans may mean that experimental models do not perfectly duplicate human physiology, which could affect how broadly study results can be applied	Studies on inflammatory bowel disease (IBD) may not be as relevant if people and animals have different gut microbiota compositions	(130,131)
Genetic Variability	The genetic geography of human inflammatory intestinal illnesses is heterogeneous, whereas the genetic uniformity observed in experimental models may restrict our comprehension of the genetic elements involved	It's possible that the genetic complexity seen in IBD-affected human populations cannot be fully captured by examining the effect of a particular gene on inflammation in a homogeneous mouse group	(132)
Environmental Factors	The influence of several environmental elements prevalent in human lifestyles may be oversimplified under controlled experimental conditions, potentially resulting in incomplete findings	The study of the effects of nutrition on inflammation may be impacted by the possibility that the standardized diets given to animals do not fairly reflect the dietary variability seen in human populations	(133)
Acute vs. Chronic Inflammation	Acute inflammatory induction is the main emphasis of many experimental models, which may overlook the chronic character of conditions like Crohn's and ulcerative colitis and restrict their long-term applicability	The intricate and continual processes implicated in chronic inflammatory bowel illnesses, such as ongoing tissue injury and healing, may not be fully captured in a short-term study that induces acute inflammation	(134)
Immune System Complexity	The complex and diverse character of human immune mechanisms in gastrointestinal inflammation may be neglected by oversimplifications of immune system interconnections in experimental models	An insufficient understanding of immunological processes in IBD may result from ignoring particular immune cell interaction or the intricacy of cytokine signaling networks	(135)
Drug Response Discrepancies	The predictive capacity of experimental models may be limited by differences between the reactions of animals and humans to therapeutic interventions, which could affect the transfer of results to clinical settings	When examined in human clinical trials, a medication that shows promise in a preclinical study utilizing animal models may reveal different responses or adverse consequences	(136)
Tissue-Culture Challenges	The physiological significance of isolated cell investigations may be endangered by *in-vitro* models, despite their value, as they lacked the intricacy of the milieu of a real organism	Conventional cell culture techniques might not adequately simulate the complex *in-vivo* circumstances of the gastrointestinal tract since they lack elements like immune cell interactions and blood flow	(137,138)
Ethical and Practical Considerations	The scope of study on the onset of disease is impacted by practical challenges and ethical limitations on animal experimentation, which limit the depth and intensity of experiments	Because of ethical regulations that forbid using animals in research, researchers might not be able to conduct the lengthy studies necessary to fully understand the development of chronic gastrointestinal illnesses	(139)

outcomes and raise the quality of life for those who suffer from these devastating illnesses (126).

13.8 LIMITATIONS OF EXPERIMENTAL MODELS

Experimental models are essential for expanding our knowledge of gastrointestinal inflammation because they offer important insights into the intricate mechanisms underlying a variety of diseases, including inflammatory bowel disease (IBD). To assure a correct translation of observations to human illness, these models do have some intrinsic limitations that need to be taken into account (127). The fundamental distinctions between human beings and animal species constitute one of the main drawbacks of experimental models. Despite the fact that rodent models—mice and rats, for example—are frequently employed because of their genetic and anatomical, or physiological parallels to humans, there are still important distinctions. For example, there are significant differences amongst species in the gut microbiota's composition, which is a critical component of gastrointestinal health. This difference casts doubt on the applicability of findings from animal studies to human circumstances, possibly restricting their direct use in therapeutic contexts (128). An important additional problem in experimental models of gastrointestinal inflammation is genetic heterogeneity. Regarding humans, IBD is a group of diseases that are predisposed to by a variety of hereditary variables. Comparatively, biologically homogeneous populations are frequently used in experimental models, which simplifies the genetic environment being studied. Because of this oversimplification, the complicated genetic interconnections behind gastrointestinal inflammation may not be fully understood, which would restrict the model's capacity to fully represent the complexity of clinical disorders as demonstrated in Table 13.4 (129).

The lives of persons who experience gastrointestinal issues will eventually be improved if we can narrow the gap between experimental data and clinical trials through a thorough and comprehensive scholarly approach (84). Tissue morphology, which includes blood circulation, and immune cell interactions may be excluded from traditional cell culture techniques. In the modern life sciences, animal cultured cells technology has become essential, serving as a foundation for research on regulation, proliferation, differentiation, including genetic manipulation. To do it successfully, certain technical skills are needed. The fundamental methods of animal cell culture and their uses are covered in this chapter (140). Technological advancements have a significant impact on our understanding of the variety and function of the gut microbiota. Innovations in molecular technologies, including DNA-based methods like high-throughput sequencing, have enabled the visualization of the diversity, function, and dynamics of gut microbial populations in close proximity to the gut lining. But given the difficulty of the interactions and the variety of microbial species present in the various microbiota, it seems necessary to create simplified models that enable the cultivation of specific model microcosms and the investigation, via these models, of the connections and the emergence of exogenous molecules manufactured by the agri-food or pharmaceutical industries with the human microbiota (141).

It is advantageous for prebiotic as well as probiotic research to combine *in-vitro* and *in-vivo* studies. Every strategy, though, has benefits and drawbacks. Researchers are able to examine host-microbe interactions in extremely controlled situations by using animal models, such as rats, mice, and piglets. Additionally, they offer immediate access to colonic fluids, which makes them a better option than research utilizing excrement for examining the metabolic activity of the gut microbiota. It is still difficult to apply the results from animal models to human situations since animal models are costly, labor-intensive, physiologically misrepresentative of human metabolisms, and different from the human microbiome (142).

13.9 CONCLUSION

Hundreds of different bacterial species make up the gut microbiota, and they have been stable over time. Certain disturbances, such as the administration of medicines or changes in food, can upset this stability and impact the variety and function of the gut microbiota. To analyze and forecast the behavior of the bacterial community in an *in-vitro* as well as *in-vivo* human gut model, one must employ a stable microbiota. Experimental animal models are a useful tool for researching the systemic effects of gastrointestinal inflammation because they faithfully mimic the complexity of the *in-vivo* microenvironment. Additionally, the use of animal models enhances the connection between preclinical research as well as clinical trials by enabling a deeper analysis of potential therapeutic methods. In conclusion, a significant gain in our knowledge of gastrointestinal inflammation has resulted from the application of both in vitro and animal investigations. These models have helped identify specific targets for intervention, create new therapeutic approaches, and deepen our understanding of the intricate mechanisms underlying inflammatory reactions in the gastrointestinal tract. By combining *in-vitro* and *in-vivo* models, researchers have been able to obtain insight into the complex linkages and mechanisms underlying gastrointestinal inflammation. This has made it easier to apply scientific discoveries in domains that are relevant to medicine.

13.10 FUTURE PROSPECTS

This will open the door to the development of tailored pharmaceuticals and therapies based on each patient's distinct characteristics. Additionally, it will be essential to develop new products that more closely resemble the human digestive system and to enhance the current experimental models. It is vital to continuously optimize *in-vitro* systems in order to represent the dynamic interactions between various cell types in the gut environment. Additionally, animal models must be improved in order to better mimic human physiology. To further our comprehension of gastrointestinal inflammation, scientists from a variety of fields, particularly immunology or genomics, as well as computational biology, must continue to collaborate. Individualized medicine and tailored therapeutics for gastrointestinal illnesses can be developed on the basis of the complete approach to investigating the intricacies of inflammation provided by the combined efforts of *in-vitro* or experimental animal models. Upcoming gastrointestinal inflammation research will surely be influenced by the incorporation of cutting-edge techniques and a multidisciplinary strategy, eventually assisting patients with enhanced treatment approaches and diagnostics.

REFERENCES

1. Jonathan, P., Katz, J. P., & Lichtenstein, G. R. (1998). Rheumatologic manifestations of gastrointestinal diseases. *Gastroenterology Clinics of North America*, 27(3), 533–562. https://doi.org/10.1016/s0889-8553(05)70020-0
2. Ahluwalia, B., Magnusson, M. K., & Öhman, L. (2017). Mucosal immune system of the gastrointestinal tract: Maintaining balance between the good and the bad. *Scandinavian Journal of Gastroenterology*, 52(11), 1185–1193. https://doi.org/10.1080/00365521.2017.1349173
3. Iddir, M., Brito, A., Dingeo, G., Fernandez Del Campo, S. S., Samouda, H., La Frano, M. R., & Bohn, T. (2020). Strengthening the immune system and reducing inflammation and oxidative stress through diet and nutrition: Considerations during the COVID-19 crisis. *Nutrients*, 12(6), 1562. https://doi.org/10.3390/nu12061562
4. Bates, C. M., & Plevris, J. N. (2013). Clinical evaluation of abdominal pain in adults. *Medicine*, 41(2), 81–86. https://doi.org/10.1016/j.mpmed.2012.11.005
5. Rogler, G., Singh, A., Kavanaugh, A., & Rubin, D. T. (2021). Extraintestinal manifestations of inflammatory bowel disease: Current concepts, treatment, and implications for disease management. *Gastroenterology*, 161(4), 1118–1132. https://doi.org/10.1053/j.gastro.2021.07.042
6. Sharma, R., Kannourakis, G., Prithviraj, P., & Ahmed, N. (2022). Precision medicine: An optimal approach to patient care in renal cell carcinoma. *Frontiers in Medicine*, 9, 766869. https://doi.org/10.3389/fmed.2022.766869
7. Elgazzar, A. H., & Alenezi, S. A. (2022). Digestive system. In Abdelhamid H. Elgazzar (ed.) *The Pathophysiologic Basis of Nuclear Medicine* (pp. 385–443). Springer. https://doi.org/10.1007/978-3-030-96252-4_10
8. Rubin, D. C., Shaker, A., & Levin, M. S. (2012). Chronic intestinal inflammation: Inflammatory bowel disease and colitis-associated colon cancer. *Frontiers in Immunology*, 3, 107. https://doi.org/10.3389/fimmu.2012.00107
9. Mosli, M., Al Beshir, M., Al-Judaibi, B., Al-Ameel, T., Saleem, A., Bessissow, T., Ghosh, S., & Almadi, M. (2014). Advances in the diagnosis and management of inflammatory bowel disease: Challenges and uncertainties. *Saudi Journal of Gastroenterology: Official Journal of the Saudi Gastroenterology Association*, 20(2), 81. https://doi.org/10.4103/1319-3767.129473
10. Dumic, I., Nordin, T., Jecmenica, M., Stojkovic Lalosevic, M., Milosavljevic, T., & Milovanovic, T. (2019). Gastrointestinal tract disorders in older age. *Canadian Journal of Gastroenterology and Hepatology*, 2019, 1–19. https://doi.org/10.1155/2019/6757524
11. Shamoon, M., Martin, N. M., & O'Brien, C. L. (2019). Recent advances in gut Microbiota mediated therapeutic targets in inflammatory bowel diseases: Emerging modalities for future pharmacological implications. *Pharmacological Research*, 148, 104344. https://doi.org/10.1016/j.phrs.2019.104344
12. Pastorelli, L., De Salvo, C., Mercado, J. R., Vecchi, M., & Pizarro, T. T. (2013). Central role of the gut epithelial barrier in the pathogenesis of chronic intestinal inflammation: Lessons learned from animal models and human genetics. *Frontiers in Immunology*, 4, 280. https://doi.org/10.3389/fimmu.2013.00280
13. DeGruttola, A. K., Low, D., Mizoguchi, A., & Mizoguchi, E. (2016). Current understanding of dysbiosis in disease in human and animal models. *Inflammatory Bowel Diseases*, 22(5), 1137–1150. https://doi.org/10.1097/mib.0000000000000750
14. Sharma, R., Young, C., & Neu, J. (2010). Molecular modulation of intestinal epithelial barrier: Contribution of microbiota. *BioMed Research International*, 2010, 1–15. https://doi.org/10.1155/2010/305879
15. Webb, D. R. (2014). Animal models of human disease: Inflammation. *Biochemical Pharmacology*, 87(1), 121–130. https://doi.org/10.1016/j.bcp.2013.06.014
16. Jiminez, J. A., Uwiera, T. C., Douglas Inglis, G., & Uwiera, R. R. E. (2015). Animal models to study acute and chronic intestinal inflammation in mammals. *Gut Pathogens*, 7, 1–31. https://doi.org/10.1186/s13099-015-0076-y
17. Estes, J. D., Wong, S. W., & Brenchley, J. M. (2018). Nonhuman primate models of human viral infections. *Nature Reviews Immunology*, 18(6), 390–404. https://doi.org/10.1038/s41577-018-0005-7
18. Kamareddine, L., Najjar, H., Sohail, M. U., Abdulkader, H., & Al-Asmakh, M. (2020). The microbiota and gut-related disorders: Insights from animal models. *Cells*, 9(11), 2401. https://doi.org/10.3390/cells9112401
19. Gasaly, N., De Vos, P., & Hermoso, M. A. (2021). Impact of bacterial metabolites on gut barrier function and host immunity: A focus on bacterial metabolism and its relevance for intestinal inflammation. *Frontiers in Immunology*, 12, 658354. https://doi.org/10.3389/fimmu.2021.658354
20. Mizoguchi, E., Low, D., Ezaki, Y., & Okada, T. (2020). Recent updates on the basic mechanisms and pathogenesis of inflammatory bowel diseases in experimental animal models. *Intestinal Research*, 18(2), 151–167. https://doi.org/10.5217/ir.2019.09154
21. Kosaka, S., Tamauchi, H., Terashima, M., Maruyama, H., Habu, S., & Kitasato, H. (2011). IL-10 controls Th2-type cytokine production and eosinophil infiltration in a mouse model of allergic airway inflammation. *Immunobiology*, 216(7), 811–820. https://doi.org/10.1016/j.imbio.2010.12.003
22. Randhawa, P. K., Singh, K., Singh, N., & Jaggi, A. S. (2014). A review on chemical-induced inflammatory bowel disease models in rodents. *The Korean Journal of Physiology & Pharmacology: Official Journal of the Korean Physiological Society and the Korean Society of Pharmacology*, 18(4), 279. https://doi.org/10.4196/kjpp.2014.18.4.279
23. Kim, H.-S., & Berstad, A. (1992). Experimental colitis in animal models. *Scandinavian Journal of Gastroenterology*, 27(7), 529–537. https://doi.org/10.3109/00365529209000116

24. Al-Hussaini, J. A.-A. S., Al-Hussaini, A. D. L., & Abd, A. H. (2012). A comparative study of the effect of Asacol and the ethanolic extracts of Matricaria chamomilla and Terminalia chebula on induced ulcerative colitis in rabbits. *Kufa Journal For Veterinary Medical Sciences*, 3(1), 36–54. https://doi.org/10.36326/kjvs/2012/v3i14081
25. Nielsen, T. S., Fredborg, M., Theil, P. K., Yue, Y., Bruhn, L. V, Andersen, V., & Purup, S. (2020). Dietary red meat adversely affects disease severity in a pig model of DSS-induced colitis despite reduction in colonic pro-inflammatory gene expression. *Nutrients*, 12(6), 1728. https://doi.org/10.3390/nu12061728
26. Chamanza, R. (2011). Non-human primates: Cynomolgus (Macaca fascicularis) and rhesus (Macaca mulatta) macaques and the common marmoset (Callithrix jacchus). *Background Lesions in Laboratory Animals E-Book: A Color Atlas*, 1, 1–15. https://doi.org/10.1016/b978-0-7020-3519-7.00001-2
27. Yang, L., Ho, N. Y., Alshut, R., Legradi, J., Weiss, C., Reischl, M., Mikut, R., Liebel, U., Müller, F., & Strähle, U. (2009). Zebrafish embryos as models for embryotoxic and teratological effects of chemicals. *Reproductive Toxicology*, 28(2), 245–253. https://doi.org/10.1016/j.reprotox.2009.04.013
28. Yi, Y., Xu, W., Fan, Y., & Wang, H.-X. (2021). Drosophila as an emerging model organism for studies of food-derived antioxidants. *Food Research International*, 143, 110307. https://doi.org/10.1016/j.foodres.2021.110307
29. Wang, R. X., & Colgan, S. P. (2017). Special pro-resolving mediator (SPM) actions in regulating gastro-intestinal inflammation and gut mucosal immune responses. *Molecular Aspects of Medicine*, 58, 93–101. https://doi.org/10.1016/j.mam.2017.02.002
30. Yadav, V., Varum, F., Bravo, R., Furrer, E., Bojic, D., & Basit, A. W. (2016). Inflammatory bowel disease: Exploring gut pathophysiology for novel therapeutic targets. *Translational Research*, 176, 38–68. https://doi.org/10.1016/j.trsl.2016.04.009
31. Eichele, D. D., & Kharbanda, K. K. (2017). Dextran sodium sulfate colitis murine model: An indispensable tool for advancing our understanding of inflammatory bowel diseases pathogenesis. *World Journal of Gastroenterology*, 23(33), 6016. https://doi.org/10.3748/wjg.v23.i33.6016
32. Zhou, G. X., & Liu, Z. J. (2017). Potential roles of neutrophils in regulating intestinal mucosal inflammation of inflammatory bowel disease. *Journal of Digestive Diseases*, 18(9), 495–503. https://doi.org/10.1111/1751-2980.12540
33. Antoniou, E., Margonis, G. A., Angelou, A., Pikouli, A., Argiri, P., Karavokyros, I., Papalois, A., & Pikoulis, E. (2016). The TNBS-induced colitis animal model: An overview. *Annals of Medicine and Surgery*, 11, 9–15. https://doi.org/10.1016/j.amsu.2016.07.019
34. Saha, B., Prasanna, S. J., Chandrasekar, B., & Nandi, D. (2010). Gene modulation and immunoregulatory roles of Interferonγ. *Cytokine*, 50(1), 1–14. https://doi.org/10.1016/j.cyto.2009.11.021
35. Vahabi, S., Sattari, M., Taheraslani, M., & Bagheban, A. A. (2011). Correlation between interleukin-1β, interleukin-6 and tumor necrosis factor-α and clinical parameters in chronic and aggressive periodontal disease. *Journal of Advanced Periodontology & Implant Dentistry*, 3(2), 51–56. https://doi.org/10.1111/j.1600-0765.2012.01500.x
36. Mortensen, J. H., Lindholm, M., Langholm, L. L., Kjeldsen, J., Bay-Jensen, A. C., Karsdal, M. A., & Manon-Jensen, T. (2019). The intestinal tissue homeostasis–The role of extracellular matrix remodeling in inflammatory bowel disease. *Expert Review of Gastroenterology & Hepatology*, 13(10), 977–993. https://doi.org/10.1080/17474124.2019.1673729
37. Rieder, F., Kessler, S., Sans, M., & Fiocchi, C. (2012). Animal models of intestinal fibrosis: New tools for the understanding of pathogenesis and therapy of human disease. *American Journal of Physiology-Gastrointestinal and Liver Physiology*, 303(7), G786–G801. https://doi.org/10.1152/ajpgi.00059.2012
38. Messer, J. S., & Chang, E. B. (2018). Microbial physiology of the digestive tract and its role in inflammatory bowel diseases. In *Physiology of the Gastrointestinal Tract* (pp. 795–810). Elsevier. https://doi.org/10.1016/b978-0-12-809954-4.00036-0
39. Zhang, E., Fang, M., Jones, C., Minze, L. J., Xing, J., & Zhang, Z. (2022). Mechanisms involved in controlling RNA virus-induced intestinal inflammation. *Cellular and Molecular Life Sciences*, 79(6), 313. https://doi.org/10.1007/s00018-022-04332-z
40. Zha, L., Garrett, S., & Sun, J. (2019). Salmonella infection in chronic inflammation and gastrointestinal cancer. *Diseases*, 7(1), 28. https://doi.org/10.3390/diseases7010028
41. Yousefi, Y., Haq, S., Banskota, S., Kwon, Y. H., & Khan, W. I. (2021). Trichuris muris model: Role in understanding intestinal immune response, inflammation and host defense. *Pathogens*, 10(8), 925. https://doi.org/10.3390/pathogens10080925
42. Goldring, M. B. (1999). The role of cytokines as inflammatory mediators in osteoarthritis: Lessons from animal models. *Connective Tissue Research*, 40(1), 1–11. https://doi.org/10.3109/03008209909005273
43. Yan, X., Pan, Q., Xin, H., Chen, Y., & Ping, Y. (2021). Genome-editing prodrug: Targeted delivery and conditional stabilization of CRISPR-Cas9 for precision therapy of inflammatory disease. *Science Advances*, 7(50), eabj0624. https://doi.org/10.1126/sciadv.abj0624
44. Ewart, D. T., Peterson, E. J., & Steer, C. J. (2018). Gene editing for inflammatory disorders. *Annals of the Rheumatic Diseases*, 78(1), 6-15. https://doi.org/10.1136/annrheumdis-2018-213454
45. MacFarlane, A. J., & Stover, P. J. (2007). Convergence of genetic, nutritional and inflammatory factors in gastrointestinal cancers. *Nutrition Reviews*, 65(suppl_3), S157–S166. https://doi.org/10.1111/j.1753-4887.2007.tb00355.x
46. Vaarala, O., Atkinson, M. A., & Neu, J. (2008). The "perfect storm" for type 1 diabetes: The complex interplay between intestinal microbiota, gut permeability, and mucosal immunity. *Diabetes*, 57(10), 2555–2562. https://doi.org/10.2337/db08-0331
47. Jenner, R. G., & Young, R. A. (2005). Insights into host responses against pathogens from transcriptional profiling. *Nature Reviews Microbiology*, 3(4), 281–294. https://doi.org/10.1038/nrmicro1126
48. Amoroso, C., Perillo, F., Strati, F., Fantini, M., Caprioli, F., & Facciotti, F. (2020). The role of gut microbiota

biomodulators on mucosal immunity and intestinal inflammation. *Cells*, 9(5), 1234. https://doi.org/10.3390/cells9051234
49. Schirbel, A., & Fiocchi, C. (2010). Inflammatory bowel disease: Established and evolving considerations on its etiopathogenesis and therapy. *Journal of Digestive Diseases*, 11(5), 266–276. https://doi.org/10.1111/j.1751-2980.2010.00449.x
50. Tabas, I., & Glass, C. K. (2013). Anti-inflammatory therapy in chronic disease: Challenges and opportunities. *Science*, 339(6116), 166–172. https://doi.org/10.1126/science.1230720
51. Wallace, K. L., Zheng, L.-B., Kanazawa, Y., & Shih, D. Q. (2014). Immunopathology of inflammatory bowel disease. *World Journal of Gastroenterology: WJG*, 20(1), 6. https://doi.org/10.3748/wjg.v20.i1.6
52. Matricon, J., Barnich, N., & Ardid, D. (2010). Immunopathogenesis of inflammatory bowel disease. *Self/Nonself*, 1(4), 299–309. https://doi.org/10.4161/self.1.4.13560
53. Aguilera, M., Darby, T., & Melgar, S. (2014). The complex role of inflammasomes in the pathogenesis of inflammatory bowel diseases–lessons learned from experimental models. *Cytokine & Growth Factor Reviews*, 25(6), 715–730. https://doi.org/10.1016/j.cytogfr.2014.04.003
54. Iismaa, S. E., Mearns, B. M., Lorand, L., & Graham, R. M. (2009). Transglutaminases and disease: Lessons from genetically engineered mouse models and inherited disorders. *Physiological Reviews*, 89(3), 991–1023. https://doi.org/10.1152/physrev.00044.2008
55. Gerondakis, S., Grumont, R., Gugasyan, R., Wong, L., Isomura, I., Ho, W., & Banerjee, A. (2006). Unravelling the complexities of the NF-κB signalling pathway using mouse knockout and transgenic models. *Oncogene*, 25(51), 6781–6799. https://doi.org/10.1038/sj.onc.1209944
56. Limanskiy, V., Vyas, A., Chaturvedi, L. S., & Vyas, D. (2019). Harnessing the potential of gene editing technology using CRISPR in inflammatory bowel disease. *World Journal of Gastroenterology*, 25(18), 2177. https://doi.org/10.3748/wjg.v25.i18.2177
57. Singer, M. (2010). Pathogen-pathogen interaction: A syndemic model of complex biosocial processes in disease. *Virulence*, 1(1), 10–18. https://doi.org/10.4161/viru.1.1.9933
58. Fung, T. C., Olson, C. A., & Hsiao, E. Y. (2017). Interactions between the microbiota, immune and nervous systems in health and disease. *Nature Neuroscience*, 20(2), 145–155. https://doi.org/10.1038/nn.4476
59. Tomlin, H., & Piccinini, A. M. (2018). A complex interplay between the extracellular matrix and the innate immune response to microbial pathogens. *Immunology*, 155(2), 186–201. https://doi.org/10.1111/imm.12972
60. Lee, Y. T., Tan, Y. J., & Oon, C. E. (2018). Molecular targeted therapy: Treating cancer with specificity. *European Journal of Pharmacology*, 834, 188–196. https://doi.org/10.1016/j.ejphar.2018.07.034
61. Kumari, P. V, & Rao, Y. S. (2019). Personalized medicine-A review. *Research Journal of Pharmacy and Technology*, 12(8), 3989–3992. https://doi.org/10.5958/0974-360x.2019.00687.5
62. Günther, C., Winner, B., Neurath, M. F., & Stappenbeck, T. S. (2022). Organoids in gastrointestinal diseases: From experimental models to clinical translation. *Gut*, 71(9), 1892–1908. https://doi.org/10.1136/gutjnl-2021-326560
63. Shah, P., Fritz, J. V, Glaab, E., Desai, M. S., Greenhalgh, K., Frachet, A., Niegowska, M., Estes, M., Jäger, C., & Seguin-Devaux, C. (2016). A microfluidics-based in vitro model of the gastrointestinal human–microbe interface. *Nature Communications*, 7(1), 11535. https://doi.org/10.1038/ncomms11535
64. Di Vincenzo, F., Del Gaudio, A., Petito, V., Lopetuso, L. R., & Scaldaferri, F. (2023). Gut microbiota, intestinal permeability, and systemic inflammation: A narrative review. *Internal and Emergency Medicine*, 19(2), 275–293. https://doi.org/10.1007/s11739-023-03374-w
65. Carter, S. R., Zahs, A., Palmer, J. L., Wang, L., Ramirez, L., Gamelli, R. L., & Kovacs, E. J. (2013). Intestinal barrier disruption as a cause of mortality in combined radiation and burn injury. *Shock (Augusta, Ga.)*, 40(4), 281. https://doi.org/10.1097/shk.0b013e3182a2c5b5
66. Kim, K. S., Jung, H., Shin, I. K., Choi, B., & Kim, D. H. (2015). Induction of interleukin-1 beta (IL-1β) is a critical component of lung inflammation during influenza A (H1N1) virus infection. *Journal of Medical Virology*, 87(7), 1104–1112. https://doi.org/10.1002/jmv.24138
67. Valkenburg, K. C., De Groot, A. E., & Pienta, K. J. (2018). Targeting the tumour stroma to improve cancer therapy. *Nature Reviews Clinical Oncology*, 15(6), 366–381. https://doi.org/10.1038/s41571-018-0007-1
68. Kapoor, S., & Padwad, Y. S. (2023). Phloretin suppresses intestinal inflammation and maintained epithelial tight junction integrity by modulating cytokines secretion in in vitro model of gut inflammation. *Cellular Immunology*, 391, 104754. https://doi.org/10.1016/j.cellimm.2023.104754
69. Papoutsopoulou, S., & Campbell, B. J. (2021). Epigenetic modifications of the nuclear factor kappa B signalling pathway and its impact on inflammatory bowel disease. *Current Pharmaceutical Design*, 27(35), 3702–3713. https://doi.org/10.2174/1381612827666210218141847
70. Yan, H., Shao, D., Lao, Y., Li, M., Hu, H., & Leong, K. W. (2019). Engineering cell membrane-based nanotherapeutics to target inflammation. *Advanced Science*, 6(15), 1900605. https://doi.org/10.1002/advs.201970088
71. Schulte, L., Hohwieler, M., Mueller, M., & Klaus, J. (2019). Intestinal organoids as a novel complementary model to dissect inflammatory bowel disease. *Stem Cells International*, 2019, 1–15. https://doi.org/10.1155/2019/8010645
72. Kaminska, B. (2005). MAPK signalling pathways as molecular targets for anti-inflammatory therapy—from molecular mechanisms to therapeutic benefits. *Biochimica et Biophysica Acta (BBA)-Proteins and Proteomics*, 1754(1–2), 253–262. https://doi.org/10.1155/2019/8010645
73. Schoultz, I., & Keita, Å. V. (2019). Cellular and molecular therapeutic targets in inflammatory bowel disease—focusing on intestinal barrier function. *Cells*, 8(2), 193. https://doi.org/10.3390/cells8020193
74. Bein, A., Shin, W., Jalili-Firoozinezhad, S., Park, M. H., Sontheimer-Phelps, A., Tovaglieri, A., Chalkiadaki, A., Kim, H. J., & Ingber, D. E. (2018). Microfluidic organ-on-a-chip models of human intestine. *Cellular and Molecular Gastroenterology and Hepatology*, 5(4), 659–668. https://doi.org/10.1016/j.jcmgh.2017.12.010

75. Ambrosini, Y. M., Shin, W., Min, S., & Kim, H. J. (2020). Microphysiological engineering of immune responses in intestinal inflammation. *Immune Network*, 20(2), 1–15. https://doi.org/10.4110/in.2020.20.e13
76. Liu, C., Chu, D., Kalantar-Zadeh, K., George, J., Young, H. A., & Liu, G. (2021). Cytokines: From clinical significance to quantification. *Advanced Science*, 8(15), 2004433. https://doi.org/10.1002/advs.202004433
77. Marrero, D., Pujol-Vila, F., Vera, D., Gabriel, G., Illa, X., Elizalde-Torrent, A., Alvarez, M., & Villa, R. (2021). Gut-on-a-chip: Mimicking and monitoring the human intestine. *Biosensors and Bioelectronics*, 181, 113156. https://doi.org/10.1016/j.bios.2021.113156
78. Thomas, D. P., Zhang, J., Nguyen, N.-T., & Ta, H. T. (2023). Microfluidic gut-on-a-chip: Fundamentals and challenges. *Biosensors*, 13(1), 136. https://doi.org/10.3390/bios13010136
79. Marzorati, M., Van den Abbeele, P., Possemiers, S., Benner, J., Verstraete, W., & Van de Wiele, T. (2011). Studying the host-microbiota interaction in the human gastrointestinal tract: Basic concepts and in vitro approaches. *Annals of Microbiology*, 61, 709–715. https://doi.org/10.1007/s13213-011-0242-5
80. Spiljar, M., Merkler, D., & Trajkovski, M. (2017). The immune system bridges the gut microbiota with systemic energy homeostasis: Focus on TLRs, mucosal barrier, and SCFAs. *Frontiers in Immunology*, 8, 1353. https://doi.org/10.3389/fimmu.2017.01353
81. Yu, L. C.-H. (2018). Microbiota dysbiosis and barrier dysfunction in inflammatory bowel disease and colorectal cancers: Exploring a common ground hypothesis. *Journal of Biomedical Science*, 25(1), 1–14. https://doi.org/10.1186/s12929-018-0483-8
82. Basson, A., Trotter, A., Rodriguez-Palacios, A., & Cominelli, F. (2016). Mucosal interactions between genetics, diet, and microbiome in inflammatory bowel disease. *Frontiers in Immunology*, 7, 290. https://doi.org/10.3389/fimmu.2016.00290
83. Noel, G., Baetz, N. W., Staab, J. F., Donowitz, M., Kovbasnjuk, O., Pasetti, M. F., & Zachos, N. C. (2017). A primary human macrophage-enteroid co-culture model to investigate mucosal gut physiology and host-pathogen interactions. *Scientific Reports*, 7(1), 45270. https://doi.org/10.1038/srep46790
84. Ponce de León-Rodríguez, M. del C., Guyot, J.-P., & Laurent-Babot, C. (2019). Intestinal in vitro cell culture models and their potential to study the effect of food components on intestinal inflammation. *Critical Reviews in Food Science and Nutrition*, 59(22), 3648–3666. https://doi.org/10.1080/10408398.2018.1506734
85. Beutler, B. (2003). Innate immune responses to microbial poisons: Discovery and function of the Toll-like receptors. *Annual Review of Pharmacology and Toxicology*, 43(1), 609–628. https://doi.org/10.1146/annurev.pharmtox.43.100901.135729
86. Geboes, K., Dewit, O., Moreels, T. G., Faa, G., & Jouret-Mourin, A. (2018). Inflammatory bowel diseases. In Anne Jouret-Mourin, Gavino Faa, Karel Geboes (eds.) *Colitis: A Practical Approach to Colon and Ileum Biopsy Interpretation*, 2nd edition, 107–140. Springer International Publishing AG. https://doi.org/10.1007/978-3-319-89503-1_8
87. Pedersen, J., Coskun, M., Soendergaard, C., Salem, M., & Nielsen, O. H. (2014). Inflammatory pathways of importance for management of inflammatory bowel disease. *World Journal of Gastroenterology: WJG*, 20(1), 64. https://doi.org/10.3748/wjg.v20.i1.64
88. Zhang, H., Shi, N., Diao, Z., Chen, Y., & Zhang, Y. (2021). Therapeutic potential of TNFα inhibitors in chronic inflammatory disorders: Past and future. *Genes & Diseases*, 8(1), 38–47. https://doi.org/10.1016/j.gendis.2020.02.004
89. Malemud, C. J. (2018). The role of the JAK/STAT signal pathway in rheumatoid arthritis. *Therapeutic Advances in Musculoskeletal Disease*, 10(5–6), 117–127. https://doi.org/10.1177/1759720x18776224
90. Olivera, P., Danese, S., & Peyrin-Biroulet, L. (2017). Next generation of small molecules in inflammatory bowel disease. *Gut*, 66(2), 199–209. https://doi.org/10.1136/gutjnl-2016-312912
91. Danese, S., Grisham, M., Hodge, J., & Telliez, J.-B. (2016). JAK inhibition using tofacitinib for inflammatory bowel disease treatment: A hub for multiple inflammatory cytokines. *American Journal of Physiology-Gastrointestinal and Liver Physiology*, 310(3), G155–G162. https://doi.org/10.1152/ajpgi.00311.2015
92. Shivaji, U. N., Nardone, O. M., Cannatelli, R., Smith, S. C., Ghosh, S., & Iacucci, M. (2020). Small molecule oral targeted therapies in ulcerative colitis. *The Lancet Gastroenterology & Hepatology*, 5(9), 850–861. https://doi.org/10.1016/s2468-1253(19)30414-5
93. Chang, C., & Lin, H. (2016). Dysbiosis in gastrointestinal disorders. *Best Practice & Research Clinical Gastroenterology*, 30(1), 3–15. https://doi.org/10.1016/j.bpg.2016.02.001
94. Biazzo, M., & Deidda, G. (2022). Fecal microbiota transplantation as new therapeutic avenue for human diseases. *Journal of Clinical Medicine*, 11(14), 4119. https://doi.org/10.3390/jcm11144119
95. Zimmermann, M., Patil, K. R., Typas, A., & Maier, L. (2021). Towards a mechanistic understanding of reciprocal drug–microbiome interactions. *Molecular Systems Biology*, 17(3), e10116. https://doi.org/10.15252/msb.202010116
96. Singh, S., Sarma, D. K., Verma, V., Nagpal, R., & Kumar, M. (2023). Unveiling the future of metabolic medicine: Omics technologies driving personalized solutions for precision treatment of metabolic disorders. *Biochemical and Biophysical Research Communications,* 682, 1–20. https://doi.org/10.1016/j.bbrc.2023.09.064
97. Torres, J., Caprioli, F., Katsanos, K. H., Lobatón, T., Micic, D., Zerôncio, M., Van Assche, G., Lee, J. C., Lindsay, J. O., & Rubin, D. T. (2016). Predicting outcomes to optimize disease management in inflammatory bowel diseases. *Journal of Crohn's and Colitis*, 10(12), 1385–1394. https://doi.org/10.1093/ecco-jcc/jjw116
98. Al Mijan, M., & Lim, B. O. (2018). Diets, functional foods, and nutraceuticals as alternative therapies for inflammatory bowel disease: Present status and future trends. *World Journal of Gastroenterology*, 24(25), 2673. https://doi.org/10.3748/wjg.v24.i25.2673
99. Yang, N., Sampathkumar, K., & Loo, S. C. J. (2017). Recent advances in complementary and replacement therapy with nutraceuticals in combating gastrointestinal

illnesses. *Clinical Nutrition*, 36(4), 968–979. https://doi.org/10.1016/j.clnu.2016.08.020

100. Chu, J. N., & Traverso, G. (2022). Foundations of gastrointestinal-based drug delivery and future developments. *Nature Reviews Gastroenterology & Hepatology*, 19(4), 219–238. https://doi.org/10.1038/s41575-021-00539-w
101. Mittal, R., Patel, A. P., Jhaveri, V. M., Kay, S.-I. S., Debs, L. H., Parrish, J. M., Pan, D. R., Nguyen, D., Mittal, J., & Jayant, R. D. (2018). Recent advancements in nanoparticle based drug delivery for gastrointestinal disorders. *Expert Opinion on Drug Delivery*, 15(3), 301–318. https://doi.org/10.1080/17425247.2018.1420055
102. Placha, D., & Jampilek, J. (2021). Chronic inflammatory diseases, anti-inflammatory agents and their delivery nanosystems. *Pharmaceutics*, 13(1), 64. https://doi.org/10.3390/pharmaceutics13010064
103. Oka, A., & Sartor, R. B. (2020). Microbial-based and microbial-targeted therapies for inflammatory bowel diseases. *Digestive Diseases and Sciences*, 65(3), 757–788. https://doi.org/10.1007/s10620-020-06090-z
104. Haskey, N., & Gibson, D. L. (2017). An examination of diet for the maintenance of remission in inflammatory bowel disease. *Nutrients*, 9(3), 259. https://doi.org/10.3390/nu9030259
105. Ruemmele, F. M., Hyams, J. S., Otley, A., Griffiths, A., Kolho, K.-L., Dias, J. A., Levine, A., Escher, J. C., Taminiau, J., & Veres, G. (2015). Outcome measures for clinical trials in paediatric IBD: An evidence-based, expert-driven practical statement paper of the paediatric ECCO committee. *Gut*, 64(3), 438–446. https://doi.org/10.1136/gutjnl-2014-307008
106. Verstockt, B., Van Assche, G., Vermeire, S., & Ferrante, M. (2017). Biological therapy targeting the IL-23/IL-17 axis in inflammatory bowel disease. *Expert Opinion on Biological Therapy*, 17(1), 31–47. https://doi.org/10.1080/14712598.2017.1258399
107. Pérez-Jeldres, T., Tyler, C. J., Boyer, J. D., Karuppuchamy, T., Yarur, A., Giles, D. A., Yeasmin, S., Lundborg, L., Sandborn, W. J., & Patel, D. R. (2019). Targeting cytokine signaling and lymphocyte traffic via small molecules in inflammatory bowel disease: JAK inhibitors and S1PR agonists. *Frontiers in Pharmacology*, 10, 212. https://doi.org/10.3389/fphar.2019.00212
108. Quigley, E. M. M., & Gajula, P. (2020). Recent advances in modulating the microbiome. *F1000Research*, 9, 1–12. https://doi.org/10.12688/f1000research.20204.1
109. Ortega, V. E., & Meyers, D. A. (2014). Pharmacogenetics: Implications of race and ethnicity on defining genetic profiles for personalized medicine. *Journal of Allergy and Clinical Immunology*, 133(1), 16–26. https://doi.org/10.1016/j.jaci.2013.10.040
110. Damas, O. M., Garces, L., & Abreu, M. T. (2019). Diet as adjunctive treatment for inflammatory bowel disease: Review and update of the latest literature. *Current Treatment Options in Gastroenterology*, 17, 313–325. https://doi.org/10.1007/s11938-019-00231-8
111. Vashist, A., Vashist, A., Gupta, Y. K., & Ahmad, S. (2014). Recent advances in hydrogel based drug delivery systems for the human body. *Journal of Materials Chemistry B*, 2(2), 147–166. https://doi.org/10.1039/c3tb21016b
112. Hu, C., & Dignam, J. J. (2019). Biomarker-driven oncology clinical trials: Key design elements, types, features, and practical considerations. *JCO Precision Oncology*, 1, 1–12. https://doi.org/10.1200/po.19.00086
113. Ahmed, Z., Mohamed, K., Zeeshan, S., & Dong, X. (2020). Artificial intelligence with multi-functional machine learning platform development for better healthcare and precision medicine. *Database*, 2020, baaa010. https://doi.org/10.1093/database/baaa010
114. Simoens, S. (2021). How do biosimilars sustain value, affordability, and access to oncology care? *Expert Review of Pharmacoeconomics & Outcomes Research*, 21(3), 327–329. https://doi.org/10.1080/14737167.2020.1813570
115. Levit, L., Balogh, E., Nass, S., & Ganz, P. A. (2013). Patient-centered communication and shared decision making. In Laura A. Levit, Erin P. Balogh, Sharyl J. Nass, and Patricia A. Ganz (eds.) *Delivering High-Quality Cancer Care: Charting a New Course for a System in Crisis*. National Academies Press (US). https://doi.org/10.17226/18359
116. Vodovotz, Y., Constantine, G., Faeder, J., Mi, Q., Rubin, J., Bartels, J., Sarkar, J., Squires Jr, R. H., Okonkwo, D. O., & Gerlach, J. (2010). Translational systems approaches to the biology of inflammation and healing. *Immunopharmacology and Immunotoxicology*, 32(2), 181–195. https://doi.org/10.3109/08923970903369867
117. Keely, S., Walker, M. M., Marks, E., & Talley, N. J. (2015). Immune dysregulation in the functional gastrointestinal disorders. *European Journal of Clinical Investigation*, 45(12), 1350–1359. https://doi.org/10.1111/eci.12548
118. Mehrzadi, S., Sheibani, M., Koosha, F., Alinaghian, N., Pourhanifeh, M. H., Tabaeian, S. A. P., Reiter, R. J., & Hosseinzadeh, A. (2023). Protective and therapeutic potential of melatonin against intestinal diseases: Updated review of current data based on molecular mechanisms. *Expert Review of Gastroenterology & Hepatology*, 17(10), 1011–1029. https://doi.org/10.1080/17474124.2023.2267439
119. Wilkins, B. J., Kelsen, J. R., & Conrad, M. A. (2022). A pattern-based pathology approach to very early-onset inflammatory bowel disease: Thinking beyond crohn disease and ulcerative colitis. *Advances in Anatomic Pathology*, 29(1), 62. https://doi.org/10.1097/pap.0000000000000327
120. Fakhoury, M., Negrulj, R., Mooranian, A., & Al-Salami, H. (2014). Inflammatory bowel disease: Clinical aspects and treatments. *Journal of Inflammation Research*, 7, 113–120. https://doi.org/10.2147/jir.s65979
121. Schreyer, A. G., Seitz, J., Feuerbach, S., Rogler, G., & Herfarth, H. (2004). Modern imaging using computer tomography and magnetic resonance imaging for inflammatory bowel disease (IBD) AU1. *Inflammatory Bowel Diseases*, 10(1), 45–54. https://doi.org/10.1097/00054725-200401000-00008
122. Makarev, E., Cantor, C., Zhavoronkov, A., Buzdin, A., Aliper, A., & Csoka, A. B. (2014). Pathway activation profiling reveals new insights into age-related macular degeneration and provides avenues for therapeutic interventions. *Aging (Albany NY)*, 6(12), 1064. https://doi.org/10.18632/aging.100711

123. Ohland, C. L., & Jobin, C. (2015). Microbial activities and intestinal homeostasis: A delicate balance between health and disease. *Cellular and Molecular Gastroenterology and Hepatology*, 1(1), 28–40. https://doi.org/10.1016/j.jcmgh.2014.11.004
124. Singh, R., Zogg, H., Wei, L., Bartlett, A., Ghoshal, U. C., Rajender, S., & Ro, S. (2021). Gut microbial dysbiosis in the pathogenesis of gastrointestinal dysmotility and metabolic disorders. *Journal of Neurogastroenterology and Motility*, 27(1), 19. https://doi.org/10.5056/jnm20149
125. Sanders, M. E., Merenstein, D. J., Reid, G., Gibson, G. R., & Rastall, R. A. (2019). Probiotics and prebiotics in intestinal health and disease: From biology to the clinic. *Nature Reviews Gastroenterology & Hepatology*, 16(10), 605–616. https://doi.org/10.1038/s41575-019-0173-3
126. Bakshi, N., Hart, A. L., Lee, M. C., Williams, A. C. de C., Lackner, J. M., Norton, C., & Croft, P. (2021). Chronic pain in patients with inflammatory bowel disease. *Pain*, 162(10), 2466. https://doi.org/10.1097/j.pain.0000000000002304
127. Stafford, I. S., Gosink, M. M., Mossotto, E., Ennis, S., & Hauben, M. (2022). A systematic review of artificial intelligence and machine learning applications to inflammatory bowel disease, with practical guidelines for interpretation. *Inflammatory Bowel Diseases*, 28(10), 1573–1583. https://doi.org/10.1093/ibd/izac115
128. Weber, B., Lackner, I., Haffner-Luntzer, M., Palmer, A., Pressmar, J., Scharffetter-Kochanek, K., Knöll, B., Schrezenemeier, H., Relja, B., & Kalbitz, M. (2019). Modeling trauma in rats: Similarities to humans and potential pitfalls to consider. *Journal of Translational Medicine*, 17, 1–19. https://doi.org/10.1186/s12967-019-2052-7
129. Loddo, I., & Romano, C. (2015). Inflammatory bowel disease: Genetics, epigenetics, and pathogenesis. *Frontiers in Immunology*, 6, 551. https://doi.org/10.3389/fimmu.2015.00551
130. Gkouskou, K. K., Deligianni, C., Tsatsanis, C., & Eliopoulos, A. G. (2014). The gut microbiota in mouse models of inflammatory bowel disease. *Frontiers in Cellular and Infection Microbiology*, 4, 28. https://doi.org/10.3389/fcimb.2014.00028
131. Peloquin, J. M., & Nguyen, D. D. (2013). The microbiota and inflammatory bowel disease: Insights from animal models. *Anaerobe*, 24, 102–106. https://doi.org/10.1016/j.anaerobe.2013.04.006
132. Lees, C. W., & Satsangi, J. (2009). Genetics of inflammatory bowel disease: Implications for disease pathogenesis and natural history. *Expert Review of Gastroenterology & Hepatology*, 3(5), 513–534. https://doi.org/10.1586/egh.09.45
133. Ramos-Lopez, O., Milagro, F. I., Riezu-Boj, J. I., & Martinez, J. A. (2021). Epigenetic signatures underlying inflammation: An interplay of nutrition, physical activity, metabolic diseases, and environmental factors for personalized nutrition. *Inflammation Research*, 70, 29–49. https://doi.org/10.1007/s00011-020-01425-y
134. Lamb, C. A., Saifuddin, A., Powell, N., & Rieder, F. (2022). The future of precision medicine to predict outcomes and control tissue remodeling in inflammatory bowel disease. *Gastroenterology*, 162(5), 1525–1542. https://doi.org/10.1053/j.gastro.2021.09.077
135. Maturo, M. G., Soligo, M., Gibson, G., Manni, L., & Nardini, C. (2020). The greater inflammatory pathway—High clinical potential by innovative predictive, preventive, and personalized medical approach. *EPMA Journal*, 11, 1–16. https://doi.org/10.1007/s13167-019-00195-w
136. Denayer, T., Stöhr, T., & Van Roy, M. (2014). Animal models in translational medicine: Validation and prediction. *New Horizons in Translational Medicine*, 2(1), 5–11. https://doi.org/10.1016/j.nhtm.2014.08.001
137. Pacheco, D. P., Vargas, N. S., Visentin, S., & Petrini, P. (2021). From tissue engineering to engineering tissues: The role and application of in vitro models. *Biomaterials Science*, 9(1), 70–83. https://doi.org/10.1039/d0bm01097a
138. Urzì, O., Gasparro, R., Costanzo, E., De Luca, A., Giavaresi, G., Fontana, S., & Alessandro, R. (2023). Three-dimensional cell cultures: The bridge between in vitro and in vivo models. *International Journal of Molecular Sciences*, 24(15), 12046. https://doi.org/10.3390/ijms241512046
139. Nugent, A. C., Ballard, E. D., Park, L. T., & Zarate, C. A. (2019). Research on the pathophysiology, treatment, and prevention of suicide: Practical and ethical issues. *BMC Psychiatry*, 19(1), 1–12. https://doi.org/10.1186/s12888-019-2301-6
140. Bailoo, J. D., Reichlin, T. S., & Würbel, H. (2014). Refinement of experimental design and conduct in laboratory animal research. *ILAR Journal*, 55(3), 383–391. https://doi.org/10.1093/ilar/ilu037
141. Holmes, E., Wijeyesekera, A., Taylor-Robinson, S. D., & Nicholson, J. K. (2015). The promise of metabolic phenotyping in gastroenterology and hepatology. *Nature Reviews Gastroenterology & Hepatology*, 12(8), 458–471. https://doi.org/10.1038/nrgastro.2015.114
142. Hernandez, E., Fawcett, A., Brouwer, E., Rau, J., & Turner, P. V. (2018). Speaking up: Veterinary ethical responsibilities and animal welfare issues in everyday practice. *Animals*, 8(1), 15. https://doi.org/10.3390/ani8010015

14 The Neuroinflammation Enigma

Cracking the Code of the Gut-Brain Connection

Abhimanyu Thakur[1,2]*, *Dipanjan Ghosh*[3], *Rajesh Sarkar*[4], *Sera Averbek*[5], *Ashok Iyaswamy*[6,7], *Prabhat Upadhyay*[8], *Sarika Gupta*[9], *Vikas Kumar*[10], and *Amrendra K. Ajay*[11]

[1]Department of Pharmacology, Delhi Pharmaceutical Science and Research University, New Delhi, India

[2]Department of Neurosurgery, Massachusetts General Hospital, Harvard Medical School, Boston, MA, USA

[3]Department of Biotechnology and Dr. B. C. Guha Centre for Genetic Engineering and Biotechnology, University of Calcutta, Kolkata, West Bengal, India

[4]Department of Medicine, Division of Biological Sciences, University of Chicago, Chicago, Illinois, USA

[5]Ludwig Center for Metastasis Research, Department of Molecular Genetics and Cell Biology, University of Chicago, Illinois, USA

[6]School of Chinese Medicine, Hong Kong Baptist University, Kowloon Tong, Hong Kong SAR

[7]Department of Biochemistry, Karpagam Academy of Higher Education, Coimbatore, India

[8]Wellman Center for Photomedicine, Massachusetts General Hospital, Harvard Medical School, Boston, USA

[9]Molecular Science Laboratory, National Institute of Immunology, New Delhi, India

[10]Center for Genomic Medicine, Massachusetts General Hospital, Harvard Medical School, Boston, USA

[11]Division of Renal Medicine, Department of Medicine, Brigham and Women's Hospital, Boston, MA, USA

*Correspondence at: athakur.dipsar@gmail.com

14.1 INTRODUCTION

The intricate communication network between the gut and the brain, often termed the "gut-brain axis (GBA)," represents one of the most fascinating areas of contemporary research, highlighting a complex, bidirectional communication system that not only ensures proper maintenance of gastrointestinal homeostasis but also influences the state of cognitive health, mood, and overall brain function (Martin, Osadchiy, Kalani, & Mayer, 2018). This axis comprises an array of cells and molecules, including neural, endocrine, and immune mediators, which together facilitate the interaction between the central nervous system (CNS) and the enteric nervous system (ENS) of the gut (Figure 14.1) (Socała et al., 2021). Among the plethora of factors influencing this communication, the gut microbiota—a diverse community of microorganisms residing in the gastrointestinal tract—plays a pivotal role (Thursby & Juge, 2017). Emerging evidence suggests that disruptions in the gut microbial composition are closely linked to neuroinflammation, potentially leading to various brain diseases, including neurodegenerative disorders, psychiatric conditions, and beyond (Kandpal et al., 2022; Stolzer et al., 2023).

The gut microbiota influences brain health through several mechanisms, with neuroinflammation being a critical mediator of disease pathology (Solanki, Karande, & Ranganathan, 2023). Neuroinflammation refers to the inflammatory response within the brain or spinal cord, which, when chronic, can lead to the progression of neurological diseases (DiSabato, Quan, & Godbout, 2016). The gut microbiota can modulate this process via several pathways: the immune system, the vagus nerve, and microbial metabolites. The gut microbiota plays a crucial role in the development and function of the host immune system (Han et al., 2022). It helps in the maturation of immune cells and the production of anti-inflammatory molecules, which can suppress neuroinflammation. However, dysbiosis—a disruption in the balance of gut microbial

DOI: 10.1201/9781003493143-14

communities—can lead to an exaggerated immune response. This imbalance can cause an increase in the permeability of the gut barrier (leaky gut), allowing pathogens and pro-inflammatory bacterial products, such as lipopolysaccharides (LPS), to enter the circulation and eventually reach the brain, contributing to neuroinflammation (Anand, Gorantla, & Chidambaram, 2023).

The vagus nerve, a principal component of the parasympathetic nervous system, serves as a major communication pathway between the gut and the brain. Gut bacteria can influence brain function by modulating the activity of the vagus nerve (Breit et al., 2018). For instance, certain probiotics have been shown to exert anti-inflammatory effects on the brain by activating the vagus nerve, thereby reducing neuroinflammation and potentially mitigating the risk of neurodegenerative diseases.

Microbial metabolites, such as short-chain fatty acids (SCFAs)—including butyrate, propionate, and acetate—are produced through the fermentation of dietary fibers by gut microbiota. These metabolites can cross the blood-brain barrier (BBB) and influence brain health by modulating immune responses, reducing oxidative stress, and directly affecting neural pathways associated with inflammation. SCFAs have been shown to possess anti-inflammatory properties and may protect against the development of diseases characterized by neuroinflammation, such as Alzheimer's disease and multiple sclerosis (MS) (Guo, Huo, Li, Han, & Zhou, 2022; Lan, Tang, Lu, Hu, & Tang, 2024; Mirzaei et al., 2021; Qian, Xie, Liu, Chen, & Tang, 2022).

The interplay between the gut microbiota and neuroinflammation has significant implications for various brain diseases. For example, in Alzheimer's disease (AD), dysbiosis has been linked to increased levels of amyloid plaques and tau phosphorylation, hallmarks of the disease, partially through the mechanism of neuroinflammation (Chen & Yu, 2023; J. de Oliveira et al., 2021). Similarly, in conditions like Parkinson's disease (PD), changes in gut microbiota composition can influence the aggregation of α-synuclein, a protein associated with the disease's pathology. Moreover, psychiatric disorders such as depression and anxiety have also been connected to alterations in the gut microbiome, with neuroinflammation acting as a possible bridge between gut dysbiosis and the manifestation of psychiatric symptoms (Ahmed, Ramadan, Elbeh, & Haridy, 2024; Góralczyk-Bińkowska, Szmajda-Krygier, & Kozłowska, 2022; Rust et al., 2023).

14.2 DIFFERENT ROUTES CONNECTING GUT AND BRAIN

14.2.1 The Vagus Nerve: A Direct Connection

The vagus nerve, also known as the tenth cranial nerve, is the longest nerve of the autonomic nervous system in the human body and serves as the primary channel for gut-brain communication. It extends from the brainstem through the neck and thorax down to the abdomen, innervating various organs, including the heart, lungs, and digestive tract. The vagus nerve is bidirectional; it sends sensory information from the gut to the brain and motor information from the brain to the gut (Yu, Xu, & Chang, 2020). This nerve plays a critical role in regulating many bodily functions, including heart rate, gastrointestinal peristalsis, sweating, and certain reflex actions such as coughing, sneezing, and vomiting. Its importance in the GBA cannot be overstated, as it directly affects gut inflammation, the release of digestive enzymes, and the gut's motility (Breit et al., 2018; Clapp et al., 2017).

14.2.2 ENS: The Second Brain

The ENS, often referred to as the "second brain," is a vast network of neurons lining the gastrointestinal tract from the esophagus to the rectum. It operates independently but in communication with the CNS. The ENS contains more neurons than the spinal cord and uses more than 30 neurotransmitters, including serotonin, which plays a significant role in regulating mood (Breit et al., 2018). This extensive network is responsible for the complex processes of digestion and has its reflexes and control mechanisms. It can function autonomously but is influenced by signals from the CNS, mediated by the vagus nerve, highlighting a crucial anatomical connection between the brain and the gut (Fülling, Dinan, & Cryan, 2019; Montanari, Imbriani, Bonsi, Martella, & Peppe, 2023; Spencer & Hu, 2020).

14.2.3 Immune System: The Inflammatory Pathway

The gut is a significant site of immune activity, housing approximately 70% of the body's immune cells. The immune system provides another critical link between the gut and the brain. Immune cells in the gut can send signals to the brain through the bloodstream, mediated by cytokines—proteins involved in cell signaling. These cytokines can cross the BBB and interact with the CNS, influencing brain function and behavior. This pathway is particularly relevant in the context of inflammation, where an immune response in the gut can lead to inflammatory cytokines affecting the brain, potentially contributing to mood disorders and neurodegeneration (Doney, Cadoret, Dion-Albert, Lebel, & Menard, 2022; Zhang, Xiao, Mao, & Xia, 2023).

14.2.4 Hormonal Routes: Endocrine Connections

The gut and brain communicate also through the endocrine system, utilizing hormones released into the bloodstream. The gut produces a significant number of hormones and neurotransmitters that can affect brain function. For instance, about 95% of the body's serotonin, a neurotransmitter crucial for mood regulation, is produced in the gut. These gut-derived hormones can influence brain function by affecting areas involved in mood regulation, appetite, and stress response (Appleton, 2018; Siba, Martynhak, & Pereira, 2023; Sun, Li, & Nie, 2020).

14.3 GUT MICROBIOTA AND BRAIN COMMUNICATION

The human gut microbiota, a complex community of microorganisms residing in our gastrointestinal tract, plays a pivotal role in health and disease, affecting not just the digestive system but also the brain and its pathology (Thursby & Juge, 2017). This diverse microbial ecosystem comprises both beneficial ("good") and potentially harmful ("bad") bacteria, each exerting significant impacts on the brain through the GBA (Sasso et al., 2023). Communication between gut microbiota and the brain through the GBA is a dynamic process that can influence neuroinflammation and neurodegeneration. Understanding how specific gut microorganisms reach the brain and cause neuroinflammation is crucial in developing targeted interventions for neurological disorders associated with dysregulated gut-brain communication, such as AD, PD, and MS (Hong, Zhang, Wu, Lu, & Zhang, 2023).

The gut microbiota can influence the brain through multiple pathways, including the endocrine, neural, and immune pathways. One significant route is via systemic circulation, where molecules produced by gut microorganisms can cross the BBB and affect brain function. This interaction is crucial in maintaining CNS homeostasis and can lead to neuroinflammation and neurodegeneration when dysregulated (Mitra, Dash, Nishan, Habiba, & Moon, 2023).

14.4 GOOD GUT BACTERIA AND THEIR IMPACT ON THE BRAIN

Beneficial gut bacteria contribute to brain health through various mechanisms, including the production of SCFAs like butyrate, propionate, and acetate, derived from the fermentation of dietary fibers. SCFAs can cross the BBB and have been shown to possess anti-inflammatory properties, reduce oxidative stress, and enhance the integrity of the BBB itself. For example, butyrate has been associated with the reduction of neuroinflammation and the improvement of memory function, showing potential as a therapeutic agent for neurodegenerative diseases like AD (Cavaliere et al., 2023). Probiotics, which are live beneficial bacteria found in certain foods and supplements, such as *Lactobacillus* and *Bifidobacterium*, have been shown to positively impact mental health (Kechagia et al., 2013; Tegegne & Kebede, 2022). They can modulate the activity of the vagus nerve, influence neurotransmitter levels, and reduce the symptoms of anxiety and depression through their anti-inflammatory effects. This psychobiotic effect highlights the potential of good gut bacteria in managing psychiatric disorders and enhancing cognitive function (Sarkar et al., 2016).

14.5 BAD GUT BACTERIA AND THEIR IMPACT ON BRAIN AND NEUROPATHOLOGY

Conversely, dysbiosis—a disruption in the balance of the gut microbiota with an overgrowth of harmful bacteria—can have detrimental effects on the brain. Certain pathogenic bacteria can produce toxins or pro-inflammatory substances that can breach the gut barrier and, potentially, the blood-brain barrier, leading to neuroinflammation, which is a risk factor for various neuropathological conditions (Tran, Lee, & Cho, 2022). For instance, species such as certain strains of *Escherichia coli* and *Clostridium* have been associated with increased levels of neuroinflammation and have been found in higher concentrations in individuals with PD and AD. These bacteria can produce LPS and other inflammatory compounds that, once they cross into the brain, can contribute to the pathogenesis of these diseases by promoting the aggregation of pathological proteins like alpha-synuclein in PD and amyloid-beta in AD (Kalyan et al., 2022). Furthermore, bad gut bacteria can influence the development of MS, an autoimmune disease affecting the central nervous system. Certain gut bacteria can alter immune system function, leading to autoimmune attacks on the myelin sheath of neurons, exacerbating the condition.

14.6 ROLE OF GUT MICROBIOTA-INDUCED NEUROINFLAMMATION

Bacterial cell wall components, specifically LPS derived from gut bacteria, have the capacity to initiate neuroinflammation and neurodegeneration by inducing inflammatory responses in the brain. These components play a significant role in the pathogenesis of neurodegenerative disorders by promoting inflammation within the brain (Tran et al., 2022). Additionally, bacterial extracellular vesicles released by bacteria serve as carriers of bioactive molecules that impact brain function and contribute to neuroinflammation, thereby playing a role in the development of neurodegenerative diseases. Specific gut microbiota has been identified as key players in causing neuroinflammation and contributing to neurodegenerative disorders (B. Sun, Sawant, Borthakur, & Bihl, 2023). *Escherichia/Shigella*, known for their pro-inflammatory properties, have been associated with neuroinflammation and AD pathogenesis. Alterations in the gut microbiome composition, such as an increase in these bacteria, can influence brain function and contribute to neurodegenerative conditions (Chandra, Sisodia, & Vassar, 2023). *Porphyromonas gingivalis*, another gut bacterium, has been linked to active invasion and infection-induced complement activation in the brains of mice, highlighting its role in neuroinflammation and its potential contribution to neurodegenerative diseases like AD (Poole et al., 2015). On the other hand, *Lactobacillus plantarum* has shown promise in improving cognitive function by restoring neurotransmitter levels and reducing amyloid-beta plaque formation in AD models, indicating its potential to alleviate neuroinflammation and its impact on neurodegenerative disorders. These examples underscore the intricate relationship between gut microbiota, neuroinflammation, and neurodegeneration, shedding light on the significant role of specific bacteria in influencing brain health and the development of neurological conditions (Sultana, Hia, & Reddy, 2024).

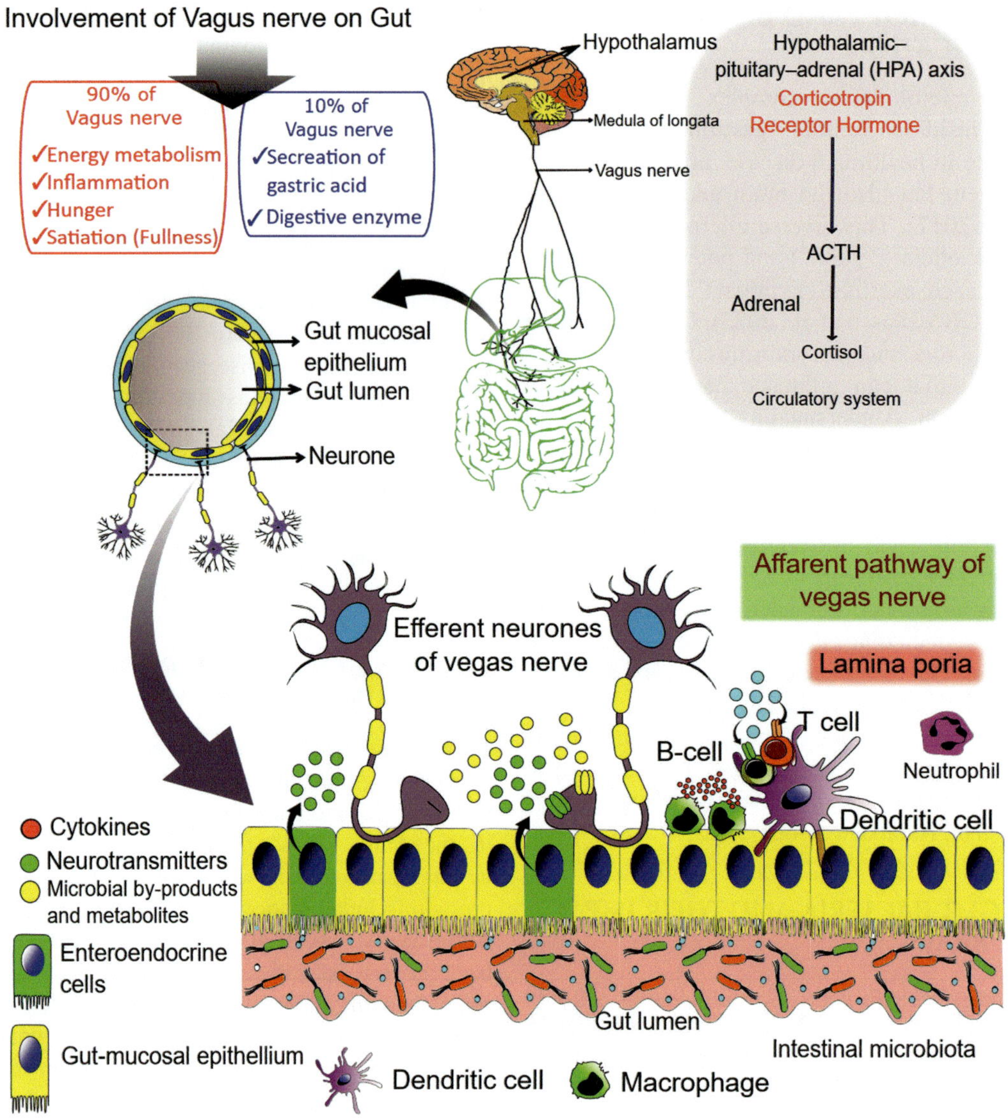

FIGURE 14.1 Communication architecture of the gut-brain axis. A schematic diagram showing the different routes of communication between the gut and brain, including the involvement of the vagus nerve via efferent and afferent pathways. Afferent pathways of the vagus nerve involve sensory information transmission from the gut to the brain. Receptors in the stomach wall send signals to the brain through afferent vagal fibers, providing information about sensations like fullness or hunger. These signals are then processed in the brain, influencing responses related to digestion and other physiological functions. On the other hand, efferent pathways of the vagus nerve involve motor responses from the brain to the gut. Once sensory signals are processed in the brain, efferent vagal signaling is sent back to the gastrointestinal smooth muscle, modulating muscle movement, and aiding in digestion. Additionally, the hypothalamus-pituitary-adrenal (HPA) axis regulates interactions between the gut microbiota, gut, and brain, impacting cognitive functions, stress-induced gastrointestinal responses, and immune modulation through neural and endocrine pathways.

14.7 UNVEILING THE DIALOGUE OF GUT MICROBIOTA WITH THE BRAIN VIA NEUROTRANSMITTER

There are two chemical neurotransmitter entities, such as butyrate and dopamine, play a pivotal role in gut-brain communication. The gut microbiota plays a role in regulating levels of neurotransmitters in experimental setups, suggesting that microorganisms serve as agents for traditional signaling molecules utilized by the nervous system (Barrett, Ross, O'Toole, Fitzgerald, & Stanton, 2012; Poutahidis et al., 2013; Yano et al., 2015). Microbes residing in the gut can manufacture neurotransmitters and stimulate their production in their host animals. The presence of neurotransmitter receptors in bacteria is necessary for communication between CNS effectors and bacteria. It has been found that bacteria have binding sites for enteric neurotransmitters that the host produces. These sites can affect how some microbiota components function and raise susceptibility to inflammatory and infectious stimuli (Hughes & Sperandio, 2008). It has been shown that *Pseudomonas fluorescens* has been shown

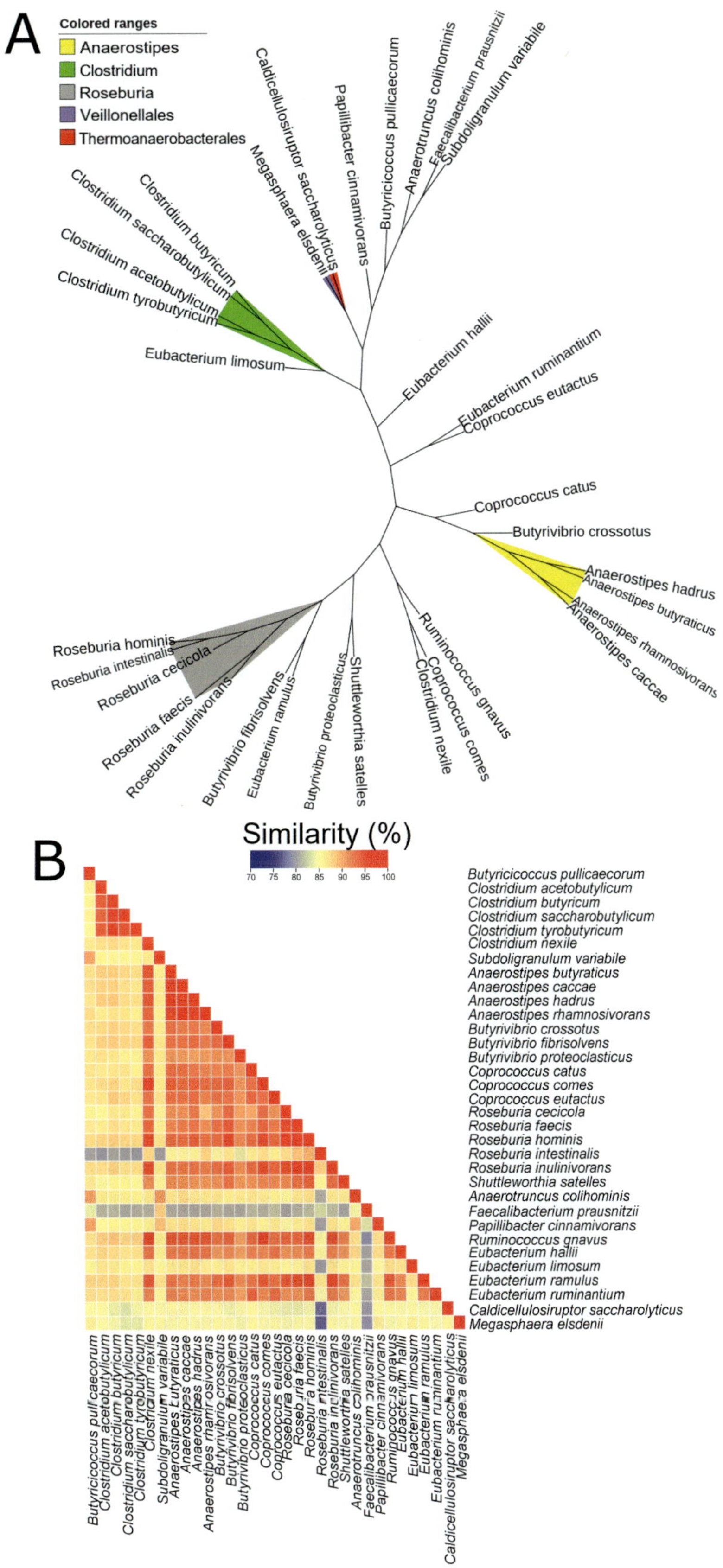

FIGURE 14.2 Landscape of the butyrate-producing bacteria. (A) A phylogenetic tree depicting butyrate-producing bacteria species based on 16S rRNA sequences is illustrated, while (B) a correlation heat map showcases the similarity index among these species. The 16S rRNA sequences utilized in this analysis were sourced from RNA Central (https://rnacentral.org/), with a total of 33 distinct butyrate-producing bacteria species included in the study.

to have a high affinity for the GABA system, with binding characteristics akin to those of a brain receptor (Guthrie & Nicholson-Guthrie, 1989). Adrenergic antagonists have the ability to selectively block the host-derived adrenaline/norepinephrine receptor found in *Escherichia coli* O157:H7 (Clarke, Hughes, Zhu, Boedeker, & Sperandio, 2006). Butyrate plays a crucial role in gastrointestinal health, neurotransmitter levels, and communication between the gut and the brain. Bacteria utilize fermentation processes to produce metabolites such as lactates and SCFAs, including acetate (C_2), propionate (C_3), and butyrate (C_4) (Silva, Bernardi, & Frozza, 2020). These substances, originating in the gastrointestinal tract, can enter circulation through the portal vein and then travel via the bloodstream to peripheral tissues, organs, and even the brain (Lerner, Neidhöfer, & Matthias, 2017). Butyrate possesses anti-inflammatory properties and positively impacts the intestinal epithelium, potentially benefiting the gut microbiota by increasing the number of ENS cholinergic neurons and modulating appetite via vagal and hypothalamic stimulation (Liu et al., 2018). For instance, various microorganisms including *Bacteroides, Bifidobacterium, Parabacteroides, and Escherichia spp.* have been identified as producers of the neurotransmitter γ-aminobutyric acid (GABA) (Strandwitz et al., 2019). Gamma-aminobutyric acid (GABA) functions as a neurotransmitter, serving as a chemical messenger within the brain. Its role involves reducing brain activity by inhibiting certain signals in the central nervous system, encompassing both the brain and spinal cord. GABA is renowned for inducing a calming effect. As evidence, it has been found that colonic cells utilize butyrate, derived from dietary fiber by intestinal microbes, as their primary energy source (Rivera-Chávez et al., 2016). Among the major producers of butyrate are Firmicutes, with genera such as *Clostridium, Ruminococcus, Eubacterium,* and *Faecalibacterium spp.* playing significant roles in its synthesis (Mokhtari, Gibson, & Hekmatdoost, 2017). The presence of neurotransmitter receptors in bacteria is necessary for communication between CNS effectors and bacteria. According to several studies, bacteria have binding sites for enteric neurotransmitters that the host produces. These sites can affect how some microbiota components function and raise susceptibility to inflammatory and infectious stimuli (Hughes & Sperandio, 2008). It has been shown that *Pseudomonas fluorescens* has been shown to have a high affinity for the GABA system, with binding characteristics akin to those of a brain receptor (Guthrie & Nicholson-Guthrie, 1989). Adrenergic antagonists have the ability to selectively block the host-derived adrenaline/norepinephrine receptor found in *Escherichia coli* O157:H7 (Clarke et al., 2006). However, an un-rooted phylogenetic tree was constructed with the corresponding butyrate-producing microbiome species with 16S rRNA data (Figure 14.2A). Figure 14.2B depicts the percentages of similarity present between the corresponding butyrate-producing microbiome species. Butyrate's natural inhibition of histone deacetylase (HDAC) affects neurotransmitter levels and is thus integral to gut-brain signaling. For instance, the HDAC inhibitory properties of sodium butyrate have been demonstrated to offer beneficial effects in rats with neurotoxicity-induced symptoms, including improved locomotor function and increased levels of dopamine in the striatum (Sharma, Taliyan, & Singh, 2015). It was observed that the increase in H3 histone acetylation corresponded with the improvement in locomotor function and dopamine levels in the striatum. HDACs typically remove acetyl groups from histones, leading to DNA compaction into heterochromatin, which reduces accessibility to transcription factors and other gene regulatory proteins. Butyrate, however, acts as a potent inhibitor of heterochromatin formation, thereby relaxing chromatin and playing significant roles in epigenetic regulation. Furthermore, the HDAC inhibitory activity of sodium butyrate has been shown to mitigate oxidative stress, alleviate α-synuclein damage, and rescue dopaminergic cells in models of Parkinson's disease (PD) (Paiva et al., 2017). Recently, HDAC inhibitors have also been explored as neuropharmacological agents to alleviate symptoms associated with glioblastomas. Moreover, infusion of sodium butyrate has been found to attenuate neuronal apoptosis in mice subjected to middle cerebral artery occlusions (Zhou et al., 2021). Together, these studies highlight how butyrate's ability to inhibit HDACs could be valuable in treating neurological disorders and adjusting neurotransmitter levels, offering promising therapeutic possibilities.

14.8 TARGETING GBA FOR DISEASE THERAPEUTICS

Given the role of GBA in health and disease, targeting this communication pathway offers promising therapeutic avenues as depicted in Figure 14.3.

14.8.1 Probiotics and Prebiotics

These strategies are designed to cultivate a gut environment enriched with beneficial bacteria, enhancing mental health outcomes. Specific probiotic strains, such as *Lactobacillus rhamnosus*, have demonstrated the ability to alleviate stress-induced markers and anxiety behaviors in mice, highlighting the role of the GABAergic system and the vagus nerve in this process. Similarly, human studies, including one where participants took *Lactobacillus helveticus* and *Bifidobacterium longum*, have shown promising results in reducing stress levels, underscoring the potential of probiotics in managing mood and stress (Partrick et al., 2021). By feeding beneficial bacteria, prebiotics enhance a healthy gut microbiome, ultimately supporting improved brain health through indirect means (De Oliveira et al., 2023; Huang, Wang, & Hu, 2016).

14.8.2 Dietary Modifications

Dietary choices play a pivotal role in the composition and health of our gut microbiota. Embracing a diet abundant in whole foods, fibers, and a variety of plant-based ingredients nurtures the growth of beneficial bacteria. These microbes, including *Bifidobacteria* and *Lactobacillus*, thrive on dietary fibers found in fruits, vegetables, legumes, and whole grains,

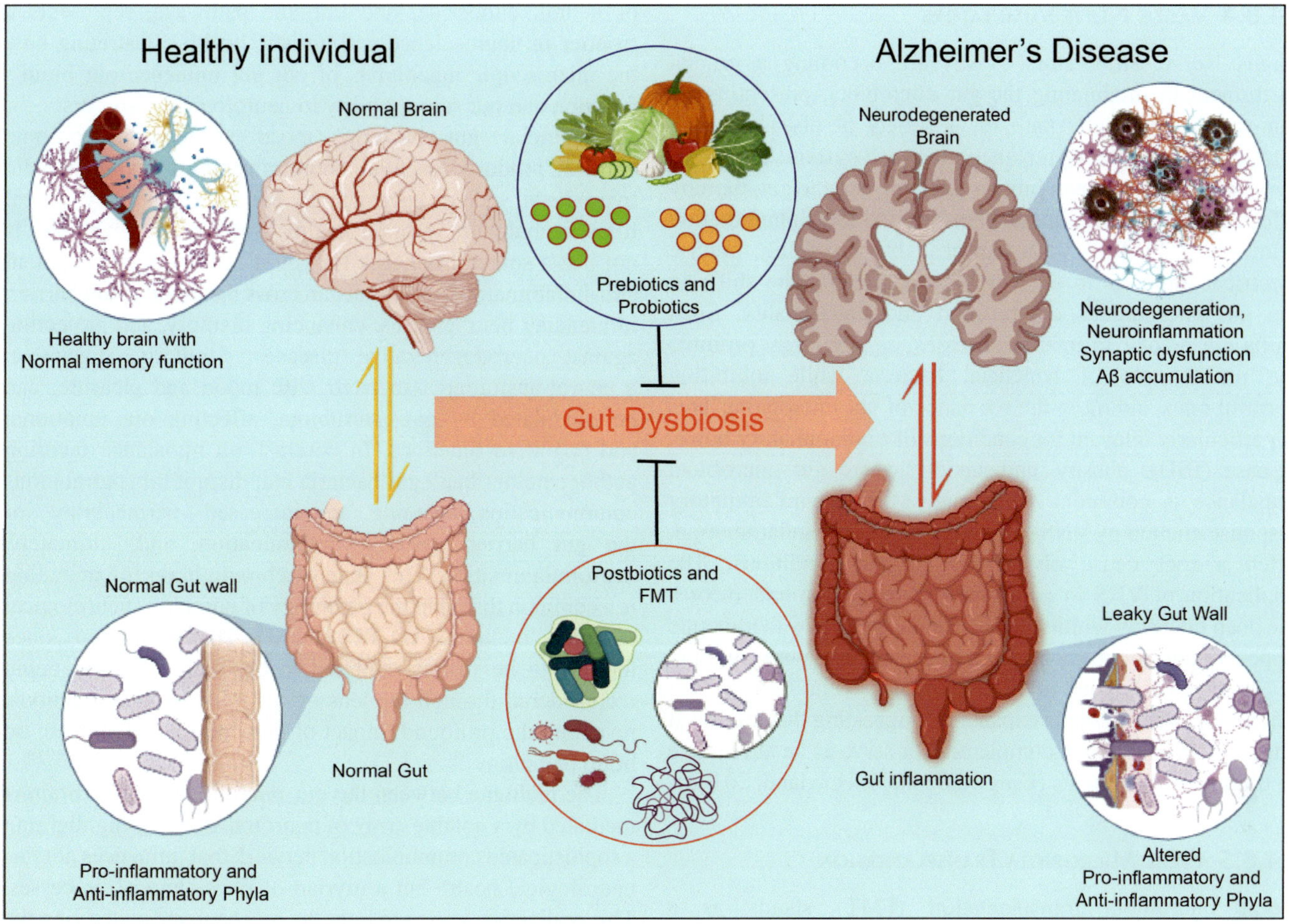

FIGURE 14.3 The intricate relationship between gut dysbiosis, Alzheimer's disease, and therapeutic interventions targeting the gut microbiota. The schematic diagram illustrates gut dysbiosis and neurodegenerative disease like Alzheimer's disease. Therapeutic interventions include postbiotics, fecal microbiota transplantation (FMT), and the role of pro-and anti-inflammatory phyla in modulating gut health. Additionally, it visualizes the concept of a leaky gut wall in the context of altered microbiota associated with Alzheimer's disease.

converting them into SCFAs like butyrate, propionate, and acetate (Holscher, 2017; Portincasa et al., 2022). SCFAs are celebrated for their anti-inflammatory effects, capable of crossing the blood-brain barrier to mitigate neuroinflammation, modulate immune responses, and enhance memory functions (Holscher, 2017).

Contrastingly, diets laden with processed foods and added sugars favor the proliferation of detrimental bacteria, which can lead to inflammation and various diseases. In the pursuit of dietary interventions that target the GBA for enhanced well-being, the Mediterranean diet emerges as a sterling example (Del Chierico, Vernocchi, Dallapiccola, & Putignani, 2014). Characterized by its rich variety of fruits, vegetables, nuts, whole grains, and olive oil, this diet is linked to a decreased risk of cognitive decline and improved mental health. Its high levels of polyphenols, antioxidants, and unsaturated fatty acids contribute to a gut environment conducive to brain health, exemplifying the profound impact of dietary choices on the symbiotic relationship between our gut microbiota and the brain (Barber, Kabisch, Pfeiffer, & Weickert, 2023; Nagpal, Shively, Register, Craft, & Yadav, 2019; Vázquez-Cuesta et al., 2024).

14.8.3 Pharmacological Interventions

Researchers are exploring pharmaceuticals that precisely target the GBA to alleviate a range of conditions. Antibiotics, for instance, are being used to recalibrate the gut microbiota, offering significant relief from irritable bowel syndrome (IBS)—a condition frequently accompanied by anxiety and depression. Rifaximin stands out as a specific, non-absorbable antibiotic that effectively targets the gastrointestinal tract, significantly improving IBS symptoms by altering the gut's bacterial composition Additionally, medications aimed at regulating the activity of the vagus nerve are under investigation for their promise in treating various psychiatric and neurological disorders, showcasing the breadth of potential in modulating this critical communication pathway between the gut and the brain (Chey, Shah, & DuPont, 2020; Saadi & McCallum, 2013).

14.8.4 Vagus Nerve Stimulators

Vagus Nerve Stimulators (VNS) are becoming a crucial instrument in modulating the gut microbiota, presenting an innovative therapeutic path for a variety of diseases. As a vital part of the parasympathetic nervous system, the vagus nerve plays a key role in linking the brain and the gut, thereby affecting gut health, immune responses, and inflammation (Howland, 2014b). VNS devices, by providing precise electrical impulses to the vagus nerve, can initiate shifts in the gut microbiota's composition and functionality. Such shifts are pivotal in treating diseases, as VNS may promote the proliferation of beneficial bacteria while inhibiting harmful ones, aiding in the rebalance of gut microbiota. This is particularly relevant for conditions like inflammatory bowel disease (IBD), obesity, and anxiety where gut microbiota imbalance is common. Additionally, the anti-inflammatory response elicited by VNS can mitigate systemic inflammation, often a root cause of various chronic conditions. The application of VNS in adjusting gut microbiota underscores the deep interplay within the GBS and unveils new therapeutic opportunities for microbiota-related diseases. With ongoing research, VNS's role in disease management through gut microbiota adjustment is expanding, suggesting that electrical stimulation might soon complement or serve as an alternative to traditional treatments (Fang et al., 2023; Howland, 2014a).

14.8.5 Fecal Microbiota Transplantation

Fecal microbiota transplantation (FMT) stands as a pioneering treatment that harnesses the gut-brain connection to address a variety of conditions by transferring stool from a healthy donor to a patient, aiming to recalibrate the gut's microbial balance. Initially celebrated for its effectiveness against *Clostridium difficile* infections, the scope of FMT has broadened to explore its potential in combating dysbiosis-related diseases, particularly neuropsychiatric disorders (Belvoncikova, Maronek, & Gardlik, 2022; Wortelboer, Nieuwdorp, & Herrema, 2019). Research indicates promising avenues for FMT in managing depression and anxiety, where alterations in gut microbiota may influence neurotransmitter levels and mood. Similarly, in ASD and neurodegenerative diseases like Parkinson's and Alzheimer's, FMT is being examined for its capacity to modify gut bacteria profiles associated with these conditions, showing potential for symptom improvement and disease progression delay. This expansion of FMT's applications underscores the therapeutic possibilities inherent in targeting the gut microbiota for brain health and disease management (Gupta, Allen-Vercoe, & Petrof, 2016).

14.9 CONCLUSIONS AND FUTURE DIRECTIONS

The intricate web of connections between the gut and brain unveils a complex and dynamic relationship that underscores the pivotal role of gut microbiota in brain communication and overall health. This relationship, facilitated by an array of neural, endocrine, and immune pathways, represents a frontier in neuroscience and holistic health, illustrating how the microscopic inhabitants of our gut influence our brain's function and our susceptibility to neurological disorders.

Beneficial gut bacteria, those virtuous microscopic entities, produce essential neurotransmitters and metabolic byproducts, such as butyrate and dopamine, that play critical roles in maintaining brain health. Butyrate, for example, is not just a source of energy for colon cells; it is also a potent anti-inflammatory agent that can cross the blood-brain barrier, influencing neurogenesis, enhancing memory, and protecting against neurodegenerative diseases. Similarly, dopamine, a neurotransmitter associated with mood and pleasure, can be modulated by gut microbiota, affecting our emotional and cognitive functions. In contrast, an imbalance favoring pathogenic or "bad" gut bacteria can disrupt this harmonious communication, leading to increased permeability of the gut barrier, systemic inflammation, and, ultimately, neuroinflammation. This state of chronic brain inflammation is a common thread in the pathology of numerous neurological disorders, including AD, PD, and MS. These conditions, once thought to be strictly confined to the brain, are now being reconsidered through the lens of gut microbiota interactions, revealing the profound impact of our intestinal ecosystem on brain pathology.

The dialogue between the gut microbiota and the brain is mediated by a notable array of neurotransmitters, highlighting a sophisticated communication network that influences not just neurological health but a myriad of physiological processes. This realization brings to light the mechanisms underlying the influence on neurological disorders and opens new avenues for targeted therapeutic interventions.

Looking forward, the exploration of the GBA calls for a deeper understanding of the specific mechanisms by which gut microbiota exert their influence on brain function. This knowledge is crucial for the development of novel therapeutic strategies aimed at modulating this complex system. The potential of microbiome-based interventions to improve mental health and treat neurological conditions is immense. By manipulating the gut microbiota through diet, probiotics, prebiotics, or even fecal microbiota transplantation, we could usher in a new era of treatments that address the root cause of neurological disorders rather than merely managing symptoms.

ACKNOWLEDGMENT

This study was partly supported by the Research Committee of Hong Kong Baptist University (HKBU-MPCF-003-2022-23) awarded to AI and AT. AT received the Ramalingaswamy Fellowship by the Department of Biotechnology, Ministry of Science and Technology, Government of India.

AUTHORS' CONTRIBUTIONS

AT conceived the initial idea and outline. AT, DG, RS, SA, contributed to the initial draft. AT, DG, and AI prepared the figures. PU, SG, VK, and AY contributed to the revision and gathered information and resources.

REFERENCES

Ahmed, G. K., Ramadan, H. K.-A., Elbeh, K., & Haridy, N. A. (2024). Bridging the gap: Associations between gut microbiota and psychiatric disorders. *Middle East Current Psychiatry*, *31*(1), 2. doi:10.1186/s43045-024-00395-9

Anand, N., Gorantla, V. R., & Chidambaram, S. B. (2023). The role of gut dysbiosis in the pathophysiology of neuropsychiatric disorders. *Cells*, *12*(1), 54.

Appleton, J. (2018). The gut-brain axis: Influence of microbiota on mood and mental health. *Integrative Medicine (Encinitas)*, *17*(4), 28–32.

Barber, T. M., Kabisch, S., Pfeiffer, A. F. H., & Weickert, M. O. (2023). The effects of the mediterranean diet on health and gut microbiota. *Nutrients*, *15*(9), 2150. doi:10.3390/nu15092150

Barrett, E., Ross, R. P., O'Toole, P. W., Fitzgerald, G. F., & Stanton, C. (2012). γ-Aminobutyric acid production by culturable bacteria from the human intestine. *Journal of Applied Microbiology*, *113*(2), 411–417. doi:10.1111/j.1365-2672.2012.05344.x

Belvoncikova, P., Maronek, M., & Gardlik, R. (2022). Gut dysbiosis and fecal microbiota transplantation in autoimmune diseases. *International Journal of Molecular Science*, *23*(18), 10729. doi:10.3390/ijms231810729

Breit, S., Kupferberg, A., Rogler, G., & Hasler, G. (2018). Vagus nerve as modulator of the brain-gut axis in psychiatric and inflammatory disorders. *Frontiers in Psychiatry*, *9*, 44. doi:10.3389/fpsyt.2018.00044

Cavaliere, G., Catapano, A., Trinchese, G., Cimmino, F., Penna, E., Pizzella, A., … & Mollica, M. P. (2023). Butyrate improves neuroinflammation and mitochondrial impairment in cerebral cortex and synaptic fraction in an animal model of diet-induced obesity. *Antioxidants*, *12*(1), 4.

Chandra, S., Sisodia, S. S., & Vassar, R. J. (2023). The gut microbiome in Alzheimer's disease: What we know and what remains to be explored. *Molecular Neurodegeneration*, *18*(1), 9. doi:10.1186/s13024-023-00595-7

Chen, Y., & Yu, Y. (2023). Tau and neuroinflammation in Alzheimer's disease: Interplay mechanisms and clinical translation. *Journal of Neuroinflammation*, *20*(1), 165. doi:10.1186/s12974-023-02853-3

Chey, W. D., Shah, E. D., & DuPont, H. L. (2020). Mechanism of action and therapeutic benefit of rifaximin in patients with irritable bowel syndrome: A narrative review. *Therapy Advances Gastroenterology*, *13*, 1756284819897531. doi:10.1177/1756284819897531

Clapp, M., Aurora, N., Herrera, L., Bhatia, M., Wilen, E., & Wakefield, S. (2017). Gut microbiota's effect on mental health: The gut-brain axis. *Clinical Practices*, *7*(4), 987. doi:10.4081/cp.2017.987

Clarke, M. B., Hughes, D. T., Zhu, C., Boedeker, E. C., & Sperandio, V. (2006). The QseC sensor kinase: A bacterial adrenergic receptor. *Proceedings of the National Academy Science U S A*, *103*(27), 10420–10425. doi:10.1073/pnas.0604343103

De Oliveira, F. L., Salgaço, M. K., de Oliveira, M. T., Mesa, V., Sartoratto, A., Peregrino, A. M., … & Sivieri, K. (2023). Exploring the potential of Lactobacillus helveticus R0052 and bifidobacterium longum R0175 as promising psychobiotics using SHIME. *Nutrients*, *15*(6), 1521. doi:10.3390/nu15061521

de Oliveira, J., Kucharska, E., Garcez, M. L., Rodrigues, M. S., Quevedo, J., Moreno-Gonzalez, I., & Budni, J. (2021). Inflammatory cascade in Alzheimer's disease pathogenesis: A review of experimental findings. *Cells*, *10*(10), 2581. doi:10.3390/cells10102581

Del Chierico, F., Vernocchi, P., Dallapiccola, B., & Putignani, L. (2014). Mediterranean diet and health: Food effects on gut microbiota and disease control. *International Journal of the Molecular Science*, *15*(7), 11678–11699. doi:10.3390/ijms150711678

DiSabato, D. J., Quan, N., & Godbout, J. P. (2016). Neuroinflammation: The devil is in the details. *Journal of Neurochemistry, 139*(S2), 136–153. doi:10.1111/jnc.13607

Doney, E., Cadoret, A., Dion-Albert, L., Lebel, M., & Menard, C. (2022). Inflammation-driven brain and gut barrier dysfunction in stress and mood disorders. *European Journal of the Neuroscience*, *55*(9–10), 2851–2894. doi:10.1111/ejn.15239

Fang, Y. T., Lin, Y. T., Tseng, W. L., Tseng, P., Hua, G. L., Chao, Y. J., & Wu, Y. J. (2023). Neuroimmunomodulation of vagus nerve stimulation and the therapeutic implications. *Frontiers in Aging Neuroscience*, *15*, 1173987. doi:10.3389/fnagi.2023.1173987

Fülling, C., Dinan, T. G., & Cryan, J. F. (2019). Gut microbe to brain signaling: What happens in vagus…. *Neuron, 101*(6), 998–1002. doi:10.1016/j.neuron.2019.02.008

Góralczyk-Bińkowska, A., Szmajda-Krygier, D., & Kozłowska, E. (2022). The microbiota-gut-brain axis in psychiatric disorders. *International Journal of Molecular Science*, *23*(19). doi:10.3390/ijms231911245

Guo, C., Huo, Y. J., Li, Y., Han, Y., & Zhou, D. (2022). Gut-brain axis: Focus on gut metabolites short-chain fatty acids. *World Journal of Clinical Cases*, *10*(6), 1754–1763. doi:10.12998/wjcc.v10.i6.1754

Gupta, S., Allen-Vercoe, E., & Petrof, E. O. (2016). Fecal microbiota transplantation: In perspective. *Therapy Advances Gastroenterology*, *9*(2), 229–239. doi:10.1177/1756283x15607414

Guthrie, G. D., & Nicholson-Guthrie, C. S. (1989). gamma-Aminobutyric acid uptake by a bacterial system with neurotransmitter binding characteristics. *Proceedings of the National Academy of Sciences of the United States of America*, *86*(19), 7378–7381. doi:10.1073/pnas.86.19.7378

Han, Y., Wang, B., Gao, H., He, C., Hua, R., Liang, C., …Xu, J. (2022). Vagus nerve and underlying impact on the gut microbiota-brain axis in behavior and neurodegenerative diseases. *Journal of Inflammation Research*, *15*, 6213–6230. doi:10.2147/jir.s384949

Holscher, H. D. (2017). Dietary fiber and prebiotics and the gastrointestinal microbiota. *Gut Microbes*, *8*(2), 172–184. doi:10.1080/19490976.2017.1290756

Hong, D., Zhang, C., Wu, W., Lu, X., & Zhang, L. (2023). Modulation of the gut–brain axis via the gut microbiota: A new era in treatment of amyotrophic lateral sclerosis. *Frontiers in Neurology*, *14*. doi:10.3389/fneur.2023.1133546

Howland, R. H. (2014a). New developments with vagus nerve stimulation therapy. *Journal of Psychosocial Nursing and Mental Health Services, 52*(3), 11–14. doi:10.3928/02793695-20140218-01

Howland, R. H. (2014b). Vagus nerve stimulation. *Current Behaviour Neuroscience Report*, *1*(2), 64–73. doi:10.1007/s40473-014-0010-5

Huang, R., Wang, K., & Hu, J. (2016). Effect of probiotics on depression: A systematic review and meta-analysis of randomized controlled trials. Nutrients, *8*(8), 483. doi:10.3390/nu8080483

Hughes, D. T., & Sperandio, V. (2008). Inter-kingdom signalling: Communication between bacteria and their hosts. *Nature Review Microbiology*, *6*(2), 111–120. doi:10.1038/nrmicro1836

Kalyan, M., Tousif, A. H., Sonali, S., Vichitra, C., Sunanda, T., Praveenraj, S. S., …Chidambaram, S. B. (2022). Role of endogenous lipopolysaccharides in neurological disorders. *Cells*, *11*(24), 4038. doi:10.3390/cells11244038

Kandpal, M., Indari, O., Baral, B., Jakhmola, S., Tiwari, D., Bhandari, V., … & Jha, H. C. (2022). Dysbiosis of gut microbiota from the perspective of the gut–brain axis: Role in the provocation of neurological disorders. *Metabolites, 12*(11), 1064.

Kechagia, M., Basoulis, D., Konstantopoulou, S., Dimitriadi, D., Gyftopoulou, K., Skarmoutsou, N., & Fakiri, E. M. (2013). Health benefits of probiotics: A review. *ISRN Nutrition, 2013*, 481651. doi:10.5402/2013/481651

Lan, Z., Tang, X., Lu, M., Hu, Z., & Tang, Z. (2024). The role of short-chain fatty acids in central nervous system diseases: A bibliometric and visualized analysis with future directions. *Heliyon, 10*(4), e26377. doi:10.1016/j.heliyon.2024.e26377

Lerner, A., Neidhöfer, S., & Matthias, T. (2017). The gut microbiome feelings of the brain: A perspective for non-microbiologists. Microorganisms, *5*(4), 66. doi:10.3390/microorganisms5040066

Liu, H., Wang, J., He, T., Becker, S., Zhang, G., Li, D., & Ma, X. (2018). Butyrate: A double-edged sword for health? *Advances Nutrition*, *9*(1), 21–29. doi:10.1093/advances/nmx009

Martin, C. R., Osadchiy, V., Kalani, A., & Mayer, E. A. (2018). The brain-gut-microbiome axis. *Cellular and Molecular Gastroenterology and Hepatology*, *6*(2), 133–148. doi:10.1016/j.jcmgh.2018.04.003

Mirzaei, R., Bouzari, B., Hosseini-Fard, S. R., Mazaheri, M., Ahmadyousefi, Y., Abdi, M., … & Karampoor, S. (2021). Role of microbiota-derived short-chain fatty acids in nervous system disorders. *Biomedicine & Pharmacotherapy*, *139*, 111661. doi:10.1016/j.biopha.2021.111661

Mitra, S., Dash, R., Nishan, A. A., Habiba, S. U., & Moon, I. S. (2023). Brain modulation by the gut microbiota: From disease to therapy. *Journal of Advanced Research*, *53*, 153–173. doi:10.1016/j.jare.2022.12.001

Mokhtari, Z., Gibson, D. L., & Hekmatdoost, A. (2017). Nonalcoholic fatty liver disease, the gut microbiome, and diet. *Advances Nutrition*, *8*(2), 240–252. doi:10.3945/an.116.013151

Montanari, M., Imbriani, P., Bonsi, P., Martella, G., & Peppe, A. (2023). Beyond the microbiota: Understanding the role of the enteric nervous system in Parkinson's disease from mice to human. *Biomedicines*, *11*(6), 1560. doi:10.3390/biomedicines11061560

Nagpal, R., Shively, C. A., Register, T. C., Craft, S., & Yadav, H. (2019). Gut microbiome-Mediterranean diet interactions in improving host health. *F1000Res*, *8*, 699. doi:10.12688/f1000research.18992.1

Paiva, I., Pinho, R., Pavlou, M. A., Hennion, M., Wales, P., Schütz, A. L., …Outeiro, T. F. (2017). Sodium butyrate rescues dopaminergic cells from alpha-synuclein-induced transcriptional deregulation and DNA damage. *Human Molecular Genetics*, *26*(12), 2231–2246. doi:10.1093/hmg/ddx114

Partrick, K. A., Rosenhauer, A. M., Auger, J., Arnold, A. R., Ronczkowski, N. M., Jackson, L. M., … & Huhman, K. L. (2021). Ingestion of probiotic (Lactobacillus helveticus and Bifidobacterium longum) alters intestinal microbial structure and behavioral expression following social defeat stress. *Scientific Reports*, *11*(1), 3763. doi:10.1038/s41598-021-83284-z

Poole, S., Singhrao, S. K., Chukkapalli, S., Rivera, M., Velsko, I., Kesavalu, L., & Crean, S. (2015). Active invasion of porphyromonas gingivalis and infection-induced complement activation in ApoE -/- mice brains. *Journal of Alzheimer's Disease*, *43*, 67–80. doi:10.3233/JAD-140315

Portincasa, P., Bonfrate, L., Vacca, M., De Angelis, M., Farella, I., Lanza, E., … & Di Ciaula, A. (2022). Gut microbiota and short chain fatty acids: Implications in glucose homeostasis. *International Journal of Molecular Sciences, 23*(3), 1105. doi:10.3390/ijms23031105

Poutahidis, T., Kearney, S. M., Levkovich, T., Qi, P., Varian, B. J., Lakritz, J. R., …Erdman, S. E. (2013). Microbial symbionts accelerate wound healing via the neuropeptide hormone oxytocin. *PLoS One*, *8*(10), e78898. doi:10.1371/journal.pone.0078898

Qian, X. H., Xie, R. Y., Liu, X. L., Chen, S. D., & Tang, H. D. (2022). Mechanisms of short-chain fatty acids derived from gut microbiota in Alzheimer's disease. *Aging Disease*, *13*(4), 1252–1266. doi:10.14336/ad.2021.1215

Rivera-Chávez, F., Zhang, L. F., Faber, F., Lopez, C. A., Byndloss, M. X., Olsan, E. E., … & Bäumler, A. J. (2016). Depletion of butyrate-producing clostridia from the gut microbiota drives an aerobic luminal expansion of Salmonella. *Cell Host Microbe*, *19*(4), 443–454. doi:10.1016/j.chom.2016.03.004

Rust, C., Malan-Muller, S., van den Heuvel, L. L., Tonge, D., Seedat, S., Pretorius, E., & Hemmings, S. M. J. (2023). Platelets bridging the gap between gut dysbiosis and neuroinflammation in stress-linked disorders: A narrative review. *Journal of Neuroimmunology, 382*, 578155. doi:10.1016/j.jneuroim.2023.578155

Saadi, M., & McCallum, R. W. (2013). Rifaximin in irritable bowel syndrome: Rationale, evidence and clinical use. *Therapeutic Advances Chronic Disease*, *4*(2), 71–75. doi:10.1177/2040622312472008

Sarkar, A., Lehto, S. M., Harty, S., Dinan, T. G., Cryan, J. F., & Burnet, P. W. J. (2016). Psychobiotics and the manipulation of bacteria-gut-brain signals. *Trends in Neurosciences*, *39*(11), 763–781. doi:10.1016/j.tins.2016.09.002

Sasso, J. M., Ammar, R. M., Tenchov, R., Lemmel, S., Kelber, O., Grieswelle, M., & Zhou, Q. A. (2023). Gut microbiome–brain alliance: A landscape view into mental and gastrointestinal health and disorders. *ACS Chemical Neuroscience*, *14*(10), 1717–1763. doi:10.1021/acschemneuro.3c00127

Sharma, S., Taliyan, R., & Singh, S. (2015). Beneficial effects of sodium butyrate in 6-OHDA induced neurotoxicity and behavioral abnormalities: Modulation of histone deacetylase activity. *Behaviour Brain Research*, *291*, 306–314. doi:10.1016/j.bbr.2015.05.052

Siba, I. P., Martynhak, B. J., & Pereira, M. (2023). When gut hormones influence brain function in depression. *Applied Biosciences, 2*(1), 31–51.

Silva, Y. P., Bernardi, A., & Frozza, R. L. (2020). The role of short-chain fatty acids from gut microbiota in gut-brain communication. *Frontiers in Endocrinology (Lausanne)*, *11*, 25. doi:10.3389/fendo.2020.00025

Socała, K., Doboszewska, U., Szopa, A., Serefko, A., Włodarczyk, M., Zielińska, A., … & Wlaź, P. (2021). The role of microbiota-gut-brain axis in neuropsychiatric and neurological disorders. *Pharmacological Research, 172*, 105840. doi:10.1016/j.phrs.2021.105840

Solanki, R., Karande, A., & Ranganathan, P. (2023). Emerging role of gut microbiota dysbiosis in neuroinflammation and neurodegeneration. *Frontiers in Neurology, 14*. doi:10.3389/fneur.2023.1149618

Spencer, N. J., & Hu, H. (2020). Enteric nervous system: Sensory transduction, neural circuits and gastrointestinal motility. *Nature Review Gastroenterology and Hepatology, 17*(6), 338–351. doi:10.1038/s41575-020-0271-2

Stolzer, I., Scherer, E., Süß, P., Rothhammer, V., Winner, B., Neurath, M. F., & Günther, C. (2023). Impact of microbiome–brain communication on neuroinflammation and neurodegeneration. *International Journal of Molecular Sciences, 24*(19), 14925.

Strandwitz, P., Kim, K. H., Terekhova, D., Liu, J. K., Sharma, A., Levering, J., ... & Lewis, K. (2019). GABA-modulating bacteria of the human gut microbiota. *Nature Microbiology, 4*(3), 396–403. doi:10.1038/s41564-018-0307-3

Sultana, O. F., Hia, R. A., & Reddy, P. H. (2024). A combinational therapy for preventing and delaying the onset of Alzheimer's disease: A focus on probiotic and vitamin co-supplementation. *Antioxidants, 13*(2), 202.

Sun, B., Sawant, H., Borthakur, A., & Bihl, J. C. (2023). Emerging therapeutic role of gut microbial extracellular vesicles in neurological disorders. Frontiers in Neuroscience, 17, 1241418. doi:10.3389/fnins.2023.1241418

Sun, L. J., Li, J. N., & Nie, Y. Z. (2020). Gut hormones in microbiota-gut-brain cross-talk. *Chinese Medicine Journal (England), 133*(7), 826–833. doi:10.1097/cm9.0000000000000706

Tegegne, B. A., & Kebede, B. (2022). Probiotics, their prophylactic and therapeutic applications in human health development: A review of the literature. *Heliyon, 8*(6), e09725. doi:10.1016/j.heliyon.2022.e09725

Thursby, E., & Juge, N. (2017). Introduction to the human gut microbiota. *Biochemical Journal, 474*(11), 1823–1836. doi:10.1042/bcj20160510

Tran, V. T. A., Lee, L. P., & Cho, H. (2022). Neuroinflammation in neurodegeneration via microbial infections. *Frontiers in Immunology, 13,* 907804. doi:10.3389/fimmu.2022.907804

Vázquez-Cuesta, S., Lozano García, N., Rodríguez-Fernández, S., Fernández-Avila, A. I., Bermejo, J., Fernández-Avilés, F., ... & Reigadas, E. (2024). Impact of the Mediterranean diet on the gut microbiome of a well-defined cohort of healthy individuals. *Nutrients, 16*(6), 793.

Wortelboer, K., Nieuwdorp, M., & Herrema, H. (2019). Fecal microbiota transplantation beyond Clostridioides difficile infections. *eBioMedicine, 44*, 716–729. doi:10.1016/j.ebiom.2019.05.066

Yano, J. M., Yu, K., Donaldson, G. P., Shastri, G. G., Ann, P., Ma, L., ...Hsiao, E. Y. (2015). Indigenous bacteria from the gut microbiota regulate host serotonin biosynthesis. *Cell, 161*(2), 264–276. doi:10.1016/j.cell.2015.02.047

Yu, C. D., Xu, Q. J., & Chang, R. B. (2020). Vagal sensory neurons and gut-brain signaling. *Current Opinion Neurobiology, 62*, 133–140. doi:10.1016/j.conb.2020.03.006

Zhang, W., Xiao, D., Mao, Q., & Xia, H. (2023). Role of neuroinflammation in neurodegeneration development. *Signal Transduction and Targeted Therapy, 8*(1), 267. doi:10.1038/s41392-023-01486-5

Zhou, Z., Xu, N., Matei, N., McBride, D. W., Ding, Y., Liang, H., ...Zhang, J. H. (2021). Sodium butyrate attenuated neuronal apoptosis via GPR41/Gβγ/PI3K/Akt pathway after MCAO in rats. *Journal of Cerebral Blood Flow & Metabolism, 41*(2), 267–281. doi:10.1177/0271678x20910533

15 Functional Foods and Nanocarriers for the Management of Gastrointestinal Inflammations

Amit Anand[1], Kenganora Mruthunjaya[2], Santhepete Nanjundiah Manjula[3], Neeraj Kumar Fuloria[4], and Hemanth P.R. Vikram[5]*

[1,2]Department of Pharmacognosy, JSS College of Pharmacy, Mysuru, JSS Academy of Higher Education and Research, Mysuru, Karnataka, India

[3]Department of Pharmacology, JSS College of Pharmacy Mysuru, Karnataka, India

[4]Department of Pharmaceutical Chemistry, Faculty of Pharmacy, AIMST University, Semeling Campus, Bedong, Kedah, Malaysia

[5]Department of Pharmaceutical Chemistry, JSS College of Pharmacy Mysuru, JSS Academy of Higher Education and Research, Mysuru, Karnataka, India

15.1 INTRODUCTION

15.1.1 Background

When combined with the creative use of nanocarriers, functional foods offer a novel strategy for gastrointestinal health that focuses on the control of inflammation. Inflammations of the gastrointestinal tract, which include disorders such as irritable bowel syndrome (IBS) and inflammatory bowel disease (IBD), present serious obstacles to world health and call for innovative and practical treatments. Functional foods have become a viable way to support gastrointestinal health since they are customized to give therapeutic effects and are supplemented with bioactive substances (Martínez-Ballesta et al., 2018). These foods actively support the promotion of health and the avoidance of disease. Functional foods include bioactive components such as probiotics, omega-3 fatty acids, and polyphenols that have anti-inflammatory, antioxidant, and immunomodulatory effects. Assuring the precise and effective transport of these bioactive substances to the site of inflammation is the difficult part. Because of their tiny size and adaptable characteristics, nanocarriers are the perfect means of delivering bioactive substances in a precise and regulated way. By preventing these substances from breaking down, enabling controlled release, and increasing their bioavailability, nanoencapsulation maximizes their potential as medicines (Jampilek et al., 2019). Furthermore, nanocarriers have the ability to precisely transport medication to inflammatory tissues by navigating the gastrointestinal tract's complex and dynamic environment and overcoming obstacles. The combination of functional meals with nanocarriers has a lot of potential to help treat gastrointestinal inflammations. A new era in the management of gastrointestinal health is being ushered in by this multidisciplinary approach, which not only tackles the problems caused by inflammation but also creates opportunities for effective and customized therapeutic approaches. The creation of functional foods with customized nanocarriers is set to transform the gastrointestinal care landscape as research in this area progresses, providing hope for better patient outcomes and increased quality of life (Vieira da Silva et al., 2016).

15.1.2 Significance of Gastrointestinal Inflammations

Numerous illnesses, such as infections, autoimmune diseases, and food allergies, can cause inflammation of the gastrointestinal tract, which includes the stomach and small and large intestines (Rothenberg, 2004). Gastrointestinal tract inflammation can cause a number of symptoms, such as diarrhea, weight loss, and stomach pain. Chronic inflammation may make serious conditions worse, such as inflammatory bowel diseases (IBD), which include Crohn's disease and ulcerative colitis. Additionally, the immune system depends heavily on the stomach, and gastrointestinal inflammation can upset this delicate balance (Karrasch et al., 2014). Dysbiosis, or imbalances in the gut microbiota, frequently coexists with inflammation and can exacerbate immunological dysfunction. This link between immune function and gastrointestinal health highlights the effects of gastrointestinal inflammation on the body as a whole, not just the digestive tract. The importance of gastrointestinal inflammations stems from their effects on health, which include immune system modulation, potential associations with mental health, and symptoms ranging from digestive to severe chronic illnesses (Phillips et al., 2015).

15.1.3 Emerging Trends in Functional Foods and Nanocarriers

New breakthroughs in the nexus of food science, technology, and health include emerging trends in functional foods and

DOI: 10.1201/9781003493143-15

nanocarriers. The significance of gut health for general health has drawn a lot of attention. Probiotic and prebiotic-rich functional foods are gaining popularity. By bolstering the gut flora, these components enhance immunological response, digestive health, and even mental wellness (Shin et al., 2015). The creation of functional plant-based meals has been fueled by the growing popularity of plant-based diets. These goods have additional health advantages in addition to tasting and feeling similar to traditional animal-based foods. Omega-3 fatty acids, plant-based proteins, and other useful ingredients are being added to a variety of food products. Food-grade nanocarriers, like liposomes and nanoparticles, are employed to encapsulate bioactive ingredients. This improves the body's ability to absorb minerals, vitamins, and antioxidants by increasing their bioavailability. Sensitive bioactive chemicals are shielded by nanocarriers from environmental variables like heat, light, and oxygen. Food goods have a longer shelf life and retain their nutritious content over time because of this increased stability (Singh, 2016). Targeted delivery of particular substances to particular cells or targets is made possible by nanocarriers. This is especially helpful for creating functional foods that target particular health concerns (P. M. Silva et al., 2023).

15.2 GASTROINTESTINAL INFLAMMATIONS: CAUSES AND CONSEQUENCES

15.2.1 Overview of Gastrointestinal Inflammatory Conditions

A collection of illnesses involving inflammation of the digestive tract are referred to as gastrointestinal inflammatory diseases. The stomach, small, large, and other related organs, as well as the oesophagus, can all be impacted by these disorders that affect the gastrointestinal (GI) system. Numerous symptoms and problems can result from gastrointestinal tract inflammation (Marusawa & Jenkins, 2014). Adverse microenvironmental variables, including pH variations, enzymes, and bacterial flora, pose a persistent challenge to the cells that line the mucosa of the human gastrointestinal system. These microenvironments, which do not include the oral cavity, also include leftover mixtures of bacteria and release enzymes from upper organs along the route. The GI epithelia respond to the hostile microenvironment by accelerating cell exfoliation, boosting peristalsis, eradicating bacteria by secreting plasma cell-immunoglobulins, and enhancing the synthesis of naturally occurring antibacterial substances including lysozyme and defensin-5 (Ramos De Mattos et al., 2015). It was discovered very recently that lysozyme was up-regulated in Crohn's colitis, collagenous colitis, atrophic duodenitis brought on by gluten, Barrett's oesophagitis, and chronic gastritis. This up-regulation is a reaction to the unique strains of bacteria that have been shown to be associated with certain illnesses recently. The purpose of upregulating lysozyme is to shield specific mucosal segments against long-term inflammation (Holtmann et al., 2018). Excessively high concentrations of oxygen free radicals can be chemically reactive and have a deleterious effect on living things by inducing "oxidative stress," which can lead to significant cellular components being damaged. Nitric oxide, superoxide anion and related ROS (reactive oxygen species) play a significant role in mediating the signaling processes. Ironically, many reactive oxygen-mediated responses can incite "redox homeostasis," which protects the cells against oxidative stress (Parker et al., 2018). Numerous gastrointestinal (GI) disorders, such as pancreatitis and liver cirrhosis, as well as gastritis, enteritis, colitis, and related malignancies, have been linked to the development of free radicals. Although the lower oesophageal sphincter (LES) is the primary cause of gastroesophageal reflux, other factors may also play a role in its development. Gastroesophageal reflux disease (GERD) is influenced by both pathologic and physiological variables. Transient relaxations of the lower oesophageal sphincter are the most frequent cause. TLESRs are brief, independent of swallowing episodes of suppression of the lower oesophageal sphincter tone. Although these are physiological in origin, they occur more frequently in the postprandial phase and significantly aggravate acid reflux in GERD patients (Hostetter & Uzal, 2022). Hiatal hernias, decreased lower oesophageal sphincter (LES) pressure, poor oesophageal clearance, and delayed stomach emptying are further problems. Other dyspeptic symptoms like fullness, bloating, early satiety, and nausea may also be present. The hallmark of Crohn's disease and ulcerative colitis is chronic and reiterative intestinal inflammation. This condition is also known as inflammatory bowel disease (IBD). IBD is caused by a complicated combination between immune responses, environmental, or microbial factors, and genetics, even if the exact cause of the disease is yet unknown. A variety of extraintestinal symptoms, such as arthritis, erythema nodosum, pyoderma gangrenosum, sclerosing cholangitis, uveitis, and iritis, are linked to inflammatory bowel disease (IBD). IBD is most likely caused by a combination of immunological, environmental, and genetic factors. In the intestinal mucosa, macrophages are primarily responsible for the development of noncaseous epithelioid granulomas. TNF- and other cytokines, as well as interleukins (IL-6, IL-8), are produced by activated macrophages. Regularly used medications include corticosteroids, amino salicylates, and immunosuppressive drugs such azathioprine. In certain situations, other medications like metronidazole and broad-spectrum antibiotics, as well as salts containing arsenical, sodium cromoglycate, bismuth, and colestyramine, can be beneficial. Because these antibody treatments alter the impacted biochemical inflammatory pathways, a novel strategy for treating IBD utilizing humanized monoclonal antibody preparations has shown promising results and may offer an alternative to conventional treatments. Other immunosuppressants, including azathioprine, are better for long-term maintenance but have a limited use in the acute setting because it takes several weeks for them to have a therapeutic impact. The most prevalent gastrointestinal ailment that gastroenterologists diagnose and treat in general practice is irritable bowel syndrome, or IBS (Sajadinejad et al., 2012).

Three main mechanisms, psychosocial variables, altered intestinal motility, and/or increased intestinal sensory function interact in IBS, a biopsychosocial condition. Novel strategies for treating the difficult symptoms of pain, constipation, and bloating include alosetron, a 5-HT3 antagonist, tegaserod, a partial 5-HT4 agonist, κ-opioid agonists, and neurokinin antagonists. A sensitivity to gluten is the cause of the chronic intestinal illness known as celiac disease. Immune-mediated enteropathy, which is linked to poor digestion and vitamin and nutrient loss, is its defining feature. These include diarrhea, weight loss, anemia, edema, and discomfort in the bones or joints in addition to stomach pain, gas, and bloating. The majority of patients with celiac disease are either misdiagnosed or present with unusual symptoms because the condition might be silent. Moreover, there may be severe small bowel inflammation in the absence of any gastrointestinal symptoms. Early identification is crucial because untreated celiac disease results in growth retardation in children and unusual symptoms such as neurological or infertility. Though the precise origin of celiac disease is uncertain, immune-mediated factors (tissue-transglutaminase autoantigen) are likely to be the main contributing factor; the condition is frequently hereditary. The management involves eliminating gluten from the diet for the rest of one's life, which significantly improves both the histology and the clinical results. A common gastrointestinal ailment that is linked to considerable morbidity and medical expenses is diverticulitis. With successive episodes, diverticulitis complications like abdominal sepsis are less likely to happen. Numerous risk factors, including nutrition, physical inactivity, and obesity, have been discovered but many of them are controllable (Babu et al., 2015). Diverticulitis can also occur due to some changes in the neuromuscular structure of the colon as well as genetic factors. One of the most prevalent, dangerous, and sneaky chronic disorders affecting humans is chronic gastritis. It is possible to assume that over half of the world's population suffers from this illness to some degree, meaning that hundreds of millions of individuals could be affected by chronic gastritis globally. Although the involvement of gastritis in the pathophysiology of common peptic ulcers and stomach malignancies is evident, the relevance of chronic gastritis as a dangerous disease is usually underestimated in clinical practice. One of the most prevalent and deadly pandemic infections, persistent gastritis can have life-threatening consequences similar to peptic ulcers (Wilkins et al., 2013).

15.2.2 Factors Contributing to Gastrointestinal Inflammations

Different things can induce gastrointestinal inflammations, and each individual may have different underlying causes. With an incidence rate of 18% worldwide, irritable bowel syndrome (IBS) is a chronic and incapacitating functional gastrointestinal illness. Although the etiology of IBS is poorly understood, a number of potential processes, including aberrant brain-gut interactions, visceral hypersensitivity, irregular gut motility, and the involvement of infectious agents, are thought to play a role in the onset and progression of the illness (Jia et al., 2018). Various investigations have shown changes in the concentrations of cytokines, mast cells, T-lymphocytes, and B-lymphocytes in the intestinal mucosa or systemic circulation, which are likely to have a role in the development of IBS. Thus, those with a hereditary tendency may acquire IBS. Researchers have looked at the role infections play in the onset and aggravation of IBS, as well as the potential involvement of certain pathogens in the disease's development and progression. The current study finds that the most prevalent bacterial, viral, and parasite pathogens contribute to the development and the progression of irritable bowel syndrome (IBS) (Ullman & Itzkowitz, 2011). Advances in the immune systems linked to persistent inflammatory disorders have resulted in an enhanced comprehension of the origins of autoimmune disorders affecting the gastrointestinal tract, such as autoimmune atrophic gastritis, celiac disease, autoimmune enteropathy, and ulcerative colitis. The tissue damage that results in the atrophy of the stomach mucosa in autoimmune atrophic gastritis, villous flattening of the small intestine in autoimmune enteropathy and celiac disease, and mucosal ulceration of the colon in ulcerative colitis is recognized as one of the main mechanisms of T-cell-mediated inflammation. These mechanisms include an increase in pro-inflammatory cytokines from T-helper cell type, abnormal regulatory T-cell function loss, or self-tolerance. One of the processes resulting in anti-inflammatory responses is vitamin D-modulated T-cell proliferation, which is mediated through the enhanced proliferation of the CD8α T-cell subgroup (Alam et al., 2017). Unlike CD8+ T cells, these T-cell types are not cytotoxic, yet they might still be useful in reducing gastrointestinal inflammation. Stress is characterized as an acute danger to homeostasis, which can affect the function of the gastrointestinal system both short and long-term and cause an adaptive or allostatic response (Brenchley & Douek, 2008). The paraventricular nucleus of the hypothalamus, the amygdala, and the periaqueductal gray comprise the brain's neuronal network, known as the central stress circuitry, which is responsible for producing the stress response. It gets information from the visceral motor cortex, which includes the medial prefrontal, anterior cingulate, and insular cortex, as well as from the somatic and visceral afferent pathways. The limbic system, the hypothalamus-pituitary-adrenal axis, and the pain modulatory systems are all part of the emotional motor system, which is the result of this central stress circuit. One important modulator of the central stress response is corticotropin releasing factor (CRF). There are two known CRF receptor subtypes: R1 and R2. Through receptors in the medial prefrontal cortex and hippocampus, circulating glucocorticoids dampen the stress response. There are other more neurotransmitters and neuroimmunomodulators under evaluation. Intestinal permeability to big antigenic compounds is increased under stress. Degranulation, depletion of colonic mucin, and activation of mast cells are possible outcomes. Additionally, stress raises the risk of colonic inflammation, which is adaptively transferred from rats to other rats by CD4+ cells that have become sensitized. Research is now being conducted to determine whether stress is linked to any of the gastrointestinal problems, such as peptic ulcer disease, inflammatory bowel disease, functional bowel disorders, and gastroesophageal reflux disease. With opposing effects in CD and UC, tobacco is the most well-established environmental factor influencing the propensity to acquire IBD and maybe its clinical outcome. However, evaluating how smoking affects the long-term clinical outcomes of IBD is done retrospectively, which frequently yields unpopular conclusions. Dietary components are the most prevalent luminal antigens in the intestine, along with bacteria, and they may have an impact on intestinal inflammation. A direct antigenic effect, modifications to gene expression, regulation of inflammatory mediators (e.g., eicosanoids), variations in the enteric flora's composition, and effects on gut permeability are a few examples of potential pathways (Ptak-Belowska et al., 2012).

15.2.3 Impact on Health and Quality of Life

An individual's quality of life can be greatly impacted by intestinal inflammations. The particular ailment, its chronicity, and the quality of its management all influence how severe the effects are (Drossman et al., 1989). Gastrointestinal tract inflammation frequently results in pain and discomfort. Inflammations of the stomach can impact the digestion and absorption of nutrients. Nutritional deficits and weight loss can result from conditions such as celiac disease and inflammatory bowel disease (IBD), which cause malabsorption of vital nutrients. Dietary restrictions are necessary for many gastrointestinal diseases (Graff et al., 2006). Long-term digestive disorders might cause lower productivity or absenteeism at work. Gastrointestinal pain, discomfort, and worry can cause sleep patterns to be disturbed, which can result in exhaustion and worsen general health. It is crucial to note that improvements in lifestyle management techniques and medical therapies can greatly enhance the quality of life for those suffering from gastrointestinal inflammations. (Chang, 2004)

15.3 FUNCTIONAL FOODS: CONCEPT AND MECHANISMS

15.3.1 Definition and Classification of Functional Foods

The science of nutrition itself has altered, leading to the introduction of functional food as a novel idea. The twentieth century saw the discovery of necessary nutrients and the establishment of nutrient standards, dietary guidelines, and food guides by the nutritionists. These efforts were mostly, if not completely, directed towards preventing deficiencies and promoting the growth, maintenance, and development of the body (Doyon & Labrecque, 2008). Beyond having adequate nutritional benefits, a food can be deemed functional if it can be demonstrated with sufficient evidence to change one or more target bodily processes in a way that significantly impacts one's overall health and well-being or lowers their risk of disease. Functional foods are anticipated to be approved for use in health claims based on either increased function or decreased risk of disease (Martirosyan et al., 2022). Their growth presents a special chance to support the enhancement of the food quality that consumers can choose for the sake of their health and well-being. But the only thing that can ensure its effectiveness is a rigorous scientific method yielding reliable facts (Granato et al., 2020). The inulin-type fructans, which are naturally occurring food components present in a variety of edible plants, are the topic of discussion regarding functional food components. These are dietary fibers, which are nondigestible oligosaccharides. Reduction of the risk of certain disorders such as constipation, type-2 diabetes, obesity, osteoporosis, or colon cancer is another potential health advantage (Figure 15.1) (Sanders, 1998).

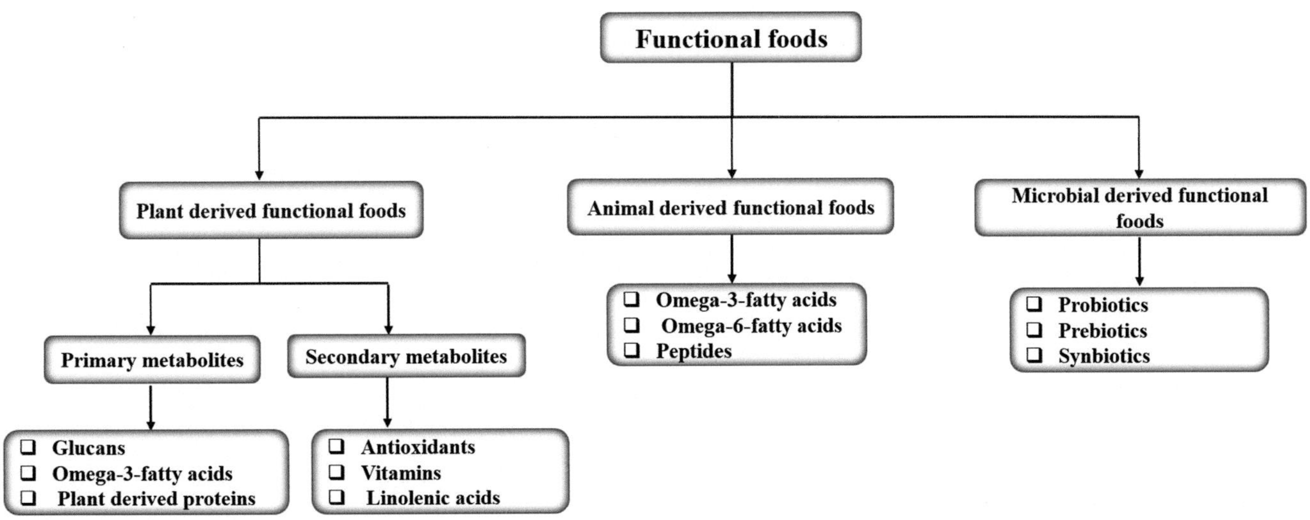

FIGURE 15.1 Schematic representation of classification of various functional foods.

15.3.2 Bioactive Compounds in Functional Foods

The naturally occurring chemical compounds found in or obtained from plants, animals, or marine sources that have the desired health or well-being effects are known as bioactive compounds. There is a belief that the nutrients found in food assist in maintaining our systems healthy and nourish our bodies (Martirosyan & Miller, 2018). Foods, particularly functional foods, include bioactive chemicals, which have positive effects on health. These advantages, which include anti-inflammatory, antifungal, antioxidant, and other preventive qualities, show how bioactive chemicals are the actual embodiment of Hippocrates' notion, "let thy food be thy medicine." Because they are the components that give functional foods their unique qualities, bioactive chemicals must be included in the functional food category (Konstantinidi & Koutelidakis, 2019). Bioactive substances found in functional foods have the potential to improve health because they target the systems involved in the management, prevention, and/or treatment of disease (Soumya et al., 2021).

15.3.3 Mechanisms of Action in Gastrointestinal Health

Prebiotics have been linked to numerous health benefits in the last few decades. Studies on the gastrointestinal (GI) tract have demonstrated these advantages through the production of short-chain fatty acids, the reduction of harmful bacterial populations, the prevention of immune system modification or injury, and the enhancement of gut barrier function (Mózsik et al., 2001). With respect to possible impacts on the GI system, nondigestible carbohydrates, like prebiotics, have the ability to significantly alter the makeup and function of intestinal microbiota. Because the enzymes needed to hydrolyze the polymer bonds in these prebiotics are typically absent from the human gut, the probiotics are able to withstand small-intestinal digestion and make it all the way to the colon, where they are fermented by good bacteria like Bifidobacteria and Lactobacilli (Yang et al., 2020). Dietary approaches to modify gut microbiota are becoming more and more popular because they have been linked to the pathophysiology of a number of GI illnesses. Because many of these polysaccharides can be metabolized by intestinal bacteria and result in the generation of SCFAs (Short-Chain Fatty Acids), studies have been concentrated on the use of prebiotics (Dupont, 2016). Prebiotics have the unique benefit of fostering the growth of target microorganisms, which in turn drive out harmful species by competing with them and safeguarding or enhancing the production of beneficial fermentation products, like SCFAs, which have immunomodulatory qualities and affect the production of pro-inflammatory cytokines and toll-like receptor-4 signaling. Lipids and other food ingredients can be made to oxidize more quickly by the release of gastric acid. Research suggests that dietary antioxidants, such as inulin, might help stop lipid peroxidation in the stomach. The addition of inulin or oligofructose supplements to the diet helps shield the body from oxidative stress and, as a result, reduces oxidative stress-related inflammatory responses. Probiotics work primarily through enhancing the epithelial barrier, increasing adhesion to intestinal mucosa and simultaneously inhibiting pathogen adhesion, producing anti-microorganism substances, competitively excluding pathogenic microorganisms, and modulating the immune system (Zhao et al., 2019). The intestinal barrier is a crucial defense mechanism that protects the organism from the environment and maintains the integrity of the epithelium. The mucous layer, secretory IgA, antimicrobial peptides, and the epithelial junction adhesion complex are the intestinal barrier's defenses. When this barrier function is compromised, food and bacterial antigens can enter the submucosa and trigger inflammatory reactions that can lead to intestinal conditions including inflammatory bowel disease. The function of the intestinal barrier may be influenced by the consumption of non-pathogenic bacteria, and probiotic bacteria have been the subject of much research due to their role in barrier maintenance. Probiotics improve the function of the intestinal barrier; however, the exact mechanisms are still unknown. Enhancing the expression of genes involved in tight junction signaling may be one way to support intestinal barrier integrity, for example, in a T84 cell barrier model, lactobacilli regulate many genes encoding adherence junction proteins, including E-cadherin and catenin. Additionally, the phosphorylation of adherence junction proteins and the quantity of protein kinase C (PKC) isoforms, are significantly influenced by the incubation of intestinal cells with lactobacilli, which positively modulates the function of the epithelial barrier. Cholesterol is the starting point for the synthesis of primary bile acids in the liver, such as chenodeoxycholic acid and cholic acid. The intestinal bacteria transform them to produce lithocholic acid, urso-deoxycholic acid (UDCA), deoxycholic acid (DCA), and secondary bile acids. In reaction to cholecystokinin (CCK), which is produced postprandially by duodenal cells, bile is secreted into the intestinal lumen. While the saturated digestive enzymes are often removed and expelled, the bile acids are reabsorbed together with dietary fats and returned to the liver through the enterohepatic circulation. Bile acids, however, have the potential to allow proteases to cross the intestinal barrier and trigger systemic inflammatory reactions (Johnson, 2001). The exocrine pancreas' acinar cells generate a range of enzymes that can break down proteins (trypsinogen, chymotrypsinogen, carboxypeptidase, elastases), lipids (pancreatic lipase, phospholipase), and carbohydrates (pancreatic amylase) when they become activated in the duodenal lumen. However, proteases may play a role in the breakdown of the epithelial tight junctions and protective mucus layer during intestinal ischemia, which may increase gut permeability and allow serine proteases of the gut wall to penetrate the wall. Reducing the input of pancreatic enzymes into the intestinal lumen prevented mucus layer breakdown, endothelial cell toxicity, and circulating neutrophil activation (Dimidi et al., 2019). The health benefits of prebiotics and probiotics in the human gut are explained through a mechanism depicted in Figure 15.2.

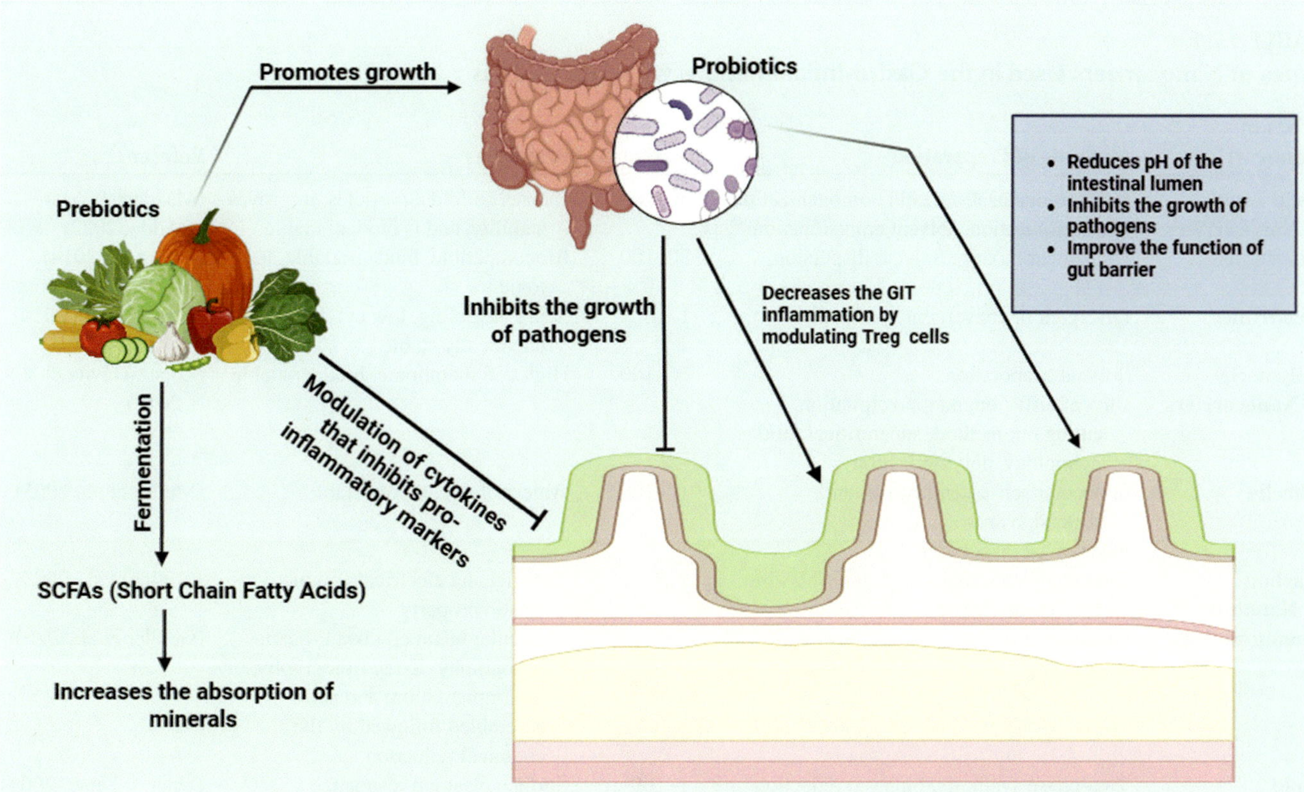

FIGURE 15.2 Mechanism of action of probiotics and prebiotics in the management of gastrointestinal inflammations.

15.4 NANOCARRIERS: DELIVERY SYSTEMS FOR ENHANCED EFFICACY

15.4.1 Introduction to Nanocarriers

Recent studies on nanobiotechnology have produced a number of fundamental and practical advancements in the medical field. Drug delivery and diagnostic instruments have been made more effective with the use of nanocarriers (Wang et al., 2019). Many types of nanostructures, including nanoparticles, nanocomposites, nanotubes, and nano-fibers, are useful for the identification and management of a wide range of illnesses. These nanostructures are also used as medications, vaccines, genes, proteins, and enzymes' carrier molecules or transportation agents (Gardikis et al., 2006). Simply nanocarriers are colloidal nanoparticles that are frequently utilized to deliver medicinal agents or other substances to a specific place. The diameter of the nanocarriers ranges from 1 to 100 nanometers (nm). Since these nanocarriers are typically thought to be harmless and inactive, they offer good biocompatibility. Nanocarriers' physiochemical characteristics, such as their surface, composition, and shape, can be changed to increase their activity while reducing their side effects. As a result, it has a wide range of effects on the delivery of drugs (Farah, 2019).

15.4.2 Types of Nanocarriers

Three types of nanocarriers can be distinguished based on their high surface-to-volume ratio: organic, inorganic, and hybrid nanocarriers (RR, 2017). The different types of nanocarriers used as a drug delivery system in gastro inflammations have been illustrated in Table 15.1 and its structure is depicted in Figure 15.3. Among the nanoparticles categorized as organic nanocarriers are solid lipid nanocarriers, liposomes, dendrimers, polymeric nanocarriers, and micelles. These more versatile, less hazardous organic nanocarriers can combine ligands with a variety of drugs to improve drug delivery (Mishra et al., 2010).

Micelles and liposome mediated drug delivery are two types of organic nanocarriers that can accumulate at the intended region due to their increased permeability and retention properties. Melted solid lipids are dispersed in water to create the solid lipid nanocarriers; they are subsequently stabilized by adding emulsifiers using micro-emulsification or high-pressure homogenization. Solid lipid nanocarriers are frequently made from room-temperature lipids, such as triglycerides, monohydric, di hydric or polyhydric alcohols, waxes, or free fatty acids (R. K. Mishra et al., 2018). A polymer-lipid hybrid nanocarrier, has been investigated as a potential oral medication delivery method. The disadvantages of traditional solid lipid nanocarriers are found to be overcome by a few next-generation nanocarriers, such as lipid drug conjugates and nanostructured lipid carriers. Any therapeutic molecule can be delivered at a specific location using a solid lipid nanocarrier, which can serve as the "perfect tailor-made carrier." Liposomes are aqueous core-encased spherical vesicles composed of lipid bilayers. They

TABLE 15.1
Types of Nanocarriers Used in the Gastro-Inflammations with Applications

Types of Nanocarriers	Methods of Preparation	Size(nm)	Properties	References
Solid Lipid Nanocarriers	Hot homogenization, cold homogenization, ultrasonication, solvent emulsification	50–100	Carries colloidal particles, improves stability, and is biodegradable	(Mazdaei & Asare-Addo, 2022)
Liposome	High shear mixing, Solvent dispersion,	50–100	Biocompatible, biodegradable, less toxic	(Sun et al., 2014)
Dendrimer	Divergent or convergent synthesis	1–10	Self-assembling, low cytotoxicity, chemically stable	(Abbasi et al., 2014)
Polymeric Nanocarriers	Solvent evaporation, solvent diffusion, nanoprecipitation, salting out method, supercritical fluid technology, polymerization	10–100	High cell membrane, biodegradable	(López-Dávila et al., 2012)
Micelle	molecular self-assembly, solvent dispersion and mechanical dispersion	10–100	Water soluble, thermostable	(Muller et al., 2005)
Carbon Nanotubes	Chemical vapor deposition, laser blasting, carbon arc discharge	1–5	Thermal and electrical conductivity, elastic property	(Muller et al., 2005)
Quantum Dots	Quantum dot		Molecular beam epitaxy, e-beam lithography, X-ray lithography, ion implantation, and self-assembled followed by the chemical reduction	(Geißler et al., 2010)
Gold Nanocarriers	Two-phase synthesis, biphasic reduction	1–100	Surface plasmon resonance property, good biocompatibility, versatility, low toxicity, and multi-surface functionality	(Üner & Yener, 2007)
Magnetic Nanocarriers	Metal alkoxide hydrolysis, coprecipitation in microemulsion hydroxide coprecipitation, glycol-thermal synthesis, citrate gel	1–100	Super para-magnetism, chemical stability, high colloidal stability	(Hallan et al., 2016)

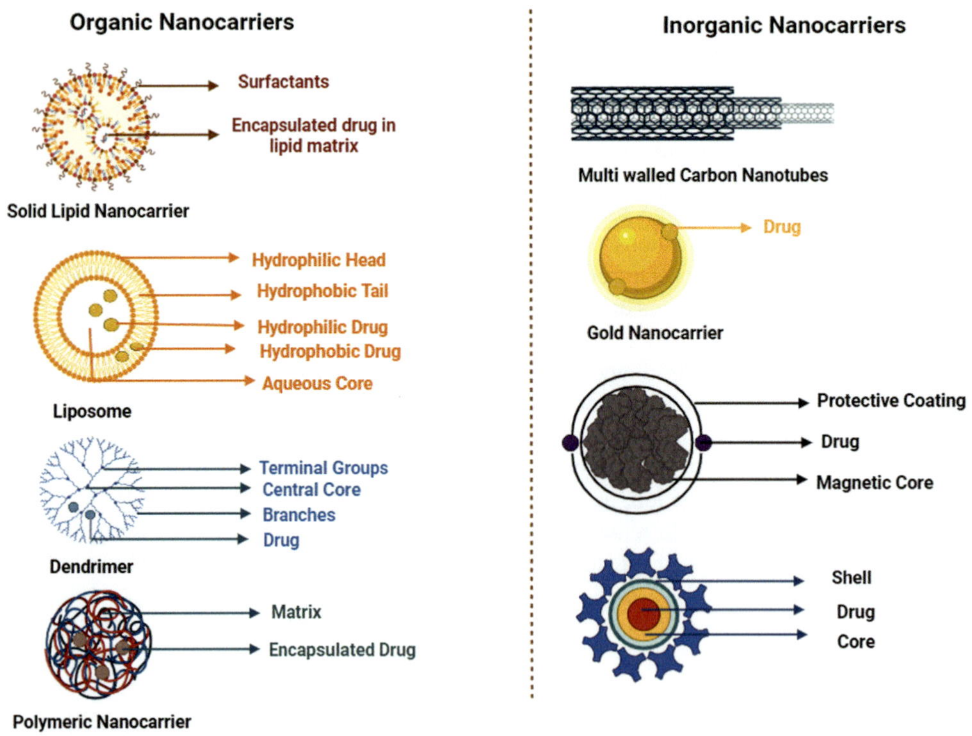

FIGURE 15.3 Various types of organic and inorganic nanocarriers as a drug delivery system for the gastrointestinal inflammations.

are helpful in getting lipophilic and hydrophilic medications to the right places. The bilayer distinguishes between a unilamellar vesicle, or one bilayer, and a multilamellar vesicle, or multiple bilayers. Physiologically active chemicals are transported by this vesicle to their intended site. Thus, PEGylated liposomes, also known as stealth liposomes, can be created by coating liposomes with polymeric molecules such as polyethylene glycol. Because it avoids the clearance of the Reticule Endothelial System, this stealth liposome has a lengthy half-life in the blood and high stability, resulting in a prolonged release of medication (Khan, 2010). Branched macromolecules called dendrimers have an initiator core in the center that gives birth to different arms, or terminal active groups. Amino acids, sugar compounds, and nucleotides can all be converted into dendrimers. Their unusual molecular weight, multivalent nature, and extensive branching structure, along with a variety of peripheral groups, make them an exceptional drug delivery vehicle. The drug molecules can be contained in the core cavities via hydrogen bonds with enhanced surface functionality, hydrophobic bonding, or chemical interactions. Covalent bonds can also be used to bind the medication molecules to terminal active groups (Kowalczuk et al., 2014). The creation of chemical and physical bonds is the main method via which drugs interact with dendrimers. Colloidal, solid polymeric nanoparticles can be made from any biodegradable polymer. The drug molecules can be ensnared in the polymer matrix by a matrix type (nanospheres) or dispersed in the polymer core by a reservoir type (nanocapsules). Natural polymers including chitosan, gelatin, albumin, collagen, alginate, etc., as well as synthetic polymers like polyethylene glycol, polyglutamic acid, and polycaprolactone, can be used to create polymeric nanocarriers. This polymeric nanocarrier has advantages over other ones in terms of stability, drug payload, half-life in systemic circulation, and extended drug release (Khizar et al., 2022). The physiochemical characteristics of the polymeric source can be altered to improve the controlled release of the medication. Additionally, advances have been made in the manufacturing of polymeric nanoparticles for the targeted administration of drugs using stimuli-sensitive polymers. When exposed to internal stimuli (low pH, redox, enzyme) as well as exterior stimuli (temperature, light, magnetic field, and electric field), these smart polymers release the medication. Smart polymer design problems include scalability, biocompatibility/toxicity, and stimuli sensitivity. Mesoporous silica, magnetic nanocarriers, quantum dots, and gold are examples of inorganic nanocarriers (Lin et al., 2011). Effective applications for inorganic nanocarriers include biosensing, cell labeling, targeting, imaging, and diagnostics. Furthermore, plasmonic, and magnetic capabilities can be achieved by varying the size or composition of the inorganic nanocarriers. On the other hand, using heavy metals as inorganic nanocarriers would cause long-term health problems. Because of their unique physicochemical and biological characteristics, carbon nanotubes are a perfect and promising material for drug delivery. These are hollow, tube-like structures with graphene sheets coiled at precise angles inside of them. Depending on how many graphene sheets are wrapped together, the carbon nanotubes can have one or many walls. These tubes have a cross-section that ranges in diameter from 0.4 to 100 nm, and their lengths extend thousands of times their diameters. The functionalized nanotubes have a prolonged half-life in the serum and are soluble in water. Non-functionalized carbon nanotubes, on the other hand, are poisonous and insoluble in water. Its flexibility, surface modification, and structural stability make it an effective agent for targeting cancer cells. Because of their inherent qualities, carbon nanotubes are a perfect source for numerous industrial applications in addition to their use in biomedicine. Another significant carbon-based nanocarrier that works well for drug delivery is graphene. Atoms in the periodic table's II–VI or III–V element groups are examples of quantum dots. Large-size quantum dots produce red fluorescence, whereas smaller ones produce blue fluorescence (Hillaireau & Couvreur, 2009). The light emission between UV-near IR wavelengths is impacted by this difference in quantum dot size. It is also known that these quantum dots work well as reporter and delivery systems. The term "micelle" was first used by McBain in 1913 to describe the colloidal aggregates that are created when detergent and water are mixed (liquid colloid). These molecules are amphiphilic, with a polar head that is hydrophilic and in touch with the external solvent and a nonpolar tail that is hydrophobic and facing the center. Similarly, an amphiphilic or detergent molecule in a nonpolar solvent can yield an inverse micelle, which has its head towards the center and its tail facing the outside. The size and shape of the generated micelle nanoparticles are determined by the amphiphilic molecule and the solution conditions (pH, ionic strength, and temperature) (Sadiq et al., 2021).

15.5 FUNCTIONAL FOODS IN THE MANAGEMENT OF GASTROINTESTINAL INFLAMMATIONS

15.5.1 Probiotics and Prebiotics

There are germs in our stomach that are genuinely helpful. These "friendly" bacteria aid in preventing the growth of yeast and harmful bacteria in the digestive system. Additionally, bacteria support healthy immune system function and the production of vitamin K. Friendly bacteria predominate in our bodies normally, but antibiotic treatment, stress, and unhealthful diets can all lead to intestinal dysbiosis, a bacterial imbalance that promotes the growth of yeast and harmful bacteria (Thomas et al., 2010). Treatment with antibiotics is a common cause of dysbiosis. The beneficial bacteria in our digestive tracts are eradicated by the antibiotics which are taken to treat infections. The balance of bacteria in our digestive tract can be restored with the use of prebiotics and probiotics. Probiotics are good microorganisms that are present in a variety of diets. These beneficial bacteria are introduced into our digestive tract when we consume probiotics. The bacterial families Lactobacillus and Bifidobacterium have common strains. Fermented foods like yogurt and sauerkraut naturally contain probiotic bacteria called lactobacilli. Prebiotics are indigestible meals that pass through the digestive system and support the growth and health of beneficial bacteria. Prebiotics maintain the health of good bacteria (Figueroa-González et al., 2011). The majority of the prebiotics that nourish the good bacteria in the stomach are derived from oligosaccharides, which are carbohydrate fibers. Whole grains, legumes, and fruits are sources of oligosaccharides. Numerous investigations have exhibited the potentially broad influence of prebiotics on the gut microbiota's composition, potentially inducing beneficial gut commensals apart from lactic acid effects on the gut, either directly or indirectly. Through the clarification of probiosis and prebiosis mechanisms, scientists can create functional meals that are optimized to boost host health. Probiotics and a prebiotic found in symbiotic fermented milk may help to strengthen intestinal health as well as the host's humoral and cell-mediated immunity (Cunningham et al., 2021).

15.5.2 Dietary Fiber and Antioxidants

Dietary fiber is necessary for preserving a healthy digestive tract and can be found in plant foods including fruit, vegetables, and whole grains. Food contains two different forms of fiber: soluble and insoluble (Das et al., 2020). Fruits, beans, and oat products contain soluble fiber, which dissolves in water and can help control blood sugar and reduce blood fat. Because insoluble fiber does not dissolve in water, it goes straight through the digestive tract. It speeds up the rate at which food moves through the digestive system and is present in vegetables and whole-grain products. In addition to aiding with digestion, fiber can help decrease blood cholesterol. Fiber helps lower the incidence of diverticular disease, constipation, and gastro-inflammation by promoting bowel regularity and maintaining a clean gastrointestinal tract. Consuming plant-based fibers in the diet is essential for preserving gut health and lowering the absorption of glucose, both of which can help those with diabetes (Eskicioglu et al., 2015). Insoluble fibers, such as cellulose and hemicelluloses, which are present in leafy vegetables, bran, and fruits act as roughage and lower the caloric content of food. Whole grains like barley and oats, as well as fruits like ripe strawberries and bananas contain soluble fiber also known as gums and pectin forms a viscous, indigestible mass that aids in trapping digestive enzymes. A diet rich in carbohydrates in grains, fruits, leaves, and cellular tubers has been shown to develop a gut microbiome. The human small intestine's microbiota is maintained by the quick uptake and fermentation of accessible carbohydrates. Antioxidants can be divided into two primary classes based on their manner of action. The first group consists of chemicals that stop the spread of the free radical chain by stabilizing electrons that have been moved to radicals or donating hydrogen to them (Goni & Hervert-Hernandez, 2011). Tocopherols, gallusans, and hydrochinons all exhibit this method of action. The second set of individuals exhibits a synergistic way of operation. It contains chelators that bind ions involved in the production of free radicals as well as oxygen scavengers. Their main action is to supply hydrogen to phenoxy radicals, which causes antioxidants to revert to their original role. The immune system needs to interact with different cell types, antigens, and inflammatory mediators in order for it to work properly. IBD-related inflammatory lesions are believed to result from immune system dysregulation, particularly with regard to T-helper cell populations. As a result, neutrophils, eosinophils, and lymphocytes invade the body in a mixed manner. This lists the impacts of PUFAs that are known to affect each of the different intestinal immune system components (Wilson et al., 2017).

15.5.3 Role of Phytochemicals

In order to defend themselves against insects, herbivorous animals, and other environmental stressors, plants create phytochemicals, which are non-nutrients, through primary and secondary metabolism (Sharma et al., 2019). The term "phenolic compounds" refers to a broad class of substances found in many fruits and vegetables, such as apples, grapes, strawberries, celery, carrots, etc. These compounds include flavonoids, phenolic acids, stilbenes, lignans, curcuminoids, and tannins. The largest family of phenolic chemicals consists of nearly 4,000 low molecular-weight molecules, known as flavonoids. Their basic structure is a fifteen-carbon skeleton, which is further subdivided into anthocyanins, iso-flavonoids, neo-flavonoids, bioflavonoids, flavanols, and flavones (Bloch & Thomson, 1997). A previous investigation into the bioactivities of flavonoids demonstrated their potential as antioxidants with anti-inflammatory properties and as modulators of gene expression linked to inflammation. Phenylpropanoid moieties combine to generate phenolic chemicals known as lignans. Major components of lignans include lignanolides, cyclolignaolides, neolignans, and bisepoxylignans. Intestinal bacteria metabolize lignans to produce mammalian lignans known as enterolignans in addition to their parent forms. They reduce oxidative stress, inhibit the expression of inflammatory cytokines, and prevent the synthesis of inflammatory genes. In addition, there are numerous more varieties of phenolics, such as coumarins, curcuminoids, hydrolyzable tannins, and condensed tannins. By preventing NF-jB activation and perhaps having antioxidative effects, it decreased neutrophil infiltration and prevented mucosal damage to the intestinal epithelial layer (Krishnan, 2017). A significant class of molecules known as terpenoids is mostly produced by flowering plants. They are divided into hemiterpenoids, monoterpenoids, sesquiterpenoids, etc. according to the quantity of isoprene units they contain. Terpenoids have antioxidant properties, lessen tissue damage, and suppress the expression of certain cytokines. Terpenoids work by downregulating the expression of NF-jB to suppress inflammation. Colic erosion is lessened with crocetin. This reveals the potential role of terpenoids in decreasing permeability, inflammation, and colonic damage in the treatment of colitis (Chandrasekara & Kumar, 2016). The effect of phytochemicals on gastrointestinal inflammation with their mechanism of action is illustrated in Table 15.2.

TABLE 15.2
Effects of Phytochemicals on Gastrointestinal Inflammation with their Potential Mechanism of Action

Phytochemicals	Source	Mechanism of Action	Effects on Gastro Intestinal Inflammation	References
Curcumin	Turmeric	Anti-inflammatory, antioxidant	Inhibits inflammatory pathways, reduces oxidative stress	(Sandur et al., 2007)
Quercetin	Fruits, vegetables	Anti-inflammatory, antioxidant	Suppresses inflammatory mediators, scavenges free radicals	(Anand David et al., 2016)
Resveratrol	Grapes, red wine	Anti-inflammatory, antioxidant	Modulates inflammatory signaling, protects against oxidative damage	(Magrone et al., 2020)
Gingerol	Ginger	Anti-inflammatory, antioxidative	Inhibits inflammatory pathways, reduces oxidative stress	(Mashhadi et al., 2012)
Allicin	Garlic	Anti-inflammatory, antimicrobial	Suppresses inflammation, fights against microbial infections	(Hussein et al., 2017)
Epigallocatechin gallate (EGCG)	Green tea	Anti-inflammatory, antioxidant	Modulates inflammatory responses, scavenges free radicals	(Mokra et al., 2023)
Berberine	Various plants, including barberry	Anti-inflammatory, antimicrobial	Suppresses inflammation, exhibits antimicrobial effects	(El-Zahar et al., 2022)

15.6 NANOCARRIERS FOR TARGETED DRUG DELIVERY IN GASTROINTESTINAL INFLAMMATIONS

15.6.1 Challenges in Conventional Drug Delivery

The cornerstone of traditional treatment consists of regular prescriptions for high doses of immune-suppressive or anti-inflammatory drugs, many of which have serious adverse effects. The best way to apply topically active drugs depends on where the inflammation is located (Kratz & Warnecke, 2012). Rectal formulations like foams, suppositories, or enemas are advised for ulcerative colitis because they efficiently reach the rectum and lower sections of the descending colon. However, in the case of pancolitis, which also affects the transverse and ascending colon, they are ineffective. In this case, as well as in Crohn's disease, which mostly affects the distal ileum and is dispersed throughout the GI tract, oral dose forms are utilized. Patients are also more compliant with these forms. However, oral medications frequently pass through first-pass metabolic inactivation, are absorbed from the stomach, and become systemically accessible, with only a limited amount of the active ingredient actually reaching the inflamed intestinal cells (Wen et al., 2015). Table 15.3 summarizes the challenges of conventional drug delivery systems in gastrointestinal inflammation.

15.6.2 Nanoparticles and Micelles

Nanoparticle (NP)-based drug delivery systems have garnered significant attention due to their favorable small size, strong drug-loading capability, and versatile surface form. Research has demonstrated that smaller nanoparticles (less than 200 nm) have the ability to more deeply penetrate the mucus barrier and enter intestinal tissues than larger submicron particles. As a drug delivery method, some researchers have suggested using NPs in the shapes of capsules, nano-emulsions, nanotubes, lipid-based nanocarriers, solid lipid nanoparticles, and nanospheres. These nanoparticles may be derived from natural materials or chemical compounds used in traditional Chinese medicine (Nicolai et al., 2010). A new formulation strategy to treat rat model experimental colitis. In an attempt to better treat inflammatory bowel disease (IBD), solid lipid microparticles (SLMs) containing curcumin were created as a novel medication delivery method for localized colon treatment. Despite its well-known anti-inflammatory qualities, curcumin has problems with solubility and absorption. The difficulty of using tailored nanoparticles (NPs) to deliver an anti-inflammatory tripeptide (KPV) to the colon in order to treat inflammation in inflammatory bowel disease (IBD). Encapsulated in a polysaccharide gel, the synthesized nanoparticles successfully released KPV in the colon, exhibiting decreased inflammatory reactions *in vitro* and guarding against colitis in a rodent model. By using NPs, therapeutic efficacy of KPV can be achieved at much lower doses, offering a unique drug delivery strategy for the treatment of IBD. Due to its low solubility and stability, resveratrol, which is well-known for its anti-inflammatory qualities in the treatment of inflammatory bowel diseases (IBD) faces difficulties. In order to circumvent this, researchers encapsulated resveratrol within β-lactoglobulin nanospheres, resulting in an approximate 1.7-fold increase in solubility. Micelles are colloidal particles that typically range in size from 5 to 100 nm and feature a core-shell configuration. Micelle production is caused by amphiphilic unimers, which frequently obstruct copolymers with hydrophilic and hydrophobic groups. These amphiphilic unimers self-assemble to form micelles with a hydrophilic corona and a hydrophobic core at critical micelle concentrations. These micelles demonstrated stable properties, a smaller particle size, and a high entrapment rate (Kabanov & Vinogradov, 2009).

15.6.3 Liposomes and Nanogels

Made of phospholipids, the same molecules that make up cell membranes, liposomes are small, spherical vesicles.

TABLE 15.3
Challenges of Conventional Drug Delivery System in Gastrointestinal Inflammations

S No.	Challenge	Description	References
1.	Variable pH Levels	The gastrointestinal tract exhibits a wide range of pH levels, from highly acidic in the stomach to more neutral in the small intestine. Designing drug delivery systems that can withstand these variations is challenging	(Homayun et al., 2019)
2.	Enzymatic Degradation	Gastrointestinal fluids contain enzymes that can break down pharmaceutical compounds, reducing drug efficacy. Formulating drugs to resist enzymatic degradation is crucial for sustained therapeutic effects	(Hua, 2020)
3.	Limited Drug Absorption	Many drugs have poor bioavailability due to challenges in absorption across the intestinal epithelium. Enhancing drug absorption and ensuring an effective therapeutic concentration is a significant hurdle	(Fasinu et al., 2011)
4.	Mucosal Barrier	The mucus layer and epithelial cells lining the gastrointestinal tract act as barriers, preventing efficient drug delivery. Overcoming these barriers to ensure drug penetration is a key consideration	(C. O. Silva et al., 2014)
5.	Short Residence Time	Rapid transit of drugs through the gastrointestinal tract limits the time available for absorption. Prolonging residence time is essential for optimal drug absorption and therapeutic impact	(Vinarov et al., 2021)
6.	First-Pass Metabolism	First-pass metabolism results from drugs that are taken from the gastrointestinal tract passing via the liver before going into the systemic circulation. This may lessen the amount of active medication that reaches the intended location	(Abuhelwa et al., 2017)
7.	Patient Compliance	Conventional oral drug delivery systems may require frequent dosing, impacting patient compliance. Developing formulations with extended-release capabilities can address this challenge	(Jones et al., 2016)
8.	Targeted Drug Delivery	Achieving targeted drug delivery to specific areas of inflammation within the gastrointestinal tract is challenging. Designing delivery systems that can release drugs at the intended site enhances therapeutic efficacy	(Bhutani et al., 2021)
9.	Toxicity and Side Effects	Conventional drug delivery systems may result in off-target effects, causing systemic toxicity. Minimizing side effects while maximizing therapeutic benefits is a constant concern	(Manzari et al., 2021)
10.	Disease Flare-ups	Gastrointestinal inflammations often involve unpredictable flare-ups. Designing drug delivery systems capable of adapting to dynamic changes in disease activity is a persistent challenge	(Husseini & Pitt, 2008)

Liposomes are becoming more and more useful in the treatment of gastrointestinal (GI) inflammations, and they have a number of advantages over conventional drug delivery techniques (Xing et al., 2016). Derived from the Chinese plant *Sophora flavescens*, oxymatrine (OM) has anti-inflammatory qualities that are advantageous for ulcerative colitis (UC). Furthermore, by combining a nitric oxide (NO) donor with the liposomes, they were able to effectively co-deliver NO, which significantly reduced inflammation in a model of ulcerative colitis (UC) in mice by downregulating pro-inflammatory cytokines, reducing macrophage infiltration, and restoring the balance between antioxidants and oxidants. For macrophage-targeted IBD therapy, researchers synthesized biomimetic liposomes (LF-lipo) using patchouli alcohol, derived from the traditional Chinese herb patchouli. These LF-lipo showed improved therapeutic efficacy in a murine colitis model, reduced inflammatory markers, suppressed inflammatory pathways, and demonstrated specific binding to activated colonic macrophages. These results point to a promising approach for addressing the unmet medical needs in the management of IBD. As a pseudo-pH-sensitive nanoparticle system, the curcumin-loaded anionic liposomes showed excellent stability and sustained release in simulated stomach and intestinal fluids. In an animal model of ulcerative colitis, the formulation demonstrated encouraging therapeutic effects, reducing clinical symptoms and averting damage to colon tissue. This suggests that liposomes could be used as delivery vehicles for a curcumin-based colon-specific drug delivery system. All things considered, liposomes offer a bright future for treating GI inflammations due to their better drug delivery, increased effectiveness, and less side effects (Kabanov & Vinogradov, 2009). For the treatment of gastrointestinal (GI) inflammations, a class of diseases marked by persistent inflammation of the digestive tract, nanogels hold great promise. Compared to traditional treatments, these minuscule hydrogel-based nanoparticles provide a number of benefits. Oral delivery of nucleic acids to an inflammatory colon, however, is hampered by the gastrointestinal system. These nanogels demonstrated a promising oral delivery system for the treatment of inflammatory bowel disease (IBD) by effectively reducing gut inflammation in a mouse colitis model. All things considered, nano gels are a formidable new treatment option for GI inflammations, with the potential to provide patients with these crippling illnesses with more individualized, targeted, and successful therapy (Yallapu et al., 2011).

15.6.4 Smart Drug Delivery Systems

By delivering medications only to the inflammatory site, these systems aim to maximize therapeutic benefit and reduce adverse effects. These are frequently biocompatible substances functionalized with certain mechanisms, such as polymers or nanoparticles. The drug release in the micelles was induced by ROS and pH, and they demonstrated regulated characteristics and stability. Micelles were shown to accumulate in inflammatory areas in *in vivo* tests, which resulted in increased therapeutic efficacy against colitis in mice. These findings raise the possibility of using this nano therapy for targeted delivery in the treatment of inflammatory bowel disease. Ali et al. Studied Budesonide-loaded nanocarriers' therapeutic efficacy in treating inflammatory bowel disease (IBD). Budesonide was encapsulated in oil-in-water emulsion-prepared poly(lactic-co-glycolic) acid (PLGA) nanoparticles, while a second batch was coated in a pH-sensitive methyl-methacrylate copolymer. In neutral to slightly alkaline pH ranges, the coated nanoparticles demonstrated controlled drug release, avoiding premature release in acidic environments. When compared to plain PLGA particles and the free drug solution, coated PLGA nanoparticles were more effective in relieving colitis in mouse models. This indicates the potential of polymeric nanocarriers for targeted drug delivery to the inflammatory intestinal mucosa, particularly when incorporating pH-dependent release characteristics for oral administration (Amin et al., 2021). Due to their inability to penetrate mucus and their unregulated release of medication, oral nanotherapeutics are not as effective in treating ulcerative colitis. In order to address this, Huang et al. modified poly (lactic-co-glycolic acid) (PLGA) nanoparticles using pluronic (PF127) and co-loaded them with catalase and curcumin. The most encouraging therapeutic results for UC were seen with oral administration of P-CAT/CUR-NPs, demonstrating their potential as potent nanotherapeutics. In order to treat ulcerative colitis caused by dextran sulfate sodium, galactosylated trimethyl chitosan–cysteine nanoparticles were developed for the oral administration of mitogen-activated protein kinase kinase 4 (MKK4) siRNA to activate macrophages. These nanoparticles demonstrated unique physicochemical properties by using anionic crosslinkers such as hyaluronic acid or tripolyphosphate. Because of their smaller size and lower zeta potential, GTC/TPP nanoparticles showed improved stability in the gastrointestinal tract, made it easier for galactose receptor-mediated endocytosis, and successfully reduced the production of TNF-α both *in vitro* and *in vivo* (Huang et al., 2023). The promising potential of oral siRNA administration for the management of inflammatory bowel disorders is highlighted by these findings. In general, intelligent medication delivery systems represent a substantial advancement in the management of gastrointestinal inflammations. These technologies have the potential to significantly enhance patient outcomes by precisely administering medication to the inflammatory site, reducing adverse effects, and enhancing therapeutic benefits (Chevalier, 2019).

15.7 COMBINED APPROACHES: FUNCTIONAL FOODS WITH NANOCARRIERS

15.7.1 Rationale for Combining Functional Foods and Nanocarriers

Utilizing nanotechnology in the creation and distribution of functional food ingredients is known as combining functional foods with nanocarriers. The rationale behind this combination is to address issues with bioavailability, stability, and targeted administration of bioactive chemicals, all of which present opportunities for several benefits (Jabr-Milane et al., 2008). Numerous beneficial food ingredients, including vitamins, antioxidants, and phytochemicals, may have low bioavailability, which means that the body is unable to absorb a substantial amount of these substances. By making the active chemicals more soluble and shielding them from digestion, nanocarriers can increase bioavailability and make it easier for the gastrointestinal tract to absorb them. Functional foods frequently include delicate bioactive ingredients that are susceptible to deterioration from heat, light, and oxygen. During the food processing and storage stages, these chemicals can be shielded from degradation and maintained in their stable state by nanocarriers. Targeted delivery of bioactive substances to particular bodily tissues or cells is made possible by nanocarriers. Targeted distribution can reduce negative effects while increasing these chemicals' medicinal effectiveness. It is possible to engineer nanocarriers such that bioactive substances are released gradually and under control. When functional foods and nanocarriers are combined, there may be synergistic effects that increase the health advantages of the bioactive substances overall. When combined, the potential health effects may be greater than when employing functional foods or nanocarriers separately. The potential to overcome obstacles related to bioavailability, stability, and targeted distribution is the driving force behind the combination of functional foods and nanocarriers, ultimately improving the health-promoting qualities of functional food items (Mcclements, 2018).

15.7.2 Synergistic Effects on Gastrointestinal Health

Functional foods are ones whose physiologically active ingredients offer additional health benefits over and beyond basic nourishment. When various bioactive components found in functional foods work together to produce better results, it's referred to as having synergistic benefits on gastrointestinal health. The potential of a number of functional foods to support gastrointestinal health via diverse pathways has been investigated. These are beneficial bacteria that are alive and can provide health advantages when consumed in sufficient quantity. Probiotics have the ability to positively impact the gut microbiota by encouraging the growth of advantageous bacteria, including Lactobacilli and Bifidobacteria. This may lead to greater digestive health by restoring the equilibrium of the gut microbiota. These are indigestible substances that encourage the growth and activity of good bacteria in the digestive system (Cencic & Chingwaru, 2010). Prebiotics like oligosaccharides and inulin feed probiotics, ensuring their survival and proper functioning. For the digestive tract to remain healthy, fiber is necessary. Foods high in soluble fiber, such as barley, lentils, and oats, can support the growth of good bacteria and help control bowel motions. Antioxidants can help lessen the detrimental effects of chronic inflammation on gastrointestinal health by lowering oxidative stress. Consuming foods high in polyphenols, such as dark chocolate, berries, and green tea, has been linked to anti-inflammatory and antioxidant benefits that may improve gastrointestinal health. A dual strategy for boosting gastrointestinal health may be achieved by combining probiotics and prebiotics (symbiotics), which may have a synergistic effect on encouraging the growth and activity of beneficial gut flora (Laparra & Sanz, n.d.).

15.8 CLINICAL APPLICATIONS AND STUDIES

15.8.1 Overview of Clinical Trials

Strong experimental evidence can be established to establish causal links between functional dietary components and health and disease/disease risk through data from randomized clinical trials. However, to maximize the quality of the data they give, clinical studies need to be carefully planned. There are two common experimental designs: parallel and crossover. Every one of these designs has benefits and drawbacks. The choice of a suitable control arm can be simple for some functional ingredients and complex for others. Studies should be long enough to guarantee biological efficacy yet short enough to maximize subject compliance, be economical, and prevent high subject dropout rates (Venkatakrishnan et al., 2019). Care must be taken while determining the active food ingredient's dosage, frequency, and timing of intake throughout the day. Randomized clinical trials evaluating the effectiveness of functional foods should utilize appropriate statistical methods for data processing and may use both established and newly developed surrogate endpoints (AbuMweis et al., 2010).

15.8.2 Future Directions in the Research

The ineffectiveness of current medication therapies, the rising incidence of IBD in East Asian and other Westernized nations, and the disease's recent globalization have all contributed to a great deal of research into the development of safe, highly effective alternative therapies based on natural substances. In the future, a concerted effort focused on determining and resolving the dietary and environmental risk factors for IBD will be given top importance. Given that nutrition is one of the primary etiological elements of the illness, it is imperative to implement a comprehensive dietary intervention that includes both the addition of food components that can target the underlying causes of IBD and the avoidance of particular foods (Streets et al., 2020). Probiotic bacterial manipulation of the gut ecology is an intriguing area of study for IBD treatment. Probiotics are still not recommended for chronic metabolic diseases, nevertheless, based on available data. "Specific targeting," which enables nutrients or bioactive substances to target and reach the precise site of inflammation to exhibit therapeutic effects, has emerged as a hot topic in IBD control research in recent years and will play a major role in future studies (Silva et al., 2020). One promising avenue for future study in the hunt for natural alternative therapeutics for gastrointestinal inflammations is the creation of innovative cell models that can replicate the GI tract. More importantly, a thorough and accurate understanding of the pathogenic pathways underlying IBD is essential because it will enable researchers to use current and potential natural therapeutic agents to develop appropriate and effective treatments (Aziz et al., 2013).

15.9 REGULATORY AND SAFETY CONSIDERATIONS

15.9.1 Regulatory Landscape for Functional Foods and Nanocarriers

Nanotechnology is used extensively in the food industry, ranging from food product research to packaging and ingredient development. Despite their negative effects on both plants and animals, traditional regulatory requirements have not yet been devised to regulate the use of nano-formulated commodities in the food and agri-sector. Thus, the utilization of nanoparticles in the food business requires efficient laws and regulations. Food Standards New Zealand and Australia (FSANZ), the supervisory agency of the Food Standards Code, actively participates in the education of nanofood components and adjuncts in Australia (Nile et al., 2020). Regulations in the European Union stipulate that food ingredients derived from nanotechnology must undergo safety testing prior to human use approval. Nanofoods, often known as food adjuncts, are thoroughly covered by the European Union New Foods Regulation (EC 258–97) (Sahoo et al., 2021).

15.9.2 Safety Profiles and Risk Assessment

The more study is done on the application of nanotechnology in the food sector, the more promise it has in food science and industry. However, a few studies have examined the ingredients used as food additives in order to examine the possible toxicity of the presence of nanomaterials in foods. There is a lack of knowledge on the bioavailability, biodistribution, pathways, and ultimate toxicity of nanomaterials when exposed to them. Most notably, when used as food additives, nanomaterials inadvertently come into touch with human organs. An increasing number of public and government sectors are paying close attention to the usage of nanomaterial compounds as taste or color additives in food (Madkour, 2021). Furthermore, by means of oral ingestion, nanoencapsulation permits direct interaction between humans and nanomaterials. One of the most popular food nanoparticles, SiO_2, has been investigated for usage in food items as a flavor or aroma carrier. Additionally, lipid incorporated nanoencapsulation has been developed to improve the solubility and bioavailability of antioxidants and to entrap bio-actives for efficient absorption and tailored site-specific delivery. However, there is still a lack of research on the safety of nanoencapsulation, which necessitates additional risk analysis, especially with regard to long-term toxicity (Momin et al., 2013).

15.10 CONCLUSION

In conclusion, the synergistic approach of utilizing functional foods and nanocarriers presents a promising avenue for the management of gastrointestinal inflammations. The integration of bioactive compounds from functional foods, known for their anti-inflammatory properties, with the advanced delivery capabilities of nanocarriers enhances bioavailability and targeted delivery to the inflamed gastrointestinal tissues. It is evident that this innovative combination holds great potential in mitigating inflammation, alleviating symptoms, and improving the overall well-being of individuals suffering from gastrointestinal disorders. The multifaceted benefits of functional foods, enriched with bioactive compounds such as polyphenols, omega-3 fatty acids, and prebiotics, can exert anti-inflammatory effects and support gut health. Nanocarriers, on the other hand, offer improved stability, controlled release, and enhanced absorption of these bioactive components, ensuring their optimal therapeutic impact. The potential to tailor nanocarriers for specific needs, such as site-specific drug delivery and sustained release, adds another layer of sophistication to the therapeutic strategies explored in the realm of gastrointestinal inflammation. However, it is essential to acknowledge the challenges and uncertainties that lie ahead, including long-term safety, regulatory considerations, and the need for more extensive clinical trials. Future research should aim to address these gaps and provide a more comprehensive understanding of the mechanisms involved, paving the way for the development of effective, targeted, and safe therapeutic interventions. In summary, the integration of functional foods and nanocarriers represents an exciting frontier in the management of gastrointestinal inflammations. As our understanding deepens and research progresses, this innovative approach holds the prospect of revolutionizing treatment strategies for individuals grappling with inflammatory conditions of the gastrointestinal tract, ushering in a new era of personalized and effective therapeutic interventions.

LIST OF ABBREVIATIONS

CD	Crohn's Disease
FSANZ	Food Standards New Zealand and Australia
GERD	Gastroesophageal reflux disease
GI	Gastrointestinal
GTC	Galactosylated trimethyl chitosan cysteine
IBD	Inflammatory bowel disease
IBS	Irritable bowel syndrome
IL	Interleukins
LES	Lower oesophageal sphincter
MKK4	Mitogen-activated protein kinase kinase 4
NP	Nanoparticle
PLGA	Poly(lactic-co-glycolic) Acid
PKC	Protein Kinase C
ROS	Reactive oxygen species
SCFAs	Short-chain fatty acids
SLMs	Solid lipid microparticles
TLERs	Transient lower oesophageal sphincter
TPP	Tripoly phosphate
TNF	Tumor necrosis factor
UC	Ulcerative colitis

ACKNOWLEDGMENTS

The authors express heartfelt gratitude towards the leadership and management of JSS College of Pharmacy, JSS Academy of Higher Education & Research (JSS AHER), Mysuru, Karnataka, India

COMPETING INTEREST STATEMENT

The authors declare no conflict of interest.

DISCLAIMER

None

REFERENCES

Abbasi, E., Aval, S. F., Akbarzadeh, A., Milani, M., Nasrabadi, H. T., Joo, S. W., Hanifehpour, Y., Nejati-Koshki, K., & Pashaei-Asl, R. (2014). Dendrimers: Synthesis, applications, and properties. *Nanoscale Research Letters*, 9(1), 1–10. https://doi.org/10.1186/1556-276X-9-247

Abuhelwa, A. Y., Williams, D. B., Upton, R. N., & Foster, D. J. R. (2017). Food, gastrointestinal pH, and models of oral drug absorption. *European Journal of Pharmaceutics and Biopharmaceutics*, 112, 234–248. https://doi.org/10.1016/j.ejpb.2016.11.034

AbuMweis, S. S., Jew, S., & Jones, P. J. H. (2010). Optimizing clinical trial design for assessing the efficacy of functional foods. *Nutrition Reviews*, 68(8), 485–499. https://doi.org/10.1111/j.1753-4887.2010.00308.x

Alam, R., Abdolmaleky, H. M., & Zhou, J. R. (2017). Microbiome, inflammation, epigenetic alterations, and mental diseases. *American Journal of Medical Genetics, Part B: Neuropsychiatric Genetics*, 174(6), 651–660. https://doi.org/10.1002/ajmg.b.32567

Amin, M. U., Ali, S., Tariq, I., Ali, M. Y., Pinnapreddy, S. R., Preis, E., Wölk, C., Harvey, R. D., Hause, G., Brüßler, J., & Bakowsky, U. (2021). Ultrasound-responsive smart drug delivery system of lipid coated mesoporous silica nanoparticles. *Pharmaceutics*, 13(9). https://doi.org/10.3390/pharmaceutics13091396

Anand David, A. V., Arulmoli, R., & Parasuraman, S. (2016). Overviews of biological importance of quercetin: A bioactive flavonoid. *Pharmacognosy Reviews*, 10(20), 84–89. https://doi.org/10.4103/0973-7847.194044

Aziz, Q., Doré, J., Emmanuel, A., Guarner, F., & Quigley, E. M. M. (2013). Gut microbiota and gastrointestinal health: Current concepts and future directions. *Neurogastroenterology and Motility*, 25(1), 4–15. https://doi.org/10.1111/nmo.12046

Babu, D., Motterlini, R., Lefebvre, R. A., & Lefebvre, R. A. (2015). Themed section: Pharmacology of the gasotransmitters REVIEW CO and CO-releasing molecules (CO-RMs) in acute gastrointestinal inflammation LINKED ARTICLES. *British Journal of Pharmacology*, 172, 1557. https://doi.org/10.1111/bph.2015.172.issue-6

Bhutani, U., Basu, T., & Majumdar, S. (2021). Oral drug delivery: Conventional to long acting new-age designs. *European Journal of Pharmaceutics and Biopharmaceutics*, 162, 23–42. https://doi.org/10.1016/j.ejpb.2021.02.008

Bloch, A., & Thomson, C. A. (1997). Position of the American dietetic association: Phytochemicals and functional foods. *Journal of Nutraceuticals, Functional and Medical Foods*, 1(1), 33–45. https://doi.org/10.1300/J133v01n01_04

Brenchley, J. M., & Douek, D. C. (2008). HIV infection and the gastrointestinal immune system. *Mucosal Immunology*, 1(1), 23–30. https://doi.org/10.1038/mi.2007.1

Cencic, A., & Chingwaru, W. (2010). The role of functional foods, nutraceuticals, and food supplements in intestinal health. *Nutrients*, 2(6), 611–625. https://doi.org/10.3390/nu2060611

Chandrasekara, A., & Josheph Kumar, T. (2016). Roots and tuber crops as functional foods: A review on phytochemical constituents and their potential health benefits. *International Journal of Food Science*, 2016. https://doi.org/10.1155/2016/3631647

Chang, L. (2004). Review article: Epidemiology and quality of life in functional gastrointestinal disorders. *Alimentary Pharmacology and Therapeutics*, 20(7), 31–39. https://doi.org/10.1111/j.1365-2036.2004.02183.x

Chevalier, R. (2019). siRNA targeting and treatment of gastrointestinal diseases. *Clinical and Translational Science*, 12(6), 573–585. https://doi.org/10.1111/cts.12668

Cunningham, M., Azcarate-Peril, M. A., Barnard, A., Benoit, V., Grimaldi, R., Guyonnet, D., Holscher, H. D., Hunter, K., Manurung, S., Obis, D., Petrova, M. I., Steinert, R. E., Swanson, K. S., van Sinderen, D., Vulevic, J., & Gibson, G. R. (2021). Shaping the future of probiotics and prebiotics. *Trends in Microbiology*, 29(8), 667–685. https://doi.org/10.1016/j.tim.2021.01.003

Das, A. K., Nanda, P. K., Madane, P., Biswas, S., Das, A., Zhang, W., & Lorenzo, J. M. (2020). A comprehensive review on antioxidant dietary fibre enriched meat-based functional foods. *Trends in Food Science and Technology*, 99, 323–336. https://doi.org/10.1016/j.tifs.2020.03.010

Dimidi, E., Cox, S. R., Rossi, M., & Whelan, K. (2019). Fermented foods: Definitions and characteristics, impact on the gut microbiota and effects on gastrointestinal health and disease. *Nutrients*, 11(8). https://doi.org/10.3390/nu11081806

Doyon, M., & Labrecque, J. A. (2008). Functional foods: A conceptual definition. *British Food Journal*, 110(11), 1133–1149. https://doi.org/10.1108/00070700810918036

Drossman, D. A., Patrick, D. L., Mitchell, C. M., Zagami, E. A., & Appelbaum, M. I. (1989). Health-related quality of life in inflammatory bowel disease functional status and patient worries and concerns. *Digestive Diseases and Sciences*, 34(9).

Dupont, H. L. (2016). Introduction: Understanding mechanisms of the actions of rifaximin in selected gastrointestinal diseases. *Alimentary Pharmacology and Therapeutics,* 43, 1–2. https://doi.org/10.1111/apt.13406

El-Zahar, K. M., Al-Jamaan, M. E., Al-Mutairi, F. R., Al-Hudiab, A. M., Al-Einzi, M. S., & Mohamed, A. A. Z. (2022). Antioxidant, antibacterial, and antifungal activities of the ethanolic extract obtained from Berberis vulgaris roots and leaves. *Molecules*, 27(18). https://doi.org/10.3390/molecules27186114

Eskicioglu, V., Kamiloglu, S., & Nilufer-Erdil, D. (2015). Antioxidant dietary fibres: Potential functional food ingredients from plant processing by-products. *Czech Journal of Food Sciences*, 33(6), 487–499. https://doi.org/10.17221/42/2015-CJFS

Farah, F. H. (2019). Nanocarriers as delivery as systems for therapeutics agents. *International Journal of Pharmaceutical Sciences and Research*, 10(8), 3487. https://doi.org/10.13040/IJPSR.0975-8232.10(8).3487-07

Fasinu, P., Pillay, V., Ndesendo, V. M. K., Du Toit, L. C., & Choonara, Y. E. (2011). Diverse approaches for the enhancement of oral drug bioavailability. *Biopharmaceutics and Drug Disposition*, 32(4), 185–209. https://doi.org/10.1002/bdd.750

Figueroa-González, I., Quijano, G., Ramírez, G., & Cruz-Guerrero, A. (2011). Probiotics and prebiotics-perspectives and challenges. *Journal of the Science of Food and Agriculture*, 91(8), 1341–1348. https://doi.org/10.1002/jsfa.4367

Gardikis, K., Hatziantoniou, S., Viras, K., Wagner, M., & Demetzos, C. (2006). Interaction of dendrimers with model lipid membranes assessed by DSC and raman spectroscopy. *Nanocarrier Technologies: Frontiers of Nanotherapy*, 9781402050411, 207–220. https://doi.org/10.1007/978-1-4020-5041-1_12

Geißler, D., Charbonnière, L. J., Ziessel, R. F., Butlin, N. G., Löhmannsröben, H. G., & Hildebrandt, N. (2010). Quantum dot biosensors for ultrasensitive multiplexed diagnostic. *Angewandte Chemie – International Edition*, 49(8), 1396–1401. https://doi.org/10.1002/anie.200906399

Goni, I., & Hervert-Hernandez, D. (2011). By-products from plant foods are sources of dietary fibre and antioxidants. *Phytochemicals – Bioactivities and Impact on Health*. InTech. https://doi.org/10.5772/27923

Graff, L. A., Walker, J. R., Lix, L., Clara, I., Rawsthorne, P., Rogala, L., Miller, N., Jakul, L., McPhail, C., Ediger, J., & Bernstein, C. N. (2006). The relationship of inflammatory bowel disease type and activity to psychological functioning and quality of life. *Clinical Gastroenterology and Hepatology*, 4(12). https://doi.org/10.1016/j.cgh.2006.09.027

Granato, D., Barba, F. J., Bursa´cbursa´c, D., Kovačevi´c, K., Lorenzo, J. M., Cruz, A. G., & Putnik, P. (2020). Functional foods: Product development, technological trends, efficacy testing, and safety. Annual Review of Food Science and Technology. https://doi.org/10.1146/annurev-food-032519

Hallan, S. S., Kaur, P., Kaur, V., Mishra, N., & Vaidya, B. (2016). Lipid polymer hybrid as emerging tool in nanocarriers for oral drug delivery. *Artificial Cells, Nanomedicine and Biotechnology*, 44(1), 334–349. https://doi.org/10.3109/21691401.2014.951721

Hillaireau, H., & Couvreur, P. (2009). Nanocarriers' entry into the cell: Relevance to drug delivery. *Cellular and Molecular Life Sciences*, 66(17), 2873–2896. https://doi.org/10.1007/s00018-009-0053-z

Holtmann, G., Shah, A., & Morrison, M. (2018). Pathophysiology of functional gastrointestinal disorders: A holistic overview. *Digestive Diseases*, 35(1), 5–13. https://doi.org/10.1159/000485409

Homayun, B., Lin, X., & Choi, H. J. (2019). Challenges and recent progress in oral drug delivery systems for biopharmaceuticals. *Pharmaceutics*, 11(3). https://doi.org/10.3390/pharmaceutics11030129

Hostetter, J. M., & Uzal, F. A. (2022). Gastrointestinal biopsy in the horse: Overview of collection, interpretation, and applications. *Journal of Veterinary Diagnostic Investigation*, 34(3), 376–388. https://doi.org/10.1177/10406387221085584

Huang, Y., Du, Z., Bao, G., Fang, G., Cappadona, M., McClements, L., Tuch, B. E., Lu, H., & Xu, X. (2023). Smart drug-delivery system of upconversion nanoparticles coated with mesoporous silica for controlled release. *Pharmaceutics*, 15(1). https://doi.org/10.3390/pharmaceutics15010089

Hua, S. (2020). Advances in oral drug delivery for regional targeting in the gastrointestinal tract – Influence of physiological, pathophysiological and pharmaceutical factors. *Frontiers in Pharmacology*, 11. https://doi.org/10.3389/fphar.2020.00524

Hussein, H. J., Hameed, I. H., & Hadi, M. Y. (2017). A review: Anti-microbial, anti-inflammatory effect and cardiovascular effects of garlic: Allium sativum. *Research Journal of Pharmacy and Technology*, 10(11), 4069–4078. https://doi.org/10.5958/0974-360X.2017.00738.7

Husseini, G. A., & Pitt, W. G. (2008). Micelles and nanoparticles for ultrasonic drug and gene delivery. *Advanced Drug Delivery Reviews*, 60(10), 1137–1152. https://doi.org/10.1016/j.addr.2008.03.008

Jabr-Milane, L., van Vlerken, L., Devalapally, H., Shenoy, D., Komareddy, S., Bhavsar, M., & Amiji, M. (2008). Multi-functional nanocarriers for targeted delivery of drugs and genes. *Journal of Controlled Release*, 130(2), 121–128. https://doi.org/10.1016/j.jconrel.2008.04.016

Jampilek, J., Kos, J., & Kralova, K. (2019). Potential of nanomaterial applications in dietary supplements and foods for special medical purposes. *Nanomaterials*, 9(2). https://doi.org/10.3390/nano9020296

Jia, W., Xie, G., & Jia, W. (2018). Bile acid–microbiota crosstalk in gastrointestinal inflammation and carcinogenesis. *Nature Reviews Gastroenterology and Hepatology,* 15(2), 111–128. https://doi.org/10.1038/nrgastro.2017.119

Johnson, I. T. (2001). New food components and gastrointestinal health. *Proceedings of the Nutrition Society*, 60(4), 481–488. https://doi.org/10.1079/pns2001106

Jones, C. R., Hatley, O. J. D., Ungell, A. L., Hilgendorf, C., Annie, S., & Rostami-Hodjegan, A. (2016). Gut wall metabolism. Application of pre-clinical models for the prediction of human drug absorption and first-pass elimination. *AAPS Journal*, 18(3), 589–604. https://doi.org/10.1208/s12248-016-9889-y

Kabanov, A. V., & Vinogradov, S. V. (2009). Nanogels as pharmaceutical carriers: Finite networks of infinite capabilities. *Angewandte Chemie – International Edition*, 48(30), 5418–5429. https://doi.org/10.1002/anie.200900441

Karrasch, T., Obermeier, F., & Straub, R. H. (2014). Systemic metabolic signaling in acute and chronic gastrointestinal inflammation of inflammatory bowel diseases. *Hormone and Metabolic Research*, 46(6), 445–451. https://doi.org/10.1055/s-0034-1374587

Khan, D. R. (2010). The use of nanocarriers for drug delivery in cancer therapy. *Journal of Cancer Science and Therapy*, 2(3), 58–62. https://doi.org/10.4172/1948-5956.1000024

Khizar, S., Alrushaid, N., Khan, A., Zine, N., Jaffrezic-Renault, N., Errachid, A., & Elaissari, A. (2022). Nanocarriers based novel and effective drug delivery system. *International Journal of Pharmaceutics,* 632, 122570.

Konstantinidi, M., & Koutelidakis, A. E. (2019). Functional foods and bioactive compounds: A review of its possible role on weight management and obesity's metabolic consequences. *Medicines*, 6(3), 94. https://doi.org/10.3390/medicines6030094

Kowalczuk, A., Trzcinska, R., Trzebicka, B., Müller, A. H. E., Dworak, A., & Tsvetanov, C. B. (2014). Loading of polymer nanocarriers: Factors, mechanisms and applications dedicated to Prof. Stanislaw Penczek on the occasion of his 80th birthday. *Progress in Polymer Science*, 39(1), 43–86. https://doi.org/10.1016/j.progpolymsci.2013.10.004

Kratz, F., & Warnecke, A. (2012). Finding the optimal balance: Challenges of improving conventional cancer chemotherapy using suitable combinations with nano-sized drug delivery systems. *Journal of Controlled Release*, 164(2), 221–235. https://doi.org/10.1016/j.jconrel.2012.05.045

Krishnan, V. (2017). Phytochemistry and functional food: The needs of healthy life. *Journal of Phytochemistry and Biochemistry*, 1, 103 www.researchgate.net/publication/321805692

Laparra, J. M., & Sanz, Y. (n.d.). Interactions of gut microbiota with functional food components and nutraceuticals. In *Advances in Processing Technology* (pp. 123–141). CRC Press.

Lin, S. Y., Hsu, W. H., Lo, J. M., Tsai, H. C., & Hsiue, G. H. (2011). Novel geometry type of nanocarriers mitigated the phagocytosis for drug delivery. *Journal of Controlled Release*, 154(1), 84–92. https://doi.org/10.1016/j.jconrel.2011.04.023

López-Dávila, V., Seifalian, A. M., & Loizidou, M. (2012). Organic nanocarriers for cancer drug delivery. *Current Opinion in Pharmacology*, 12(4), 414–419. https://doi.org/10.1016/j.coph.2012.02.011

Madkour, L. H. (2021). A review: Metal nanoparticles and their safety processing in functional foods. *Journal Chemical Science Chemical Engineering*, 2(1), 18–36. www.journalofchemistry.net

Magrone, T., Magrone, M., Russo, M. A., & Jirillo, E. (2020). Recent advances on the anti-inflammatory and antioxidant properties of red grape polyphenols: In vitro and in vivo studies. *Antioxidants*, 9(1) 35. https://doi.org/10.3390/antiox9010035

Manzari, M. T., Shamay, Y., Kiguchi, H., Rosen, N., Scaltriti, M., & Heller, D. A. (2021). Targeted drug delivery strategies for precision medicines. *Nature Reviews Materials*, 6(4), 351–370. https://doi.org/10.1038/s41578-020-00269-6

Martínez-Ballesta, Mc., Gil-Izquierdo, Á., García-Viguera, C., & Domínguez-Perles, R. (2018). Nanoparticles and controlled delivery for bioactive compounds: Outlining challenges for new "smart-foods" for health. *Foods*, 7(5). https://doi.org/10.3390/foods7050072

Martirosyan, D., Lampert, T., & Ekblad, M. (2022). Classification and regulation of functional food proposed by the functional food center. *Functional Food Science*, 2(2), 25. https://doi.org/10.31989/ffs.v2i2.890

Martirosyan, D., & Miller, E. (2018). Bioactive compounds: The key to functional foods. *Bioactive Compounds in Health and Disease*, 1(3), 36–39. https://doi.org/10.31989/bchd.v1i3.539

Marusawa, H., & Jenkins, B. J. (2014). Inflammation and gastrointestinal cancer: An overview. *Cancer Letters*, 345(2), 153–156. https://doi.org/10.1016/j.canlet.2013.08.025

Mashhadi, N. S., Ghiasvand, R., Askari, G., Hariri, M., Darvishi, L., & Mofid, M. R. (n.d.). Anti-oxidative and anti-inflammatory effects of ginger in health and physical activity: Review of current Evidence. *Original Article International Journal of Preventive Medicine,* (Suppl 1), S36. www.ijpm.ir

Mazdaei, M., & Asare-Addo, K. (2022). A mini-review of Nanocarriers in drug delivery systems. *British Journal of Pharmacy*, 7(1), 1–3. https://doi.org/10.5920/bjpharm.780

Mcclements, D. J. (2018). Encapsulation, protection, and delivery of bioactive proteins and peptides using nanoparticle and microparticle systems: A review. *Advances in Colloid and Interface Science*, 253, 1–22.

Mishra, B., Patel, B. B., & Tiwari, S. (2010). Colloidal nanocarriers: A review on formulation technology, types and applications toward targeted drug delivery. *Nanomedicine: Nanotechnology, Biology, and Medicine*, 6(1), 9–24. https://doi.org/10.1016/j.nano.2009.04.008

Mishra, R. K., Tiwari, S. K., Mohapatra, S., & Thomas, S. (2018). Efficient nanocarriers for drug-delivery systems: Types and fabrication. Nanocarriers for Drug Delivery: Nanoscience and Nanotechnology in Drug Delivery, 1–41. https://doi.org/10.1016/B978-0-12-814033-8.00001-1

Mokra, D., Joskova, M., & Mokry, J. (2023). Therapeutic effects of green tea polyphenol (–)-Epigallocatechin-3-Gallate (EGCG) in relation to molecular pathways controlling inflammation, oxidative stress, and apoptosis. *International Journal of Molecular Sciences*, 24(1), 340. https://doi.org/10.3390/ijms24010340

Momin, J. K., Jayakumar, C., & Prajapati, J. B. (2013). Potential of nanotechnology in functional foods. *Emirates Journal of Food and Agriculture*, 25(1), 10–19. https://doi.org/10.9755/ejfa.v25i1.9368

Mózsik, G., Bódis, B., Figler, M., Király, Á., Karádi, O., Pár, A., Rumi, G., Süt, G., Tóth, G., & Vincze, Á. (2001). Mechanisms of action of retinoids in gastrointestinal mucosal protection in animals, human healthy subjects and patients. *Life Sciences*, 69(25–26), 3103–3112.

Muller, J., Huaux, F., Moreau, N., Misson, P., Heilier, J. F., Delos, M., Arras, M., Fonseca, A., Nagy, J. B., & Lison, D. (2005). Respiratory toxicity of multi-wall carbon nanotubes. *Toxicology and Applied Pharmacology*, 207(3), 221–231. https://doi.org/10.1016/j.taap.2005.01.008

Nicolai, T., Colombani, O., & Chassenieux, C. (2010). Dynamic polymeric micelles versus frozen nanoparticles formed by block copolymers. *Soft Matter*, 6(14), 3111–3118. https://doi.org/10.1039/b925666k

Nile, S. H., Baskar, V., Selvaraj, D., Nile, A., Xiao, J., & Kai, G. (2020). Nanotechnologies in food science: Applications, recent trends, and future perspectives. *Nano-Micro Letters*, 12(1). https://doi.org/10.1007/s40820-020-0383-9

Parker, E. A., Roy, T., D'Adamo, C. R., & Wieland, L. S. (2018). Probiotics and gastrointestinal conditions: An overview of evidence from the Cochrane Collaboration. *Nutrition*, 45, 125–134.e11. https://doi.org/10.1016/j.nut.2017.06.024

Phillips, J. T., Selmaj, K., Gold, R., Fox, R. J., Havrdova, E., Giovannoni, G., Abourjaily, H., Pace, A., Novas, M., Hotermans, C., Viglietta, V., & Meltzer, L. (2015). Clinical significance of gastrointestinal and flushing events in patients with multiple sclerosis treated with delayed-release dimethyl fumarate. *International Journal of MS Care*, 17(5), 236–243. https://doi.org/10.7224/1537-2073.2014-069

Ptak-Belowska, A., Targosz, A., & Brzozowski, T. (2012). MicroRNAs and chronic inflammation contribution to gastrointestinal integrity. *Current Medicinal Chemistry*, 19(1), 133–136.

Ramos De Mattos, B. R., Pereira, M., Garcia, G., Nogueira, J. B., Paiatto, L. N., Albuquerque, C. G., Souza, C. L., Gustavo, L., Fernandes, R., Maria, W., Cunha Tamashiro, S., & Simioni, P. U. (2015). Inflammatory bowel disease: An overview of immune mechanisms and biological treatments. *Mediators of Inflammation,* (1), 493012

Rothenberg, M. E. (2004). Eosinophilic gastrointestinal disorders (EGID). *Journal of Allergy and Clinical Immunology*, 113(1), 11–28. https://doi.org/10.1016/j.jaci.2003.10.047

Rajesh, W. (2017). Types of nanocarriers–Formulation method and applications. *Journal of Bioequivalence & Bioavailability*, 09(03). https://doi.org/10.4172/jbb.10000e77

Sadiq, U., Gill, H., & Chandrapala, J. (2021). Casein micelles as an emerging delivery system for bioactive food components. *Foods*, 10(8), 1965. https://doi.org/10.3390/foods10081965

Sandur, S. K., Ichikawa, H., Pandey, M. K., Kunnumakkara, A. B., Sung, B., Sethi, G., & Aggarwal, B. B. (2007). Role of pro-oxidants and antioxidants in the anti-inflammatory and apoptotic effects of curcumin (diferuloylmethane). *Free Radical Biology and Medicine*, 43(4), 568–580. https://doi.org/10.1016/j.freeradbiomed.2007.05.009

Sahoo, M., Vishwakarma, S., Panigrahi, C., & Kumar, J. (2021). Nanotechnology: Current applications and future scope in food. *Food Frontiers*, 2(1), 3–22. https://doi.org/10.1002/fft2.58

Sajadinejad, M. S., Asgari, K., Molavi, H., Kalantari, M., & Adibi, P. (2012). Psychological issues in inflammatory bowel disease: An overview. *Gastroenterology Research and Practice,* 2(1), 106502. https://doi.org/10.1155/2012/106502

Sanders, M. E. (1998). Overview of functional foods: Emphasis on probiotic bacteria. *International Dairy Journal*, 8(5–6), 341–347.

Sandur, S. K., Ichikawa, H., Pandey, M. K., Kunnumakkara, A. B., Sung, B., Sethi, G., & Aggarwal, B. B. (2007). Role of pro-oxidants and antioxidants in the anti-inflammatory and apoptotic effects of curcumin (diferuloylmethane). *Free Radical Biology and Medicine*, 43(4), 568–580. https://doi.org/10.1016/j.freeradbiomed.2007.05.009

Sharma, Dr. R., Kumar, S., Kumar, V., & Thakur, A. (2019). Comprehensive review on nutraceutical significance of phytochemicals as functional food ingredients for human health management. *Journal of Pharmacognosy and Phytochemistry*, 8(5), 385–395. https://doi.org/10.22271/phyto.2019.v8.i5h.9589

Shin, G. H., Kim, J. T., & Park, H. J. (2015). Recent developments in nanoformulations of lipophilic functional foods. *Trends in Food Science and Technology*, 46(1), 144–157. https://doi.org/10.1016/j.tifs.2015.07.005

Silva, C. O., Sarmento, B., & Reis, C. P. (2014). Oral delivery of biopharmaceuticals. In Mucosal Delivery of Biopharmaceuticals: Biology, Challenges and Strategies, (Vol. 9781461495246, pp. 125–147). Springer US. https://doi.org/10.1007/978-1-4614-9524-6_5

Silva, D. R., Sardi, J. de C. O., Pitangui, N. de S., Roque, S. M., Silva, A. C. B. da, & Rosalen, P. L. (2020). Probiotics as an alternative antimicrobial therapy: Current reality and future directions. Journal of Functional Foods, 73. https://doi.org/10.1016/j.jff.2020.104080

Silva, P. M., Gonçalves, C., Pastrana, L. M., Coimbra, M. A., Vicente, A. A., & Cerqueira, M. A. (2023). Recent advances in oral delivery systems of resveratrol: Foreseeing their use in functional foods. *Food and Function*, 14(23), 10286–10313. https://doi.org/10.1039/d3fo03065b

Singh, H. (2016). Nanotechnology applications in functional foods; Opportunities and challenges. *Preventive Nutrition and Food Science*, 21(1), 1–8. https://doi.org/10.3746/pnf.2016.21.1.1

Soumya, N. P. P., Mini, S., Sivan, S. K., & Mondal, S. (2021). Bioactive compounds in functional food and their role as therapeutics. *Bioactive Compounds in Health and Disease*, 4(3), 24–39. https://doi.org/10.31989/bchd.v4i3.786

Streets, J., Bhatt, P., Bhatia, D., & Sutariya, V. (2020). Sunitinib-loaded MPEG-PCL micelles for the treatment of age-related macular degeneration. *Scientia Pharmaceutica*, 88(3), 1–12. https://doi.org/10.3390/scipharm88030030

Sun, M., Qu, S., Hao, Z., Ji, W., Jing, P., Zhang, H., Zhang, L., Zhao, J., & Shen, D. (2014). Towards efficient solid-state photoluminescence based on carbon-nanodots and starch composites. *Nanoscale*, 6(21), 13076–13081. https://doi.org/10.1039/c4nr04034a

Thomas, D. W., Greer, F. R., Bhatia, J. J. S., Abrams, S. A., Daniels, S. R., Schneider, M. B., Silverstein, J., Stettler, N., Heitlinger, L. A., Baker, R. D., Gremse, D. A., Gugig, R., & Lightdale, J. R. (2010). Clinical report – Probiotics and prebiotics in pediatrics. *Pediatrics*, 126(6), 1217–1231. https://doi.org/10.1542/peds.2010-2548

Ullman, T. A., & Itzkowitz, S. H. (2011). Intestinal inflammation and cancer. *Gastroenterology*, 140(6), 1807–1816.e1. https://doi.org/10.1053/j.gastro.2011.01.057

Üner, M., & Yener, G. (2007). Importance of solid lipid nanoparticles (SLN) in various administration routes and future perspectives. *International Journal of Nanomedicine*, 2(3), 289–300. https://doi.org/10.2147/IJN.S2.3.289

Venkatakrishnan, K., Chiu, H. F., & Wang, C. K. (2019). Popular functional foods and herbs for the management of type-2-diabetes mellitus: A comprehensive review with special reference to clinical trials and its proposed mechanism. *Journal of Functional Foods*, 57, 425–438. https://doi.org/10.1016/j.jff.2019.04.039

Vieira da Silva, B., Barreira, J. C. M., & Oliveira, M. B. P. P. (2016). Natural phytochemicals and probiotics as bioactive ingredients for functional foods: Extraction, biochemistry and protected-delivery technologies. *Trends in Food Science and Technology*, 50, 144–158. https://doi.org/10.1016/j.tifs.2015.12.007

Vinarov, Z., Abdallah, M., Agundez, J. A. G., Allegaert, K., Basit, A. W., Braeckmans, M., Ceulemans, J., Corsetti, M., Griffin, B. T., Grimm, M., Keszthelyi, D., Koziolek, M., Madla, C. M., Matthys, C., McCoubrey, L. E., Mitra, A., Reppas, C., Stappaerts, J., Steenackers, N., … Augustijns, P. (2021). Impact of gastrointestinal tract variability on oral drug absorption and pharmacokinetics: An UNGAP review. *European Journal of Pharmaceutical Sciences*, 162. https://doi.org/10.1016/j.ejps.2021.105812

Wang, N., Cheng, X., Li, N., Wang, H., & Chen, H. (2019). Nanocarriers and their loading strategies. *Advanced Healthcare Materials*, 8(6), 1801002. https://doi.org/10.1002/adhm.201801002

Wen, H., Jung, H., & Li, X. (2015). Drug delivery approaches in addressing clinical pharmacology-related issues: Opportunities and challenges. *AAPS Journal*, 17(6), 1327–1340. https://doi.org/10.1208/s12248-015-9814-9

Wilkins, T., Embry, K., & George, R. (2013, May). *Diagnosis and management of acute diverticulitis. American Family Physician*, 87(9), 612–620. www.aafp.org/afp.

Wilson, D. W., Nash, P., Singh, H., Griffiths, K., Singh, R., De Meester, F., Horiuchi, R., & Takahashi, T. (2017). The role of food antioxidants, benefits of functional foods, and influence of feeding habits on the health of the older person: An overview. *Antioxidants*, 6(4), 81. https://doi.org/10.3390/antiox6040081

Xing, H., Hwang, K., & Lu, Y. (2016). Recent developments of liposomes as nanocarriers for theranostic applications. *Theranostics*, 6(9), 1336–1352. https://doi.org/10.7150/thno.15464

Yallapu, M. M., Jaggi, M., & Chauhan, S. C. (2011). Design and engineering of nanogels for cancer treatment. *Drug Discovery Today*, 16(9–10), 457–463. https://doi.org/10.1016/j.drudis.2011.03.004

Yang, K., Zhang, L., Liao, P., Xiao, Z., Zhang, F., Sindaye, D., Xin, Z., Tan, C., Deng, J., Yin, Y., & Deng, B. (2020). Impact of gallic acid on gut health: Focus on the gut microbiome, immune response, and mechanisms of action. *Frontiers in Immunology*, 11, 580208. https://doi.org/10.3389/fimmu.2020.580208

Zhao, M., Chen, Y., Wang, C., Xiao, W., Chen, S., Zhang, S., Yang, L., & Li, Y. (2019). Systems pharmacology dissection of multi-scale mechanisms of action of Huo-Xiang-Zheng-qi formula for the treatment of gastrointestinal diseases. *Frontiers in Pharmacology*, 9(JAN). https://doi.org/10.3389/fphar.2018.01448

16 Polysaccharide-Based Nanoparticles for the Management of Gastrointestinal Inflammations

Jitu Halder, Tushar Kanti Rajwar, Ajit Mishra, Guru Prasanna Sahoo, Biswakanth Kar, Goutam Ghosh, and Goutam Rath

School of Pharmaceutical Sciences, Sikhsa 'O' Anusandhan (Deemed to be University), Bhubaneswar, Odisha, India

16.1 INTRODUCTION

Gastrointestinal inflammation is a complex biological response to injury as a result of different stimuli such as pathogens, damaged cells, or irritants(1). This generally accompanied by alterations in vascular structure and function. Multiple diseases, including celiac disease and inflammatory bowel syndrome (IBS), may be the source of this inflammation in the stomach(2). IBD typically manifests in a person's teens or 20s, while it can also strike an older person. Roughly 15% of newly diagnosed cases of IBS are in people over 60(3). In individuals with postinfectious irritable bowel syndrome, many stressors can compromise the immune system and raise the chance of developing infectious disorders like gastritis. Conversely, people with IBS are more likely to experience infectious gastroenteritis. Many microbial (Campylobacter jejuni, Clostridium, Mycobacterium), viral (norovirus, for example), and parasitic (Giardia, helminthes) agents are known to play a role in the development of chronic inflammatory bowel syndromes(4). It is believed that the symbiotic microbiome in the digestive tract protects the gut by reducing toxic stimuli and speeds up the healing process. Changes in the microbiome or its absence can worsen type 2 immunity and cause inflammatory, allergy, and infectious consequences, including parasite illnesses. In order to eliminate inflammatory triggers and support the restoration of tissue homeostasis, the signals at the sites of inflammation facilitate quick cell recruitment and differentiation(5). Chronic inflammation, however, can be brought on by recurrent inflammatory triggers or dysregulation of the restoration systems. Furthermore, GI immunity is directly and indirectly modulated by nutritional components such as probiotics, prebiotics, antioxidants, and n-3 polyunsaturated fatty acids(6). High-fat diets alter satiety and brain signaling, as well as the gut's natural microbiome populations, which can result in inflammation and obesity(7).

The range of treatments for gastrointestinal inflammation has increased within the last several years. Drugs such as aminosalicylates, corticosteroids (CSs), immunomodulators, and biologics are used in combination with other general interventions and, in certain cases, surgical resection to control symptoms(8). The discovery of certain inhibitors of tumor necrosis factor (TNF) is a ground-breaking development that has allowed for long-term remission and altered the course of IBD in a sizable portion of patients(9). New therapeutic approaches are necessary because primary non-response to TNF inhibitors was noted in as many as 40% of patients in clinical trials and 10–15% of patients in clinical series; secondary loss of response happened in around 23–46% of patients after a year of treatment(10). Small molecules, apheresis therapy, enhanced intestinal microecology, cell therapy, and exosome therapy are among the novel therapeutic approaches that are being developed. Recent advancements in therapeutic modalities, particularly the introduction of biologics, have not only aided in the modification of the manner of treatment for gastric inflammation but have also altered the outlook on treatment. Traditionally, a clinical symptom score has been used to assess the therapy outcomes. These days, objective signs like specific biomarkers, new bioactive, nanoparticulate systems can also be used in treatment for gastric inflammation.

In contrast to different approaches, nanotherapy approaches are anticipated to gain popularity as treatment modalities(11). Generally speaking, nanomedicines are a subfield of nanotechnology that seeks to use the characteristics of nanoparticles in the detection and treatment of medical conditions(12, 13). Particular physicochemical characteristics of a nanoparticulate system include good biocompatibility, bioavailability, and possible antibacterial action(14). The ability to target the gut and treat intestinal illnesses is made possible by the notable ease with which the surfaces of nanomedicines can be changed with various ligands(15). Additionally, gut-targeting nanoparticulate system has reached a pinnacle with the introduction of oral nanomedicines. In addition to being a handy drug that directly interacts with the gut microbiota and remains in the intestine for extended periods of time provides a stronger defense against illnesses by controlling the gut-X axis(16). Additionally, it has the ability to display dual-targeting by accumulating at areas of inflammation, anomalous blood arteries, and anomalous lymphatic networks(17). Nanoparticulates are extremely tiny particles, usually less than 100 nm in size. Because of their dimensions falling into the nanoscale region, nanoparticulates are also called "zero-dimensional" nanomaterials(18).

DOI: 10.1201/9781003493143-16

Nanoparticles are the most sought-after drug delivery vehicles due to their small size, high surface area-to-volume ratio, and capacity to encapsulate a wide range of medicines(19). The effectiveness of the active ingredient in pharmaceuticals and the release kinetics in drug delivery systems are greatly influenced by the shape of nanoparticles. Targetability and circulation times can be controlled because of the way that nanoparticulates size and shape affect how they travel through the blood(20).

Currently, considerable research efforts have been focused on developing biocompatible and biodegradable nanoparticles and to promote their applications in various areas and biomedicine in particular. Polysaccharides, are natural polymer with unique characteristics such as biocompatibility, biodegradability, and tunable properties, have made them appealing choices among different natural polymers(21). Polysaccharides are linked to an extensive range of biological and pharmacological actions. These include those that are neuroprotective, antitumor, anticancer, immunological modulating, wound healing, and others(22). Primarily, the biological activities are determined by the molecular weight, type, and linkage of monosaccharides(23). Polysaccharides appear to have some potential anti-SARS-CoV-2 action, according to recent research. Numerous routes have been proposed to explain the inhibitory impact of polysaccharides on SARS-CoV-2, such as N-(2-hydroxypropyl)−3-trimethylammonium chitosan chloride. These include changing the structure of the receptor binding domain, preventing the SARS-CoV-2 virus from being transcribed and replicating, and boosting the immune system following immunization(24).

Given their capability to coat many different substances, including small molecules like proteins peptides and nucleic acid, polysaccharides offer a flexible foundation for drug delivery(25). Moreover, polysaccharides can be polysaccharides extracted from natural sources have attracted extensive attention in recent years owing to their safety, accessibility and potent anti-inflammatory activities(26). A great deal of attention has been paid to the production of nanoparticles from sugars as a means to tackle stomach inflammation including such diseases as inflammatory bowel disease (IBD), colorectal cancer and others in the gastrointestinal tract(27). For example, citosan oligosaccharides (COS) have anti-inflammatory properties because they prevent macrophage responses brought on by microbial debris. The pathophysiology, onset, and progression of ulcerative colitis (UC) are significantly influenced by the large amounts of pro-inflammatory cytokines produced by macrophages, including tumor necrosis factor-α (TNFα), IL-6, inducible nitric oxide synthase (iNOS), cyclooxygenase-2 (COX-2), prostaglandin E2 (PG-E2), and nitric oxide (NO). It has been demonstrated that COS inhibits the expression and release of these pro-inflammatory mediators(28). In a recent study Mahami et al., design a 5-aminosalicylic acid (5-ASA) and berberine (BBR) loaded chitosan nanoparticles cross-linked by HPMCP (hydroxypropyl methylcellulose phthalate) for inflammation in the colon mucosa(29). The new research on the targeted drugs delivery of polysaccharide-based nanoparticles hyaluronic acid, pectin, dextran, inuline, sodium alginate, and chitosan—against gastrointestinal inflammation is included in this chapter. Numerous methods have also been discussed, such as microbial activation, pH-responsive systems, and enzymatic breakdown. Next, we provide recent findings on polysaccharide-based nanoparticulate drug delivery strategies, highlighting its salient features, advantages, and disadvantages with regard to treating gastrointestinal inflammations. Lastly, we outline future prospects for the field's advancement and the therapeutic use of nanoparticles based on polysaccharides to alleviate inflammation in the gastrointestinal tract.

16.2 ROLE OF POLYSACCHARIDES IN GASTROINTESTINAL DISORDER

The human gut is home to a variety of species of bacteria up to 100 trillion in counts. Various physiological systems such as function of immune system, inflammatory activity, homeostasis etc. are influenced by gut microbiota. Multiple studies highlighted that the action of probiotics or good bacteria affects the intestinal health and disturbances in gut microbiota is known as dysbiosis. In addition to prevent related diseases such as inflammatory bowel diseases (IBD), colorectal cancer (CRC), healthy microbial communities can also reduce inflammation, modulate immune responses, suppress tumorigenesis, and inhibit the growth of infectious bacterial species(30).

These microorganisms present in the human gut use dietary carbohydrates comprised of both polysaccharides as their growth factors. The non-starch polysaccharides (NPS) and resistant starch (RS) which are not broken down by human digestive enzymes, undergo fermentation by commensal microorganisms inside the large intestine. In the large intestine these gut microbiotas use carbohydrates as their source of energy and eliminate short-chain fatty acids as their byproducts, which have an impact on the intestinal microorganisms(31). Further, different factors like enzymes, pH, and microbial growth play a vital role in the effectiveness of gut microbiota. According to a study of Kanner et al., gastric acid secretion affects the oxidation of lipids and other food ingredients. Dietary polysaccharides such as inulin play a major role in prevention of peroxidation of lipid inside stomach. Moreover, dietary polysaccharides such as inulin/ oligofructose play a major role in the protection of gut microorganisms by reducing the oxidative stress and associated inflammation(32). A cross-over control study was carried out in between 59 volunteers divided into two groups and administered with β-galactooligosaccharides (β-GOS) (divided into two groups; i. GOS obtaining from probiotic, ii. GOS obtaining from industrial β-GOS) along at a dose of 7 g/ day which have an effect on the growth of *Bifidobacterium*. The investigation showed that after 10 days a significant increase in *Bifido* bacteria in fecal samples. Further study demonstrates that β-GOS mixture enhances the immunity of the test population by increasing the immunomodulatory cytokine IL-10(33). Acid reflux and heartburn symptoms of Gastroesophageal reflux disease, a long-term disease caused by acid flowing backward to esophagus mechanism. Polysaccharides such as chitosan has higher biocompatibility, biodegradability and targeted drug delivery mechanism. Chitosan in the form of nanoparticles serves as a drug delivery system with more efficacy as compared to the other conventional formulation. Various techniques for deliberation of nanoparticles such as coatings, microspheres and gels have been created to improve the medication targeting, controlled release, retention time in the esophagus. Chitosan nanoparticles have potential to target particular biochemical pathways linked to modulation of inflammation, protection of esophageal tissue and regulation of acid(34). Dietary fibers like pectins have both microbiota dependent and independent impacts on the gastrointestinal immune barrier. Modulation of immunological responses, encouragement of epithelial integrity, and enhancement of the mucus layer are examples of direct impacts on microbiota. Also, pectins can interact with toll-like receptors (TLRs) or galectins to direct affect these molecules. The generation of short-chain fatty acids (SCFAs), the development of commensal adhesion to epithelial cells, the stimulation of microbial diversity and the anti-adhesive properies of pathogens to epithelial cells are the examples of indirect impacts(35).

16.3 DIFFERENT POLYSACCHARIDE-BASED NANOPARTICLES

16.3.1 Chitosan

Chitosan, a cationic amino polysaccharide derived from chitin, is made up of glucosamine and N-acetyl glucosamine. A biopolymer called chitin is present in the cell structures of "fungi (2–44%), prawn cuticles (30–40%), algae (diatoms, brown algae, green algae), insect cuticles (5–25%), mollusks, and the exoskeletons of crustaceans (20–30%)". Chitosan, which is chitin in its deacetylated form, may be modified chemically and is soluble in a wide range of organic and aqueous solvents(36). Enzymes, like chitotriodidase, di-N-acetylchitobiase, N-acetyl-β-D-glucosaminidase, and lysozymes, mediate the degradation of chitosan, just like they do cellulose. Studies have been conducted on this process(37). Moreover, it has been demonstrated that this substance possesses hemostatic, analgesic, mucoadhesive, antibacterial, and antifungal qualities, all of which are crucial for tissue healing and medication delivery. In addition to enabling the polymer to dissolve in an aqueous media, free amino groups produced during chitin deacetylation also take part in chemical changes. It is a special kind of polymer that finds several uses in scaffolds, medication administration, and dressings for wound healing(38). However, because of its low solubility, thermal stability, and challenging processability, chitin has limited application potential despite its many benefits, including biodegradability, biocompatibility, and strong mechanical strength. Chitin can be modified to chitosan to lessen the majority of these disadvantages. For a range of tissue healing and drug delivery applications, chitosan has been processed into hydrogels, two-dimensional films, and three-dimensional porous structures(39). Chitosan is classified into three categories based on molecular weight (Mw): "medium molecular weight (MMWC) with a Mw of 50 to 250 kDa, low molecular weight (LMWC) with a Mw of less than 50 kDa, and high molecular weight (HMWC) with a Mw of >250 kDa". As of late, even shorter chitosan chains (less than 20 kDa), also known as oligomers or chitooligosaccharides (COS), have attracted interest in the pharmaceutical research field because of unique properties such exceptional solubility in water, which makes material processing easier(40). Citosan oligosaccharides (COS) inhibit macrophage reactions triggered by microbial debris, hence their anti-inflammatory characteristics. The high levels of pro-inflammatory cytokines produced by macrophages, such as "nitric oxide (NO), cyclooxygenase-2 (COX-2), prostaglandin E2 (PG-E2), iNOS, TNFα, and IL-6" have a significant impact on the pathophysiology, onset, and progression of ulcerative colitis (UC). It has been shown that COS prevents these pro-inflammatory mediators from being produced and released(41). The following few examples show the application of chitosan and its derivatives for the management of gastrointestinal inflammation. Oral chitosan nanoparticles was investigated for the anti-inflammatory effects on "LPS-inflamed Caco-2 cells and the relate mechanisms and found that by reducing the permeability of the intestinal epithelial monolayer and the release of pro-inflammatory cytokines in Caco-2 cells—both of which were largely regulated by the NF-κB signaling pathway". It can decrease the inflammatory response generated by LPS(42). In another study Soni et al., synthesized a melatonin-loaded chitosan nanoparticles to improve the drug biopharmaceutical profiles and improve the therapeutic efficacy against inflammatory bowel diseases. They found significant reduction in nitric oxide (NO), IL-1β and IL-6 expression via NF-kB p65 inhibition in vitro. Also their histopathological and immune-histological data indicated protective effect of melatonin loaded chitosan nanoparticles against dextran sodium sulfate induced insults and anti-inflammatory action with reduction of inflammatory markers, Nitric Oxide Synthase-2 (NOS2) and Nitro-tyrosine(43). Different colon-targeted and controlled drugs delivery systems are available for the treatment, although their success rate is not particularly high. Because of their smaller size, distinct physicochemical features, and potential for targeting, colon-targeted polymers like chitosan can be used to make nanoparticles that may reduce gastrointenstinal inflammation. A colon-targeted chitosan nanoparticles of rifaximin was prepared by Jantinder and team for colon targeting and improve drug solubility. The study result demonstrates the control release profile of drug in the upper GI tract with colonic conditions(44). Several studies demonstrate chitosan can improve the intestinal absorption of the drugs. Based on this Sun et al., developed a berberine hydrochloride loaded mucus-penetrating sodium alginate-chitosan nanoparticles for penetration of the drug into the inflammatory site of the colon. The developed NPs' in vivo therapeutic efficiency showed that they could reduce the spleen index, which would enhance immunologic function, as well as the disease activity index and colon mucosa damage index, which would lessen colonic injury. Moreover, they increase mucus permeability and transport medications to the site of intestinal inflammation(45).

16.3.2 Pectin

Pectin is a soluble fiber (polysaccharide) found in fruits. It is used for flavoring and cooking. It is also sometimes used medicinally. Pectin is a complex polysaccharide found in the main cell walls of plants and abundant in the green parts of plants. Plants are the source of pectin, a heterogeneous, negatively charged, and water-soluble polysaccharide. Pectin is the main substance in the middle of the lamella, connecting the plant cells. It is deposited by exocytosis into the cell wall by vesicles produced in the Golgi apparatus. Linear 1,4-linked α-D-galacturonic acid units compose its structure. It has attracted much interest in drug delivery applications because it is abundant, biocompatible, biodegradable, and has low toxicity. Considerable research has been done in the food and pharmaceutical industries on the gelling, emulsifying, and thickening qualities. The solubility and gelation of pectin are also significantly influenced by the size of its molecules. Strong gels are produced by pectins whose molecular weight (Mw) is more than 300 kDa, but pectins whose Mw is less than 10 kDa do not gel. In the food business, it is specifically utilized as an additive agent. Its use as an excipient in the pharmaceutical sector appears to have increased recently. It can be completely broken down by the enzymes called pectinase, which are only created by gut flora, but it is resistant to human digestive enzymes. The physicochemical properties of pectins are determined by the esterification [degree of esterification (DE)] and amidation [degree of amidation (DA)] ratio of carboxyl groups. Many scientists successfully developed colonic delivery systems by loading various medications into pectin-based nanoparticles (NPs) via the ionotropic gelation process(46,47). Pectins may have an anti-inflammatory effect through inhibiting macrophages' excessive pro-inflammatory activity. Macrophages are cells that control the degree of inflammation and the regeneration of tissue. This effect is shown to be present in pectins that have a low degree of methyl esterification of the carboxyl groups of the galacturonan core of the macromolecule. The pathophysiology of inflammatory bowel illnesses involves intestinal epithelial cells in addition to leukocytes. Markov and team evaluated the effect of the degree of pectin methyl esterification on inflammation of the colon in mice after oral administration and on the inflammatory response of human colon epithelium cells of the Caco-2 line in vitro with male BALB/c mice. The study revealed that mice's colon inflammation was suppressed by low-methyl esterified pectin AU701, which has over 70% free carboxyl groups(48). S Yener et.al prepared and investigated the melatonin-loaded pectin-based nanoparticles for inflammatory bowel disease (IBD). A modified ionotropic gelation method was used to load melatonin into nanoparticles. Colonic fibrosis, oxidative stress, and colonic inflammatory markers following histologic damage all showed a significant reduction(49). Ruifeng Luo and co-workers developed calcium pectinate and hyaluronic acid-modified lactoferrin nanoparticles loaded rhein with dual-targeting for ulcerative colitis treatment. The assessment of cellular uptake verified that NPs could precisely target and increase the rate of uptake through HA and LF ligands. Through the inhibition of the "TLR4/MyD88/NF-κB signaling pathway", in vivo tests demonstrated that CP/HA/RH-NPs greatly reduced inflammation and expedited colonic healing(50).

16.3.3 Sodium Alginate

Alginates are "1,4C-linked β-D-mannuronic and α-L-guluronic acid" linear polyanionic unbranched polysaccharides. Typically, alginates are derived from brown seaweeds such *Macrocystis sp.*, *Lessonia sp.*, and *Laminaria sp*(51). They can also be made by fermentation using microorganisms. Using ionic or covalent cross-linkers, alginates have been presented in a variety of formulations for tissue engineering and drug delivery applications, such as hydrogels, injectable solutions, sponges, and fibers(52). It is a water-soluble polymer made up of an alternating block of β-D-mannuronic acid and α-L-guluronic acid that are linked by one to four linkages(53). Through internal and external gelation, alginate is cross-linked with polyvalent cations. Due to their flexible structure, alginate-Ca2+gels lose medication throughout the gelation process(54). For targeting, the gelation property has been investigated. The pH dependent solubility of sodium alginate is another characteristic that makes it appropriate for targeting. It is a great option for the development of site-specific drug delivery systems based on its solubility in the alkaline pH of the colon and insoluble in the acidic pH of the stomach. For site specific drug delivery in inflammatory bowel disease Arif et al., designed a disulfide linked sodium alginate nanoparticlesand result showed a remarkable increase in drug release in pH 7.4 buffer with glutathione from drug-loaded nanoparticles, indicating that the nanoparticles could be used for colon-specific drug delivery(55). Alginate nanoparticles'-controlled distribution of drugs has grown in interest for the stomach organ. Certain therapeutic substances, including proteins, are intolerable in cases of acute gastrointestinal distress. IgY was encapsulated in 0.2% w/v sodium alginate nanoparticles using the ionic gelation process as a preventive measure against E. Coli O157:H7. In simulated gastric fluid (pH 1.2) and simulated intestinal fluid (pH 6.8), respectively, an in vitro release investigation revealed a highly significant difference in the amounts of released IgY at 10% and 99.84%. IgY may be effectively and safely delivered through the gastrointestinal system using prepared nanoparticles made from a 0.2%w/v stock solution(56). However, alginate degradation is also mediated by enzymes, just like other polysaccharides, and frequently occurs through bioerosion, a very sluggish process in the body's environment. Their breakdown could be sped up by adding enzymes like alginate lyase. Alginate has been used to treat gastrointestinal inflammation in a number of published studies in conjunction with other proteins and/or polysaccharides. As previously indicated, oppositely charged protein-polysaccharide molecules interact to produce polyelectrolyte complexes, which encapsulate a greater concentration of medication. Curcumin nanoparticles (NPs) based on alginate-gelatin polyelectrolyte complex were created by Sarika and James(57). Below its isoelectric point, gelatin cations, forming nanohybrid complexes with anionic polysaccharides such as sodium alginate. The very hydrophobic compound curcumin is enclosed in the center of this structure. Despite a burst release effect in acidic buffers, they reported a 60% release in 48 hours at pH 7.4 and 69% encapsulation effectiveness. Studies on cellular absorption revealed that, in contrast to free curcumin, the nanohybrid complex loaded with curcumin was shielded from degradation. Protein-polysaccharide complexes may be useful as transporters of bioactives that are sensitive or unstable(57). Chang et al., prepared a "pH-sensitive and reduction-responsive sodium alginate nanospheres cross-linked with disulfide linkages for colon-specific drug delivery for inflammation. The drug release in pH 6.0 buffer with GSH from drug-loaded nanospheres exhibited a marked increase, indicating that the nanospheres may be used for colon-specific drug delivery"(58). In order to improve biostability and promote prolonged release in the colon, Mirpoor et al. investigated into the use of beta-lactoglobulin as a carrier for quercetin, encasing it in a sodium alginate shell. The alginate coating on spherical core-shell structures was verified by morphological investigation. Due to pancreatic hydrolysis and poor interactions at neutral pH, release studies revealed ~77% of quercetin released in SGF and <5% at pH 7.4. Alginate and beta-lactoglobulin are proposed in the study as a potentially useful nanohybrid for hydrophobic bioactive protection(59).

16.3.4 Hyaluronic Acid

Hyaluronic acid, commonly known as hyaluronan, is a non-sulfated GAG that exists in a polyanionic form. This important structural element of soft connective tissue extracellular matrix (ECM) aids in controlling synovial fluid lubrication, water content, structural characteristics, intercellular and intracellular communication, restoration, pathological processes, and shock absorption(60). The units of these linear polysaccharides are β-1,3-N-acetyl-D-glucosamine and α-1,4-D-glucuronic acid, which are alternately linked by β(1 →3) linkages(61). The active enzyme hyaluronidase breaks down HA molecules, which typically have molecular weights greater than one million Da. The molecular weight of HA controls a number of biological processes. For example, high molecular weight HA is known to decrease cell proliferation and have anti-inflammatory, immunosuppressive, and anti-angiogenic effects(62). Shorter degradation fragments of HA facilitate angiogenesis and cell migration(63). Because HA has hydroxyl and carboxylic groups on it, it can be chemically modified in a number of ways to adjust the parameters related to mechanical and degradation without losing its bioactivity. Due to its ability to healing process of wounds, facilitate surface lubrication of the eye, osteoarthritis, rheumatoid arthritis, and dry eye syndrome. HA also has the potential to be used as a dermal filler, an anti-wrinkle agent, and a skin-rejuvenating agent. High levels of HA bind to CD44 receptors, which are overexpressed in many types of cancer cells. Drug delivery locations have been the focus of research pertaining to this characteristic(64). Moreover, HA is a nontoxic, non-inflammatory, non-immunogenic, biodegradable, and biocompatible polymer. Consequently, it has found use in a multitude of applications, including tissue engineering, gene-drug delivery, and drug delivery systems. It is also very accessible to chemical changes because to the presence of functional groups like hydroxyl, carboxyl, and N-acetyl groups. HA and its derivatives have been studied as drug carriers to improve drug targeting, extend release, and facilitate transdermal absorption(65). In another study, Vafaei et al., develop budesonide loaed amphiphilic hyaluronic acid (HA) nanoparticles for targeting of the inflamed intestinal mucosa. Budesonide-loaded HA nanocarriers were investigated for their potential as a treatment using the model of inflammatory CACO-2 cells. Using flow cytometry, it was discovered that inflamed CACO-2 cells had the highest expression of CD44 receptors. When compared to the same dose of free drug, drug-loaded nanoparticles showed a greater anti-inflammatory effect on the secretion of TNF-α and IL-8 in an inflammatory cell model(66). Using a one-pot synthesis approach, Kim et al. (2023) created sulfasalazine-loaded gold nanoparticles (AuNPs) coated with hyaluronic acid (HA) to improve bioavailability and control drug release for the treatment of ulcerative colitis (UC). According to the study, HA-coated gold NPs are a potentially effective colon targeting agent(67). Using the thin film hydration approach, Vafaei et al. (2022) produced budesonide-loaded hyaluronic acid (HA) nanoparticles (NPs) and evaluated their therapeutic potential in the treatment of inflammatory bowel disease (IBD). Compared to free budesonide, treatment with budesonide-loaded NPs decreased the inflammatory score in rats with acetic acid-induced colitis. Rats with IBD-induced elevated MPO activity compared to healthy rats by a large margin; however, the treatment of NPs loaded with budesonide reduced MPO activity by a factor of 1.5. Additionally, animals given budesonide-loaded HA NPs showed reduced levels of tumor necrosis factor-alpha (TNF-α). Budesonide-loaded self-assembled nanoparticles (NPs) were effectively targeted to colonic mucosa in the study, reducing inflammation and encouraging tissue regeneration(68).

16.3.5 Dextran

Dextran is a polysaccharide with glucoside linkages that is produced by microorganisms using natural glucose sources. It can be found in a range of molecular weights, including the widely used dextran 40 (40 kDa) and dextran 70 (70 kDa) (69). The branching pattern of dextran determines its physical characteristics, such as its solubility. High water solubility dextran is produced by linear linking (>95%). Dexttran becomes insoluble at higher branching percentages (>43%) (70). Through chemical or enzymatic derivatization events, the hydroxyl groups in dextran can be modified to create dextran with specific qualities. Dextran's qualities, such as its biocompatibility, biodegradability, and capacity to be enzymatically broken down by colonic bacteria, have prompted research into its potential for colon targeting(71). Dextran, as a colloid, replenishes blood plasma lost due to severe bleeding, osmotically expanding plasma volume. Its distribution volume is distributed throughout the body's blood volume. Dextran improves cardiac output and blood flow by increasing blood volume, which enables red blood cells to more efficiently travel to end-organ microvasculature. Zhang et al. developed pH-sensitive nanoparticles by modifying dex-trans with glycidyl methacrylate and acrylic acid (dex-GMA/AAC). These were meant as a model carrier for drugs designed to be delivered selectively in the colon. In this study, the susceptible stomach degradation drug erythromycin was loaded into these pH-sensitive dextran NPs. "The results showed that the NPs have some degradability towards dextranase though dex-GMA/AAC NPs were about three times more stable than simple dextran NPs. Simple NPs exhibited limited (less than 5%) drug release at pH 1.2 while cross-linked-NPs released ≈40% of the drug within the same time period". However Dex-GMA/AAC NPs have got reduced swelling under acidic conditions thus providing protection against gastric acid effects on a drug which increases with increase in pH due to carboxylic group ionization leading to further swelling and partial degradation during incubation with a dextranase causing high drug release (>75%) at pH 7.4. In conclusion, authors found that carriers based on pH-sensitive dex-GMA/AAC could effectively deliver drugs into the colon when used properly(72). Saboktakin et al. prepared dextran sulfate-chitosan DS-CS hydrogels cipro-chitosan complexes as nanoparticles for colonic drug delivery of 5-ASA. DS-CS hydrogel complexes were formed by developing a complex coacervation method. A small percentage of the loaded drug about 20%) was released after and nearly 90% of the drug was released after due to the hydrolysis of the hydrogel complexes inside the colon. As a result, the DS-CS hydrogel complex effectively alleviates the hydrolysis of the complex and the release rate of the drug in the stomach, ensuring controlled drug release(73).

16.3.6 Inulin

A natural fructose polymer discovered mainly in plants, inulin, is largely present in several plants such as Jerusalem artichoke, chicory root, and asparagus root, but even in commonly consumed vegetables such as onion, banana, leek, garlic, rye, barley, and wheat(74). Structurally, "inulin comprises linear units of D-fructose linked together by β-(2-1) glycosidic bonds with a terminal residue of glucose". Inulin composed of only ten units of fructose is therefore sometimes known as an oligofructose. Inulin's recognition of endogenous properties, including anti-inflammatory effect and stimulation of beneficial intestinal bacteria growth such as *Bifidobacterium* and *Lactobacillus*, maintains intimal microbial balance and human health(75). Functional changes in the gut microbiota were also induced by orally administered inulin-type fructans in a preliminary study to the patients with active ulcerative colitis, although the composition of the microbiota was not significantly affected. Inulin-type fructans administration at high doses also promoted the abundance of *Bifidobacteriaceae* and *Lachnospiraceae* and increased scattered short-chain fatty acids levels. Inulin is a potential candidate as a colon-targeted drug carrier because it has β-(2-1)-linked glycosidic bonds that are not degraded in the stomach before arriving in the colon, where they are broken down by colonic enzymes and bacteria that are *Bifidobacteria*(76). In vivo studies demonstrated that sulfobutyl ether carboxymethylcellulose nanoparticles of budesonide were localized in inflamed colons. These systems were shown to be susceptible to the reactions of enzymes, redox, and pH which were publicized, thereby letting the drugs be released only in the area of inflamed colon(77, 78). "Apremilast-loaded inulin nanoparticles, coupled with mannose as a macrophage-targeting ligand, showed enhanced uptake in inflamed macrophages compared to untreated macrophages and mannose receptor-negative cell lines"(79). In mice with ulcerative colitis (UC), analysis of in vivo distribution indicated that 60% of nanoparticles with mannose accumulation were found in the inflamed colon. Besides its capacity to target the colon and its anti-inflammatory effects, the potential of inulin for delivering drugs specifically to macrophages presents an encouraging prospect for treating inflammatory bowel disease (IBD)(80).

16.4 RECENT ADVANCEMENTS AND DRAWBACKS

In polysaccharide-based nanoparticulate systems, natural polysaccharides play a significant role as biopolymers made up of repeating units of mono- or disaccharides connected by glycosidic linkages(81). Biologically safe, biocompatible, biodegradable, and possessing a multitude of biological activities are natural polysaccharides derived from a wide range of sources. By modifying gut microbiota and inflammation-related signaling pathways, some naturally occurring polysaccharides may also act as nanocarriers to aid in the remission of gastrointestinal inflammation. This can be achieved in conjunction with medication to lessen the dosage and side effects of medication while also enhancing the symptoms of this condition. For example, oral Dictyophora indusiata polysaccharide administration has been reported to modulate the composition of the intestinal flora, enhance the expression of mucin and tight junction proteins, inhibit the secretion of inflammatory factors, regulate oxidative stress, and improve the function of the intestinal barrier to alleviate the symptoms of colitis induced by dextran sulfate sodium(82). Natural polysaccharides are not digested by the gastrointestinal system since they are prebiotics; nonetheless, the gut bacteria in the intestinal tract can ferment and make use of them. The intestinal barrier may interact with some charged natural polysaccharides, perhaps extending the duration of stay in the inflammatory gut. Its outstanding qualities make it a potentially significant ally in the delivery of nanomedicine. In the medical field, DNSs are mostly utilized for drug delivery. In contrast to conventional drug delivery methods, polysaccharide-based DNSs let medications enter epithelial and inflammatory cells, allowing the treatments to act on the affected areas at higher concentrations and for longer periods of time. A crucial element of DNSs are nanocarriers based on polysaccharides. Polysaccharide-based nanocarriers have the potential to greatly increase drug efficacy, improve adherence to treatment, and decrease systemic adverse effects. This could lead to the creation of targeted medications with low side effects and sustained, long-term remissions(83). Therefore, it is acknowledged that one of the most appealing biopolymers for inflammatory bowel diseases treatment in nanomedicine is natural polysaccharides. Natural polysaccharides have drawn a lot of attention lately for their potential use as nanocarriers in nanomedicine delivery systems. Thus, we examined the processes of polysaccharide-based nanocarriers in the synergistic treatment of gastrointestinal inflammation as well as the transport and breakdown process of DNSs built by polysaccharide-based nanocarriers following oral administration. The various polysaccharide nanoparticle-based delivery systems for gastrointenstinal inflammation are displayed in Table 16.1. The remedial care of due to the high level of intricacy of the GI system and the fact that the colon and rectum are the most afflicted gastrointestinal sections given their location, treating gastrointestinal inflammation therapeutically is a difficult undertaking. Numerous investigations have been conducted in the area of colon-targeted drug delivery systems, with differing degrees of success, particularly in the areas of micro- and nanoparticulate systems for the treatment of gastrointestinal inflammation. The methods and most recent developments in orally delivered micro- and nanoparticulate systems for successful therapy in gastrointestinal inflammation have been covered in the above section. Numerous recent approaches utilizing natural polysaccharide-based colonic systems have demonstrated their significant potential to reach the colonic region. These approaches primarily focus on examining the process of selective degradability by the colonic microbiota. Additional characteristics of particular polysaccharides, like mucoadhesion, have also been optimized for these strategies in order to improve contact with biological tissue and get past the intestinal barrier(84).

TABLE 16.1
Polysaccharide-based Nanoformulations for Inflammatory Bowel Disease

Polymer	Drug	Target Mechanism	Method	Outcome	Reference
Hyaluronic acid	Budesonide	Enzymatic degradation	Thin film hydration method	Show greater efficacy in reducing colitis as compared to free medication. suggested that nanoparticles might be used to deliver drugs specifically to the inflammatory colon mucosa.	(85)
	Curcumin	Enzymatic degradation	Emulsification and solvent evaporation method	Improve anti-inflammatory Effect.	(86)
	Apigenin Mn(II)	Enzymatic degradation	Ultrasonication	Potentially moderate inflammatory factors, so efficiently repairing the intestinal barrier and greatly enhancing the injured colon tissue.	(87)
	Lysine proline-valine	Enzymatic degradation	Water-in-oil-in-water (W/O/W) double emulsion-solvent evaporation method	It works in tandem to reduce inflammation and expedite mucosal repair in the fight against ulcerative colitis.	(88)
Chitosan	Rifaximin	Enzymatic degradation	Ionic gelation method	Increased rifaximin solubility. demonstrated excellent formulation stability and regulated drug distribution in the intended area.	(89)
	5-aminosalicylic acid and berberine	pH dependent	Ionic gelation	They extended the colon's length, reduced its moist weight, and enhanced the ulcer index and disease activity metrics.	(29)
	Berberine	High concentration of ROS	Dialysis method	The study investigated the in vivo anti-inflammatory impact and the resulting modulation of gut microbiota in mice that were induced to develop colitis using dextran sodium sulfate.	(90)
	Bilirubin	High concentration of ROS	Ultrasonication	Accumulate permanently in the stomach and exert powerful antioxidant and anti-inflammatory effects.	(91)

TABLE 16.1 (Continued)
Polysaccharide-based Nanoformulations for Inflammatory Bowel Disease

Polymer	Drug	Target Mechanism	Method	Outcome	Reference
	Curcumin	pH-responsive systems	Ionotropic gelation method	In order to overcome inadequate CN absorption and availability, coating CS-NPs with Eudragit FS 30D enabled the colon to receive CN in a pH-stimulated and sustained manner.	(92)
	Diclofenac sodium	Enzymatic degradation	Spinning disk processor	The drug's uptake into Caco-2 cells upon formulation was three times more than the drug's absorption from a control solution.	(93)
Dextran	Cerium Oxide	High concentration of ROS	Precipitation method	Produced a significant CT contrast and gathered in the large intestine's colitis area.	(94)
Inulin	ornidazole	Enzymatic degradation	Dialysis method	Broken down in the colon by inulinase, it is quickly released in the enzymatic intestinal environment.	(78)
	4-aminothiophenol	Enzymatic degradation	Chemical reduction	Higher accumulation in the inflamed colon.	(77)
	Cannabidiol	Enzymatic degradation	Dialysis method	considerably enhanced the number, species composition, and abundance of short chain fatty acids in the intestinal flora of mice, thereby successfully preserving the intestinal flora's equilibrium.	(95)
Cyclodextrin	Rhein	Enzymatic degradation	Ultrasonication	The strongest anti-inflammatory and antioxidant benefits were obtained by the on-demand drug release technique, and this suggests that it will be a promising carrier in the treatment of ulcerative colitis.	(96)
Heparin	Human serum albumin	Enzymatic degradation	Desolvation	The nanoparticles show the ability to load both medications at the same time; in vitro, they have a greater anti-inflammatory effect than the single-drug-loaded nanoparticles; and in a mouse model of colitis, they selectively bind to inflamed bowel following enema treatment.	(97)
Starch	Curcumin	Enzymatic degradation	Dialysis method	The inflammation-related opening of the tight junctions better allowed to cross the epithelium by paracellular transport.	(98)
Pectin	Melatonin	Enzymatic degradation	Modified polyelectrolyte complexation method	Melatonin treatment reduced the damage score by 67.1% when taken intracolonally and by 73.2% when taken orally.	(99)
	Taurine	Enzymatic degradation	Ionic gelation method	Animals treated with taurine-loaded chitosan pectin nanoparticles had the best therapeutic outcome.	(100)
Alginate	Peptides (lysine-proline-valine)	Enzymatic degradation	Solvent evaporation	Pre-treated mice exhibited a decreased anti-inflammatory response when administered drugs via nanoparticles.	(101)
	Prednisolone and inulin	Enzymatic degradation	Spray freeze-drying	Alginate nanoparticles effectively encapsulated active substances and were biodegradable.	(102)

16.5 FUTURE DEVELOPMENT AND CONCLUSION

A number of benefits of polysaccharides, including as pH, bacterial enzyme sensitivity, and biocompatibility, make them indispensable for the management of gastrointestinal disease. Creating and perfecting nanoparticles from polysaccharides that are only degradable in the colon and resistant to the digestive system is one possibility. The polymer can be changed by chemical engineering to generate nanoparticles without changing the polymer's primary physical and chemical properties. The GI tract is seeing the emergence of new uses for nanoparticles treatment. Several physiological obstacles must be overcome by conventional therapy for inflammatory gastric disorders, including pH changes in the GI tract, enzymatic degradation, systemic absorption, and metabolism, which reduces the medications' bioavailability in the colon. High pay load, protection against GI degradation, and prolonged delivery are some of the novel drug delivery technologies, such as micro/nano formulations, that can lead to improved therapeutic response. NP medication administration is, however, severely constrained by the potential toxicity concern connected to synthetic polymers. With their biodegradability, biocompatibility, and FDA GRAS classification, polysaccharides may lessen the drawbacks of synthetic polymers and present a number of benefits, including the potential for targeted drug administration in the treatment of gastric inflammation. Due to their poor intestinal barrier crossing effectiveness and inadequate resilience to gastrointestinal injury, drug delivery systems built using single polysaccharides as nanocarriers may not achieve optimal targeted delivery efficiency. The efficiency of drug delivery to the gastrointestinal tract can be significantly increased by modifying polysaccharides to suit specific needs or combining them with compounds with different properties to create delivery systems that maximize their stability in the gastrointestinal tract and allow them to accumulate in the areas of intestinal lesions. While a lot of research is now being done on the development of polysaccharide-based delivery systems for drugs that target the colon, not as much is being done on these systems to treat stomach inflammation. To find or chemically alter polysaccharides that are better suited to treating stomach inflammation and can be utilized to build delivery systems with colon-specific characteristics, more study will be needed. Additionally, it remains a crucial area of research to figure out how to get beyond the intestinal mucosal barrier and intestinal pathogen barrier to get to the target cells as much as possible when concentrated in the lesion regions. A few researchers have experimented with several successful techniques, including the alteration of mucinolytic and hydrophilic enzymes. They're still not flawless, though. Further research may concentrate on improving the efficiency of colon-targeted delivery through the modification of the nanoparticulate delivery system as needed. This will be based on the unique microenvironment of the gastrointestinal tract and the receptors that are specifically expressed in the lesions of patients with inflammatory bowel disease.

REFERENCES

1. Verma G, Marella A, Shaquiquzzaman M, Alam MM. Immunoinflammatory responses in gastrointestinal tract injury and recovery. *Acta Biochimica Polonica.* 2013;60(2):143–149.
2. Kvietys PR, Granger DN. The vascular endothelium in gastrointestinal inflammation. In Clive Page, John L. Wallace (eds.) *Immunopharmacology of the Gastrointestinal System*: Elsevier; 1993. p. 69–94.
3. Cai Z, Wang S, Li J. Treatment of inflammatory bowel disease: a comprehensive review. *Frontiers in Medicine.* 2021;8:765474.
4. Shariati A, Fallah F, Pormohammad A, Taghipour A, Safari H, Chirani AS, et al. The possible role of bacteria, viruses, and parasites in initiation and exacerbation of irritable bowel syndrome. *Journal of Cellular Physiology.* 2019;234(6):8550–69.
5. Zheng D, Liwinski T, Elinav E. Interaction between microbiota and immunity in health and disease. *Cell Research.* 2020;30(6):492–506.
6. Gu B-H, Kim M, Yun C-HJN. Regulation of gastrointestinal immunity by metabolites. *Nutrients.* 2021;13(1):167.
7. Jamar G, Ribeiro DA, Pisani LP. High-fat or high-sugar diets as trigger inflammation in the microbiota-gut-brain axis. *Critical Reviews in Food Science and Nutrition.* 2021;61(5):836–54.
8. Cai Z, Wang S, Li J. Treatment of inflammatory bowel disease: a comprehensive review. *Frontiers in Medicine.* 2021;8:765474.
9. Zundler S, Becker E, Weidinger C, Siegmund B. Anti-adhesion therapies in inflammatory bowel disease—molecular and clinical aspects. *Frontiers in Immunology.* 2017;8:283799.
10. Pouillon L, Bossuyt P, Peyrin-Biroulet L. Considerations, challenges and future of anti-TNF therapy in treating inflammatory bowel disease. *Expert Opinion on Biological Therapy.* 2016;16(10):1277–90.
11. Prasad M, Lambe UP, Brar B, Shah I, Manimegalai J, Ranjan K, et al. Nanotherapeutics: an insight into healthcare and multi-dimensional applications in medical sector of the modern world. *Biomedicine & Pharmacotherapy.* 2018;97:1521–37.
12. Chowdhury R. *A Review on the Application of Nanotechnology in Pharmaceutical Science*: East West University; 2015.
13. Uskoković V, Uskoković DP. *Nanotechnologies in preventive and regenerative medicine: Quo Vadis, Domine? Nanotechnologies in preventive and regenerative medicine*: Elsevier; 2018. p. 513–66.
14. Yusuf A, Almotairy ARZ, Henidi H, Alshehri OY, Aldughaim MS. Nanoparticles as drug delivery systems: a review of the implication of nanoparticles' physicochemical properties on responses in biological systems. *Polymers.* 2023;15(7):1596.
15. Liu W, Dong Z, Liu K, Lu Y, Wu W, Qi J, et al. Targeting strategies of oral nano-delivery systems for treating inflammatory bowel disease. *International Journal of Pharmaceutics.* 2021;600:120461.
16. Sun A, Liu H, Sun M, Yang W, Liu J, Lin Y, et al. Emerging nanotherapeutic strategies targeting gut-X axis against diseases. *Biomedicine & Pharmacotherapy.* 2023;167:115577.
17. Giordo R, Wehbe Z, Paliogiannis P, Eid AH, Mangoni AA, Pintus G, editors. Nano-targeting vascular remodeling in cancer: recent developments and future directions. *Seminars in Cancer Biology*; 2022: Elsevier. 10.1016/j.semcancer.2022.03.001
18. Oake A, Bhatt P, Pathak YV. Understanding surface characteristics of nanoparticles. *Surface Modification of Nanoparticles for Targeted Drug Delivery.* 2019:1–17. doi.org/10.1007/978-3-030-06115-9_1
19. Khan AU, Khan M, Cho MH, Khan MM. Selected nanotechnologies and nanostructures for drug delivery, nanomedicine and cure. *Bioprocess and Biosystems Engineering.* 2020;43(8):1339–57.

20. Toy R, Peiris PM, Ghaghada KB, Karathanasis E. Shaping cancer nanomedicine: the effect of particle shape on the in vivo journey of nanoparticles. *Nanomedicine*. 2014;9(1):121–34.
21. Bhatia S, Bhatia S. Natural polymers vs synthetic polymer. *Natural Polymer Drug Delivery Systems: Nanoparticles, Plants, and Algae*. 2016:95–118. doi.org/10.1007/978-3-319-41129-3_3
22. Yu Y, Shen M, Song Q, Xie J. Biological activities and pharmaceutical applications of polysaccharide from natural resources: a review. *Carbohydrate Polymers*. 2018;183:91–101.
23. Gong P, Wang S, Liu M, Chen F, Yang W, Chang X, et al. Extraction methods, chemical characterizations and biological activities of mushroom polysaccharides: a mini-review. *Carbohydrate Research*. 2020;494:108037.
24. Tan RSL, Hassandarvish P, Chee CF, Chan LW, Wong TW. Chitosan and its derivatives as polymeric anti-viral therapeutics and potential anti-SARS-CoV-2 nanomedicine. *Carbohydrate Polymers*. 2022;290:119500.
25. Ngwuluka NC. Responsive polysaccharides and polysaccharides-based nanoparticles for drug delivery. In Abdel Salam Hamdy Makhlouf and Nedal Y. Abu-Thabit (eds.) *Stimuli Responsive Polymeric Nanocarriers for Drug Delivery Applications*, Volume 1: Elsevier; 2018. p. 531–54.
26. Hou C, Chen L, Yang L, Ji X. An insight into anti-inflammatory effects of natural polysaccharides. *International Journal of Biological Macromolecules*. 2020;153:248–55.
27. Dos Santos AM, Carvalho SG, Meneguin AB, Sábio RM, Gremião MPD, Chorilli M. Oral delivery of micro/nanoparticulate systems based on natural polysaccharides for intestinal diseases therapy: challenges, advances and future perspectives. *Journal of Controlled Release*. 2021;334:353–66.
28. Muanprasat C, Chatsudthipong VJP, therapeutics. Chitosan oligosaccharide: biological activities and potential therapeutic applications. *Pharmacology & Therapeutics*. 2017;170:80–97.
29. Mahami S, Salehi M, Mehrabi M, Vahedi H, Hassani MS, Bitaraf FS, et al. pH-sensitive HPMCP-chitosan nanoparticles containing 5-aminosalicylic acid and berberine for oral colon delivery in a rat model of ulcerative colitis. *International Journal of Biological Macromolecules*. 2023;244:125332.
30. Álvarez-Mercado AI, Plaza-Diaz J. Dietary polysaccharides as modulators of the gut microbiota ecosystem: an update on their impact on health. *Nutrients*. 2022;14(19):4116.
31. Ho Do M, Seo YS, Park H-Y. Polysaccharides: bowel health and gut microbiota. *Critical Reviews in Food Science and Nutrition*. 2021;61(7):1212–24.
32. Kanner J, Lapidot T. The stomach as a bioreactor: dietary lipid peroxidation in the gastric fluid and the effects of plant-derived antioxidants. *Free Radical Biology and Medicine*. 2001;31(11):1388–95.
33. Guarino MPL, Altomare A, Emerenziani S, Di Rosa C, Ribolsi M, Balestrieri P, et al. Mechanisms of action of prebiotics and their effects on gastro-intestinal disorders in adults. *Nutrients*. 2020;12(4):1037.
34. Herdiana Y. Chitosan nanoparticles for gastroesophageal reflux disease treatment. *Polymers*. 2023;15(16):3485.
35. Beukema M, Faas MM, de Vos P. The effects of different dietary fiber pectin structures on the gastrointestinal immune barrier: impact via gut microbiota and direct effects on immune cells. *Experimental & Molecular Medicine*. 2020;52(9):1364–76.
36. Pellis A, Guebitz GM, Nyanhongo GS. Chitosan: sources, processing and modification techniques. *Gels*. 2022;8(7):393.
37. Kaczmarek MB, Struszczyk-Swita K, Li X, Szczęsna-Antczak M, Daroch MJFib. Enzymatic modifications of chitin, chitosan, and chitooligosaccharides. *Frontiers in Bioengineering Biotechnology*. 2019;7:243.
38. Harugade A, Sherje AP, Pethe A. Chitosan: a review on properties, biological activities and recent progress in biomedical applications. *Reactive Functional Polymers*. 2023;2023:105634.
39. Wang W, Xue C, Mao X. Chitosan: structural modification, biological activity and application. *International Journal of Biological Macromolecules*. 2020;164:4532–46.
40. Ferreira J, Coelho A, Moutinho L. Dynamic capabilities, creativity and innovation capability and their impact on competitive advantage and firm performance: the moderating role of entrepreneurial orientation. *Technovation*. 2020;92:102061.
41. Yang E-J, Kim J-G, Kim J-Y, Kim SC, Lee NH, Hyun C-G. Anti-inflammatory effect of chitosan oligosaccharides in RAW 264.7 cells. *Central European Journal of Biology*. 2010;5:95–102.
42. Tu J, Xu Y, Xu J, Ling Y, Cai Y. Chitosan nanoparticles reduce LPS-induced inflammatory reaction via inhibition of NF-κB pathway in Caco-2 cells. *International Journal of Biological Macromolecules*. 2016;86:848–56.
43. Soni JM, Sardoiwala MN, Choudhury SR, Sharma SS, Karmakar S. Melatonin-loaded chitosan nanoparticles endows nitric oxide synthase 2 mediated anti-inflammatory activity in inflammatory bowel disease model. *Materials Science Engineering: C*. 2021;124:112038.
44. Kumar J, Newton AM. Rifaximin-chitosan nanoparticles for inflammatory bowel disease (IBD). *Recent Patents on Inflammation Allergy Drug Discovery*. 2017;11(1):41–52.
45. Sun L, Nie X, Lu W, Zhang Q, Fang W, Gao S, et al. Correction: mucus-penetrating alginate-Chitosan nanoparticles loaded with berberine hydrochloride for oral delivery to the inflammation site of ulcerative colitis. *AAPS PharmSciTech*. 2022;23(6):218.
46. Das S. Pectin based multi-particulate carriers for colon-specific delivery of therapeutic agents. *International Journal of Pharmaceutics*. 2021;605:120814.
47. Ashford M, Fell J, Attwood D, Sharma H, Woodhead P. An evaluation of pectin as a carrier for drug targeting to the colon. *Journal of Controlled Release*. 1993;26(3):213–20.
48. Markov PA, Volkova MV, Khasanshina ZR, Martinson EA, Popov SV. Anti-inflammatory activity of high and low methoxylated apple pectins, in vivo and in vitro. *Voprosy Pitaniia*. 2021;90(6):92–100.
49. Yener S, Akbulut KG, Karakuş R, Erdoğan D, Acartürk F. Development of melatonin loaded pectin nanoparticles for the treatment of inflammatory bowel disease: in vitro and in vivo studies. *Journal of Drug Delivery Science Technology*. 2022;67:102861.
50. Luo R, Lin M, Fu C, Zhang J, Chen Q, Zhang C, et al. Calcium pectinate and hyaluronic acid modified lactoferrin

nanoparticles loaded rhein with dual-targeting for ulcerative colitis treatment. *Carbohydrate Polymers*. 2021;263:117998.

51. Kaidi S, Bentiss F, Jama C, Khaya K, Belattmania Z, Reani A, et al. Isolation and structural characterization of alginates from the Kelp species Laminaria ochroleuca and Saccorhiza polyschides from the Atlantic Coast of Morocco. *Colloids and Interfaces*. 2022;6(4):51.
52. Milivojevic M, Pajic-Lijakovic I, Bugarski B. Recent advances in alginates as material for biomedical applications. *Alginates*. 2019;2019:25–88.
53. Kabir SF, Rahman A, Yeasmin F, Sultana S, Masud RA, Kanak NA, et al. Occurrence, distribution, and structure of natural polysaccharides. In M. Naeem, Tariq Aftab, M. Masroor A. Khan (eds.) *Radiation-Processed Polysaccharides*: Elsevier; 2022. p. 1–27.
54. De'Nobili MD, Curto LM, Delfino JM, Pérez CD, Bernhardt D, Gerschenson LN, et al. Alginate utility in edible and nonedible film development and the influence of its macromolecular structure in the antioxidant activity of a pharmaceutical/food interface. *Alginic Acid: Chemical Structure, Uses and Health Benefits Moore, A, Ed.* 2015;1:119–69.
55. Arif M, Chi Z, Liu Y-J, Liu C-G. Preparation, characterization, and in vitro drug release behavior of thiolated alginate nanoparticles loaded budesonide as a potential drug delivery system toward inflammatory bowel diseases. *Journal of Biomaterials Science, Polymer Edition*. 2020;31(18):2299–317.
56. Bakhshi M, Ebrahimi F, Nazarian S, Zargan J, Behzadi F, Gariz DS. Nano-encapsulation of chicken immunoglobulin (IgY) in sodium alginate nanoparticles: in vitro characterization. *Biologicals*. 2017;49:69–75.
57. Sarika P, James NR. Polyelectrolyte complex nanoparticles from cationised gelatin and sodium alginate for curcumin delivery. *Carbohydrate Polymers*. 2016;148:354–61.
58. Chang D, Lei J, Cui H, Lu N, Sun Y, Zhang X, et al. Disulfide cross-linked nanospheres from sodium alginate derivative for inflammatory bowel disease: preparation, characterization, and in vitro drug release behavior. *Carbohydrate Polymers*. 2012;88(2):663–9.
59. Mirpoor SF, Hosseini SMH, Nekoei A-R. Efficient delivery of quercetin after binding to beta-lactoglobulin followed by formation soft-condensed core-shell nanostructures. *Food Chemistry*. 2017;233:282–9.
60. Marinho A, Nunes C, Reis S. Hyaluronic acid: a key ingredient in the therapy of inflammation. *Biomolecules*. 2021;11(10):1518.
61. Salave S, Rana D, Sharma A, Bharathi K, Gupta R, Khode S, et al. Polysaccharide based implantable drug delivery: development strategies, regulatory requirements, and future perspectives. *Polysaccharides*. 2022;3(3):625–54.
62. Lierova A, Kasparova J, Filipova A, Cizkova J, Pekarova L, Korecka L, et al. Hyaluronic acid: known for almost a century, but still in vogue. *Pharmaceutics*. 2022;14(4):838.
63. Giraudo MV, Di Francesco D, Catoira MC, Cotella D, Fusaro L, Boccafoschi F. Angiogenic potential in biological hydrogels. *Biomedicines*. 2020;8(10):436.
64. Yasin A, Ren Y, Li J, Sheng Y, Cao C, Zhang K, et al. Advances in hyaluronic acid for biomedical applications. *Frontiers in Bioengineering Biotechnology*. 2022;10:910290.
65. Huang G, Huang H. Application of hyaluronic acid as carriers in drug delivery. *Drug Delivery*. 2018;25(1):766–72.
66. Vafaei SY, Esmaeili M, Amini M, Atyabi F, Ostad SN, Dinarvand R. Self assembled hyaluronic acid nanoparticles as a potential carrier for targeting the inflamed intestinal mucosa. *Carbohydrate Polymers*. 2016;144:371–81.
67. Kim H-M, Park JH, Choi YJ, Oh J-M, Park J. Hyaluronic acid-coated gold nanoparticles as a controlled drug delivery system for poorly water-soluble drugs. *RSC Advances*. 2023;13(8):5529–37.
68. Vafaei SY, Abdolghaffari AH, Mahjub R, Eslami SM, Esmaeili M, Abdollahi M, et al. Budesonide-loaded hyaluronic acid nanoparticles for targeted delivery to the inflamed intestinal mucosa in a rodent model of colitis. *BioMed Research International*. 2022;2022: 7776092.
69. Naessens M, Cerdobbel A, Soetaert W, Vandamme EJ. Leuconostoc dextransucrase and dextran: production, properties and applications. *Journal of Chemical Technology Biotechnology: International Research in Process, Environmental & Clean Technology*. 2005;80(8): 845–60.
70. Neely WB. Dextran: structure and synthesis. In M.L. Wolfrom, R. Stuart Tipson (eds.) *Advances in Carbohydrate Chemistry*, Vol. 15: Elsevier; 1961. p. 341–69.
71. Petrovici AR, Pinteala M, Simionescu N. Dextran formulations as effective delivery systems of therapeutic agents. *Molecules (Basel, Switzerland)*. 2023;28(3):1086.
72. Zhang H, Wu H, Fan L, Li F, Gu Ch, Jia M. Preparation and characteristics of pH-sensitive derivated dextran hydrogel nanoparticles. *Polymer Composites*. 2009;30(9):1243–50.
73. Saboktakin MR, Tabatabaie RM, Maharramov A, Ramazanov MA. Synthesis and characterization of chitosan hydrogels containing 5-aminosalicylic acid nanopendents for colon: specific drug delivery. *Journal of Pharmaceutical Sciences*. 2010;99(12):4955–61.
74. Mudannayake DC, Jayasena DD, Wimalasiri KM, Ranadheera CS, Ajlouni S. Inulin fructans–food applications and alternative plant sources: a review. *International Journal of Food Science Technology*. 2022;57(9):5764–80.
75. Tremaroli V, Bäckhed F. Functional interactions between the gut microbiota and host metabolism. *Nature*. 2012;489(7415):242–9.
76. Roberfroid MB. Introducing inulin-type fructans. *British Journal of Nutrition*. 2005;93(S1):S13-S25.
77. Sun Q, Luan L, Arif M, Li J, Dong Q-J, Gao Y, et al. Redox-sensitive nanoparticles based on 4-aminothiophenol-carboxymethyl inulin conjugate for budesonide delivery in inflammatory bowel diseases. *Carbohydrate Polymers*. 2018;189:352–9.
78. Shivhare K, Garg C, Priyam A, Gupta A, Sharma AK, Kumar P. Enzyme sensitive smart inulin-dehydropeptide conjugate self-assembles into nanostructures useful for targeted delivery of ornidazole. *International Journal of Biological Macromolecules*. 2018;106:775–83.
79. Laird ME, Tong LX, Sicco KIL, Kim RH, Meehan SA, Franks AG. Novel use of apremilast for adjunctive treatment of recalcitrant pyoderma gangrenosum. *JAAD Case Reports*. 2017;3(3):228–9.

80. Sun Q, Arif M, Chi Z, Li G, Liu C-G. Macrophages-targeting mannosylated nanoparticles based on inulin for the treatment of inflammatory bowel disease (IBD). *International Journal of Biological Macromolecules*. 2021;169:206–15.
81. Barclay TG, Day CM, Petrovsky N, Garg S. Review of polysaccharide particle-based functional drug delivery. *Carbohydrate Polymers*. 2019;221:94–112.
82. Kanwal S, Joseph TP, Aliya S, Song S, Saleem MZ, Nisar MA, et al. Attenuation of DSS induced colitis by Dictyophora indusiata polysaccharide (DIP) via modulation of gut microbiota and inflammatory related signaling pathways. *Journal of Functional Foods*. 2020;64:103641.
83. Viscido A, Capannolo A, Latella G, Caprilli R, Frieri G. Nanotechnology in the treatment of inflammatory bowel diseases. *Journal of Crohn's Colitis*. 2014;8(9):903–18.
84. Boateng J, Okeke O, Khan S. Polysaccharide based formulations for mucosal drug delivery: a review. *Current Pharmaceutical Design*. 2015;21(33):4798–821.
85. Vafaei SY, Abdolghaffari AH, Mahjub R, Eslami SM, Esmaeili M, Abdollahi M, et al. Budesonide-loaded hyaluronic acid nanoparticles for targeted delivery to the inflamed intestinal mucosa in a rodent model of colitis. *BioMed Research International*. 2022;2022: 7776092.
86. Hlaing SP, Cao J, Lee J, Kim J, Saparbayeva A, Kwak D, et al. Hyaluronic acid-conjugated PLGA nanoparticles alleviate ulcerative colitis via CD44-mediated dual targeting to inflamed colitis tissue and macrophages. *Pharmaceutics*. 2022;14(10):2118.
87. Lv F, Zhang Y, Peng Q, Zhao X, Hu D, Wen J, et al. Apigenin-Mn (II) loaded hyaluronic acid nanoparticles for ulcerative colitis therapy in mice. *Frontiers in Chemistry*. 2022;10:969962.
88. Xiao B, Xu Z, Viennois E, Zhang Y, Zhang Z, Zhang M, et al. Orally targeted delivery of tripeptide KPV via hyaluronic acid-functionalized nanoparticles efficiently alleviates ulcerative colitis. *Molecular Therapy*. 2017;25(7):1628–40.
89. Kumar J, Newton AM. Rifaximin-chitosan nanoparticles for inflammatory bowel disease (IBD). *Recent Patents on Inflammation & Allergy Drug Discovery*. 2017;11(1):41–52.
90. Zhao L, Du X, Tian J, Kang X, Li Y, Dai W, et al. Berberine-loaded carboxylmethyl chitosan nanoparticles ameliorate DSS-induced colitis and remodel gut microbiota in mice. *Frontiers in Pharmacology*. 2021;12:644387.
91. Huang Z, Shi Y, Wang H, Chun C, Chen L, Wang K, et al. Protective effects of chitosan-bilirubin nanoparticles against ethanol-induced gastric ulcers. *International Journal of Nanomedicine*. 2021:8235–50.
92. Raj PM, Raj R, Kaul A, Mishra AK, Ram A. Biodistribution and targeting potential assessment of mucoadhesive chitosan nanoparticles designed for ulcerative colitis via scintigraphy. *RSC Advances*. 2018;8(37):20809–21.
93. Huanbutta K, Sriamornsak P, Luangtana-Anan M, Limmatvapirat S, Puttipipatkhachorn S, Lim L-Y, et al. Application of multiple stepwise spinning disk processing for the synthesis of poly (methyl acrylates) coated chitosan–diclofenac sodium nanoparticles for colonic drug delivery. *European Journal of Pharmaceutical Sciences*. 2013;50(3-4):303–11.
94. Naha PC, Hsu JC, Kim J, Shah S, Bouché M, Si-Mohamed S, et al. Dextran-coated cerium oxide nanoparticles: a computed tomography contrast agent for imaging the gastrointestinal tract and inflammatory bowel disease. *ACS Nano*. 2020;14(8):10187–97.
95. Zhang X, Gao X, Yi X, Yu H, Shao M, Li Y, et al. Multi-targeting inulin-based nanoparticles with cannabidiol for effective prevention of ulcerative colitis. *Materials Today Bio*. 2024;25:100965.
96. Xu H, Dai L, Nie W, Luo R, Pu X, Dong L, et al. Oral cyclodextrin supramolecular for ulcerative colitis treatment through macrophages targeting/ROS response-based accurate on-demand drug release strategy. *Materials Design*. 2023;226:111606.
97. Zhang S, Cho WJ, Jin AT, Kok LY, Shi Y, Heller DE, et al. Heparin-coated albumin nanoparticles for drug combination in targeting inflamed intestine. *Advanced Healthcare Materials*. 2020;9(16):2000536.
98. Salah N, Dubuquoy L, Carpentier R, Betbeder D. Starch nanoparticles improve curcumin-induced production of anti-inflammatory cytokines in intestinal epithelial cells. *International Journal of Pharmaceutics: X*. 2022;4:100114.
99. Yener S, Akbulut KG, Karakuş R, Erdoğan D, Acartürk F. In vitro and in vivo evaluation of pectin-based melatonin nanoparticles for the treatment of inflammatory bowel disease. Review. 2021;3. https://doi. org/10.21203/rs.
100. Ahmed O, Abdel-Halim M, Farid A, Elamir A. Taurine loaded chitosan-pectin nanoparticle shows curative effect against acetic acid-induced colitis in rats. *Chemico-Biological Interactions*. 2022;351:109715.
101. Laroui H, Dalmasso G, Nguyen HTT, Yan Y, Sitaraman SV, Merlin D. Drug-loaded nanoparticles targeted to the colon with polysaccharide hydrogel reduce colitis in a mouse model. *Gastroenterology*. 2010;138(3):843–53.e2.
102. Gamboa A, Araujo V, Caro N, Gotteland M, Abugoch L, Tapia C. Spray freeze-drying as an alternative to the ionic gelation method to produce chitosan and alginate nano-particles targeted to the colon. *Journal of Pharmaceutical Sciences*. 2015;104(12):4373–85.

17 Lipid Nanocarriers for the Treatment of Gastrointestinal Inflammations

Anas Islam[1,2]*, Ambareen Fatima Ahmed*[2]*, Asad Ahmad*[1]*,*
Usama Ahmad[1]**, and Yusuf Asad*[3]

[1]Department of Pharmacy, Integral University, Lucknow, Uttar Pradesh, India

[2]School of Pharmaceutical Sciences, Integral University, Lucknow, Uttar Pradesh, India

[3]Biomedical Engineering Program, Department of Electrical and Computer Engineering, North Dakota State University, Fargo, North Dakota, USA

17.1 INTRODUCTION

The definition of inflammation has evolved from its four cardinal signs of calor, rubor, tumor, and dolor as proposed by Celsus during the first century A.D. to the addition of the fifth sign of inflammation by Virchow in 1871 as the loss of function. With the advancement of cell-based techniques and molecular biology in the 19th century, the term inflammation has been defined more elaborately. The body's defense response to injury caused by any agent—be it microbial, immunological, physical, or chemical—is called inflammation. Clinically, this definition encompasses a multitude of biological processes (Scott et al., 2004).

Inflammation helps to alleviate harmful substances, flush them out, and promote tissue repair. Two fundamental processes—the early inflammatory response and the subsequent healing process—are involved in inflammation. While both of these processes typically serve as defenses against harmful agents, inflammation and healing can also have serious negative effects on the body (Carton et al., 2006).

The body's initial reaction is known as acute inflammation, which usually goes away quickly after the underlying cause is treated and is followed by healing. It lasts for two weeks or longer. It starts following a particular injury that triggers the migration of neutrophils and macrophages to the site of inflammation by the release of soluble mediators such as cytokines, acute phase proteins, and chemokines. Chronic inflammation develops either immediately after the acute inflammatory stimulation or after a prolonged period of time has elapsed. Chronic inflammation is characterized by its prolonged duration. In chronic inflammation, T lymphocytes and plasma cells go to the site of inflammation (Hannoodee & Nasuruddin, 2023).

Gastrointestinal inflammations are a type of disease that affects the gastrointestinal tract. This tract runs from the oral cavity to the anal cavity. Several conditions can cause inflammation of the gastrointestinal tract, including gastritis, periodontal disease, peptic ulcer disease, Crohn's disease, and ulcerative colitis. All of these conditions involve chronic inflammation in various parts of the gastrointestinal mucosa, which can result in epithelial compromises and tissue damage. There are also autoimmune components associated with the inflammatory bowel diseases along with systemic effects (Bamford, 1999).

17.1.1 Major Types of Gastrointestinal Inflammations

Of the gastrointestinal tract inflammatory conditions associated with inflammatory bowel disease (IBD), the most frequently occurring conditions are Crohn's disease and ulcerative colitis. The pathological mechanisms involve complex interaction of immunological, microbial, environmental, and genetic variables.

Genetic predisposition is just one of the numerous factors causing immune system dysregulation in inflammatory bowel disease (IBD) (Geremia et al., 2014) as well as dietary habits (Keewan et al., 2020) and the gut microbiome (Matsuoka & Kanai, 2015). Potentially, such mechanisms can take into account cytokine modulation, microbiome therapies, or dietary readjustments to manage and treat IBD.

Gastritis is the other condition of the gastrointestinal inflammation. Gastritis is the inflamed, irritated, or eroded stomach lining as a pathological condition that may manifest abruptly (acute) or develop gradually (chronic). Overindulgence in alcohol or long-term use of nonsteroidal anti-inflammatory medicines (NSAIDs), including ibuprofen or aspirin, are common causes of the illness. bacterium infection, especially Helicobacter pylori (HP), persistent bile reflux, stress, and some autoimmune diseases (Elseweidy, 2017).

Eosinophilic esophagitis (EoE) refers to a persistent inflammatory condition affecting the esophagus, primarily observed among the pediatric demographic. EoE is characterized by clinical symptoms of esophageal dysfunction and eosinophilic infiltration of the esophageal mucosa (Sutton et al., 2015).

In genetically susceptible individuals, duodenal inflammation is the hallmark of celiac disease (CD), a multiorgan autoimmune condition of the chronic intestinal disease group that is triggered by gluten consumption. Today, there is a great deal of research on the pathophysiology of celiac disease, which expands the concept beyond autoimmune alone and explains why the condition is hereditary. The

DOI: 10.1201/9781003493143-17

identification of multiple genes implicated in interleukin signaling and immune-related pathways has resulted from the genomic profiling of this illness (Ivanova et al., 2023).

Other types of gastrointestinal inflammations include inflammation of the parts of the small intestine such as duodenitis, jejunitis, and ileitis. These are all caused by infections, Crohn's disease, or autoimmune disorders.

17.1.2 Current Treatment, Challenges, and Limitations

Current drugs or therapies available for the management of IBD focus on reducing inflammation and improving patient outcomes. IBD is frequently treated with anti-inflammatory and immunosuppressive medications, although these treatments have unfavorable side effects and have limited success. In the early times, IBD was treated with, thiopurines, mesalamine, and methotrexate as well as corticosteroids. Newer therapies such as monoclonal antibody therapy were later introduced. Drugs of the category that are target-specific for gut inflammation such as anti-integrins, IL-23 inhibitors, JAK/STAT inhibitors, and S1P receptor modulators are undergoing clinical trials. Although extensive research has been carried out in this regard, there are limitations regarding the accessibility and affordability of these therapies. Moreover, such therapies may not show the same effect among different individuals (Nadpara et al., 2020). The effectiveness and safety of medications for inflammatory bowel disease (IBD) presently pose a substantial clinical obstacle, thereby underscoring the necessity for sophisticated pharmaceutical formulations that enable precise drug administration to the sites of inflammation (Luo et al., 2022).

Gastritis of the non-infectious type can be treated by histamine H2 receptor inhibitors, proton pump inhibitors (PPI), and antacids. However, their use is linked to various side effects. Though traditional medicines are available, with good safety profiles, the outcomes need to be extensively evaluated and clinically explored (Duque-Buitrago et al., 2023). Treatment of gastritis mediated through H. pylori follows a complex regime that includes PPI, antibiotics, bismuth, and probiotics as adjuvants (García García & Gómez Rodríguez, 2023).

Treatment of EoE includes corticosteroids, some biologic therapies, and PPI with dietary modifications. Currently, there exists a restricted number of authorized compositions for the management of eosinophilic esophagitis (EoE), including the orodispersible tablet of budesonide and the oral suspension of budesonide (Miehlke et al., 2020).

17.1.3 Promise of Nanotechnology in Gastrointestinal Drug Selivery

17.1.3.1 Advantages of Nanocarriers for Targeted Therapy

For a drug to elicit a pharmacological action, it must reach its target site in an appreciable amount. Of the many drug delivery systems, nanocarriers offer a promising tool for targeted therapy owing to their distinct form, size, content, and surface characteristics. A broad variety of nanocarriers, including liposomes, metallic nanoparticles, silica compounds, carbon nanotubes, nano-pattern surfaces, and nano-hydrogels, are being extensively studied (Heidari & Mehdi Ommati, 2023).

The oral route is, therefore, recommended for drug administration by the virtue of the convenience of use, its appropriateness, and flexibility it presents in designing a dosing regimen, and most importantly, patient compliance. That pathway offers a prospective platform for drug delivery, since the great surface area of the gut and the different cell types along the tract (Pridgen et al., 2015). Therefore, nanocarrier use would result in an increase in the efficacy and decrease in the systemic side effects that come with increasing the quantity of medication at the colonic location.

Moreover, more of the drug can be delivered to the colonic site, thus enhancing the efficacy of the therapy while reducing its systemic side effects. Treatment of inflammatory bowel disease using anti-inflammatory drug-loaded nanocarriers presents as a promising alternative due to the increased accumulation of medication within the inflammatory tissues and the reduced systemic adverse effects (Beloqui et al., 2016). Research has pointed out that the eminent performance of the microencapsulated nanogels for the delivery of the therapeutic TNF-siRNA across the pH of the stomach into the inflamed colon is a potential application in the oral management of the inflammatory bowel disease (IBD), a very fascinating approach for therapy (Knipe et al., 2016). Nanotechnology bears a great promise in enhancing the delivery of medications to Eosinophilic esophagitis (EoE). The study has optimized the manufacturing of isoliquiritigenin (ILQ) loaded into self-nano emulsion, hence increasing the bioavailability of the drug for treating endophthalmitis (EOE). Excellent biophysical properties were demonstrated by this formulation, which included a small droplet size, a suitable negative zeta potential, a significant drug content, and consistent stability (Cao et al., 2022).

17.1.3.2 Overcoming Traditional Drug Delivery Hurdles

Obstacles with the traditional drug delivery systems have been grouped into physicochemical, biopharmaceutical, physiological, and clinical (Date et al., 2016).

Physicochemical barriers include the poor water solubility, pH-dependent solubility, extensive ionization at gastric pH range, probability of the degradation. Biopharmaceutical parameters that need to be overcome via traditional methods are the poor drug permeability, pre-systemic metabolism, drug efflux transporters, variation in the pH across the GI range. Physiological barriers that pose a problem are the protective mucus barrier that hinders drug absorption, gastric- emptying, peristaltic movement of the stomach and intestines, presence of gastric and intestinal enzymes, gut microbes. Clinical hindrances include the patient-to-patient variation in disease state, fed and fasted state of individuals.

17.1.4 Lipid Nanocarriers as a Promising Platform

The bulk of lipid nanoparticles are usually composed of lipid components such as phospholipids, cholesterol, PEGylated lipids, triglycerides, and other lipids (Copland et al., 2005). The lipid structure and hydrophilic surfactant present in lipid nanocarriers make them attractive carriers for the encapsulation of medicinal compounds. In contrast to other current technologies, nanoencapsulation improves the contact surface, physical stability, and protection, which makes such chemicals easier to dissolve and absorb.

Transferrin is a protein that is overexpressed in inflammatory tissue, and solid lipid nanoparticles (SLNPs) with a negative surface charge have been designed to preferentially cling to it (Yang et al., 2022). Anti-inflammatory drugs can be encapsulated by these SLNPs and protected from being broken down by stomach and intestinal fluids. They have demonstrated cyto-compatibility and effective anti-inflammatory properties in vitro (Sunnap et al., 2022).

Gemini lipid nanoparticles (GLNPs), which are derived from gemini cationic lipids, have proven to be successful in delivering siRNA therapeutics to inflamed gastrointestinal tissues. This delivery method reduces gene expression and alleviates gut inflammation in a murine model (Verma et al., 2022).

Oral liposomes, such as budsomes, have been developed to encapsulate corticosteroid anti-inflammatory drugs and release them within inflamed intestinal tissues, resulting in favorable anti-colitis effects with minimal side effects (Xian et al., 2023).

These innovative lipid nanoparticle-based drug delivery systems offer the potential for targeted and efficacious treatment of gastrointestinal inflammations (Figure 17.1).

17.2 LIPID NANOCARRIERS FOR GASTROINTESTINAL DELIVERY

17.2.1 Classification of Lipid Nanocarriers

Lipid nanocarriers have been classified as per their composition and structure into following types (da Costa de Quadros et al., 2023).

Liposomes are spherical vesicles composed of lipid bilayers that can enclose substances that are hydrophilic or hydrophobic. Liposomes have been intensively studied as the possible drug carriers, because of their proven biocompatibility and efficient delivery and protection of bioactive substances.

Thus, liposomes may fall into the following classes according to size and the quantity of lipid bilayers: small unilamellar vesicles (SUVs), large unilamellar vesicles (LUVs), multilamellar vesicles (MLVs), and multivesicular vesicles (MVVs). Their features are given in Table 17.1 (Yingchoncharoen et al., 2016).

Solid lipid nanoparticles (SLNs) are those formed of a lipid-based nanoparticle that has a solid lipid core matrix. Benefits such as controlled release, enhanced stability, and higher bioavailability of the bioactive being encapsulated are just some of the myriad of benefits these nanoparticles offer. It is then no wonder that SLNs have generated this much interest towards them in the drug delivery field.

The other class of lipid-based nanoparticles, which is comprised of both solid and liquid lipids, is referred to as nanostructured lipid carriers or NLCs. NLCs have special composition and hence have better stability along with a

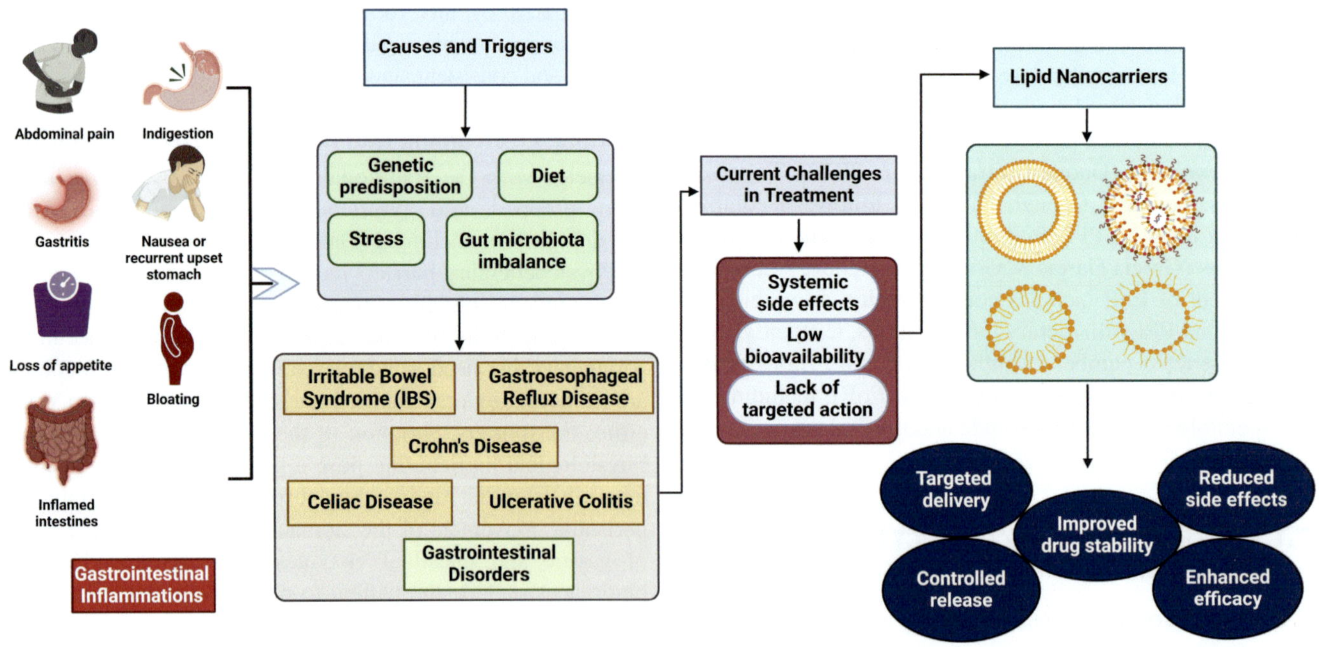

FIGURE 17.1 Overview of the causes, consequences, and current challenges in treating gastrointestinal inflammation, highlighting the innovative role of lipid nanocarrier technology.

TABLE 17.1
Type of Liposome with Characteristic Features

SN	Type	Characteristic Feature	Size
1	SUV	Single lipid bilayer	20–100 nm
2	LUV	Single lipid bilayer	> 100 nm
3	MLV	Multiple lipid bilayer	0.5–10 μm
4	MVV	One bigger liposome encapsulating several smaller liposomes	--

controlled release and higher drug loading capacity compared to SLNs. Consequently, NLCs have found application in medication delivery to be attractive.

Lipid-polymer hybrid nanoparticles are a kind of nanoparticle that combine the merits of both lipids and polymers. In the face of all these advantages, lipid-polymer hybrid nanoparticles have the potential to deliver higher encapsulation efficiency, controlled release, and improved stability of bioactive substances. Therefore, they represent a very promising option in medication delivery.

Ethosomes are lipid-based vesicular carriers with high contents of ethanol. Ethosomes are considered potential drug delivery systems for transdermal applications, as these carriers possess the ability to augment the transcutaneous permeation of actives. Thus, the interest towards etheromes was drawn because of its potential usage in drug delivery.

Amphiphilic lipids self-assemble into micellar structures in water. Micelles can increase the bioavailability of any hydrophobic bioactive compound by way of solubilization. In this way, lipid-based micelles have been explored for potential application in drug delivery.

Lipids, on a colloidal dispersing in water, form lipid emulsions. They can serve as a potential vehicle for bioactive chemicals and hence can be frequently incorporated into parenteral nutrition. The subsequent to it, lipid emulsions are considered a promising route of delivery for bioactive substances.

17.2.2 Fabrication and Characterization of Lipid Nanocarriers

17.2.2.1 Key Components and Formulation Strategies

Lipids: Typically, lipids have been employed in most of the lipid nanocarriers. They include the phospholipid bilayer within the liposomes and the core matrices present within the solid lipid nanoparticles (SLNs) and nanostructured lipid carriers (NLCs). Hence, their presence brings about stability and improved encapsulation capacity to the carriers and also is responsible for their controlled drug release characteristics.

Surfactants: Surfactants are commonly applied to lipid nanocarriers for stabilization of lipid bilayers and improvements of dispersibility. Hydrophilic surfactants in the lipid nanocarriers, for example, liposomes, will be able to increase the solubility and bioavailability of bioactive compounds.

Solid lipids: In the formulation of SLN and NLC, solid lipids are incorporated. This would give them structure and add to providing controlled release properties.

Liquid lipids: Liquid lipids in NLCs are generally formulated with solid lipids to provide a more flexible lipid matrix that enhances the drug loading capacity and stability of the carriers.

Polymers: Lipid-polymer hybrid nanoparticles add on the polymers to lipids, and its major goal is to enhance stability, raise encapsulation efficiency, and permit controlled release of bioactive compounds.

Ethanol: Ethosomes belong to high concentrations of ethanol of a lipid nanocarrier. An example is ethanol that enhances the permeation of bioactive compounds through the skin for transdermal drug delivery.

Water: Lipid emulsions are known as colloidal dispersions of lipids in water. Water in this form of emulsion represents the continuous phase that provides opportunity for the dispersion and delivery of bioactive compounds.

17.2.2.2 Techniques for Preparation and Characterization

Film hydration and then extrusion is one of the most favorite methods in which liposomes are made. In this method, the phospholipids are treated with an organic solvent. With a rotary evaporator, the solution is evaporated in order to form a thin lipid layer. These thin films are then hydrated with water media to form liposomes. Most of the time, those liposomes prepared using this method are MLVs. One can still reduce the particle size of these vesicles using extrusion, ultrasonication, or high-pressure homogenization (Zhang, 2023). Reverse-phase evaporation is another method for the production of liposomes.

This involves the dissolution of a combination of lipids in solvents, after which the lipid solvents are given a chance to evaporate under low pressure. This leads to the formation of a lipid film with secondary reconstitution in the organic phase. As the organic solvent further evaporates, the resulting dispersion gets thicker, and one is finally left with an aqueous suspension containing liposomes (Shi & Qi, 2021).

The liposome preparation process can be accurately and consistently carried out using the microfluidic technology. Liquid streams are divided through grooves in the channels to agitate the aggressive blending of an aqueous phase and an organic phase, usually composed of lipids, on a microchip. Lipids unite to produce liposomes when the organic phase in the lipid stream diffuses and dilutes into the water stream. Eventually, when the mixture achieves an equilibrium condition, stable liposomes are produced. With the use of a mixing cartridge that enables quick mixing of aqueous and organic solvents, this system permits regulated, bottom-up, molecular self-assembly of liposomes in the context of drug delivery applications (Shah et al., 2019). Nanocarriers can be characterized by morphological evaluations by microscopic techniques, nanocarrier lamellarity, size determination by light scattering techniques, and size exclusion chromatography (SEC), determining nanocarrier surface charge (zeta potential) (Fan et al., 2021).

17.2.2.3 Factors Influencing Stability and Release Profile

Many parameters, including the composition, size, surface charge, encapsulation effectiveness, and lipophilicity of the chemical being integrated, affect the stability and release profile of lipid nanocarriers. These parameters are of vital importance to determine how good the lipid nanocarriers work in delivering drugs, especially when it comes to their ability to increase bioavailability, permit controlled release, and ensure stability of the incorporated medication. Besides, the composition of the matrix of the nanocarrier comprises the selection of lipids and surfactants, which have a big effect on both stability and release characteristics. For instance, lipid nanocarriers loaded with resveratrol were made stable through the DODAB:MO liposomal system, which promoted drug release and cellular uptake through endocytosis in yeast (Célia Barbosa et al., 2019). Furthermore, the choice of solid lipids, for example, the combination of beeswax and carnauba wax, is a factor that has some impact on the release rate of the drug. With changing beeswax content in carriers, drug release rates changed and influenced the stability and release dynamics of flurbiprofen-loaded carriers (D. Baviskar et al., 2012). Also, the size and surface characteristics of the particles are of great importance. Less-sized particles promote an enhanced uptake of drugs by the cells, and the increase in positive charge facilitates interaction with negatively charged cell membranes. The surface modification of lipid-based nanocarriers has been explored as a means for targeted drug delivery, and it was observed that it improved penetration efficiency and controlled drug release (S. Priya et al., 2022).

Manufacturing processes also influence lipid nanocarrier stability and release profile. Preparation methodologies for the production of solid lipid nanoparticles (SLNs) and nanostructured lipid carriers (NLCs) influence the physical stability and the release behavior of the resulting materials. For example, using high-pressure homogenization, omega-3 fatty acid nanoemulsions containing curcumin and quercetin were successfully produced (G. Vaz et al., 2022). Such biocompatible and biodegradable nanocarriers, along with extended-release and targeting ability, are of great importance for cancer therapies. The excellent temporal and thermal stability, in addition to the ability to release the drug slowly over time and in a controlled manner, has enabled the wide adoption of lipid-based nanocarriers in oncology, such as liposomes, solid lipid nanoparticles, and NLCs (Buddhadev Layek et al., 2020).

Some factors affecting the performance of lipid nanocarriers have been discussed below:

a. **Composition of Lipid Nanocarriers:** A large proportion of the lipids used for preparation, their phase behavior, and the presence of additional molecules (like drugs or targeting ligands) can significantly affect the nanocarrier's stability and drug release profile. For instance, inclusion of cholesterol enhances stability, while choice between solid lipids and liquid lipids influences the release kinetics of encapsulated drugs (Heurtault et al., 2003).
b. **Size and Polydispersity Index (PDI):** The particle size and uniformity distribution, which could be from ~30 nm to 3–50 μm for most nanocarriers, are determinants of nanoparticles' behavior. Particularly, nanoparticles of a small size can interact with biological membranes and influence drug release rates, while, conversely, higher PDI indicates variable particle size distribution, which should be good for predictable behavior in biological systems (Danaei et al., 2018).
c. **Surface Modification and Functionalization:** Changes in the surface properties of lipid nanocarriers can be made, such that they could be used for specific interactions with biological systems, affecting their stability and release profile. For example, PEGylation can enhance the circulation time by reducing opsonization (Ishida & Kiwada, 2008).
d. **Environmental Conditions:** pH, ionic strength, and the presence of enzymes or other biological molecules can affect the stability and drug-release behavior of lipid nanocarriers. For instance, enzyme-responsive liposomes are formulated to release their cargo in the presence of specific enzymes, which can be elevated in certain diseases or physiological conditions (Fouladi et al., 2017). Storage conditions, such as temperature, humidity, and light exposure, can affect the physical stability of lipid nanocarriers, potentially leading to changes in particle size, zeta potential, and even drug leakage. For example, G. Ioele et al., in the composition of solid lipids and liposomes, found that their formulations could significantly reduce photodegradation of encapsulated drugs, leading to effective retention and prolonged efficacy (G. Ioele et al., 2021).
e. **Encapsulation Efficiency and Drug Loading:** The rate at which drugs are encapsulated and their loading within the nanocarrier affects both stability and release. High drug loading may compromise the structural integrity of nanocarriers, while low encapsulation efficiency can promote premature drug release (Fathi et al., 2013).
f. **Release Mechanisms:** Various lipid nanocarriers use mechanisms of release, from diffusion to erosion and triggered release (e.g., pH-sensitive or thermosensitive release), which can be tailored according to the application and the target site (Son et al., 2017).
g. **Lipid Composition and Ratio:** The selection of particular types of lipids and their proportion in the formulation of nanocarriers can have a great effect on their stability and release properties. This is because, for instance, some solid lipids, compared to liquid lipids or oil (if applicable), can alter the crystallinity and rigidity of the nanocarriers, influence drug encapsulation efficiency, and release rates.
h. **Manufacturing Process:** The method used in producing lipid nanocarriers influences size, morphology, and

stability. Different techniques, such as high-pressure homogenization, ultrasonication, and microemulsion, can produce nanocarriers with varying characteristics. Makoni et al. (2019) pointed out that this was successful in manufacturing efavirenz-loaded solid lipid nanoparticles and nanostructured lipid carriers without the use of toxic organic solvents, indicating that the manufacturing process impacts the stability and release profile of nanocarriers (Makoni et al., 2019). However, for example, the phase-inversion temperature method used in preparing solid lipid nanoparticles with volatile oil-loaded quercetin showed that the method could produce carriers exhibiting good stability and small particle sizes important for effective drug delivery (Yotsanan Weerapol et al., 2022).

i. **Temperature Sensitivity:** The stability and release profile of lipid nanocarriers can be sensitive to temperature changes, which can induce phase transitions in the lipid matrix affecting drug encapsulation and release. This characteristic can be used to create temperature-responsive lipid nanocarriers that will release a controlled amount of material in response to the temperature at the target site. Cardoso et al. (2022) showed that temperature sensitivity is another factor in the generation of multi-stimuli-responsive vesicles for targeted drug administration. Their work, therefore, involves the building of solid magnetoliposomes that regulate the release of doxorubicin under thermal, magnetic, and pH stimuli (Cardoso et al., 2022). These factors collectively determine the effectiveness of lipid nanocarriers in drug delivery applications. The design and optimization of lipid nanocarriers thus require careful consideration of these parameters to achieve desired therapeutic outcomes.

17.2.3 Advantages and Challenges of Lipid Nanocarriers for Gastrointestinal Applications

17.2.3.1 Enhanced Mucosal Adhesion and Permeability

Lipid nanocarriers can enhance the mucosal adhesion of drugs and thus the absorption through the intestinal epithelial lining. Nanostructured lipid carriers (NLC) and solid lipid nanoparticles (SLN) have exhibited improved mucosal adhesion, which can significantly increase the bioavailability of poorly soluble drugs (Danaei et al., 2018). This is especially advantageous for drugs that need to be present at the site of absorption for a long time to exert their therapeutic effect.

17.2.3.2 Protection from Enzymatic Degradation

Lipid-based nanocarriers provide a protective environment for the encapsulated drugs, protecting the drugs from the enzymatic environment of the GI tract. This is very crucial for the delivery of biologically active molecules, such as peptides and proteins, which are prone to enzymatic degradation (Plaza-Oliver et al., 2021).

17.2.3.3 Targeted Delivery to Inflamed Tissues

Lipid nanocarriers can be modified with specific ligands or antibodies to make them targeted to inflamed GI tissues. The use of targeted therapy can, therefore, be considered in the treatment of inflammatory bowel diseases by reducing systemic side effects and improving therapeutic efficacy (Salvi & Pawar, 2019).

17.2.3.4 Controlled Drug Release and Improved Bioavailability

The various lipid nanocarriers are offering controlled drug release, which can be designed for desired release profiles (sustained or targeted) from the preparation. This ultimately results in improved bioavailability and improved outcomes for therapeutics. For example, nanoemulsions and SLNs have been developed for the oral delivery of vitamins due to significant enhancement in their solubility and bioavailability (Hsu et al., 2019). The controlled pace at which the encapsulated medication releases helps in achieving improved bioavailability and lowering the frequency of dosages through maintaining drug levels in the systemic circulation. This is especially beneficial for those medications having poor therapeutic windows.

17.2.3.5 Challenges for Scaling up Production and Potential Safety Concerns

The most common challenges of scaling up the production of lipid nanocarriers to industrial levels include formulation complexity, the need for precise control of particle size and distribution, and scalable production methods. Tackling these challenges requires innovations in manufacturing technologies and quality control for reproducibility and affordability. These barrier entry issues in expanding manufacturing may translate into a change in the stability and biological activity of nanocarriers at an industrial scale, despite the systems featuring such intricate interactions between their constituents.

Although lipid nanocarriers are generally viewed as safe and biocompatible, potential safety concerns remain, including long-term use, toxicity, and immunogenicity. Given their use in clinical applications, careful safety profiling must include the constituents of the formulated nanocarriers, associated degradation products, and chronic exposure effects. Long-term safety evaluations should also be carried out to ensure that immunogenicity, carrier material toxicity, and bioaccumulation are all part of the requisites for safety assessment. It is thus important that safety be assessed for each individual nanosystem in operation

17.3 MECHANISM OF ACTION OF LIPID NANOCARRIERS IN GASTROINTESTINAL INFLAMMATIONS

Lipid nanocarriers have emerged as a promising drug delivery system for the treatment of gastrointestinal inflammations. Their unique properties, such as high stability, biocompatibility, and ability to encapsulate a wide range of drugs, make them an attractive option for targeting inflamed gastrointestinal tissues. In this section, we will explore the mechanism of action of lipid nanocarriers in the context of gastrointestinal inflammations, shedding light on their potential applications and the latest advancements in this field (Montoto et al., 2020).

17.3.1 Inflammation Process

17.3.1.1 Overview of Gastrointestinal Tract Functionality

The gastrointestinal (GI) tract plays a crucial role in waste removal, nutrient absorption, and digestion. It features a physical barrier composed of villi, finger-like projections that assist in nutrient absorption from food. Despite being home to trillions of microorganisms, this epithelial barrier prevents their entry into the body. The gut's natural microbiota, unique to each individual, significantly influences the GI tract's mucosal immune responses, aimed at thwarting pathogen colonization and limiting opportunistic infections from commensal bacteria. Chronic inflammatory conditions in the GI tract can arise from various immune system disruptions, including infections, damage, barrier function abnormalities, autoimmune mechanisms, and shifts in commensal microorganism composition. These persistent inflammatory processes can lead to diseases such as food allergies, celiac disease, and inflammatory bowel disease (Mathias et al., 2020).

The inflammation involves a complex interplay of immune cells, cytokines, and other inflammatory mediators. Initial triggers can be toxins, tissue injury, or microbial infections, leading to the activation of T-cells, dendritic cells, macrophages, and others. These cells release pro-inflammatory cytokines like tumor necrosis factor-alpha and interleukin-6, amplifying the inflammatory response. Increased vascular permeability facilitates further immune cell activation and infiltration into the GI system, causing local tissue injury and dysfunction. Over time, chronic inflammation can induce anatomical changes in the GI tract's mucosa, potentially resulting in conditions like inflammatory bowel disease (Frosali et al., 2015).

Various factors, including immunological, endocrine, neurological conditions, or mechanical, chemical, and microbial damage, can trigger intestinal inflammation. Persistent gut inflammation can lead to dysbiosis or an imbalance in the gut microbiota, impacting overall health and disease states. The gut microbiota is shaped by chronic inflammation, which is in turn driven by gut bacteria to induce immunological activation. The immune system, gut bacteria, smooth muscle cells, and epithelium interact with the enteric nervous system, which is housed within the gastrointestinal wall, to preserve mucosal barrier function and control motor and secretory processes. Inflammatory bowel disease (IBD) is a long-term immune-mediated illness that causes inflammation and digestive problems. Transmembrane glycoprotein CD34 is aberrantly elevated in the intestinal tracts of individuals with inflammatory bowel disease (IBD) and has a role in facilitating immune cell migration to the site of inflammation (Zheng et al., 2023).

17.3.1.2 Initiation of the Inflammatory Response

5-hydroxytryptamine (5-HT) and substance P are critical mediators in gastrointestinal (GI) inflammation. 5-HT, also known as serotonin, is involved in various physiological processes, including inflammation in the GI tract. Substance P is a neuropeptide that plays a significant role in neurogenic inflammation and contributes to the development of gastrointestinal (GI) inflammation. It triggers the release of pro-inflammatory cytokines and chemokines, thereby promoting inflammation within the GI tract through the induction of histamine, prostaglandins, and other inflammatory mediators. During GI inflammation, cytokines such as tumor necrosis factor-alpha (TNF-α), interleukin-6 (IL-6), and interleukin-1 (IL-1) are produced. Additionally, chemokines like CCL2 (also known as monocyte chemoattractant protein-1, MCP-1) play a crucial role in attracting immune cells to the inflamed GI areas (Chen et al., 2022).

Histamine, a powerful inflammatory mediator, is also produced during GI inflammation, leading to further inflammation. The GI tract inflammation is exacerbated by the production and release of prostaglandins, such as prostaglandin E2 (PGE2). Moreover, other inflammatory mediators like leukotrienes and reactive oxygen species (ROS) are produced during GI inflammation. These pro-inflammatory molecules serve as effector molecules in GI inflammation, influencing the inflammatory response within the gastrointestinal tract (Chen et al., 2022).

17.3.1.3 Chronic Inflammation and GI Disorders

Chronic low-grade inflammation is a key factor predisposing individuals to a range of gastrointestinal disorders associated with obesity. Directly linked to this inflammatory state are conditions such as gallbladder disease, gastroesophageal reflux disease (GERD), and non-alcoholic fatty liver disease (NAFLD). Obesity can alter immunological regulatory systems, thereby predisposing individuals to additional gastrointestinal disorders. This includes inflammatory bowel diseases (IBD), such as Crohn's disease and ulcerative colitis. The inflammatory state associated with obesity is characterized by increased levels of pro-inflammatory cytokines like interleukin-6 (IL-6) and tumor necrosis factor-alpha (TNF-α), along with the activation of immune cells such as macrophages. These chronic inflammatory conditions can lead to symptoms like abdominal pain, diarrhea, and malabsorption, affecting the normal operation of the digestive tract (Tillman et al., 2023).

Moreover, gut inflammation is associated with an increased risk of developing gastrointestinal cancers in various chronic inflammatory gastrointestinal disorders. Conditions such as autoimmune chronic gastritis caused by Helicobacter pylori, celiac disease, GERD, and IBDs have been linked to a heightened risk of neoplasms. Inflammation in the gastrointestinal tract can lead to the development of neoplasms, including gastric marginal zone lymphomas and enteropathy-associated T cell lymphoma, as well as epithelial tumors like esophageal squamous cell carcinoma and adenocarcinoma, gastric adenocarcinoma and neuroendocrine tumors, small bowel adenocarcinoma and neuroendocrine tumors, and colorectal cancer. A histologically identifiable inflammation-dysplasia-carcinoma cancerogenic sequence can eventually lead to carcinogenesis through pathogenetic processes and microenvironmental/microbiome changes brought on by inflammation. Long-term ileostomy, celiac disease, and inflammatory bowel diseases (IBDs) have all been repeatedly linked to an increased risk of small intestine cancer. Intestinal T-cell lymphomas can occur in patients with non-celiac enteropathies, such as autoimmune enteropathy. Individuals who have common variable immunodeficiency are more susceptible to intestine lymphomas and stomach adenocarcinomas, which typically develop in conjunction with nodular lymphoid hyperplasia (Vanoli et al., 2023).

17.3.1.4 Molecular Mediators of Inflammation

Chronic inflammation in the gastrointestinal tract is a hallmark of inflammatory bowel disease (IBD), which is a major risk factor for colorectal cancer. Connexin 43 (Cx43) proteins generate gap junctions (GJs), which are intercellular channels that are involved in intercellular communication. The etiology of inflammatory and malignant disorders has been linked to epigenetic mechanisms including DNA methylation. The TET-2 enzyme, also known as the ten-eleven translocation-2 enzyme, plays a role in controlling the activity of certain genes that can either suppress tumors or promote cancer by demethylating DNA. In inflammatory conditions and colorectal cancer, TET-2 levels decrease while Cx43 levels increase. The presence of DSS-induced inflammation or an inflammatory environment in a laboratory model of colon epithelium was found to regulate the expression of Cx43 and TET-2. Furthermore, reduced levels of both TET-2 and Cx43 were observed in tissue samples from patients with ulcerative colitis or sporadic colorectal cancer. This may have contributed to healthy intestinal tissues becoming malignant growths due to inflammation (Khan et al., 2022).

One of the proteins that helps cells communicate with each other through gap junctions is Cx43. These gap junctions, which contain Cx43 proteins, facilitate the passage of chemicals and signals between cells, enabling coordinated cellular responses and intercellular communication. Changes in TET-2 expression have been associated with colorectal cancer and inflammatory bowel disease. Studies on intestinal inflammation have shown that TET-2 expression levels are affected, suggesting a role for TET-2 in both inflammation and the development of cancer (Khan et al., 2022).

Immunological tolerance to benign microorganisms can be weakened when the GI tract barrier fails, leading to chronic inflammation. This inflammation further weakens the barrier and allows germs to move from the intestines to other tissues, causing systemic inflammation that may affect organ function. Environmental stressors, such as certain food components, play a significant role in initiating gastrointestinal inflammation and breaking down immunological tolerance. When the barrier malfunctions, it compromises the mucosal immune system's ability to defend against gut bacteria properly, resulting in pro-inflammatory cytokine release and leukocyte migration. In inflammatory conditions, dendritic cells express co-stimulatory molecules more frequently, leading to non-tolerogenic adaptive immune responses. To restore tolerance in chronic inflammatory disorders, strategies aiming at influencing DC maturation status and producing regulatory T-cells may be used (Jensen et al., 2022).

17.3.1.5 Consequences of Unresolved Inflammation

Western individuals often experience a chronic inflammatory pattern in their gastrointestinal tract, which is characterized by high levels of C-reactive protein and a lack of typical organ-related symptoms. This inflammation can gradually and severely affect various bodily systems such as the skin, joints, and neurological functions. The gut serves as an important barrier to protect overall bodily health, with ongoing submucosal inflammation acting as surveillance against external threats. The gut microbiome—an extensive community of cells within the gut—is being studied for its role in influencing local and distant signs of inflammation. Rapid changes in diet can prompt the release of pro-inflammatory signals by the microbiome, leading to further gut inflammation. Addressing this link between the microbiome and GI tract inflammation requires collaborative efforts across multiple disciplines to effectively manage its impact on human health (Ribaldone et al., 2019).

How the microbiome interacts with the host, as well as what is happening in this relationship, is still not clear. However, it has come to light that inflammation is one of the most contributing pathologies in various conditions, including metabolic disorders, obesity, irritable bowel syndrome (IBS), inflammatory bowel disease (IBD), allergy, asthma, cardiovascular diseases, and infections. The inflammatory process in the GI tract lacks the acute inflammatory signs of the past century and is now considered a chronic, creeping phenomenon. It is also a prominent feature in hyperimmune diseases, such as rheumatoid arthritis and degenerative neuropathies. Furthermore, understanding the relationship of inflammation in gastrointestinal disorders has implications for disease classification and therapeutic options. Therefore, further research is needed to explore the complex relationship between the gut microbiome, inflammation, and disease progression (Ribaldone et al., 2019).

In the digestive system, the symbiotic microbiome has a protective role in eliminating pathogens and enhancing the healing response. Type 2 immune responses are regulated by the microbiota and are crucial for maintaining balance between immune responses at mucosal surfaces. Rapid cell recruitment and differentiation to eliminate inflammatory inducers and support the restoration of tissue homeostasis is mediated by signals at the sites of inflammation. Chronic inflammation may result from deregulation of the repair mechanisms or persistent inflammatory stressors. Numerous stressors have the potential to compromise immune function and raise the risk of infectious disorders, including postinfectious irritable bowel syndrome and gastritis. Patients with IBS are also more likely to get infectious gastroenteritis. Both directly and indirectly, nutritional components such as probiotics, prebiotics, antioxidants, and n-3 polyunsaturated fatty acids affect gastrointestinal immunity (Oz et al., 2016).

17.3.2 Role of Lipid Nanocarriers in Modulating Inflammation

Lipid nanocarriers have shown promising roles in modulating gastrointestinal inflammation through various mechanisms, including modulation of pro-inflammatory gene expression, improvement of intestinal barrier function, and targeted delivery of therapeutic agents.

17.3.2.1 Modulation of pro-inflammatory gene expression

Lipoxin A4 analogs, acting through lipid nanocarriers, can attenuate the induction of intestinal epithelial pro-inflammatory gene expression, suggesting a therapeutic potential for diseases characterized by mucosal inflammation by down-regulating IL-8 secretion and neutrophil recruitment, influenced through the NF-κB pathway (Gewirtz et al., 2002).

17.3.2.2 Improvement of intestinal barrier function

Chitosan and fucoidan-based nanocarriers for oral delivery of berberine have shown protective effects against intestinal epithelia tight junction disruption induced by lipopolysaccharides, suggesting their usefulness for ameliorating LPS-induced intestinal injuries (Wu et al., 2014).

17.3.2.3 Targeted delivery of therapeutics

Lipid nanoparticles have the ability to deliver payloads to inflammatory gut-homing leukocytes selectively, which can silence interferon-γ in the gut and improve therapeutic outcomes. This was illustrated in a study that uses targeted lipid nanocarriers, demonstrating that conformation-sensitive targeting of lipid nanocarriers for RNA therapeutics in a murine model of colitis is possible (Dammes et al., 2021). For instance, by changing the surface of nanocarriers using ligands or antibodies that bind specifically to indicators of inflammatory tissue or cells (Yang et al., 2023), targeting options will be given, such as changing the surface of nanocarriers. Cationic lipid nanocarriers have been shown to elicit inflammatory response-related cellular signaling pathways, such as the NLRP3 inflammasome and Toll-like receptor 2 (TLR2) pathways. This suggests that they could be used as a novel class of safe and synthetic adjuvant for vaccines (Lonez et al., 2014).

17.3.2.4 Activation of inflammatory cascades

Cationic lipid nanocarriers may trigger a pro-inflammatory response that involves the activation of inflammatory response-related cellular signaling pathways, including the NLRP3 inflammasome and Toll-like receptor 2 (TLR2) pathways (Yang et al., 2023).

17.3.2.5 Comparison of nanocarriers in treatment efficacy

Variable in vitro and in vivo efficacy was shown in treating inflammatory bowel disease (IBD) among a variety of lipid-based nanocarriers containing curcumin, with the solubility and stability of hydrophobic medicinal drugs being increased through the lipid nanocarriers and thus improved its bioavailability, particularly beneficial to treat conditions such as gastrointestinal inflammation where the efficacy of treatment relies on effective transport and absorption to achieve a good therapeutic response. These lipid nanocarriers have been shown to enhance vitamin bioavailability and oral delivery, which are important aspects of reducing inflammation (Hsu et al., 2019).

17.3.2.6 Enhanced bioavailability of therapeutic agents

The solubility and stability of hydrophobic medicinal drugs are greatly improved by lipid nanocarriers, which leads to higher bioavailability. This is especially advantageous in medicines for the treatment of gastrointestinal inflammation since it determines the success of the treatment through good transport and uptake, improving the adherence of patients (Niu et al., 2016).

17.3.2.7 Modulation of immune response

Lipid nanocarriers can either directly interact with immune cells or deliver immunomodulatory agents. For instance, cationic lipid nanocarriers have been shown to activate the Toll-like receptor 2 (TLR2) and NLRP3 inflammasome pathways, depending on the context, which can have both pro- and anti-inflammatory effects (Lonez et al., 2014).

17.3.2.8 Protection against gastrointestinal enzymes

The therapeutic agent with lipid nanocarriers protects them from degradation by gastrointestinal enzymes, ensuring that a greater fraction of the drug reaches its target site in an active form. This enhances the lifespan of the drug and cuts the dosage requirement. This also increases compliance of patients (Niu et al., 2016).

17.3.2.9 Decrease in inflammatory mediators

Lipid nanocarriers also deliver anti-inflammatory agents that specifically target decreasing or altering the levels of mediators of inflammation, such as cytokines, chemokines, and eicosanoids within the gastrointestinal tract. Specific delivery in this way reduces the inflammation response, enhances healing of the intestinal mucosa, and provides relief from symptoms associated with conditions such as inflammatory bowel disease (IBD) (Fumk, 2001).

The above findings suggest that lipid nanocarriers would be a potential tool in the treatment of gastrointestinal inflammation by modulation of pro-inflammatory responses, strengthening of intestinal barrier functions, and delivery of the therapeutic agent in a targeted manner. Further development and research in this area may result in innovative treatment approaches for inflammatory bowel disease and related conditions.

17.3.3 Evidence Supporting the Efficacy of Lipid Nanocarriers in Gastrointestinal Inflammations

Beloqui et al. (2016) assessed the efficacy of the intestinal delivery of three distinct lipid-based nanocarriers containing curcumin. Results of the study indicate that the nanostructured lipid carriers (NLCs) may ameliorate inflammation and are useful in the current use for lowering such inflammation (Beloqui et al., 2016). In a separate study, the impact of oleuropein, derived from olive oil, and high siRNA nanolipid carriers (NLCs) on intestinal inflammation were evaluated by Huguet-Casquero et al. (2020). This showed the promising role played by these nanocarriers in increasing the efficacy of antioxidant and anti-inflammatory treatment of oxidative stress in acute colitis patients (Huguet-Casquero et al., 2020).

Mishra et al. (2020) discovered that a novel formulation for the treatment of ulcerative colitis (UC) using lipid-based nanocarriers that contained celecoxib, a nonsteroidal anti-inflammatory drug, was able to moderate the severity of the disease. The approach demonstrated effective control of inflammation, with implications for improved therapeutic outcomes (Mishra et al., 2020). Gastrointestinal and efficacy analysis of treatment by siRNA lipid nanoparticles was shown to have been studied by these researchers, with an emphasis on their fate at different stages along the gastrointestinal (GI) tract and the effectiveness of the treatment for such diseases. It is established that the efficacy of LNP siRNA entrapment is not solely dependent upon the pH levels, suggesting its versatile potential in the treatment of persistent inflammation (Ball et al., 2018).

In developing a novel 1:1 binary lipid nanoparticle formulation, Sharma et al. (2019) aimed to treat an inflammatory bowel disorder by oral delivery of curcumin. It reported the results of this study indicate that the nanocarriers increased the uptake and therapeutic intervention provided by these nanocarriers, hence providing possible betterment for the treatment of intestinal inflammation (Sharma et al., 2019). Magnolol (Mag) is a natural anti-inflammatory compound derived from traditional Chinese medicine, having potential in the treatment of ulcerative colitis (UC). Its oral delivery, however, is a challenge due to poor solubility, stability, and bioavailability. This study proposed preparing hyaluronic acid (HA)-functionalized lipid-polymer hybrid nanoparticles (LPNs) that encapsulate Mag and target inflamed colonic lesions. The HA-LPNs showed similar physicochemical properties and were more in adequate cellular uptake and therapeutic intervention than the Gly-LPNs in vitro and in vivo experiments. The results indicate that HA-LPNs can be a potential oral delivery system for Mag in the treatment of UC (Li et al., 2024).

The CRISPR/Cas9 system is a promising gene editing tool for the treatment of colon diseases, such as ulcerative colitis and colitis-associated colorectal cancer. However, the oral delivery faces challenges such as instability, low transfection efficiency, and off-target effects. A new oral delivery system targeting CD98, a major player in colon inflammation and carcinogenesis, with lipid nanoparticles (LNPs) derived from mulberry leaves was described in the work. LNPs showed enhanced stability and penetration toward the colon, and effectively alleviated UC and CAC animal models (Ma et al., 2024).

6-Shogaol, a compound derived from ginger known for its anti-inflammatory properties, faces challenges with oral absorption and stability. Researchers utilized a natural-lipid nanoparticle drug delivery system (NP-DDS) to encapsulate and release 6-shogaol into the colon, where it transformed into more potent molecules, M2 and M13. This approach significantly improved the biodistribution, anti-inflammatory efficacy, and wound-healing activity of 6-shogaol and its metabolites in a mouse model of ulcerative colitis. The study highlights the potential of natural lipid nanoparticles as a novel treatment method for inflammatory bowel disease (Yang et al., 2020). In a separate study, extremely small solid archaeolipid nanoparticles (SAN) were developed, loaded with the potent anti-inflammatory drug dexamethasone (Dex). SAN-Dex demonstrated superior anti-inflammatory efficiency and stability compared to traditional solid lipid nanoparticles (SLN), suggesting its potential as an oral treatment option for inflammatory bowel disease (Higa et al., 2017).

A novel oral delivery strategy involving IL-22 mRNA encapsulated in lipid nanoparticles (nLNPs), which specifically target the injured colonic mucosa, was introduced. These nLNPs comprise three primary lipids found in ginger, a natural colon-targeting agent. The study showed that oral administration of IL-22/nLNPs increased IL-22 mRNA expression in the colon and alleviated symptoms of ulcerative colitis (UC) in a mouse model, offering a new therapeutic avenue for UC patients (Sung et al., 2022). Furthermore, researchers developed lipid-polymer hybrid nanoparticles to encapsulate the antioxidant enzyme superoxide dismutase (SOD), creating a novel local drug delivery method for inflammatory bowel disease (IBD). This technique protected SOD from degradation, enhanced its penetration and targeting of inflamed tissues, and reduced inflammatory cytokines in a colitis mouse model, suggesting a promising enzyme therapy approach for IBD (Le et al., 2021).

17.4 CURRENT RESEARCH AND APPLICATIONS OF LIPID NANOCARRIERS IN GASTROINTESTINAL INFLAMMATIONS

Recent advancements in the understanding and application of lipid nanocarrier systems for therapeutic purposes have significantly enriched the field of gastrointestinal (GI) inflammation treatment. The exploration of lipid nanocarriers, including liposomes, solid lipid nanoparticles (SLNs), and nanostructured lipid carriers (NLCs), has revealed their potential to enhance therapeutic delivery to areas of GI tract inflammation. Their biocompatibility, ability to target specific inflammation regions, and capacity to encapsulate a diverse range of therapeutic molecules position them as promising candidates for treating inflammatory bowel diseases (IBD) such as Crohn's disease and ulcerative colitis.

17.4.1 Preclinical Models and Proof-of-Concept Studies

Preclinical models, notably murine models of colitis, have played a crucial role in assessing the efficacy and safety of therapies based on lipid nanocarriers for managing GI inflammations. These models closely mimic human IBD pathology, allowing for an accurate evaluation of therapeutic outcomes, drug targeting efficiency, and potential side effects. The use of dextran sulfate sodium (DSS)-induced colitis models has provided valuable insights into the mechanisms of action of lipid nanocarriers, such as their role in modulating immune responses, enhancing mucosal healing, and decreasing levels of pro-inflammatory cytokines.

Proof-of-concept studies utilizing these models have effectively demonstrated the potential of lipid nanocarriers for targeted drug delivery in the treatment of IBD. Research has shown that lipid nanocarriers can efficiently transport anti-inflammatory medications, such as budesonide, directly to inflamed tissues. This targeted delivery minimizes the drug's systemic exposure and associated adverse effects. Moreover, the development of theranostic lipid nanocarriers, which combine therapeutic and diagnostic capabilities, represents a significant advancement in personalized medicine for IBD treatment.

17.4.1.1 Evaluation of Drug Loading and Release In Vitro

Investigations into lipid nanoparticle (LNP) drug loading and release, especially concerning gastrointestinal inflammation, have shown promising results for enhancing the administration and efficacy of anti-inflammatory drugs. Drug loading is defined as the amount of drug encapsulated relative to the total mass of the nanoparticle carrier, a key factor affecting the dosage and therapeutic effectiveness of the delivery system. High drug loading efficiency is desired to reduce the quantity of carrier material and minimize potential side effects. Techniques such as UV-Vis spectroscopy, high-performance liquid chromatography (HPLC), and liquid chromatography-mass spectrometry (LC-MS) are commonly utilized to determine the drug content within the carriers after separation of the free drug from the loaded nanoparticles, typically through centrifugation or filtration methods. In vitro release studies are crucial for understanding how the drug is released from LNPs under simulated physiological conditions, offering insights into the release kinetics and mechanisms that may predict the nanocarrier system's in vivo behavior. Techniques like dialysis bag diffusion, reverse dialysis, and sample-and-separate methods are employed, with the choice depending on the physicochemical properties of the drug and nanocarrier, as well as the intended action site. The release medium often simulates physiological fluids, including phosphate-buffered saline (PBS), gastric, or intestinal fluid, with enzymes added to mimic metabolic conditions (Müller et al., 2000; Tiwari et al., 2012; Zhang and Zhang, 2010).

One study explored nanostructured lipid carriers (NLCs) loaded with budesonide for treating inflammatory bowel disease (IBD). These NLCs, characterized by their 200 nm size and negative zeta potential, were produced via high-pressure homogenization and demonstrated improved efficacy in reducing TNF-α release from activated macrophages and alleviating symptoms in a dextran sulfate-induced mouse model of colitis. This indicates the potential of NLCs in decreasing pro-inflammatory cytokines and neutrophil infiltration in the colon, suggesting a precise drug delivery approach for IBD treatment (Beloqui et al., 2013). Another investigation focused on the topical delivery of lipid nanoparticles, with implications for gastrointestinal applications due to their targeted effects and controlled release in various body layers. Lipid nanoparticles loaded with ketoprofen and naproxen showed reduced drug penetration but increased permeability and accumulation in the horny layer, presenting a promising delivery system for topical medication administration. This approach may offer targeted medication administration and sustained anti-inflammatory effects, potentially benefiting the treatment of gastrointestinal inflammations (Puglia et al., 2008).

Resveratrol-loaded silk fibroin nanoparticles have shown significant intestinal anti-inflammatory effects in a rat model of colitis, as reported by Perez et al. (2014). These nanoparticles exhibited immunomodulatory effects both in vitro and in vivo, suggesting that silk fibroin nanoparticles could serve as an effective vehicle for the controlled release of medicinal agents like resveratrol. This approach presents a novel treatment strategy for inflammatory bowel diseases (Lozano-Pérez et al., 2014). Additionally, studies on solid lipid nanoparticles (SLNs) and nanostructured lipid carriers (NLCs) loaded with amphotericin B have explored their biopharmaceutical characteristics, including drug loading, release patterns, and in vitro hemolytic and antifungal properties. These lipid formulations demonstrated sustained release profiles and reduced toxicity compared to traditional formulations, indicating their potential as promising candidates for treating gastrointestinal infections and associated inflammations (Jansook et al., 2018).

Researchers explored SLNs loaded with etoposide aimed at treating gastric cancer. These nanoparticles showcased a sustained release profile, high drug loading, and an optimal size range. They induced higher rates of apoptosis and increased cytotoxicity against SGC7901 cells compared to the free drug, suggesting their effectiveness in enhancing therapeutic outcomes for gastric cancer through a mechanism involving mitochondrial-mediated apoptosis (Wang et al., 2014). Scientists developed a nanocarrier composed of chitosan and fucoidan to orally deliver berberine, targeting the damaged intestinal epithelium tight junction barrier. These nanoparticles exhibited controlled release profiles—slow in simulated stomach fluid and fast in simulated intestinal fluid, alongside efficient drug loading. In vitro and in vivo studies demonstrated their capability to repair tight junction breakdown in intestinal epithelia caused by lipopolysaccharide, highlighting their potential in treating inflammatory bowel disease (Wu et al., 2014).

Zhang et al. (2013) aimed to mitigate the gastrointestinal discomfort caused by oral triptolide (TP) administration by developing TP-SLNs using a central composite design. These optimized nanoparticles showed prolonged in vitro drug release behavior, a narrow size distribution, and significant drug loading. They exhibited stability in artificial stomach fluids, indicating their potential to reduce gastrointestinal discomfort and improve the oral bioavailability of triptolide for medical applications (Zhang et al., 2013). Researchers investigated SLNs loaded with sulpiride, using a unique ultrasonic melt-emulsification method to enhance its intestinal permeability. These SLNs displayed high drug loading and a limited size distribution, significantly increasing sulpiride's permeability in a rat everted sac intestinal model. This suggests that SLNs could enhance the oral absorption of drugs with generally poor absorption (Ibrahim et al., 2013). Researchers utilized tailored nanoparticles to deliver an anti-inflammatory tripeptide to the colon, targeting inflammation sites in inflammatory bowel disease (IBD). These nanoparticles, prepared via double-emulsion/solvent evaporation, provided enhanced in vivo protection against colitis' histologic and inflammatory aspects compared to previous treatments, offering a novel therapeutic approach for IBD through targeted drug delivery to inflamed gastrointestinal regions (Laroui et al., 2010).

A novel liposome formulation (ST-H2S lipo) specifically targets the spleen to deliver hydrogen sulfide (H2S), showing greater success in reducing the severity of colitis in a dextran sulfate sodium (DSS) induced colitis model compared to both long-circulating liposomes loaded with H2S donors (LC-H2S lipo) and unloaded H2S donors. The superior therapeutic efficacy of ST-H2S lipo is attributed to its enhanced systemic immunomodulatory effect, making it a promising therapeutic approach for IBD treatment (Oh et al., 2023). To improve the cellular absorption and bioavailability of tripterine, a challenging insoluble and enterotoxic drug, researchers developed selenized polymer-lipid hybrid nanoparticles (Se@Tri-PLNs). These nanoparticles exhibited significantly higher cellular uptake in Caco-2 cells, increased stability in digestive fluids, and markedly enhanced oral bioavailability of tripterine. In addition, Se@Tri-PLNs demonstrated potent anti-enteritis activity in vivo, suggesting that such polymer-lipid hybrid nanoparticles, especially when modified with selenium, offer a viable treatment strategy for managing hard-to-treat inflammatory diseases like IBD by amplifying the therapeutic benefits of phytomedicine (Ren et al., 2023).

17.4.1.2 Efficacy and Safety Assessment in Animal Models

The evaluation of lipid nanocarriers in animal models for gastrointestinal (GI) inflammation is a crucial step towards developing targeted treatments for inflammatory bowel disease (IBD), including Crohn's disease and ulcerative colitis. These animal models, often involving rodents like mice and rats with chemically induced colitis through agents such as dextran sulfate sodium (DSS) or trinitrobenzene sulfonic acid (TNBS), effectively replicate the human condition of IBD, showcasing pathological features like inflammation, ulceration, and dysbiosis. This allows for an in-depth assessment of the therapeutic potential of nanoparticle treatments. Studies have shown that lipid nanocarriers can significantly mitigate inflammation, reduce mucosal damage, and restore intestinal barrier function in these models. For example, Gemini lipid nanoparticles (GLNPs) have been demonstrated to alleviate gut inflammation by delivering TNF-α siRNA directly to the inflamed tissues, thereby inhibiting the differentiation of pro-inflammatory CD4+ T-cells and reducing cytokine production (Verma et al., 2022). Additionally, plant-derived exosomal nanoparticles have been investigated for their natural anti-inflammatory properties, showing promise in easing IBD symptoms (Li et al., 2023). These findings underscore the potential of lipid nanoparticles as a modality for delivering medications, particularly for the treatment of gastrointestinal inflammation and IBD.

One of the primary efficacy endpoints in these models is the reduction in inflammatory markers, including cytokines (TNF-α, IL-6, IL-1β) and other biomarkers of inflammation (CRP, SAA). ELISA, Western blotting, and RT-PCR are used to quantify these markers in tissue samples. Samples are histologically analyzed with hematoxylin and eosin (H&E) staining of colon tissue; the extent of inflammation, ulcer presence, and the degree of regeneration of epithelial cells are assessed. Improvement in those parameters evidences the healing efficacy of lipid nanocarriers. Other parameters such as improvement in clinical symptoms, such as weight loss, stool consistency, blood in feces, are also used to assess the effectiveness of lipid nanocarriers. These symptoms are scored to provide a quantitative measure of disease severity and therapeutic response.

Assessing the safety of lipid nanocarriers involves its evaluation through biochemical assays (such as liver and kidney function tests), histological examination of major organs, and observation of any adverse effects or behavioral changes in animals. Evaluating the potential immunogenicity and allergic reactions generated by the nanocarriers is crucial. The immunological response markers are monitored for indications of immune reactions, and if there are any reports of hypersensitivity or anaphylaxis, these should be observed. Long-term studies are carried out to estimate whether the lipid nanocarriers have resulted in any chronic effects, including carcinogenicity, reproductive toxicity, and influence on the gut microbiota. The safety profile of lipid nanocarriers is equally important and is evaluated in acute and chronic toxicity studies using animal models. Parameters such as mortality, behavioral changes, weight loss, histopathological changes in major organs, hematologic, and biochemical markers are monitored to establish the potential toxicity. Lipid nanocarriers generally have a favorable safety profile, thanks to their biocompatibility and biodegradation. The composition of the lipid matrix, size, surface charge, and presence of targeting ligands can influence biodistribution and clearance, thus potentially affecting toxicity. Thus, for instance, the in vivo safety of zingerone-encapsulated solid lipid nanoparticles was evaluated in a rodent model and no significant toxicity or side effect at therapeutic doses was found. These findings lead to the conclusion that such lipid nanocarriers safely deliver anti-inflammatory compounds to the site of inflammation in the GI tract (Sunnapu et al., 2022).

Using an innovative methodology known as quality by design (QbD), a groundbreaking investigation devised transferrin anchored PLGA nanocarriers loaded with Tofacitinib citrate (Tofa) (Tofa-P/tfr NCs). These cutting-edge nanocarriers have been ingeniously designed to improve the retention of medication at the intended site and extend the effectiveness of treatment in models of ulcerative colitis (UC) by selectively binding to the transferrin receptor (TFR-1/CD71), which is excessively expressed on macrophages and colon epithelial cells. This remarkable study unveiled the tremendous potential of the nanocarriers in reducing inflammation of the mucosal lining by specifically targeting surface receptors. In addition, the study demonstrated the nanocarriers' remarkable physicochemical properties, extraordinary compatibility with biological systems, substantial therapeutic efficacy against UC in models induced by dextran sulfate sodium (DSS), and notable uptake by the desired cells (Zeeshan et al., 2023).

17.4.2 Emerging Perspectives and Research Opportunities

While preclinical investigations have showcased the capability of lipid nanocarriers to enhance the administration of drugs and specifically target inflamed tissues within the gastrointestinal tract, the translation of these discoveries into clinical trials remains a pivotal avenue for inquiry. The absence of clinical trials delving into the application of lipid nanocarriers in gastrointestinal inflammations presents a significant opportunity for researchers to bridge this gap and propel our comprehension of their potential utilities in the management of inflammatory gastrointestinal disorders. The forthcoming trajectories and research prospects concerning lipid nanocarriers for the treatment of gastrointestinal inflammations center around augmenting the efficacy of drug administration, mitigating toxicity, and amplifying bioavailability to enhance treatment outcomes.

17.4.2.1 Enhanced Drug Delivery to Inflamed Tissues

It has been shown that combining lipids and anti-inflammatory medications into a nanocarrier is a helpful strategy for treating inflammatory bowel disease (IBD). Researchers have explored the effectiveness of different carriers made of lipids in improving drug absorption and reducing inflammation. These carriers include tiny capsules with a lipid core and shell, nanostructured carriers made of lipids, and self-nanoemulsifying drug delivery systems. Interestingly, having better drug absorption does not always mean better treatment for inflammatory bowel disease (IBD); instead, carriers that can hold onto the drug at the location in the intestines may be more effective treatments for IBD (Beloqui et al., 2016).

17.4.2.2 Overcoming Physiological Barriers

To surpass the challenges posed by acid-induced hydrolysis in the stomach, enzymatic breakdown in the gastrointestinal (GI) tract, and limited mucosa permeability barriers, scientists are creating lipid-based nanocarriers to bolster the oral absorption of large molecules. These innovative carriers hold promise in optimizing intestinal absorption mechanisms, dissolving capacity, and robustness of biomolecules, as evidenced by recent advancements (Karamanidou et al., 2016).

17.4.2.3 Targeted Drug Delivery Systems

There is exciting progress in the realm of research focused on creating lipid nanocarriers capable of specifically targeting inflamed tissues. This cutting-edge advancement holds great potential in improving treatment outcomes and minimizing unwanted side effects. In fact, the use of nanoformulations such as budesonide-loaded NLCs has already exhibited promising results in reducing inflammation, surpassing the efficacy of conventional formulations. This highlights the potential for lipid nanocarriers to serve as a promising alternative to polymeric nanoparticles for precisely targeting and treating inflammatory bowel disease (Beloqui et al., 2013).

17.4.2.4 Improving Stability and Bioavailability

The therapeutic effectiveness of lipid-based nanocarriers depends on their stability and storage times. The focus of research is on improving the physicochemical properties of these nanocarriers, which include, particle size, polydispersity index (PDI) and drug encapsulation efficiency (EE) in order to improve the bioavailability and therapeutic efficacy of these nanocarriers (Danaei et al., 2018).

17.4.2.5 PEGylation and Surface Modification of Lipid Nanocarriers

Research is necessary on the PEGylation and surface modification of lipid nanocarriers. They can improve the circulation time of these carriers in blood, avoid their uptake by the mononuclear phagocyte system and also increase their potential to accumulate at inflammation sites. Another research showed that mucin binding was reduced while lipolysis was inhibited in PEGylated lipid nanoparticles leading to high oral absorption of fenofibrate as a model for Biopharmaceutics Classification System (BCS) II drugs (Zhang et al., 2014).

17.4.2.6 Lipid Nanocarriers for Improved Diagnosis and Precision Medicine

A significant research opportunity is represented by the development of lipid nanocarriers that can be used for diagnosing and improving the precision medicine in gastrointestinal inflammations. By avoiding destruction through this complex microenvironment, functional nanocarriers enhance drug bioavailability and therapeutic efficacy. In addition to targeting therapy strategies, it may also provide a means for precise diagnostics of inflammatory bowel diseases (IBD) (Wang et al., 2021).

17.4.2.7 Nanoformulations for pH-Sensitive Drug Release

Another fascinating idea is to make nanoformulations with pH-sensitive drug release capabilities. They ensure that therapeutic medications are released only at the inflamed site since during the inflammation process, there is usually a change in pH. Such targeted release improves treatment efficiency while minimizing systemic adverse reactions. As an effective way of reducing systemic side effects, the use of pH-sensitive nanocarriers administered orally may really allow for focused and localized action against IBD (Zeeshan et al., 2019).

17.4.2.8 Lipid Nanocarriers for Co-Delivery of Drugs and Nutraceuticals

In complementary ways, lipid nanocarriers are useful in concurrent delivery of drugs and nutraceuticals that enhance therapeutic outcomes. Such an approach may exploit both the specific therapeutic effects of medicines as well as the anti-inflammatory and reparative properties of some nutraceuticals. New generations of bioactive lipid nanocarriers loaded with synthetic and natural anti-inflammatory actives have great potential to reduce inflammation, elevate antioxidant activity and provide controlled release mechanisms (Lacatusu et al., 2019).

17.4.2.9 Lipid Nanocarriers for Enhancing Mucosal Vaccination

A novel prophylactic measure against gastrointestinal inflammations could be made available through investigations into the employment of lipid nanocarriers for mucosal vaccination against pathogens causing these infections Exploiting such carriers enhances antigen delivery to the mucosal immune system with concomitant maximal presentation that leads to protective immunity. The development of lipid-based nanocarriers as adjuvants and delivery systems for vaccines targeting the gastrointestinal tract could pave the way for innovative treatments and preventive measures for inflammatory diseases of the GI tract.

17.4.2.10 Lipid Nanocarriers for the oral delivery of Peptides and Proteins

Lipid-based nanocarriers could offer an oral delivery approach to treating gastrointestinal diseases that is less invasive. Peptides and proteins can be packed into these systems to shield them from enzymatic degradation within the gastrointestinal tract and facilitate absorption across the intestinal mucosa. A review on lipid-based nanocarriers for oral peptide delivery lays out the importance of the nanocarriers' composition and physicochemical properties in dictating the prospects of these carriers for interacting with the gastrointestinal environment and improving the pharmacokinetic/pharmacodynamic (PK/PD) profile of their drug cargoes. This could have a significant impact on the future of oral administration of peptide-based therapeutics for gastrointestinal inflammations, potentially improving the ease of patient compliance and therapeutic outcomes (Niu et al., 2016).

17.4.2.11 Nanocarriers for Targeted Delivery of Vitamins and Antioxidants

Antioxidants and vitamins play a pivotal role in the management of inflammation and wound healing in the gastrointestinal tract. However, in many cases, the solubility, stability, and bioavailability of these therapeutic nutrients are low, which limits their efficiency in the treatment of gastrointestinal diseases. The use of lipid nanocarriers to improve the oral delivery of different vitamins has clearly generated an improved solubility, chemical stability, and bioavailability of pretty much any type of lipophilic molecule such as vitamins and also quite different families of biologically active molecules such as vitamin derivatives and antioxidants in suppressing gastrointestinal inflammation (Hsu et al., 2019).

17.4.2.12 Innovations in Lipid Nanocarriers for Enhanced Mucosal Healing

Another novel approach to the treatment of gastrointestinal inflammations is the development of lipid nanocarriers that target specifically the process of mucosal healing. The delivery of the therapeutic agent directly to the damaged mucosa could facilitate better mucosal healing by promoting an efficient response to the same, from the site. The development and evaluation studies of lipid nanocarriers, such as solid lipid nanoparticles (SLN), nanostructured lipid carriers (NLC), and lipid nanoemulsions (LNE), for quercetin delivery as a therapeutic agent, state the potential of enhancing bioaccessibility and giving controlled release. Controlled release via matrices is particularly useful for chronic conditions such as IBD, where the therapeutic effect has to be prolonged to control inflammation and foster mucosal healing (Aditya et al., 2014).

17.5 CONCLUSION

This chapter underscores the potential that lipid nanocarriers have to transform the management of complex problems associated with the treatment of gastrointestinal inflammations. Therefore, through the use of nanotechnology, the lipid nanocarrier provides an efficient and targeted drug delivery system that can circumvent the setbacks arising from poor bioavailability and systemic toxicity as are displayed by the conventional treatments. The overall outlook for lipid nanocarriers—from fabrication and characterization to clinical applications—only mirrors their flexibility and biocompatibility. These challenges include but are not limited to scalability and gaining approval through regulatory processes. Therefore, future studies should be aimed at optimization of such carriers with regard to safety, efficacy, and patient acceptability to create space for new kinds of treatments that can serve to improve patients' quality of life while battling gastrointestinal inflammations.

REFERENCES

Aditya, N., Macedo, A., Doktorovova, S., Souto, E., Kim, S., Chang, P., & Ko, S. (2014). Development and evaluation of lipid nanocarriers for quercetin delivery: A comparative study of solid lipid nanoparticles (SLN), nanostructured lipid carriers (NLC), and lipid nanoemulsions (LNE). *Lwt – Food Science and Technology*, *59*, 115–121. https://doi.org/10.1016/j.lwt.2014.04.058

Ball, R. L., Bajaj, P., & Whitehead, K. A. (2018). Oral delivery of siRNA lipid nanoparticles: Fate in the GI tract. *Scientific Reports*, *8*, Article number: 2175. https://doi.org/10.1038/s41598-018-20632-6

Bamford, K. B. (1999). Chronic gastrointestinal inflammation. *FEMS Immunology & Medical Microbiology*, *24*(2), 161–168. https://doi.org/10.1111/j.1574-695x.1999.tb01277.x

Barbosa, C., Santos-Pereira, C., Soares, I., Martins, V., Terra-Matos, J., Côrte-Real, M., ... & Gerós, H. (2019). Resveratrol-loaded lipid nanocarriers are internalized by endocytosis in yeast. *Journal of Natural Products*, *82*(5), 1240–1249. https://doi.org/10.1021/acs.jnatprod.8b01003

Baviskar, D. T., Amritkar, A. S., Chaudhari, H. S., & Jain, D. K. (2012). Modulation of drug release from nanocarriers loaded with a poorly water soluble drug (flurbiprofen) comprising natural waxes. *Die Pharmazie-An International Journal of Pharmaceutical Sciences*, *67*(8), 701–705. https://doi.org/10.1691/ph.2012.1146

Beloqui, A., Coco, R., Alhouayek, M., Solinís, M., Rodríguez-Gascón, A., Muccioli, G., & Préat, V. (2013). Budesonide-loaded nanostructured lipid carriers reduce inflammation in murine DSS-induced colitis. *International Journal of Pharmaceutics*, *454*(2), 775–83. https://doi.org/10.1016/j.ijpharm.2013.05.017

Beloqui, A., Memvanga, P. B., Coco, R., et al. (2016). A comparative study of curcumin-loaded lipid-based nanocarriers in the treatment of inflammatory bowel disease. *Colloids and Surfaces B: Biointerfaces*, *143*, 274–282. https://doi.org/10.1016/j.colsurfb.2016.03.038

Cao, M., Wang, Y., Jing, H., Wang, Z., Meng, Y., Geng, Y., Miao, M., & Li, X.-M. (2022). Development of an oral isoliquiritigenin self-nano-emulsifying drug delivery system (ILQ-SNEDDS) for effective treatment of eosinophilic esophagitis induced by food allergy. *Pharmaceuticals*, *15*(12), 1587. https://doi.org/10.3390/ph15121587

Cardoso, B. D., Cardoso, V. F., Lanceros-Méndez, S., & Castanheira, E. M. (2022). Solid magnetoliposomes as multi-stimuli-responsive systems for controlled release of doxorubicin: Assessment of lipid formulations. *Biomedicines*, *10*(5), 1207. https://doi.org/10.3390/biomedicines10051207

Carton, J., Daly, R., & Ramani, P. (2006). Inflammation. In Carton, J., Daly, R., Ramani, P., (eds.), *Clinical Pathology* (pp. 11–18). Oxford University PressOxford. https://doi.org/10.1093/oso/9780198569466.003.0002

Chen, W., Zhao, Y., Dai, Y., & Nie, K. (2022). Gastrointestinal inflammation plays a critical role in chemotherapy-induced nausea and vomiting. *European Journal of Pharmacology*, *936*, 175379. https://doi.org/10.1016/j.ejphar.2022.175379

Copland, M. J., Rades, T., Davies, N. M., & Baird, M. A. (2005). Lipid based particulate formulations for the delivery of antigen. *Immunology & Cell Biology*, *83*(2), 97–105. https://doi.org/10.1111/j.1440-1711.2005.01315.x

da Costa de Quadros, C., de Sousa Araujo, A. C., Latorres, J. M., Michelon, M., & de las Mercedes Salas-Mellado, M. (2023). Lipid nanocarriers as an alternative for the delivery of bioactive compounds beneficial to health. *Current Bioactive Compounds*, *19*(8), 1–14. https://doi.org/10.2174/1573407219666230306142421

Dammes, N., Goldsmith, M., Ramishetti, S., Dearling, J., Veiga, N., Packard, A., & Peer, D. (2021). Conformation sensitive targeting of lipid nanoparticles for RNA therapeutics. *Nature Nanotechnology*, *16*, 1030–1038. https://doi.org/10.1038/s41565-021-00928-x

Danaei, M., Dehghankhold, M., Ataei, S., Hasanzadeh Davarani, F., Javanmard, R., Dokhani, A., ... & Mozafari, M. R. (2018). Impact of particle size and polydispersity index on the clinical applications of lipidic nanocarrier systems. *Pharmaceutics*, *10*(2), 57. https://doi.org/10.3390/pharmaceutics10020057

Date, A. A., Hanes, J., & Ensign, L. M. (2016). Nanoparticles for oral delivery: Design, evaluation and state-of-the-art. *Journal of Controlled Release*, *240*, 504–526. https://doi.org/10.1016/j.jconrel.2016.06.016

Duque-Buitrago, L. F., Tornero-Martínez, A., Loera-Castañeda, V., & Mora-Escobedo, R. (2023). Use of food and food-derived products in the treatment of gastritis: A systematic review. *Critical Reviews in Food Science and Nutrition*, *63*(22), 5771–5782. https://doi.org/10.1080/10408398.2021.2024131

Elseweidy, M. M. (2017). Brief review on the causes, diagnosis and therapeutic treatment of gastritis disease. *Alternative & Integrative Medicine*, *6*(1), 1–6. https://doi.org/10.4172/2327-5162.1000231

Fan, Y., Marioli, M., & Zhang, K. (2021). Analytical characterization of liposomes and other lipid nanoparticles for drug delivery. *Journal of Pharmaceutical and Biomedical Analysis*, *192*, 113642. https://doi.org/10.1016/j.jpba.2020.113642

Fathi, M., & Varshosaz, J. (2013). Novel hesperetin loaded nanocarriers for food fortification: Production and characterization. *Journal of Functional Foods*, *5*(3), 1382–1391.

Fouladi, F., Steffen, K. J., & Mallik, S. Enzyme-responsive liposomes for the delivery of anticancer drugs. *Bioconjugate Chemistry*, *28*(4), 857–868.

Frosali, S., Pagliari, D., Gambassi, G., Landolfi, R., Pandolfi, F., & Cianci, R. (2015). How the intricate interaction among toll-like receptors, microbiota, and intestinal immunity can influence gastrointestinal pathology. *Journal of Immunology Research*. *2015*(1), 489821. https://doi.org/10.1155/2015/489821

Funk, C. (2001). Prostaglandins and leukotrienes: Advances in eicosanoid biology. *Science*, *294*(5548), 1871–1875. https://doi.org/10.1126/science.294.5548.1871

García García, D., & Gómez Rodríguez, B. (2023). Eradication treatment on Helicobacter Pylori infection. *Revista Andaluza de Patología Digestiva*, *46*(1), 16–25. https://doi.org/10.37352/2023461.2

Geremia, A., Biancheri, P., Allan, P., Corazza, G. R., & Di Sabatino, A. (2014). Innate and adaptive immunity in inflammatory bowel disease. *Autoimmunity Reviews*, *13*(1), 3–10. https://doi.org/10.1016/j.autrev.2013.06.004

Gewirtz, A. T., Collier-Hyams, L. S., Young, A. N., Kucharzik, T., Guilford, W. J., Parkinson, J. F., ... & Madara, J. L. (2002). Lipoxin a4 analogs attenuate induction of intestinal epithelial proinflammatory gene expression and reduce the severity of dextran sodium sulfate-induced colitis. *The Journal of Immunology*, *168*(10), 5260–5267. https://doi.org/10.4049/jimmunol.168.10.5260

Hannoodee, S., & Nasuruddin, D. N. (2023). *Acute Inflammatory Response*. StatPearls Publishing. http://europepmc.org/abstract/MED/32310543

Heidari, R., & Mehdi Ommati, M. (2023). Gastroenterological aspects of nanopharmacology. In Ahmadian, E., Cucchiarini, M., Eftekhari, A. (eds.), *Nanopharmacology and Nanotoxicology: Clinical Implications and Methods* (pp. 139–167). Bentham Science Publishers. https://doi.org/10.2174/9789815079692123010009

Heurtault, B., Saulnier, P., Pech, B., Venier-Julienne, M. C., Proust, J. E., Phan-Tan-Luu, R., & Benoît, J. P. (2003). The influence of lipid nanocapsule composition on their size distribution. *European Journal of Pharmaceutical Sciences*, *18*(1), 55–61.

Higa, L. H., Jerez, H. E., De Farias, M. A., Portugal, R. V., Romero, E. L., & Morilla, M. J. (2017). Ultra-small solid archaeolipid nanoparticles for active targeting to macrophages of the inflamed mucosa. *Nanomedicine*, *12*(10), 1165–1175. https://doi.org/10.2217/nnm-2016-0437

Hsu, C. Y., Wang, P. W., Alalaiwe, A., Lin, Z. C., & Fang, J. Y. (2019). Use of lipid nanocarriers to improve oral delivery of vitamins. *Nutrients*, *11*(1), 68. https://doi.org/10.3390/nu11010068

Huguet-Casquero, A., Xu, Y., Gainza, E., Pedraz, J. L., et al. (2020). Oral delivery of oleuropein-loaded lipid nanocarriers alleviates inflammation and oxidative stress in acute colitis. *International Journal of Pharmaceutics*, *579*, 119166. https://doi.org/10.1016/j.ijpharm.2020.119515

Ibrahim, W., Alomrani, A., & Yassin, A. (2013). Novel sulpiride-loaded solid lipid nanoparticles with enhanced intestinal permeability. *International Journal of Nanomedicine*, *9*, 129–144. https://doi.org/10.2147/IJN.S54413

Ioele, G., Grande, F., De Luca, M., Occhiuzzi, M. A., Garofalo, A., & Ragno, G. (2021). Photodegradation of anti-inflammatory drugs: Stability tests and lipid nanocarriers for their photoprotection. *Molecules*, *26*(19), 5989. https://doi.org/10.3390/molecules26195989

Ishida, T., & Kiwada, H. (2008). Accelerated blood clearance (ABC) phenomenon upon repeated injection of PEGylated liposomes. *International Journal of Pharmaceutics*, *354*(1-2), 56–62.

Ivanova, M., Bottiglieri, L., Sajjadi, E., Venetis, K., & Fusco, N. (2023). Malignancies in patients with celiac disease: Diagnostic challenges and molecular advances. *Genes*, *14*(2), 376. https://doi.org/10.3390/genes14020376

Jansook, P., Pichayakorn, W., & Ritthidej, G. (2018). Amphotericin B-loaded solid lipid nanoparticles (SLNs) and nanostructured lipid carrier (NLCs): Effect of drug loading and biopharmaceutical characterizations. *Drug Development and Industrial Pharmacy*, *44*, 1693–1700. https://doi.org/10.1080/03639045.2019.1569023

Jensen, S. K., Pærregaard, S. I., Brandum, E. P., Jørgensen, A. S., Hjortø, G. M., & Jensen, B. A. (2022). Rewiring host–microbe interactions and barrier function during gastrointestinal inflammation, *Gastroenterology Report*, *10*, goac008. https://doi.org/10.1093/gastro/goac008

Karamanidou, T., Bourganis, V., Kammona, O., & Kiparissides, C. (2016). Lipid-based nanocarriers for the oral administration of biopharmaceutics. *Nanomedicine*, *11*(22), 3009–3032. https://doi.org/10.2217/nnm-2016-0265

Keewan, E., Narasimhulu, C. A., Rohr, M., Hamid, S., & Parthasarathy, S. (2020). Are fried foods unhealthy? The dietary peroxidized fatty acid, 13-HPODE, induces intestinal inflammation in vitro and in vivo. *Antioxidants*, *9*(10), 926. https://doi.org/10.3390/antiox9100926

Khan, M. A., Ashraf, G., Ashraf, H., Hassan, Y., Rawal, A., Ashraf, I., Shahab, Q. A., Hasimoglu, Z.. (2022). Gastroenterology. In. Khan, M. A., Ashraf, G., Ashraf, H. (eds.), *Medical Analogies for Clinician-Patient Communication*. Springer. https://doi.org/10.1007/978-3-030-87293-9_5

Knipe, J. M., Strong, L. E., & Peppas, N. A. (2016). Enzyme- and pH-responsive microencapsulated nanogels for oral delivery of siRNA to induce TNF-α knockdown in the intestine. *Biomacromolecules*, *17*(3), 788–797. https://doi.org/10.1021/acs.biomac.5b01518

Lacatusu, I., Badea, N., Badea, G., Mihaila, M., Ott, C., Stan, R., & Meghea, A. (2019). Advanced bioactive lipid nanocarriers loaded with natural and synthetic anti-inflammatory actives. *Chemical Engineering Science*, *200*, 113–126. https://doi.org/10.1016/j.ces.2019.01.044

Laroui, H., Dalmasso, G., Nguyen, H. T. T., Yan, Y., Sitaraman, S. V., & Merlin, D. (2010). Drug-loaded nanoparticles targeted to the colon with polysaccharide hydrogel reduce colitis in a mouse model. *Gastroenterology*, *138*(3), 843–853. https://doi.org/10.1053/j.gastro.2009.11.003

Layek, B., Gidwani, B., Tiwari, S., Joshi, V., Jain, V., & Vyas, A. (2020). Recent advances in lipid-based nanodrug delivery systems in cancer therapy. *Current Pharmaceutical Design*, *26*(27), 3218–3233. https://doi.org/10.2174/1381612826666200622133407

Le, Z., He, Z., Liu, H., Liu, L., Liu, Z., & Chen, Y. (2021). Antioxidant enzymes sequestered within lipid–polymer hybrid nanoparticles for the local treatment of inflammatory bowel disease. *ACS Applied Materials & Interfaces*, *13*(47), 55966–55977. https://doi.org/10.1021/acsami.1c19457

Li, D. F., Tang, Q., Yang, M. F., Xu, H. M., Zhu, M. Z., Zhang, Y., ... & Yao, J. (2023). Plant-derived exosomal nanoparticles: Potential therapeutic for inflammatory bowel disease. *Nanoscale Advances*, *5*(14), 3575–3588. https://doi.org/10.1039/D3NA00093A

Li, W., Lin, J., Zhou, J., He, S., Wang, A., Hu, Y., Li, H., et al. (2024). Hyaluronic acid-functionalized DDAB/PLGA nanoparticles for improved oral delivery of magnolol in the treatment of ulcerative colitis. *International Journal of Pharmaceutics*, 653, 123878. https://doi.org/10.1016/j.ijpharm.2024.123878

Lonez, C., Bessodes, M., Scherman, D., Vandenbranden, M., Escriou, V., & Ruysschaert, J. (2014). Cationic lipid nanocarriers activate toll-like Receptor 2 and NLRP3 inflammasome pathways. *Nanomedicine: Nanotechnology, Biology, and Medicine*, *10*(4), 775–782. https://doi.org/10.1016/j.nano.2013.12.003

Lozano-Pérez, A. A., Rodriguez-Nogales, A., Ortiz-Cullera, V., Algieri, F., Garrido-Mesa, J., Zorrilla, P., ... & Galvez, J. (2014). Silk fibroin nanoparticles constitute a vector for controlled release of resveratrol in an experimental model of inflammatory bowel disease in rats. *International Journal of Nanomedicine*, *9*(1), 4507–4520. https://doi.org/10.2147/IJN.S68526

Luo, H., Cao, G., Luo, C., Tan, D., Vong, C. T., Xu, Y., Wang, S., Lu, H., Wang, Y., & Jing, W. (2022). Emerging pharmacotherapy for inflammatory bowel diseases. *Pharmacological Research*, *178*, 106146. https://doi.org/10.1016/j.phrs.2022.106146

Ma, L., Ma, Y., Gao, Q., Liu, S., Zhu, Z., Shi, X., Dai, F., Reis, R. L., et al. (2024). Mulberry leaf lipid nanoparticles: A naturally targeted CRISPR/Cas9 oral delivery platform for alleviation of colon diseases. *Small*, 20, 2307247. https://doi.org/10.1002/smll.202307247

Makoni, P. A., Wa Kasongo, K., & Walker, R. B. (2019). Short term stability testing of efavirenz-loaded solid lipid nanoparticle (SLN) and nanostructured lipid carrier (NLC) dispersions. *Pharmaceutics*, *11*(8), 397. https://doi.org/10.3390/pharmaceutics11080397

Mathias, C. B., McAleer, J. P., & Szollosi, D. E. (2020). Inflammatory diseases of the gastrointestinal tract and pharmacological treatments. *Pharmacology of Immunotherapeutic Drugs*, 175–205. https://doi.org/10.1007/978-3-030-19922-7_6

Matsuoka, K., & Kanai, T. (2015). The gut microbiota and inflammatory bowel disease. *Seminars in Immunopathology*, *37*(1), 47–55. https://doi.org/10.1007/s00281-014-0454-4

Miehlke, S., Lucendo, A. J., Straumann, A., Jan Bredenoord, A., & Attwood, S. (2020). Orodispersible budesonide tablets for the treatment of eosinophilic esophagitis: A review of the latest evidence. *Therapeutic Advances in Gastroenterology*, *13*, 175628482092728. https://doi.org/10.1177/1756284820927282

Mishra, R. K., Ahmad, A., Kumar, A., Vyawahare, A., et al. (2020). Lipid-based nanocarrier-mediated targeted delivery of celecoxib attenuate severity of ulcerative colitis. *Materials Science and Engineering: C*, *110*, 110698. https://doi.org/10.1016/j.msec.2020.111103

Montoto, S S., Muraca, G., & Ruiz, M E. (2020, October 30). Solid lipid nanoparticles for drug delivery: Pharmacological and biopharmaceutical aspects. *Frontiers in Molecular Biosciences*, *7*. https://doi.org/10.3389/fmolb.2020.587997

Müller, R. H., Mäder, K., Gohla, S. (2000). Solid lipid nanoparticles (SLN) for controlled drug delivery – A review of the state of the art. *European Journal of Pharmaceutics and Biopharmaceutics*, *50*(1), 161–177. https://doi.org/10.1016/S0939-6411(00)00087-4

Nadpara, N., Reichenbach, Z. W., Ehrlich, A. C., & Friedenberg, F. (2020). Current status of medical therapy for inflammatory bowel disease: The wealth of medications. *Digestive Diseases and Sciences*, *65*(10), 2769–2779. https://doi.org/10.1007/s10620-020-06471-4

Niu, Z., Conejos-Sánchez, I., Griffin, B. T., O'Driscoll, C. M., & Alonso, M. J. (2016). Lipid-based nanocarriers for oral peptide delivery. *Advanced Drug Delivery Reviews*, *106*, 337–354. https://doi.org/10.1016/j.addr.2016.04.001

Oh, C., Lee, W., Park, J., Choi, J., Lee, S., Li, S., ... & Im, H. J. (2023). Development of spleen targeting H2S donor loaded liposome for the effective systemic immunomodulation and treatment of inflammatory bowel disease. *ACS Nano*, *17*(5), 4327–4345. https://doi.org/10.1021/acsnano.2c08898

Oz, H. S., Yeh, S. L., & Neuman, M. G. (2016). Gastrointestinal inflammation and repair: Role of microbiome, infection, and nutrition. *Gastroenterology Research and Practice*, *2016*. https://doi.org/10.1155/2016/6516708

Tiwari, G., Tiwari, R., Sriwastawa, B., Bhati, L., Pandey, S., Pandey, P., & Bannerjee, S. K. (2012). Drug delivery systems: An updated review. *International Journal of Pharmaceutical Investigation*, *2*(1), 2. https://doi.org/10.4103%2F2230-973X.96920

Plaza-Oliver, M., Santander-Ortega, M. J., & Lozano, M. V. (2021). Current approaches in lipid-based nanocarriers for oral drug delivery. *Drug Delivery and Translational Research*, *11*, 471–497. https://doi.org/10.1007/s13346-021-00908-7

Pridgen, E. M., Alexis, F., & Farokhzad, O. C. (2015). Polymeric nanoparticle drug delivery technologies for oral delivery applications. *Expert Opinion on Drug Delivery*, *12*(9), 1459–1473. https://doi.org/10.1517/17425247.2015.1018175

Priya, S., Desai, V. M., & Singhvi, G. (2022). Surface modification of lipid-based nanocarriers: A potential approach to enhance targeted drug delivery. *ACS Omega*, *8*(1), 74–86. https://doi.org/10.1021/acsomega.2c05976

Puglia, C., Blasi, P., Rizza, L., Schoubben, A., Bonina, F., Róssi, C., & Ricci, M. (2008). Lipid nanoparticles for prolonged topical delivery: An in vitro and in vivo investigation. *International Journal of Pharmaceutics*, *357*(1–2), 295–304. https://doi.org/10.1016/j.ijpharm.2008.01.045

Ren, Y., Qi, C., Ruan, S., Cao, G., Ma, Z., & Zhang, X. (2023). Selenized polymer-lipid hybrid nanoparticles for oral delivery of tripterine with ameliorative oral anti-enteritis activity and bioavailability. *Pharmaceutics*, *15*(3), 821. https://doi.org/10.3390/pharmaceutics15030821

Ribaldone, D. G., Pellicano, R., & Actis, G. C. (2019). Inflammation in gastrointestinal disorders: Prevalent socioeconomic factors. *Clinical and Experimental Gastroenterology*, *12*, 321–329. https://doi.org/10.2147/CEG.S210844

Salvi, V. R., & Pawar, P. (2019). Nanostructured lipid carriers (NLC) system: A novel drug targeting carrier. *Journal of Drug Delivery Science and Technology*, *51*, 255–267. https://doi.org/10.1016/j.jddst.2019.02.017

Scott, A., Khan, K. M., Cook, J. L., & Duronio, V. (2004). What is "inflammation"? Are we ready to move beyond Celsus?. *British Journal of Sports Medicine*, *38*(3), 248–249. https://doi.org/10.1136/bjsm.2003.011221

Shah, V. M., Nguyen, D. X., Patel, P., Cote, B., Al-Fatease, A., Pham, Y., Huynh, M. G., Woo, Y., & Alani, A. W. (2019). Liposomes produced by microfluidics and extrusion: A comparison for scale-up purposes. *Nanomedicine: Nanotechnology, Biology and Medicine*, *18*, 146–156. https://doi.org/10.1016/j.nano.2019.02.019

Sharma, M., Sharma, S., Wadhwa, J., et al. (2019). Improved uptake and therapeutic intervention of curcumin via designing binary lipid nanoparticulate formulation for oral delivery in inflammatory bowel disorder. *Artificial Cells, Nanomedicine, and Biotechnology*, *47*(1), 3642–3655. https://doi.org/10.1080/21691401.2018.1543191

Shi, N. Q., & Qi, X. R. (2021). Preparation of drug liposomes by reverse-phase evaporation. *Liposome-Based Drug Delivery Systems*, 37–46. https://doi.org/10.1007/978-3-662-49231-4_3-1

Son, G. H., Lee, B. J., Cho, C. W. (2017). Mechanisms of drug release from advanced drug formulations such as polymeric-based drug-delivery systems and lipid nanoparticles. *Journal of Pharmaceutical Investigation*, 47, 287–296.

Sung, J., Alghoul, Z., Long, D., Yang, C., & Merlin, D. (2022). Oral delivery of IL-22 mRNA-loaded lipid nanoparticles targeting the injured intestinal mucosa: A novel therapeutic solution to treat ulcerative colitis. *Biomaterials*, *288*, 121707. https://doi.org/10.1016/j.biomaterials.2022.121707

Sunnap, O., Subramanian, S., Vemula, P. K., & Karuppannan, S. (2022). Zingerone-encapsulated solid lipid nanoparticles as oral drug-delivery systems to potentially target inflammatory diseases. *ChemNanoMat*, *8*(12). https://doi.org/10.1002/cnma.202200388

Sutton, A. G., Mir, S., & Steiner, M. J. (2015). Esophagitis: Allergic and eosinophilic. *Pediatrics in Review*, *36*(8), 375–376. https://doi.org/10.1542/pir.36-8-375

Tillman, A. C., & Giorgi, M. (2023). Obesity, inflammation, and diseases of the gastrointestinal tract. In Mehrzad, R., (ed.), *Inflammation and Obesity* (pp. 101–118). Academic Press. https://doi.org/10.1016/B978-0-323-90960-0.00008-4

Vanoli, A., Parente, P., Fassan, M., Mastracci, L., & Grillo, F. (2023). Gut inflammation and tumorigenesis: Every site has a different tale to tell. *Internal and Emergency Medicine*, *18*(8), 2169–2179. https://doi.org/10.1007/s11739-023-03320-w

Vaz, G. R., Carrasco, M. C. F., Batista, M. M., Barros, P. A. B., Oliveira, M. D. C., Muccillo-Baisch, A. L., ... & Dora, C. L. (2022). Curcumin and quercetin-loaded lipid nanocarriers: Development of omega-3 mucoadhesive nanoemulsions for intranasal administration. *Nanomaterials*, *12*(7), 1073. https://doi.org/10.3390/nano12071073

Verma, P., Arora, A., Rana, K., Mehta, D., Kar, R., Verma, V., ... & Bajaj, A. (2022). Gemini lipid nanoparticle (GLNP)-mediated oral delivery of TNF-α siRNA mitigates gut inflammation via inhibiting the differentiation of CD4+ T cells. *Nanoscale*, *14*(39), 14717–14731. https://doi.org/10.1039/D1NR05644A

Verma, P., Arora, A., Rana, K., Mehta, D., Kar, R., Verma, V., Srikanth, C. V., Patil, V. S., & Bajaj, A. (2022). Gemini lipid nanoparticle (GLNP)-mediated oral delivery of TNF-α siRNA mitigates gut inflammation *via* inhibiting the differentiation of CD4 + T cells. *Nanoscale*, *14*(39), 14717–14731. https://doi.org/10.1039/D1NR05644A

Wang, J., Zhu, R., Sun, X., Zhu, Y., Liu, H., & Wang, S. (2014). Intracellular uptake of etoposide-loaded solid lipid nanoparticles induces an enhancing inhibitory effect on gastric cancer through mitochondria-mediated apoptosis pathway. *International Journal of Nanomedicine*, *9*, 3987–3998. https://doi.org/10.2147/IJN.S64103

Wang, L., Yu, M., & Yang, H. (2021). Recent progress in the diagnosis and precise nanocarrier-mediated therapy of inflammatory bowel disease. *Journal of Inflammation Research*, *14*, 1701–1716. https://doi.org/10.2147/JIR.S304101

Weerapol, Y., Manmuan, S., Chaothanaphat, N., Limmatvapirat, S., Sirirak, J., Tamdee, P., & Tubtimsri, S. (2022). New approach for preparing solid lipid nanoparticles with volatile oil-loaded quercetin using the phase-inversion temperature method. *Pharmaceutics*, *14*(10), 1984. https://doi.org/10.3390/pharmaceutics14101984

Wu, S., Don, T., Lin, C., & Mi, F. (2014). Delivery of berberine using chitosan/fucoidan-taurine conjugate nanoparticles for treatment of defective intestinal epithelial tight junction barrier. *Marine Drugs*, *12*, 5677–5697. https://doi.org/10.3390/md12115677

Xian, S., Zhu, J., Wang, Y., Song, H., & Wang, H. (2023). Oral liposomal delivery of an activatable budesonide prodrug reduces colitis in experimental mice. *Drug Delivery*, *30*(1). https://doi.org/10.1080/10717544.2023.2183821

Yang, C., Alpini, G., Glaser, S., & Merlin, D. (2022). Lipid nanoparticles for oral delivery of nucleic acids for treating inflammatory bowel disease. *Nanomedicine*, *17*(21), 1501–1504. https://doi.org/10.2217/nnm-2022-0222

Yang, C., Zhang, M., Lama, S., Wang, L., & Merlin, D. (2020). Natural-lipid nanoparticle-based therapeutic approach to deliver 6-shogaol and its metabolites M2 and M13 to the colon to treat ulcerative colitis. *Journal of Controlled Release*, *323*, 293–310. https://doi.org/10.1016/j.jconrel.2020.04.032

Yang, T. M., Miao, M., Yu, W. Q., Wang, X., Xia, F. J., Li, Y. J., & Guo, S. D. (2023). Targeting macrophages in atherosclerosis using nanocarriers loaded with liver X receptor agonists: A narrow review. *Frontiers in Molecular Biosciences*, *10*, 1147699. https://doi.org/10.3389/fmolb.2023.1147699

Yingchoncharoen, P., Kalinowski, D. S., & Richardson, D. R. (2016). Lipid-based drug delivery systems in cancer therapy: What is available and what is yet to come. *Pharmacological Reviews*, *68*(3), 701–787. https://doi.org/10.1124/pr.115.012070

Zeeshan, M., Ain, Q. U., Sunny, A., Raza, F., Mohsin, M., Khan, S., ... & Ali, H. (2023). QbD-based fabrication of transferrin-anchored nanocarriers for targeted drug delivery to macrophages and colon cells for mucosal inflammation healing. *Biomaterials Science*, *11*(4), 1373–1397. https://doi.org/10.1039/D2BM01719A

Zeeshan, M., Ali, H., Khan, S., Khan, S.A., & Weigmann, B. (2019). Advances in orally-delivered pH-sensitive nanocarrier systems; an optimistic approach for the treatment of inflammatory bowel disease. *International Journal of Pharmaceutics*, *558*, 201–214.

Zhang, C., Gu, C., Peng, F., Liu, W., Wan, J., Xu, H., Lam, C., & Yang, X. (2013). Preparation and optimization of triptolide-loaded solid lipid nanoparticles for oral delivery with reduced gastric irritation. *Molecules*, *18*, 13340–13356. https://doi.org/10.3390/molecules181113340

Zhang, H. (2023). *Thin-Film Hydration Followed by Extrusion Method for Liposome Preparation* (pp. 57–63). Humana. https://doi.org/10.1007/978-1-0716-2954-3_4

Zhang, L. I., & Zhang, L. (2010). Lipid–polymer hybrid nanoparticles: Synthesis, characterization and applications. *Nano Life*, *1*(1–2), 163–173. https://doi.org/10.1142/S179398441000016X

Zhang, X., Chen, G., Zhang, T., Ma, Z., & Wu, B. (2014). Effects of PEGylated lipid nanoparticles on the oral absorption of one BCS II drug: A mechanistic investigation. *International Journal of Nanomedicine*, *9*, 5503–5514. https://doi.org/10.2147/IJN.S73340

Zheng, L. F., Liu, S., & Zhu, J. X. (2023). Advances in research on the role of inflammation in gut functional disorders. *Frontiers in Physiology*, *14*, 1248333. https://doi.org/10.3389/fphys.2023.1248333

18 Exosomes Mediated Nanocarriers for Gastrointestinal Inflammation Management

Sakshi Soni, Aashi Jain, Priyanshu Nema, Vandana Soni, and Sushil K. Kashaw*
Department of Pharmaceutical Sciences, Dr. Harisingh Gour University, Sagar, Madhya Pradesh, India

18.1 INTRODUCTION

Cells communicate by releasing tiny membrane-bound carriers known as exosomes, which contain a rich cargo of diverse biomolecules such as DNA, proteins, and lipids. Within their lumen, exosomes predominantly contain major regulatory molecules including nucleic acids like DNA, messenger RNA (mRNA), and microRNA (miRNA), alongside proteins and lipids. Certain proteins, such as CD9, CD63, and CD81 from the tetraspanin family, are consistently found in exosomes across various cell types, often serving as marker proteins for their positive isolation [1–4]. These proteins play crucial roles in adhesion, interaction with other proteins of the MHC complex and integrins, and are essential for the formation and maintenance of large macromolecular complexes. Additionally, protein families with diverse members (e.g., ANNEXINS, RAB, SNARE), with endosomal origins, have been identified in exosomes from different cells, contributing to their membrane assembly, fusion, and trafficking. Exosomes, tiny packages carrying cellular messages, belong to a group of extracellular vesicles secreted by various cells. These packages typically range from 30 to 100 nanometers in size. EVs encompass diverse classes such as membrane particles, microvesicles, ectosomes, apoptotic bodies, and exosomes [5, 6]. These vesicles are classified based on their biogenesis pathway, size, their membrane lipid composition, the force needed to sediment them, and the proteins they express. Notably, exosomes are primarily generated through the endolysosomal pathway, distinguishing them from microvesicles which bud from cytoplasmic membrane. The activation of growth factor receptors on the cell surface sets off a chain reaction, leading to the active release of exosomes through exocytosis. This mechanism is vital for both cell communication and receptor removal. Exosomes possess the ability to release their cargo via direct membrane fusion, establishing a more intimate communication with recipient cells facilitating the exchange of transmembrane proteins and lipids. Alternatively, they can be internalized by recipient cells "grab" exosomes in different ways, like using tiny "hands" (pinocytosis), "vacuums" (phagocytosis), specific "locks and keys" (receptor-mediated endocytosis), and clathrin-mediated endocytosis. Once internalized, the intraluminal contents of exosomes deposit their contents within the cell's main compartment, modulating intercellular activities within both physiological and pathological frameworks [7]. Exosomes play a complex role in the immune system, acting as double-edged swords that can both activate and suppress immune responses.

While exosomes are severely augmented in phosphatidylserine, their membrane composition differs from macrovesicles, reflecting more similarity to their parent cells. Imagine exosomes as custom-built cars: the same chassis, but specific modifications create different functionalities.

Tiny messengers called exosomes start their journey inside special compartments called endosomes. These compartments mature and fill with smaller bubbles called intraluminal vesicles, kind of like packing boxes inside a truck. Eventually, the truck fuses with the cell's outer membrane, releasing the packing boxes (exosomes) to travel around the body. So, exosomes are dynamic membrane vesicles that play diverse roles in intercellular communication and modulation, with implications for various physiological and pathological processes [8]. Figure 18.1 shows the biogenesis of exosomes.

The properties and benefits of exosomes as nanocarriers are manifold. These minuscule extracellular vesicles shuttle proteins, lipids, and nucleic acids across cells, facilitating intercellular communication. Their potential as nanocarriers has sparked significant interest, particularly in fields like drug delivery, diagnostics, and regenerative medicine. Exosomes exhibit diverse characteristics and offer several advantages in this regard as shown in Figure 18.2.

Gastrointestinal inflammation, affecting organs such as the mouth, gastrointestinal tract, stomach, small intestine, and large intestine (colon), encompasses a range of inflammatory processes. In particular, the term inflammatory bowel disease (IBD) denotes a cluster of chronic autoimmune disorders affecting the gastrointestinal tract, notably Crohn's disease (CD) and ulcerative colitis (UC). These conditions involve immunological dysregulation and persistent inflammation, resulting in damage to the gastrointestinal tract. While the precise etiology of IBD remains elusive, factors such as immune response variations, environmental influences, genetic predisposition, and alterations in gut microbiota are believed

DOI: 10.1201/9781003493143-18

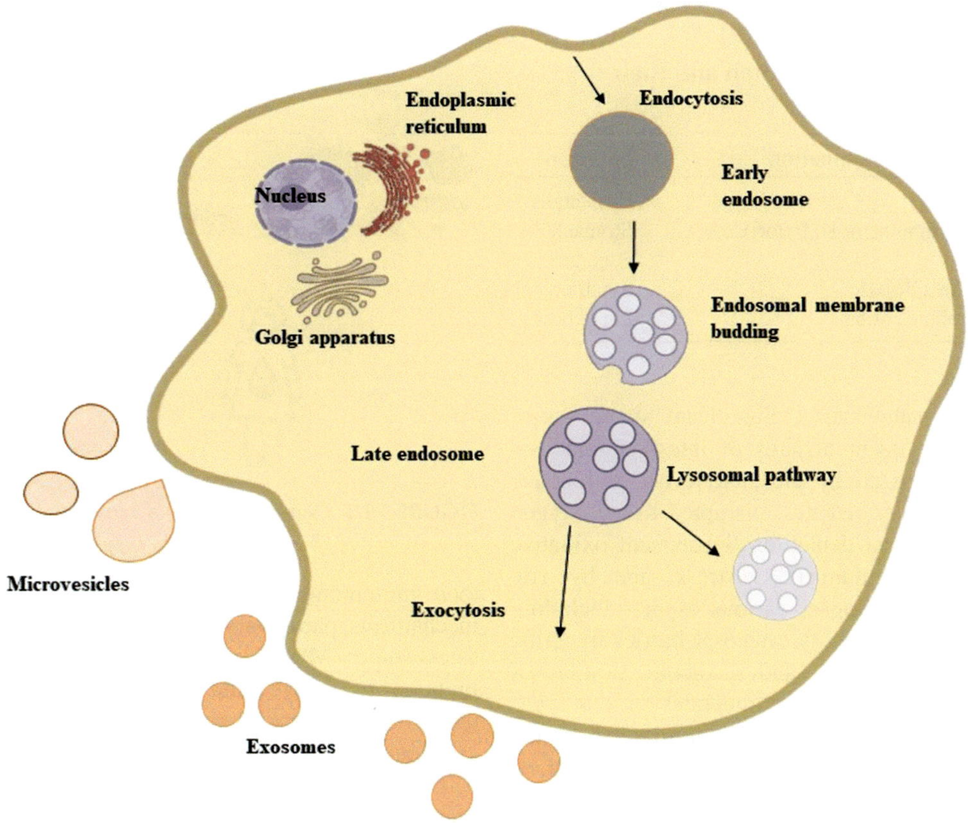

FIGURE 18.1 Exosomes biogenesis. During the process of endocytosis, early endosomes mature into late endosomes, which comprise a group of intraluminal vesicles known as multivesicular bodies (MVBs). MVBs have the ability to degrade when they fuse with lysosomes or release as exosomes when they interact with the plasma membrane.

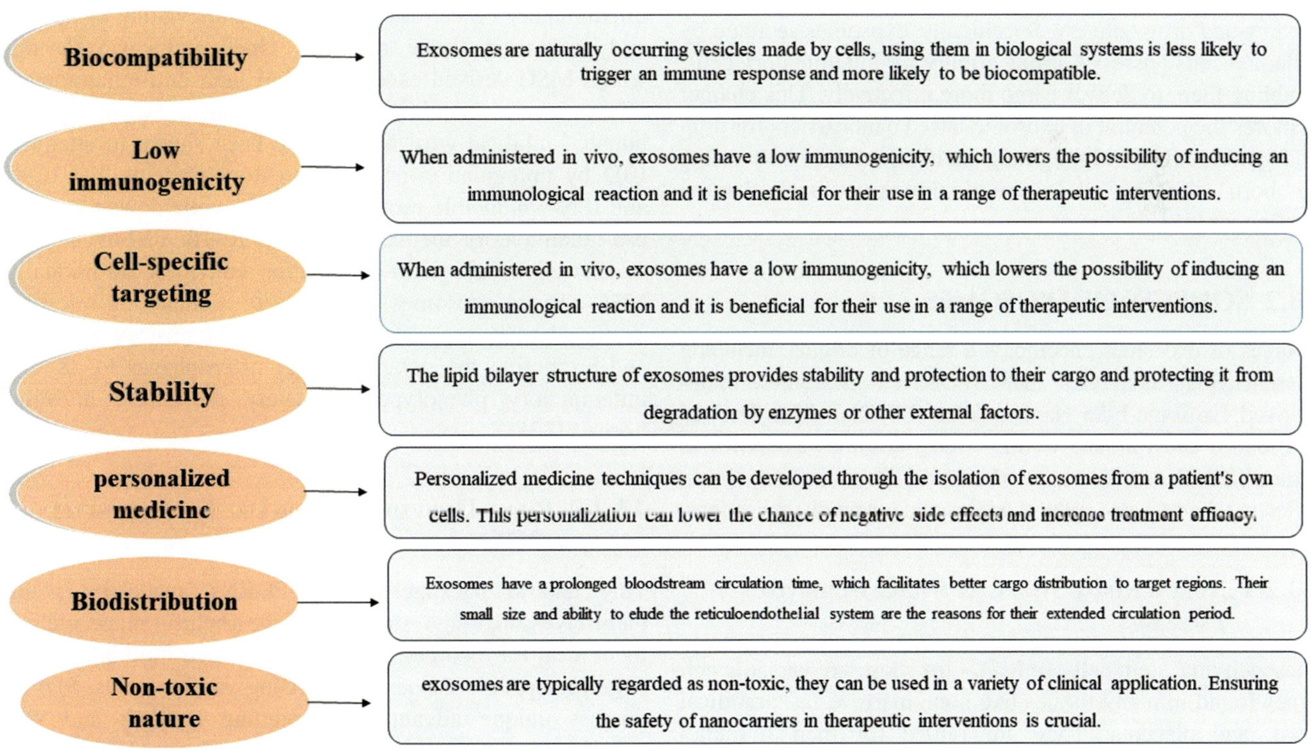

FIGURE 18.2 Various properties of exosomes.

TABLE 18.1
Type of Gastrointestinal Inflammation and Their Location [24]

Type of Gastrointestinal Inflammation	Location
Gastro-esophageal reflux	Esophagus
Chronic gastritis (autoimmune or H. Pylori)	Stomach
Chronic pancreatitis	Pancreas
Coeliac disease, Crohn's disease	Small bowel
Ulcerative colitis, Crohn's colitis	Large bowel

to be pivotal in its pathogenesis. Significant alterations in the microenvironment occur at sites of intestinal mucosal inflammation, characterized by elevated levels of reactive nitrogen and oxygen intermediates, disrupted tissue oxygen balance, and local nutrient depletion. Subsequent oxidative stress, endoplasmic reticulum (ER) stress, and hypoxic stress response signaling impact various aspects including immunomodulation, cell viability, mucosal barrier integrity, and the composition, diversity, and metabolic profiles of the gut microbiota [9]. Table 18.1 enumerates various types of gastrointestinal inflammation, each contributing to the complex landscape of gastrointestinal inflammatory disorders. Traditional treatments for inflammation often involve high doses of medications, but their long-term efficacy is limited by accumulating toxicities. Nanoparticle-based drugs offer a promising alternative, and exosomes, tiny lipid-bilayer enclosed vesicles secreted by cells, have emerged as versatile drug carriers. Exosomes have a natural ability to transport proteins, lipids, and genes between cells, making them ideal for targeted drug delivery. Specifically, exosomes secreted by inflamed cells possess a strong affinity for inflammatory cells, enabling them to deliver cargo more effectively. This chapter explores the potential of exosome-based nanocarriers for drug and gene delivery, highlighting their advantages in treating stubborn diseases while minimizing traditional therapy challenges [10].

18.2 SOURCES OF EXOSOMES

Sources of exosomes encompass a range of origins, including Mesenchymal Stem Cell (MSCs)-Derived Exosomes, Plant-Derived Exosome-Like Nanoparticles (PDENs), Milk-Derived Exosomes, and Immune Cell-Derived Exosomes as shown in Figure 18.3. Exosomes can also be found in various body fluids, reflecting their diverse cellular origins and potential applications.

18.2.1 Mesenchymal Stem Cell (MSCs)-Derived Exosomes

Mesenchymal stem cells, or MSCs for short, are versatile cell types found in many tissues like bone marrow, fat, umbilical cord, and placenta. They are valued for their immune-suppressive and anti-inflammatory properties, along with their ability to expand in vitro without ethical concerns. MSCs exert their immunomodulatory effects mainly through paracrine mechanisms, particularly via extracellular vesicles (EVs) such as exosomes. MSC-derived exosomes have shown promising immunosuppressive and anti-inflammatory properties in numerous in vitro and animal model studies, particularly in gastrointestinal diseases. These exosomes serve as effective carriers for therapeutic molecules like genes, drugs, and RNA, protecting them from degradation and facilitating their uptake by target cells through endocytosis [11, 12].

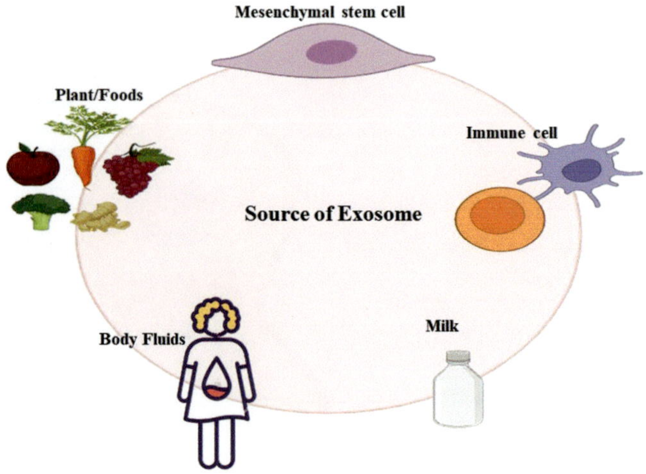

FIGURE 18.3 Various sources of exosomes.

Stem cell-derived exosomes, including those from MSCs, play crucial roles in self-renewal, immunomodulation, expansion, and tissue repair. They have demonstrated therapeutic benefits in various diseases, including inflammatory bowel disease (IBD), by alleviating symptoms and promoting tissue recovery. One mechanism through which MSC-derived exosomes exert their anti-inflammatory effects is via ubiquitination. For instance, exosomes from human umbilical cord MSCs have been shown to attenuate IBD by upregulating anti-inflammatory proteins like IL-10 and IFN-γ inducible protein-10 (IP10) while downregulating pro-inflammatory mediators such as IL-1β, TNF-α, and IL-6, as well as ubiquitin-conjugating enzymes. Additionally, MSC-derived exosomes contribute to maintaining intestinal mucosal barrier integrity, suppressing inflammatory responses, and promoting the polarization of macrophages to an anti-inflammatory phenotype, ultimately preventing intestinal fibrosis [13].

18.2.2 Plant-Derived Exosome-Like Nanoparticles (PDENs)

Tiny natural packages called PDENs, made by plants themselves, are tough, stable, and have a helpful shape, making them ideal for incorporating drugs into their lipid bilayer and targeting specific tissues. Comprising various lipids, PDENs possess unique advantages, including strength and site-specific targeting, making them attractive for nanotechnology-based drug development [14]. PDENs, commonly released by consumable vegetation such as vine [15], citrus [16], zingiber

[17], citron [18], Brassica oleracea [19], Daucus carota [20], Cocos nucifera [21], and Malus domestica [22], capitalize on beneficial healing properties derived from the innate biochemicals found within these botanical sources. Exosome-like nanoparticles derived from zingiber (GDENs) have demonstrated efficacy in suppressing cancerous cell growth and ameliorating inflammatory bowel conditions [23].

The benefits of PDENs as biomolecules include their natural origin, facilitating intercellular communication, and their phospholipid-rich characteristics, protecting cargo from external agents. These tiny messengers have a built-in ability to enter cells, even bypassing tough roadblocks like the blood-brain barrier and the placenta. Additionally, PDENs are immune-tolerated, scalable for industrial applications, non-toxic, and derived from organic fruits and vegetables [24].

PDENs harbor diverse bioactive molecules essential for plant communication and inter-kingdom signaling, demonstrating promising therapeutic potential in areas such as cancer suppression, modulation of inflammatory responses, and regulation of intestinal dysbiosis. PDENs exhibit favorable safety profiles characterized by minimal toxicity, low immunogenicity, and exceptional biocompatibility, have positioned PDENs as promising candidates for applications in treating conditions like IBD [25].

18.2.3 Milk-Derived Exosomes

Mammalian milk, a complex fluid rich in biological compounds like proteins, antibodies, and peptides, provides crucial protection against diseases such as diabetes, inflammatory bowel disease, and obesity. Within mammalian milk, including milk obtained from various animals or humans, exists a bioactive extracellular vesicle called milk-derived exosomes (MDEs). However, characterizing MDEs proves challenging due to the lack of standardized methods for milk processing, storage, and exosome isolation. Despite this, MDEs offer numerous advantages over other delivery vectors, including higher yield, additional therapeutic benefits, and oral delivery capabilities. They exhibit high biocompatibility and remain intact after gastrointestinal absorption, suggesting excellent stability and rendering them suitable as drug carriers [26].

Studies have highlighted the potential of MDEs in treating various intestinal diseases, such as inflammatory bowel disease, necrotizing enterocolitis, colorectal cancer, and intestinal ischemia-reperfusion injury. MDEs regulate intestinal immune homeostasis, restore gut microbiota composition, improve intestinal structure and integrity, and alleviate conditions such as oxidative stress, cell apoptosis, and inflammation. These properties, combined with their nanoparticle nature, make MDEs promising tools for medical and research applications across different mammalian species, including rat [27], horse [28], buffalo [29], donkey [30], goat [31], sheep [28], bovine [32], human, and camel milk [33]. Breast milk-derived exosomes exert multifaceted effects on the neonatal intestinal system, including promoting development, modulating inflammation, and maintaining epithelial integrity. They facilitate the delivery of proteins and nucleic acids, thereby protecting cargo from degradation and enhancing intestinal structure and absorptive function. Porcine milk-derived exosomal miRNAs alleviate intestinal mucosal damage induced by deoxynivalenol (DON) and inhibit inflammation and apoptosis in response to lipopolysaccharide (LPS). Furthermore, exosomes extracted from yak milk stimulate the growth and endurance of intestinal epithelial cells (IECs) in low-oxygen environments, whereas exosomes sourced from goat milk function as organic sensors for identifying inflammatory responses. Exosomes derived from rat milk augment the activation of genes associated with cellular growth and amplify the functionality of IECs [34, 35].

18.2.4 Immune Cell-Derived Exosomes

Exosomes originating from immune cells have attracted considerable research interest and have been utilized in studies related to vaccine development and drug delivery, resulting in noteworthy therapeutic advancements. Primary immune cells like macrophages, monocytes, and dendritic cells (DCs) are key contributors of these exosomes in therapeutic research. Immune cell-derived exosomes have demonstrated the ability to evade immune surveillance and prolong circulation, enhancing their therapeutic impact. Particularly, exosomes from macrophages and monocytes offer a unique opportunity to evade mononuclear phagocyte control, thereby improving drug delivery to target sites and enhancing therapeutic efficacy [36].

Macrophages can polarize into two main phenotypes: M1 macrophages, associated with pro-inflammatory responses, and M2 macrophages, linked to anti-inflammatory actions. Extracellular vesicles originating from various subtypes of M2 macrophages, such as M2a, M2b, and M2c, have been investigated in mouse models of experimental colitis. Notably, exosomes from M2b macrophages significantly alleviate colitis severity by upregulating regulatory T cells (Treg) and interleukin-4 (IL-4), these vesicles decrease the expression of cytokines linked to colitis (IL-6, IL-1β, and IL-17A) via the C-C motif chemokine ligand 1/8 (CCL1/CCR8) pathway. These findings underscore the therapeutic potential of immune cell-derived exosomes, particularly those from M2 macrophages, in modulating inflammatory responses and treating colitis [37].

18.2.5 Exosomes in Body Fluids

Our bodies are full of microscopic communication channels! Various cell types like red blood cells, immune cells, blood vessel cells, fat cells, and skin cells release tiny packages called extracellular vesicles (EVs). These EVs, which include exosomes, travel through fluids like blood, saliva, milk, and tears, delivering messages and performing important tasks such as immune response modulation, antigen presentation, intercellular communication, and non-classical protein secretion [38]. Exosomes are usually extracted from various bodily fluids employing ultracentrifugation, relying on the principle of sedimentation to produce a highly refined exosomal fraction deemed the benchmark. Alternatively, size exclusion filtration or chromatography entails passing through pores with diameters less than 100 nm, followed by centrifugation to concentrate exosomes, thus maintaining their biological functionality and integrity. [39].

18.3 EXOSOMAL COMPONENTS AS BIOMARKERS FOR GASTROINTESTINAL INFLAMMATION

Extracellular vesicles and their constituents have been implicated in the pathogenesis of various gastrointestinal disorders, including inflammatory bowel disease (IBD) (see Table 18.2). Ongoing research in this domain continues to yield encouraging findings, with certain investigations pinpointing extracellular vesicles from the placenta as early indicators of preeclampsia [40]. Additionally, specific microRNAs within exosomes and circulating extracellular vesicles have emerged as potential biomarkers for gastrointestinal malignancies, while the mRNA content of these vesicles suggests their utility as fecal biomarkers reflecting mucosal inflammation [41, 42]. Although several biomarkers have been identified as relevant to IBD, none have yet been recognized as singular, highly specific, and sensitive markers for its diagnosis or for distinguishing between ulcerative colitis (UC) and Crohn's disease (CD) patients. Some of the more promising biomarkers include antibodies against Saccharomyces cerevisiae (ASCA), perinuclear anti-neutrophil cytoplasmic antibodies (pANCA), C-reactive protein (CRP), lactoferrin, and calprotectin [43]. Salivary exosomal proteasome subunit alpha type 7 (PSMA7) has also exhibited potential as a biomarker for inflammatory bowel disease (IBD), demonstrating higher expression levels in patients with Crohn's disease (CD) compared to those with ulcerative colitis (UC) [44]. Furthermore, Leoni et al. (2015) observe elevated levels of endogenous extracellular vesicles (EVs) containing ANXA1 in IBD patients compared to healthy individuals. Since ANXA1-containing EVs increase in circulation in response to inflammation, they could potentially serve as a marker for intestinal mucosal inflammation [45]. Identifying a biomarker that is highly specific and sensitive for IBD could not only facilitate early and accurate diagnosis but also enhance treatment strategies, prognosis, and potentially uncover biological pathways for the development of novel therapies [46].

18.4 ROLE OF EXOSOMES IN INFLAMMATORY RESPONSES

Exosomes, tiny vesicles released by cells, emerge as fascinating players in the complex dance of inflammation. These intricate messengers are believed to possess immune-modulating properties, carrying information and molecules that influence the inflammatory pathway. Within the liver, a key battleground for inflammatory processes, resident immune cells like Kupffer cells, NK cells, T cells, and B cells orchestrate the response as shown in Figure 18.4. Exosomes released from these cells, brimming with specific proteins like alix, Tsg101, and annexin, reflect their origin and even differ based on the cell's location (e.g., basolateral vs. apical surfaces of epithelial cells) [47]. Their impact on the immune system is diverse. Antigen-presenting cell (APC)-derived exosomes, equipped with MHC-I/II and costimulatory molecules, can stimulate T cell proliferation, fueling the immune response [48]. Similarly, exosomes from B cells carry BCR, while those from T cells showcase TCR and costimulatory molecules. This personalized packaging allows targeted communication between cells. Beyond these basic building blocks, exosomes can harbor disease-specific cargo. Cancer cell-derived exosomes might express MHC class II-peptide complexes, while platelet-derived exosomes carry coagulation factors. This unique content underscores the potential of exosomes as biomarkers for various diseases.

TABLE 18.2
Exosome Components and their Effects in Studies of Inflammatory Bowel Disease (IBD) As Biomarker

Source	Exosome Component	Effects/Potential	Ref.
Bone marrow-derived MSCs	Metallothionein-2	Maintenance of intestinal barrier integrity Polarization of M2b macrophages Induction of IL-10 from macrophages	72
IECs	Annexin A1	Activates mucosal wound repair Potential biomarker of intestinal mucosal inflammation	45
Hookworm	81 proteins including tetraspanin 14–3-3 protein, enolase and heat shock proteins, together with 52 miRNA species	Protection against colitis by significantly suppressing IFNγ, IL-6, IL-1β, and IL-17a and upregulating anti-inflammatory cytokine IL-10	73, 74
Ginger	mdo-miR7267-3	Shapes gut microbiota	75
Porcine milk	miR-4334, miR-219 and miR-338	Prevents LPS-induced intestinal inflammation, apoptosis and damage via inhibiting TLR4/NF-κB and p53 pathways	76
Grapes	GELNs lipids	Induces Lgr5+ stem cells Enhances in vivo targeting of intestinal stem cells Remodels and protects intestinal tissue against DSS-induced colitis	77
Intestinal epithelial cell line Caco-2	miR-34c and PlncRNA1	Modulates ZO-1, MAZ and occluding expression Modulates intestinal barrier integrity	78

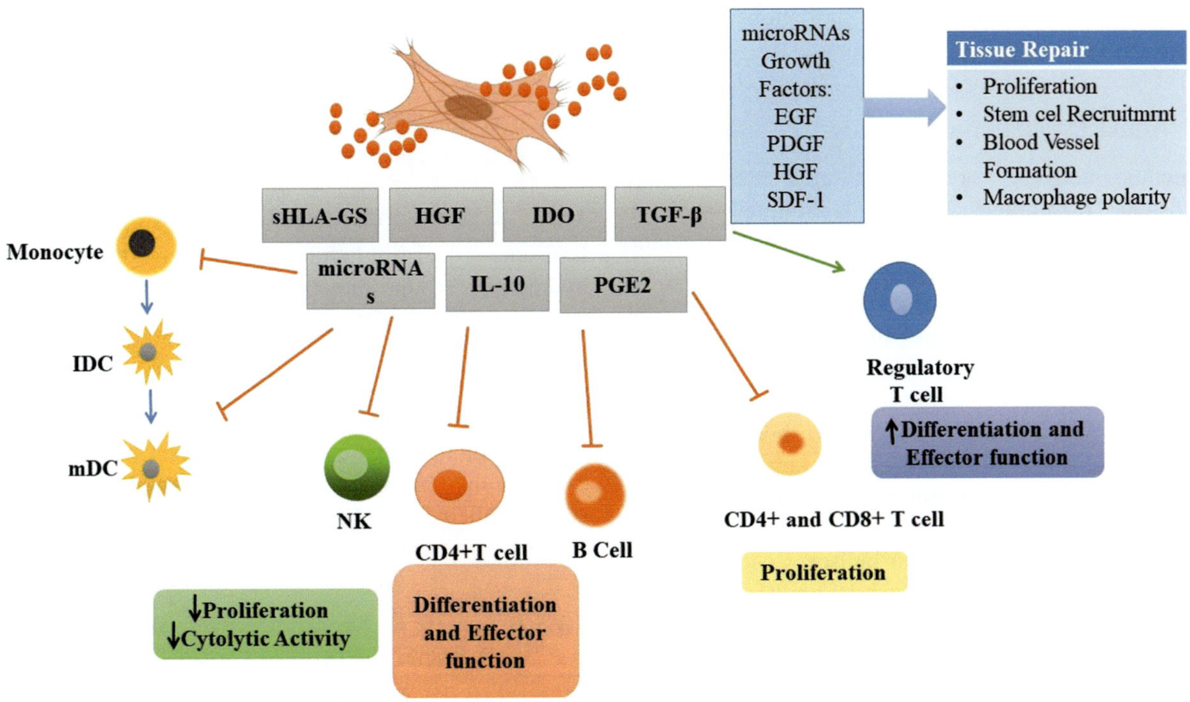

FIGURE 18.4 The stimulatory and inhibitory impact of immune cell-derived exosomes.

Note: Abbreviations: CD: Crohn disease; DC: dendritic cell; EGF: epidermal growth factor; FGF: fibroblast growth factor; IDO: indoleamine 2,3 dioxygenase; IGF: insulin-like growth factor; IL-10: interleukin-10; NK: natural killer; PDGF: platelet-derived growth factor; PGE2: prostaglandin E2; VEGF: vascular endothelial growth factor; TGF-β: transforming growth factor-β; Shla-g5: soluble human leukocyte antigen-G5.

But how do these tiny messengers exert their influence? It seems a complex dance of receptors and adhesion molecules plays a crucial role. Exosomes interact with various receptors on recipient cells, including MHC molecules, costimulatory molecules, cytokine receptors, and tetraspanin-associated proteins. This intricate interaction facilitates binding and initiates immunomodulatory processes. One intriguing player in this dance is galectin-5, a molecule found on exosomes that modulates cell adhesion and triggers intracellular signals [49]. This highlights the intricate mechanisms employed by exosomes to communicate with their targets. While the exact mode of action after attachment remains a mystery, the delivery of antigens and MHC complexes by exosomes suggests their role in shaping the immune response. Studies have shown that exosomes from tumor cells can activate CD8+ T lymphocytes, but only in the presence of recipient DCs expressing the correct MHC haplotype. This hints at the complex interplay between exosomes and the existing immune landscape. Unraveling the secrets of exosomes and their influence on inflammation promises exciting breakthroughs in disease diagnosis and treatment. Their diverse functionalities and targeted communication capabilities hold immense potential for therapeutic interventions. Research continues to shed light on the intricate dance of exosomes in the inflammatory response, paving the way for a deeper understanding of their role in health and disease [50, 51].

18.5 EXOSOME-MEDIATED NANOCARRIERS FOR TARGETED DELIVERY

Exosomes, tiny extracellular vesicles brimming with information, are emerging as revolutionary nanocarriers for targeted drug delivery. Their inherent ability to transfer molecules between cells and low immunogenicity make them ideal candidates for overcoming the limitations of traditional drug delivery methods. Exosome engineering offers exciting possibilities to improve safety, effectiveness, and specificity. Natural exosomes are increasingly employed in immunomodulatory, regenerative, and anticancer medicine. However, recent advancements like tailored loading schemes, membrane modifications, and even hybridization with artificial nanoparticles further enhance their efficacy. These tailored techniques demonstrate incredible potential for improved drug utilization, precise targeting, and controlled release. Two main methods exist for loading therapeutic compounds into exosomes: passive and active encapsulation [52].

18.5.1 Passive Encapsulation

This method involves incubating drugs with isolated exosomes or the cells that produce them. Drugs passively diffuse into the exosomes, resulting in loaded vesicles released by the cells. Studies using this approach have successfully loaded Paclitaxel and Curcumin into exosomes, demonstrating improved drug effectiveness and target specificity. However, this method has limitations in loading efficiency and stability.

18.5.2 Active Encapsulation

This method actively forces drug molecules through the exosome membrane. Various techniques such as sonication, electroporation, and the use of surfactants achieve this without compromising the exosome's structure or function. Additionally, transfection, freezing-thawing, and membrane fusion techniques offer alternative avenues for loading various cargoes, including nucleic acids and nanoparticles [77].

Exosomes, nature's tiny messengers, boast immense potential for delivering drugs directly to target cells. With their inherent ability to interact with cellular membranes and trigger specific pathways, they can be equipped with "targeting ligands"—like molecular bullseyes—to zero in on diseased cells. This targeted approach minimizes side effects and maximizes therapeutic impact.

18.5.3 Genetic Modification

This involves engineering exosomes to express specific binding molecules on their surface, such as peptides, receptors, or even antibodies. These can be attached to the exosome's natural proteins or incorporated into its membrane. Popular targets include: Lamp2b and tetraspanins. These naturally occurring exosomal proteins can be modified to include targeting sequences. Specific receptor proteins: Some exosomes already express receptors like HER2 and EGFR, which can be leveraged for targeted delivery [78].

18.5.4 Chemical Conjugation

Ligands are directly attached to the exosome surface using chemical reactions. This offers flexibility for attaching various molecules, including antibodies. While full-sized antibodies have limitations, smaller fragments like scFvs are being explored for their targeting efficiency [53]. Targeting peptides: These smaller molecules offer advantages like reduced size and immunogenicity, allowing them to target tumor-associated receptors [54]. Aptamers: These artificial DNA or RNA molecules bind specifically to target molecules with high affinity and can be easily modified for targeted delivery [55].

18.5.5 Directly Targeting the Exosome Envelope

18.5.5.1 Chemical and Mechanical Techniques

Modifying the exosome surface with proteins, peptides, or other molecules using techniques like click chemistry can enhance their targeting ability. For example, neuropilin-1 targeting peptide attached to exosomes improved their penetration of the blood-brain barrier, demonstrating the potential for glioma treatment [56].

18.5.5.2 Non-Covalent Modifications

Techniques like electrostatic interactions can also be employed. Cationized pullulan used to modify exosomes led to their accumulation in liver tissue, suggesting its potential for liver disease treatment [57].

18.5.5.3 Hydrophobic Insertion

Exosomes' dual lipid structure allows for the integration of hydrophobic drugs directly into their membranes, offering another avenue for drug loading [58].

18.5.5.4 Targeted Ligand-Modified PEG Phospholipids

This technique involves inserting these molecules into the exosome membrane for targeted delivery. For instance, folic acid-modified PEG-cholesterol enabled precise targeting of inflamed synovial tissue in a model of arthritis [59].

18.5.6 Indirectly Modifying Exosomes via Parental Cells

18.5.6.1 Genetic Engineering

Transfecting parental cells with genes for specific receptor proteins allows them to be expressed on the exosome surface, enhancing their affinity for target cells. This strategy was successfully employed to create exosomes with improved uptake by injured heart cells in a model of myocardial infarction [60].

18.5.7 Hybrid Membrane Engineering

18.5.7.1 Exosome-Liposome Hybridization

Fusing exosomes with liposomes offers several advantages, including the ability to modify the exosome surface with various ligands or polyethylene glycol for enhanced targeting and stability. This approach also holds promise for delivering CRISPR-Cas9 technology for targeted gene editing [61, 62].

18.6 ADVANTAGES OF EXOSOME-MEDIATED DRUG DELIVERY

Exosomes, tiny cellular messengers, offer a revolutionary platform for targeted drug delivery thanks to their unique properties:

18.6.1 Biocompatibility and Stealth

Imagine microscopic "Trojan horses" delivering drugs directly to diseased cells. Exosomes, naturally produced by our own cells, offer this precise advantage. Their biocompatibility minimizes immune reactions, allowing them to navigate the body undetected and deliver their cargo without raising any red flags. This translates to potentially fewer side effects compared to traditional drug delivery methods.

18.6.2 Enhanced Cellular Uptake and Penetration

Unlike synthetic nanoparticles that struggle to gain entry, exosomes naturally interact with recipient cells. This facilitates efficient drug uptake and penetration into target tissues, ensuring the medication reaches the intended destination, not healthy bystander cells. This targeted delivery maximizes therapeutic impact while minimizing off-target effects.

18.6.3 Controlled Release and Targeting

Imagine equipping these "Trojan horses" with specific homing signals. By engineering exosomes with targeting ligands or surface modifications, researchers can direct them to specific cell types or tissues. This controlled release ensures drugs reach the exact location needed, minimizing leakage and optimizing treatment efficacy.

By delivering high drug concentrations directly to target cells, exosomes minimize the overall amount of drug needed, leading to potentially fewer side effects. Exosomes can encapsulate various therapeutic cargo, including drugs, genetic material, and imaging agents, allowing for diverse therapeutic applications. Their natural origin minimizes immune responses and ensures safe degradation within the body after delivering their payload.

18.7 THERAPEUTIC APPLICATIONS OF EXOSOMES IN GASTROINTESTINAL INFLAMMATION

Exosomes, small extracellular vesicles secreted by various cell types, have gained significant attention in recent years due to their therapeutic potential in various diseases, including gastrointestinal (GI) inflammation. Here are some therapeutic applications of exosomes in GI inflammation:

18.7.1 Anti-inflammatory and Immunomodulatory Effects

Exosomes possess anti-inflammatory properties by modulating immune responses. They can suppress the activation of immune cells such as T cells and macrophages, thereby reducing inflammation in the GI tract [63].

18.7.2 Suppressing Pro-inflammatory Cytokines

Exosomes can carry specific cargo, including microRNAs and proteins, that can inhibit the expression of pro-inflammatory cytokines such as TNF-α, IL-1β, and IL-6. By downregulating these cytokines, exosomes help in dampening the inflammatory response in GI conditions [64].

18.7.3 Promoting Tissue Repair and Regeneration

Exosomes contain factors that promote tissue repair and regeneration. They can stimulate epithelial cell proliferation and migration, enhance angiogenesis, and facilitate the repair of damaged tissue in the GI tract [65].

18.7.4 Exosome-Based Therapies for Specific GI Conditions

Ulcerative Colitis: Exosome-based therapies for ulcerative colitis aim to reduce inflammation, promote mucosal healing, and restore intestinal barrier function. Exosomes derived from mesenchymal stem cells (MSCs) have shown promise in preclinical studies for their immunomodulatory effects [66].

Crohn's Disease: Exosomes may mitigate inflammation in Crohn's disease by modulating immune responses and promoting tissue repair. MSC-derived exosomes have been investigated for their potential to alleviate inflammation and promote intestinal healing in Crohn's disease models [67].

Irritable Bowel Syndrome (IBS): While research on exosome-based therapies specifically for IBS is limited, the anti-inflammatory and tissue repair properties of exosomes could potentially benefit patients with IBS by alleviating symptoms associated with intestinal inflammation [68].

Exosomes have the potential to modulate the gut microbiome and its interactions with the host immune system. They can deliver bioactive molecules to gut microbiota, influencing their composition and activity. This modulation of the gut microbiome by exosomes holds promise for the treatment of various GI disorders, including inflammation [69].

18.8 CONCLUSION

In conclusion, exosomes represent a fascinating avenue in the realm of gastrointestinal inflammation therapy, offering a multifaceted approach to address the complexities of inflammatory conditions affecting the GI tract. These small extracellular vesicles possess remarkable anti-inflammatory and immunomodulatory effects, effectively suppressing pro-inflammatory cytokines while promoting tissue repair and regeneration. Exosome-based therapies show promise across a spectrum of GI disorders, including ulcerative colitis, Crohn's disease, and potentially irritable bowel syndrome. By harnessing the therapeutic potential of exosomes derived from various sources such as mesenchymal stem cells, plants, milk, and immune cells, researchers are exploring innovative strategies for targeted drug delivery and modulation of the gut microbiome. The versatility and biocompatibility of exosomes as nanocarriers offer distinct advantages over traditional drug delivery methods, paving the way for precise and effective treatment approaches with minimized side effects. Through active encapsulation techniques and surface modifications, exosomes can be tailored to target specific cells or tissues, enhancing therapeutic efficacy while minimizing off-target effects. Overall, the therapeutic applications of exosomes in gastrointestinal inflammation hold immense promise, presenting novel opportunities for the development of more effective and personalized treatment modalities. Further research and clinical trials are warranted to fully elucidate the therapeutic potential of exosomes and translate these findings into tangible benefits for patients suffering from GI inflammatory disorders.

REFERENCES

1. Adamcik, J., Klinov, D. V., Witz, G., Sekatskii, S. K., & Dietler, G. (2006). Observation of single-stranded DNA on mica and highly oriented pyrolytic graphite by atomic force microscopy. *FEBS Letters*, 580(24), 5671–5675. https://doi.org/10.1016/j.febslet.2006.09.017
2. Valadi, H., Ekström, K., Bossios, A., Sjöstrand, M., Lee, J. J., & Lötvall, J. O. (2007). Exosome-mediated transfer of mRNAs and microRNAs is a novel mechanism of genetic exchange between cells. *Nature Cell Biology*, 9(6), 654–659. https://doi.org/10.1038/ncb1596
3. Colombo, M., Raposo, G., & Théry, C. (2014). Biogenesis, secretion, and intercellular interactions of exosomes and other extracellular vesicles. *Annual Review of Cell and Developmental Biology*, 30, 255–289. https://doi.org/10.1146/annurev-cellbio-101512-122326
4. Thakur, B. K., Zhang, H., Becker, A., Matei, I., Huang, Y., Costa-Silva, B., & Lyden, D. (2014). Double-stranded DNA in exosomes: A novel biomarker in cancer detection. *Cell Research*, 24(6), 766–769. https://doi.org/10.1038/cr.2014.44
5. Théry, C., Zitvogel, L., & Amigorena, S. (2002). Exosomes: Composition, biogenesis and function. *Nature Reviews Immunology*, 2(8), 569–579. https://doi.org/10.1038/nri855
6. Zhang, Y., Bi, J., Huang, J., Tang, Y., Du, S., & Li, P. (2020). Exosome: A review of its classification, isolation techniques, storage, diagnostic and targeted therapy applications. *International Journal of Nanomedicine*, 15, 6917–6934. https://doi.org/10.2147/IJN.S264498
7. Tan, L., Wu, H., Liu, Y., Zhao, M., Li, D., & Lu, Q. (2016). Recent advances of exosomes in immune modulation and autoimmune diseases. *Autoimmunity*, 49(6), 357–365. https://doi.org/10.1080/08916934.2016.1191477
8. Oberemko, A. V., & Popandopulo, A. G. (2014). Extracellular vesicles: Classification, functions and clinical relevance. *Biotechnologia Acta*, 7(6), 102–108. https://doi.org/10.1124/pr.112.005983
9. Ocansey, D. K. W., Wang, L., Wang, J., Yan, Y., Qian, H., Zhang, X., & Mao, F. (2019). Mesenchymal stem cell–gut microbiota interaction in the repair of inflammatory bowel disease: An enhanced therapeutic effect. *Clinical and Translational Medicine*, 8, 1–17. https://doi.org/10.1186/s40169-019-0251-8
10. Zhang, M., Zang, X., Wang, M., Li, Z., Qiao, M., Hu, H., & Chen, D. (2019). Exosome-based nanocarriers as bio-inspired and versatile vehicles for drug delivery: Recent advances and challenges. *Journal of Materials Chemistry B*, 7(15), 2421–2433. https://doi.org/10.1039/C9TB00170K
11. Pittenger, M. F., Mackay, A. M., Beck, S. C., Jaiswal, R. K., Douglas, R., Mosca, J. D., & Marshak, D. R. (1999). Multilineage potential of adult human mesenchymal stem cells. *Science*, 284(5411), 143–147. https://doi.org/10.1126/science.284.5411.143
12. Bagno, L., Hatzistergos, K. E., Balkan, W., & Hare, J. M. (2018). Mesenchymal stem cell-based therapy for cardiovascular disease: Progress and challenges. *Molecular Therapy*, 26(7), 1610–1623. https://doi.org/10.1016/j.ymthe.2018.05.009
13. Li, T., Yan, Y., Wang, B., Qian, H., Zhang, X., Shen, L., & Xu, W. (2013). Exosomes derived from human umbilical cord mesenchymal stem cells alleviate liver fibrosis. *Stem Cells and Development*, 22(6), 845–854. https://doi.org/10.1089/scd.2012.0395
14. Teng, Y., Ren, Y. I., Sayed, M., Hu, X., Lei, C., Kumar, A., & Zhang, H. G. (2018). Plant-derived exosomal microRNAs shape the gut microbiota. *Cell Host & Microbe*, 24(5), 637–652. https://doi.org/10.1016/j.chom.2018.10.001
15. Ju, S., Mu, J., Dokland, T., Zhuang, X., Wang, Q., Jiang, H., & Zhang, H. G. (2013). Grape exosome-like nanoparticles induce intestinal stem cells and protect mice from DSS-induced colitis. *Molecular Therapy*, 21(7), 1345–1357. https://doi.org/10.1038/mt.2013.64
16. Wang, B., Zhuang, X., Deng, Z. B., Jiang, H., Mu, J., Wang, Q., & Zhang, H. G. (2014). Targeted drug delivery to intestinal macrophages by bioactive nanovesicles released from grapefruit. *Molecular Therapy*, 22(3), 522–534. https://doi.org/10.1038/mt.2013.190
17. Brahmbhatt, M., Gundala, S. R., Asif, G., Shamsi, S. A., & Aneja, R. (2013). Ginger phytochemicals exhibit synergy to inhibit prostate cancer cell proliferation. *Nutrition and Cancer*, 65(2), 263–272. https://doi.org/10.1080/01635581.2013.749925
18. Raimondo, S., Naselli, F., Fontana, S., Monteleone, F., Dico, A. L., Saieva, L., & Alessandro, R. (2015). Citrus limon-derived nanovesicles inhibit cancer cell proliferation and suppress CML xenograft growth by inducing TRAIL-mediated cell death. *Oncotarget*, 6(23), 19514. https://doi.org/10.18632%2Foncotarget.4004
19. Deng, Z., Rong, Y., Teng, Y., Mu, J., Zhuang, X., Tseng, M., & Zhang, H. G. (2017). Broccoli-derived nanoparticle inhibits mouse colitis by activating dendritic cell AMP-activated protein kinase. *Molecular Therapy*, 25(7), 1641–1654. https://doi.org/10.1016/j.ymthe.2017.01.025
20. Mu, J., Zhuang, X., Wang, Q., Jiang, H., Deng, Z. B., Wang, B., & Zhang, H. G. (2014). Interspecies communication between plant and mouse gut host cells through edible plant derived exosome-like nanoparticles. *Molecular Nutrition & Food Research*, 58(7), 1561–1573. https://doi.org/10.1002/mnfr.201300729
21. Zhao, Z. (2018). Isolation of exosome-like nanoparticles and analysis of microRNAs derived from coconut water based on small RNA high-throughput sequencing. *Journal of Agricultural and Food Chemistry*, 66(11), 2749–2757. https://doi.org/10.1021/acs.jafc.7b05614
22. Yu, S., Zhao, Z., Xu, X., Li, M., & Li, P. (2019). Characterization of three different types of extracellular vesicles and their impact on bacterial growth. *Food Chemistry*, 272, 372–378. https://doi.org/10.1016/j.foodchem.2018.08.059
23. Fujita, D., Arai, T., Komori, H., Shirasaki, Y., Wakayama, T., Nakanishi, T., & Tamai, I. (2018). Apple-derived nanoparticles modulate expression of

organic-anion-transporting polypeptide (OATP) 2B1 in Caco-2 cells. *Molecular Pharmaceutics*, 15(12), 5772–5780. https://doi.org/10.1021/acs.molpharmaceut.8b00921

24. Logozzi, M., Di Raimo, R., Mizzoni, D., & Fais, S. (2022). The potentiality of plant-derived nanovesicles in human health—A comparison with human exosomes and artificial nanoparticles. *International Journal of Molecular Sciences*, 23(9), 4919. https://doi.org/10.3390/ijms23094919
25. Cong, M., Tan, S., Li, S., Gao, L., Huang, L., Zhang, H. G., & Qiao, H. (2022). Technology insight: Plant-derived vesicles—How far from the clinical biotherapeutics and therapeutic drug carriers?. *Advanced Drug Delivery Reviews*, 182, 114108. https://doi.org/10.1016/j.addr.2021.114108
26. Munagala, R., Aqil, F., Jeyabalan, J., & Gupta, R. C. (2016). Bovine milk-derived exosomes for drug delivery. *Cancer Letters*, 371(1), 48–61. https://doi.org/10.1016/j.canlet.2015.10.020
27. Alvarez, M. L., Khosroheidari, M., Ravi, R. K., & DiStefano, J. K. (2012). Comparison of protein, microRNA, and mRNA yields using different methods of urinary exosome isolation for the discovery of kidney disease biomarkers. *Kidney International*, 82(9), 1024–1032. https://doi.org/10.1038/ki.2012.256
28. Le Doare, K., Holder, B., Bassett, A., & Pannaraj, P. S. (2018). Mother's milk: A purposeful contribution to the development of the infant microbiota and immunity. *Frontiers in Immunology*, 9, 325188. https://doi.org/10.3389/fimmu.2018.00361
29. Victora, C. G., Bahl, R., Barros, A. J., França, G. V., Horton, S., Krasevec, J., & Rollins, N. C. (2016). Breastfeeding in the 21st century: Epidemiology, mechanisms, and lifelong effect. *The Lancet*, 387(10017), 475–490. https://doi.org/10.1016/S0140-6736(15)01024-7
30. Hock, A., Miyake, H., Li, B., Lee, C., Ermini, L., Koike, Y., & Pierro, A. (2017). Breast milk-derived exosomes promote intestinal epithelial cell growth. *Journal of Pediatric Surgery*, 52(5), 755–759. https://doi.org/10.1016/j.jpedsurg.2017.01.032
31. Santos-Coquillat, A., González, M. I., Clemente-Moragón, A., González-Arjona, M., Albaladejo-García, V., Peinado, H., ... & Salinas, B. (2022). Goat milk exosomes as natural nanoparticles for detecting inflammatory processes by optical imaging. *Small*, 18(6), 2105421. https://doi.org/10.1002/smll.202105421
32. Samuel, M., Chisanga, D., Liem, M., Keerthikumar, S., Anand, S., Ang, C. S., & Mathivanan, S. (2017). Bovine milk-derived exosomes from colostrum are enriched with proteins implicated in immune response and growth. *Scientific Reports*, 7(1), 5933. https://doi.org/10.1038/s41598-017-06288-8
33. Ibrahim, H. M., Mohammed-Geba, K., Tawfic, A. A., & El-Magd, M. A. (2019). Camel milk exosomes modulate cyclophosphamide-induced oxidative stress and immuno-toxicity in rats. *Food & Function*, 10(11), 7523–7532. https://doi.org/10.1039/C9FO01914F
34. Zempleni, J., Sukreet, S., Zhou, F., Wu, D., & Mutai, E. (2019). Milk-derived exosomes and metabolic regulation. *Annual Review of Animal Biosciences*, 7, 245–262. https://doi.org/10.1146/annurev-animal-020518-115300
35. Garcia, C., Duan, R. D., Brévaut-Malaty, V., Gire, C., Millet, V., Simeoni, U., & Armand, M. (2013). Bioactive compounds in human milk and intestinal health and maturity in preterm newborn: An overview. *Cellular and Molecular Biology*, 59, 108–131. https://doi.org/10.3390%2Fnu10020164
36. Haney, M. J., Klyachko, N. L., Zhao, Y., Gupta, R., Plotnikova, E. G., He, Z., & Batrakova, E. V. (2015). Exosomes as drug delivery vehicles for Parkinson's disease therapy. *Journal of Controlled Release*, 207, 18–30. https://doi.org/10.1016/j.jconrel.2015.03.033
37. Yang, R., Liao, Y., Wang, L., Hu, Y., Wu, Z., & Sun, X. (2019). Exosomes derived from M2b macrophages attenuate DSS-induced colitis. *Frontiers in Immunology*, 10, 476632. https://doi.org/10.3389/fimmu.2019.02346
38. Zebrowska, A., Skowronek, A., Wojakowska, A., Widlak, P., & Pietrowska, M. (2019). Metabolome of exosomes: Focus on vesicles released by cancer cells and present in human body fluids. *International Journal of Molecular Sciences*, 20(14), 3461. https://doi.org/10.3390/ijms20143461
39. Muller, L., Hong, C. S., Stolz, D. B., Watkins, S. C., & Whiteside, T. L. (2014). Isolation of biologically-active exosomes from human plasma. *Journal of Immunological Methods*, 411, 55–65. https://doi.org/10.1016/j.jim.2014.06.007
40. Salomon, C., Guanzon, D., Scholz-Romero, K., Longo, S., Correa, P., Illanes, S. E., & Rice, G. E. (2017). Placental exosomes as early biomarker of preeclampsia: Potential role of exosomal microRNAs across gestation. *The Journal of Clinical Endocrinology & Metabolism*, 102(9), 3182–3194. https://doi.org/10.1210/jc.2017-00672
41. Nedaeinia, R., Manian, M., Jazayeri, M. H., Ranjbar, M., Salehi, R., Sharifi, M., & Ghayour-Mobarhan, M. (2017). Circulating exosomes and exosomal microRNAs as biomarkers in gastrointestinal cancer. *Cancer Gene Therapy*, 24(2), 48–56. https://doi.org/10.1038/cgt.2016.77
42. Mitsuhashi, S., Feldbrügge, L., Csizmadia, E., Mitsuhashi, M., Robson, S. C., & Moss, A. C. (2016). Luminal extracellular vesicles (EVs) in inflammatory bowel disease (IBD) exhibit proinflammatory effects on epithelial cells and macrophages. *Inflammatory Bowel Diseases*, 22(7), 1587–1595. https://doi.org/10.1097/MIB.0000000000000840
43. Bennike, T., Birkelund, S., Stensballe, A., & Andersen, V. (2014). Biomarkers in inflammatory bowel diseases: Current status and proteomics identification strategies. *World Journal of Gastroenterology: WJG*, 20(12), 3231. https://doi.org/10.3748%2Fwjg.v20.i12.3231
44. Zheng, X., Chen, F., Zhang, Q., Liu, Y., You, P., Sun, S., & Chen, N. (2017). Salivary exosomal PSMA7: A promising biomarker of inflammatory bowel disease. *Protein & Cell*, 8(9), 686–695. https://doi.org/10.1007/s13238-017-0413-7
45. Leoni, G., Neumann, P. A., Kamaly, N., Quiros, M., Nishio, H., Jones, H. R., & Nusrat, A. (2015). Annexin A1–containing extracellular vesicles and polymeric nanoparticles promote epithelial wound repair. *The Journal of Clinical Investigation*, 125(3), 1215–1227. https://doi.org/10.1172/jci76693
46. Liu, R., Tang, A., Wang, X., Chen, X., Zhao, L., Xiao, Z., & Shen, S. (2018). Inhibition of lncRNA NEAT1 suppresses the inflammatory response in IBD by modulating the intestinal epithelial barrier and by exosome-mediated polarization of macrophages. *International Journal of*

Molecular Medicine, 42(5), 2903–2913. https://doi.org/10.3892/ijmm.2018.3829
47. Seki, E., & Brenner, D. A. (2008). Toll-like receptors and adaptor molecules in liver disease: Update. *Hepatology*, 48(1), 322–335. https://doi.org/10.1002/hep.26717
48. Segura, E., Guérin, C., Hogg, N., Amigorena, S., & Théry, C. (2007). CD8+ dendritic cells use LFA-1 to capture MHC-peptide complexes from exosomes in vivo. *The Journal of Immunology*, 179(3), 1489–1496. https://doi.org/10.4049/jimmunol.179.3.1489
49. Morelli, A. E., Larregina, A. T., Shufesky, W. J., Sullivan, M. L., Stolz, D. B., Papworth, G. D., ... & Thomson, A. W. (2004). Endocytosis, intracellular sorting, and processing of exosomes by dendritic cells. *Blood*, 104(10), 3257–3266. https://doi.org/10.1182/blood-2004-03-0824
50. Zhou, Q., Li, M., Wang, X., Li, Q., Wang, T., Zhu, Q., & Li, X. (2012). Immune-related microRNAs are abundant in breast milk exosomes. *International Journal of Biological Sciences*, 8(1), 118. https://doi.org/10.7150%2Fijbs.8.118
51. Théry, C., Ostrowski, M., & Segura, E. (2009). Membrane vesicles as conveyors of immune responses. *Nature Reviews Immunology*, 9(8), 581–593. https://doi.org/10.1038/nri2567
52. Li, Y. J., Wu, J. Y., Liu, J., Xu, W., Qiu, X., Huang, S., & Xiang, D. X. (2021). Artificial exosomes for translational nanomedicine. *Journal of Nanobiotechnology*, 19, 1–20. https://doi.org/10.1186/s12951-021-00986-2
53. Zheng, Y., Li, M., Weng, B., Mao, H., & Zhao, J. (2022). Exosome-based delivery nanoplatforms: Next-generation theranostic platforms for breast cancer. *Biomaterials Science*, 10(7), 1607–1625. https://doi.org/10.1039/D2BM00062H
54. Ellert-Miklaszewska, A., Poleszak, K., & Kaminska, B. (2017). Short peptides interfering with signaling pathways as new therapeutic tools for cancer treatment. *Future Medicinal Chemistry*, 9(2), 199–221. https://doi.org/10.4155/fmc-2016-0189
55. Kar, R., Dhar, R., Mukherjee, S., Nag, S., Gorai, S., Mukerjee, N., & Thorat, N. D. (2023). Exosome-based smart drug delivery tool for cancer theranostics. *ACS Biomaterials Science & Engineering*, 9(2), 577–594. https://doi.org/10.1021/acsbiomaterials.2c01329
56. Raza, F., Zafar, H., Jiang, L., Su, J., Yuan, W., Qiu, M., & Paiva-Santos, A. C. (2024). Progress of cell membrane-derived biomimetic nanovesicles for cancer phototherapy. *Biomaterials Science*, 1, 1–13. https://doi.org/10.1039/D3BM01170D
57. Tamura, R., Uemoto, S., & Tabata, Y. (2017). Augmented liver targeting of exosomes by surface modification with cationized pullulan. *Acta Biomaterialia*, 57, 274–284. https://doi.org/10.1016/j.actbio.2017.05.013
58. Alvarez-Erviti, L., Seow, Y., Yin, H., Betts, C., Lakhal, S., & Wood, M. J. (2011). Delivery of siRNA to the mouse brain by systemic injection of targeted exosomes. *Nature Biotechnology*, 29(4), 341–345. http://www.nature.com/doifinder/10.1038/nbt.1807
59. Zheng, Y., Li, M., Weng, B., Mao, H., & Zhao, J. (2022). Exosome-based delivery nanoplatforms: Next-generation theranostic platforms for breast cancer. *Biomaterials Science*, 10(7), 1607–1625. https://doi.org/10.1039/D2BM00062H
60. Sun, B., Wang, L., Guo, W., Chen, S., Ma, Y., & Wang, D. (2023). New treatment methods for myocardial infarction. *Frontiers in Cardiovascular Medicine*, 10, 1–15. https://doi.org/10.3389%2Ffcvm.2023.1251669
61. Alvarez-Erviti, L., Seow, Y., Yin, H., Betts, C., Lakhal, S., & Wood, M. J. (2011). Delivery of siRNA to the mouse brain by systemic injection of targeted exosomes. *Nature Biotechnology*, 29(4), 341–345. http://www.nature.com/doifinder/10.1038/nbt.1807
62. Wang, L., Ouyang, M., Xing, S., Zhao, S., Liu, S., Sun, L., & Yu, H. (2022). Mesenchymal stem cells and their derived exosomes promote malignant phenotype of polyploid non-small-cell lung cancer cells through AMPK signaling pathway. *Analytical Cellular Pathology*, 2022, 1–15. https://doi.org/10.1155/2022/8708202
63. Seo, Y., Kim, H. S., & Hong, I. S. (2019). Stem cell-derived extracellular vesicles as immunomodulatory therapeutics. *Stem Cells International*, 2019, 1–10. https://doi.org/10.1155/2019/5126156
64. Li, T., Yan, Y., Wang, B., Qian, H., Zhang, X., Shen, L., & Xu, W. (2013). Exosomes derived from human umbilical cord mesenchymal stem cells alleviate liver fibrosis. *Stem Cells and Development*, 22(6), 845–854. https://doi.org/10.1089/scd.2012.0395
65. Attia, N., & Mashal, M. (2020). Mesenchymal stem cells: The past present and future. In Turksen, K., (ed.), *Cell Biology and Translational Medicine, Volume 11: Stem Cell Therapy-Potential and Challenges* (pp. 107–129). Springerhttps://doi.org/10.1007/978-3-030-71925-8
66. Zheng, X., Zhu, Y., & Zhao, Z. (2023). The role of amino acid metabolism in inflammatory bowel disease and other inflammatory diseases. *Frontiers in Immunology*, 14, 1284133. https://doi.org/10.3389/fimmu.2023.1284133
67. Li, D. F., Yang, M. F., Xu, J., Xu, H. M., Zhu, M. Z., Liang, Y. J., & Yao, J. (2022). Extracellular vesicles: The next generation theranostic nanomedicine for inflammatory bowel disease. *International Journal of Nanomedicine*, 17, 3893. https://doi.org/10.2147%2FIJN.S370784
68. Li, D. F., Yang, M. F., Xu, J., Xu, H. M., Zhu, M. Z., Liang, Y. J., & Yao, J. (2022). Extracellular vesicles: the next generation theranostic nanomedicine for inflammatory bowel disease. *International Journal of Nanomedicine*, 17, 3893. https://doi.org/10.2147%2FIJN.S370784
69. Yan, Y., Li, K., Jiang, J., Jiang, L., Ma, X., Ai, F., & Si, W. (2023). Perinatal tissue-derived exosomes ameliorate colitis in mice by regulating the Foxp3+ Treg cells and gut microbiota. *Stem Cell Research & Therapy*, 14(1), 43. https://doi.org/10.1186%2Fs13287-023-03263-1
70. Liu, H., Liang, Z., Wang, F., Zhou, C., Zheng, X., Hu, T., & Lan, P. (2019). Exosomes from mesenchymal stromal cells reduce murine colonic inflammation via a macrophage-dependent mechanism. *JCI Insight*, 4(24), 1–13. https://doi.org/10.1172%2Fjci.insight.131273
71. Eichenberger, R. M., Giacomin, P. R., Dent, L. A., Sotillo, J., & Loukas, A. (2018). Hookworm secreted extracellular vesicles interact with host cells and prevent inducible colitis in mice. *Frontiers in Immunology*, 9, 357235. https://doi.org/10.3389/fimmu.2018.00850
72. Ocansey, D. K., Zhang, L., Wang, Y., Yan, Y., Qian, H., Zhang, X., & Mao, F. (2020). Exosome-mediated effects

and applications in inflammatory bowel disease. *Biological Reviews*, 95(5), 1287–1307. https://doi.org/10.1111/brv.12608

73. Rani, P., Onteru, S. K., & Singh, D. (2020). Genome-wide profiling and analysis of microRNA expression in buffalo milk exosomes. *Food Bioscience*, 38, 100769. https://doi.org/10.1016/j.fbio.2020.100769
74. Urzì, O., Raimondo, S., & Alessandro, R. (2021). Extracellular vesicles from plants: Current knowledge and open questions. *International Journal of Molecular Sciences*, 22(10), 5366. https://doi.org/10.3390/ijms22105366
75. Pérez-Bermúdez, P., Blesa, J., Soriano, J. M., & Marcilla, A. (2017). Extracellular vesicles in food: Experimental evidence of their secretion in grape fruits. *European Journal of Pharmaceutical Sciences*, 98, 40–50. https://doi.org/10.1016/j.ejps.2016.09.022
76. Chen, S., Zhang, C., He, B., He, R., Xu, L., & Zhang, S. (2021). The role of lncRNAs in regulating the intestinal mucosal mechanical barrier. *BioMed Research International*, 2021, 1–13. https://doi.org/10.1155/2021/2294942
77. Zeng, H., Guo, S., Ren, X., Wu, Z., Liu, S., & Yao, X. (2023). Current strategies for exosome cargo loading and targeting delivery. *Cells*, 12(10), 1416.
78. Liang, Y., Duan, L., Lu, J., & Xia, J. (2021). Engineering exosomes for targeted drug delivery. *Theranostics*, 11(7), 3183.

19 Plant-Based Bio-Actives Encapsulated Nanocarriers for Gastrointestinal Inflammations Management

*Subhasundar Maji[1], Sandipan Dasgupta[2], Ria Das[3], and Slim Smaoui[4]**

[1]Charnock Health Care Institute, Kolkata, India

[2]Maulana Abul Kalam Azad University of Technology, Haringhata, India

[3]Narapatipara Simurali, Chakdah, Kolkata, West Bengal, India

[4]Center of Biotechnology of Sfax (CBS), University of Sfax, Road of Sidi Mansour Km 6, Sfax-Tunisia

19.1 INTRODUCTION

In recent years, gastrointestinal (GI) inflammatory diseases have emerged as significant health concerns worldwide, imposing a substantial burden on affected individuals and healthcare systems. These conditions encompass a broad spectrum of disorders affecting the digestive tract, including inflammatory bowel disease (IBD), gastritis, and enteritis, among others. The prevalence of GI inflammatory diseases is on the rise, attributed to various factors such as changes in lifestyle, diet, and environmental influences. Despite advancements in medical science, managing GI inflammation remains challenging, often requiring long-term therapy and presenting risks of adverse effects from conventional treatments. The traditional pharmacological approaches to GI inflammation management, including corticosteroids, immunosuppressants, and biologics, are associated with limitations such as systemic side effects, diminished efficacy over time, and high costs. Additionally, the heterogeneity of GI inflammatory diseases and the variable responses to existing treatments underscore the need for innovative therapeutic strategies that can provide targeted and personalized interventions. In this context, the intersection of plant-based bio-actives and nanocarrier technology represents a promising frontier in GI inflammation management. Plant-derived constituents have long been recognized for their potential anti-inflammatory, antioxidant, and immunomodulatory characteristics, offering potential as natural alternatives or adjuncts to conventional therapies. However, challenges related to their bioavailability, stability, and targeted delivery have hindered their clinical translation. Nanocarrier technology, on the other hand, offers a sophisticated platform for the encapsulation, protection, and controlled release of bioactive compounds. Nanocarriers, such as liposomes, nanoparticles made of polymers, and hydrogels, can enhance the solubility, stability, and bioavailability of plant-based bio-actives while enabling their precise delivery to inflamed GI tissues. By harnessing the synergistic benefits of plant-based bio-actives and nanocarriers, researchers aim to overcome the limitations of current therapies and usher in a new era of precision medicine for GI inflammation. The rationale for innovative approaches in gastrointestinal inflammation management lies in the intricate molecular mechanisms driving these conditions. By understanding these pathways, researchers can target specific points for intervention, paving the way for novel therapeutic strategies. Plant-based bio-actives, including polyphenols and flavonoids from sources like turmeric and green tea, offer promising anti-inflammatory and immunomodulatory properties. Encapsulating these bio-actives within nanocarriers, such as liposomes and nanoparticles made of polymers, enhances their stability and enables controlled release in the gastrointestinal tract, maximizing their therapeutic efficacy. Through in vitro and in vivo studies, the effectiveness of these formulations in alleviating gastrointestinal inflammations is validated, providing insights into their pharmacokinetics and safety profiles. Considerations such as biocompatibility and scalability underscore the translational potential of these nanotechnological interventions, offering hope for improved management of GI inflammatory disorders. This chapter aims to unravel the molecular complexities underlying gastrointestinal inflammations, delving into the intricate interplay of signaling pathways and immune responses driving the pathogenesis of these conditions. It further explores a diverse spectrum of plant-based bio-actives, renowned for their anti-inflammatory, antioxidant, and immunomodulatory properties. Emphasis is placed on polyphenols, flavonoids, and terpenoids sourced from plants like turmeric, green tea, and aloe vera. The core focus is on elucidating the principles of nanocarrier design and fabrication, ranging from liposomes to nanoparticles made of polymers. These nanocarriers act as protective vehicles, ensuring the stability and controlled release of plant-derived bio-actives within the gastrointestinal tract. The chapter provides insights from in vitro and in vivo studies, validating the potency of plant-based bio-active-loaded nanocarriers in alleviating gastrointestinal inflammations. Additionally, it comprehensively addresses pharmacokinetics, safety profiles,

DOI: 10.1201/9781003493143-19

and mechanisms of action, offering a nuanced understanding of the therapeutic potential of these formulations. Considerations such as biocompatibility, biodegradability, and scalability are explored, underscoring the translational feasibility of these nanotechnological interventions.

19.2 GASTROINTESTINAL INFLAMMATIONS: MOLECULAR INSIGHTS

Gastrointestinal (GI) inflammatory processes encompass a diverse array of conditions affecting the digestive tract, ranging from mild discomfort to severe diseases with significant morbidity and mortality (McDowell et al., 2023; Vuyyuru et al., 2022). Inflammation within the GI tract can arise from various etiologies, including infections, autoimmune disorders, dietary factors, and environmental triggers. Recognizing the fundamental mechanics driving GI inflammation is essential for developing effective management strategies, including novel approaches utilizing plant-based bio-actives encapsulated in nanocarriers (Burns et al., 2019; Iwata et al., 2021; Oz et al., 2016).

19.2.1 Anatomy and Physiology of the Gastrointestinal Tract

The GI tract is a complex system comprising the mouth, esophagus, stomach, small intestine, and large intestine, each with specialized functions in digestion and absorption. The mucosal lining of the GI tract serves as a barrier against harmful pathogens and toxins while facilitating nutrient uptake. Key components of GI immunity, such as gut-associated lymphoid tissue (GALT) and resident immune cells, play crucial roles in maintaining homeostasis and mounting appropriate immune responses to foreign antigens (Azzouz & Sharma, 2023; Mörbe et al., 2021).

19.2.2 Inflammatory Pathways in Gastrointestinal Diseases

Gastrointestinal inflammatory processes can manifest through various mechanisms, including innate and adaptive immune responses, dysregulated cytokine signaling, and disruptions in the gut microbiota. In conditions like inflammatory bowel disease (IBD), comprising Crohn's disease and ulcerative colitis, aberrant immune activation against commensal gut bacteria leads to chronic inflammation and tissue damage. Similarly, infections with pathogens such as Helicobacter pylori can trigger inflammatory cascades, contributing to gastritis and peptic ulcer disease (Axelrad et al., 2021; Yoo et al., 1587).

19.2.3 Molecular Pathways and Key Factors

The molecular pathways and key factors involved in gastrointestinal (GI) inflammation are essential for the development of targeted interventions for managing GI inflammatory diseases.

19.3 INFLAMMATORY SIGNALING PATHWAYS IN GASTROINTESTINAL INFLAMMATION

The GI tract's inflammatory response is coordinated by multiple signaling pathways, including those for mitogen-activated protein kinase (MAPK), nuclear factor kappa-light-chain-enhancer of activated B cells (NF-κB), and Janus kinase/signal transducer and activator of transcription (JAK/STAT) (Zhao et al., 2021). Dysregulation of these pathways can lead to excessive production of pro-inflammatory cytokines, such as tumor necrosis factor-alpha (TNF-α), interleukin-1β (IL-1β), and interleukin-6 (IL-6), perpetuating tissue damage and chronic inflammation (Chen et al., 2018).

19.4 CELLULAR PLAYERS IN GASTROINTESTINAL INFLAMMATION

Various immune cells, epithelial cells, and resident microbiota within the GI mucosa actively participate in the inflammatory process. Macrophages, dendritic cells, and T lymphocytes play pivotal roles in initiating and propagating immune responses against luminal antigens. (Yoo et al., 2020). Dysbiosis, characterized by alterations throughout the composition and function of gut microbiota, has been linked to the etiology of inflammatory gastrointestinal disorders, highlighting the intricate interplay between host immunity and microbial communities (Hrncir, 2022; Kandpal et al., 2022).

19.5 KEY FACTORS CONTRIBUTING TO GASTROINTESTINAL INFLAMMATION

The onset and development of GI inflammation are influenced by several variables, including genetic predisposition, environmental triggers, dysregulated immune responses, and alterations in the gut microbiome. (Virolainen et al., n.d.) Genetic polymorphisms associated with genes involved in immune regulation, barrier function, and microbial sensing can increase susceptibility to GI inflammatory diseases. Environmental factors, such as diet, stress, and microbial exposures, can modulate the gut immune milieu and influence disease outcomes (Bander et al., 2020).

19.6 MODULATION OF MOLECULAR PATHWAYS BY PLANT-BASED BIO-ACTIVES

Plant-derived bioactive compounds possess a diverse array of molecular targets and mechanisms of action that can modulate inflammatory pathways within the GI tract. Polyphenols, such as resveratrol, curcumin, and quercetin, have anti-inflammatory properties through suppression of NF-κB activation, suppressing cytokine production, and enhancing antioxidant defenses (Yahfoufi et al., 2018; H. Zhang & Tsao, 2016). Flavonoids, terpenoids, and alkaloids found in various plant sources exhibit similar immunomodulatory properties, making them promising candidates for GI inflammation management (Dias et al., 2021; Huang et al., 2022).

19.7 SIGNIFICANCE OF TARGETING MOLECULAR MECHANISMS

Targeting molecular mechanisms underlying GI inflammation represents a promising strategy for developing effective therapies for GI inflammatory diseases. The intricate molecular mechanisms underlying gastrointestinal (GI) inflammation offer valuable targets for therapeutic intervention in the management of GI inflammatory diseases (see Table 19.1).

19.8 NANOCARRIERS IN DRUG DELIVERY

Nanocarriers play a pivotal role in facilitating the delivery of plant-based bio-actives for the management of gastrointestinal inflammations. This section highlights the design principles of nanocarriers, discusses the challenges associated with gastrointestinal drug delivery, and explores how nanocarriers enhance the bioavailability of bioactive compounds.

19.9 NANOCARRIER DESIGN PRINCIPLES

Bioactive chemicals are designed to be encapsulated and protected by nanocarriers, facilitating their delivery to target tissues within the gastrointestinal tract. Several design principles govern the development of nanocarriers for effective drug delivery:

19.9.1 Particle Size and Surface Properties

Nanocarriers are designed with precise control over particle size and surface characteristics to optimize interactions with biological barriers and target cells. Small particle sizes enhance cellular uptake and tissue penetration, while surface modifications, such as PEGylation or ligand conjugation, facilitate targeted delivery and minimize immune recognition (Martínez-Ballesta et al., 2018).

19.9.2 Efficiency of Encapsulation and Drug Loading

Nanocarriers must exhibit high encapsulation efficiency and drug-loading capacity to ensure the efficient delivery of bioactive compounds. Various encapsulation techniques, including nanoprecipitation, emulsification, and solvent evaporation, are employed to encapsulate hydrophobic or hydrophilic compounds within the nanocarrier matrix while maintaining stability and uniformity (Alshawwa et al., 2022a).

19.9.3 Controlled Release Kinetics

In order to facilitate the continued and extended distribution of bioactive substances to the target location, nanocarriers are designed with regulated release kinetics in mind. This controlled release profile is achieved through the selection of biocompatible polymers, lipid bilayers, or stimuli-responsive materials that modulate drug release in response to physiological cues or external stimuli (Hadidi et al., 2023a; see Table 19.2).

19.10 CHALLENGES IN GASTROINTESTINAL DRUG DELIVERY

Gastrointestinal drug delivery poses several challenges that must be addressed to ensure the efficacy of therapeutic interventions:

19.10.1 Harsh Gastrointestinal Environment

The acidic pH, enzymatic activity, and rapid transit time within the gastrointestinal tract pose challenges for drug stability and bioavailability. Nanocarriers must withstand these harsh conditions and protect encapsulated bioactive compounds from degradation until they reach the target site (Alqahtani et al., 2021; Hua, 2019; Vinarov et al., 2021).

19.10.2 Mucosal Barrier and Absorption Limitations

The gastrointestinal mucosa serves as a formidable barrier to drug absorption, limiting the bioavailability of orally

TABLE 19.1
Targeting Molecular Mechanisms in GI Inflammation Management

Name	Targeting Molecular Mechanisms in GI Inflammation Management	References
Precision Therapy	Targeting specific molecular pathways allows for personalized treatment tailored to individual disease pathogenesis.	(Neurath, 2019; Xavier & Podolsky, 2007)
Modulation of Immune Responses	Targeting key immune signaling pathways can dampen inflammation while preserving immune homeostasis.	(Abraham & Cho, 2009; Kaser et al., 2010)
Restoration of Gut Barrier Function	Targeting mechanisms involved in maintaining gut barrier integrity is crucial for preventing microbial translocation and immune activation.	(Marchiando et al., 2010; Turner, 2009)
Reduction of Oxidative Stress and Tissue Damage	Targeting oxidative stress pathways mitigates tissue damage and promotes tissue repair.	(Grisham, 1994; Rezaie et al., 2007)
Enhanced Bioavailability and Targeted Delivery	Utilizing nanocarriers ensures optimal delivery of bioactive compounds to inflamed gastrointestinal mucosa, maximizing therapeutic efficacy.	(Anselmo & Mitragotri, 2019; Paiva-Santos et al., 2021a)

TABLE 19.2
Advantages for Nanocarriers Drug Delivery

Name	Advantages	References
Encapsulation and Protection:	Nanocarriers protect bioactive compounds from degradation and ensure their safe delivery to target tissues within the gastrointestinal tract.	(Alshawwa et al., 2022a; Hadidi et al., 2023a; Martínez-Ballesta et al., 2018)
Enhanced Bioavailability:	Nanocarriers improve the bioavailability of bioactive compounds by enhancing solubility, stability, and absorption, leading to increased therapeutic efficacy.	
Targeted Delivery:	Nanocarriers enable targeted delivery of bioactive compounds to specific tissues or cells within the gastrointestinal tract, minimizing systemic exposure and off-target effects.	

TABLE 19.3
Types of Bioavailability Enhancing Process

Type	Explanation	References
Improved Solubility and Stability	Encapsulation within nanocarriers improves the solubility and stability of hydrophobic bioactive compounds, overcoming limitations associated with poor aqueous solubility and chemical instability. This enhancement in solubility ensures uniform dispersion and prolonged release of bioactive compounds within the gastrointestinal lumen.	(Gao et al., 2013; P. Liu et al., 2021; X. Liu et al., 2023)
Targeted Delivery and Tissue Penetration	Nanocarriers enable targeted delivery of bioactive compounds to inflamed tissues within the gastrointestinal tract, minimizing systemic exposure and off-target effects. Surface modifications and ligand conjugation facilitate specific interactions with target cells or receptors, enhancing tissue penetration and therapeutic efficacy.	(Hua et al., 2015; Kotla et al., 2022; X. Li et al., 2020; P. Liu et al., 2021)
Sustained Release Profile	Nanocarriers provide a sustained release profile, prolonging the exposure of inflamed tissues to bioactive compounds and minimizing fluctuations in drug concentration. This sustained release kinetics ensures continuous therapeutic effects, reducing the frequency of dosing and improving patient compliance.	(Pridgen et al., 2014, 2015; Raval et al., 2013; Teruel et al., 2020a)

administered therapeutics. Nanocarriers are designed to overcome mucosal barriers through enhanced permeation, mucoadhesion, or receptor-mediated transport mechanisms, facilitating efficient drug delivery to inflamed tissues (Azman et al., 2022; Luo et al., 2021; Vancamelbeke & Vermeire, 2017).

19.10.3 Immune Response and Clearance

Nanocarriers may elicit immune responses or undergo rapid clearance by the reticuloendothelial system, reducing their efficacy in delivering bioactive compounds. Surface modifications and stealth coatings are employed to minimize immunogenicity and prolong circulation time, enhancing the accumulation of nanocarriers at the target site (Ko et al., 2022; Lôbo et al., 2021; Rana et al., 2023; Su & Kang, 2020a).

19.10.4 Enhancing Bioavailability Through Nanocarriers

Nanocarriers offer unique advantages for enhancing the bioavailability of bioactive compounds within the gastrointestinal tract (see Table 19.3).

19.11 PLANT-BASED BIO-ACTIVES AS ANTI-INFLAMMATORY AGENTS

This section focuses on the most current advancements in the development and assessment of vectors that have the ability to release anti-inflammatory phytochemicals. In this way, in accordance to natural compounds issued from plants, several delivery techs will be examined. Notably, functionalized nanovectors loaded with polyphenol- vegetal plant, phytocannabinoids, phytosterols, terpenoids and essential oils will be investigated.

19.11.1 Polyphenols

This huge family of diverse molecules was presented in numerous vascular plants as secondary metabolites. Polyphenols (PPs) structure, range of strength from basic to complicated structures, mutually possess in benzene rings bearing one or numerous hydroxyl functions. By virtue of their aptitude to interplay with enzymes, proteins and membrane receptors, PPs are powerful effectors of biological processes (Festa et al., 2023). For instance, some isolated biomolecules such as quercetin, resveratrol, and tannins are acknowledged as painkillers (Mostafa et al., 2023). It should be mentioned that the effectiveness of PPs is based on their bioavailability and metabolism, which is defined as the quantity of administered medication that may enter the bloodstream. PPs are recognized for their limited bioavailability due to two main extrinsic variables: low stability in gastrointestinal circumstances and intrinsic factors such as chemical structure, molecular weight, and low hydrosolubility (Li et al., 2024). Nano-delivery systems have been invented to work around these challenges (Chen & Hu, 2024). As an illustration, the quercetin, semi-lipophilic flavonol omnipresent in plants, was encapsulated in nanosized vectors (Bandakinda et al., 2024). Interestingly, compared to the free form the developed vectors could enhance the relative oral bioavailability till 6 folds. Similar trends were reported by Barras et al. (2009). These authors noted an enhancement of solubility up to 100 times, without product degradation. To attain efficient antioxidant activity against lipid peroxidation caused by peroxyl radicals, Pool et al. (2012) shaped quercetin-charged polylactic-co-glycolide. The newly developed product displayed an efficient anti-inflammatory agent. Similar trends were examined by Chakraborty et al. (2012); however, in contrast to 20% of free quercetin, the oral potential of given quercetin-PLGA NP in a rat model could maintain till ninety percent of this inflammation. Another PPs, the ellagitannin and ellagic acid were deeply discussed. For instance, Bala et al. (2006) created ellagic acid-loaded PLGA NP with high loading efficacy, stability, cytotoxic activity and a high AIA. Bertrand et al. (2014) studied the AIA of curcumin and concluded that only a minor % was detected in blood plasma. Several in vitro studies (Naghsh et al., 2023; Chou et al., 2023; Bhojwani et al., 2023), have recommended the AIA of curcumin. This may be connected to its capacity to decrease the generation of inflammatory mediators, stop chondrocyte apoptosis and extracellular matrix degradation (Patnaik et al., 2023), and function as an antioxidant by reducing the excessive generation of reactive O2 species (García-Santos et al., 2023). Zhang et al. (2016) proved the *in vivo* curcumin NP. The hydrogel/glass-based NP could slow the osteoarthritis disease progression, and pain relief, and conserve the chondroprotective activity of curcumin, and may enhance its bioavailability.

19.11.2 Phytocannabinoids

Assigned as lipophilic substances isolated from *Cannabis sativa* L., (Cannabaceae), cannabis extracts comprises over 460 compounds of which ~80 are phytocannabinoids (Zandkarimi et al., 2023). On the phytocannabinoids case, The AIA may be regulated by receptors such as the T-type Ca2+ channel, deorphanized G protein-coupled receptor, peroxisome proliferator-activated receptors, and vanilloid type-1 receptor (Gao et al., 2024). Esposito et al. (2016) encapsulated phytocannabinoids in nanostructured lipid carriers. Similarly, Duran Lobato et al. (2016) produced lipid NP formulations as an oral delivery vehicle for cannabis derivatives against chronic pain conditions. De La Ossa et al. (2012) developed cannabidiol – PCL particles as a suitable dosage form. This material was prepared with O/W emulsion and presented high entrapment efficiency (~100%). PLGA nanoparticles loaded with Δ9-THC were synthesized by Martín-Banderas et al. (2015).

19.11.3 Phytosterols

Phytosterols have various bioactive properties like AIA, hoverer their bioavailability is partial. Based on the nanotechnology employment, NP- phytosterols could maintain and/or enhance the AIA. Leong et al. (2011) created nanodispersions of phytosterols on the resulting vectors by homogenization at high pressure. Using antisolvent precipitation in the presence of a non-ionic surfactant, Rossi et al. (2010) created phytosterol colloidal particles. Because of the surface charge, the manufactured capsules exhibited high stability. This was the outcome of steric stability brought on by the use of a non-ionic stabilizer, as well as the presence of water and hydroxyl groups on the particle surface. (Türk & Lietzow 2004) rapidly expanded a supercritical solution into an aqueous solution to create continuous suspensions of phytosterol submicron particles. After 12 months, the resulting material demonstrated excellent stability.

19.11.4 Carbohydrates

Numerous polysaccharides obtained from plants have a robust AIA. For instance, *Echinacea purpurea* and *Echinacea angustifolia* are potential candidates (Geszke-Moritz et al., 2023). Aqueous fractions isolated from the roots of these asteraceae contain echinacin. This latter had good AIA. From the aerial parts of *Comarum palustre*, the comaruman was confirmed to own AIA. Specifically, the fragments of galacturonan were able to lessen neutrophil adherence to fibronectin in vitro more so than the original pectin (Singh et al., 2023). As a consequence of their aptitude to modify neutrophil count, the fixation of complement function, and function of macrophages, sulfated polysaccharides from *Artemisia tripartite* showed good AIA (Xie et al., 2008). Takahashi et al. (2009) produced liposomes containing aloe vera gel in relation to previous work on the nano-delivery of AIA. The generated gel may promote the production of skin cell lines and collagen in the skin. Aloe vera and curcumin were co-encapsulated, which enhanced the phytocompounds' antioxidant activity and delivery (Kitture et al., 2015).

19.11.5 Essential Oils

They are used in food preservation (Meenu et al., 2023) and medicine as analgesics, sedatives, anti-inflammatory, spasmolytics, and local anesthetics due to their well-known antibacterial and fungicidal action, scent, and taste (Liang et al., 2023). The desired qualities of EOs have been explored in great detail, and there is a wealth of literature on this subject. However, due to their extreme sensitivity to light, oxygen, humidity, and high temperatures, EOs have restricted uses (Ali et al., 2023). In terms of enhanced solubility, bioavailability, defense against toxic effects, pharmacological action, stability, improved tissue macrophage distribution, sustained delivery, and protection against physical and chemical degradation, novel technologies for the administration of plant extracts may present significant advantages over traditional formulations. According to Parris et al., encapsulating pure EOs of oregano and cassia in maize zein NC using phase separation technique yields a quick and high-yield process (Parris et al., 2005). According to reports, particles laden with EOs are less digestible in the gastrointestinal tract, release in the small intestine gradually and in the large intestine more quickly. It is noteworthy that minimal to nonexistent interaction between the EO and other feed constituents was seen in their encapsulated form. Thymol and carvacrol EOs have also been nanoencapsulated using the liquid–liquid dispersion approach by zein, a prolamine protein found in corn that is highly recognized for having remarkable biodegradability and biological compatibility (Wu et al., 2012). Using cyclodextrins (CD) as carriers is among the most successful strategies for durability and dissolution of natural active substances (Pinho et al., 2014). Water molecules occupy the somewhat polar CD cavity in aqueous solution, making it easy to replace them with suitable "guest molecules" that are less polar than water (Loftsson & Brewster, 1996). Research on the encapsulation and release of thymol and cinnamon aldehyde from CD was conducted by Cevallos et al. (2010). The experiments examining the impact of CDs' water adsorption on the release of encapsulated chemicals are what make this work significant. The significance of choosing suitable storage settings for hydrophobic tastes encapsulated in CD or for forecasting the shelf life of functional items made with nanoencapsulated chemicals was demonstrated by the results. According to De Olivera et al. (2014), alginate/cashew gum NP was prepared by spray-drying with the goal of creating a biopolymer mix for the encapsulation of Lippiasidoides essential oil, which is mostly made up of thymol (50–70%). Over the last few decades, novel regulated and triggered release methods have often accompanied encapsulation approaches. Therefore, stimuli-responsive materials have been used as drug and other active agent vectors as well as capsule shell material (Fleige et al., 2012). The fabrication of polyamide capsules loaded with cumin and basil oil that may release their cargo oil when exposed to UV light has been studied by Bizzarro et al. (2016). The literature has evaluated a great deal of polymeric capsules that are photoresponsive as well as responsive to other stimuli. A few nanoscale delivery methods for phytochemicals with anti-inflammatory properties are included in Table 19.4.

19.12 MECHANISMS OF ACTION IN INFLAMMATION MODULATION

19.12.1 Natural plant products-derived anti-inflammatory impact Modulators of NRF2

It is believed that phytochemicals are the most important factor Nrf2-activators (Baralić et al., 2024). In fact, the Nrf2 pathway is thought to be the mechanism by which many phytochemicals produce AIA (Kakarla et al., 2023). This route is also thought to function as an auxiliary to improve phase 2 enzymes, GSH production and turnover [GCL and glutathione reductase (GR)], the development of HO-1 and NQO1 levels, and increased ROS levels (Qin et al., 2023). It has been discovered that ROS, such as superoxide anion, activate NF-κB to cause inflammatory pain (Qin et al., 2023), increase of TNFα, IL-1β and endothelin levels. Consequently, phytochemicals employment could induce Nrf2 pathway and reduce ROS production is a possible method for treating inflammatory and painful conditions. Nrf2 inducer phytochemicals can be categorized: According to Pouremamali et al. (2022) chemical categories, Michael acceptors include oxidizable phenols and quinones, isothiocyanates, dithiolethiones, polyenes, and vicinal dimercaptans. Their capacity to react by alkylation, oxidation, or reduction with SH groups is what unites them. The majority of phytochemicals are conjugate addition products of nucleophilic amino acids bound in electrophile-sensitive proteins and are Michael acceptors (olefins or acetylenes conjugated with electron-withdrawing groups) (Dinkova-Kostova et al., 2024). Additionally, phytochemicals can interact with SH groups to function as nucleophiles in the signaling pathway of the Michael reaction, which leads to the upregulation of phase 2 enzymes.

19.12.1.1 Flavonoids

Two major molecules, quercetin and naringenin, belonging to flavonoids are deeply studied. These latter inhibited the generation of inflammatory pain in mice caused by carrageenan and superoxide anion by lowering oxidative stress, IL-33, TNFα, IL-6 secretion, and NF-κB activation (Ferraz et al., 2020). The flavanone naringenin decreased TiO2/prosthesis-like-induced chronic arthritis by enhancing leukocyte recruitment, cartilage degradation, bone resorption, oxidative strain, and production of inflammatory cytokines (IL-33, TNF-α, IL-1β, and IL-6), as well as NF-κB activation (Manchope et al., 2018). Additionally, quercetin has been shown to lower chronic inflammatory response in mice with Ehrlich tumor-induced chronic arthritis and TiO2/prosthesis-like chronic arthritis (Calixto-Campos et al., 2015). The following information was noted: a decrease in phase 2 enzymes HO-1, GPx1, and GR as well as a decrease in Nrf2 and leukocyte recruitment, oxidative stress, and the generation of inflammatory cytokines (TNF-α, IL-1β, and IL-6).

19.12.1.2 Curcumin

These substances could prevent leukocyte recruitment, oxidative stress, TNF-α, and IL-1β production, and also increase Nrf2 activation and HO-1 expression, hence reducing superoxide anion-induced inflammatory pain (Fattori et al., 2016). In a mouse model of ischemia/reperfusion, curcumin-induced Nrf2 expression and its downstream enzymes required the activation of the PI3K/Akt signaling pathway (Wu et al., 2013). Research showed that via binding to GSK-3β (Bustanji et al., 2009), Cyclooxygenase-1, and Cyclooxygenase -2 (Selvam et al., 2005), curcumin may also have an effect on therapeutic capacity as a Michael acceptor. Certain phenolic

TABLE 19.4
Anti-Inflammatory Nanoparticle-Mediated Delivery of Some Phytochemical Compounds

Phytoconstituents	Molecules	Nanovector Type	Deliver System Kind	Experimental Model	Main Findings	References
Polyphenols	Quercetin	solid lipid nanoparticles	soya lecithin, Tween-80 and PEG 400	Rats	↗F_{rel} of QT-SLNs to quercetin suspension till 570% - Tmax and MRT for quercetin in plasma were both delayed.	Li et al. (2009)
	Quercetin and catechin	Polymeric nanoparticles	poly (dl-lactide-co-glycolide) (PLGA)	In vitro	- inhibition of the action of free radicals and chelating properties	Pool et al., (2012)
	Quercetin	nanoparticles	Eudragit® E and polyvinyl alcohol	In vitro	-↗Antioxidant activity	Wu et al., (2012)
	Quercetin	Nanoparticles	Poly (D,L-lactic-co-glycolic acid)	Rats	MMP-9 upregulation throughout prevention of gastric inflammation	Chakraborty et al., (2012)
	Quercetin and epigallocatechin gallate	Nanocapsules	LabrafacTM Lipophile WL 1349 Phospholipon® 90G and Solutol® HS15	In vitro	Efficient delivery for inti-inflammatory effect	Barras et al., (2009)
	Resveratrol	Nanoparticles	PLGA	Wistar male rats	↗in absorptivity and permeability	Singh & Pai, (2014a)
	Resveratrol	Nanoparticles	Eudragit RL 100	In vitro and in vivo	↗in absorptivity and permeability	Singh & Pai, (2014b)
	Resveratrol	Nanoparticles	carboxymethyl chitosan	in vitro/in vivo	↗in vivo absorption, duration of action ↗ of relative bioavailability	Zu et al., (2016)
	Resveratrol	Nanoparticles	Lipid Compritol 888ATO and Phospholipon 80H	In vitro	↗solubility, stability and intracellular delivery	Teskač & Kristl, (2010)
	Resveratrol	Nanoparticles	stearic acid based SLNs	Wistar male and rats	↗ bioavailability for resveratrol	Pandita et al., (2014)
	Ellagic acid	Nanoparticles	PLGA	In vitro	↗ stability	Bala et al. (2006)
	Ellagic acid	Nanoparticles	LGA-PCL	Rats	↗oral bioavailability of Ellagic acid	Sonaje et al. (2007)
	Curcumin	Nanoparticles	Hydrogel	Rats	Suppression of mRNA expression of pro-inflammatory mediators IL-1β and TNF-α, MMPs 1, 3, and 13, and aggrecanase ADAMTS5.	Zhang et al., (2016)
	Curcumin	Lipid nanoparticules	Lecithin liposomes	Rats	↗ stability	Takahashi et al., (2009)
Phytocannabinoids	Cannabinoid drugs	NLC: nanostructured lipid carriers	URB597, AM251 and rimonabant	In vivo	↗recovery, morphology and dimensional distribution.	Esposito et al. (2016)
	cannabinoid derivative CB13	Nanoparticles	Lecithin	In vitro	↗oral delivery	Durán-Lobato et al., (2016)
	Cannabinoid	Nanoparticles	PLGA	In vitro/In vivo	↗ stability	Durán-Lobato et al., (2016)
	Cannabidiol	Nanoparticles	PCL	In vitro	↗ stability	De La Ossa et al., (2012)

	Δ9-tetrahid-ocannabinol	Nanoparticles	PLGA	In vivo	↗ stability	Martin-Banderas et al., (2015)
Phytosterol	Vagapure® FTE (contain minimum 99% free phytosterols)	Nanodispersions	Tween 20	In vitro	↗ stability	Leong et al. (2011)
	Phytosterol	Nanodispersions	Tween 20	In vitro	↗ bioaccessibility	Rossi et al., (2010)
Essential oils	Oregano and Cassia EO	Nanoparticles	Corn zein	In vitro	↗ stability	Parris et al., (2005)
	Thymol and carvacrol	Nanoparticles	Corn zein	In vitro	↗ biological potential	Wu, and Luo (2012)
	Cinnamon and thyme EO	Inclusion	Inclusion in cyclodextrins	In vitro	↗ stability	Cevallos et al., (2010)
	Lippiasidoides EO	Nanoparticles	Alginate/cashew gum	In vitro	↗ stability	de Oliveira et al., (2014)
	Cumin and basil EO	Nanoparticles	Polyamide	In vitro	↗ stability	Fleige et al., (2012)
Terpenoids	Squalene	Nanocapsules (polyelectrolyte multilayer adsorption)	polycation –poly and polyanion: polysodium	In vitro	↗ stability	Adamczak et al., (2014)
	Lycopene	Nanoparticles	Inclusion in cyclodextrins	In vitro	↗ stability	Butnariu et al., (2011)
	Carvacrol	Cyclodextrins	Inclusion in cyclodextrins	In vivo	↗ stability	Guimarães et al. (2015)
	Linalool	Cyclodextrins	Inclusion in cyclodextrins	In vivo	↗ stability	Menezes eta al. (2014)

acids, such as caffeic acid, raised hepatic GSH levels without causing damage from APAP (Subramanya et al., 2018). Nrf2 stimulation and related enzymes, HO-1 and NQO1, by caffeic acid, as seen in human liver L-02 cells, was attributed to these effects. The H atom of the OH at position 3 in caffeic acid formed water H_2O-mediated H interactions with Ser508, while the carbonyl group of CA may establish H-bonds with Gly603 and Ser363, enhancing the affinity of adhesion between caffeic acid and Keap1 (Pang and others, 2016).

19.12.1.3 Terpenoids

As an appropriate illustration, consider the inhibition of inflammatory pain caused by carrageenan, TNF-α and IL-1β production in mice, and kaurenoic acid, a diterpene present in Sphagneticola trilobata with AIA (Mizokami et al., 2012). It has been demonstrated to have an analgesic impact on both Nrf2 activation and expression of inflammatory cytokine genes (TNFα and IL-1β) via activating Nrf2 and regulating the synthesis of phase 2 enzymes NQO-1, HO-1, and glutamate-cysteine ligase catalytic subunits (Kim et al., 2017).

19.12.1.4 Oxidizable Phenols and Quinones

These compounds were categorized as phase 2 inducers prior to the finding of the Nrf2/ARE pathway (Staurengo-Ferrari et al., 2019). Catechol was an in vivo inducer of Nrf2 that oxidizes cytochrome p450. When Quinone included necessary cysteine residues, it interacted with Keap1, activating Nrf2 and elevating phase 2 enzyme expressions (Bensasson et al., 2008). It has been demonstrated that rosmarinic acid, either before or after therapy, reduces the neuropathic pain caused by chronic constriction injury (CCI) by preventing the generation of TNFα and mechanical and thermal allodynia (Anantharaju et al., 2016). These outcomes were linked to Nrf2 activation. Furthermore, via lowering the synthesis of TNFα and COX-2 proteins, oxidative and nitrosative stress, and NF-κB activation, this chemical can suppress C tetrachloride-induced liver poisoning in mice (Domitrović et al., 2013). Conversely, relationships with Keap1 were shown for carsonic acid. Carsonic acid interacts with Keap1 via alkylating essential cysteine residues. The fact that this effect is absent in dominant negative Nrf2 cortical neurons highlights the significance of Nrf2 activation in carsonic acid action (Staurengo-Ferrari et al., 2019).

19.12.1.5 NRF2 Modulation by Synergic Effects and Analgesic Drugs

By interrelating with μ/δ/κ opioid receptors (OR), opioids could provoke their pain-mitigating acts (Staurengo-Ferrari et al., 2019), and to attain better analgesic effects, evidences has shown that Nrf2-activators can accord with OR. Sulforaphane, for example, was reported to enhance the antihyperalgesic and morphine antiallodynic effects by activating the Nrf2 transcription factor. According to Redondo et al. (2017), in the spinal cord sulforaphane increases the synthesis of HO-1 and NQO1, which are produced through an Nrf2 antioxidant-mediated pathway. In combination with butorphanol, fentanyl (an opioid drug) may increase the genes' activity NQO1 and HO-1 downstream by activating Nrf2-ARE signaling through k-OR. According to Zhang et al. (2016), this fact may be able to prevent oxidative stress in the cardiac ischemia-reperfusion damage model. Likewise, in diabetic neuropathy associated with type 2 diabetes, Nrf2 activation may enhance the antinociceptive effects of Δ-OR (McDonnell et al., 2017). Some molecular effects of cannabis depend on Nrf2, and these effects were studied in a Parkinson's disease model (Staurengo-Ferrari et al., 2019). For instance, in MCF-7 and MDA-MB-231 breast cancer cell lines, anandamide causes Nrf2 initiation and subsequent HO-1 mRNA expression (Staurengo-Ferrari et al., 2019). Additionally, by increasing Nrf2 and ATF4 transcription factors in a method involving Nrf2-Hmox1 and the Nrf2/ATF4 pathways, cannabidiol may be able to regulate LPS-induced oxidative stress and inflammation in BV2 cells (Juknat et al., 2013).

19.13 SYNERGIES: PLANT-BASED BIO-ACTIVES AND NANOCARRIER TECHNOLOGY

19.13.1 Rationale for Combining Bio-actives and Nanocarriers

The rationale for combining bio-actives and nanocarriers lies in the synergistic benefits that arise from their integration, particularly in the context of managing gastrointestinal inflammations. Bioactive compounds derived from plants possess inherent therapeutic properties, including anti-inflammatory, antioxidant, and immunomodulatory effects. However, their efficacy is often limited by poor solubility, stability, and bioavailability within the gastrointestinal tract (Bergonzi et al., 2022; Haruna & Yahaya, 2021). Nanocarriers, such as liposomes, nanoparticles made of polymers, and lipid-based formulations, offer solutions to these challenges by encapsulating bioactive compounds, protecting them from degradation, and facilitating targeted delivery to inflamed tissues (Alshawwa et al., 2022b; Rehman et al., 2020). The combination of bio-actives and nanocarriers enhances therapeutic outcomes through several mechanisms. Firstly, nanocarriers improve the solubility and stability of bioactive compounds, ensuring their sustained release and prolonged efficacy within the GI tract (Hadidi et al., 2023b). Secondly, nanocarriers enable site-specific delivery of bioactive compounds to inflamed mucosal tissues, minimizing systemic exposure and reducing off-target effects. Additionally, nanocarriers can be engineered to incorporate surface modifications or ligands that enhance cellular uptake and tissue penetration, further augmenting the therapeutic potential of encapsulated bio-actives.

19.13.2 Enhancement of Stability and Efficacy

Encapsulation of plant-based bio-actives within nanocarriers contributes significantly to the enhancement of stability and efficacy, overcoming challenges related to bioavailability and degradation.

19.13.2.1 Enhanced Stability Through Encapsulation

Plant-derived bioactive compounds, such as polyphenols and flavonoids, are often susceptible to environmental factors, including heat, light, and oxygen, leading to degradation and reduced efficacy. Nanocarriers act as protective shields, preventing premature degradation of bio-actives during transit through the harsh conditions of the gastrointestinal tract. The encapsulation provides a stable microenvironment, safeguarding the integrity of bioactive compounds and ensuring their availability at the target site (Mozafari et al., 2008).

19.13.2.2 Improved Bioavailability

Nanocarriers facilitate the efficient delivery of plant-based bio-actives to the gastrointestinal mucosa, addressing the challenge of limited bioavailability associated with these compounds. The encapsulation process enhances solubility, stability, and absorption, promoting sustained release and prolonged exposure of bioactive compounds to the inflamed tissues. This improved bioavailability ensures that therapeutic concentrations are achieved, maximizing the efficacy of the plant-derived compounds. (Torchilin, 2005a).

19.13.2.3 Controlled Release for Prolonged Efficacy

Nanocarriers enable controlled release kinetics, allowing for distribution of bioactive substances to the target location over an extended period of time. This controlled release not only enhances therapeutic efficacy but also minimizes potential side effects associated with high concentrations of bio-actives. By modulating the release profile, nanocarriers ensure a more stable and prolonged interaction between plant-based bio-actives and the inflamed gastrointestinal tissues (Uhrich et al., 1999).

19.14 PRESERVATION OF BIOLOGICAL ACTIVITY

Encapsulation in nanocarriers helps preserve the biological activity of plant-based bio-actives by shielding them from enzymatic degradation in the gastrointestinal environment. This protection ensures that the bio-actives reach the target tissues in their active form, maintaining their ability to modulate molecular pathways, regulate immune responses, and exert anti-inflammatory effects (Muller & Keck, 2004; see Table 19.5).

19.15 CONTROLLED RELEASE MECHANISMS

Varios controlled release mechanism pathways are shown in Table 19.6.

19.16 EXPERIMENTAL EVIDENCE AND CASE STUDIES

19.16.1 Experimental Studies Demonstrating Effectiveness

Remarkable characteristics of lipid-based nanocarriers include a high surface-to-mass ratio and prolonged stability over time, the ability to modify and conjugate, and the potential to encapsulate nutraceutical compounds. Poorly water-soluble substances can be greatly improved in intestinal absorption by lipid-based formulations. Three primary mechanisms have been identified as to how lipophilic excipients improve absorption, they are as follows: Nutraceuticals must dissolve in order to be absorbed by enterocytes, which are intestinal absorptive cells, from the small intestine lumen. The GI tract's solubilization is improved by lipid-based nanocarriers. by keeping the bioactive compounds soluble and avoiding precipitation in the gut environment.

TABLE 19.5
Synergistic Effects in Gastrointestinal Inflammation Management

Synergistic Effects in Gastrointestinal Inflammation Management	References
Enhanced Anti-Inflammatory Action: Combining the inherent anti-inflammatory properties of plant-based bio-actives (e.g., curcumin, quercetin) with the targeted delivery capabilities of nanocarriers results in a synergistic effect, maximizing the suppression of inflammatory pathways in the gastrointestinal mucosa.	(Boots et al., 2008; Pescosolido et al., 2014; Schaffer et al., 2015)
Improved Bioavailability and Stability: The encapsulation of plant-derived compounds within nanocarriers enhances their stability and bioavailability, ensuring a sustained and efficient release at the inflamed site. This synergy contributes to a prolonged and intensified therapeutic impact on gastrointestinal inflammation.	(C. H. Liu et al., 2010; Paiva-Santos et al., 2021b)
Modulation of Immune Responses: The combination of immunomodulatory properties inherent in certain plant bio-actives and the controlled release facilitated by nanocarriers leads to a synergistic modulation of immune responses. This dual-action approach helps restore immune homeostasis within the gastrointestinal tract.	(Du et al., 2014; Fulton & Najahi-Missaoui, 2023; Heng et al., 2022; Hwang et al., 2023; Iyer et al., 2013; Jadhav et al., 2012; Kshirsagar et al., 2023; Sun et al., 2014)
Preservation of Gut Barrier Integrity: Nanocarriers protect plant-based bio-actives from premature degradation, preserving their ability to regulate tight junction proteins and mucins. This synergy contributes to the restoration and maintenance of gut barrier integrity in inflamed tissues.	(Akhtar et al., 2012)

TABLE 19.6
Controlled Release Mechanism Pathway

Controlled Release Mechanisms	Details	References
Various Controlled Release Strategies	- Polymeric Nanoparticles: Utilizing biodegradable polymers, such as PLGA or chitosan, to encapsulate plant-based bio-actives and control their release kinetics through diffusion or degradation of the polymer matrix.	(Begines et al., 2020; Danhier et al., 2012; Gagliardi et al., 2021; Zielinska et al., 2020)
	- Liposomes: Employing lipid-based vesicles to encapsulate bioactive compounds and modulate release profiles through lipid composition, size, and surface modifications.	(Lombardo & Kiselev, 2022; Nsairat et al., 2022; Torchilin, 2005b)
	- Hydrogels: Incorporating bio-actives into hydrogel matrices to achieve sustained release and localized delivery in response to environmental cues, such as pH or temperature.	(Bashir et al., 2020; Cao et al., 2021; Lee & Mooney, 2001)
Sustained and Targeted Delivery Approaches	- pH-Sensitive Nanocarriers: Designing nanocarriers that respond to pH changes along the gastrointestinal tract, releasing bio-actives selectively in inflamed regions with altered pH levels.	(Karimi et al., 2016; Pandita et al., 2023; Sah & Sah, 2015)
	- Ligand-Functionalized Nanocarriers: Surface modification of nanocarriers with targeting ligands, such as antibodies or peptides, to facilitate specific binding to receptors overexpressed on inflamed tissues, enabling targeted delivery.	(Miyazaki et al., 2021; Ying et al., 2022)
Examples of Controlled Release Systems	- Curcumin-Loaded PLGA Nanoparticles: PLGA nanoparticles encapsulating curcumin demonstrate sustained release kinetics, enhancing bioavailability and anti-inflammatory efficacy in gastrointestinal inflammation models.	(Feltrin et al., 2022; Mohanty et al., 2012)
	- Resveratrol-Loaded Liposomes: Liposomal formulations of resveratrol exhibit controlled release characteristics, prolonging exposure to inflamed tissues and mitigating oxidative stress and inflammation in the GI tract.	(Balanč et al., 2015; Isailović et al., 2013)
	- Alginate Hydrogel Encapsulated Quercetin: Quercetin-loaded alginate hydrogels enable sustained release of the bioactive compound, offering prolonged protection against oxidative damage and inflammatory responses in the gut mucosa.	(Cadena-Velandia et al., 2020; Selvaraj et al., 2021)

The intestine lymphatic system is capable of carrying highly lipophilic substances into the systemic circulation. Normal circumstances involve the transfer of chemicals via the portal vein, into the systemic circulation (by first-pass metabolism). Lipid-based systems safeguard molecules from first-pass metabolism by enlisting the help of both internal and external lipids transporters and promoting lipophilic bioactive compounds co-administered and their intestinal lymphatic transit. Through interacting with lipid transporters in the apical membrane, lipophilic excipients and lipids can alter the lipid pool inside the enterocyte as well as the manifestation of lipid-binding proteins inside cells (Ezzati Nazhad Dolatabadi & Omidi, 2016; Porter et al., 2007; Subramanian et al., 2021). Two main types of inflammatory bowel illness are Crohn's disease (CD) and ulcerative colitis (UC)/ inflammatory bowel disease (IBD), both of which are linked to an imbalance in the gut microbiome, commonly referred to as a chronic relapsing GI disorder. Currently, there is no permanent cure for either of these two types. Severe gastrointestinal (GI) symptoms, such as diarrhea, anemia, bleeding, abdominal discomfort, and weight loss, are brought on by this morbid illness (Fiocchi, 2012; Pithadia & Jain, 2011a). It can be challenging to distinguish between a steady gut's symptoms and persons in a chronic condition given that clinical symptoms in IBD patients are not necessarily connected with how serious the sickness is. Professionals who treat mucosal lesions may show minor signs and symptoms or none at all. To assess and monitor the illness's development, accurate evaluation is needed. Even though a set procedure to identify IBD does not yet exist, evaluating the symptoms of the illness or how the body responds to medication or treatment can be useful. To determine the occurrence of a disease and to guide treatment choices, clinicians employ a variety of tools including endoscopy, radiation monitoring, laboratory indicators, clinical symptoms, and histological examination of tissue samples. Several of the imaging techniques listed below are also employed in the clinical diagnosis of IBD. MRI, PET, CT, and SPECT are the most often used techniques for imaging IBD. Mucosal lesions imaged in situ can be facilitated by endoscopy in the diagnosis of IBD. However, it is expensive, unpleasant, time-consuming, and has a chance of puncture. No, endoscopy can't be used to evaluate transmural inflammation (Fernandes et al., 2017; Peyrin-Biroulet et al., 2014a, 2014b). In order to evaluate and image IBD, imaging techniques can provide promising features. With the most recent advancements in multifunctional nanoparticles, IBD diagnosis is guided by a multidisciplinary approach. Utilizing nanotechnology in conjunction with imaging techniques can enable cellular or molecular applications, as well as early identification of IBD and monitoring of disease severity (Rieber et al., 2000a, 2000b; Saisho & Yamaguchi, 2004). A great deal of research has been done about non-invasive serum indicators to find indicators that are beneficial for diagnosing illnesses, division into smaller groups, tracking illness manifestation, and estimating comorbidities and the course of the patient because proper management of IBD is crucial for the disease's prognosis (Arasaradnam et al., 2016; Monasta et al., 2017).

19.16.2 Case Studies Highlighting Positive Outcome

Using macrophages labeled with indium (^{111}In)/iron oxide nanoparticles, Wu et al. improved the SPECT/CT and MRI imaging quality. Superparamagnetic iron oxide (SPIO) nanoparticle absorption was shown to be high (7.4 pg iron per cell) using ex vivo mass spectrometry. After 24 hours, every tissue associated with the reticuloendothelial region or the system of mononuclear phagocytes contained in-labeled cells. In evaluating IBD patients' pain throughout their entire gut, Basirat et al. assessed the precision of fecal calprotectin and Ileocolonoscopy combined with 99mTc(V)-dimercaptosuccinic acid (DMSA) as new methods for inflammatory location as a result of colonoscopy limitations. There were significant differences in the calprotectin level: 90%, 0.90, and 100% in terms of sensitivity, positive likelihood ratio, and positive predictive value, in identifying active illness by colonoscopy (Wu et al., 2014).

According to Naha et al., using a dextran coating ceriadoxine NPs would encourage assembling at IBD inflammatory sites. Cerium oxide NPs coated with dextran offered a strong CT contrast and were able to be located at the intestinal location of colitis. Zhou et al. created an in vivo fluorescence cascade amplifier (HCFA) that is powered by cytoplasmic proteins and hypoxia-activatable to visualize hypoxia associated with IBD. Mesoporous silica nanoparticles (MSN) doped with 4-aminobenzoic acid (azo) were used in their architecture as a carrier to load cytoplasmic protein-binding squarylium dye (SQ) and black hole quencher 2 (BHQ2). Afterward, the β-cyclodextrin polymer (β-CDP) joined in a host-guest exchange with azo to produce HCFA. According to the results of the fluorescence imaging, HCFA was able to detect and distinguish between different levels of cellular hypoxia in vivo (Basirat et al., 2016; Naha et al., 2020; Zhou et al., 2020). Biological research, especially that related to inflammatory bowel disease (IBD), has been using nanosensors more and more in the past few decades due to their rapid development. Using nanostructures to diagnose inflammatory bowel disease has become more common in recent years. The subsequent paragraphs will address the majority of recent research on IBD nano diagnosis studies. Irritable bowel syndrome (IBS) and inflammatory bowel disease (IBD) can be distinguished using chemical sensors, as described by Karban et al., built on programmable molecularly modified Au NPs. 75% of instances of Crohn's and colitis and 81% of cases of IBD and IBS were correctly classified as having distinct discriminative powers, according to the findings (Karban et al., 2016; Ruckh & Clark, 2014). Shepherd et al. used chemiresistive metal oxide gas sensor in conjunction with headspace gas chromatography to construct a stool analysis technique. For the purpose of detecting IBD, data on the fluctuation in resistivity across time from many samples was loaded into an artificial neural network program (Shepherd et al., 2014).

With a 76% average prediction precision overall, the machine successfully distinguished samples from IBD and IBS with 88% specificity and 76% sensitivity, respectively. In the past, silver nanoparticles and miRNAs were utilized as fluorescence quenching diagnostic probes for inflammation. These probes' low limits of detection (LODs) are caused by biomatrix interference, though. In order to tackle this issue, Fang et al. created a novel DNA/AgNC-cDNA nanosensor that operated by increasing fluorescence. To combat IBD infection, they employed the IBD biomarker miR-223. In comparison to the traditional fluorescence method, the newly designed probe demonstrated excellent specificity and sensitivity ten times. (Fang et al., 2020).

19.16.3 Lessons Learned from Experimental Models

IBD can be brought on by oxidative stress, genetic susceptibility, chronic illnesses, and alterations in the microbiota, despite the fact the accurate MOA of the disease is yet unknown. (Gersemann et al., 2012; Lu et al., 2018; Neuman, 2007). Adverse side effects of traditional IBD medications include an increased risk of infections and several types of malignancies (Bonovas et al., 2016). Furthermore, individuals need to take medications for an extended period of time to keep the illness from relapsing (Pithadia & Jain, 2011b; Rahimi et al., 2010). Finding a medication that is able to be delivered locally to the inflammatory tissues while avoiding systemic side effects will be beneficial for people with inflammatory bowel disease. Nanomedicine is a therapeutic approach that uses nanoparticles or nanocarriers to deliver medication to specific targets (Sargazi, Afzali, et al., 2019; Sargazi, Khajeh Ebrahimi, et al., 2019). In the experimental UC, Miroliaee and colleagues revealed that freshly synthesized selenium-NPs coupled with silymarin had an ameliorative effect (Miroliaee et al., 2011). Laroui et al. decided to employ modified NPs to administer KPV to the colon, a tripeptide with anti-inflammatory properties in a UC mice model. Their findings indicate that this tripeptide can be given at a concentration much less than tripeptide alone to the site of inflammation by using NP-based DDS (Laroui et al., 2010).

Using an encapsulated NP/hydrogel combination, Theiss and colleagues were able to successfully deliver prohibiting, a protein that is widely expressed, to the colon. They also noted the encouraging reduction of the impact of NMs when administering medication (Theiss et al., 2011). It was suggested by Ocansey et al. that NPs that resemble exosomes be used inside the IBD framework to carry proteins, functional RNAs, and synthetic compounds to the colon (Ocansey et al., 2020). Similarly, in a UC model in mice, NPs that resemble exosomes generated from grapes produced favorable protection.

Research has also focused on the application of NPs in the creation of innovative oral DDS, in which the pH or reactive oxygen species or the existence of enzymes primarily triggers the release of medications in the GI tract (Mittal et al., 2018; M. Zhang & Merlin, 2018). Effective medication delivery methods for the treatment of IBD have also been introduced, including RNAi-based NPs with a variety of clinical uses. Excellent biocompatibility has also been demonstrated by manufactured chondroitin sulfate-curcumin NPs, which may be used as a desirable therapeutic option for IBD (Gou et al., 2019). Fascinatingly, intestine organoids that include poly (lactide-co-glycolide) (PLGA) NPs have been considered as a possible DDS to transport traditional IBD drugs to the inflammatory regions (Davoudi et al., 2018).

19.17 SAFETY CONSIDERATIONS AND REGULATORY PERSPECTIVES

19.17.1 Addressing Safety Concerns

The phrase "nutraceutical" which means "a food or components of a food that have therapeutic or health advantages, such as the ability to prevent or treat disease," has become more popular lately. The terms ``pharmaceutical" and "nutrition" are the origin of this phrase. Many innovative nano-formulations and multiple patents, including β-carotene, fish oil, resveratrol, quercetin, epigallocatechin-3-gallate, curcumin, and gallic acid, were summarized in a recent paper. The anti-inflammatory and antioxidant properties of these innovative formulations are completely characterized by their effectiveness and toxicity in laboratory specimens. However, colitis research has never examined these components. It is necessary to do more research utilizing these exclusive patents and nano-formulations (Helal et al., 2019). A new approach to treating IBD is combination therapy, which is the subject of extensive pre-clinical research. Whether used alone or in conjunction with other molecules or biologics, natural macromolecules and nutritional supplements have demonstrated positive results over the period of treating inflammatory bowel disease (IBD) in recent years. In this sense, mesalazine (5-ASA) and zinc oxide nanoparticles together improve the medication's effectiveness in treating colitis. One useful technique to target cells for colitis treatment is the simultaneous administration of CD98siRNA and curcumin via hyaluronic acid (HA)-functionalized polymeric nanoparticles. Curcumin and celecoxib together are delivered using pH-sensitive nanoparticles as a possible colitis treatment. When silymarin and nano selenium are administered together, NF-κB is inhibited, which helps manage inflammatory bowel disease. In IBD patients, combination therapy for migraines using ω-3 fatty acids and nano-curcumin has to be investigated as well. Likewise, ω-3 polyunsaturated fatty acids contained in solid lipid nanoparticles based on resveratrol, which inhibits colorectal cancer in vitro models, can potentially lessen the risk of IBD complications (Gugulothu et al., 2014; J. Li et al., 2017; Serini et al., 2018; Varshosaz et al., 2015; Xiao et al., 2016).

Phytochemicals have the ability to influence multiple inflammatory and pathogenic targets. Owing to their enormous pharmacological activity, natural molecules currently face significant restrictions and impediments in IBD therapy, chief among them being a compromised therapeutic activity prior to reaching the colon that is inflamed. Different synthetic polymer-based nanoparticles that are hazardous to biological systems are being used to solve these difficulties. Future research endeavors should focus on encapsulating these chemicals into dose delivery process based on natural proteins and/or polysaccharides in order to distribute and/or target them with less chance of adverse effects than synthetic polymer-based nanocarriers. A biodegradable carrier system like this could be the perfect formulation for colon targeting in the future because it can maximize colon targeting, protect active phytochemicals from gastrointestinal instability, and degrade naturally by the colonic microflora without changing their composition. To absorb and disperse 6-shogaol, a molecule that is biologically active and present in ginger, throughout the colon, a medication delivery device utilizing natural lipid nanoparticles was created. The possibility exists for Ulva lactuca polysaccharide-selenium nanoparticles to be used therapeutically to lessen acute colitis symptoms by acting as an anti-inflammatory. By triggering dendritic cell activation of protein kinase triggered by adenosine monophosphate, broccoli extract-derived nanoparticles protect against colitis. In colitis models, TNF-α was reduced by liposomes including krill oil and entrapped with beneside. This natural delivery system has a lot of promise as an oral nanovehicle for IBD medication delivery (Deng et al., 2017; Yang et al., 2020; Zhu et al., 2017).

19.17.2 Regulatory Landscape for Innovative Therapies

Nowadays, there is a concept called "nanofoods," which is centered on using nanotechnologies to prepare safe foods that have better nutritional and sensory qualities and can offer a variety of health benefits. Aiming to enhance the bioavailable nature of bioactive substances used in food fortification, the development of novel nanobiosystems has been the attention of food scientists, adhering to the general principles of this concept. The primary determinants of oral bioavailability, such as the capacity of bioactive compounds to be absorbed, transformed, and bioavailable, as well as bioactive-loaded nanocarriers, have been the subject of numerous research projects. The ways in which bioactive substances and nanoparticles are transported via intestinal epithelial cells, the release, dissolution, and communication of these substances with food digestion using nanocarriers, and phase I and phase II reaction changes in terms of chemistry and biology are just a few of the well-studied physicochemical processes that contribute to the factors limiting oral bioavailability. Plant-based nanocarriers are useful in numerous biological contexts, such as tissue engineering and medication delivery. They are the perfect choice for healthcare applications because of their special qualities, which include simplicity of surface functionalization, natural abundance, high stability, and cost efficiency. Because of their structural diversity, chemical diversity, and biodiversity, plants are a significant source of various compounds that have been shown to offer a variety of medicinal benefits. These days, there is no denying that natural ingredients have contributed significantly to the growth of cutting-edge pharmaceutical formulations and continue to do so. Progress in nanotechnology has enabled the development of novel colloidal formulations capable of modifying the pharmacological and biological characteristics of drugs. The special physical, chemical and technological aspects of treatments utilizing nanoparticles have made numerous effective uses in the treatment of cancer possible. To enhance and maximize cell and tissue contact, studying and modifying the dimensions, charge, shape, and patterning of medicinal molecules at the nanoscale is essential (Dima et al., 2021).

19.18 FUTURE DIRECTIONS AND CHALLENGES

Despite advancements in utilizing plant-based bio-actives encapsulated in nanocarriers for gastrointestinal inflammation management, several future directions and challenges lie ahead, including emerging trends, technologies on the horizon, and addressing current obstacles.

19.18.1 Emerging Trends in the Field

As research in the field of plant-based bio-actives and nanocarrier technology continues to evolve, several emerging trends are becoming apparent. These include:

19.18.2 Personalized Medicine

Tailoring treatments to individual patient profiles, including genetic predisposition, immune status, and gut microbiota composition, to optimize therapeutic outcomes (Wang & Wang, 2023).

19.18.3 Multi-Target Approaches

Developing formulations that target multiple molecular pathways simultaneously to achieve synergistic effects and enhanced efficacy in managing gastrointestinal inflammations (Chauhan et al., 2020; C. Zhang et al., 2014).

19.18.4 Integration of Biomaterials

Exploring novel biomaterials and hybrid systems that combine the advantages of different nanocarriers, such as liposomes, polymeric nanoparticles, and lipid-based formulations, to improve stability, targeting, and bioavailability (Alshawwa et al., 2022c; Ferreira Soares et al., 2020; Shah et al., 2022).

19.18.5 Technologies on the Horizon

Several cutting-edge technologies hold promise for revolutionizing the field of gastrointestinal inflammation management using plant-based bio-actives encapsulated in nanocarriers. These include:

19.18.6 Targeted Delivery Systems

Advancements in nanotechnology, such as ligand-mediated targeting and stimuli-responsive nanocarriers, enable precise delivery of bioactive compounds to inflamed gastrointestinal tissues while minimizing systemic exposure (Chen et al., 2021; Su & Kang, 2020b).

19.18.7 Gene Editing and Therapeutics

Utilizing gene editing technologies, such as CRISPR-Cas9, to modify immune cell behavior and molecular pathways implicated in gastrointestinal inflammation, offering potential long-term therapeutic solutions (Kim et al., 2021; H. Li et al., 2020; Mengstie & Wondimu, 2021).

19.18.8 Microbiome Engineering

Engineering gut microbiota using probiotics, prebiotics, and fecal microbiota transplantation to restore microbial balance and modulate immune responses in gastrointestinal inflammatory diseases (Airola et al., 2023; Bai et al., 2023; Hu et al., 2022).

19.18.9 Addressing Current Challenges

Despite significant progress, several challenges must be addressed to realize the full potential of plant-based bio-actives encapsulated in nanocarriers for gastrointestinal inflammation management. These include:

19.18.9.1 Optimizing Formulation Stability

Improving the stability of nanocarrier formulations, particularly during storage and gastrointestinal transit, to ensure consistent therapeutic efficacy and shelf-life (Din et al., 2017).

19.18.9.2 Enhancing Targeting Efficiency

Creating plans to reduce off-target effects on healthy tissues while increasing the targeting efficacy of nanocarriers to inflammatory gastrointestinal tissues (Min et al., 2021; Su & Kang, 2020c).

19.18.9.3 Clinical Translation and Regulatory Approval

Navigating the regulatory landscape and conducting rigorous clinical trials to validate the safety, efficacy, and long-term outcomes of nanocarrier-based therapies for gastrointestinal inflammations (Gogtay et al., 2017).

19.19 CONCLUSION

This book chapter has explored a groundbreaking approach to managing gastrointestinal inflammations through the innovative utilization of plant-based bio-actives encapsulated within nanocarriers. By exploring the major molecular pathways underlying gastrointestinal inflammations and the diverse array of plant-derived compounds with therapeutic potential, we have uncovered promising avenues for intervention. Furthermore, insights into the design and fabrication of nanocarriers have highlighted the importance of controlled release and targeted delivery for enhancing the efficacy of these bioactive compounds. The integration of plant-based bio-actives with nanocarriers holds significant implications for the management of gastrointestinal inflammations. By harnessing the synergistic properties of these compounds and the unique capabilities of nanocarriers, targeted and effective therapies can be developed. This approach offers the potential to minimize systemic side effects, improve patient outcomes, and address the complex challenges associated with current treatment modalities for conditions such as inflammatory bowel disease and gastritis. Finally, the integration of plant-based bio-actives encapsulated within nanocarriers provides an achievable strategy for the management of gastrointestinal inflammations. By combining the principles of traditional herbal medicine with cutting-edge nanotechnology, this approach paves the way for the development of targeted, efficient, and minimally invasive therapies. Moving forward, continued research and innovation in this field will be crucial for optimizing formulations, enhancing efficacy, and translating these advancements into clinical practice, ultimately improving the lives of patients with gastrointestinal inflammatory disorders.

ABBREVIATIONS

IBD - Inflammatory Bowel Disease
GI - Gastrointestinal tract
TNF-α - Tumor necrosis factor
PPs - polyphenols
Eos - Essential oils
CD - Cyclodextrins
CD - Crohn's Disease
NPs - Nanoparticles
MOA - Mechanism of action

REFERENCES

Abraham, C., & Cho, J. H. (2009). Inflammatory bowel disease. *The New England Journal of Medicine*, *361*(21), 2066–2078. https://doi.org/10.1056/NEJMRA0804647

Adamczak, M., Kupiec, A., Jarek, E., Szczepanowicz, K., & Warszyński, P. (2014). Preparation of the squalene-based capsules by membrane emulsification method and polyelectrolyte multilayer adsorption. *Colloids and Surfaces A: Physicochemical and Engineering Aspects*, *462*, 147–152.

Airola, C., Severino, A., Porcari, S., Fusco, W., Mullish, B. H., Gasbarrini, A., Cammarota, G., Ponziani, F. R., & Ianiro, G. (2023). Future modulation of gut microbiota: From eubiotics to FMT, engineered bacteria, and phage therapy. *Antibiotics*, *12*(5). https://doi.org/10.3390/ANTIBIOTICS12050868

Akhtar, F., Rizvi, M. M. A., & Kar, S. K. (2012). Oral delivery of curcumin bound to chitosan nanoparticles cured Plasmodium yoelii infected mice. *Biotechnology Advances*, *30*(1), 310–320. https://doi.org/10.1016/J.BIOTECHADV.2011.05.009

Alqahtani, M. S., Kazi, M., Alsenaidy, M. A., & Ahmad, M. Z. (2021). Advances in oral drug delivery. *Frontiers in Pharmacology*, *12*, 618411. https://doi.org/10.3389/FPHAR.2021.618411

Alshawwa, S. Z., Kassem, A. A., Farid, R. M., Mostafa, S. K., & Labib, G. S. (2022a). Nanocarrier drug delivery systems: Characterization, limitations, future perspectives and implementation of artificial intelligence. *Pharmaceutics*, *14*(4), 883. https://doi.org/10.3390/PHARMACEUTICS14040883

Alshawwa, S. Z., Kassem, A. A., Farid, R. M., Mostafa, S. K., & Labib, G. S. (2022b). Nanocarrier drug delivery systems: Characterization, limitations, future perspectives and implementation of artificial intelligence. *Pharmaceutics*, *14*(4). https://doi.org/10.3390/PHARMACEUTICS14040883

Alshawwa, S. Z., Kassem, A. A., Farid, R. M., Mostafa, S. K., & Labib, G. S. (2022c). Nanocarrier drug delivery systems: Characterization, limitations, future perspectives and implementation of artificial intelligence. *Pharmaceutics*, *14*(4). https://doi.org/10.3390/PHARMACEUTICS14040883

Ali, S., Ekbbal, R., Salar, S., Yasheshwar, Ali, S. A., Jaiswal, A. K., ... & Gaurav. (2023). Quality standards and pharmacological interventions of natural oils: Current scenario and future perspectives. *ACS Omega*, *8*(43), 39945–39963.

Anantharaju, P. G., Gowda, P. C., Vimalambike, M. G., & Madhunapantula, S. V. (2016). An overview on the role of dietary phenolics for the treatment of cancers. *Nutrition Journal*, *15*(1), 1–16.

Ansari, K. A., Vavia, P. R., Trotta, F., & Cavalli, R. (2011). Cyclodextrin-based nanosponges for delivery of resveratrol: In vitro characterisation, stability, cytotoxicity and permeation study. *Aaps Pharmscitech*, *12*, 279–286.

Anselmo, A. C., & Mitragotri, S. (2019). Nanoparticles in the clinic: An update. *Bioengineering & Translational Medicine*, *4*(3). https://doi.org/10.1002/BTM2.10143

Arasaradnam, R. P., McFarlane, M., Daulton, E., Skinner, J., O'Connell, N., Wurie, S., Chambers, S., Nwokolo, C., Bardhan, K., Savage, R., & Covington, J. (2016). Non-invasive exhaled volatile organic biomarker analysis to detect inflammatory bowel disease (IBD). *Digestive and Liver Disease: Official Journal of the Italian Society of Gastroenterology and the Italian Association for the Study of the Liver*, *48*(2), 148–153. https://doi.org/10.1016/J.DLD.2015.10.013

Axelrad, J. E., Cadwell, K. H., Colombel, J. F., & Shah, S. C. (2021). The role of gastrointestinal pathogens in inflammatory bowel disease: A systematic review. *Therapeutic Advances in Gastroenterology*, *14*. https://doi.org/10.1177/17562848211004493/SUPPL_FILE/SJ-DOCX-1-TAG-10.1177_17562848211004493.DOC

Azman, M., Sabri, A. H., Anjani, Q. K., Mustaffa, M. F., & Hamid, K. A. (2022). Intestinal absorption study: Challenges and absorption enhancement strategies in improving oral drug delivery. *Pharmaceuticals*, *15*(8), 975. https://doi.org/10.3390/PH15080975

Azzouz, L. L., & Sharma, S. (2023). *Physiology, Large Intestine*. StatPearls. www.ncbi.nlm.nih.gov/books/NBK507857/

Bai, X., Huang, Z., Duraj-Thatte, A. M., Ebert, M. P., Zhang, F., Burgermeister, E., Liu, X., Scott, B. M., Li, G., & Zuo, T. (2023). Engineering the gut microbiome. *Nature Reviews*

Bioengineering, *1*(9), 665–679. https://doi.org/10.1038/s44222-023-00072-2

Bandakinda, M., Kumar, A., & Mishra, A. (2024). Toxicity issues of nanoparticles in the delivery of phytoconstituents and cosmeceuticals. In Pooja, D. & Kulhari, H. (eds.), *Nanotechnology Based Delivery of Phytoconstituents and Cosmeceuticals* (pp. 329–342). Springer Nature Singapore.

Balanč, B. D., Ota, A., Djordjević, V. B., Šentjurc, M., Nedović, V. A., Bugarski, B. M., & Ulrih, N. P. (2015). Resveratrol-loaded liposomes: Interaction of resveratrol with phospholipids. *European Journal of Lipid Science and Technology*, *117*(10), 1615–1626. https://doi.org/10.1002/EJLT.201400481

Bala, I., Bhardwaj, V., Hariharan, S., Kharade, S. V., Roy, N., & Ravi Kumar, M. N. V. (2006). Sustained release nanoparticulate formulation containing antioxidant-ellagic acid as potential prophylaxis system for oral administration. *Journal of Drug Targeting*, *14*(1), 27–34.

Bander, Z. Al, Nitert, M. D., Mousa, A., & Naderpoor, N. (2020). The gut microbiota and inflammation: An overview. *International Journal of Environmental Research and Public Health*, *17*(20), 1–22. https://doi.org/10.3390/ijerph17207618

Barnes, M., Flaherty, J., Caron, M., Naqvee, A., & Bierer, B. (n.d.). The evolving regulatory landscape for clinical trials in India. Retrieved February 10, 2024, from www.thehindu.com/sci-tech/health/policy-and-issues/post-stringent-norms-clinical-trials-in-india

Baralić, K., Živanović, J., Marić, Đ., Bozic, D., Grahovac, L., Antonijević Miljaković, E., ... & Đukić-Ćosić, D. (2024). Sulforaphane—A compound with potential health benefits for disease prevention and treatment: Insights from pharmacological and toxicological experimental studies. *Antioxidants*, *13*(2), 147.

Barras, A., Mezzetti, A., Richard, A., Lazzaroni, S., Roux, S., Melnyk, P., ... & Monfilliette-Dupont, N. (2009). Formulation and characterization of polyphenol-loaded lipid nanocapsules. *International Journal of Pharmaceutics*, *379*(2), 270–277.

Bashir, S., Hina, M., Iqbal, J., Rajpar, A. H., Mujtaba, M. A., Alghamdi, N. A., Wageh, S., Ramesh, K., & Ramesh, S. (2020). Fundamental concepts of hydrogels: Synthesis, properties, and their applications. *Polymers*, *12*(11), 1–60. https://doi.org/10.3390/POLYM12112702

Basirat, V., Azizi, Z., Anbardan, S. J., Asl, M. T., Farbod, Y., Teimouri, A., Daryani, E., & Daryani, N. E. (2016). Accuracy of 99mTc (V)-dimercaptosuccinic acid scintigraphy and fecal calprotectin compared with colonoscopy in localizing active lesions in inflammatory bowel disease. *Middle East Journal of Digestive Diseases*, *8*(3), 212–218. https://doi.org/10.15171/MEJDD.2016.23

Begines, B., Ortiz, T., Pérez-Aranda, M., Martínez, G., Merinero, M., Argüelles-Arias, F., & Alcudia, A. (2020). Polymeric nanoparticles for drug delivery: Recent developments and future prospects. *Nanomaterials*, *10*(7), 1–41. https://doi.org/10.3390/NANO10071403

Bensasson, R. V., Zoete, V., Dinkova-Kostova, A. T., & Talalay, P. (2008). Two-step mechanism of induction of the gene expression of a prototypic cancer-protective enzyme by diphenols. *Chemical Research in Toxicology*, *21*(4), 805–812.

Bergonzi, M. C., Heard, C. M., & Garcia-Pardo, J. (2022). Bioactive molecules from plants: Discovery and pharmaceutical applications. *Pharmaceutics*, *14*(10), 2116. https://doi.org/10.3390/PHARMACEUTICS14102116

Bhojwani, H., Begwani, K., Bhor, V., Bedi, P., Balasinor, N., Raut, S., & Joshi, U. (2023). Synthesis and biological evaluation of benzamide-chalcone hybrids as potential c-Met kinase and COX-2 inhibitors. *Archiv der Pharmazie*, e2200405.

Bertrand, N., Wu, J., Xu, X., Kamaly, N., & Farokhzad, O. C. (2014). Cancer nanotechnology: the impact of passive and active targeting in the era of modern cancer biology. *Advanced Drug Delivery Reviews*, *66*, 2–25.

Bonovas, S., Fiorino, G., Allocca, M., Lytras, T., Nikolopoulos, G. K., Peyrin-Biroulet, L., & Danese, S. (2016). Biologic therapies and risk of infection and malignancy in patients with inflammatory bowel disease: A systematic review and network meta-analysis. *Clinical Gastroenterology and Hepatology: The Official Clinical Practice Journal of the American Gastroenterological Association*, *14*(10), 1385–1397.e10. https://doi.org/10.1016/J.CGH.2016.04.039

Boots, A. W., Haenen, G. R. M. M., & Bast, A. (2008). Health effects of quercetin: From antioxidant to nutraceutical. *European Journal of Pharmacology*, *585*(2–3), 325–337. https://doi.org/10.1016/J.EJPHAR.2008.03.008

Burns, G., Pryor, J., Holtmann, G., Walker, M. M., Talley, N. J., & Keely, S. (2019). Immune activation in functional gastrointestinal disorders. *Gastroenterology & Hepatology*, *15*, 539.

Bizzarro, V., Carfagna, C., Cerruti, P., Marturano, V., & Ambrogi, V. (2016, May). Light-responsive polymer microcapsules as delivery systems for natural active agents. In D'Amore, A., Acierno, D., & Grassia, L. (eds.), *AIP Conference Proceedings* (Vol. 1736, No. 1). AIP Publishing.

Bustanji, Y., Taha, M. O., Almasri, I. M., Al-Ghussein, M. A., Mohammad, M. K., & Alkhatib, H. S. (2009). Inhibition of glycogen synthase kinase by curcumin: Investigation by simulated molecular docking and subsequent in vitro/in vivo evaluation. *Journal of Enzyme Inhibition and Medicinal Chemistry*, *24*(3), 771–778.

Butnariu, M. V., & Giuchici, C. V. (2011). The use of some nanoemulsions based on aqueous propolis and lycopene extract in the skin's protective mechanisms against UVA radiation. *Journal of nanobiotechnology*, *9*, 1–9.

Cadena-Velandia, Z. G., Montenegro-Alarcón, J. C., Marquínez-Casas, X., & Mora-Huertas, C. E. (2020). Quercetin-loaded alginate microparticles: A contribution on the particle structure. *Journal of Drug Delivery Science and Technology*, *56*, 101558. https://doi.org/10.1016/J.JDDST.2020.101558

Calixto-Campos, C., Corrêa, M. P., Carvalho, T. T., Zarpelon, A. C., Hohmann, M. S., Rossaneis, A. C., ... & Verri, W. A. (2015). Quercetin reduces Ehrlich tumor-induced cancer pain in mice. *Analytical cellular pathology*, *2015*.

Cao, H., Duan, L., Zhang, Y., Cao, J., & Zhang, K. (2021). Current hydrogel advances in physicochemical and biological response-driven biomedical application diversity. *Signal Transduction and Targeted Therapy*, *6*(1), 1–31. https://doi.org/10.1038/s41392-021-00830-x

Cevallos, P. A. P., Buera, M. P., & Elizalde, B. E. (2010). Encapsulation of cinnamon and thyme essential oils components (cinnamaldehyde and thymol) in β-cyclodextrin: Effect of interactions with water on complex stability. *Journal of Food Engineering*, *99*(1), 70–75.

Chakraborty, S., Stalin, S., Das, N., Choudhury, S. T., Ghosh, S., & Swarnakar, S. (2012). The use of nano-quercetin to arrest mitochondrial damage and MMP-9 upregulation during prevention of gastric inflammation induced by ethanol in rat. *Biomaterials*, *33*(10), 2991–3001.

Chauhan, V. M., Zhang, H., Dalby, P. A., & Aylott, J. W. (2020). Advancements in the co-formulation of biologic therapeutics. *Journal of Controlled Release*, *327*, 397–405. https://doi.org/10.1016/J.JCONREL.2020.08.013

Chen, L., Deng, H., Cui, H., Fang, J., Zuo, Z., Deng, J., Li, Y., Wang, X., & Zhao, L. (2018). Inflammatory responses and inflammation-associated diseases in organs. *Oncotarget*, *9*(6), 7204–7218. www.impactjournals.com/oncotarget/

Chen, L. H., & Hu, J. N. (2024). Development of nano-delivery systems for loaded bioactive compounds: Using molecular dynamics simulations. *Critical Reviews in Food Science and Nutrition*, 1–22.

Chen, L., Hong, W., Ren, W., Xu, T., Qian, Z., & He, Z. (2021). Recent progress in targeted delivery vectors based on biomimetic nanoparticles. *Signal Transduction and Targeted Therapy*, *6*(1), 1–25. https://doi.org/10.1038/s41392-021-00631-2

Chou, A. H., Lee, H. C., Liao, C. C., Yu, H. P., & Liu, F. C. (2023). ERK/NF-kB/COX-2 signaling pathway plays a key role in curcumin protection against acetaminophen-induced liver injury. *Life*, *13*(11), 2150.

Danhier, F., Ansorena, E., Silva, J. M., Coco, R., Le Breton, A., & Préat, V. (2012). PLGA-based nanoparticles: An overview of biomedical applications. *Journal of Controlled Release: Official Journal of the Controlled Release Society*, *161*(2), 505–522. https://doi.org/10.1016/J.JCONREL.2012.01.043

Davoudi, Z., Peroutka-Bigus, N., Bellaire, B., Wannemuehler, M., Barrett, T. A., Narasimhan, B., & Wang, Q. (2018). Intestinal organoids containing poly(lactic-co-glycolic acid) nanoparticles for the treatment of inflammatory bowel diseases. *Journal of Biomedical Materials Research. Part A*, *106*(4), 876–886. https://doi.org/10.1002/JBM.A.36305

Deng, Z., Rong, Y., Teng, Y., Mu, J., Zhuang, X., Tseng, M., Samykutty, A., Zhang, L., Yan, J., Miller, D., Suttles, J., & Zhang, H. G. (2017). Broccoli-derived nanoparticle inhibits mouse colitis by activating dendritic cell AMP-activated protein kinase. *Molecular Therapy: The Journal of the American Society of Gene Therapy*, *25*(7), 1641–1654. https://doi.org/10.1016/J.YMTHE.2017.01.025

de Oliveira, E. F., Paula, H. C., & de Paula, R. C. (2014). Alginate/cashew gum nanoparticles for essential oil encapsulation. *Colloids and Surfaces B: Biointerfaces*, *113*, 146–151.

De La Ossa, D. H. P., Ligresti, A., Gil-Alegre, M. E., Aberturas, M. R., Molpeceres, J., Di Marzo, V., & Suárez, A. T. (2012). Poly-ε-caprolactone microspheres as a drug delivery system for cannabinoid administration: Development, characterization and in vitro evaluation of their antitumoral efficacy. *Journal of Controlled Release*, *161*(3), 927–932.

Dias, M. C., Pinto, D. C. G. A., & Silva, A. M. S. (2021). Plant flavonoids: Chemical characteristics and biological activity. *Molecules*, *26*, 17, 5377. https://doi.org/10.3390/molecules26175377

Dima, C., Assadpour, E., Dima, S., & Jafari, S. M. (2021). Nutraceutical nanodelivery; an insight into the bioaccessibility/bioavailability of different bioactive compounds loaded within nanocarriers. *Critical Reviews in Food Science and Nutrition*, *61*(18), 1–35. https://doi.org/10.1080/10408398.2020.1792409

Din, F. U., Aman, W., Ullah, I., Qureshi, O. S., Mustapha, O., Shafique, S., & Zeb, A. (2017). Effective use of nanocarriers as drug delivery systems for the treatment of selected tumors. *International Journal of Nanomedicine*, *12*, 7291. https://doi.org/10.2147/IJN.S146315

Dinkova-Kostova, A. T., Hakomäki, H., & Levonen, A. L. (2024). Electrophilic metabolites targeting the KEAP1/NRF2 partnership. *Current Opinion in Chemical Biology*, *78*, 102425.

Domitrović, R., Škoda, M., Marchesi, V. V., Cvijanović, O., Pugel, E. P., & Štefan, M. B. (2013). Rosmarinic acid ameliorates acute liver damage and fibrogenesis in carbon tetrachloride-intoxicated mice. *Food and Chemical Toxicology*, *51*, 370–378.

Du, J. Z., Mao, C. Q., Yuan, Y. Y., Yang, X. Z., & Wang, J. (2014). Tumor extracellular acidity-activated nanoparticles as drug delivery systems for enhanced cancer therapy. *Biotechnology Advances*, *32*(4), 789–803. https://doi.org/10.1016/j.biotechadv.2013.08.002

Durán-Lobato, M., Martín-Banderas, L., Lopes, R., Gonçalves, L. M. D., Fernández-Arévalo, M., & Almeida, A. J. (2016). Lipid nanoparticles as an emerging platform for cannabinoid delivery: Physicochemical optimization and biocompatibility. *Drug development and Industrial Pharmacy*, *42*(2), 190–198.

Ezzati Nazhad Dolatabadi, J., & Omidi, Y. (2016). Solid lipid-based nanocarriers as efficient targeted drug and gene delivery systems. *TrAC Trends in Analytical Chemistry*, *77*, 100–108. https://doi.org/10.1016/J.TRAC.2015.12.016

Esposito, E., Drechsler, M., Cortesi, R., & Nastruzzi, C. (2016). Encapsulation of cannabinoid drugs in nanostructured lipid carriers. *European Journal of Pharmaceutics and Biopharmaceutics*, *102*, 87–91.

Fang, B., Guo, P., Yang, M., Ma, Y., Yan, X., Jia, Z., Gao, W., Ahmad, S., Xu, C., Liu, C., & Zhang, M. (2020). A novel fluorescent enhancing platform based on DNA-scaffolded silver nanoclusters for potential inflammatory bowel disease-associated microRNA detection. *Talanta*, *218*, 121122. https://doi.org/10.1016/J.TALANTA.2020.121122

Fattori, V., Hohmann, M. S., Rossaneis, A. C., Pinho-Ribeiro, F. A., & Verri Jr, W. A. (2016). Capsaicin: current understanding of its mechanisms and therapy of pain and other pre-clinical and clinical uses. *Molecules*, *21*(7), 844.

Feltrin, F. da S., Agner, T., Sayer, C., & Lona, L. M. F. (2022). Curcumin encapsulation in functional PLGA nanoparticles: A promising strategy for cancer therapies. *Advances in Colloid and Interface Science*, *300*, 102582. https://doi.org/10.1016/J.CIS.2021.102582

Fernandes, S. R., Rodrigues, R. V., Bernardo, S., Cortez-Pinto, J., Rosa, I., Da Silva, J. P., Gonçalves, A. R., Valente, A., Baldaia, C., Santos, P. M., Correia, L., Venâncio, J., Campos, P., Pereira, A. D., & Velosa, J. (2017). Transmural healing is associated with improved long-term outcomes of patients with Crohn's disease. *Inflammatory Bowel Diseases*, *23*(8), 1403–1409. https://doi.org/10.1097/MIB.0000000000001143

Ferraz, C. R., Carvalho, T. T., Manchope, M. F., Artero, N. A., Rasquel-Oliveira, F. S., Fattori, V., ... & Verri Jr, W. A. (2020). Therapeutic potential of flavonoids in pain and inflammation: Mechanisms of action, pre-clinical and clinical data, and pharmaceutical development. *Molecules*, *25*(3), 762.

Ferreira Soares, D. C., Domingues, S. C., Viana, D. B., & Tebaldi, M. L. (2020). Polymer-hybrid nanoparticles: Current advances in biomedical applications. *Biomedicine & Pharmacotherapy*, *131*, 110695. https://doi.org/10.1016/J.BIOPHA.2020.110695

Festa, J., Hussain, A., Al-Hareth, Z., Singh, H., & Da Boit, M. (2023). Anthocyanins and vascular health: A matter of metabolites. *Foods*, *12*(9), 1796.

Fiocchi, C. (2012). Towards a "cure" for IBD. *Digestive Diseases (Basel, Switzerland)*, *30*(4), 428–433. https://doi.org/10.1159/000338148

Fleige, E., Quadir, M. A., & Haag, R. (2012). Stimuli-responsive polymeric nanocarriers for the controlled transport of active compounds: Concepts and applications. *Advanced drug delivery reviews*, *64*(9), 866–884.

Fulton, M. D., & Najahi-Missaoui, W. (2023). Liposomes in cancer therapy: How did we start and where are we now. *International Journal of Molecular Sciences*, *24*(7), 102582. https://doi.org/10.3390/ijms24076615

Gagliardi, A., Giuliano, E., Venkateswararao, E., Fresta, M., Bulotta, S., Awasthi, V., & Cosco, D. (2021). Biodegradable polymeric nanoparticles for drug delivery to solid tumors. *Frontiers in Pharmacology*, *12*, 601626. https://doi.org/10.3389/FPHAR.2021.601626/FULL

Gao, H., Li, Y., Zhang, X., Zhang, H., Tian, Y., & Li, B. (2024). Unraveling the G protein-coupled receptor superfamily in aphids: Contractions and duplications linked to phloem feeding. *General and Comparative Endocrinology*, *347*, 114435.

Gao, L., Liu, G., Ma, J., Wang, X., Zhou, L., Li, X., & Wang, F. (2013). Application of drug nanocrystal technologies on oral drug delivery of poorly soluble drugs. *Pharmaceutical Research*, *30*(2), 307–324. https://doi.org/10.1007/S11095-012-0889-Z/METRICS

García-Santos, W. H., Mayorquín-Torres, M. C., Maldonado-Domínguez, M., Medina-Campos, O. N., Pedraza-Chaverri, J., & Iglesias-Arteaga, M. A. (2023). Mechanistic insights on Pd-catalyzed three-component reactions of alkynols, methyl orthoformate, and salicylaldehyde derivatives. Application to the synthesis of steroid chroman ketals and spiroketals with antioxidant activity. *The Journal of Organic Chemistry*, *88*(21), 14860–14873.

Gersemann, M., Wehkamp, J., & Stange, E. F. (2012). Innate immune dysfunction in inflammatory bowel disease. *Journal of Internal Medicine*, *271*(5), 421–428. https://doi.org/10.1111/J.1365-2796.2012.02515.X

Geszke-Moritz, M., Nowak, G., & Moritz, M. (2023). Pharmacological properties and safe use of 12 medicinal plant species and their bioactive compounds affecting the immune system. *Applied Sciences*, *13*(11), 6477.

Gogtay, N. J., Ravi, R., & Thatte, U. M. (2017). Regulatory requirements for clinical trials in India: What academicians need to know. *Indian Journal of Anaesthesia*, *61*(3), 192. https://doi.org/10.4103/IJA.IJA_143_17

Gou, S., Huang, Y., Wan, Y., Ma, Y., Zhou, X., Tong, X., Huang, J., Kang, Y., Pan, G., Dai, F., & Xiao, B. (2019). Multi-bioresponsive silk fibroin-based nanoparticles with on-demand cytoplasmic drug release capacity for CD44-targeted alleviation of ulcerative colitis. *Biomaterials*, *212*, 39–54. https://doi.org/10.1016/J.BIOMATERIALS.2019.05.012

Grisham, M. B. (1994). Oxidants and free radicals in inflammatory bowel disease. *Lancet (London, England)*, *344*(8926), 859–861. https://doi.org/10.1016/S0140-6736(94)92831-2

Gugulothu, D., Kulkarni, A., Patravale, V., & Dandekar, P. (2014). pH-sensitive nanoparticles of curcumin–celecoxib combination: Evaluating drug synergy in ulcerative colitis model. *Journal of Pharmaceutical Sciences*, *103*(2), 687–696. https://doi.org/10.1002/JPS.23828

Guimarães, A. G., Oliveira, M. A., dos Santos Alves, R., dos Passos Menezes, P., Serafini, M. R., de Souza Araújo, A. A., ... & Júnior, L. J. Q. (2015). Encapsulation of carvacrol, a monoterpene present in the essential oil of oregano, with β-cyclodextrin, improves the pharmacological response on cancer pain experimental protocols. *Chemico-Biological Interactions*, *227*, 69–76.

Hadidi, M., Tan, C., Assadpour, E., Kharazmi, M. S., & Jafari, S. M. (2023a). Emerging plant proteins as nanocarriers of bioactive compounds. *Journal of Controlled Release*, *355*, 327–342. https://doi.org/10.1016/J.JCONREL.2023.01.069

Hadidi, M., Tan, C., Assadpour, E., Kharazmi, M. S., & Jafari, S. M. (2023b). Emerging plant proteins as nanocarriers of bioactive compounds. *Journal of Controlled Release*, *355*, 327–342. https://doi.org/10.1016/J.JCONREL.2023.01.069

Haruna, A., & Yahaya, S. M. (2021). Recent advances in the chemistry of bioactive compounds from plants and soil microbes: A review. *Chemistry Africa*, *4*(2), 231. https://doi.org/10.1007/S42250-020-00213-9

Helal, N. A., Eassa, H. A., Amer, A. M., Eltokhy, M. A., Edafiogho, I., & Nounou, M. I. (2019). Nutraceuticals' novel formulations: The good, the bad, the unknown and patents involved. *Recent Patents on Drug Delivery & Formulation*, *13*(2), 105–156. https://doi.org/10.2174/1872211313666190503112040

Heng, W. T., Yew, J. S., & Poh, C. L. (2022). Nanovaccines against viral infectious diseases. *Pharmaceutics*, *14*(12), 2554. https://doi.org/10.3390/pharmaceutics14122554

Hrncir, T. (2022). Gut microbiota dysbiosis: Triggers, consequences, diagnostic and therapeutic options. *Microorganisms*, *10*(3), 578. https://doi.org/10.3390/MICROORGANISMS10030578

Hu, Y., Chen, Y., Xu, J., Wang, X., Luo, S., Mao, B., Zhou, Q., & Li, W. (2022). Metagenomic discovery of novel CRISPR-Cas13 systems. *Cell Discovery*, *8*(1), 1–4. https://doi.org/10.1038/s41421-022-00464-5

Hua, S. (2019). Physiological and pharmaceutical considerations for rectal drug formulations. *Frontiers in Pharmacology*, *10*, 489933. https://doi.org/10.3389/FPHAR.2019.01196

Hua, S., Marks, E., Schneider, J. J., & Keely, S. (2015). Advances in oral nano-delivery systems for colon targeted drug delivery in inflammatory bowel disease: selective targeting to diseased versus healthy tissue. *Nanomedicine: Nanotechnology, Biology, and Medicine*, *11*(5), 1117–1132. https://doi.org/10.1016/J.NANO.2015.02.018

Huang, W., Wang, Y., Tian, W., Cui, X., Tu, P., Li, J., Shi, S., & Liu, X. (2022). Biosynthesis investigations of terpenoid, alkaloid, and flavonoid antimicrobial agents derived from medicinal plants. *Antibiotics*, *11*(10), 1380. https://doi.org/10.3390/ANTIBIOTICS11101380

Hwang, J., Huang, H., Sullivan, M. O., & Kiick, K. L. (2023). Controlled delivery of vancomycin from collagen-tethered peptide vehicles for the treatment of wound infections. *Molecular Pharmaceutics*, *20*(3), 1696–1708. https://doi.org/10.1021/acs.molpharmaceut.2c00898

Isailović, B. D., Kostić, I. T., Zvonar, A., Dordević, V. B., Gašperlin, M., Nedović, V. A., & Bugarski, B. M. (2013). Resveratrol loaded liposomes produced by different techniques. *Innovative Food Science & Emerging Technologies*, *19*, 181–189. https://doi.org/10.1016/J.IFSET.2013.03.006

Iwata, K., Mikami, Y., Kato, M., Yahagi, N., & Kanai, T. (2021). Pathogenesis and management of gastrointestinal inflammation and fibrosis: From inflammatory bowel diseases to endoscopic surgery. *Inflammation and Regeneration*, *41*(1), 1–8. https://doi.org/10.1186/S41232-021-00174-7

Iyer, A. K., Singh, A., Ganta, S., & Amiji, M. M. (2013). Role of integrated cancer nanomedicine in overcoming drug resistance. *Advanced Drug Delivery Reviews*, *65*(13–14), 1784–1802. https://doi.org/10.1016/j.addr.2013.07.012

Jadhav, M. P., Shinde, V. M., Chandrakala, S., Jijina, F., Menon, H., Arora, B., Kurkure, P. A., Parikh, P. M., & Kshirsagar, N. A. (2012). A randomized comparative trial evaluating the safety and efficacy of liposomal amphotericin B (Fungisome TM) versus conventional amphotericin B in the empirical treatment of febrile neutropenia in India. *Indian Journal of Cancer*, *49*(1), 107–113. https://doi.org/10.4103/0019-509X.98933

Juknat, A., Pietr, M., Kozela, E., Rimmerman, N., Levy, R., Gao, F., ... & Vogel, Z. (2013). Microarray and pathway analysis reveal distinct mechanisms underlying cannabinoid-mediated modulation of LPS-induced activation of BV-2 microglial cells. *PloS One*, *8*(4), e61462.

Kakarla, R., Karuturi, P., Siakabinga, Q., Kasi Viswanath, M., Dumala, N., Guntupalli, C., ... & Gujjari, L. (2024). Current understanding and future directions of cruciferous vegetables and their phytochemicals to combat neurological diseases. *Phytotherapy Research*, *38*, 1381–1399.

Kandpal, M., Indari, O., Baral, B., Jakhmola, S., Tiwari, D., Bhandari, V., Pandey, R. K., Bala, K., Sonawane, A., & Jha, H. C. (2022). Dysbiosis of gut microbiota from the perspective of the gut–brain axis: Role in the provocation of neurological disorders. *Metabolites*, *12*(11). https://doi.org/10.3390/METABO12111064

Karban, A., Nakhleh, M. K., Cancilla, J. C., Vishinkin, R., Rainis, T., Koifman, E., Jeries, R., Ivgi, H., Torrecilla, J. S., & Haick, H. (2016). Programmed nanoparticles for tailoring the detection of inflammatory bowel diseases and irritable bowel syndrome disease via breathprint. *Advanced Healthcare Materials*, *5*(18), 2339–2344. https://doi.org/10.1002/ADHM.201600588

Karimi, M., Eslami, M., Sahandi-Zangabad, P., Mirab, F., Farajisafiloo, N., Shafaei, Z., Ghosh, D., Bozorgomid, M., Dashkhaneh, F., & Hamblin, M. R. (2016). pH-Sensitive stimulus-responsive nanocarriers for targeted delivery of therapeutic agents. *Wiley Interdisciplinary Reviews. Nanomedicine and Nanobiotechnology*, *8*(5), 696. https://doi.org/10.1002/WNAN.1389

Kaser, A., Zeissig, S., & Blumberg, R. S. (2010). Inflammatory bowel disease. *Annual Review Immunology*, *28*, 573–621. https://doi.org/10.1146/ANNUREV-IMMUNOL-030409-101225

Kim, S., Hupperetz, C., Lim, S., & Kim, C. H. (2021). Genome editing of immune cells using CRISPR/Cas9. *BMB Reports*, *54*(1), 59. https://doi.org/10.5483/BMBREP.2021.54.1.245

Kim, W., Kim, B. G., Lee, J. S., Lee, C. K., Yeon, J. E., Chang, M. S., ... & Kim, Y. J. (2017). Randomised clinical trial: The efficacy and safety of oltipraz, a liver X receptor alpha-inhibitory dithiolethione in patients with non-alcoholic fatty liver disease. *Alimentary Pharmacology & Therapeutics*, *45*(8), 1073–1083.

Kitture, R., Ghosh, S., More, P. A., Gaware, S., Datar, S., Chopade, B. A., & Kale, S. N. (2015). Curcumin-loaded, self-assembled aloevera template for superior antioxidant activity and trans-membrane drug release. *Journal of Nanoscience and Nanotechnology*, *15*(6), 4039–4045.

Ko, C. N., Zang, S., Zhou, Y., Zhong, Z., & Yang, C. (2022). Nanocarriers for effective delivery: modulation of innate immunity for the management of infections and the associated complications. *Journal of Nanobiotechnology 2022*, *20*(1), 1–31. https://doi.org/10.1186/S12951-022-01582-8

Kotla, N. G., Singh, R., Baby, B. V., Rasala, S., Rasool, J., Hynes, S. O., Martin, D., Egan, L. J., Vemula, P. K., Jala, V. R., Rochev, Y., & Pandit, A. (2022). Inflammation-specific targeted carriers for local drug delivery to inflammatory bowel disease. *Biomaterials*, *281*, 121364. https://doi.org/10.1016/J.BIOMATERIALS.2022.121364

Kshirsagar, N., Pahuja, M., Chatterjee, N. S., & Kamboj, V. P. (2023). Gaps in translating basic science research from bench to bedside. *Indian Journal of Medical Research*, *158*(3), 228–232. https://doi.org/10.4103/ijmr.ijmr_1665_22

Laroui, H., Dalmasso, G., Nguyen, H. T. T., Yan, Y., Sitaraman, S. V., & Merlin, D. (2010). Drug-loaded nanoparticles targeted to the colon with polysaccharide hydrogel reduce colitis in a mouse model. *Gastroenterology*, *138*(3), 914–924. https://doi.org/10.1053/J.GASTRO.2009.11.003

Lee, K. Y., & Mooney, D. J. (2001). Hydrogels for tissue engineering. *Chemical Reviews*, *101*(7), 1869–1879. https://doi.org/10.1021/CR000108X

Leong, W. F., Lai, O. M., Long, K., Man, Y. B. C., Misran, M., & Tan, C. P. (2011). Preparation and characterisation of water-soluble phytosterol nanodispersions. *Food Chemistry*, *129*(1), 77–83.

Li, H., Yang, Y., Hong, W., Huang, M., Wu, M., & Zhao, X. (2020). Applications of genome editing technology in the targeted therapy of human diseases: mechanisms, advances and prospects. *Signal Transduction and Targeted Therapy*, *5*(1), 1–23. https://doi.org/10.1038/s41392-019-0089-y

Li, J., Chen, H., Wang, B., Cai, C., Yang, X., Chai, Z., & Feng, W. (2017). ZnO nanoparticles act as supportive therapy in DSS-induced ulcerative colitis in mice by maintaining gut homeostasis and activating Nrf2 signaling. *Scientific Reports*, *7*(1), 1–11. https://doi.org/10.1038/srep43126

Li, H., Zhao, X., Ma, Y., Zhai, G., Li, L., & Lou, H. (2009). Enhancement of gastrointestinal absorption of quercetin by solid lipid nanoparticles. *Journal of Controlled Release*, *133*(3), 238–244.

Li, M., Luo, J., Nawaz, M. A., Stockmann, R., Buckow, R., Barrow, C., ... & Rasul Suleria, H. A. (2024). Phytochemistry, bioaccessibility, and bioactivities of sesame seeds: An overview. *Food Reviews International*, *40*(1), 309–335.

Li, X., Lu, C., Yang, Y., Yu, C., & Rao, Y. (2020). Site-specific targeted drug delivery systems for the treatment of inflammatory bowel disease. *Biomedicine & Pharmacotherapy*, *129*, 110486. https://doi.org/10.1016/J.BIOPHA.2020.110486

Liang, J., Zhang, Y., Chi, P., Liu, H., Jing, Z., Cao, H., ... & Kong, D. (2023). Essential oils: Chemical constituents, potential neuropharmacological effects and aromatherapy-A review. *Pharmacological Research-Modern Chinese Medicine*, *6*, 100210.

Liu, C. H., Wu, C. T., & Fang, J. Y. (2010). Characterization and formulation optimization of solid lipid nanoparticles in vitamin K1 delivery. *Drug Development and Industrial Pharmacy*, *36*(7), 751–761. https://doi.org/10.3109/03639040903460453

Liu, P., Gao, C., Chen, H., Vong, C. T., Wu, X., Tang, X., Wang, S., & Wang, Y. (2021). Receptor-mediated targeted drug delivery systems for treatment of inflammatory bowel disease: Opportunities and emerging strategies. *Acta Pharmaceutica Sinica. B*, *11*(9), 2798. https://doi.org/10.1016/J.APSB.2020.11.003

Liu, X., Zhao, L., Wu, B., & Chen, F. (2023). Improving solubility of poorly water-soluble drugs by protein-based strategy: A review. *International Journal of Pharmaceutics*, *634*, 122704. https://doi.org/10.1016/J.IJPHARM.2023.122704

Lôbo, G. C. N. B., Paiva, K. L. R., Silva, A. L. G., Simões, M. M., Báo, S. N., & Radicchi, M. A. (2021). Nanocarriers used in drug delivery to enhance immune system in cancer therapy. *Pharmaceutics*, *13*(8), 1167. https://doi.org/10.3390/PHARMACEUTICS13081167

Loftsson, T., & Brewster, M. E. (1996). Pharmaceutical applications of cyclodextrins. 1. Drug solubilization and stabilization. *Journal of Pharmaceutical Sciences*, *85*(10), 1017–1025.

Lombardo, D., & Kiselev, M. A. (2022). Methods of liposomes preparation: Formation and control factors of versatile nanocarriers for biomedical and nanomedicine application. *Pharmaceutics*, *14*(3), 543. https://doi.org/10.3390/PHARMACEUTICS14030543

Lu, Y., Li, X., Liu, S., Zhang, Y., & Zhang, D. (2018). Toll-like receptors and inflammatory bowel disease. *Frontiers in Immunology*, *9*(JAN), 326971. https://doi.org/10.3389/FIMMU.2018.00072

Luo, Z., Paunović, N., & Leroux, J. C. (2021). Physical methods for enhancing drug absorption from the gastrointestinal tract. *Advanced Drug Delivery Reviews*, *175*, 1–38. https://doi.org/10.1016/J.ADDR.2021.05.024

Manchope, M. F., Artero, N. A., Fattori, V., Mizokami, S. S., Pitol, D. L., Issa, J. P., ... & Verri, W. A. (2018). Naringenin mitigates titanium dioxide (TiO 2)-induced chronic arthritis in mice: Role of oxidative stress, cytokines, and NFκB. *Inflammation Research*, *67*, 997–1012.

Marchiando, A. M., Graham, W. V., & Turner, J. R. (2010). Epithelial barriers in homeostasis and disease. *Annual Review of Pathology*, *5*, 119–144. https://doi.org/10.1146/ANNUREV.PATHOL.4.110807.092135

Martin-Banderas, L., Muñoz-Rubio, I., Prados, J., Álvarez-Fuentes, J., Calderón-Montaño, J. M., López-Lázaro, M., ... & Fernández-Arévalo, M. (2015). In vitro and in vivo evaluation of Δ9-tetrahidrocannabinol/PLGA nanoparticles for cancer chemotherapy. *International Journal of Pharmaceutics*, *487*(1–2), 205–212.

Martínez-Ballesta, Mc., Gil-Izquierdo, Á., García-Viguera, C., & Domínguez-Perles, R. (2018). Nanoparticles and controlled delivery for bioactive compounds: Outlining challenges for new "Smart-foods" for health. *Foods*, *7*(5), 72. https://doi.org/10.3390/FOODS7050072

McDowell, C., Farooq, U., & Haseeb, M. (2023). *Inflammatory Bowel Disease*. StatPearls. www.ncbi.nlm.nih.gov/books/NBK470312/

McDonnell, C., Leánez, S., & Pol, O. (2017). The induction of the transcription factor Nrf2 enhances the antinociceptive effects of delta-opioid receptors in diabetic mice. *PLoS One*, *12*(7), e0180998.

Menezes, P. P., Serafini, M. R., Quintans-Júnior, L. J., Silva, G. F., Oliveira, J. F., Carvalho, F. M. D. S., ... & Araújo, A. A. S. (2014). Inclusion complex of (−)-linalool and β-cyclodextrin. *Journal of Thermal Analysis and Calorimetry*, *115*, 2429–2437.

Mengstie, M. A., & Wondimu, B. Z. (2021). Mechanism and applications of CRISPR/Cas-9-mediated genome editing. *Biologics: Targets & Therapy*, *15*, 353. https://doi.org/10.2147/BTT.S326422

Meenu, M., Padhan, B., Patel, M., Patel, R., & Xu, B. (2023). Antibacterial activity of essential oils from different parts of plants against Salmonella and Listeria spp. *Food Chemistry*, *404*, 134723.

Min, K., Yoo, S., Han, M. S., & Tae, G. (2021). Effective and prolonged targeting of a nanocarrier to the inflammation site by functionalization with ZnBPMP and chitosan. *Materials Science and Engineering: C*, *131*, 112521. https://doi.org/10.1016/J.MSEC.2021.112521

Miroliaee, A. E., Esmaily, H., Vaziri-Bami, A., Baeeri, M., Shahverdi, A. R., & Abdollahi, M. (2011). Amelioration of experimental colitis by a novel nanoselenium-silymarin mixture. *Toxicology Mechanisms and Methods*, *21*(3), 200–208. https://doi.org/10.3109/15376516.2010.547887

Mittal, R., Patel, A. P., Jhaveri, V. M., Kay, S. I. S., Debs, L. H., Parrish, J. M., Pan, D. R., Nguyen, D., Mittal, J., & Jayant, R. D. (2018). Recent advancements in nanoparticle based drug delivery for gastrointestinal disorders. *Expert Opinion on Drug Delivery*, *15*(3), 301–318. https://doi.org/10.1080/17425247.2018.1420055

Miyazaki, T., Nakagawa, Y., & Cabral, H. (2021). Strategies for ligand-installed nanocarriers. *Handbook of Nanotechnology Applications: Environment, Energy, Agriculture and Medicine*, 633–655. https://doi.org/10.1016/B978-0-12-821506-7.00024-7

Mizokami, S. S., Arakawa, N. S., Ambrosio, S. R., Zarpelon, A. C., Casagrande, R., Cunha, T. M., ... & Verri Jr, W. A. (2012). Kaurenoic acid from Sphagneticola trilobata inhibits inflammatory pain: effect on cytokine production and activation of the NO–cyclic GMP–protein kinase G–ATP-sensitive potassium channel signaling pathway. *Journal of Natural Products*, *75*(5), 896–904.

Mohanty, C., Das, M., & Sahoo, S. K. (2012). Sustained wound healing activity of curcumin loaded oleic acid based polymeric bandage in a rat model. *Molecular Pharmaceutics*, *9*(10), 2801–2811. https://doi.org/10.1021/MP300075U

Monasta, L., Pierobon, C., Princivalle, A., Martelossi, S., Marcuzzi, A., Pasini, F., & Perbellini, L. (2017). Inflammatory bowel disease and patterns of volatile organic compounds in the exhaled breath of children: A case-control study using Ion Molecule Reaction-Mass Spectrometry. *PloS One*, *12*(8). https://doi.org/10.1371/JOURNAL.PONE.0184118

Mörbe, U. M., Jørgensen, P. B., Fenton, T. M., von Burg, N., Riis, L. B., Spencer, J., & Agace, W. W. (2021). Human gut-associated lymphoid tissues (GALT); diversity, structure, and function. *Mucosal Immunology*, *14*(4), 793–802. https://doi.org/10.1038/s41385-021-00389-4

Mostafa, H., Cheok, A., Meroño, T., Andres-Lacueva, C., & Rodriguez-Mateos, A. (2023). Biomarkers of berry intake: Systematic review update. *Journal of Agricultural and Food Chemistry*, *71*(31), 11789–11805.

Mozafari, M. R., Johnson, C., Hatziantoniou, S., & Demetzos, C. (2008). Nanoliposomes and their applications in food nanotechnology. *Journal of Liposome Research*, *18*(4), 309–327. https://doi.org/10.1080/08982100802465941

Muller, R. H., & Keck, C. M. (2004). Challenges and solutions for the delivery of biotech drugs–A review of drug nanocrystal technology and lipid nanoparticles. *Journal of Biotechnology*, *113*(1–3), 151–170. https://doi.org/10.1016/J.JBIOTEC.2004.06.007

Naghsh, N., Musazadeh, V., Nikpayam, O., Kavyani, Z., Moridpour, A. H., Golandam, F., ... & Ostadrahimi, A. (2023). Profiling inflammatory biomarkers following curcumin supplementation: An umbrella meta-analysis of randomized clinical trials. *Evidence-Based Complementary and Alternative Medicine*, *2023*, 1–12.

Naha, P. C., Hsu, J. C., Kim, J., Shah, S., Bouché, M., Si-Mohamed, S., Rosario-Berrios, D. N., Douek, P., Hajfathalian, M., Yasini, P., Singh, S., Rosen, M. A., Morgan, M. A., & Cormode, D. P. (2020). Dextran-coated cerium oxide nanoparticles: A computed tomography contrast agent for imaging the gastrointestinal tract and inflammatory bowel disease. *ACS Nano*, *14*(8), 10187–10197. https://doi.org/10.1021/acsnano.0c03457

Neuman, M. G. (2007). Immune dysfunction in inflammatory bowel disease. *Translational Research*, *149*(4), 173–186. https://doi.org/10.1016/J.TRSL.2006.11.009

Neurath, M. F. (2019). Targeting immune cell circuits and trafficking in inflammatory bowel disease. *Nature Immunology*, *20*(8), 970–979. https://doi.org/10.1038/S41590-019-0415-0

Nsairat, H., Khater, D., Sayed, U., Odeh, F., Al Bawab, A., & Alshaer, W. (2022). Liposomes: Structure, composition, types, and clinical applications. *Heliyon*, *8*(5), 9394. https://doi.org/10.1016/J.HELIYON.2022.E09394

Ocansey, D. K. W., Zhang, L., Wang, Y., Yan, Y., Qian, H., Zhang, X., Xu, W., & Mao, F. (2020). Exosome-mediated effects and applications in inflammatory bowel disease. *Biological Reviews of the Cambridge Philosophical Society*, *95*(5), 1287–1307. https://doi.org/10.1111/BRV.12608

Oz, H. S., Yeh, S.-L., & Neuman, M. G. (2016). *Editorial Gastrointestinal Inflammation and Repair: Role of Microbiome, Infection, and Nutrition*. Wiley. https://doi.org/10.1155/2016/6516708

Qin, S., Chen, Z., Wen, Y., Yi, Y., Lv, C., Zeng, C., ... & Shi, M. (2023). Phytochemical activators of Nrf2: A review of therapeutic strategies in diabetes: Phytochemical activators of Nrf2. *Acta Biochimica et Biophysica Sinica*, *55*(1), 11.

Paiva-Santos, A. C., Herdade, A. M., Guerra, C., Peixoto, D., Pereira-Silva, M., Zeinali, M., Mascarenhas-Melo, F., Paranhos, A., & Veiga, F. (2021a). Plant-mediated green synthesis of metal-based nanoparticles for dermopharmaceutical and cosmetic applications. *International Journal of Pharmaceutics*, *597*, 120311. https://doi.org/10.1016/J.IJPHARM.2021.120311

Paiva-Santos, A. C., Herdade, A. M., Guerra, C., Peixoto, D., Pereira-Silva, M., Zeinali, M., Mascarenhas-Melo, F., Paranhos, A., & Veiga, F. (2021b). Plant-mediated green synthesis of metal-based nanoparticles for dermopharmaceutical and cosmetic applications. *International Journal of Pharmaceutics*, *597*. https://doi.org/10.1016/J.IJPHARM.2021.120311

Pandita, D., Kumar, S., Poonia, N., & Lather, V. (2014). Solid lipid nanoparticles enhance oral bioavailability of resveratrol, a natural polyphenol. *Food Research International*, *62*, 1165–1174.

Pandita, D., Vakar, Poonia, N., Chaudhary, G., Jain, G. K., Lather, V., & Khar, R. K. (2023). pH-sensitive polymeric nanocarriers for enhanced intracellular drug delivery. *Smart Polymeric Nano-Constructs in Drug Delivery*, 65–107. https://doi.org/10.1016/B978-0-323-91248-8.00004-0

Pang, C., Zheng, Z., Shi, L., Sheng, Y., Wei, H., Wang, Z., & Ji, L. (2016). Caffeic acid prevents acetaminophen-induced liver injury by activating the Keap1-Nrf2 antioxidative defense system. *Free Radical Biology and Medicine*, *91*, 236–246.

Parris, N., Cooke, P. H., & Hicks, K. B. (2005). Encapsulation of essential oils in zein nanospherical particles. *Journal of Agricultural and Food Chemistry*, *53*(12), 4788–4792.

Patnaik, R., Riaz, S., Sivani, B. M., Faisal, S., Naidoo, N., Rizzo, M., & Banerjee, Y. (2023). Evaluating the potential of Vitamin D and curcumin to alleviate inflammation and mitigate the progression of osteoarthritis through their effects on human chondrocytes: A proof-of-concept investigation. *Plos One*, *18*(12), e0290739.

Pescosolido, N., Giannotti, R., Plateroti, A. M., Pascarella, A., & Nebbioso, M. (2014). Curcumin: Therapeutical potential in ophthalmology. *Planta Medica*, *80*(4), 249–254. https://doi.org/10.1055/s-0033-1351074

Peyrin-Biroulet, L., Reinisch, W., Colombel, J. F., Mantzaris, G. J., Kornbluth, A., Diamond, R., Rutgeerts, P., Tang, L. K., Cornillie, F. J., & Sandborn, W. J. (2014a). Clinical disease activity, C-reactive protein normalisation and mucosal healing in Crohn's disease in the SONIC trial. *Gut*, *63*(1), 88–95. https://doi.org/10.1136/GUTJNL-2013-304984

Peyrin-Biroulet, L., Reinisch, W., Colombel, J. F., Mantzaris, G. J., Kornbluth, A., Diamond, R., Rutgeerts, P., Tang, L. K., Cornillie, F. J., & Sandborn, W. J. (2014b). Clinical disease activity, C-reactive protein normalisation and mucosal healing in Crohn's disease in the SONIC trial. *Gut*, *63*(1), 88–95. https://doi.org/10.1136/GUTJNL-2013-304984

Pinho, E., Grootveld, M., Soares, G., & Henriques, M. (2014). Cyclodextrins as encapsulation agents for plant bioactive compounds. *Carbohydrate polymers*, *101*, 121–135.

Pithadia, A. B., & Jain, S. (2011a). Treatment of inflammatory bowel disease (IBD). *Pharmacological Reports: PR*, *63*(3), 629–642. https://doi.org/10.1016/S1734-1140(11)70575-8

Pithadia, A. B., & Jain, S. (2011b). Treatment of inflammatory bowel disease (IBD). *Pharmacological Reports: PR*, *63*(3), 629–642. https://doi.org/10.1016/S1734-1140(11)70575-8

Pool, H., Quintanar, D., de Dios Figueroa, J., Mano, C. M., Bechara, J. E. H., Godínez, L. A., & Mendoza, S. (2012). Antioxidant effects of quercetin and catechin encapsulated into PLGA nanoparticles. *Journal of Nanomaterials*, *2012*, 86–86.

Popov, S. V., Ovodova, R. G., Markov, P. A., Nikitina, I. R., & Ovodov, Y. S. (2006). Protective effect of comaruman, a pectin of cinquefoil Comarum palustre L., on acetic acid-induced colitis in mice. *Digestive diseases and sciences*, *51*, 1532–1537.

Pouremamali, F., Pouremamali, A., Dadashpour, M., Soozangar, N., & Jeddi, F. (2022). An update of Nrf2 activators and inhibitors in cancer prevention/promotion. *Cell Communication and Signaling*, *20*(1), 100

Porter, C. J. H., Trevaskis, N. L., & Charman, W. N. (2007). Lipids and lipid-based formulations: optimizing the oral delivery of lipophilic drugs. *Nature Reviews Drug Discovery*, *6*(3), 231–248. https://doi.org/10.1038/nrd2197

Pridgen, E. M., Alexis, F., & Farokhzad, O. C. (2014). Polymeric nanoparticle technologies for oral drug delivery. *Clinical Gastroenterology and Hepatology: The Official Clinical Practice Journal of the American Gastroenterological Association*, *12*(10), 1605–1610. https://doi.org/10.1016/J.CGH.2014.06.018

Pridgen, E. M., Alexis, F., & Farokhzad, O. C. (2015). Polymeric nanoparticle drug delivery technologies for oral delivery applications. *Expert Opinion on Drug Delivery*, *12*(9), 1459–1473. https://doi.org/10.1517/17425247.2015.1018175

Rahimi, R., Shams-Ardekani, M. R., & Abdollahi, M. (2010). A review of the efficacy of traditional Iranian medicine for inflammatory bowel disease. *World Journal of Gastroenterology*, *16*(36), 4504–4514. https://doi.org/10.3748/WJG.V16.I36.4504

Rana, I., Oh, J., Baig, J., Moon, J. H., Son, S., & Nam, J. (2023). Nanocarriers for cancer nano-immunotherapy. *Drug Delivery and Translational Research*, *13*(7), 1936–1954. https://doi.org/10.1007/S13346-022-01241-3/FIGURES/5

Raval, M. K., Ramani, R. V, & Sheth, N. R. (2013). Formulation and evaluation of sustained release enteric-coated pellets of budesonide for intestinal delivery. *International Journal of Pharmaceutical Investigation*, *3*(4), 203. https://doi.org/10.4103/2230-973X.121294

Redondo, A., Chamorro, P. A. F., Riego, G., Leánez, S., & Pol, O. (2017). Treatment with sulforaphane produces antinociception and improves morphine effects during inflammatory pain in mice. *Journal of Pharmacology and Experimental Therapeutics*, *363*(3), 293–302.

Rehman, A. U., Akram, S., Seralin, A., Vandamme, T., & Anton, N. (2020). Lipid nanocarriers: Formulation, properties, and applications. *Smart Nanocontainers: Micro and Nano Technologies*, 355–382. https://doi.org/10.1016/B978-0-12-816770-0.00021-6

Rezaie, A., Parker, R. D., & Abdollahi, M. (2007). Oxidative stress and pathogenesis of inflammatory bowel disease: An epiphenomenon or the cause? *Digestive Diseases and Sciences*, *52*(9), 2015–2021. https://doi.org/10.1007/S10620-006-9622-2

Rieber, A., Wruk, D., Potthast, S., Nüssle, K., Reinshagen, M., Adler, G., & Brambs, H. J. (2000a). Diagnostic imaging in Crohn's disease: Comparison of magnetic resonance imaging and conventional imaging methods. *International Journal of Colorectal Disease*, *15*(3), 176–181. https://doi.org/10.1007/S003840000219

Rieber, A., Wruk, D., Potthast, S., Nüssle, K., Reinshagen, M., Adler, G., & Brambs, H. J. (2000b). Diagnostic imaging in Crohn's disease: Comparison of magnetic resonance imaging and conventional imaging methods. *International Journal of Colorectal Disease*, *15*(3), 176–181. https://doi.org/10.1007/S003840000219

Rossi, L., ten Hoorn, J. W. S., Melnikov, S. M., & Velikov, K. P. (2010). Colloidal phytosterols: Synthesis, characterization and bioaccessibility. *Soft Matter*, *6*(5), 928–936.

Ruckh, T. T., & Clark, H. A. (2014). Implantable nanosensors: Towards continuous physiologic monitoring. *Analytical Chemistry*, *86*(3), 1314. https://doi.org/10.1021/AC402688K

Sah, E., & Sah, H. (2015). Recent trends in preparation of poly(lactide-co-glycolide) nanoparticles by mixing polymeric organic solution with antisolvent. *Journal of Nanomaterials*, *2015*, 794601. https://doi.org/10.1155/2015/794601

Saisho, H., & Yamaguchi, T. (2004). Diagnostic imaging for pancreatic cancer: Computed tomography, magnetic resonance imaging, and positron emission tomography. *Pancreas*, *28*(3), 273–278. https://doi.org/10.1097/00006676-200404000-00011

Sargazi, G., Afzali, D., Mostafavi, A., Shadman, A., Rezaee, B., Zarrintaj, P., Saeb, M. R., Ramakrishna, S., & Mozafari, M. (2019). Chitosan/polyvinyl alcohol nanofibrous membranes: Towards green super-adsorbents for toxic gases. *Heliyon*, *5*(4), 5. https://doi.org/10.1016/J.HELIYON.2019.E01527

Sargazi, G., Khajeh Ebrahimi, A., Afzali, D., Badoei-dalfard, A., Malekabadi, S., & Karami, Z. (2019). Fabrication of PVA/ZnO fibrous composite polymer as a novel sorbent for arsenic removal: Design and a systematic study. *Polymer Bulletin*, *76*(11), 5661–5682. https://doi.org/10.1007/S00289-019-02677-3/METRICS

Schaffer, M., Schaffer, P. M., & Bar-Sela, G. (2015). An update on Curcuma as a functional food in the control of cancer and inflammation. *Current Opinion in Clinical Nutrition and Metabolic Care*, *18*(6), 605–611. https://doi.org/10.1097/MCO.0000000000000227

Selvam, C., Jachak, S. M., Thilagavathi, R., & Chakraborti, A. K. (2005). Design, synthesis, biological evaluation and molecular docking of curcumin analogues as antioxidant, cyclooxygenase inhibitory and anti-inflammatory agents. *Bioorganic & Medicinal Chemistry Letters*, *15*(7), 1793–1797.

Selvaraj, S., Shanmugasundaram, S., Maruthamuthu, M., Venkidasamy, B., & Shanmugasundaram, S. (2021). Facile synthesis and characterization of quercetin-loaded alginate nanoparticles for enhanced in vitro anticancer effect against human leukemic cancer U937 cells. *Journal of Cluster Science*, *32*(6), 1507–1518. https://doi.org/10.1007/S10876-020-01913-5/METRICS

Serini, S., Cassano, R., Corsetto, P. A., Rizzo, A. M., Calviello, G., & Trombino, S. (2018). Omega-3 PUFA loaded in resveratrol-based solid lipid nanoparticles: Physicochemical properties and antineoplastic activities in human colorectal cancer cells in vitro. *International Journal of Molecular Sciences*, *19*(2). https://doi.org/10.3390/IJMS19020586

Shah, S., Famta, P., Raghuvanshi, R. S., Singh, S. B., & Srivastava, S. (2022). Lipid polymer hybrid nanocarriers: Insights into synthesis aspects, characterization, release mechanisms, surface functionalization and potential implications. *Colloid and Interface Science Communications*, *46*, 100570. https://doi.org/10.1016/J.COLCOM.2021.100570

Shepherd, S. F., McGuire, N. D., De Lacy Costello, B. P. J., Ewen, R. J., Jayasena, D. H., Vaughan, K., Ahmed, I., Probert, C. S., & Ratcliffe, N. M. (2014). The use of a gas chromatograph coupled to a metal oxide sensor for rapid assessment of stool samples from irritable bowel syndrome and inflammatory bowel disease patients. *Journal of Breath Research*, *8*(2), 026001. https://doi.org/10.1088/1752-7155/8/2/026001

Sethuram, L., & Thomas, J. (2023). Therapeutic applications of electrospun nanofibers impregnated with various biological macromolecules for effective wound healing strategy–A review. *Biomedicine & Pharmacotherapy*, *157*, 113996.

Sonaje, K., Italia, J. L., Sharma, G., Bhardwaj, V., Tikoo, K., & Kumar, M. R. (2007). Development of biodegradable nanoparticles for oral delivery of ellagic acid and evaluation of their antioxidant efficacy against cyclosporine A-induced nephrotoxicity in rats. *Pharmaceutical Research*, *24*, 899–908.

Singh, G., & Pai, R. S. (2014a). Optimized PLGA nanoparticle platform for orally dosed trans-resveratrol with enhanced bioavailability potential. *Expert Opinion on Drug Delivery*, *11*(5), 647–659.

Singh, G., & Pai, R. S. (2014b). In-vitro/in-vivo characterization of trans-resveratrol-loaded nanoparticulate drug delivery system for oral administration. *Journal of Pharmacy and Pharmacology*, *66*(8), 1062–1076.

Singh, S., Semwal, B. C., Sharma, H., & Sharma, D. (2023). Impact of phytomolecules with nanotechnology on the treatment of inflammation. *Current Bioactive Compounds*, *19*(10), 122–148.

Staurengo-Ferrari, L., Badaro-Garcia, S., Hohmann, M. S., Manchope, M. F., Zaninelli, T. H., Casagrande, R., & Verri Jr, W. A. (2019). Contribution of Nrf2 modulation to the mechanism of action of analgesic and anti-inflammatory drugs in pre-clinical and clinical stages. *Frontiers in Pharmacology*, *9*, 1536.

Su, S., & Kang, P. M. (2020a). Recent advances in nanocarrier-assisted therapeutics delivery systems. *Pharmaceutics*, *12*(9), 1–27. https://doi.org/10.3390/PHARMACEUTICS12090837

Su, S., & Kang, P. M. (2020b). Recent advances in nanocarrier-assisted therapeutics delivery systems. *Pharmaceutics*, *12*(9), 1–27. https://doi.org/10.3390/PHARMACEUTICS12090837

Su, S., & Kang, P. M. (2020c). Recent advances in nanocarrier-assisted therapeutics delivery systems. *Pharmaceutics*, *12*(9), 1–27. https://doi.org/10.3390/PHARMACEUTICS12090837

Subramanian, P., Kurek, M. A., & Jensen, I.-J. (2021). Lipid-based nanocarrier system for the effective delivery of nutraceuticals. *Molecules*, *26*(18), 5510. https://doi.org/10.3390/MOLECULES26185510

Subramanya, S. B., Venkataraman, B., Meeran, M. F. N., Goyal, S. N., Patil, C. R., & Ojha, S. (2018). Therapeutic potential of plants and plant derived phytochemicals against acetaminophen-induced liver injury. *International Journal of Molecular Sciences*, *19*(12), 3776.

Sun, T., Zhang, Y. S., Pang, B., Hyun, D. C., Yang, M., & Xia, Y. (2014). Engineered nanoparticles for drug delivery in cancer therapy. *Angewandte Chemie (International Ed. in English)*, *53*(46), 12320–12364. https://doi.org/10.1002/ANIE.201403036

Teruel, A. H., Gonzalez-Alvarez, I., Bermejo, M., Merino, V., Marcos, M. D., Sancenon, F., Gonzalez-Alvarez, M., & Martinez-Mañez, R. (2020a). New insights of oral colonic drug delivery systems for inflammatory bowel disease therapy. *International Journal of Molecular Sciences*, *21*(18), 1–30. https://doi.org/10.3390/IJMS21186502

Teskač, K., & Kristl, J. (2010). The evidence for solid lipid nanoparticles mediated cell uptake of resveratrol. *International Journal of Pharmaceutics*, *390*(1), 61–69.

Theiss, A. L., Laroui, H., Obertone, T. S., Chowdhury, I., Thompson, W. E., Merlin, D., & Sitaraman, S. V. (2011). Nanoparticle-based therapeutic delivery of prohibitin to the colonic epithelial cells ameliorates acute murine colitis. *Inflammatory Bowel Diseases*, *17*(5), 1163–1176. https://doi.org/10.1002/IBD.21469

Torchilin, V. P. (2005a). Recent advances with liposomes as pharmaceutical carriers. *Nature Reviews. Drug Discovery*, *4*(2), 145–160. https://doi.org/10.1038/NRD1632

Torchilin, V. P. (2005b). Recent advances with liposomes as pharmaceutical carriers. *Nature Reviews. Drug Discovery*, *4*(2), 145–160. https://doi.org/10.1038/NRD1632

Turner, J. R. (2009). Intestinal mucosal barrier function in health and disease. *Nature Reviews. Immunology*, *9*(11), 799–809. https://doi.org/10.1038/NRI2653

Türk, M., & Lietzow, R. (2004). Stabilized nanoparticles of phytosterol by rapid expansion from supercritical solution into aqueous solution. *Aaps Pharmscitech*, *5*, 36–45.

Uhrich, K. E., Cannizzaro, S. M., Langer, R. S., & Shakesheff, K. M. (1999). Polymeric systems for controlled drug release. *Chemical Reviews*, *99*(11), 3181–3198. https://doi.org/10.1021/CR940351U

Vancamelbeke, M., & Vermeire, S. (2017). The intestinal barrier: A fundamental role in health and disease. *Expert Review of Gastroenterology & Hepatology*, *11*(9), 821. https://doi.org/10.1080/17474124.2017.1343143

Varshosaz, J., Minaiyan, M., & Khaleghi, N. (2015). Eudragit nanoparticles loaded with silybin: A detailed study of preparation, freeze-drying condition and in vitro/in vivo evaluation. *Journal of Microencapsulation*, *32*(3), 211–223. https://doi.org/10.3109/02652048.2014.995728

Vinarov, Z., Abdallah, M., Agundez, J. A. G., Allegaert, K., Basit, A. W., Braeckmans, M., Ceulemans, J., Corsetti, M., Griffin, B. T., Grimm, M., Keszthelyi, D., Koziolek, M., Madla, C. M., Matthys, C., McCoubrey, L. E., Mitra, A., Reppas, C., Stappaerts, J., Steenackers, N., … & Augustijns, P. (2021). Impact of gastrointestinal tract variability on oral drug absorption and pharmacokinetics: An UNGAP review. *European Journal of Pharmaceutical Sciences*, *162*, 105812. https://doi.org/10.1016/J.EJPS.2021.105812

Virolainen, S. J., Vonhandorf, A., Viel, K. C. M. F., Weirauch, M. T., & Kottyan, L. C. (n.d.). Gene-environment interactions and their impact on human health, *Genes and Immunity*, *24(1*), 1–11.https://doi.org/10.1038/s41435-022-00192-6

Vuyyuru, S. K., Kedia, S., Sahu, P., & Ahuja, V. (2022). *Immune-Mediated Inflammatory Diseases of the Gastrointestinal Tract: Beyond Crohn's Disease and Ulcerative Colitis*, *Journal of Gastroenterology and Hepatology*, 6(2), 100–111. https://doi.org/10.1002/jgh3.12706

Wang, R. C., & Wang, Z. (2023). Precision medicine: Disease subtyping and tailored treatment. *Cancers*, *15*(15), 3837. https://doi.org/10.3390/CANCERS15153837

Wu, Y., Briley-Saebo, K., Xie, J., Zhang, R., Wang, Z., He, C., Tang, C. Y., & Tao, X. (2014). Inflammatory bowel disease: MR- and SPECT/CT-based macrophage imaging for monitoring and evaluating disease activity in experimental mouse model--Pilot study. *Radiology*, *271*(2), 400–407. https://doi.org/10.1148/RADIOL.13122254

Xavier, R. J., & Podolsky, D. K. (2007). Unravelling the pathogenesis of inflammatory bowel disease. *Nature*, *448*(7152), 427–434. https://doi.org/10.1038/NATURE06005

Xiao, B., Zhang, Z., Viennois, E., Kang, Y., Zhang, M., Han, M. K., Chen, J., & Merlin, D. (2016). Combination therapy for ulcerative colitis: Orally targeted nanoparticles prevent mucosal damage and relieve inflammation. *Theranostics*, *6*(12), 2250–2266. https://doi.org/10.7150/THNO.15710

Xie, G., Schepetkin, I. A., Siemsen, D. W., Kirpotina, L. N., Wiley, J. A., & Quinn, M. T. (2008). Fractionation and characterization of biologically-active polysaccharides from Artemisia tripartita. *Phytochemistry*, *69*(6), 1359–1371.

Wu, J., Li, Q., Wang, X., Yu, S., Li, L., Wu, X., ... & Zhao, Y. (2013). Neuroprotection by curcumin in ischemic brain injury involves the Akt/Nrf2 pathway. *PloS One*, *8*(3), e59843.

Wu, Y., Luo, Y., & Wang, Q. (2012). Antioxidant and antimicrobial properties of essential oils encapsulated in zein nanoparticles prepared by liquid–liquid dispersion method. *LWT-Food Science and Technology*, *48*(2), 283–290.

Yahfoufi, N., Alsadi, N., Jambi, M., & Matar, C. (2018). The immunomodulatory and anti-inflammatory role of polyphenols. *Nutrients*, *10*(11), 1618. https://doi.org/10.3390/NU10111618

Yang, C., Zhang, M., Lama, S., Wang, L., & Merlin, D. (2020). Natural-lipid nanoparticle-based therapeutic approach to deliver 6-shogaol and its metabolites M2 and M13 to the colon to treat ulcerative colitis. *Journal of Controlled Release: Official Journal of the Controlled Release Society*, *323*, 293–310. https://doi.org/10.1016/J.JCONREL.2020.04.032

Ying, N., Lin, X., Xie, M., & Zeng, D. (2022). Effect of surface ligand modification on the properties of anti-tumor nanocarrier. *Colloids and Surfaces B: Biointerfaces*, *220*, 112944. https://doi.org/10.1016/J.COLSURFB.2022.112944

Yoo, J. Y., Groer, M., Dutra, S. V. O., Sarkar, A., & McSkimming, D. I. (2020). Gut microbiota and immune system interactions.

Microorganisms, *8*(10), 1–22. https://doi.org/10.3390/MICROORGANISMS8101587

Yoo, J. Y., Groer, M., Valeria, S., Dutra, O., Sarkar, A., & Mcskimming, D. I. (1587). Gut microbiota and immune system interactions. *Microorganisms, 8*, 1587. https://doi.org/10.3390/microorganisms8101587

Zhang, C., Zhai, S., Li, X., Zhang, Q., Wu, L., Liu, Y., Jiang, C., Zhou, H., Li, F., Zhang, S., Su, G., Zhang, B., & Yan, B. (2014). Synergistic action by multi-targeting compounds produces a potent compound combination for human NSCLC both in vitro and in vivo. *Cell Death & Disease*, *5*(3), e1138. https://doi.org/10.1038/CDDIS.2014.76

Zhang, H., & Tsao, R. (2016). Dietary polyphenols, oxidative stress and antioxidant and anti-inflammatory effects. *Current Opinion in Food Science*, *8*, 33–42. https://doi.org/10.1016/J.COFS.2016.02.002

Zhang, M., & Merlin, D. (2018). Nanoparticle-based oral drug delivery systems targeting the colon for treatment of ulcerative colitis. *Inflammatory Bowel Diseases*, *24*(7), 1401–1415. https://doi.org/10.1093/IBD/IZY123

Zhang, X., Li, J., Yang, J., & Huang, Z. (2016). Fentanyl combined with butorphanol protects myocardial ischemia/reperfusion injury via κ-opioid receptor-mediated Nrf2-ARE signaling. *International Journal Clinical Experiment Medicine*, *9*, 2500–2506.

Zhao, H., Wu, L., Yan, G., Chen, Y., Zhou, M., Wu, Y., & Li, Y. (2021). Inflammation and tumor progression: Signaling pathways and targeted intervention. *Signal Transduction and Targeted Therapy*, 6(1), 263. https://doi.org/10.1038/s41392-021-00658-5

Zhou, Y., Yang, S., Guo, J., Dong, H., Yin, K., Huang, W. T., & Yang, R. (2020). In vivo imaging of hypoxia associated with inflammatory bowel disease by a cytoplasmic protein-powered fluorescence cascade amplifier. *Analytical Chemistry*, *92*(8), 5787–5794. https://doi.org/10.1021/ACS.ANALCHEM.9B05278/SUPPL_FILE/AC9B05278_SI_001.PDF

Zhu, C., Zhang, S., Song, C., Zhang, Y., Ling, Q., Hoffmann, P. R., Li, J., Chen, T., Zheng, W., & Huang, Z. (2017). Selenium nanoparticles decorated with Ulva lactuca polysaccharide potentially attenuate colitis by inhibiting NF-KB mediated hyper inflammation. *Journal of Nanobiotechnology*, *15*(1), 1–15. https://doi.org/10.1186/S12951-017-0252-Y/FIGURES/8

Zielinska, A., Carreiró, F., Oliveira, A. M., Neves, A., Pires, B., Nagasamy Venkatesh, D., Durazzo, A., Lucarini, M., Eder, P., Silva, A. M., Santini, A., & Souto, E. B. (2020). Polymeric nanoparticles: Production, characterization, toxicology and ecotoxicology. *Molecules*, *25*(16), 3731. https://doi.org/10.3390/MOLECULES25163731

Zu, Y., Zhang, Y., Wang, W., Zhao, X., Han, X., Wang, K., & Ge, Y. (2016). Preparation and in vitro/in vivo evaluation of resveratrol-loaded carboxymethyl chitosan nanoparticles. *Drug Delivery*, *23*(3), 971–981.

20 Future Prospects and Clinical Studies for the Treatment of Gastrointestinal Inflammations

Sanjeeb Kumar Kar[1], Shubhrajit Mantry[2], Ashutosh Behera[3], Laliteshwar Pratap Singh[4], Brijyog[5], and Deepak Kumar Patra[6]*

[1]Department of Pharmaceutical Chemistry, Florence College of Pharmacy, Irba, Ranchi, Jharkhand, India

[2]Department of Pharmacy, Sarala Birla University, Ranchi, Jharkhand, India

[3]Department of Pharmaceutical Quality Assurance, Florence College of Pharmacy, Irba, Ranchi, Jharkhand, India

[4]Department of Pharmaceutical Chemistry, Narayan Institute of Pharmacy, Gopal Narayan Singh University, Sasaram, Bihar, India

[5]Department of Pharmacognosy, Institute of Pharmacy, Harish Chandra P.G. College, Varanasi, Uttar Pradesh, India

[6]Department of Pharmaceutical Chemistry, Dhanvantari College of Pharmacy, Ormanjhi, Ranchi, Jharkhand, India

20.1 BACKGROUND

Gut microbiota, evolving alongside their hosts, forms an essential component of the human organism. Microbiota obtained at birth evolves alongside the host, maintaining stability and diversity throughout adulthood and into death. Additionally, the gut microbiota plays well-being of individuals is significantly impacted by various factors, including both physical and mental influences. Several factors are crucial in ascertaining a person's general well-being and happiness. Comprising various microbes such as bacteria, archaea, eukaryote, viruses, and parasites, the human gut microbiota demonstrates significant diversity (Adak & Khan, 2019). Recognizing the crucial role of the gut microbiota, also known as the microbiome, in human health underscores the need for a deeper comprehension of the elements shaping the composition and operations of this microbial community. Among these factors, diet has emerged as a pivotal driver influencing both the composition and functionality of the microbiota (Wastyk et al., 2021). Two-way communication pathways connecting the central nervous system (CNS) and the gastrointestinal tract are collectively referred to as the gut-brain axis. In this context, the gut microbiota constitutes the ecological community of both beneficial and harmful microorganisms residing in the gut, with certain members playing the interaction between the brain and the gastrointestinal system is facilitated by an important role (Dalile et al., 2019; Nag, Bhatt, et al., 2023). The makeup of the microbiota in the digestive system can be impacted by a range of external factors, including environmental variables pH, oxygen levels, nutrient availability, water activity, and temperature. These factors create conditions for different microbial populations to thrive and engage in various activities as they interact involving their surroundings, encompassing the human host. The gut microbiota includes members from there three domains of life, namely archaea, bacteria, and eukaryote. Additionally, the gut microbiota is also inclusive of viruses. These microorganisms form complex trophic relationships among themselves and with the human host, spanning from symbiosis to parasitism. The autochthonous (indigenous) are microorganisms that are present along with allochthonous or transient microorganisms both make up the human gut microbiota (Milani et al., 2017).

The gut microbiota is an intricate assembly of microbial communities, comprising trillions of fungi, viruses, archaea, micro-eukaryotes, and predominantly bacteria. While the majority of these microbes are beneficial, some can act as pathogens. The gut microbiota has a significant role in substantial role in shaping the health of both humans and animals, and disruptions in its normal composition are observed during gut diseases. There are indications that the gastrointestinal microbiota assumes a significant role in regulating gut inflammation (Akhtar et al., 2022). The gut microbiome is a dynamic ecosystem created by a community of microorganisms inhabiting the gastrointestinal tract, along with their microbial genomes and interdependent interactions. In addition to their essential roles in host metabolic processes, safeguarding the gut barrier, and combating pathogens, the associations between the gut microbiota and immune system

DOI: 10.1201/9781003493143-20

are crucial in shaping the host's immune response (Ciernikova et al., 2023). The gut microbiota, comprising bacteria along with other microbes like viruses and fungi, along with the microbiome regulates metabolism, inflammation, haematopoiesis, and immunity. Short-chain fatty acids (SCFAs), tryptophan (TRP), and bile acid metabolites, which are generated by microbiota, have an impact on genetic and epigenetic modifications but also contribute to these processes but also play a role in modulating the metabolism of immune cells (Gholap et al., 2024; Nag, Mitra, et al., 2023). This includes effects on immunosuppressive and inflammatory cells, contributing to the maintenance of both gut and systemic homeostasis (Wang et al., 2024). The abundant population of the gut microbiota is sustained through various means, including indigestible fibers (leftovers from the diet), mucus generated by the gut, and nutrients released from dead cells. The composition of the host's diet significantly affects the growth of the microbiota and the gut environment. Functional gut bacteria have the potential to generate potentially harmful compounds, including carcinogens, immunotoxins, and neurotoxins, alongside producing biologically potential compounds, i.e., SCFAs, minerals, vitamins, and other substances with varied properties (Islam et al., 2023). The gut microbiota, often referred to considered as a novel organ system within the body, is closely implicated in various pathological conditions induced by diet, diabetes, encompassing obesity, cardiovascular complications, and inflammatory bowel disease (IBD). The gastrointestinal system plays a pivotal role in digesting and absorbing dietary oils and fats while also acting as a barrier against pathogen invasion via innate immune functions. The microbiota present in the gut is an essential component of the immune barrier, also known as the biological barrier. It performs critical functions in identifying harmful dietary habits that affect normal intestinal functions (Mohanto et al., 2023; Ye et al., 2021). The gut microbiota consists of the intestines are inhabited by trillions of microorganisms, including bacteria. Their functions encompass aiding in food digestion, synthesizing vitamins, enhancing the gut microbiome consists of trillions of microorganisms that aid in the digestion of food, support the body's feedback mechanism, which is functioned for protecting against harmful bacteria. The composition of gut microbes is different in individuals' subjects and can be altered by factors, i.e., diet, lifestyle, co-morbidities, and medication history. Imbalances in these microbes can result in various digestive and health issues. Additionally, during injury conditions, the modulation of the CNS is critically influenced by the microbiome and its potential role in the development of the CNS cannot be overstated. In animal models, changes in the gut microbiome have been observed to impact the CNS development, repair, and regeneration (Guha et al., 2023). The gastrointestinal tract constitutes a micro-environment characterized by biodiversity, hosting hundreds of bacteria, archaea, and fungi. The interplay among these diverse microbial communities contributes to the establishment of intestinal microbial homeostasis, commonly referred to as gut microbiota, with probiotics being a significant component of these intestinal microorganisms. Maintaining a balanced intestinal microecology is crucial for physiological functions such as nutrient absorption and metabolism, immune responses, and biological antagonism. Conversely, disruptions in gut microbiota balance can lead to corresponding pathological reactions, posing a threat to human health (Liu et al., 2023).

20.2 PRESENT RESEARCH GAP IN GUT MICROBIOTA AND GI DISORDERS

The mammalian gastrointestinal tract serves as a primary reservoir for diverse and intricate microbial communities. An imbalanced microbiota, also known as gut dysbiosis, is strongly linked with the development of a common inflammatory disorder affecting the gastrointestinal tract. The gut microbiota, comprising several bacterial species, undergone extensive exploration through various studies, including physiological, epidemiological, and omics-based investigations in humans. Despite these efforts, the understanding of the unclear influence of on the development, the complex microbiota in the intestine plays an important role or resistance of mucosal inflammation remains a topic of interest. To address this gap, Ikeda E et al. conducted a study that aimed to examine how the gut microbiota affects susceptibility to UC in a mouse model. To achieve this, researchers compared colitis induced by DSS was observed in inbred BALB/c mice. The susceptibility of these mice to this condition was evaluated. The assessment included clinical symptom evaluation of colitis and analysis of the fecal microbiota. The researchers utilized a cohousing approach to assess whether the digestive system is functioning properly by examining the microbiota plays a preliminary function in determining disease initiation and susceptibility (Ikeda et al., 2024). Both functional foods and acute alcohol intake can affect the gut microbiota. Considering this, the study postulated that Camellia oil (CO) might exert a significant the effect on the microorganisms in the digestive system, subsequently influencing the state of oxidative stress of the host. To explore this hypothesis, Xu Z et al. conducted a study the aim is to examine how camellia oil can protect the stomach against injury caused by ethanol in mice. The findings indicated that pretreatment with camellia oil significantly mitigated gastric ulcer injury. Notably, there was a substantial inhibition of reactive oxygen species (ROS), as evidenced by a reduction in malondialdehyde concentration, the group that received camellia oil pretreatment showed a noteworthy reduction in the levels of pro-inflammatory factors in their gastric tissue. This was accompanied by a decrease in the overall inflammation in the said tissue. As a result, The study proposed that the utilization of camellia oil could alleviate ethanol-induced gastric mucosal injury by improving antioxidant and anti-inflammatory status, along with the regulation of gut microbiota and its metabolites (Xu et al., 2023). In a study led by Bennett C et al. aim was to explore the possible link between the α-diversity and relative abundance of gastrointestinal bacterial taxa and the response magnitude of markers indicative of exercise-induced gastrointestinal syndrome during extreme-heat stress. The results showed that the relative abundance of different bacterial groups demonstrated modestly favorable associations (suggesting low perturbations) or adverse associations with the magnitude of gastrointestinal integrity perturbations and symptoms. Furthermore, these associations were noted to potentially impact changes in body temperature in response to extreme-heat stress (Bennett et al., 2020). In the human gastrointestinal tract, there exists a complex ecosystem known as the gut microbiota, comprising approximately 1,200 species. This gut microbiota possesses a critical attribute in regulating glucose and lipid metabolism. The modulation of the gut microbiota has emerged as an innovative therapeutic approach for managing metabolic syndromes, including obesity, non-alcoholic fatty liver disease (NAFLD), and diabetes. To explore this avenue, Shan-Shan S et al. investigated a polysaccharide that cannot dissolve in water, was extracted the study focused on the water-insoluble polysaccharide (WIP) derived from the sclerotium of Poria cocos, a commonly used Traditional Chinese Medicine. The research observed that WIP treatment led to an increase in butyrate levels in regulating the gut mucosa and activating the PPAR-γ pathway within the intestines are both important factors to consider for optimal health, and enhancement of the gut function were observed. This study marks the first documentation of a polysaccharide derived from the sclerotium of *P. cocos*, that is not soluble in water, is known to influence gut microbiota and has been found effective in improving conditions of hyperglycemia and hyperlipidaemia. Consequently, WIP from *P. cocos*, acting as a prebiotic, exhibits potential for preventing or treating metabolic diseases, providing new insights into the mechanisms underlying the efficacy of this traditional herbal medicine in regulating lipid and glucose metabolism (Sun et al., 2019). The GEM project incorporates genetic, environmental, and microbial factors; Turpin W et al. conducted case-control studies that disclosed distinct microbial compositions with a difference in the gut bacteria of people with Crohn's disease (CD) when compared to those who are healthy. Despite the fact that the reasons for CD are still unknown, studies on the spread of diseases comprehended that diet plays a significant possession in the likelihood of developing CD, potentially by influencing bacterial specifications/ compositions and gut inflammation. Additionally, the cohort model of 2289 the study analyzed the health of first-degree relatives (FDRs) of patients with CD and found that adhering to Mediterranean-like dietary patterns can result in a beneficial microbiome and decreased intestinal inflammation. The knowledge acquired from this research is expected to pave the way for future dietary interventions aimed at modifying microbial composition and reducing gut inflammation in order to prevent illnesses (Turpin et al., 2022). Extensive research has been conducted in understanding the gut microbiota's effect on IBD, which is a persistent inflammatory ailment that affects the digestive system. Collij V et al. underscored the significance of establishing gold standards, enhancing and broadening biobanking initiatives, and investigating various constituents of the gut microbiota to accurately elucidate the role of the gut microbiota in IBD. The conclusion drawn is that the utilization the treatment of patients with IBD can be improved by considering the role of gut microbiota is still in its early stages. It remains to be seen whether addressing a singular facet of the gut microbiota is adequate for treating IBD. It is probable that future management of IBD will entail microbiome-targeted treatments can be used alongside the existing immunosuppressive approaches. The structure of microorganisms residing in the digestive system holds significant potential for clinical the study suggests potential applications in IBD, including the development of screening tools and personalized treatments (Collij et al., 2021). The modulation of the gastrointestinal microbiota through diet is thought to be a promising approach for enhancing healthy aging and cognitive decline is one of the functional outcomes that are associated with aging. Est A et al. as part of the NU-AGE study, the relationship between cognition, gastrointestinal microbiota, and diet in healthy older adults from the Netherlands was explored. A microarray targeting the 16S ribosomal RNA gene was used to determine 452 fecal samples were analyzed for their microbiota profile collected from 226 participants, while a 7-days of food recorded to assess their dietary intake. The study found a gradient from pro-inflammatory to anti-inflammatory diets that was co-related with the type of food consumed, whether animal- or plant-based. The study also identified specific food items, i.e., nuts, peanuts, seeds, red meat, fruits, and, grain products, that were positively indicated to microbiota composition. The researchers concluded that a plant-based diet that included nuts, fresh fruits, seeds, and peanuts had anti-inflammatory potential and was reflected in a gastrointestinal it is possible to identify a microbiota profile that has a greater potential to reduce inflammation. On the other hand, a diet primarily based on animal products showed a greater tendency towards a gastrointestinal microbiota profile that was pro-inflammatory. Although inflammation plays a significant role in the decline of cognitive function, there were no definitive correlations found in the study between cognitive functioning and gastrointestinal microbiota (van Soest et al., 2020). Diet plays a role in the symptoms of *Helicobacter pylori* (*H. pylori*) infectious microorganism, and one diet-related microbial metabolite, Trimethylamine N-oxide (TMAO), inflammatory and metabolic diseases are known to be linked with the aforementioned statement. Wu D et al. conducted an investigation with the contribution of TMAO consumption on the microbiota composition and inflammation in *H. pylori*-infected mice. The urease activity and the expression of genes associated with growth and metabolism were observed to increase in vitro experiments in the presence of TMAO. Additionally, TMAO was found to promote the production of virulence factors. The study also observed increased mice infected with *H. pylori* displayed a decrease in gut microbiota diversity and richness, as well as an increase in the production of inflammatory markers due to TMAO intake. Further analysis showed that the relative abundance of *Escherichia Shigella* was significant in *H. pylori*-infected mice after

TMAO administration, which was positively correlated with the levels of CXCL1, CRP, and LPS. The findings of the study suggested that TMAO might exacerbate *H. pylori*-induced inflammation by increasing the survival and potency of *H. pylori*. Additionally, mice infected with *H. pylori* display changes in their gut microbiota may be associated with inflammation caused by TMAO (Wu et al., 2020). Recent studies suggest a connection between alterations in the association between the normal microbiome and the development of atherosclerotic burden, which can lead to cardiovascular disease, is being studied. Although various gastrointestinal diseases can cause disturbances to the normal gut microbiome, it is not clear what impact these diseases have on the risk of developing cerebrovascular disease. In an exploratory analysis, Roth W et al. investigated the research looked into the correlation between gastrointestinal disorders and ischemic stroke in individuals aged 66 years and older. The findings were based on an average of 1,725,246 beneficiaries in each analysis. According to the study, a number of types of gastrointestinal diseases were linked to a higher risk of ischemic stroke, even after accounting for established risk factors for stroke (Roth et al., 2020). Identifying and addressing research gaps in the field of gut microbiome are crucial for advancing our understanding of this complex ecosystem. Some potential research gaps includes importance to comprehend the lasting impacts of alterations in the gut microbiome, to establish disease-specific associations, interactions between different microbial species in the gut, developing personalized medicine approaches based on an individual's gut microbiome profile, exploration of the intricate interactions between the gut microbiome immune system and metabolism of the host are important factors to consider, and other physiological functions is necessary to unravel the complexities of microbiome-host crosstalk and gastro inflammatory disorders needs an extensive study and an intensive importance towards understanding the problem to overcome the future problems associated with the further growth of the disorder due to negligence. The enhancement of current knowledge by addressing the research gaps could lead to a better understanding of the gut microbiome and its implications for human health, which is complex and requires a detailed perspective.

20.3 FUTURE PROSPECTIVE AND ROADMAP IN GUT MICROBIOTA RESEARCH FOR GASTROINTESTINAL INFLAMMATORY DISEASES

The latest progress in investigating the restorative prospect of gut microbes for the treatment of inflammatory diseases has indicated that disrupting the normal microflora causes many of the human diseases mentioned earlier. Consequently, restoring the microbial community by administering suitable microbes is now considered an effective approach (Mukherjee et al., 2018). The utilization of advanced molecular microbiota analysis techniques, such as 16S rRNA sequencing using a highly efficient barcoded pyrosequencing method, enables scientists to comprehensively and effectively examine microbiota for the very first time (Hang et al., 2014). Functional metagenomics, which involves studying the connections between gut microbiota and metabolic phenotypes, can be assessed utilizing nuclear magnetic resonance (NMR-H1)- based profiling of fecal, urine, or other biological specimens (Mai & Draganov, 2009). Preliminary findings in this field, which aimed to establish a connection between microbiota and changes in its composition caused by probiotic treatment, show encouraging results (Mai & Draganov, 2009). Sequencing the internal transcribed spacer region 2 (ITS2) of the fungal 26S rDNA and bacterial 16S rDNA is a standard molecular method used to identify and analyze the phylogeny of the gut microbiota. The shotgun sequencing method is commonly used to analyze metagenomes and meta-transcriptomes of microbes in fecal or biopsy specimens (Jovel et al., 2016). This approach is especially useful for identifying individual-specific and community-level variations within gut microbiomes. The utilization of mass spectrometric (MS) analysis and NMR spectroscopic investigation in metabolomic profiling and metaproteomic has yielded supplementary benefits in examining alterations in the gut microbiome, particularly in pathogenic or sick states (Vernocchi et al., 2016). Upon examining the gut virome, it was uncovered that 16S or 18S rRNA genes are insufficient. As a result, empirical metataxonomic methods are used to study virus-like particles (VLPs) by extracting and sequencing DNA from VLPs and employing traditional techniques, i.e., microscopy and culturing. These are obtained directly from the human microbiota rather than sewage or water sources to characterize gut-associated phages accurately. This methodology demonstrates a significant possibility for extending our acquaintance of the virome and can function as a vital tool for metagenomic analyses that bring to light on the human gut (Mukherjee et al., 2018).

Investigating unknown biomarkers and comprehending their clinical significance is essential for the progress of personalized medicine, enhancement of diagnostics, and the creation of targeted medicines (Goetz & Schork, 2018). Biomarkers are quantifiable indications of biological processes or circumstances within the body, encompassing substances such as proteins and DNA and characteristics like imaging features or behavioral traits. Comprehending biomarkers facilitates the formulation of individualized treatment approaches. Through the process of categorizing patients according to their distinct biomarker profiles, healthcare practitioners can provide treatments that are highly likely to be efficacious for each person, resulting in enhanced treatment results and diminished healthcare expenses (Ahmad et al., 2023).

Contemporary research endeavors have been aimed at exploring the intricate realm of gut microbiota, which constitutes a multifaceted ecosystem of microorganisms that inhabit the gastrointestinal tract. These microbes facilitate digestion and play an integral role in sustaining overall health and well-being (Kho & Lal, 2018; Thursby & Juge, 2017). Presently, scholars are scrutinizing numerous techniques that endeavor to precisely regulate the composition and functionality of gut microbiota, which entails comprehending the intricate interplay between microbial species and the human body. By manipulating the composition and activity of these microorganisms, researchers aim to enhance the immune system, improve digestion, abate inflammation, and prevent or treat medical conditions such as obesity, diabetes, and IBD (Ma et al., 2023; Zhang et al., 2022). Some techniques being explored include probiotics, prebiotics, fecal microbiota transplantation, and dietary interventions. These methods seek to introduce beneficial microbes into the gut, promote their growth and activity, and eliminate harmful bacteria (Cunningham et al., 2021). However, the challenge lies in identifying the appropriate balance of microorganisms and comprehending their interactions with individuals' unique biology and lifestyle (Altveş et al., 2020). Although much remains to be learned about gut microbiota and its potential affect in human health, the prospective benefits of modulating this complex ecosystem are vast and awe-inspiring. Ongoing research is investigating the influence of nutrition on the gut microbiome and how dietary interventions can alter the makeup and activity of microorganisms to enhance overall well-being. This involves examining the impact of particular foods, dietary patterns (i.e., Mediterranean or a high-fiber diet), and dietary supplements (i.e., polyphenols or omega-3 fatty acids) on the variety and functioning of gut microbiota (Bianchetti et al., 2023; Horn et al., 2022; Riaz Rajoka et al., 2017; Singh et al., 2017).

20.3.1 Understanding the Exploration of Unidentified Biomarkers and Clinical Importance

Biomarkers are quantifiable chemicals obtained from tissue or biofluid samples. This comprehensive explanation encompasses typical laboratory indices, such as the content of hemoglobin, biological methodologies, and the expression profile of genes in blood cell samples. They are increasingly crucial as minimally invasive, resource-efficient, and cost-effective methods for treating gastrointestinal inflammation with the goal of remedy (Norouzinia et al., 2017). An investigation into the connections between different components and elements, such as biomarkers and cytokines, in the immunoregulatory pathways related to IBD would facilitate the development of novel approaches for diagnosing, predicting the course of the disease and designing treatment strategies for IBD patients (Barani et al., 2021; Norouzinia et al., 2017). At present, numerous biomarkers are widely employed in clinical settings. It is important to justify that although there are various biomarkers available for diagnosing and classifying IBD patients (Garg et al., 2023) into subtypes, i.e., CD and UC, none of these biomarkers can be used in isolation to achieve a precise diagnosis (Chen et al., 2020; Cioffi et al., 2015; Liu et al., 2023; Norouzinia et al., 2017). Instead, they should be used as supplementary tools for assessing procedures or monitoring diseases (Mendoza & Abreu, 2009). In addition to clinical and endoscopic assessments, serological and immunologic biomarkers are widely employed to characterize disease processes. When compared to intrusive procedures, biomarkers provide the benefit of being safe (Tang et al., 2022). However, they are not necessarily directly related to the existence of endoscopic abnormalities (Khalilpour et al., 2016; Tang et al., 2022).

During the arachidonic acid cascade, prostaglandin E2 (PGE2) is produced as a result of a sequence of processes that cause inflammation. This appears when cyclooxygenase and PGE2 synthase function on arachidonic acid, discharged from cell membrane phospholipids via phospholipase A2 in response to inflammation (Ricciotti & FitzGerald, 2011; Wang et al., 2021). The PGE2 chemical mediator is a crucial player that has a fascinating ability to both promote and suppress inflammation, further, observed to elevate in the bloodstream, due to mucosal inflammation (Chinen et al., 2011; Goodwin & Ceuppens, 1983; Nataraj et al., 2001). The PGE2 are eliminated from the body through enzymatic reactions and oxidation mechanisms. The tetranor-PGEM is a substance represented by the chemical formula 9,15-dioxo-11a-hydroxy-13,14-dihydro-2,3,4,5-tetranor-prostan-1,20-dioic acid. The main constituent of these metabolites is tetranor-PGEM, also depicted as prostaglandin major urinary metabolite (PGE-MUM) (An et al., 2021; Hankin et al., 1997).

The PGE-MUM is a potentially stabilized condition that does not follow a specific sleep-wake pattern, commonly called a "chronotype." However, smokers are more susceptible to this condition, especially if their Brinkman index exceeds 400 (An et al., 2021; Hankin et al., 1997; Saito et al., 2012). Studies suggest that this could be due to the damage smoking causes to the lungs, leading to impaired breathing and oxygen flow. Therefore, smokers must be aware of the risks associated with this condition and take necessary measures to prevent its onset (Okayasu et al., 2014). Oral administration of nonsteroidal anti-inflammatory drugs (NSAIDs) or therapeutics reduces the magnitude, while PGE mediators augment the magnitude (Maseda & Ricciotti, 2020). Various studies on PGE-MUM as an inflammation indicator exhibit increased levels in UC. It has been observed that extremely elevated levels of PGE-MUM are associated with chronic enteropathy, which is linked to SLCO2AI (Sakurai et al., 2022; Sakurai & Saruta, 2023). Furthermore, there have been reports indicating that the levels of PGE-MUM are elevated in patients suffering from interstitial pneumonia, in addition to intestinal disorders (Hagiwara et al., 2023).

MiRNA, a small noncoding RNA molecule consisting of 17-25 nucleotides, is widely distributed throughout the body, and its expression is known to be disrupted in various diseases, including ulcerative colitis (UC) (Zhou et al., 2021). In fact, it plays a crucial role in the progression of UC to colorectal neoplasms and regulates the immune response in UC patients. Several studies demonstrate that miRNAs can serve as valuable biomarkers for diagnosing UC (Ashique et al., 2024; Shahnazari et al., 2023; Zhou et al., 2021). The individuals with active UC have considerably elevated expression levels of miR-29a and miR-21 compared to those with and without IBS, with 92% and 100% specificities, respectively. Moreover, the levels of circulating miR-375 are significantly higher in patients with UC compared to those with CD and non-IBD (Ahmed Hassan et al., 2020). This highlights the diagnostic potential of miR-375 as a marker for UC. Multiple research studies have indicated the potential for predicting the response to treatment and the need to improve treatment efficacy (Zhou et al., 2021).

The IBD is a chronic condition that causes digestive tract inflammation. Recent studies have found that individuals diagnosed with IBD often have high levels of a group of cytokines and growth modulators called Oncostatin M (OSM) (Bani Ahluwalia Luiza Moraes & Öhman, 2018; Mohammad Y. Alshahrani Ali Gaithan Alkhathami & Wahab, 2023). OSM has been shown to bind to specific receptors on the surface of cancer cells, which can inhibit their growth and induce programmed cell death (apoptosis). This suggests that OSM could be a potential therapeutic agent for treating cancer (Gholap et al., 2024; Jiang et al., 2019; Singh et al., 2024). However, further research is needed to determine how OSM inhibits cancer cell growth (Nag et al., 2024) and to develop effective treatment protocols (Wolf et al., 2023). According to recent reports, OSM can be utilized to anticipate the outcomes of anti-TNFα (anti-tumor Necrosis Factor Alpha) antibodies (Guo et al., 2022; Nag et al., 2024). In addition, patients with low levels of OSM in their blood (Guo et al., 2022) before introducing the treatment had a significantly higher probabilities of achieving clinical remission via TNF therapy (Minar et al., 2019; Mohammad Y. Alshahrani Ali Gaithan Alkhathami & Wahab, 2023). Patients who have high levels of OSM in their bloodstream are more likely to discontinue using anti-TNFα antibodies. In addition, patients with elevated OSM levels in their colonic mucosa are less likely to demonstrated to anti-TNF-α antibodies and vedolizumab (Sakurai & Saruta, 2023; Verstockt et al., 2019). In contrast, low levels of serum OSM indicated the effectiveness of anti-TNF-α antibody treatment in the absence of vedolizumab. While there may be variations in findings between the lining of the colon and the blood (Ashique, Garg, Singh, et al., 2023), OSM has been linked to patients who have developed resistance to anti-TNFα antibodies and is considered a promising biomarker for treating IBD (W. M. Kim et al., 2017; Wolf et al., 2023).

Integrins are specific molecules that help cells stick to each other. They are present on the surface of cells and made up of two subunits, i.e., α and β chain. As of now, twenty-four distinct permutations of these chains have been delineated (Takada et al., 2007). By acting as a receptor within the extracellular matrix network of epithelial cells, the αvβ6 protein supports the integrity of the epithelial barrier and maintains tissue structure. The autoantibody against epithelial adhesion molecules in the pathophysiology of UC was identified as an anti-integrin αvβ6 antibody (Kuwada et al., 2021; Livanos et al., 2023). In individuals with UC, the positive rate for this antibody was significantly higher at 92.0% compared to only 5.2% in healthy people. The diagnostic accuracy of UC was 92.0% for sensitivity and 94.8% for specificity (Livanos et al., 2023; Sakurai & Saruta, 2023). Moreover, a direct relationship was observed between the antibody titer of αvβ6 and Mayo score. Therefore, it was conclusively determined to be beneficial as a diagnostic indicator (Kuwada et al., 2021).

A considerable endeavor has been undertaken to explore non-invasive serum biomarkers for the appropriate management of IBD and enhancement of the condition's prognosis. These biomarkers are designed to aid in diagnosing, classifying, monitoring disease development, and predicting treatment outcomes and repercussions (Chen et al., 2020). Despite substantial research, the current biomarkers for IBD need improvement due to their need for more specificity and sensitivity. The incorporation of many biomarkers, such as the CD classifier, MHI, and UCRI, can improve the evaluation of disease progression. Further examination is necessary to discover new economical and readily obtainable indicators (Ahmad et al., 2023). Prioritize the prediction of complex medical conditions before they worsen and assess the likelihood of patients being readmitted or experiencing a recurrence after surgery. Moreover, it is crucial to establish rigorous and standardized processes for discovering new biomarkers and determining the endpoints of clinical trials (Ahmad et al., 2023; Califf, 2018). Utilizing objective measures is crucial for assessing the disease activity level and determining the therapy's efficacy. Before implementation in clinical practice, any recently discovered indicators must be validated through multicenter international collaborations.

20.3.2 Considerations in Microbiota Modulations for Gastrointestinal Inflammatory Disorders

When evaluating interventions that aim to improve the microbiome, it is crucial to consider the complex and ever-changing environment described in the introduction. The microbiome is an ecosystem of microorganisms that live within and on a host organism, and it plays a significant role in maintaining the host's health. However, the microbiome is a complex system influenced by various factors, such as diet, lifestyle, genetics, and environmental factors (Thursby & Juge, 2017), further summarized in Figure 20.1. The overly simplistic notions regarding the impact of a certain supplement or medication on the interaction between the microbiota and the host have generated much enthusiasm and disappointment (Quigley & Gajula, 2020). A thorough understanding of the potential interactions that occurred between the diet, genome, immune system, resident commensals of the host and an invention should make one cognizant of the challenges that will be encountered in the future.

20.3.2.1 Lifestyle Modification

The correlation between nutrition and the gut flora is now unequivocally proven. The gut microbiota is greatly influenced by nutrition in the short and long term. This is a natural conclusion as the microbiota predominantly depends on the nutrients from our meals for sustenance (Conlon & Bird, 2014). Evidence from comparative studies on groups or individuals with different dietary patterns supports the long-term effects of food. These inequalities suggest long-term dietary patterns maintained consistently, or at least for a substantial duration

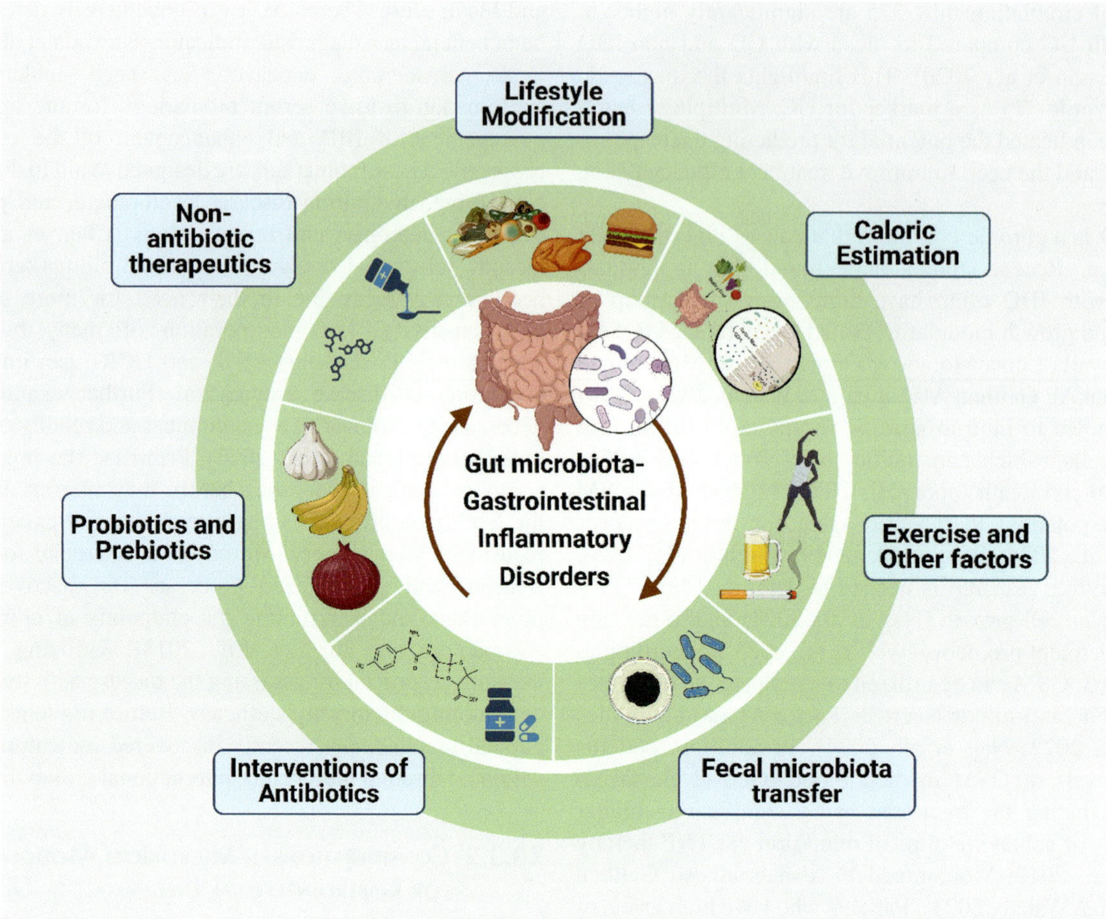

FIGURE 20.1 Summarization of various factors modulated microbiome to treat gastrointestinal inflammatory disorders (*created via Biorender.com*).

(Bianchetti et al., 2023). In the foreseeable future, altering one's diet in major ways, i.e., reduction in fiber intake, removing gluten or fermentable poly-saccharides and polyols or intake of high protein, can also impact the composition of the microbiome. Diets that are rich in carbohydrates promote the proliferation of *Lachnospiraceae, Ruminococcaceae*, and *Clostridium* cluster XVIII bacteria while inhibiting the growth of *Bifidobacteria, Enterobacteriaceae*, and *Bacteroides* (Conlon & Bird, 2014). Conversely, diets that contain a high amount of fat promote the growth of bile-tolerant bacteria, i.e., *Bacteroides, Bilophila*, and *Alistipes*. Conversely, protein-rich meals promote the proliferation of butyrate bacteria, i.e., *Faecalbacterium prausnitzii*, *Bacteroides*, *Eubacterium rectale*, *Lactobacilli*, and *Roseburia* (Huang & Liu, 2019; Shortt et al., 2018).

20.3.2.2 Caloric Estimation

The implementation of extreme diet methods such as fasting exemplifies the challenges in dietary studies. While there have been observations of alterations in the variety and makeup of microbiota in individuals with anorexia nervosa and similar eating disorders, distinguishing between cause and effect has been challenging (Navarro-Tapia et al., 2021). However, it is yet unclear whether specific patterns of gut microorganisms are associated with eating disorders and contribute to the development of these diseases. Due to the significant interest in the possible impact of the gut microbiota on obesity, several studies have examined the microbiome's involvement in several calorie-reducing techniques. The alterations in the microbiota caused by fasting have been linked to advantageous metabolic outcomes and have shown favorable effects on gut inflammation and CNS complications (Desai et al., 2016). Scientists have researched the potential impact of the microbiota on weight loss and the beneficial metabolic effects of bariatric surgery. A recent comprehensive review detailed the proven alterations in the fecal microbiota that occur after gastric bypass and other bariatric operations. A possible negative outcome of a restricted or insufficient diet is that bacteria utilize host glycans in the mucus membrane as replacements for dietary glycans, thus disrupting the stability of the mucus layer. Specific interactions between bacteria and nutrients normally maintain this stability (Desai et al., 2016). The impairment of the mucus layer is particularly prone to happen in diets lacking in fiber and can make the host more vulnerable to viruses. However, it should be noted that the breakdown of host glycans is not always detrimental, as exhibited by the associations observed between *Akkermansia muciniphilia* and enhanced health necessities (Geerlings et al., 2018; Ouyang et al., 2020).

20.3.2.3 Exercise and Other Factors

The impact assessment of physical exercise on the gut microbiome has similar difficulties, as physical exercise and nutritional habits are virtually always connected as part of a "healthy lifestyle." Professional athletes often consume significantly larger quantities of protein, affecting the microbiome's composition. Several additional factors further complicated the outcomes of exercise, i.e., the body's physical characteristics (i.e., lean or obese) significantly influence the effects of six weeks of endurance exercise on microbial diversity and the concentration of SCFAs (Ashique et al., 2022; Aya et al., 2023). This observation may be attributed to diet-related alterations in the microbiota before exercise. However, data suggests that exercise has a distinct impact on the microbiota. For instance, researchers have seen a rise in the number of Veillonella bacteria in marathon runners. They found that introducing these specific bacteria into mice improved endurance, as they converted the lactate produced during exercise into propionate (Boytar et al., 2023; Wegierska et al., 2022). Additional lifestyle factors, e.g., tobacco smoking, alcohol use, and recreational drug usage, have also been impacted with alterations in the microbiome (Fatima et al., 2024; Huang & Shi, 2019; Prabhu et al., 2014). The connection between cigarette smoking and oral cancer has sparked significant attention towards the oral microbiota.

20.3.2.4 Fecal Microbiota Transfer

Fecal microbiota transplantation (FMT) has gained scientific credibility in the past decade while being used empirically for decades if not millennia. The most notable outcomes have been documented in recurrent *clostridioides difficile*-associated illness (CDAD), with reported approximately 90% recovery rate (Gupta et al., 2016; Kim & Gluck, 2019). Several preparations and delivery techniques have been used, showing variations in effectiveness. However, it is still unknown which specific components of the FMTs are necessary for effectiveness. Based on empirical evidence, FMT has been extensively utilized in various medical conditions. Online resources now provide guidance on how to execute FMT at home (Gupta et al., 2016). Engaging in this technique is not recommended: recent instances of serious systemic infections and even fatalities resulting from FMT serve as a reminder of the potential risks associated with this therapy. Significantly, outcomes achieved through FMT in other medical conditions are considerably less remarkable than those documented in CDAD (Kelly et al., 2021). Unsurprisingly, the studies have shifted from studying an illness caused by a single organism to one involving different characteristics (Wilson et al., 2019). Despite significant efforts, we still do not fully understand the exact function of the microbiome in causing this disorder. Although, several systematic reviews and clinical trials indicated that the effectiveness of FMT in treating UC and IBS, the data from individual research present a rather ambiguous scenario. Some studies report no positive effects or even worse outcomes for FMT. The FMT is a potential technique, requires significant improvement when used outside of the context of CDAD (Kelly et al., 2021; Wilson et al., 2019). The management of intricate polygenic disorders will inevitably necessitate a highly customized and individualized strategy, which ultimately should encompass the identification of the optimal microbial combination that yields the highest efficacy for each specific phenotype. The involvement of the small intestinal microbiota in developing symptoms in hepatic encephalopathy has been well-established. Some clinical trial findings indicate that FMT may be beneficial in this context (Kelly et al., 2021; Wilson et al., 2019). The microbiome may have a significant influence on the progression of NAFLD and its more severe form, i.e., non-alcoholic steatohepatitis. Modulating the microbiome, such as through FMT, shows potential in treating these conditions. It is important to note that the gut microbiome consists of many species other than bacteria, i.e., viruses, which can be affected by FMT (Koning et al., 2023).

20.3.2.5 Interventions of Antibiotics

Without exception, antibiotics that are taken by mouth or that are eliminated by the bile and undergo circulation between the intestines and liver, regardless of how they are administered, will have some degree of impact on the gut flora (Yoon & Yoon, 2018). The presence of "innocent bystander" effects can weaken the host's ability to defend against infections, leading to CDAD or fungal overgrowth. Certain populations are particularly vulnerable to these effects. Antibiotic resistance is a significant worldwide concern, and the escalating resistance to *H. pylori*, a prevalent pathogen, is worrisome (Savoldi et al., 2018). The human gut microbiome is a "reservoir of antibiotic resistance genes." In a recent study, apparently healthy individuals in China were found to have 1,093 antibiotic-resistant genes. The existence of horizontal and vertical transfer of these genes is worrisome (Patangia, Ryan, et al., 2022).

Antibiotics have various effects on the microbiota, including their influence on inflammation, metabolism, and tumorigenesis. The overall outcome, whether harmful or advantageous, depends on the specific antibiotic used and aspects related to the microbes and the host (Patangia, Anthony Ryan, et al., 2022; Ramirez et al., 2020). It is important to acknowledge that the use of antibiotics may have significant implications for human health in the long run. The consequences of such effects are still being studied, and it is vital to consider them carefully (Ramirez et al., 2020). There is growing evidence that infants are more vulnerable to the negative effects of antibiotics. Even small amounts of antibiotics that we consume through animal products can increase the risk of developing inflammatory and metabolic disorders later in life. Therefore, it is important to raise awareness and take action to reduce the use of antibiotics worldwide (Patangia, Anthony Ryan, et al., 2022; Ramirez et al., 2020).

20.3.2.6 Probiotics and Prebiotics

Prebiotic substances are present in various cereals and plants, such as onions, garlic, bananas, chicory root, and Jerusalem artichokes. These substances are known to promote the growth of beneficial bacteria in the gut (Ashique, Faruk, et al., 2023; Ashique, Garg, Bhatt, et al., 2023). Fructo-oligosaccharides (FOSs) are encountered in many plants, including wheat, but in small quantities that may not furnish probiotic benefits in their natural forms. Additional prebiotics that exhibit higher biological activity and selectivity include oligofructose (OF), fructo-oligosaccharides (FOSs), galacto-oligosaccharides (GOSs), chicory fiber, and inulin (Ashique, Mishra, et al., 2023). Human milk oligosaccharides serve as paramount probiotics for breast milk-fed newborns, facilitating the growth of *Bifidobacteria* (Stiverson et al., 2014). Consequently, they have been associated with numerous health advantages. Breast milk contains additional constituents that benefit the infant's microbiota and immune system. Several research on prebiotics is centered around creating specialized prebiotics that selectively target specific groups of microorganisms in the gut microbiota (Ashaolu, 2020). However, some researchers need to be more skeptical about the significance and practicality of this selective approach, considering the functional redundancy that naturally exists in the gut microbiome. Recently, clinical research on prebiotics has placed significant emphasis on metabolic effects, specifically blood glucose control, calcium balance, and weight reduction (Ashaolu, 2020; Megur et al., 2022). Studies have also examined the effects of this substance on the immune system (such as boosting antibody responses due to vaccines and promotion of anti-inflammatory cytokines), the nervous system (with potential benefits for mood and cognition), the cardiovascular system (including improvements in lipid profiles), and the gastrointestinal system (with notable effects on colon transit and significant for IBS) (Megur et al., 2022).

20.3.2.7 Consequences of Non-Antibiotic Therapeutics on Gut Microbiome

The composition of the microbiota can be altered by specific interventions that modify the inherent defense mechanisms of the body against bacterial colonization (Le Bastard et al., 2018). Proton pump inhibitors (PPIs) can sometimes lead to reduced acidity in the stomach, which has been found to have an inconsistent association with an increased risk of infections such as *Clostridioides difficile* infection, enteric infections, and small intestinal bacterial overgrowth (Maier et al., 2018). Research on human feces has confirmed reduced Clostridiales and a rise in Actinomycetales, Micrococcaceae, and Streptococcaceae in individuals utilized PPIs (Le Bastard et al., 2018; Pilmis et al., 2020). These alterations have previously been linked to an increased vulnerability to the adverse effects of antibiotics (Maier et al., 2018). Likewise, several medicines that modify the movement and passage of substances through the intestines can potentially affect the composition of the microbiota. Numerous medications are expected to interact with the microbiota, leading to either increased or decreased effectiveness or side effects (Maier et al., 2018; Pilmis et al., 2020). This presents a promising area for future study on the microbiome.

20.4 CLINICAL-BASED EVIDENCE TREATMENT TARGETED GUT MICROBIOTA FOR GI DISORDERS

The GI tract is home to a complex ecosystem of microorganisms collectively known as the gut microbiota. Emerging evidence suggests that alterations in the composition and function of the gut microbiota play a pivotal role in the pathogenesis of various GI disorders, i.e., UC, IBS, IBD, and gastroenteritis. Consequently, targeting the gut microbiota has garnered significant interest as a therapeutic approach to manage these conditions. This introduction will provide an overview of the clinical evidence supporting gut microbiota-targeted treatments for GI disorders, highlighting key findings from clinical trials and observational studies. Furthermore, the introduction will underscore the challenges and opportunities associated with microbiota-based therapies, including variability in treatment response, the need for personalized approaches, and safety considerations (Guinane & Cotter, 2013; Hani et al., 2023; Nishida et al., 2022; Pramanik et al., 2021). It will also address the importance of robust clinical trial designs, standardized methodologies for microbiota analysis, and long-term follow-up to establish the efficacy and safety of these interventions. Overall, this introduction aims to set the stage for the subsequent discussion on the role of gut microbiota-targeted treatments in improving clinical outcomes and quality of life for individuals with GI disorders. By synthesizing current clinical evidence and highlighting future research directions, it seeks to inform clinicians, researchers, and stakeholders about the evolving landscape of microbiota-based therapies in gastroenterology. In summary, clinical-based evidence for treatments targeting the gut microbiota in GI disorders is essential for validating efficacy, ensuring safety, identifying optimal interventions, advancing personalized medicine, obtaining regulatory approval, guiding clinical practice, and educating stakeholders. By generating robust evidence through well-designed trials, researchers can improve outcomes and quality of life for individuals with GI disorders (Guinane & Cotter, 2013; Nishida et al., 2022). Table 20.1 summarized various clinical investigations associated gut microbiome in inflammatory disorders for therapeutic and diagnostic purposes.

20.5 CONCLUSION

Gut microbiota, evolving alongside their hosts, forms an essential component of the human organism. Microbiota obtained at birth evolves alongside the host, maintaining stability and diversity throughout adulthood and into death. Additionally, the gut microbiota plays in well-being of individuals is significantly impacted by various factors,

TABLE 20.1
Recent Clinical Investigations Associated Gut Microbiome for the Theranostics Approaches in Various Gastrointestinal Inflammatory Disorders

Sl No.	Study Title	Application/ Trial Registration No.	Sponsor/s	Current Phase/ Status	References
1.	Fructose Effect on Neuroinflammation and Feelings (MOODYFRUCTOSE)	NCT05371067	University Hospital, Rouen	Not Applicable	(University Hospital, 2021)
2.	Evaluation of the Diagnostic Potential of Artificial Intelligence-assisted Fecal Microbiome Testing for Inflammatory Bowel Disease	NCT05797207	Istanbul Medipol University Hospital	Recruiting	(Istanbul Medipol University Hospital, 2023)
3.	DIAMANTE – Diseases Infectious Analysis Microflora Antibiotic Effect	NCT00424905	Sanofi	Phase 4	(Sanofi, 2006)
4.	Study of Fecal Bacteria in Early Diagnosis of Colorectal Cancer	NCT02845973	Shanghai Jiao Tong University School of Medicine	Completed	(Shanghai Jiao Tong University School of Medicine, 2012)
5.	Gut Microbiota and Color-rectal Cancer. (CCR-microbiota)	NCT04662853	Maimónides Biomedical Research Institute of Córdoba	Not Applicable	(Maimónides Biomedical Research Institute of Córdoba, 2017)
6.	Prevalence of FODMAP Intolerance and JHS in FGID and Association with Microbiome, Dyssynergic Defecation and Dietary Intervention (PreDiMi)	NCT03460613	University Of Zurich	Not Applicable	(University of Zurich, 2017)
7.	The Gastrointestinal Microbiome of Infants with GERD and PPI Therapy: A Pilot Study	NCT02359604	Medical University of Graz	Completed	(Medical University of Graz, 2014)
8.	Analysis of Intestinal Microflora Combined with DNA Methylation in Stool to Detect Colorectal Cancer (IMMDC)	NCT04302363	Weidong Liu, Md	Not Applicable	(WEIDONG LIU, 2018)
9.	Validation of the Small Intestine Microbiome Aspiration (SIMBA) Capsule	NCT04094558	University Of Calgary	Not Applicable	(University of Calgary, 2021)
10.	Research on Gut Microbiome and Metabolomics Alterations in *C. Difficile* Infected IBD Patients	NCT05714566	Renji Hospital	Recruiting	(RenJi Hospital, 2022)
11.	Gut and Oral Microbiome of GDM Patients During Pregnancy	NCT04420507	Peking Union Medical College Hospital	Completed	(Peking Union Medical College Hospital, 2017)
12.	Microbiome Test for the Detection of Colorectal Polyps and Cancer	NCT02141945	Metabiomics Corp	Completed	(Metabiomics Corp, 2014)
13.	To Evaluate the Beneficial Effect of Probiotics on NAFLD Patients and the Role of Gut Microbiota Modulation	NCT05402449	Genmont Biotech Incorporation	Recruiting	(GenMont Biotech Incorporation, 2022)
14.	The Effect of Prebiotics on the Microbiome in Irritable Bowel Syndrome Patients: The Diet and Microbiome Study	NCT01829932	Queen's University	Completed	(Queen's University, 2013)
15.	Mucosa Adherent Intestinal Microbiome in Microscopic Colitis and Colorectal Cancer (CMBACT)	NCT06172647	Hospital Mutua De Terrassa	Recruiting	(Hospital Mutua de Terrassa, 2022)

including both physical and mental influences. The latest progress in investigating the restorative prospect of gut microbes for the treatment of inflammatory diseases has indicated that disrupting the normal microflora causes many of the human diseases mentioned earlier. Consequently, restoring the microbial community by administering suitable microbes is now considered an effective approach. By synthesizing current clinical evidence and highlighting future research directions, it seeks to inform clinicians, researchers, and stakeholders about the evolving landscape of microbiota-based therapies in gastroenterology. Clinical-based evidence for treatments targeting the gut microbiota in GI disorders is essential for validating efficacy, ensuring safety, identifying optimal interventions, advancing personalized medicine, obtaining regulatory approval, guiding clinical practice, and educating stakeholders. By generating robust evidence through well-designed trials, researchers can improve outcomes and quality of life for individuals with GI disorders.

LIST OF ABBREVIATIONS

CNS: Central nervous system
SCFAs: Short-chain fatty acids
TRP: Tryptophan
IBD: Inflammatory bowel disease
ROS: Reactive oxygen species
NAFLD: Non-alcoholic fatty liver disease
CD: Crohn's disease
H. pylori: *Helicobacter pylori*
UC: Ulcerative colitis
OSM: Oncostatin M
TNF: Tumor Necrosis Factor
FMT: Fecal microbiota transplantation
PPIs: Proton pump inhibitors

REFERENCES

Adak, A., & Khan, M. R. (2019). An insight into gut microbiota and its functionalities. *Cellular and Molecular Life Sciences: CMLS*, *76*(3), 473–493. https://doi.org/10.1007/s00018-018-2943-4

Ahmad, A., Imran, M., & Ahsan, H. (2023). Biomarkers as biomedical bioindicators: Approaches and techniques for the detection, analysis, and validation of novel biomarkers of diseases. *Pharmaceutics*, *15*(6), 1630. https://doi.org/10.3390/pharmaceutics15061630

Ahmed Hassan, E., El-Din Abd El-Rehim, A. S., Mohammed Kholef, E. F., & Abd-Elgwad Elsewify, W. (2020). Potential role of plasma miR-21 and miR-92a in distinguishing between irritable bowel syndrome, ulcerative colitis, and colorectal cancer. *Gastroenterology and Hepatology from Bed to Bench*, *13*(2), 147–154. https://doi.org/10.22037/ghfbb.v13i2.1868

Akhtar, M., Chen, Y., Ma, Z., Zhang, X., Shi, D., Khan, J. A., & Liu, H. (2022). Gut microbiota-derived short chain fatty acids are potential mediators in gut inflammation. *Animal Nutrition*, *8*, 350–360. https://doi.org/https://doi.org/10.1016/j.aninu.2021.11.005

Altveş, S., Yildiz, H. K., & Vural, H. C. (2020). Interaction of the microbiota with the human body in health and diseases. *Bioscience of Microbiota, Food and Health*, *39*(2), 23–32. https://doi.org/10.12938/bmfh.19-023

An, Y., Yao, J., & Niu, X. (2021). The signaling pathway of PGE(2) and its regulatory role in T cell differentiation. *Mediators of Inflammation*, *2021*, 9087816. https://doi.org/10.1155/2021/9087816

Ashaolu, T. J. (2020). Immune boosting functional foods and their mechanisms: A critical evaluation of probiotics and prebiotics. *Biomedicine & Pharmacotherapy*, *130*, 110625. https://doi.org/https://doi.org/10.1016/j.biopha.2020.110625

Ashique, S., Bhowmick, M., Pal, R., Khatoon, H., Kumar, P., Sharma, H., Garg, A., Kumar, S., & Das, U. (2024). Multi drug resistance in colorectal cancer- approaches to overcome, advancements and future success. *Advances in Cancer Biology – Metastasis*, *10*, 100114. https://doi.org/https://doi.org/10.1016/j.adcanc.2024.100114

Ashique, S., De Rubis, G., Sirohi, E., Mishra, N., Rihan, M., Garg, A., Reyes, R.-J., Manandhar, B., Bhatt, S., Jha, N. K., Singh, T. G., Gupta, G., Singh, S. K., Chellappan, D. K., Paudel, K. R., Hansbro, P. M., Oliver, B. G., & Dua, K. (2022). Short chain fatty acids: Fundamental mediators of the gut-lung axis and their involvement in pulmonary diseases. *Chemico-Biological Interactions*, *368*, 110231. https://doi.org/https://doi.org/10.1016/j.cbi.2022.110231

Ashique, S., Faruk, A., Ahmad, F. J., Khan, T., & Mishra, N. (2023). It is all about probiotics to control cervical cancer. *Probiotics and Antimicrobial Proteins*, *16*, 979–992. https://doi.org/10.1007/s12602-023-10183-2

Ashique, S., Garg, A., Bhatt, S., Sirohi, E., Fatima, N., Bajwa, N., Dua, K., & Mishra, N. (2023). Synbiotics in cervical cancer. In N. Mishra, S. Bhatt, K. R. Paudel, P. M Hansbro, & K. Dua (Eds.), *Synbiotics for the Management of Cancer* (pp. 135–155). Springer Nature Singapore. https://doi.org/10.1007/978-981-19-7550-9_6

Ashique, S., Garg, A., Singh, V., Rai, G., Mishra, N., Soni, M. L., Kumar, S., & Madamsetty, V. S. (2023). Role of block copolymers in colon cancer. In N. Mishra & V. Pandey (Eds.), *Block Co-Polymeric Nanocarriers: Design, Concept, and Therapeutic Applications* (pp. 181–209). Springer Nature Singapore. https://doi.org/10.1007/978-981-99-6917-3_7

Ashique, S., Mishra, N., Garg, A., Sibuh, B. Z., Taneja, P., Rai, G., Djearamane, S., Wong, L. S., Al-Dayan, N., Roychoudhury, S., Kesari, K. K., Slama, P., Roychoudhury, S., & Gupta, P. K. (2023). Recent updates on correlation between reactive oxygen species and synbiotics for effective management of ulcerative colitis. *Frontiers in Nutrition*, *10*, 1126579. https://doi.org/10.3389/fnut.2023.1126579

Aya, V., Jimenez, P., Muñoz, E., & Ramírez, J. D. (2023). Effects of exercise and physical activity on gut microbiota composition and function in older adults: A systematic review. *BMC Geriatrics*, *23*(1), 364. https://doi.org/10.1186/s12877-023-04066-y

Bani Ahluwalia Luiza Moraes, M. K. M., & Öhman, L. (2018). Immunopathogenesis of inflammatory bowel disease and mechanisms of biological therapies. *Scandinavian Journal of Gastroenterology*, *53*(4), 379–389. https://doi.org/10.1080/00365521.2018.1447597

Barani, M., Rahdar, A., Sargazi, S., Amiri, M. S., Sharma, P. K., & Bhalla, N. (2021). Nanotechnology for inflammatory bowel disease management: Detection, imaging and treatment. *Sensing and Bio-Sensing Research*, *32*, 100417. https://doi.org/https://doi.org/10.1016/j.sbsr.2021.100417

Bennett, C. J., Henry, R., Snipe, R. M. J., & Costa, R. J. S. (2020). Is the gut microbiota bacterial abundance and composition associated with intestinal epithelial injury, systemic inflammatory profile, and gastrointestinal symptoms in response to exertional-heat stress? *Journal of Science and Medicine in Sport*, *23*(12), 1141–1153. https://doi.org/https://doi.org/10.1016/j.jsams.2020.06.002

Bianchetti, G., De Maio, F., Abeltino, A., Serantoni, C., Riente, A., Santarelli, G., Sanguinetti, M., Delogu, G., Martinoli, R., Barbaresi, S., Spirito, M. De, & Maulucci, G. (2023). Unraveling the gut microbiome-diet connection: Exploring the impact of digital precision and personalized nutrition on microbiota composition and host physiology. *Nutrients*, *15*(18), 3931. https://doi.org/10.3390/nu15183931

Boytar, A. N., Skinner, T. L., Wallen, R. E., Jenkins, D. G., & Dekker Nitert, M. (2023). The effect of exercise prescription on the human gut microbiota and comparison between clinical and apparently healthy populations: A systematic review. *Nutrients*, *15*(6). https://doi.org/10.3390/nu15061534

Califf, R. M. (2018). Biomarker definitions and their applications. *Experimental Biology and Medicine (Maywood, N.J.)*, *243*(3), 213–221. https://doi.org/10.1177/1535370217750088

Chen, P., Zhou, G., Lin, J., Li, L., Zeng, Z., Chen, M., & Zhang, S. (2020). Serum biomarkers for inflammatory bowel disease. *Frontiers in Medicine*, *7*, 123. https://doi.org/10.3389/fmed.2020.00123

Chinen, T., Komai, K., Muto, G., Morita, R., Inoue, N., Yoshida, H., Sekiya, T., Yoshida, R., Nakamura, K., Takayanagi, R., & Yoshimura, A. (2011). Prostaglandin E2 and SOCS1 have a role in intestinal immune tolerance. *Nature Communications*, *2*, 190. https://doi.org/10.1038/ncomms1181

Ciernikova, S., Sevcikova, A., Drgona, L., & Mego, M. (2023). Modulating the gut microbiota by probiotics, prebiotics, postbiotics, and fecal microbiota transplantation: An emerging trend in cancer patient care. *Biochimica et Biophysica Acta (BBA) – Reviews on Cancer*, *1878*(6), 188990. https://doi.org/https://doi.org/10.1016/j.bbcan.2023.188990

Cioffi, M., Rosa, A. De, Serao, R., Picone, I., & Vietri, M. T. (2015). Laboratory markers in ulcerative colitis: Current insights and future advances. *World Journal of Gastrointestinal Pathophysiology*, *6*(1), 13–22. https://doi.org/10.4291/wjgp.v6.i1.13

Collij, V., Klaassen, M. A. Y., Weersma, R. K., & Vila, A. V. (2021). Gut microbiota in inflammatory bowel diseases: Moving from basic science to clinical applications. *Human Genetics*, *140*(5), 703–708. https://doi.org/10.1007/s00439-020-02218-3

Conlon, M. A., & Bird, A. R. (2014). The impact of diet and lifestyle on gut microbiota and human health. *Nutrients*, *7*(1), 17–44. https://doi.org/10.3390/nu7010017

Cunningham, M., Azcarate-Peril, M. A., Barnard, A., Benoit, V., Grimaldi, R., Guyonnet, D., Holscher, H. D., Hunter, K., Manurung, S., Obis, D., Petrova, M. I., Steinert, R. E., Swanson, K. S., van Sinderen, D., Vulevic, J., & Gibson, G. R. (2021). Shaping the future of probiotics and prebiotics. *Trends in Microbiology*, *29*(8), 667–685. https://doi.org/10.1016/j.tim.2021.01.003

Dalile, B., Van Oudenhove, L., Vervliet, B., & Verbeke, K. (2019). The role of short-chain fatty acids in microbiota-gut-brain communication. *Nature Reviews. Gastroenterology & Hepatology*, *16*(8), 461–478. https://doi.org/10.1038/s41575-019-0157-3

Desai, M. S., Seekatz, A. M., Koropatkin, N. M., Kamada, N., Hickey, C. A., Wolter, M., Pudlo, N. A., Kitamoto, S., Terrapon, N., Muller, A., Young, V. B., Henrissat, B., Wilmes, P., Stappenbeck, T. S., Núñez, G., & Martens, E. C. (2016). A dietary fiber-deprived gut microbiota degrades the colonic mucus barrier and enhances pathogen susceptibility. *Cell*, *167*(5), 1339–1353.e21. https://doi.org/10.1016/j.cell.2016.10.043

Fatima, N., Ashique, S., Upadhyay, A., Kumar, S., Kumar, H., Kumar, N., & Kumar, P. (2024). Current landscape of therapeutics for the management of hypertension–A review. *Current Drug Delivery*, *21*(5), 662–682. https://doi.org/10.2174/1567201820666230623121433

Garg, A., Ashique, S., Kumar, S., Soni, M. L., Shravastava, A., Pandey, V., Garud, N., Agrawal, A., Akram, W., & Mishra, N. (2023). An insight to block copolymers in inflammatory bowel disease management. In N. Mishra & V. Pandey (Eds.), *Block Co-polymeric Nanocarriers: Design, Concept, and Therapeutic Applications* (pp. 227–244). Springer Nature Singapore. https://doi.org/10.1007/978-981-99-6917-3_9

Geerlings, S. Y., Kostopoulos, I., de Vos, W. M., & Belzer, C. (2018). Akkermansia muciniphila in the human gastrointestinal tract: When, where, and how? *Microorganisms*, *6*(3), 75. https://doi.org/10.3390/microorganisms6030075

GenMont Biotech Incorporation. (2022). *To Evaluate the Beneficial Effect of Probiotics on NAFLD Patients and the Role of Gut Microbiota Modulation*. https://clinicaltrials.gov/study/NCT05402449?cond=Gastrointestinal Diseases&intr=Gut Microbiome Testing&page=2&rank=19

Gholap, A. D., Kapare, H. S., Pagar, S., Kamandar, P., Bhowmik, D., Vishwakarma, N., Raikwar, S., Garkal, A., Mehta, T. A., Rojekar, S., Hatvate, N., & Mohanto, S. (2024). Exploring modified chitosan-based gene delivery technologies for therapeutic advancements. *International Journal of Biological Macromolecules*, *260*, 129581. https://doi.org/https://doi.org/10.1016/j.ijbiomac.2024.129581

Goetz, L. H., & Schork, N. J. (2018). Personalized medicine: Motivation, challenges, and progress. *Fertility and Sterility*, *109*(6), 952–963. https://doi.org/10.1016/j.fertnstert.2018.05.006

Goodwin, J. S., & Ceuppens, J. (1983). Regulation of the immune response by prostaglandins. *Journal of Clinical Immunology*, *3*(4), 295–315. https://doi.org/10.1007/BF00915791

Guha, L., Agnihotri, T. G., Jain, A., & Kumar, H. (2023). Gut microbiota and traumatic central nervous system injuries: Insights into pathophysiology and therapeutic approaches. *Life Sciences*, *334*, 122193. https://doi.org/https://doi.org/10.1016/j.lfs.2023.122193

Guinane, C. M., & Cotter, P. D. (2013). Role of the gut microbiota in health and chronic gastrointestinal disease: Understanding a hidden metabolic organ. *Therapeutic Advances in Gastroenterology*, *6*(4), 295–308. https://doi.org/10.1177/1756283X13482996

Guo, A., Ross, C., Chande, N., Gregor, J., Ponich, T., Khanna, R., Sey, M., Beaton, M., Yan, B., Kim, R. B., & Wilson, A. (2022). High oncostatin M predicts lack of clinical remission for patients with inflammatory bowel disease on tumor necrosis factor α antagonists. *Scientific Reports*, *12*(1), 1185. https://doi.org/10.1038/s41598-022-05208-9

Gupta, S., Allen-Vercoe, E., & Petrof, E. O. (2016). Fecal microbiota transplantation: In perspective. *Therapeutic Advances in Gastroenterology*, *9*(2), 229–239. https://doi.org/10.1177/1756283X15607414

Hagiwara, S.-I., Abe, N., Hosoi, K., Hara, T., Ishige, T., Shimizu, H., Mizuochi, T., Kakiuchi, T., Kunisaki, R., Matsuoka, R., Kondou, H., Kakuta, F., Nakayama, Y., Kimura, T., Maeyama, T., Honma, H., Hirano, D., Saruta, M., Yoshida, T., … & Etani, Y. (2023). Utility of a rapid assay for prostaglandin E-major urinary metabolite as a biomarker in pediatric ulcerative colitis. *Scientific Reports*, *13*(1), 9898. https://doi.org/10.1038/s41598-023-37145-6

Hang, J., Desai, V., Zavaljevski, N., Yang, Y., Lin, X., Satya, R. V., Martinez, L. J., Blaylock, J. M., Jarman, R. G., Thomas, S. J., & Kuschner, R. A. (2014). 16S rRNA gene pyrosequencing of reference and clinical samples and investigation of the temperature stability of microbiome profiles. *Microbiome*, *2*(1), 31. https://doi.org/10.1186/2049-2618-2-31

Hani, U., Gowda, B. H. J., Haider, N., Ramesh, K., Paul, K., Ashique, S., Ahmed, M. G., Narayana, S., Mohanto, S., & Kesharwani, P. (2023). Nanoparticle-based approaches for treatment of hematological malignancies: A comprehensive review. *AAPS PharmSciTech*, *24*(8), 233. https://doi.org/10.1208/s12249-023-02670-0

Hankin, J. A., Wheelan, P., & Murphy, R. C. (1997). Identification of novel metabolites of prostaglandin E2 formed by isolated rat hepatocytes. *Archives of Biochemistry and Biophysics*, *340*(2), 317–330. https://doi.org/10.1006/abbi.1997.9921

Horn, J., Mayer, D. E., Chen, S., & Mayer, E. A. (2022). Role of diet and its effects on the gut microbiome in the pathophysiology of mental disorders. *Translational Psychiatry*, *12*(1), 164. https://doi.org/10.1038/s41398-022-01922-0

Hospital Mutua de Terrassa. (2022). *Mucosa Adherent Intestinal Microbiome in Microscopic Colitis and Colorectal Cancer (CMBACT)*. https://clinicaltrials.gov/study/NCT06172647?cond=GastrointestinalDiseases&intr=Gut Microbiome Testing&page=3&rank=21

Huang, C., & Shi, G. (2019). Smoking and microbiome in oral, airway, gut and some systemic diseases. *Journal of Translational Medicine*, *17*(1), 225. https://doi.org/10.1186/s12967-019-1971-7

Huang, P., & Liu, Y. (2019). A reasonable diet promotes balance of intestinal microbiota: Prevention of precolorectal cancer. *BioMed Research International*, *2019*, 3405278. https://doi.org/10.1155/2019/3405278

Ikeda, E., Yamaguchi, M., & Kawabata, S. (2024). Gut microbiota-mediated alleviation of dextran sulfate sodium-induced colitis in mice. *Gastro Hep Advances*, *3*, 461–470. https://doi.org/https://doi.org/10.1016/j.gastha.2024.01.016

Islam, M. M., Islam, M. M., Rahman, M. A., Ripon, M. A. R., & Hossain, M. S. (2023). Gut microbiota in obesity and related complications: Unveiling the complex interplay. *Life Sciences*, *334*, 122211. https://doi.org/https://doi.org/10.1016/j.lfs.2023.122211

Istanbul Medipol University Hospital. (2023). *Evaluation of the Diagnostic Potential of Artificial Intelligence-assisted Fecal Microbiome Testing for Inflammatory Bowel Disease*. https://clinicaltrials.gov/study/NCT05797207?cond=GastrointestinalDiseases&intr=GutMicrobiomeTesting&rank=2

Jiang, G.-M., Tan, Y., Wang, H., Peng, L., Chen, H.-T., Meng, X.-J., Li, L.-L., Liu, Y., Li, W.-F., & Shan, H. (2019). The relationship between autophagy and the immune system and its applications for tumor immunotherapy. *Molecular Cancer*, *18*(1), 17. https://doi.org/10.1186/s12943-019-0944-z

Jovel, J., Patterson, J., Wang, W., Hotte, N., O'Keefe, S., Mitchel, T., Perry, T., Kao, D., Mason, A. L., Madsen, K. L., & Wong, G. K.-S. (2016). Characterization of the gut microbiome using 16S or shotgun metagenomics. *Frontiers in Microbiology*, *7*, 459. https://doi.org/10.3389/fmicb.2016.00459

Kelly, C. R., Yen, E. F., Grinspan, A. M., Kahn, S. A., Atreja, A., Lewis, J. D., Moore, T. A., Rubin, D. T., Kim, A. M., Serra, S., Nersesova, Y., Fredell, L., Hunsicker, D., McDonald, D., Knight, R., Allegretti, J. R., Pekow, J., Absah, I., Hsu, R., … & Laine, L. (2021). Fecal microbiota transplantation is highly effective in real-world practice: Initial results from the FMT national registry. *Gastroenterology*, *160*(1), 183–192.e3. https://doi.org/10.1053/j.gastro.2020.09.038

Khalilpour, A., Kazemzadeh-Narbat, M., Tamayol, A., Oklu, R., & Khademhosseini, A. (2016). Biomarkers and diagnostic tools for detection of Helicobacter pylori. *Applied Microbiology and Biotechnology*, *100*(11), 4723–4734. https://doi.org/10.1007/s00253-016-7495-7

Kho, Z. Y., & Lal, S. K. (2018). The human gut microbiome – A potential controller of wellness and disease. *Frontiers in Microbiology*, *9* 1835. https://doi.org/10.3389/fmicb.2018.01835

Kim, K. O., & Gluck, M. (2019). Fecal microbiota transplantation: An update on clinical practice. *Clinical Endoscopy*, *52*(2), 137–143. https://doi.org/10.5946/ce.2019.009

Kim, W. M., Kaser, A., & Blumberg, R. S. (2017). A role for oncostatin M in inflammatory bowel disease. *Nature Medicine*, *23*(5), 535–536. https://doi.org/10.1038/nm.4338

Koning, M., Herrema, H., Nieuwdorp, M., & Meijnikman, A. S. (2023). Targeting nonalcoholic fatty liver disease via gut microbiome-centered therapies. *Gut Microbes*, *15*(1), 2226922. https://doi.org/10.1080/19490976.2023.2226922

Kuwada, T., Shiokawa, M., Kodama, Y., Ota, S., Kakiuchi, N., Nannya, Y., Yamazaki, H., Yoshida, H., Nakamura, T., Matsumoto, S., Muramoto, Y., Yamamoto, S., Honzawa, Y., Kuriyama, K., Okamoto, K., Hirano, T., Okada, H., Marui, S., Sogabe, Y., … & Seno, H. (2021). Identification of an anti-integrin αvβ6 autoantibody in patients with ulcerative colitis. *Gastroenterology*, *160*(7), 2383–2394.e21. https://doi.org/10.1053/j.gastro.2021.02.019

Le Bastard, Q., Al-Ghalith, G. A., Grégoire, M., Chapelet, G., Javaudin, F., Dailly, E., Batard, E., Knights, D., & Montassier, E. (2018). Systematic review: Human gut dysbiosis induced by non-antibiotic prescription medications. *Alimentary Pharmacology \& Therapeutics*, *47*(3), 332–345. https://doi.org/https://doi.org/10.1111/apt.14451

Liu, R., Li, D., Haritunians, T., Ruan, Y., Daly, M. J., Huang, H., & McGovern, D. P. B. (2023). Profiling the inflammatory bowel diseases using genetics, serum biomarkers, and smoking information. *IScience*, *26*(10), 108053. https://doi.org/https://doi.org/10.1016/j.isci.2023.108053

Liu, W., Zheng, C., Li, Q., Xu, T., Cao, W., Shi, M., Huang, F., Liu, L., Luo, Y., Zhang, W., Xiao, Q., Liu, Z., & Deng, X. (2023). Preoperative oral probiotics relieve insulin resistance and gut dysbacteriosis in patients with gastric cancer after gastrectomy. *Journal of Functional Foods*, *101*, 105426. https://doi.org/https://doi.org/10.1016/j.jff.2023.105426

Livanos, A. E., Dunn, A., Fischer, J., Ungaro, R. C., Turpin, W., Lee, S.-H., Rui, S., Del Valle, D. M., Jougon, J. J., Martinez-Delgado, G., Riddle, M. S., Murray, J. A., Laird, R. M., Torres, J., Agrawal, M., Magee, J. S., Dervieux, T., Gnjatic, S., Sheppard, D., … & Mehandru, S. (2023). Anti-integrin αvβ6 autoantibodies are a novel biomarker that antedate ulcerative colitis. *Gastroenterology*, *164*(4), 619–629. https://doi.org/10.1053/j.gastro.2022.12.042

Ma, T., Shen, X., Shi, X., Sakandar, H. A., Quan, K., Li, Y., Jin, H., Kwok, L.-Y., Zhang, H., & Sun, Z. (2023). Targeting gut microbiota and metabolism as the major probiotic mechanism – An evidence-based review. *Trends in Food Science & Technology*, *138*, 178–198. https://doi.org/https://doi.org/10.1016/j.tifs.2023.06.013

Mai, V., & Draganov, P. V. (2009). Recent advances and remaining gaps in our knowledge of associations between gut microbiota and human health. *World Journal of Gastroenterology*, *15*(1), 81–85. https://doi.org/10.3748/wjg.15.81

Maier, L., Pruteanu, M., Kuhn, M., Zeller, G., Telzerow, A., Anderson, E. E., Brochado, A. R., Fernandez, K. C., Dose, H., Mori, H., Patil, K. R., Bork, P., & Typas, A. (2018). Extensive impact of non-antibiotic drugs on human gut bacteria. *Nature*, *555*(7698), 623–628. https://doi.org/10.1038/nature25979

Maimónides Biomedical Research Institute of Córdoba. (2017). *Gut Microbiota and Color-Rectal Cancer. (CCR-microbiota).* https://clinicaltrials.gov/study/NCT04662853?cond=GastrointestinalDiseases&intr=Gut Microbiome Testing&rank=5

Maseda, D., & Ricciotti, E. (2020). NSAID-gut microbiota interactions. *Frontiers in Pharmacology*, *11*, 1153. https://doi.org/10.3389/fphar.2020.01153

Medical University of Graz. (2014). *The Gastrointestinal Microbiome of Infants With GERD and PPI Therapy: A Pilot Study.* https://clinicaltrials.gov/study/NCT02359604?cond=GastrointestinalDiseases&intr=Gut Microbiome Testing&rank=7

Megur, A., Daliri, E. B.-M., Baltriukienė, D., & Burokas, A. (2022). Prebiotics as a tool for the prevention and treatment of obesity and diabetes: Classification and ability to modulate the gut microbiota. *International Journal of Molecular Sciences*, *23*(11), 6097. https://doi.org/10.3390/ijms23116097

Mendoza, J. L., & Abreu, M. T. (2009). Biological markers in inflammatory bowel disease: Practical consideration for clinicians. *Gastroenterologie Clinique et Biologique*, *33 Suppl 3*, S158–73. https://doi.org/10.1016/S0399-8320(09)73151-3

Metabiomics Corp. (2014). *Microbiome Test for the Detection of Colorectal Polyps and Cancer.* https://clinicaltrials.gov/study/NCT02141945?cond=GastrointestinalDiseases&intr=Gut Microbiome Testing&page=2&rank=18

Milani, C., Duranti, S., Bottacini, F., Casey, E., Turroni, F., Mahony, J., Belzer, C., Delgado Palacio, S., Arboleya Montes, S., Mancabelli, L., Lugli, G. A., Rodriguez, J. M., Bode, L., de Vos, W., Gueimonde, M., Margolles, A., van Sinderen, D., & Ventura, M. (2017). The first microbial colonizers of the human gut: Composition, activities, and health implications of the infant gut microbiota. *Microbiology and Molecular Biology Reviews: MMBR*, *81*(4), e00036-17. https://doi.org/10.1128/MMBR.00036-17

Minar, P., Lehn, C., Tsai, Y.-T., Jackson, K., Rosen, M. J., & Denson, L. A. (2019). Elevated pretreatment plasma oncostatin M is associated with poor biochemical response to infliximab. *Crohn's & Colitis*, *360*(3), otz026. https://doi.org/10.1093/crocol/otz026

Mohammad Y. Alshahrani Ali Gaithan Alkhathami, M. A. A. A. M. Z. A. S. M. W. A., & Wahab, S. (2023). Phytochemicals as potential inhibitors of interleukin-8 for anticancer therapy: In silico evaluation and molecular dynamics analysis. *Journal of Biomolecular Structure and Dynamics*, *0*(0), 1–12. https://doi.org/10.1080/07391102.2023.2294387

Mohanto, S., Narayana, S., Merai, K. P., Kumar, J. A., Bhunia, A., Hani, U., Al Fatease, A., Gowda, B. H. J., Nag, S., Ahmed, M. G., Paul, K., & Vora, L. K. (2023). Advancements in gelatin-based hydrogel systems for biomedical applications: A state-of-the-art review. *International Journal of Biological Macromolecules*, *253*, 127143. https://doi.org/https://doi.org/10.1016/j.ijbiomac.2023.127143

Mukherjee, S., Joardar, N., Sengupta, S., & Sinha Babu, S. P. (2018). Gut microbes as future therapeutics in treating inflammatory and infectious diseases: Lessons from recent findings. *The Journal of Nutritional Biochemistry*, *61*, 111–128. https://doi.org/10.1016/j.jnutbio.2018.07.010

Nag, S., Bhatt, M., Ghosh, S., Dey, A., Paul, S., Dutta, S., Mohanto, S., Jaswanth Gowda, B. H., & Ahmed, M. G. (2023). Drug delivery for neurological disorders using nanotechnology. In A. Gautam & V. Chaudhary (Eds.), *Theranostic Applications of Nanotechnology in Neurological Disorders* (pp. 135–165). Springer Nature Singapore. https://doi.org/10.1007/978-981-99-9510-3_7

Nag, S., Mitra, O., Tripathi, G., Adur, I., Mohanto, S., Nama, M., Samanta, S., Gowda, B. H. J., Subramaniyan, V., Sundararajan, V., & Kumarasamy, V. (2024). Nanomaterials-assisted photothermal therapy for breast cancer: State-of-the-art advances and future perspectives. *Photodiagnosis and Photodynamic Therapy*, *45*, 103959. https://doi.org/10.1016/J.PDPDT.2023.103959

Nag, S., Mitra, O., Tripathi, G., Samanta, S., Bhattacharya, B., Chandane, P., Mohanto, S., Sundararajan, V., Malik, S., Rustagi, S., Adhikari, S., Mohanty, A., León-Figueroa, D. A., Rodriguez-Morales, A. J., Barboza, J. J., & Sah, R. (2023). Exploring the theranostic potentials of miRNA and epigenetic networks in autoimmune diseases: A comprehensive review. *Immunity, Inflammation and Disease*, *11*(12), e1121. https://doi.org/https://doi.org/10.1002/iid3.1121

Nataraj, C., Thomas, D. W., Tilley, S. L., Nguyen, M. T., Mannon, R., Koller, B. H., & Coffman, T. M. (2001). Receptors for prostaglandin E(2) that regulate cellular immune responses in the mouse. *The Journal of Clinical Investigation*, *108*(8), 1229–1235. https://doi.org/10.1172/JCI13640

Navarro-Tapia, E., Almeida-Toledano, L., Sebastiani, G., Serra-Delgado, M., García-Algar, Ó., & Andreu-Fernández, V. (2021). Effects of microbiota imbalance in anxiety and eating disorders: Probiotics as novel therapeutic approaches. *International Journal of Molecular Sciences*, *22*(5), 2351. https://doi.org/10.3390/ijms22052351

Nishida, A., Nishino, K., Ohno, M., Sakai, K., Owaki, Y., Noda, Y., & Imaeda, H. (2022). Update on gut microbiota in gastrointestinal diseases. *World Journal of Clinical Cases*, *10*(22), 7653–7664. https://doi.org/10.12998/wjcc.v10.i22.7653

Norouzinia, M., Chaleshi, V., Alizadeh, A. H. M., & Zali, M. R. (2017). Biomarkers in inflammatory bowel diseases: Insight into diagnosis, prognosis and treatment. *Gastroenterology and Hepatology from Bed to Bench*, *10*(3), 155–167.

Okayasu, I., Ohnishi, H., Sarandi, I., Shojima, J., Komatsu, J., Oritsu, M., Sasabe, M., Nanami, K. O., Matsuura, M., Azumi, J.-I., Ito, S., & Fujiwara, M. (2014). Significant increase of prostaglandin E-major urinary metabolite in male smokers: A screening study of age and gender differences using a simple radioimmunoassay. *Journal of Clinical Laboratory Analysis*, *28*(1), 32–41. https://doi.org/10.1002/jcla.21640

Ouyang, J., Lin, J., Isnard, S., Fombuena, B., Peng, X., Marette, A., Routy, B., Messaoudene, M., Chen, Y., & Routy, J.-P. (2020). The bacterium Akkermansia muciniphila: A sentinel for gut permeability and its relevance to HIV-related inflammation. *Frontiers in Immunology*, *11*, 645. https://doi.org/10.3389/fimmu.2020.00645

Patangia, D. V, Anthony Ryan, C., Dempsey, E., Paul Ross, R., & Stanton, C. (2022). Impact of antibiotics on the human microbiome and consequences for host health. *MicrobiologyOpen*, *11*(1), e1260. https://doi.org/10.1002/mbo3.1260

Patangia, D. V, Ryan, C. A., Dempsey, E., Stanton, C., & Ross, R. P. (2022). Vertical transfer of antibiotics and antibiotic resistant strains across the mother/baby axis. *Trends in Microbiology*, *30*(1), 47–56. https://doi.org/10.1016/j.tim.2021.05.006

Peking Union Medical College Hospital. (2017). *Gut and Oral Microbiome of GDM Patients During Pregnancy*. https://clinicaltrials.gov/study/NCT04420507?cond=GastrointestinalDiseases&intr=Gut Microbiome Testing&page=2&rank=14

Pilmis, B., Le Monnier, A., & Zahar, J.-R. (2020). Gut microbiota, antibiotic therapy and antimicrobial resistance: A narrative review. *Microorganisms*, *8*(2), 269. https://doi.org/10.3390/microorganisms8020269

Prabhu, P., Srinivas, R., Vishwanathan, K., & Raavi, A. (2014). Factors influencing alcohol and tobacco addiction among patients attending a de-addiction Centre, South India. *Journal of International Society of Preventive & Community Dentistry*, *4*(2), 103–107. https://doi.org/10.4103/2231-0762.137645

Pramanik, S., Mohanto, S., Manne, R., Rajendran, R. R., Deepak, A., Edapully, S. J., Patil, T., & Katari, O. (2021). Nanoparticle-based drug delivery system: The magic bullet for the treatment of chronic pulmonary diseases. *Molecular Pharmaceutics*, *18*(10), 3671–3718. https://doi.org/10.1021/acs.molpharmaceut.1c00491

Queen's University. (2013). *The Effect of Prebiotics on the Microbiome in Irritable Bowel Syndrome Patients: The Diet and Microbiome Study*. https://clinicaltrials.gov/study/NCT01829932?cond=GastrointestinalDiseases&intr=Gut Microbiome Testing&page=2&rank=20

Quigley, E. M. M., & Gajula, P. (2020). Recent advances in modulating the microbiome. *F1000Research*, *9*. https://doi.org/10.12688/f1000research.20204.1

Ramirez, J., Guarner, F., Bustos Fernandez, L., Maruy, A., Sdepanian, V. L., & Cohen, H. (2020). Antibiotics as major disruptors of gut microbiota. *Frontiers in Cellular and Infection Microbiology*, *10*, 572912. https://doi.org/10.3389/fcimb.2020.572912

RenJi Hospital. (2022). *Research on Gut Microbiome and Metabolomics Alterations in C.Difficile Infected IBD Patients*. https://clinicaltrials.gov/study/NCT05714566?cond=GastrointestinalDiseases&intr=Gut Microbiome Testing&page=2&rank=12

Riaz Rajoka, M. S., Shi, J., Mehwish, H. M., Zhu, J., Li, Q., Shao, D., Huang, Q., & Yang, H. (2017). Interaction between diet composition and gut microbiota and its impact on gastrointestinal tract health. *Food Science and Human Wellness*, *6*(3), 121–130. https://doi.org/https://doi.org/10.1016/j.fshw.2017.07.003

Ricciotti, E., & FitzGerald, G. A. (2011). Prostaglandins and inflammation. *Arteriosclerosis, Thrombosis, and Vascular Biology*, *31*(5), 986–1000. https://doi.org/10.1161/ATVBAHA.110.207449

Roth, W. H., Cai, A., Zhang, C., Chen, M. L., Merkler, A. E., & Kamel, H. (2020). Gastrointestinal disorders and risk of first-ever ischemic stroke. *Stroke*, *51*(12), 3577–3583. https://doi.org/10.1161/STROKEAHA.120.030643

Saito, T., Miyatake, N., Sakano, N., Oda, K., Katayama, A., Nishii, K., & Numata, T. (2012). Relationship between cigarette smoking and muscle strength in Japanese men. *Journal of Preventive Medicine and Public Health = Yebang Uihakhoe Chi*, *45*(6), 381–386. https://doi.org/10.3961/jpmph.2012.45.6.381

Sakurai, T., Akita, Y., Miyashita, H., Miyazaki, R., Maruyama, Y., Saito, T., Shimada, M., Yamasaki, T., Arhihiro, S., Kato, T., Matsuura, T., Ikegami, M., Okayasu, I., & Saruta, M. (2022). Prostaglandin E-major urinary metabolite diagnoses mucosal healing in patients with ulcerative colitis in remission phase. *Journal of Gastroenterology and Hepatology*, *37*(5), 847–854. https://doi.org/10.1111/jgh.15782

Sakurai, T., & Saruta, M. (2023). Positioning and usefulness of biomarkers in inflammatory bowel disease. *Digestion*, *104*(1), 30–41. https://doi.org/10.1159/000527846

Sanofi. (2006). *DIAMANTE – Diseases Infectious Analysis Microflora Antibiotic Effect*. https://clinicaltrials.gov/study/NCT00424905?cond=GastrointestinalDiseases&intr=Gut Microbiome Testing&rank=3

Savoldi, A., Carrara, E., Graham, D. Y., Conti, M., & Tacconelli, E. (2018). Prevalence of Antibiotic Resistance in Helicobacter pylori: A systematic review and meta-analysis in World Health Organization Regions. *Gastroenterology*, *155*(5), 1372–1382.e17. https://doi.org/10.1053/j.gastro.2018.07.007

Shahnazari, M., Afshar, S., Emami, M. H., Amini, R., & Jalali, A. (2023). Novel biomarkers for neoplastic progression from ulcerative colitis to colorectal cancer: A systems biology approach. *Scientific Reports*, *13*(1), 3413. https://doi.org/10.1038/s41598-023-29344-y

Shanghai Jiao Tong University School of Medicine. (2012). *Study of Fecal Bacteria in Early Diagnosis of Colorectal Cancer*. https://clinicaltrials.gov/study/NCT02845973?cond=GastrointestinalDiseases&intr=GutMicrobiomeTesting&rank=4

Shortt, C., Hasselwander, O., Meynier, A., Nauta, A., Fernández, E. N., Putz, P., Rowland, I., Swann, J., Türk, J., Vermeiren, J., & Antoine, J.-M. (2018). Systematic review of the effects of the intestinal microbiota on selected nutrients and non-nutrients. *European Journal of Nutrition*, *57*(1), 25–49. https://doi.org/10.1007/s00394-017-1546-4

Singh, H., Chopra, H., Singh, I., Mohanto, S., Ahmed, M. G., Ghumra, S., Seelan, A., Survase, M., Kumar, A., Mishra, A., Mishra, A. K., & Kamal, M. A. (2024). Molecular targeted therapies for cutaneous squamous cell carcinoma: Recent developments and clinical implications. *EXCLI Journal*, *23*(SE-Review articles), 300–334. https://doi.org/10.17179/excli2023-6489

Singh, R. K., Chang, H.-W., Yan, D., Lee, K. M., Ucmak, D., Wong, K., Abrouk, M., Farahnik, B., Nakamura, M., Zhu, T. H., Bhutani, T., & Liao, W. (2017). Influence of diet on the gut microbiome and implications for human health. *Journal of Translational Medicine*, *15*(1), 73. https://doi.org/10.1186/s12967-017-1175-y

Stiverson, J., Williams, T., Chen, J., Adams, S., Hustead, D., Price, P., Guerrieri, J., Deacon, J., & Yu, Z. (2014). Prebiotic oligosaccharides: Comparative evaluation using in vitro cultures of infants' fecal microbiomes. *Applied and Environmental Microbiology*, *80*(23), 7388–7397. https://doi.org/10.1128/AEM.02200-14

Sun, S.-S., Wang, K., Ma, K., Bao, L., & Liu, H.-W. (2019). An insoluble polysaccharide from the sclerotium of Poria cocos improves hyperglycemia, hyperlipidemia and hepatic steatosis in ob/ob mice via modulation of gut microbiota. *Chinese Journal of Natural Medicines*, *17*(1), 3–14. https://doi.org/https://doi.org/10.1016/S1875-5364(19)30003-2

Takada, Y., Ye, X., & Simon, S. (2007). The integrins. *Genome Biology*, *8*(5), 215. https://doi.org/10.1186/gb-2007-8-5-215

Tang, N., Chen, H., Chen, R., Tang, W., & Zhang, H. (2022). Combination of serological biomarkers and clinical features to predict mucosal healing in Crohn's disease: A multicenter cohort study. *BMC Gastroenterology*, *22*(1), 229. https://doi.org/10.1186/s12876-022-02304-y

Thursby, E., & Juge, N. (2017). Introduction to the human gut microbiota. *The Biochemical Journal*, *474*(11), 1823–1836. https://doi.org/10.1042/BCJ20160510

Turpin, W., Dong, M., Sasson, G., Raygoza Garay, J. A., Espin-Garcia, O., Lee, S.-H., Neustaeter, A., Smith, M. I., Leibovitzh, H., Guttman, D. S., Goethel, A., Griffiths, A. M., Huynh, H. Q., Dieleman, L. A., Panaccione, R., Steinhart, A. H., Silverberg, M. S., Aumais, G., Jacobson, K., … & Croitoru, K. (2022). Mediterranean-like dietary pattern associations with gut microbiome composition and subclinical gastrointestinal inflammation. *Gastroenterology*, *163*(3), 685–698. https://doi.org/10.1053/j.gastro.2022.05.037

University Hospital, R. (2021). *Fructose Effect on Neuroinflammation and Feelings (MOODYFRUCTOSE)*. https://clinicaltrials.gov/study/NCT05371067?cond=GastrointestinalDiseases&intr=Gut Microbiome Testing&rank=1

University of Calgary. (2021). *Validation of the Small Intestine Microbiome Aspiration (SIMBA) Capsule*. https://clinicaltrials.gov/study/NCT04094558?cond=GastrointestinalDiseases&intr=Gut Microbiome Testing&rank=9

University of Zurich. (2017). *Prevalence of FODMAP Intolerance and JHS in FGID and Association With Microbiome, Dyssynergic Defecation and Dietary Intervention (PreDiMi)*. https://clinicaltrials.gov/study/NCT03460613?cond=GastrointestinalDiseases&intr=Gut Microbiome Testing&rank=6

van Soest, A. P. M., Hermes, G. D. A., Berendsen, A. A. M., van de Rest, O., Zoetendal, E. G., Fuentes, S., Santoro, A., Franceschi, C., de Groot, L. C. P. G. M., & de Vos, W. M. (2020). Associations between pro- and anti-inflammatory gastro-intestinal microbiota, diet, and cognitive functioning in Dutch healthy older adults: The NU-AGE study. *Nutrients*, *12*(11), 3471. https://doi.org/10.3390/nu12113471

Vernocchi, P., Del Chierico, F., & Putignani, L. (2016). Gut microbiota profiling: Metabolomics based approach to unravel compounds affecting human health. *Frontiers in Microbiology*, *7*, 1144. https://doi.org/10.3389/fmicb.2016.01144

Verstockt, B., Verstockt, S., Dehairs, J., Ballet, V., Blevi, H., Wollants, W.-J., Breynaert, C., Van Assche, G., Vermeire, S., & Ferrante, M. (2019). Low TREM1 expression in whole blood predicts anti-TNF response in inflammatory bowel disease. *EBioMedicine*, *40*, 733–742. https://doi.org/https://doi.org/10.1016/j.ebiom.2019.01.027

Wang, B., Wu, L., Chen, J., Dong, L., Chen, C., Wen, Z., Hu, J., Fleming, I., & Wang, D. W. (2021). Metabolism pathways of arachidonic acids: Mechanisms and potential therapeutic targets. *Signal Transduction and Targeted Therapy*, *6*(1), 94. https://doi.org/10.1038/s41392-020-00443-w

Wang, J., Zhu, N., Su, X., & Yang, R. (2024). Gut microbiota: A double-edged sword in immune checkpoint blockade immunotherapy against tumors. *Cancer Letters*, *582*, 216582. https://doi.org/https://doi.org/10.1016/j.canlet.2023.216582

Wastyk, H. C., Fragiadakis, G. K., Perelman, D., Dahan, D., Merrill, B. D., Yu, F. B., Topf, M., Gonzalez, C. G., Van Treuren, W., Han, S., Robinson, J. L., Elias, J. E., Sonnenburg, E. D., Gardner, C. D., & Sonnenburg, J. L. (2021). Gut-microbiota-targeted diets modulate human immune status. *Cell*, *184*(16), 4137–4153.e14. https://doi.org/10.1016/j.cell.2021.06.019

Wegierska, A. E., Charitos, I. A., Topi, S., Potenza, M. A., Montagnani, M., & Santacroce, L. (2022). The connection between physical exercise and gut microbiota: Implications for competitive sports athletes. *Sports Medicine*, *52*(10), 2355–2369. https://doi.org/10.1007/s40279-022-01696-x

Weidong Liu, M. (2018). *Analysis of Intestinal Microflora Combined With DNA Methylation in Stool to Detect Colorectal Cancer (IMMDC)*. https://clinicaltrials.gov/study/NCT04302363?cond=GastrointestinalDiseases&intr=Gut Microbiome Testing&rank=8

Wilson, B. C., Vatanen, T., Cutfield, W. S., & O'Sullivan, J. M. (2019). The super-donor phenomenon in fecal microbiota transplantation. *Frontiers in Cellular and Infection Microbiology*, *9*, 1–11. https://doi.org/10.3389/fcimb.2019.00002

Wolf, C. L., Pruett, C., Lighter, D., & Jorcyk, C. L. (2023). The clinical relevance of OSM in inflammatory diseases: A comprehensive review. *Frontiers in Immunology*, *14*, 1239732. https://doi.org/10.3389/fimmu.2023.1239732

Wu, D., Cao, M., Li, N., Zhang, A., Yu, Z., Cheng, J., Xie, X., Wang, Z., Lu, S., Yan, S., Zhou, J., Peng, J., & Zhao, J. (2020). Effect of trimethylamine N-oxide on inflammation and the gut microbiota in Helicobacter pylori-infected mice. *International Immunopharmacology*, *81*, 106026. https://doi.org/https://doi.org/10.1016/j.intimp.2019.106026

Xu, Z., Zhang, H., Yang, Y., Ma, X., & Yang, C. (2023). Camellia oil (Camellia oleifera Abel.) alleviates gastric injury induced by ethanol associated with modulation of gut microbiota in mice. *Oil Crop Science*, *8*(1), 61–71. https://doi.org/https://doi.org/10.1016/j.ocsci.2023.02.006

Ye, Z., Xu, Y.-J., & Liu, Y. (2021). Influences of dietary oils and fats, and the accompanied minor content of components on the gut microbiota and gut inflammation: A review. *Trends in Food Science & Technology*, *113*, 255–276. https://doi.org/https://doi.org/10.1016/j.tifs.2021.05.001

Yoon, M. Y., & Yoon, S. S. (2018). Disruption of the gut ecosystem by antibiotics. *Yonsei Medical Journal*, *59*(1), 4–12. https://doi.org/10.3349/ymj.2018.59.1.4

Zhang, L., Liu, Y., Sun, Y., & Zhang, X. (2022). Combined physical exercise and diet: Regulation of gut microbiota to prevent and treat of metabolic disease: A review. *Nutrients*, *14*(22), 4774. https://doi.org/10.3390/nu14224774

Zhou, J., Liu, J., Gao, Y., Shen, L., Li, S., & Chen, S. (2021). miRNA-based potential biomarkers and new molecular insights in ulcerative colitis. *Frontiers in Pharmacology*, *12*, 707776. https://doi.org/10.3389/fphar.2021.707776

Index

J

K

L

M

N

O

P

Q

R

Q

R

U

V

W

X

Y

Z